W9-CHK-386

THE SCIENCE OF GENETICS

THE SCIENCE OF GENETICS

SIXTH EDITION

GEORGE W. BURNS
Emeritus, Ohio Wesleyan University

PAUL J. BOTTINO
University of Maryland

Macmillan Publishing Company
NEW YORK

Collier Macmillan Publishers
LONDON

Macmillan Publishing Company
866 Third Avenue, New York, New York 10022

Collier Macmillan Canada, Inc.

Library of Congress Cataloging in Publication Data

Burns, George W.
The science of genetics / George W. Burns, Paul J. Bottino.—6th ed.
p. cm.
Bibliography: p.
Includes index.
ISBN 0-02-317400-5
1. Genetics. I. Bottino, P. J. II. Title.
QH430.B869 1989
575.1—dc19 88-22102
CIP

Printing: 1 2 3 4 5 6 7 8 Year: 9 0 1 2 3 4 5 6 7 8

PREFACE

Genetic knowledge is expanding at an explosive rate, one that extends the boundaries of understanding by a factor of approximately eight during a four-year college career. It is important to include in any new edition of a text as much of this most recent knowledge as possible without allowing an unreasonably encyclopedic expansion. Our overall approach has been to update the book without changing the basic strengths that have appealed to users of earlier editions.

This sixth edition includes a major rearrangement of topics as well as a new introductory chapter that attempts to develop the concept of the gene through a historical approach. It has been written primarily to enhance student interest. Other chapters have been combined into single chapters; sex determination with inheritance related to sex, multiple alleles with blood group genetics, and polygenic inheritance with statistics. Totally new chapters on mutation and recombinant DNA have been added. The chapters on DNA structure and function and gene expression have been moved to a position earlier in the book so that almost every topic than Mendelian genetics can be dealt with on the molecular level. The chapters on molecular aspects remain in a block for those instructors who prefer to start a course with this material. In addition to the new material in the text, boxes of information have been added to cover new, or controversial, topics in genetics. Some new problems have been added and others retained from the fifth edition. These are designed to lead the student through necessary reasoning steps to a fuller grasp of genetic principles. A hand calculator will be useful in solving many of these problems.

A number of people have been very helpful during the course of preparing this edition. A large group of individuals are acknowledged in the figure legends for photographs which they kindly provided. A number of colleagues at the University of Maryland have been helpful over the years with stimulating discussion and information, including: Neal Barnett, Steve Wolniak, John Watson, Richard Imberski. Listed below are our reviewers who provided a great amount of constructive criticism, but as usual the final burden of responsibility falls on the authors: Alan G. Atherly, Iowa State University; Glenn C. Bewley, North Carolina State University; Darrel S. English, Northern Arizona University; John Erickson, Western Washington University; David J. Fox, Biological Consultants; Jack R. Girton, Iowa State University; Robert M. Kitchin, University of Wyoming; Joyce B. Maxwell, California

State University, Northridge; B. W. Mortimore, Graceland College; Harry Nickla, Creighton University; Gene A. Pratt, University of Wyoming; Earline Rupert, Clemson University; David Sadava, Claremont Colleges; Alvin Sarachek, Wichita State University; and Dan R. Varney, Eastern Kentucky University. In addition to the reviewers, we received valuable insight and assistance from our development editor, Barbara Branca. Two people at Macmillan were very helpful and deserve special thanks: Dora Rizzuto, Production Supervisor, who exhibited infinite patience, and Alma Orenstein, Designer, who has provided a fresh, exciting modern design to give this book a new look. Finally, we thank our wives, Hermine and Ann, for their support and encouragement.

G. W. B.
P. J. B.

BRIEF CONTENTS

DETAILED CONTENTS

CHAPTER 7

SEX DETERMINATION AND INHERITANCE RELATED TO SEX 117

CHAPTER 8

IDENTIFICATION OF THE GENETIC MATERIAL 153

CHAPTER 11

GENE STRUCTURE AND ORGANIZATION 233

CHAPTER 12

REGULATION OF GENE ACTION 247

CHAPTER 13

GENE MUTATION 261

CHAPTER 17

CHAPTER 18

CHAPTER 19

CHAPTER 20

RECOMBINANT DNA: NEW GENETICS AND THE FUTURE 407

INTRODUCTION TO GENETICS

CHAPTER

1

What is the science of **genetics?** Very simply, it is the study of two main subjects, **heredity** and **variation.** Heredity is the cause of the similarities between individuals. This is the reason that brothers and sisters with the same parents resemble each other. Variation is the cause of the differences between individuals. This is the reason that brothers and sisters who do resemble each other are still unique individuals. The science of genetics attempts to explain the mechanism and the basis for both similarities and differences between related individuals.

In addition to understanding the mechanisms of heredity and variation, genetics is also involved with other interesting subjects. The explanation for the tremendous variation one encounters in all forms of life is a major question addressed by genetics. This variation provides the raw materials on which the processes of speciation act. Another area of interest to geneticists is the question of development. All organisms begin their life cycle as a single cell. Yet, in multicellular forms, this single cell gives rise to the very complex multicellular organism, containing numerous cell types. How this process of development occurs is a major question in genetics.

From the beginning of the domestication of plants and animals by humans, genetic principles have been applied for their improvement for human consumption. At first, people chose only those plants and animals most desirable to produce the next generation. As the scientific basis of these desirable characteristics became known, specific genetic principles were applied to the breeding of crop plants and animals for use in specific environments.

The medical aspects of genetics are among the most recent. There are now about three thousand human inherited diseases. Many hospital beds are occupied by persons with these diseases, most of which do not have a cure. Understanding the nature and pattern of inheritance of these diseases allows genetic counselors to give meaningful information to young couples so that they may make intelligent decisions regarding the planning of their families.

Finally, applications from the emerging field of biotechnology are making important contributions in medicine, diagnosing and treating diseases, and in agriculture, developing new varieties of plants and animals for food.

DEVELOPMENT OF THE GENE CONCEPT

Genetics revolves around one central concept, the **gene.** Since the 1860s, geneticists have devoted their efforts to defining and understanding this central theme. To appreciate this fact, we must remember that almost since the beginning of time, humans have tried to explain the patterns of their inheritance observed in populations. Most of the incorrect theories to explain inheritance were based on one central idea, the blending or mixing of characteristics from the two parents to produce offspring, which appeared intermediate between the parents.

There were various theories advanced to explain the mechanism of this blending inheritance; however, the correct explanation came with the publishing of the work of **Gregor Mendel** (Figure 1–1) in 1866. Based on hybridization experiments with peas, Mendel proposed the concept

FIGURE 1–1. *Gregor Mendel, Ca 1860.* (Photo used exclusively with the permission of Dr. Hugh Iltis, Botany Department, University of Wisconsin.)

of hereditary units. Equal numbers of these units (factors) were inherited from each parent and determined the observable characteristics of hybrids. This was the first conceptualization of what is now referred to as *particulate* inheritance. Characteristics themselves are not inherited, but the particles, units, or factors that determine or control the observable character are transmitted from parent to offspring. The appearance of the character in the offspring is determined by the particular combination of factors inherited from the two parents. This was the beginning of the concept of a gene, the gene being the modern term for the hereditary units or particles originally described by Mendel.

The significance of Mendel's work was unappreciated and mostly unknown until 1900 when it was independently rediscovered by three biologists: **Hugo de Vries, Carl Correns,** and **Erik von Tschermak.** From 1866 to 1900, other fields of biology, particularly cytology, had developed so that in 1902, **Walter S. Sutton** proposed the *chromosome theory of inheritance,* in which he postulated that the newly rediscovered hereditary factors were physically located on chromosomes. This was based mainly on the parallel behavior between the pairs of factors and the pairs of chromosomes during meiosis. This theory was important because it provided a mechanism of transmission to explain the behavior of the newly discovered Mendelian factors.

FIGURE 1–2. *William Bateson.* (Used with permission of the Genetics Society of America.)

The chromosome theory of inheritance was just that—a theory. The race was on to convert this theory into fact. First, **William Bateson** (Figure 1–2), in 1906, actually discovered the principle of linkage, a very important concept in genetics, of several factors being associated together on each chromosome. However, he failed to explain this phenomenon correctly. Bateson receives additional credit for the development of the science of genetics. Many consider him the real founder of the science after Mendel's original discoveries. He was the first to have Mendel's paper translated into English and the first to show that Mendel's theory also applied to animals. Bateson coined several terms we use today, including giving the name *genetics* to the new field. The actual term *gene* was first used in 1909 by W. Johannsen.

Starting around 1912, the center of contribution to genetic thought shifted from Europe to the United States and the laboratory of **Thomas Hunt Morgan** at Columbia University. Initially, Morgan had been skeptical of the chromosome theory of inheritance, but through the elegant work of three Ph.D. students, **C. B. Bridges, H. J. Muller,** and **A. H. Sturtevant,** Morgan turned the chromosome theory of inheritance into the concept of genes being located in a linear array on each chromosome. By carrying out extensive breeding experiments in the fruit fly *Drosophila melanogaster,* Morgan and his students were able to show that the genes of the fly were linked together in groups on each of the chromosomes. This linkage could be broken only by the orderly process of crossing over. By carefully analyzing the results of crosses and using the crossing over data, they could construct a physical map of each chromosome, showing the relative location of each of the genes and their relative distances apart.

With the 1926 publication of Morgan's book, *The Theory of the Gene,* the concept was finally secure of inheritance due to units transmitted from parent to offspring that behaved in a regular orderly manner. This field of investigation of genes then began a new direction, the role of genes in development. This was mainly facilitated by the discovery by Muller in 1927 that X-rays cause mutations and that these gene mutations alter normal developmental patterns in flies. In 1937, **Richard Goldschmidt** attempted to define a gene on the basis of its inferred physiological action on the development of the eye in *Drosophila.* His conclusion was that genes exist as points on a chromosome and have to be arranged in a proper order to control the normal developmental process of the eye. Mutations resulting in altered eye development were caused by disruption in the correct arrangement of the genes on the chromosome. Goldschmidt tried to create a chemical interpretation for his ideas, but did not have any experimental evidence. His main contribution may have been in stimulating exploration of questions on the chemical nature of the gene.

If there is a transition from classical Mendelian genetics to modern genetics it is marked by a change in thought from the mechanics of inheritance to the chemical basis of inheritance. The modern questions of genetics centered around the chemical nature of the gene and how it acts to produce the observable characteristics of living organisms. Thought in this area actually began shortly after the rediscovery of Mendel's work with **A. E. Garrod** and his theory of *inborn errors of metabolism.* Garrod proposed that certain metabolic diseases in humans were inherited and had Mendelian factors as their basis. It took thirty-five years before his ideas were revived and considered to be important contributions in genetics.

In the early 1940s, two significant discoveries were made concerning the chemical nature of the gene. In the laboratory of **Oswald Avery** it was established that genes were composed of a specific type of nucleic acid—deoxyribonucleic acid (DNA)—and not protein. Avery and two colleagues, **C. M. MacLeod** and **M. McCarty,** were able to demonstrate that when DNA from dead pneumonia-causing (virulent) bacteria was incorporated into avirulent bacteria, it stably transformed the avirulent bacteria into virulent bacteria. Some authors consider this the beginning of what we now call **molecular genetics.** Investigations into the gene were thereafter focused on DNA, its structure and function, in producing the hereditary characteristics of the organism.

Second, while studying the biochemical basis for eye color in *Drosophila,* **George Beadle** and **E. L. Tatum** were able to show that the lack of brown color in various mutants was due to a defect in one step in the biosynthesis of the brown pigment. Eventually they showed that each of the steps in the synthesis of the brown eye color pigment was the work of a single gene. This led them to the famous **one-gene one-enzyme** hypothesis, which proposed that the action of each gene is through the synthesis of a protein (enzyme), which in turn catalyzes a single chemical reaction. They were able to test and prove this hypothesis through the use of a multitude of mutants in the fungus *Neurospora.* In most cases, each mutation was due to an alteration in a single gene. With these two discoveries a new level of investigation in genetics began—the molecular basis of inheritance.

In 1953, one of the most significant twentieth-century discoveries in biology was made by **James Watson** and **Francis Crick.** Their paper, published in the British journal *Nature,* began in the following way: "We wish to suggest a structure for the salt of deoxyribonucleic acid (D. N. A.)." In this paper and a subsequent one they propose the molecular structure of DNA and consequently the molecular composition of the gene.

The gene could then be redefined in molecular terms. To do this, the physicist-turned-biologist **Seymour Benzer** published the results of a number of genetic studies on T4, a virus of the common colon bacteria, *Escherichia coli.* By examining a large number of progeny viruses, Benzer was able to detect very rare genetic events. Benzer was able to demonstrate crossing over within a gene, to show that protein products of two different normal genes functioned together in combined mutants to produce the normal phenotype, and to attempt to map mutant sites into molecular units along the DNA molecule. Simply, Benzer was able to demonstrate that the linear array of genes on chromosomes, as shown by Morgan, extended down to the molecule of DNA making up the chromosome. Benzer's work was able to define the gene in terms of function, recombination, and mutation, and place an accurate molecular size estimate on the conceptual gene components.

Now that the chemical composition and structure of genes were known, two remaining parts of the gene story were to determine the mechanism of regulation of the activity of genes and how the information encoded in DNA became translated into proteins. While it had been known for many years that the ultimate products of most genes were proteins, it was not known how the activity of individual genes was regulated. In 1961, it was **Francois Jacob** and **Jacques Monod** who finally provided genetic evidence for a method of gene regulation in bacteria now called the **operon.** The operon consists of a series of genetic regions, interacting in such a way as to regulate whether a messenger ribonucleic acid (RNA) copy of a gene is made. If made, this messenger RNA species directs the synthesis of a protein. Watson and Crick proposed that the genetic information of an organism lies in the sequence of bases in DNA, but the nature of this information wasn't understood until 1961–1964. The **genetic code,** as it is called, was solved by **M. W. Nirenberg, J. H. Matthaei,** and **P. Leder,** biochemists at the National Institutes of Health. They synthesized small RNA molecules of known composition and observed which amino acid was incorporated into protein in a cell-free protein-synthesizing system. By testing the sixty-four possibilities of four RNA bases, taken three at a time, they succeeded in identifying the exact code word for each amino acid.

MODERN STUDIES ON GENES

It has been over a hundred years since Mendel and his original explanations of particulate inheritance to present times where the structure and functioning of genes at the molecular level is well known. Now the question to ask is, "Is there anything more to know about genes?" The answer is that the closer and the longer we look at genes, the more we find to learn. There are several new areas of investigation of genes that are providing an exciting life for geneticists.

Oncogenes

These are genes associated with cancer. Originally oncogenes were found in viruses, and were thought to be transferred to vertebrate cells, thus causing cancer. However, it was found later that these genes were already present in the cells of most vertebrates. Gene products of normal oncogenes play a normal role in cell growth. It is now thought that mutations in oncogenes bring about increased production of gene product, which causes the rapid cell proliferation characteristic of neoplastic growth.

Antibody Diversity

The standard immune response of vertebrates allows for the production of a virtually unlimited number of unique antibodies in response to an unlimited number of possible antigens. Each individual antibody is produced in response to and is specific for each antigen. The basis for this interesting complex response lies in the phenomenon of somatic recombination of segments of a relatively small number of genes and the selective production of a unique protein product from these recombined genes. The number of unique combinations of gene segments is almost infinite, allowing for the tremendous diversity of antibody types. The 1987 Nobel Prize in Medicine was awarded to S. Tonegawa for the primary discoveries in this exciting area.

Homeotic Mutations

In *Drosophila* the production of a leg in place of an antenna is a homeotic mutation. This is caused by mutations in large complexes of genes controlling the location and development of structures. These complexes of genes are under the control of a single master gene. This single master, by controlling the expression of the gene complex, determines pattern formation. Within the master gene are repeating units of genetic material called **homeoboxes,** which seem to play a role in the overall regulation of the system. Homeobox sequences are found in many organisms, including humans, but no specific role has been assigned to them yet.

Behavior

Very recently, studies of identical twins separated since birth revealed that genes play a much greater role in determining behavior than previously thought. Although these twins were not raised together, they exhibited many similar behavioral traits. These new studies indicate that in any behavioral characteristic studied, genes exert at least a 50 percent influence on the trait. The information so far obtained suggests that our environment can have only a 50 percent influence at best on the development of our personality. The conclusion is that genes play a major role in shaping almost any type of behavior including alcoholism, criminality, intelligence, political attitudes, schizophrenia, and sociability.

Recombinant DNA

The latest chapter in the gene story is still being written. Through a number of recent technical advancements it is now possible to isolate any gene virtually intact. Intact genes can be attached to DNA molecules from other organisms, such as bacteria, and reintroduced into the bacterium where the gene is replicated and fully expressed. When placed into an appropriate vector, these cloned genes can be transferred into unrelated organisms where they are also expressed. The possible applications for this technology are already being explored. They include commercial production of human insulin, and eventually other human proteins in bacteria, gene therapy, the correction of a genetic disorder by inserting the correct gene in place of the defective one, and genetic manipulation of crop plants for such things as herbicide and cold tolerance, as well as insect resistance. It has been estimated that by the year 2000 everyone will come into contact on a daily basis with some product of recombinant DNA technology.

MONOHYBRID INHERITANCE

CHAPTER

2

Early attempts to determine fundamental genetic mechanisms frequently failed because investigators tried to examine simultaneously all discernible traits. Mendel's success in preparing the groundwork of modern understanding lay in (1) concentrating on one or a few characters at a time, (2) making controlled crosses and keeping careful numerical records of results, and (3) suggesting "factors" as the particulate causes of various genetic patterns. Similarly, in order to learn something of the inheritance of *vestigial* wings in the fruit fly, *Drosophila* (Figure 2–1), an individual having normal, full-sized wings would be crossed with one having vestigial (i.e., small, nonfunctional) wings; all other traits would be ignored. The appearance of the offspring through several generations and the numbers of *normal* and *vestigial* individuals produced would be carefully recorded and evaluated.

Such a cross, involving contrasting expressions of one trait, is referred to as a **monohybrid cross.** Although this term is restricted by some geneticists to cases in which the parents *differ* detectably in a single allele pair (as in the example of the fruit flies), it may also be extended to any cross in which only a single character is being considered, whether or not the parents differ discernibly. The latter, broader definition will be used in this text.

Although Mendel knew nothing of chromosomes and their behavior in nuclear division, and had only inferred a particulate mechanism in the transmission of genetic traits, he recognized a pattern of behavior for his genetic factors, as he called them. As a result, he developed what has long been known as **Mendel's first law, the principle of segregation.** It states essentially that hereditary patterns are determined by

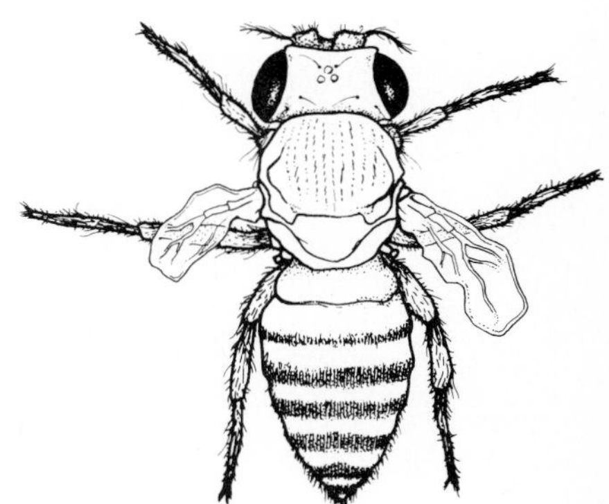

FIGURE 2–1. *Vestigal wing, resulting from the action of a recessive gene, in* Drosophila melanogaster.

factors (genes) that occur in pairs in an individual but *segregate from each other in the formation of sex cells (gametes) so that any one gamete receives only one or the other of the paired alleles.* The double number of genes is then restored in the progeny. A second Mendelian law will be explored in Chapter 3.

Let us begin our study of genetics along the general lines of Mendel's technique, but using a different organism for the sake of variety. Shortly, it will be seen that the same principles apply to human beings.

THE STANDARD MONOHYBRID CROSS

Complete Dominance

The common house plant, variegated coleus (*Coleus blumei,* of the mint family), is a frequently cultivated flowering plant that lends itself readily to simple genetic experimentation. It has many of the desirable characteristics of an organism useful for genetic experimentation. Desirable characteristics of this plant include a relatively short reproductive cycle, easily made crosses between different individuals (yet self-pollination is also possible), and appearance of many of its genetically determined traits in the seedling stage.

One such character involves the shape of the leaf margins. Some plants have shallowly crenate edges; others have rather deeply incised leaves (Figures 2–2 and 2–3). Plants displaying one or the other of these two traits may be referred to as "shallow" and "deep," respectively. Imagine two such plants, each the product of a long series of generations obtained by self-pollination, in which the same individual serves as both maternal and paternal parent. If, in such repeated breeding, deep always gave rise to deep offspring only, and shallow to shallow only, each such line of descent would be established as "pure-breeding" for this trait. The geneticist customarily uses the term **homozygous** in such cases. The precise genetic implication of this latter term will be seen shortly.

Suppose now a cross were to be made between two such homozygous individuals, one deep and the other shallow. In theory, the outcome might be any one of the following: (1) all the offspring may resemble one or the other parent only; (2) some of the progeny may

FIGURE 2–2. *The common house plant* Coleus, *showing shallow-lobed leaves.*

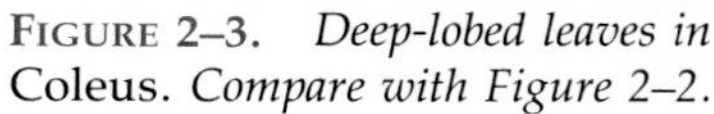

FIGURE 2–3. *Deep-lobed leaves in* Coleus. *Compare with Figure 2–2.*

resemble one parent, whereas the remainder may look like the other parent; (3) all the offspring, though like each other, may be intermediate in appearance between the two parents; (4) the offspring may look like neither parent and yet not be clearly intermediate between them; or (5) there may be a considerable range of types in the progeny, some being about as deeply lobed as one parent, some about as shallowly cut as the other, with the remainder forming a continuum of variation between the two parental types. When the offspring of this cross are examined, however, all of them are seen to be *deep,* like one parent. The *shallow* trait does not appear at all. Since the days of Mendel, a characteristic that expresses itself in all the offspring of such a monohybrid cross (as *deep* does in this case) has been termed **dominant,** and the trait that fails to be expressed (here, *shallow*) has been referred to as **recessive.**

If various individuals of this progeny are either intercrossed or self-pollinated, what will be the results? From similar work of Mendel (Table 2-1) it can be predicted that about three-fourths of the second generation should be deep and about one-fourth shallow. This predicted outcome is actually observed. (It should be noted, however, that only with quite large samples would a very close approach to a 3:1 ratio be expected.)

TABLE 2–1. **Mendel's Earliest Experiments on Peas**

Parents	*First generation*	*Second generation*				*Ratio*
Round × wrinkled seed	all round	5,474	round	1,850	wrinkled	2.96:1
Yellow × green cotyledons	all yellow	6,022	yellow	2,001	green	3.01:1
Gray-brown × white seed coats	all gray-brown	705	gray-brown	224	white	3.15:1
Inflated × constricted pods	all inflated	882	inflated	299	constricted	2.95:1
Green × yellow pods	all green	428	green	152	yellow	2.82:1
Axial × terminal flowers	all axial	651	axial	207	terminal	3.14:1
Long × short stems	all long	787	long	277	short	2.84:1
TOTALS		14,949		5,010		Av. 2.98:1

According to standard terminology, results in the *Coleus* example may be summarized as follows:

(*Parental generation*)	P	deep × shallow
(*First filial generation*)	F_1	all deep
(*Second filial generation*)	F_2	$\frac{3}{4}$ deep + $\frac{1}{4}$ shallow

F_2 progeny are obtained either by "selfing" or by interbreeding members of the preceding F_1.

Mechanism of the Monohybrid Cross

Genes and their location. It now becomes important to determine what mechanism might account for these results. Obviously, what is transmitted from parent to offspring is not the trait itself, but rather something that *determines* the later development of that trait at an appropriate time and place in a particular environment. These determiners are called *genes,* which occur in two or more structural forms, called alleles. Their actual structure and behavior will be examined later.

Because the sex cells or **gametes** constitute the only link in sexually reproducing organisms between parent and offspring, it is clear that genes must be transmitted from generation to generation via this gametic bridge. Where, then, are the genes located in the gametes? Cytological examination of the gametes of a great many sexually reproducing organisms, both plant and animal, discloses that, although the egg cell has a large amount of cytoplasm, the sperm is largely nucleus, having relatively little cytoplasm. This relationship breaks down, of course, in such isogamous forms as some of the algae, in which relative amounts of cytoplasm and nuclear material are quite similar. Though this observation does not furnish conclusive *proof* of the location of the genes within the nuclei of the sex cells, it does offer a *probability* that the genes in the *Coleus* example are more likely to be nuclear than cytoplasmic in location. The possibility of other genes being located in the cytoplasm is discussed in Chapter 16.

That sperm and egg in this kind of cross do make equal genetic contribution to the next generation is strongly indicated by the fact that, whether *deep* serves as the **pistillate** (egg-contributing) parent or as the **staminate** (sperm-contributing) parent, the result is the same: all the F_1 are deep-lobed (Figure 2–4). Such **reciprocal crosses** do not, in every case, produce the same results. In fact, whether reciprocal crosses give the same or different results provides certain important additional information concerning the operation of the genes involved. (See Chapter 16.)

Assuming that genes are situated in the nucleus, it would be useful to locate them more precisely within the nucleus. Genes should be associated with some cell organelle that is (1) quantitatively distributed with complete exactness during nuclear division and (2) contributed equally by sperm and egg at gametic union, or syngamy. Although

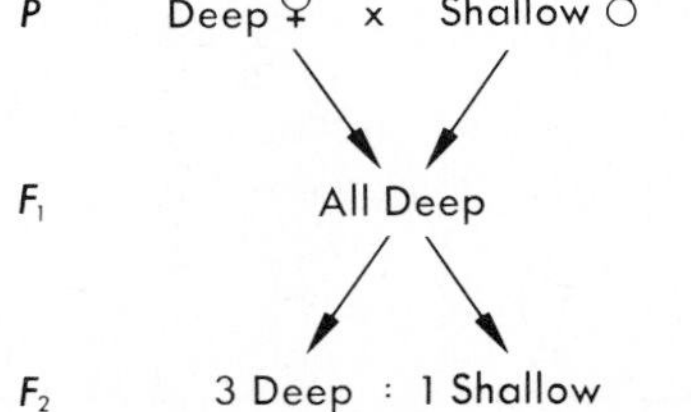

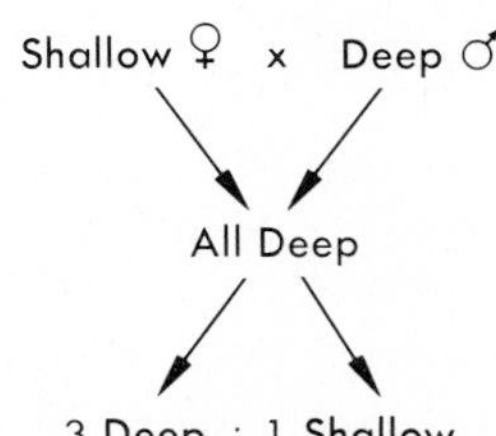

FIGURE 2–4. *Diagram of a reciprocal cross in* Coleus. *Note that, in this case, results in the F_1 and F_2 are the same regardless of which plant is used as the pistillate parent.*

Chapter 4 deals more closely with the cellular processes involved, previous work that you as a student have had in the life sciences surely suggests the chromosomes might well serve as the vehicles of gene transmission in such cases. Pending a test of that hypothesis, this is the assumption on which attempts to explain these observations in *Coleus* (and, by implication, other organisms as well) will be based. Regardless of the attractiveness of a theory, however, acceptance must rest on testing, and revision must always be considered carefully in the light of new experimental evidence. *Coleus* cells have a discrete nucleus; that is, *Coleus* is a **eukaryote.** Genetic hypotheses with regard to *Coleus* will have to be modified somewhat for **prokaryotes** (which do not have a discrete nucleus and in which chromosomes *as known in eukaryotes* are absent). Both bacteria and cyanobacteria are prokaryotic organisms, the former being of great importance to geneticists and, indeed, to the human species. Viruses, which also have genetic material, will require additional refinements or widening of hypotheses that are based on eukaryotes.

For the present, then, it will be assumed that genes are situated on the chromosomes, transmitted via the gametes from one generation to the next, and contributed equally by both parents. If D represents the dominant allele for deep lobes and d the recessive, the gametic contribution of the P generation will be designated as

Ⓓ and ⓓ

where the two alleles, one for deep (D) and one for shallow (d), are said to form an allelic pair.

Based on (1) the observation that chromosomes of both fusing gametes become incorporated within the nucleus of the zygote, each maintaining a separate identity, and (2) the assumption that chromosomes serve as likely candidates for gene location, it then follows that the zygotes (which will become F_1 individuals) can be represented as Dd. So, also, the P individuals must be represented as DD and dd, respectively. Thus each of the parental individuals possesses two identical alleles in each of their somatic cells; that is, each is **homozygous.** In the same way, the F_1 plants must have in each of their somatic cells one allele for deep (D) *and* one for shallow (d); thus they are **heterozygous** with respect to this pair of alleles. Because in the heterozygous F_1 the effect of the dominant allele appears to mask completely the presence of the recessive allele, this case illustrates **complete dominance.**

The assumption is, therefore, that there are two alleles for a given characteristic in the somatic cells and only one in the gametes. Some sort of nuclear division, prior to sex cell formation, which will reduce the allele number from two per trait to one—i.e., separate the alleles—is thus required. Chapter 4 explores this possibility further. The original *Coleus* cross may now be written as follows:

P	deep DD	×	shallow dd
P gametes	Ⓓ	×	ⓓ
F_1		deep Dd	
F_1 gametes	Ⓓ	+	ⓓ

Phenotype and genotype. *Detectable* traits of the organism, with regard to the character or characters under consideration, constitute the **phe-**

notype, which is generally designated by a descriptive word or phrase. Phenotypic traits may be either morphological or physiological although in the final analysis it is not always easy to separate the two. On the other hand, an individuals' *genetic makeup* is termed the **genotype** and is customarily given by letters of the alphabet or other convenient symbols. In this case, *deep* and *shallow* represent *phenotypes,* whereas *DD, Dd, dd, D,* and *d* represent *genotypes* (somatic and gametic).

An individual's phenotype is not always a rigidly expressed "either-or" condition but is often modified by environmental influences. For example, although a quantitative character such as height is genetically determined in many plants (and in humans), even in those cases in plants in which only a single pair of alleles can be shown to be operating, height variation in both the tall and short individuals often occurs. Such variations, usually clustering fairly closely around a mean, can be shown in suitable, controlled experiments, to be the results of environmental influences. In other words, genotype determines the phenotypic

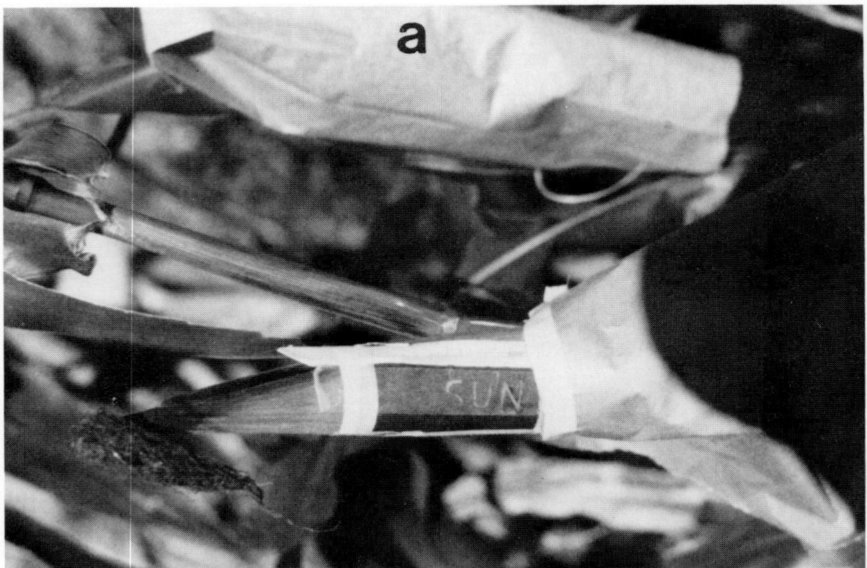

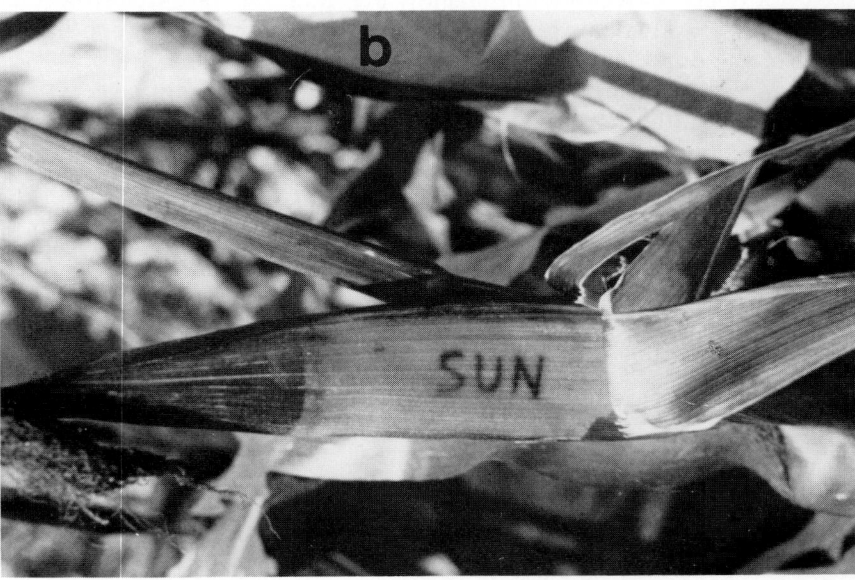

FIGURE 2–5. *"Sun-red" corn. Action of the sun-red gene is to produce red pigment wherever exposed to the light; in the absence of light, the tissue remains green. In this instance, the husks had been covered with black paper with the word* sun *cut out so as to expose the husk (a). When the cover is removed, the action of the sun-red gene can be observed (b).* (Photo courtesy of Dr. M. G. Neuffer, University of Missouri.)

range within which an individual will fall; environment determines where in that range the individual will occur. In certain environments, too, it is possible that a particular genotype will not be expressed at all. In corn, for example, one allele ("sun-red") produces red grains if the ear is exposed to light (Figure 2–5), but as long as the husks are intact the grains remain white, so that both sun-red and white genotypes remain phenotypically indistinguishable. Furthermore, as will soon be shown, phenotype may often be physiological and therefore "observable" only in the biochemical sense.

Probability method of calculating ratios. In a nuclear division, which segregates the alleles D and d during gamete formation, *one-half* of the F_1 gametes should carry the dominant D and *one-half* the recessive d. Furthermore, if syngamy is random so that a D egg has an equal chance to be fertilized by either a D or d sperm, mating in the F_1 to produce the F_2 can be represented as follows:

$$\begin{array}{rl} \textit{eggs from } \female\ F_1 & \frac{1}{2}\,\textcircled{D} + \frac{1}{2}\,\textcircled{d} \\ \textit{sperms from } \male\ F_1 & \frac{1}{2}\,\textcircled{D} + \frac{1}{2}\,\textcircled{d} \\ \hline F_2 & \frac{1}{4}\,DD + \frac{1}{4}\,Dd + \frac{1}{4}\,dD + \frac{1}{4}\,dd \end{array}$$

or $\frac{1}{4}\,DD + \frac{2}{4}\,Dd + \frac{1}{4}\,dd$ (i.e., $\frac{1}{4} + \frac{1}{2} + \frac{1}{4}$) for a 1:2:1 **monohybrid genotypic ratio.**

The basis for this kind of calculation is the **product law (rule) of probability.** Briefly stated, this rule holds that the probability of the simultaneous occurrence of two independent events equals the product of the probabilities of their separate occurrences. Thus if one-half the eggs bear the allele D and one-half the sperms also carry the D allele, and if fertilization is a completely random event, then the probability of a DD zygote is $\frac{1}{2} \times \frac{1}{2} = \frac{1}{4}$. However, to determine the total probability for traits that *do not* require simultaneous occurrence of independent events, one *adds*. For example, the heterozygote probability is $\frac{1}{4} + \frac{1}{4} = \frac{1}{2}$. Adding the two genotypes (DD and Dd) that give the dominant phenotype is $\frac{1}{4} + \frac{1}{2} = \frac{3}{4}$.

As to phenotypes, we see that DD and Dd organisms appear indistinguishable on visual bases, so that the 1:2:1 genotypic ratio gives our observed 3:1 (F_2) monohybrid phenotypic ratio. A 3:1 ratio is shown in Figure 2–6. (As will be seen, these ratios may occur in the F_1, given certain P individuals, and are, of course, not the only possible monohybrid ratios.) Thus assumptions developed so far have led to an explanation completely compatible with earlier observations. These assumptions remain to be tested further, both by cytological examination and by additional breeding experiments. Note that the so-called checkerboard or Punnett square method has not been used, but rather a *probability* method of calculating genotypic and phenotypic ratios. This latter system becomes extremely advantageous when calculating ratios involving several pairs of alleles.

FIGURE 2–6. *A 3:1 purple:white ratio in corn. This is the F_2 of homozygous purple × white.*

The Testcross

DD and *Dd* individuals in *Coleus* look alike as is true in all such instances where dominance is complete. This raises two questions. First, can homozygous dominants and heterozygous individuals be distinguished in any way? The answer is *yes,* and the *testcross* is the means whereby this can be determined. In the testcross an individual having the dominant phenotype (which therefore could be either homozygous or heterozygous, and can be represented as having the genotype *D*—) is crossed with one having the recessive phenotype. In *Coleus* the testcross of a homozygous dominant is as follows:

P	deep *DD*	×	shallow *dd*
P gametes	eggs	Ⓓ	
	sperms	ⓓ	
F_1	all *Dd* (deep)		

If the P dominant phenotype had been heterozygous, the result would be the classic **1:1 monohybrid testcross ratio:**

P	deep *Dd*	×	shallow *dd*
P gametes	*eggs*	$\frac{1}{2}$ Ⓓ + $\frac{1}{2}$ ⓓ	
	sperms	1 ⓓ	
F_1		$\frac{1}{2}$ *Dd* (deep) + $\frac{1}{2}$ *dd* (shallow)	

Note that *DD* × *dd* produces an F_1 all of the dominant phenotype, whereas *Dd* × *dd* gives rise to a 1:1 ratio in the progeny. A 1:1 testcross ratio in corn is shown in Figure 2–7.

At this point use of the expression F_1 must be explained. It will be used to designate the first generation resulting from *any* given mating regardless of the parental genotypes. Therefore the term F_1 does not necessarily imply heterozygosity. In the two testcrosses just outlined, note that the F_1 genotype depends on the P genotypes and may, of course, be either heterozygous (as in the first instance) or homozygous (as in the *dd* individuals of the second case).

The second question suggested by the visual similarity of homozygous dominants and heterozygotes is the matter of how genes operate to exert their phenotypic effects. Although a complete answer must be deferred to later considerations of the molecular nature of the gene and the chemistry of its action, a genetic trait in Mendel's peas offers a

FIGURE 2–7. *A 1:1 testcross in corn. This ear resulted from the cross white × heterozygous purple.*

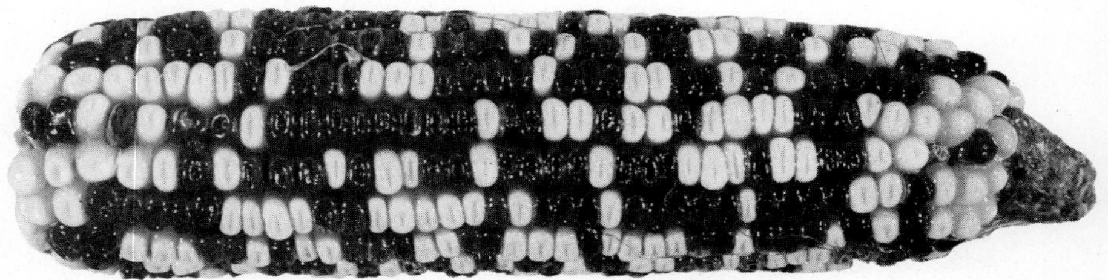

tempting suggestion. It also furnishes an introduction to a variation in the classic 3:1 phenotypic ratio seen in *Coleus*. This case will be examined in the following section.

MODIFICATIONS OF THE 3:1 PHENOTYPIC RATIO

Incomplete Dominance

The data of Table 2–1 were first reported by Mendel in a pair of originally widely ignored papers read before the Natural History Society at Brünn on February 8 and March 8, 1865. Mendel's results led him to designate round seeds as (completely) dominant to wrinkled seeds; indeed, homozygous and heterozygous round seeds cannot be differentiated macroscopically. However, when cells of the cotyledons of the three genotypes (*WW*, *Ww* round, and *ww* wrinkled) are examined *microscopically*, an abundance of well-formed starch grains is seen in the *WW* plants, and very few in *ww* individuals. Heterozygotes show an intermediate number of grains, many appearing imperfect or eroded. Moreover, *ww* embryos test higher in reducing sugars (the raw material from which starch molecules are constructed) than do those of *Ww* genotype. *WW* embryos test lowest of all for sugar. In plant cells starch is synthesized from glucose-1-phosphate under the influence of an enzyme system. Although a discussion of the chemistry of the sugar $\rightleftarrows$ starch interconversion is outside the scope of a beginning course in genetics, a plausible explanation is that gene *W* is responsible for the production of one of the enzymes required for the reaction glucose-1-phosphate $\rightarrow$ starch. Likewise, gene *w* may produce either a smaller quantity of the enzyme or, perhaps, a molecule that functions only imperfectly and produces a less efficient enzyme molecule, or is unable to function at all because of sufficient structural differences. The larger amount of starch in *WW* plants is associated with higher water retention (starch is a hydrophilic colloid) and therefore with plump, distended, spherical seeds. Homozygous recessive embryos, on the other hand, retain considerably less water upon reaching maturity and hence appear wrinkled. The starch content of heterozygotes, however, is apparently sufficient to produce embryos visually indistinguishable from those of *WW* genotype. Consistent with these differences in starch content, heterozygotes have an intermediate amount of functional enzyme. The relationship between genes and enzymes suggested here is a fundamental concern of modern genetics and offers much help in elucidating the nature and action of the gene. It will receive considerable attention in later chapters of this book.

For present purposes, however, it is clear that on the gross, macroscopic level, a case of complete dominance is operating, yielding the familiar 3:1 phenotypic and 1:2:1 genotypic ratios. But on the microscopic, chemical, or molecular level allele *W* must be considered only **incompletely dominant** to the *w* allele. Therefore, at the *physiological* level the cross *Ww* × *Ww* yields identical phenotypic and genotypic ratios of 1:2:1. Whether round versus wrinkled seeds is considered as complete dominance or as incomplete dominance depends entirely on the level of analysis. Because the physiological level more accurately reflects the actual operation of the genes involved, most geneticists would prefer to regard this as a case of incomplete dominance. It is important to note that in incomplete dominance the heterozygotes are phenotypically *intermediate* between the two homozygous types.

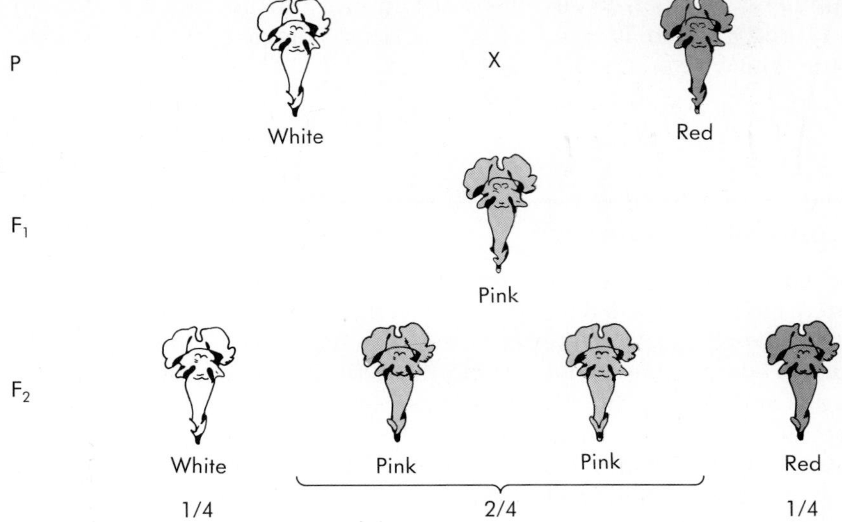

FIGURE 2–8. *A case of incomplete dominance in snapdragon. The pink F_1 hybrid is truly intermediate between the red and white parents, and the F_2 segregates in a characteristic 1:2:1 phenotypic ratio.*

Incomplete dominance in other organisms. Examples of incomplete dominance at the macroscopic level occur in both plants and animals. For example, radishes may be long, oval, or round. Crosses of long × round produce an F_1 of wholly oval phenotype:

$$
\begin{array}{lll}
\text{P} & \text{long} \quad \times & \text{round} \\
 & l^1l^1 & l^2l^2 \\
F_1 & \text{all oval} & \\
 & l^1l^2 & \\
F_1 \text{ gametes} \quad \textit{eggs} & \tfrac{1}{2}\,\text{\textcircled{l^1}} + \tfrac{1}{2}\,\text{\textcircled{l^2}} & \\
\textit{sperms} & \tfrac{1}{2}\,\text{\textcircled{l^1}} + \tfrac{1}{2}\,\text{\textcircled{l^2}} & \\
\hline
F_2 & \tfrac{1}{4}\,l^1l^1 + \tfrac{2}{4}\,l^1l^2 + \tfrac{1}{4}\,l^2l^2 &
\end{array}
$$

Practice in this text is to use lowercase letters with numerical superscripts to designate incompletely dominant alleles.

Flower color in snapdragon (Figure 2–8) is likewise associated with a pair of incompletely dominant alleles. The same is true of height in the cultivated geranium (*Pelargonium*).

Codominance

Cattle. In shorthorn cattle alleles for red and white coat color occur. Crosses between red (r^1r^1) and white (r^2r^2) produce offspring (r^1r^2) whose coats appear at a distance to be reddish gray, or *roan.* Superficially this would seem to be a case of incomplete dominance, but close examination of roan animals reveals that the coat is composed of a *mixture of red hairs and white hairs,* rather than hairs all of a color intermediate between red and white. Instances such as this, where the heterozygote exhibits a mixture of the phenotypic characters of both homozygotes, instead of a single intermediate expression, illustrate **codominance.** Genotypic and phenotypic ratios are identical in incomplete dominance and codominance, but the distinction reflects something of the way in which alleles operate, a topic that will be discussed in detail in later chapters.

Humans. Several illustrations of codominance are found in human genetics. One of these involves a pair of codominant alleles, *M* and *N*, responsible for production of antigenic substances M and N, respectively, on the surfaces of the red blood cells. Although the genetics of blood antigens is explored more fully in Chapter 17, it is interesting to note that, whereas persons of genotype *MM* produce antigen M and *NN* individuals produce the somewhat different antigen N, heterozygotes (*MN*) produce *both* M and N antigens, not a single intermediate substance. All persons belong to one of three phenotypic classes, that is, M, MN, or N. However, because antibodies to these antigens rarely occur, they are not considered in transfusion, and most people do not know their M–N blood group. Accumulated data from families in which both parents have been identified as MN show a close approximation to the expected 1:2:1 ratio in the children.

Lethal Alleles

A second major modification of the classic monohybrid ratios is produced by alleles whose effect is sufficiently drastic to kill bearers of certain genotypes.

Tay-Sachs disorder in humans. Infants normally exhibit a "startle response" in which they stiffen their arms and legs upon hearing a sudden noise. Continuation of this reaction beyond four months old to six months old *may* be symptomatic of the Tay-Sachs disorder (juvenile amaurotic idiocy), which is due to homozygosity for a **recessive** allele. Homozygous recessives fail to produce an enzyme, hexosaminidase A (hex A), with the result that a lipid, ganglioside GM_2, accumulates in the brain. Hex A is one of two enzymes in a sequence of lipid catabolic reactions (the other is hexosaminidase B) and consists of α and β chains of amino acid residues (polypeptide chains). Tay-Sachs disorder results from a change (mutation) in the allele responsible for the synthesis of the α polypeptide chain. This defect prevents the production of functional hex A.[1] Mental and motor deterioration rapidly follow the onset of symptoms, accompanied by paralysis and degeneration of the retina, the latter leading to blindness. The disorder culminates in death, generally before the age of four years. There is no treatment. The defective recessive allele responsible is, therefore, termed **lethal.**

Sickle-cell in humans. Sickle-cell disorder, involving a pair of **incompletely dominant** alleles, is particularly prevalent in blacks in the United States and in certain African peoples. Because the alleles are incompletely dominant, certain phenotypic ratios occur, differing from those of Tay-Sachs.

The hemoglobin of most persons is of a particular chemical structure and is known as hemoglobin A (for adult hemoglobin). Many chemical variants of hemoglobin A are found in relatively small numbers of people; one of these, hemoglobin S, is involved in the sickle-cell disorder. Alleles responsible for hemoglobin types here are Hb^A and Hb^S. Most persons belong to genotype Hb^AHb^A. Their erythrocytes contain only hemoglobin A and are biconcave disk-shaped (Figure 2–9). Persons with *sickle-cell anemia* are of the genotype Hb^SHb^S and are characterized by a collection of symptoms, chiefly a chronic hemolytic anemia. In the blood of such persons the erythrocytes under reduced oxygen tension become distorted and appear sickle-shaped (Figure 2–9). Sickle-shaped cells not

[1]Failure to produce *both* enzymes results in Sandhoff disease, whose symptoms are similar to Tay-Sachs. Hex B consists only of β chains, so Sandhoff disease involves defects in both α and β chains.

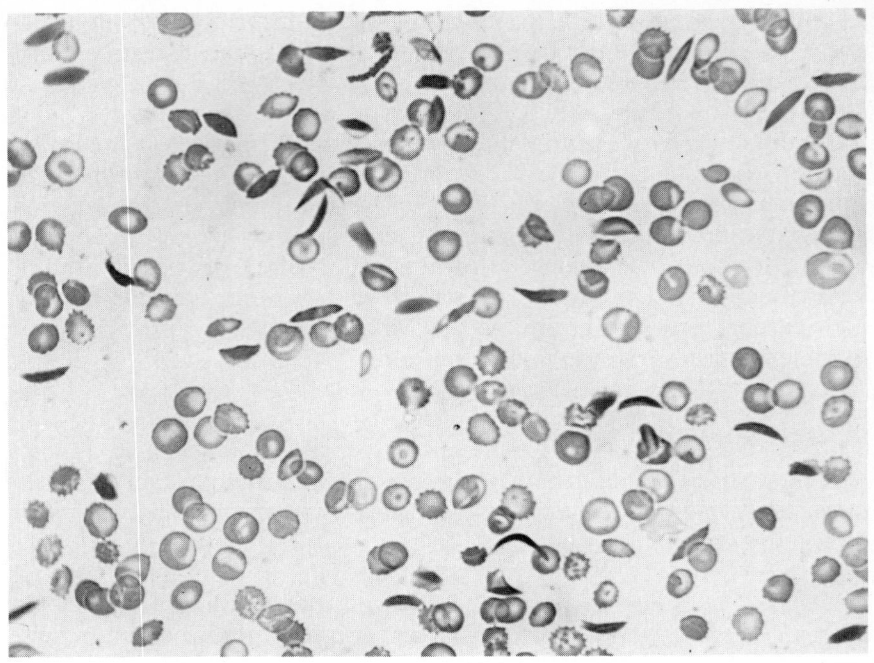

FIGURE 2–9. *Normal and sickle-cell red blood cells.*

only impede circulation by blocking capillaries, but also cannot properly perform the function of carrying oxygen and carbon dioxide to and from the tissues.

In heterozygotes, Hb^AHb^S, some red cells contain hemoglobin A, others hemoglobin S. Because both types of hemoglobin are produced, rather than a single intermediate form, this is another case of codominance. Microscopic examination of heterozygotes' blood under low oxygen tension discloses both normal and sickled erythrocytes. Under normal conditions heterozygotes manifest none of the severe symptoms of Hb^SHb^S persons, though they may suffer some periodic discomfort and even develop anemia after a time at high altitudes. Genotypes and phenotypes in this disorder are as follows:

$\boldsymbol{Hb^AHb^A}$ normal (hemoglobin A only; no sickling of red cells)

$\boldsymbol{Hb^AHb^S}$ sickle-cell trait (hemoglobins A and S; sickling under reduced oxygen tensions)

$\boldsymbol{Hb^SHb^S}$ sickle-cell anemia (hemoglobin S only; sickling under normal oxygen tension)

The chemistry of sickle-cell and other hemoglobin disorders, as well as the operation of the alleles involved, will be examined in more detail in later chapters.

If a large number of $Hb^AHb^S \times Hb^AHb^S$ marriages are studied, children are found to fall into a ratio of 1 homozygous normal:2 heterozygous sickle-cell trait:1 homozygous sickle-cell anemia. In the past most or all of those with sickle-cell anemia died, often before reaching reproductive age. With modern methods of treatment, however, these individuals often live well beyond reproductive age. When the Hb^SHb^S genotype proves fatal, the phenotypic ratio becomes 2:1. Whenever a 2:1 ratio occurs, a lethal allele may be suspected.

Corn. Most familiar plants are characteristically autotrophic; hence they are able to manufacture all required food from carbon dioxide and water in the process of photosynthesis. For this process, in all but the few autotrophic bacteria, the presence of a green, light-absorbing pigment, chlorophyll, is required. In corn (*Zea mays*) several pairs of genes affecting chlorophyll production have been described in the literature. One such gene, designated *G*, for normal chlorophyll production, is completely dominant to its allele *g*, so, as a result *G*— plants contain chlorophyll and are photosynthetic. On the other hand, *gg* plants produce no chlorophyll and are yellowish white (because they still produce the yellow carotenoid pigments). (See Figure 2–10.)

On the average, about one-fourth of the progeny of two heterozygous parents are thus without chlorophyll, and seedlings show the classic 3:1 ratio. In corn, germination and early seedling development take place at the expense of a food-storage tissue, the endosperm, in the grain. In normal green plants, by the time this food reserve has been exhausted (in about ten to fourteen days' time), the seedling has developed a sufficient root system and amount of green tissue to be physiologically independent. Therefore, in the cross of two heterozygotes, the initial 3:1 phenotypic ratio becomes what might be termed a 3:0 or a "1:0" ratio after some two weeks:

P	green	×	green		
	Gg		*Gg*		
F_1	$\frac{1}{4}$ green	+	$\frac{2}{4}$ green	+	$\frac{1}{4}$ nongreen (die)
	GG		*Gg*		*gg*

FIGURE 2–10. *A lethal gene in corn. The chlorophyll-less plants are unable to manufacture their own food and will die as soon as food stored in the grain has been consumed. Photo shows progeny plants of the cross heterozygous green × heterozygous green; there are 36 green and 12 "albino" seedlings in the flat, a perfect 3:1 ratio.*

Note that after allele *g* has exerted its lethal effect, the genotypic ratio is converted from 1:2:1 to 2:1, as in the case of sickle-cell in humans. So whereas homozygous green plants, for example, initially make up one-fourth of the F_1, they later comprise *one-third* of the surviving progeny because of the death of *gg* individuals. This 2:1 genotypic ratio, or the occurrence of only one phenotypic class where two would be expected in a 3:1 ratio, is a clear indication of a (recessive) lethal allele.

In the instance just described, the occurrence and action of the lethal allele is easily discerned because the death of the homozygous recessives takes place only after nearly two weeks of growth following germination. Yet the effect of other lethals might conceivably be produced at almost any time between syngamy (i.e., in the zygote state) on through embryogeny to a very late point in life. Obviously, the effect of lethals killing very late in life might be hard to separate from other causes of death.

Mouse. A classic case of a recessive lethal that kills early in embryo development, and one of the first to be reported in the literature, was "yellow" in mice. Early in this century L. Cuénot noted that black × black always produced black offspring, but that yellow × black produced yellow and black in a 1:1 ratio. He concluded correctly that yellow is heterozygous. Yet crosses of yellow × yellow always produced yellow and black in a ratio of 2:1, with litters of such matings being about one-fourth smaller than those from other crosses. Letting A^Y represent an allele for yellow and *a* one for black as Cuénot did, Cuénot's crosses may be represented as follows:

Testcross:

P	yellow	×	black
	A^Ya		aa
F_1	$\frac{1}{2}$ yellow	+	$\frac{1}{2}$ black
	A^Ya		aa

or, **simple monohybrid:**

P	yellow	×	yellow		
	A^Ya		A^Ya		
F_1	$\frac{1}{4}\ A^YA^Y$	+	$\frac{2}{4}\ A^Ya$	+	$\frac{1}{4}\ aa$
	die		yellow		black

Thus allele A^Y appears to be dominant with respect to coat color but recessive with respect to lethality. For some time the nature of the action of A^YA^Y was unknown. Although selective fertilization was considered a possible factor, it was suggested that A^YA^Y animals are conceived but die soon after. G. G. Robertson and later G. J. Eaton and M. M. Green were able to demonstrate that about one-fourth of the embryos of pregnant yellow (A^Ya) females that had been mated to yellow males did die soon after conception. In this case death usually occurs at gastrulation.

Fowl. The well-known "creeper" condition in fowl falls into the same category. Creeper birds have much shortened and deformed legs and wings, giving them a squatty appearance and creeping gait. Creeper × creeper always produces two creeper to one normal with the homozygous creepers having such gross deformities (greater than in heterozygotes) that death occurs during incubation, generally about the fourth day:

$$
\begin{array}{llllll}
P & \text{creeper} & \times & \text{creeper} & & \\
 & c^1c^2 & & c^1c^2 & & \\
F_1 & \tfrac{1}{4}\ \text{normal} & + & \tfrac{2}{4}\ \text{creeper} & + & \tfrac{1}{4}\ \text{(die)} \\
 & c^1c^1 & & c^1c^2 & & c^2c^2
\end{array}
$$

It has been shown that the creeper allele produces general retardation of embryo growth, with the effect being greatest at the stage of limb bud formation.

A dominant lethal in humans? Thus far, only cases in which lethality itself is recessive have been examined. Reasoning *a priori,* there is no reason not to expect lethal dominant alleles, provided death of the affected individual occurs somewhat after reproduction has taken place. Just such a situation is illustrated by Huntington's disease, a hereditary disorder in humans characterized by involuntary jerking of parts of the body and a progressive degeneration of the nervous system, accompanied by gradual mental and physical deterioration. The mean age of onset of these symptoms is about thirty-five to forty (although it has been reported to take place as early as the first decade of life and as late as sixty to seventy), by which time many afflicted persons have produced children. Affected offspring always have at least one parent who exhibits symptoms of the disease sooner or later. The variability in age of onset (which may be due to still other alleles) makes identity of persons carrying this dominant allele difficult to ascertain in some cases and impossible when parents and grandparents die at relatively early ages from other causes. Although Huntington's disease might not be considered as lethal in all cases in terms of *reproduction,* it is eventually lethal because it does cause death.

The lethals discussed in this chapter are summarized in Table 2–2.

TABLE 2–2. **Comparison of Certain Lethal Genes**

		Dominance		*F_1 phenotypic ratio of heterozygote × heterozygote*		
Organism	*Phenotype*	*Phenotype*	*Lethality*	*Before lethality occurs*	*After lethality occurs*	*Age at death*
Human	Sickle-cell	Codom.	Recessive	1 normal:2 sickle-cell trait:1 sickle-cell anemia*	1 normal:2 sickle-cell trait	Adolescence or later
Human	Tay-Sachs	Recessive	Recessive	3 normal:1 Tay-Sachs	All normal	5 years
Corn	Albinism	Recessive	Recessive	3 green:1 albino	All green ("1:0")	10–14 days
Mouse	Yellow	Inc. dom.?	Recessive	Unknown	2 yellow:1 black	Postzygote
Fowl	Creeper	Inc. dom.	Recessive	Unknown	2 creeper:1 normal	Early embryo
Human	Huntington's disease	Dominant	Dominant	3 diseased: 1 normal	All normal (1:0)	Middle age but variable

*The 1:2:1 ratio listed under F_1 phenotype for sickle-cell anemia is based on microscopic examination of blood samples.

PROBLEMS

2–1 Make a list of several phenotypic characters in your family for as many generations and individuals as possible. Consider traits singly and try to determine the kind of inheritance involved. Save any that seem not to fit patterns developed in this chapter until somewhat later on.

2–2 In human beings, a downward-pointed frontal hairline ("widow's peak") is a heritable trait. A person with widow's peak always has at least one parent who also has this trait, whereas persons with a straight frontal hairline may occur in families in which one or even both parents have widow's peaks. When both parents have a straight frontal hairline, all children also have a straight hairline. Using W and w to symbolize alleles for this trait, what is the genotype of an individual *without* widow's peak?

2–3 Some individuals have one whorl of hair on the back of the head (because of a completely dominant allele), whereas others have two. In the following pedigrees solid symbols represent one whorl, open symbols two:

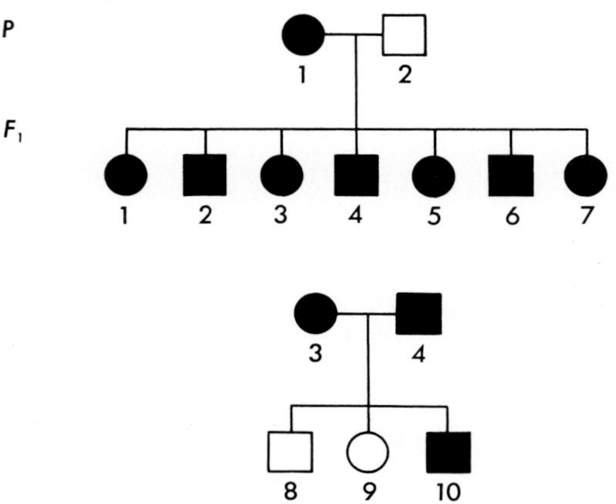

(a) Using the first letter of the alphabet, give the probable genotype of (1) P 2; (2) P 3; (3) F_1 7; (4) F_1 8; (5) F_1 10.
(b) What should be the phenotypic ratio of the progeny produced by the marriage of F_1 7 × F_1 8, assuming a large family?

2–4 Radishes may have any of three shapes: long, spherical, or ovoid. Crosses of long × spherical always produce an F_1 consisting only of ovoid radishes. In terms of the alleles discussed in this chapter, how should the pair of alleles involved here be designated?

2–5 Albinism, the total lack of pigment, is due to a recessive allele. A man and a woman plan to marry and wish to know the probability that they will have any albino children. What could you tell them if (a) both are normally pigmented, but each has one albino parent; (b) the man is an albino, the woman is normal, but her father is an albino; (c) the man is an albino and the woman's family includes no albinos for at least three generations?

2–6 Cystic fibrosis of the pancreas is an inherited condition characterized by faulty metabolism of fats. Affected individuals are homozygous for the allele responsible and ordinarily die in childhood. Such individuals also produce a much higher concentration of chlorides in their sweat than homozygous normals, whereas heterozygotes, who live a normal lifespan, have an intermediate chloride concentration. In terms of the types of alleles discussed in this chapter, how would you designate (a) this pair of alleles as to dominance; (b) the allele for cystic fibrosis?

2–7 One study has estimated the number of persons in the United States who are heterozygous for the cystic allele at about 8 million. A couple planning marriage decide to have a sweat test because a brother of the man died in infancy from the disorder. The tests show the man to be heterozygous and the woman homozygous normal. (a) What is the chance that any of their children might have cystic fibrosis? (b) Could any of the couple's grandchildren have it?

2–8 In a certain plant the cross purple × blue yields purple- and blue-flowered progeny in equal proportions, but blue × blue always gives rise only to blue. (a) What does this tell you about the genotypes of blue- and purple-flowered plants? (b) Which phenotype is dominant?

2–9 In cattle, the cross horned × hornless sometimes produces only hornless offspring, and in other crosses horned and hornless appear in equal numbers. A cattle owner has a large herd of hornless cattle in which horned progeny occasionally appear. He has red, roan, and white animals and wishes to establish a pure-breeding line of red hornless animals. How should he proceed?

2–10 In corn, resistance to a certain fungus is conferred by allele h, which is completely recessive to its allele H for susceptibility. If a resistant plant (♀) is pollinated by a homozygous susceptible plant (♂), give the genotypes for (a) the pistillate parent, (b) the staminate parent, (c) sperm, (d) egg, (e) polar nucleus, (f) F_1 embryo, (g) endosperm surrounding the F_1 embryo, (h) epidermis of kernels that contain the F_1 embryos. You may wish to consult Appendix B–6, Figure B–5, before answering.

2–11 Two curly-winged fruit flies (*Drosophila*) are mated; the F_1 consists of 341 curly and 162 normal. Explain.

2–12 Using the sixth letter of the alphabet, give the genotype of each of the following persons in Figure 1–4: I-1, I-2, II-1, II-2, II-3, II-4, III-1, and III-2.

2–13 The marriage between II-1 and II-2 in Figure 1–4 represents what sort of genetic cross?

2–14 Using the first letter of the alphabet, give the genotype of each of the following persons from Figure 1–5: I-1, II-1, III-4.

2–15 No ancestry information is given for II-1 in Figure 1–5; how do you justify your designation of her genotype?

2–16 Rh negative children (those not producing rhesus antigen D) may be born to either Rh positive or Rh negative parents, but Rh positive children always have at least one Rh positive parent. Which phenotype is due to a dominant gene?

2–17 A normal couple has five children, two of whom suffer from a somewhat uncommon genetic disorder that has, however, appeared sporadically in this familial line. (a) What kind of gene is responsible in this case: completely dominant, completely recessive, codominant, or incompletely dominant? (b) What does the occurrence of affected children in this family tell you about the parental genotypes?

2–18 Among the many antigenic substances that may occur on human red blood cells are two, known as M and N. Individuals' phenotypes are referred to as M, N, or MN depending on which antigen or antigens they produce (one, the other, or both). In the case of an M father and an N mother, children are always MN. What kind of genes are responsible for these antigens? (Choose from among the choices given in Problem 2–17a.)

2–19 **Thalassemia** is a hemoglobin defect in humans; it occurs in two forms, (a) **thalassemia minor,** in which erythrocytes are small (**microcytic**) and increased in number (**polycythemic**), but health is essentially normal, and (b) **thalassemia major,** characterized by early, severe anemia, enlargement of the spleen, microcytes, and polycythemia, among other symptoms. The latter form usually culminates in death before reproductive age. From the following hypothetical pedigree, determine the mode of inheritance:

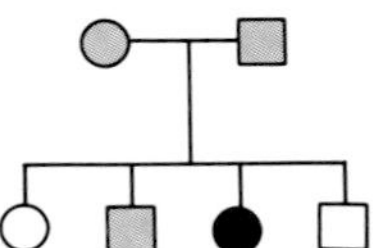

Clear symbols represent normal persons, tinted symbols thalassemia minor, and the blackened thalassemia major.

2–20 In families in which both parents have sickle-cell trait, what is the probability of their having (a) a child with sickle-cell trait, (b) a normal child?

2–21 A normal individual, Hb^AHb^A, receives a transfusion of blood from a person who has sickle-cell trait. Would this transfusion transmit the sickle-cell trait to the recipient? Explain.

2–22 Phenylketonuria (PKU) is a heritable condition in humans involving inability to metabolize the amino acid phenylalanine because of failure to produce the enzyme phenylalanine hydroxylase. If not diagnosed and treated very soon after birth, PKUs develop such severe mental retardation (among other symptoms) that they almost never reproduce. PKU children, therefore, are born to parents who are not PKUs. The normal brother of a PKU seeks the advice of a genetic counselor before a contemplated marriage. (a) What is the probability that he is heterozygous? (b) If PKU occurs once in 25,000 live births in the United States, and he contemplates marrying a normal woman in whose family no cases of PKU have occurred since her ancestors came over on the *Mayflower*, what is the probability of their having a PKU child? (c) What else might the genetic counselor consider telling this couple?

2–23 In **juvenile amaurotic idiocy,** children are normal until about age six. Subsequently there is a progressive decline in mental development, an impairment of vision leading to blindness, and muscular degeneration, culminating in death, usually before age twenty. The trait may appear in families in which both parents are completely normal. A couple, age twenty-five, planning marriage, are first cousins; siblings of both parties have died of the disorder. (a) Knowing no more than you do at this point about the genotypes of these two persons, what is the probability of *both* of them being heterozygous? (b) On the basis of your answer to part (a), what could you tell them about the chance of their having an affected child? (c) Heterozygotes can be detected by an increase in vacuolization of **lymphocytes** (a type of white blood cell). If such a test should disclose that both persons are actually heterozygotes, what then could you say about the probability of their having an affected child?

2–24 **Dentinogenesis imperfecta** (opalescent dentine) is an hereditary tooth disorder characterized by defective dentine that splits from stress due to normal biting and chewing, and occurs in about 1 in 8,000 persons. Teeth of affected persons range in color from amber to opalescent blue. Affected children occur only in families where one or both parents have opalescent dentine; children with normal dentine may have parents who are both normal, or one normal and one affected, or both affected. (a) Is the gene for this condition completely dominant, codominant, incompletely dominant, or recessive? (b) A man with opalescent dentine marries a woman who has normal teeth; one of his parents was affected, one was not, but neither parent of the woman had opalescent dentine. What ratio of affected to unaffected would be found in the children of a large number of such families? (c) What kind of cross does the marriage in part (b) represent?

2–25 In addition to the M and N antigens of human red blood cells referred to in this chapter, two more human blood antigens, A and B (among many others) may occur on the surfaces of the red blood cells. Persons belong to any one of four blood groups: A, having only antigen A; B, having only antigen B; AB, having both antigens; and O, having neither of these antigens. Examine the following blood group pedigrees (a + sign in the progeny indicates a child of that group can occur):

P	Progeny			
	A	B	AB	O
A × A	+			+
A × O	+			+
A × B	+	+	+	+
B × B		+		+
B × O		+		+
AB × A	+	+	+	
AB × B	+	+	+	
AB × O	+	+		
AB × AB	+	+	+	
O × O				+

Assuming three alleles, one for production of antigen A, another for production of antigen B, and a third that results in failure to produce either antigen, what is the dominance relationship among these three alleles? (This question deals with a genetic situation that will be taken up in a later chapter, but you might wish to try your powers of inductive reasoning at this point.)

DIHYBRID INHERITANCE

CHAPTER 3

The behavior of a single pair of alleles through several generations has been followed in the preceding chapter. It will be interesting now to examine the course of two or more independently segregating pairs of alleles through a number of generations. Behavior of nonindependently segregating alleles will be considered in Chapter 6.

CLASSIC TWO-PAIR RATIOS

Complete Dominance in Two Pairs

In addition to the leaf margin trait of *Coleus* considered in Chapter 2, another vegetative character involves the venation pattern, easily observed on lower surfaces of leaves. A typically encountered vein arrangement is the one shown in Figure 3–1. Here a single midvein branches in standard pinnate ("featherlike") fashion. An alternative expression is the highly irregular arrangement seen in Figure 3–2. For convenience let the terms *regular* and *irregular*, respectively, refer to these two phenotypes. A simple monohybrid cross involving these two characters can easily demonstrate that *irregular* is completely dominant to *regular*.

Therefore a cross of a doubly homozygous *deep irregular* with a *shallow regular* individual (using D and d again to represent the deep and shallow genotypes, and I and i to denote irregular and regular) may be represented as follows:

FIGURE 3–1. *Regular venation pattern, a trait caused by a recessive gene, in* Coleus.

P	deep irregular	×	shallow regular
	DDII		*ddii*
P gametes	ⓓ*DI*		ⓓ*di*
F_1		deep irregular *DdIi*	

The F_1 individuals in this instance are referred to as dihybrid individuals because they are heterozygous for each of the two pairs of alleles.

What will be the result of crossing two members of this F_1 to produce an F_2? The outcome of this cross can be predicted on the basis of the theses developed thus far. Recalling assumptions already made and the experimental evidence for them, when *considering one pair of alleles at a time,* the phenotypic result will be the usual 3:1 ratio:

F_1		$Dd \times Dd$
F_1 *gametes*	eggs	$\frac{1}{2}$ (D) + $\frac{1}{2}$ (d)
	sperms	$\frac{1}{2}$ (D) + $\frac{1}{2}$ (d)
F_2 *genotypes*		$\frac{1}{4}\,DD + \frac{2}{4}\,Dd + \frac{1}{4}\,dd$
F_2 *phenotypes*		$\frac{3}{4}$ deep + $\frac{1}{4}$ shallow

Likewise, *Ii* × *Ii* will produce the same $\frac{3}{4}$ irregular (*I*—): $\frac{1}{4}$ regular (*ii*) progeny. To determine the *phenotypic ratio* of the dihybrid cross *DdIi* ×

DdIi, the simplest and most logical (though not necessarily correct) assumption is that either the deep or shallow phenotype may be associated *at random* with either the irregular or regular phenotype if the two pairs of alleles involved segregate independently. That is, in terms of the product law of probability noted in Chapter 2, the phenotypic ratio expected here is the *arithmetic product* of its two component monohybrid ratios, thus

$$Dd \times Dd \text{ yields: } \tfrac{3}{4} \text{ deep} + \tfrac{1}{4} \text{ shallow}$$
$$Ii \times Ii \text{ yields: } \tfrac{3}{4} \text{ irregular} + \tfrac{1}{4} \text{ regular}$$

The F_2 result, based on random combination of phenotypic classes, is as follows: ($\frac{3}{4}$ deep $\times$ $\frac{3}{4}$ irregular) = $\frac{9}{16}$ deep irregular + ($\frac{1}{4}$ shallow $\times$ $\frac{3}{4}$ irregular) = $\frac{3}{16}$ shallow irregular + ($\frac{3}{4}$ deep $\times$ $\frac{1}{4}$ regular) = $\frac{3}{16}$ deep regular + ($\frac{1}{4}$ shallow $\times$ $\frac{1}{4}$ regular) = $\frac{1}{16}$ shallow regular.

This 9:3:3:1 ratio is the same as Mendel observed in his work with peas and is one of the classical phenotypic ratios. On this basis, the *genotypic* ratio of this cross is readily determined:

$$\begin{array}{rl} Dd \times Dd \text{ yields:} & \tfrac{1}{4}DD + \tfrac{2}{4}Dd + \tfrac{1}{4}dd \\ Ii \times Ii \text{ yields:} & \tfrac{1}{4}II + \tfrac{2}{4}Ii + \tfrac{1}{4}ii \\ \hline F_2 \text{ genotypes:} & \tfrac{1}{16}DDII + \tfrac{2}{16}DdII + \tfrac{1}{16}ddII \\ & + \tfrac{2}{16}DDIi + \tfrac{4}{16}DdIi + \tfrac{2}{16}ddIi \\ & + \tfrac{1}{16}DDii + \tfrac{2}{16}Ddii + \tfrac{1}{16}ddii \end{array}$$

FIGURE 3–2. *Irregular venation pattern, produced by the dominant allele of the gene for regular venation in* Coleus.

Closer examination of this 1:2:1:2:4:2:1:2:1 genotypic ratio reveals the 9:3:3:1 phenotypic ratio:

$$
\begin{array}{lclclcl}
\frac{1}{16}\,DDII & & \frac{1}{16}\,ddII & & \frac{1}{16}\,Ddii & & \frac{1}{16}\,ddii \\
\frac{2}{16}\,DdII & & \frac{2}{16}\,ddIi & & \frac{2}{16}\,Ddii & & \\
\frac{2}{16}\,DDIi & & & & & & \\
\frac{4}{16}\,DdIi & + & & + & & + & \\
\hline
\frac{9}{16}\,D{-}I{-} & & \frac{3}{16}\,ddI{-} & & \frac{3}{16}\,D{-}ii & & \frac{1}{16}\,ddii \\
\text{deep} & & \text{shallow} & & \text{deep} & & \text{shallow} \\
\text{irregular} & & \text{irregular} & & \text{regular} & & \text{regular}
\end{array}
$$

Alternatively, this 9:3:3:1 ratio could have been calculated more quickly in this fashion:

$$
\begin{array}{c}
Dd \times Dd \text{ yields } \frac{3}{4}\,D{-} + \frac{1}{4}\,dd \\
Ii \times Ii \text{ yields } \frac{3}{4}\,I{-} + \frac{3}{4}\,ii \\
\hline
\frac{3}{4}\,D{-} \times \frac{3}{4}\,I{-} \text{ yields } \frac{9}{16}\,D{-}\ I{-} \\
\frac{3}{4}\,D{-} \times \frac{1}{4}\,ii \text{ yields } \frac{3}{16}\,D{-}\ ii \\
\frac{1}{4}\,dd \times \frac{3}{4}\,I{-} \text{ yields } \frac{3}{16}\,ddI{-} \\
\frac{1}{4}\,dd \times \frac{1}{4}\,ii \text{ yields } \frac{1}{16}\,ddii
\end{array}
$$

to get the same result as obtained by collecting and summing the nine genotypes immediately above. *This ratio will be useful in later cases:*

$$
\begin{array}{l}
9\ D{-}I{-} \\
3\ ddI{-} \\
3\ D{-}ii \\
1\ ddii
\end{array}
$$

A 9:3:3:1 phenotypic ratio in corn is shown in Figure 3–3.

It is imperative to recognize the importance of one tacit assumption involved in this method of calculating dihybrid ratios. From the behavior of chromosomes at nuclear division, to be discussed in the next chapter, it will be clear that the calculation on page 27 is based on the expectation that, in gametes, association of *D* with *I* or *i*, and of *d* with *I* or *i*, is random. This then implies *equal* numbers of four possible gamete genotypes: *DI, dI, Di,* and *di* from a *DdIi* individual. The results of the cross *DdIi* × *DdIi* could be calculated either as on page 27 or by expected gamete genotypes:

$$
\begin{array}{ll}
DdIi \text{ eggs:} & \frac{1}{4}\,\textcircled{DI} + \frac{1}{4}\,\textcircled{dI} + \frac{1}{4}\,\textcircled{Di} + \frac{1}{4}\,\textcircled{di} \\
DdIi \text{ sperms:} & \frac{1}{4}\,\textcircled{DI} + \frac{1}{4}\,\textcircled{dI} + \frac{1}{4}\,\textcircled{Di} + \frac{1}{4}\,\textcircled{di} \\
\hline
\end{array}
$$

A useful exercise would be to verify that, if gamete union is also random, the phenotypic and genotypic ratios arrived at in this way will be identical with those calculated on page 27. Production of four kinds of

FIGURE 3–3. *A 9:3:3:1 purple–starchy: purple–sweet:white–starchy:white–sweet F_2 phenotypic ratio in corn. Sweet kernels are shriveled; starchy ones are plump.*

gametes in equal number by a doubly heterozygous individual will occur if each pair of alleles is on a different pair of chromosomes. That is, each pair of alleles here behaves exactly as in a one-pair cross. Cytological justification for this conclusion will be detailed in the next chapter.

Mendel's Dihybrid Crosses

Some of Mendel's experiments with the garden pea involved crosses between plants producing yellow round seeds and those giving rise to green wrinkled seeds. From his monohybrid crosses it was clear to Mendel that yellow and round were dominant traits. Letting *G* represent yellow and *W* round, we can represent this cross of Mendel's as

P round yellow × wrinkled green
WWGG *wwgg*

F_1 all round yellow
WwGg

F_2 315 round yellow + 108 round green + 101 wrinkled yellow + 32 wrinkled green
G—W— *W—gg* *wwG—* *wwgg*

(The dash in three of the genotypes signifies that the allele so represented may be either a dominant or a recessive.) This observed ratio closely approximates a 9:3:3:1 ratio. The cross and its results are shown in Figure 3–4. It was clear to Mendel that the allelic pair for seed color was assorting independently of the allelic pair for wrinkled/round seeds. Thus, Mendel concluded that *members of different pairs of alleles assort independently into the gametes.* This is Mendel's second law, the **principle of independent assortment.** Exceptions to this law disclose some facts about gene locations and will be discussed in a later chapter.

Gamete and Zygote Combinations

As seen earlier, a monohybrid such as *Dd* produces two kinds of gametes (*D* and *d*) in equal numbers, which can combine by syngamy to form three different zygote genotypes (*DD, Dd, dd*), from which two phenotypes (deep and shallow) will be discernible in the progeny. Likewise, a doubly heterozygous individual (*DdIi*) produces four kinds of gametes (*DI, Di, dI, di*), again in equal numbers if the two pairs of alleles are located on different chromosome pairs. With random syngamy, nine different zygote genotypes are produced:

DDII *Ddii*
DDIi *ddII*
DdII *ddIi*
DdIi *ddii*
Ddii

from which four phenotypes (deep irregular, deep regular, shallow irregular, shallow regular) will be apparent in the young seedlings.

A third pair of alleles in *Coleus* (Figure 3–1) can be symbolized as follows:

W (no white area at the base of the leaf blade)

w (white area at the base of the leaf blade)

What gamete combinations can be produced by the trihybrid *DdIiWw?* If each of the three pairs of alleles is on a different chromosome

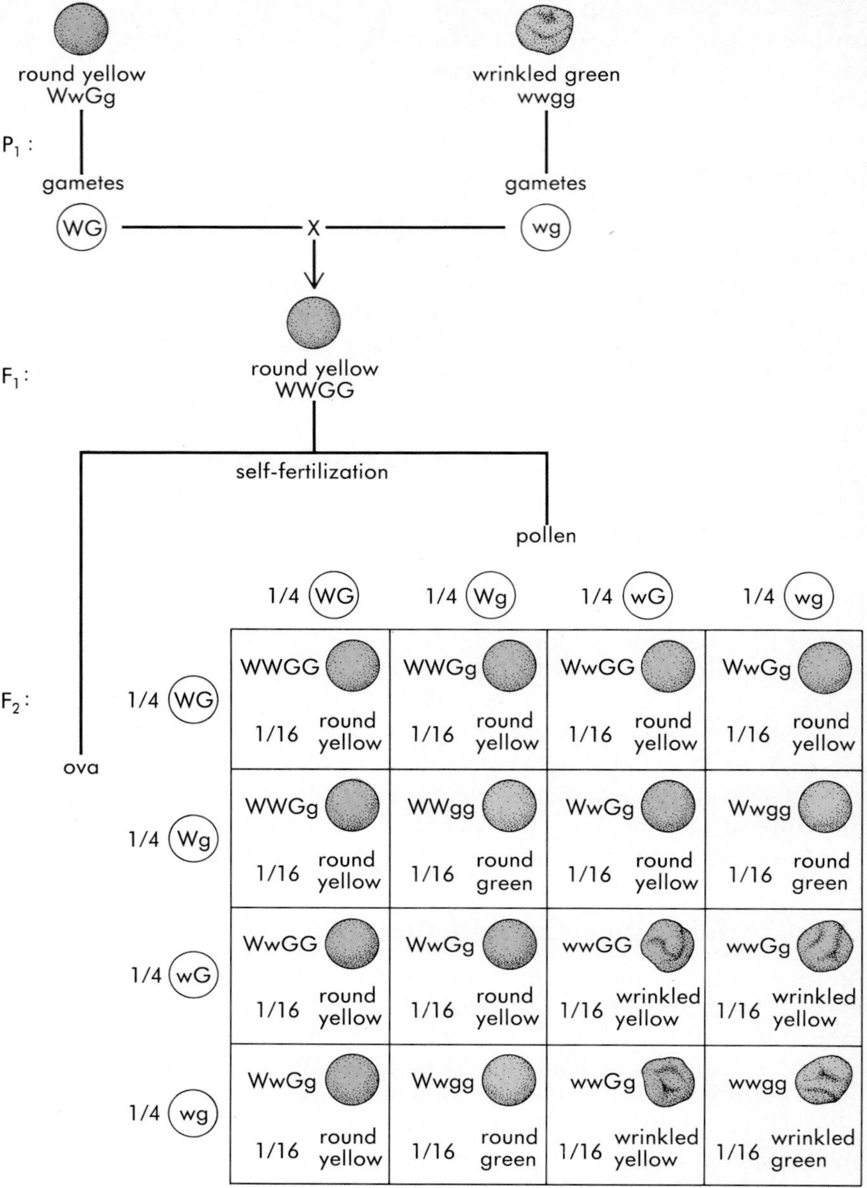

FIGURE 3–4. *Summary of Mendel's results with the dihybrid cross showing independent assortment of the two pairs of alleles resulting in the 9:3:3:1 phenotypic ratio.*

pair, then either allele of any pair can combine with either allele of any other. A simple application of the probability method indicates eight possible gamete genotypes. As just seen, the dihybrid *DdIi* produces four gamete types, $\frac{1}{4}$ *DI* + $\frac{1}{4}$ *Di* + $\frac{1}{4}$ *dI* + $\frac{1}{4}$ *di*. With the addition of the *W*, *w* pair, two additional gamete types, *W* and *w*, become possible, and may combine with any of the dihybrid gamete genotypes:

$$\frac{1}{4}\,\textcircled{\mathit{DI}} + \frac{1}{4}\,\textcircled{\mathit{Di}} + \frac{1}{4}\,\textcircled{\mathit{dI}} + \frac{1}{4}\,\textcircled{\mathit{di}}$$

$$\frac{1}{2}\,\textcircled{\mathit{W}} + \frac{1}{2}\,\textcircled{\mathit{w}}$$

$\frac{1}{8}$ each of *DIW, DiW, dIW, diW, DIw, Diw, dIw, diw*

Again applying the probability method used for monohybrid and dihybrid cases, these eight gamete genotypes may be expected to combine

TABLE 3–1. Parental Gamete Genotypes, Gamete Combinations, and Progeny Phenotypes and Genotypes (Dominance Complete in All Pairs)

Number of pairs of heterozygous alleles	*Number of gamete genotypes*	*Number of progeny phenotypes*	*Number of progeny genotypes*	*Number of possible combinations of gametes*
1	2	2	3	4
2	4	4	9	16
3	8	8	27	64
4	16	16	81	256
n	2^n	2^n	3^n	4^n

randomly into twenty-seven zygote genotypes, producing eight phenotypic classes in the progeny.

Thus, considering the number of possible gamete genotypes per pair of heterozygous alleles in which dominance is complete and that are located on different chromosome pairs, the emergence of these mathematical relationships can be detected (where n = number of pairs of chromosomes with single gene differences):

number of gamete genotypes produced by parents $= 2^n$
number of progeny phenotypic classes $= 2^n$
number of progeny genotypic classes $= 3^n$

These relationships are summarized and extended in Table 3–1.

Testcross

The testcross is a useful way of determining homozygosity or heterozygosity of a dominant phenotype in single-gene cases and can be equally valuable in determining genotypes in situations with two or more pairs of genes. In a two-pair cross, in which each allele is on a different chromosome pair (i.e., the genes are not linked), the resulting progeny ratio is the product of two one-pair ratios. Thus, as we have seen, a monohybrid testcross ratio is 1:1, and a dihybrid testcross produces a 1:1:1:1 ratio when one parent is heterozygous for both allelic pairs:

P	*DdIi* × *ddii*						
P gametes	$\frac{1}{4}$ (*DI*)	+	$\frac{1}{4}$ (*dI*)	+	$\frac{1}{4}$ (*Di*)	+	$\frac{1}{4}$ (*di*)
		×	1 (*di*)				
F_1	$\frac{1}{4}$ *DdIi* deep irregular	+	$\frac{1}{4}$ *ddIi* shallow irregular	+	$\frac{1}{4}$ *Ddii* deep regular	+	$\frac{1}{4}$ *ddii* shallow regular

The testcross is a very useful technique in mapping gene locations on chromosomes, as will be discussed in Chapter 6.

With the information arrived at thus far, it will be useful to determine testcross ratios in (1) cases in which only one of two pairs is heterozygous and (2) cases such as trihybrids or other polyhybrids. Problems at the end of this chapter explore cases such as these.

MODIFICATIONS OF THE 9:3:3:1 RATIO

Incomplete Dominance

Having examined dihybrid crosses in which dominance is complete in both pairs, we should now determine whether incomplete dominance in one or both pairs has any effect on ratios and on numbers of phenotypic classes.

In tomato, two pairs of alleles, located on different pairs of chromosomes, are

$D-$	tall plant	h^1h^1 hairless stems
dd	dwarf plant	h^1h^2 scattered short hairs
		h^2h^2 very hairy stems

Crossing two individuals of genotype Ddh^1h^2 produces progeny as follows:

$\frac{3}{16}$ tall, hairless
$\frac{6}{16}$ tall, scattered hairs
$\frac{3}{16}$ tall, very hairy
$\frac{1}{16}$ dwarf, hairless
$\frac{2}{16}$ dwarf, scattered hairs
$\frac{1}{16}$ dwarf, very hairy

Note that this is precisely what would be expected. Tall and dwarf segregate in a 3:1 ratio, and hairless, scattered hairs, and very hairy segregate in a 1:2:1 ratio, producing a dihybrid 3:6:3:1:2:1 phenotypic ratio here. In such a case as this, because $Dd \times Dd$ produces two phenotypic classes in the offspring and $h^1h^2 \times h^1h^2$ is responsible for three phenotypic classes, the cross $Ddh^1h^2 \times Ddh^1h^2$ produces offspring of $2 \times 3 = 6$ phenotypic classes. If dominance is incomplete, the number of phenotypic classes is 3^n (where again $n =$ the number of chromosome pairs with a single allele difference). In a dihybrid situation in which one pair of genes exhibits complete dominance and the other incomplete dominance, as in this example from tomato, the number of F_1 phenotypic classes from two doubly heterozygous parents is 2×3, as just noted.

Thus, again, incomplete dominance increases the number of phenotypic classes. Further possibilities of this sort are suggested in some of the problems at the end of this chapter.

Epistasis

Mouse. The laboratory mouse occurs in a number of colors and patterns. The wild type (the customary or most frequently encountered phenotype in natural populations, often used as a standard of comparison) or "agouti," is characterized by color-banded hairs in which the part nearest the skin is gray, then a yellow band, and finally the distal part is either black or brown. The wild type has rather obvious selection value in natural surroundings, enhancing concealment of the individual. Two other colors are albino and solid black. In albinos there is a total lack of pigment, producing white hair and pink eyes (the latter results when blood vessel color shows through unpigmented irises).

The cross *black* × *albino* produces a uniform F_1 of agouti that when inbred, in certain instances, results in an F_2 of 9 *agouti*:3 *black*:4 *albino*. The segregation of the F_2 into sixteenths immediately suggests two genes, and this particular ratio implies a 9:3:3:1 ratio in which the $\frac{1}{16}$

class and one of the $\frac{3}{16}$ classes are indistinguishable. These results would then indicate further that the individuals in this case are heterozygous for both alleles. Assume one of the two pairs of genes to include one allele for color production and another allele for color inhibition (the latter perhaps responsible for either a defective enzyme or the absence of a particular enzyme required for a specific intermediate biochemical step in pigment production). Further assume the other genes to include one allele for agouti and one for black. Represent the genes as follows:

A	agouti	*C*	color
a	solid color	*c*	color inhibition

Note that dominance of agouti over black is suggested by the $\frac{9}{16}$ agouti class versus the $\frac{3}{16}$ black in the F_2 (this is tantamount to a 3:1 segregation). On these assumptions, the cross in mouse may be diagramed in this way:

P	black ×	albino
	aaCC	*AAcc*
F_1	*AaCc* agouti	
F_2	$\frac{9}{16}$ $A-C-$	agouti
	$\frac{3}{16}$ $aaC-$	black
	$\frac{3}{16}$ $A-cc$	albino
	$\frac{1}{16}$ $aacc$	albino

In this particular example, allele *c* (which is recessive to its own allele *C*) actually masked the effect of either $A-$ or *aa* so that any $-cc$ individual is albino. Such a gene, which masks the effect of one or both members of a *different* pair of alleles, is said to be **epistatic;** the masked gene or genes may be termed **hypostatic.** Here *c* is epistatic to *A* and *a*, and this case illustrates recessive epistasis because the *recessive* of one is epistatic to another. Note that epistasis is quite different from dominance in that the masking operates between different pairs of alleles rather than between members of one pair.

Clover. An interesting example in white clover (*Trifolium repens*, the clover so frequently seen in lawns), reported in 1943 by S. S. Atwood and J. T. Sullivan, furnishes presumptive evidence that epistasis depends on a gene-enzyme relationship in a series of sequential biochemical steps.

Some strains of white clover test high in hydrocyanic acid (HCN), whereas others test negatively for this substance. HCN content is associated with more vigorous growth; it does not harm cattle eating such varieties. Often the cross *positive* × *negative* results in an F_1 testing uniformly positive for HCN and an F_2 segregating 3 positive:1 negative, suggesting a single pair of alleles with the *positive* dominant trait.

One series of crosses reported by Atwood and Sullivan, however, produced unexpected totals:

P	positive × negative
F_1	positive
F_2	351 positive + 256 negative

A 3:1 expectancy in the F_2 would be approximately 455 positive:152 negative; the actual results differ sufficiently from a 3:1 ratio to cast doubt on its relevance here. (Statistical tests, described in Chapter 5, give objective support to the hypothesis that, although a ratio of 351:256

FIGURE 3–5. *A 9:7 purple:yellow epistatic ratio in corn.*

could occur by chance alone in a 3:1 expectancy, this result is unlikely enough to cause one to look for a better explanation.) Note that these results are very close to a 9:7 ratio, which would be 342 positive:265 negative. A 9:7 ratio immediately suggests an epistatic expression of the 9:3:3:1, which, in turn, indicates two pairs of genes. One such 9:7 ratio (purple:yellow) in corn is illustrated in Figure 3–5.

HCN formation follows a path that may be represented thus:

$$\longrightarrow \text{(precursor)} \xrightarrow[]{\substack{\text{enzyme} \\ \text{“}\alpha\text{”}}} \text{cyanogenic glucoside} \xrightarrow[]{\substack{\text{enzyme} \\ \text{“}\beta\text{”}}} \text{HCN}$$

Each of the conversions indicated by the arrows is enzymatically controlled. Tests of F_2 individuals of the cross just described for (1) HCN, (2) enzyme "β," and (3) cyanogenic glucoside revealed four classes of individuals:

Class	*Glucoside*	*Enzyme "β"*	*HCN*
1	+	+	+
2	0	+	0
3	+	0	0
4	0	0	0

Each of these four classes was then tested for HCN after adding either enzyme "β" or glucoside to their leaf extracts, with the following results:

Class	*Control test for HCN*	*HCN test after adding enzyme "β"*	*HCN test after adding glucoside*
1	+	+	+
2	0	0	+
3	0	+	0
4	0	0	0

Results show that class 1 plants produce both glucoside and enzyme "β" (and, by inference, also enzyme "α"); class 2 plants produce enzyme "β" but no glucoside (hence, by inference, no enzyme "α"); class 3 plants produce glucoside (and, there, enzyme "α") but no enzyme "β"; class 4 plants produce neither enzyme "β" nor glucoside (therefore, by inference, no enzyme "α"). Conclusions are summarized in tabular form:

	Production			
Class	*Enzyme "α"*	*Glucoside*	*Enzyme "β"*	*Substance accumulating*
1	+	+	+	HCN
2	0	0	+	precursor
3	+	+	0	glucoside
4	0	0	0	precursor

It therefore seems highly likely that production of enzymes "α" and "β" is determined by two different genes:

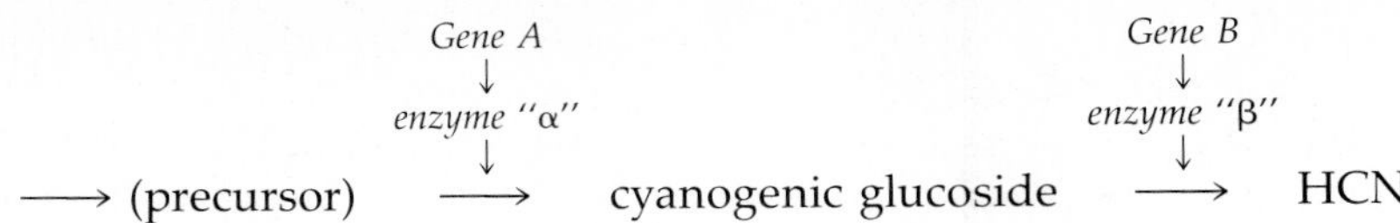

So an individual must have at least one dominant of each of two pairs of alleles, *A* and *B*, in order to carry the process from precursor to HCN. Addition of genotypes to the cross developed on page 33 produces the following:

P positive *AABB* × negative *aabb*

F_1 positive *AaBb*

F_2
- $\frac{9}{16}$ HCN positive *A–B–* ("class 1") (9)
- $\frac{3}{16}$ HCN negative *aaB–* ("class 2")
- $\frac{3}{16}$ HCN negative *A–bb* ("class 3")
- $\frac{1}{16}$ HCN negative *aabb* ("class 4")

("class 2"), ("class 3"), ("class 4") together: (7)

A little reflection will serve to indicate that this situation may be considered in any of the following ways:

9:7	HCN : no HCN
12:4	glucoside : no glucoside
9:3:4	HCN : glucoside : precursor
9:3:3:1	enzymes "α" and "β" : enzyme "β" only : enzyme "α" only : neither enzyme

The ratio chosen depends, of course, on the level of chemical analysis to which consideration is carried. Thus we are considering phenotype in terms of chemical reaction and products, or of enzyme production. Because enzymes are proteins in whole or in part, this suggests again a relationship between gene and enzyme and, therefore, between protein synthesis and phenotype. This will be a promising avenue to explore later in this book.

Incidentally, an explanation for the occurrence of all positive cyanide content progeny from crosses of negative × negative (which is also reported) works out on these bases quite readily:

P negative *aaBB* × negative *AAbb*

F_1 positive *AaBb*

The cross (P) positive × negative that produces an F_2 of 3 positive:1 negative, referred to on page 33, genotypically would be as follows:

P positive *AABB* × negative *aaBB*

F_1 positive *AaBB*

F_2 3 positive *A–BB* : 1 negative *aaBB*

Epistasis (which can operate in any cross that involves two or more genes) can be recognized *by the reduction in number of expected phenotypic classes* in which two or more of the classes become indistinguishable from each other.

Many other ratios are possible and have been reported in the literature. Additional examples are included in the problems at the end of this chapter, as well as in Table 3–2. One of these, however, requires additional explanation. In certain breeds of domestic fowl (e.g., White Leghorn), individuals are white because of a dominant color-inhibiting allele *I*. Even though birds may carry alleles for color, such alleles cannot be expressed in the presence of *I*−. On the other hand, such breeds as the White Silkie are white because of homozygosity for the recessive allele *c*, which blocks synthesis of a necessary pigment precursor.

Crosses of White Leghorn (*IICC*) × White Silkie (*iicc*) (case 6) produce an F_2 ratio of 13:3, as follows:

9 *I*−*C*− white (because of color-inhibitor *I*)
3 *iiC*− colored
3 *I*−*cc* white (because of both *I* and *cc*)
1 *iicc* white (because of *cc*)

The actual color of the *iiC*− individuals depends on the presence of additional genes for particular colors. These are not shown here.

Lethal Alleles

Lethal alleles also reduce the number of expected phenotypic classes in a two-gene cross, but the change is somewhat different from that caused

TABLE 3–2. **Summary of Dihybrid Ratios in the Cross *AABB* × *aabb***

		Case	*AABB*	*AABb*	*AaBB*	*AaBb*	*AAbb*	*Aabb*	*aaBB*	*aaBb*	*aabb*
More than four phenotypic classes	*A* and *B* both incompletely dominant	1	1	2	2	4	1	2	1	2	1
	A incompletely dominant; *B* completely dominant	2	3		6		1	2	3		1
Four phenotypic classes	*A* and *B* both completely dominant (classic ratio)	3	9				3		3		1
Fewer than four phenotypic classes	*aa* epistatic to *B* and *b* Recessive epistasis	4	9				3		4		
	A epistatic to *B* and *b* Dominant epistasis	5	12						3		1
	A epistatic to *B* and *b*; *bb* epistatic to *A* and *a* Dominant and recessive epistasis	6	13*						3		
	aa epistatic to *B* and *b*; *bb* epistatic to *A* and *a* Duplicate recessive epistasis	7	9				7				
	A epistatic to *B* and *b*; *B* epistatic to *A* and *a* Duplicate dominant epistasis	8	15								1
	Duplicate interaction	9	9				6				1

*The 13 is composed of the 12 classes immediately above, plus the one *aabb* from the last column.

by epistasis. For example, consider the case of corn, in which tall ($D-$) and dwarf (dd) phenotypes are known in addition to the green and albino condition described in Chapter 2. Note that the cross $DdGg \times DdGg$ will produce the classic 9:3:3:1 phenotypic ratio of seedlings, but because of the lethal effect of gg, this becomes 9 tall green:3 dwarf green (that is, 3 tall:1 dwarf) after gg has exerted its lethal effect. In humans, for example, consider the dominant lethal for Huntington's disease (Chapter 2) together with another pair of alleles, free ear lobes ($A-$) versus attached ear lobes (aa) (Figure 1–1). A marriage involving two double heterozygotes, $HhAa$ (where H represents the dominant lethal for Huntington's disease and h its recessive allele for "normal"), would, statistically (or actually in collections of family data), produce a 9:3:3:1 ratio initially, which would ultimately become 3 free (normal):1 attached (normal) after the diseased individuals had died. Recall also the 2:1 ratio in the case of the yellow mice trait.

Penetrance and Expressivity

With some genes not all individuals may exhibit phenotypes that are consistent with their genotypes. For example, in humans a dominant allele ($F-$) for bent little finger (camptodactyly) affects the attachment of some of the muscles to a joint of the finger. Although some persons exhibit bent little finger on both hands, many are camptodactylous on only one hand. If, in a given population, only 65 percent of $F-$ persons have bent little finger on both hands, that allele is said to exhibit 65 percent **penetrance.** The dominant for polydactyly (P) also has reduced penetrance in that heterozygotes are not always polydactylous.

Polydactyly is interesting for another reason. Some $P-$ persons have an extra digit on both hands and feet, whereas in others some of the extremities are not affected. Even in those who are polydactylous, the trait may range from complete extra digits to a mere rudiment. Thus, P exhibits variable **expressivity.** This is quite different from penetrance. In *expressivity* it is the *degree* of expression that varies, whereas *penetrance* refers to the inability of a given allele to express itself at all in some individuals. Figure 3–6 summarizes penetrance and expressivity through a hypothetical flower color example in which color is determined by the intensity of the pigment.

Many factors interact in a complex way to produce variable expressivity and reduced penetrance. Some of these causes are intracellular, others intercellular, still others depend on the overall genotype, and many reflect the influence of the environment.

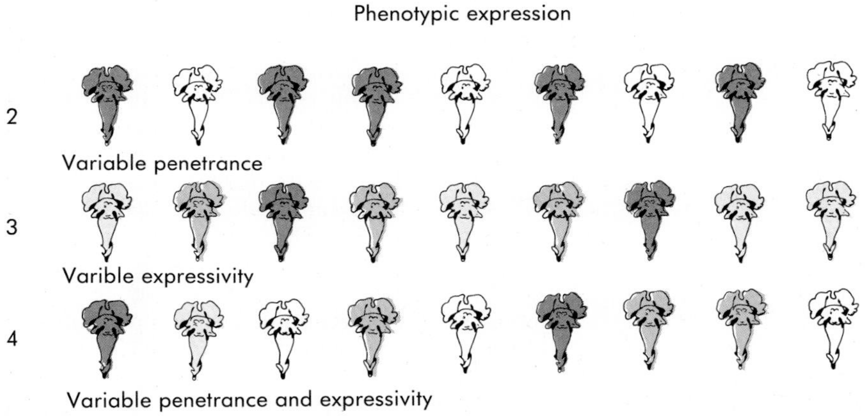

Phenotypic expression

2

Variable penetrance

3

Varible expressivity

4

Variable penetrance and expressivity

FIGURE 3–6. *Penetrance and expressivity shown in a hypothetical flower color example in which color is determined by the intensity of the pigment produced.*

PROBLEMS

3–1 How many different matings can be made in a population in which only one pair of alleles is considered?

3–2 If two *DdIi Coleus* plants are crossed, what fraction of the offspring will be (a) shallow irregular, (b) deep regular, (c) *DDIi*, (d) *ddII*?

3–3 How many progeny phenotypic classes result if two *Coleus* plants (a) of genotype *DdIiWw* are crossed; (b) heterozygous for one pair of alleles (showing complete dominance) on each of its pairs of chromosomes are crossed?

3–4 What is the phenotypic ratio of the testcross (a) *DdII* × *ddii* in *Coleus*, (b) *DdIiWw* × *ddiiww* in *Coleus*?

3–5 How many different gamete classes are produced by the tetrahybrid *AaBbCcDd*?

3–6 If two tetrahybrids like that of the preceding problem are crossed, how many of each of the following can be expected in the progeny: (a) phenotypic classes, (b) genotypic classes?

3–7 In how many ways can gametes of two tetrahybrids (*AaBbCcDd*) be combined to form zygotes?

3–8 How many phenotypic classes are produced by a testcross in which one parent is heterozygous for (a) two genes, (b) three genes, (c) four genes, (d) n genes?

3–9 Suggest a mathematical formula for determining the probability of a completely homozygous recessive progeny individual resulting from selfing a plant heterozygous for n genes, all of which exhibit complete dominance.

3–10 Assume the following sets of genes in human beings:

A–	free ear lobes	h^1h^1	straight hair
aa	attached ear lobes	h^1h^2	wavy hair
		h^2h^2	curly hair
R–	nonred hair	*P–*	polydactylous
rr	red hair	*pp*	nonpolydactylous

(a) A husband and wife, Aah^1h^2PpRr × Aah^1h^2ppRr, want to know the probability of their having a child with attached earlobes, wavy red hair, and is nonpolydactylous. What would you tell them? (b) What is the chance that they might have a child with curly red hair (without regard to the other traits)?

3–11 A few of the many known genes (each on a different chromosome pair) in tomato are

P	smooth-skinned fruit	*p*	"peach" (pubescent fruit)
W	yellow flowers	*w*	white flowers
h^1	hairless stems and leaves	h^2	hairy stems and leaves
C	cut leaf margins	*c*	leaves not cut ("potato")
Cc shows complete dominance of *C*.			

The h^1, h^2 pair shows incomplete dominance. A tetrahybrid smooth, yellow, cut, scattered-hair plant is self-pollinated. (a) How many phenotypic classes can occur in the progeny? (b) What fraction of the offspring can be expected to be peach, white, potato, hairless? (c) What fraction of the progeny can be expected to be peach, white, potato, with scattered hairs?

3–12 You raise 100 tomato plants from seed received from a friend and find 37 red-fruited plants with scattered short hairs on stems and leaves, 19 red hairless, 18 red very hairy, 13 yellow-fruited with scattered short hairs, 7 yellow very hairy, and 6 yellow hairless. Suggest genotypes and phenotypes for the unknown parent plants from which the 100 seeds were obtained.

3–13 Coat in guinea pigs may be either long or short; matings of short × short may produce long-haired progeny, but long × long gives rise only to long. Additionally, coat color may be yellow, cream, or white. The mating cream × cream produces progeny of each of the three colors. Given the following incomplete pedigree:

P	long yellow × short white
F_1	all short –

(a) What is the coat color in the F_1? (b) If members of the F_1 were interbred, what fraction of their progeny would be long cream?

3–14 What progeny phenotypic ratio results from the cross $AaBbc^1c^2$ × $AaBbc^1c^2$ if *bb* individuals die during an early embryo stage?

3–15 In addition to the genes for flower color in snapdragon described in Chapter 2, leaves in this plant may be broad, narrow, or intermediate. From the cross red broad × white narrow this F_2 was obtained: 10 red broad, 20 red intermediate, 10 red narrow, 20 pink broad, 40 pink intermediate, 20 pink narrow, 10 white broad, 20 white intermediate, and 10 white narrow. (a) How many genes are involved and what kind of dominance is demonstrated by this case? (b) Which of these F_2 phenotypic classes can you recognize as homozygous?

3–16 The Christmas poinsettia (*Euphorbia pulcherrima*) produces colored modified leaves (bracts) below the clusters of small flowers. These bracts may be red, pink, or white. Work of Steward and Arisumi (1966) shows that red pigmentation results from a two-step, enzymatically controlled biochemical process from a colorless precursor via an intermediate pink pigment that is converted to a red pigment in the second step. Let *W* represent the completely dominant allele that is responsible for production of the enzyme catalyzing the first step, from colorless to pink, and *P* the completely dominant allele for producing the enzyme bringing about conversion of the pink pigment to a red one. If a *wwpp* white-bracted plant is crossed to a doubly homozygous red-bracted one, what phenotypic ratio can be expected in the F_2?

3–17 Normal hearing depends on the presence of at least one dominant of each of two genes, *D* and *E*. If you examined the collective progeny of a large number of *DdEe* × *DdEe* marriages, what phenotypic ratio would you expect to find?

3–18 In sweet pea the cross white flowers × white flowers produced an F_1 of all purple flowers. An F_2 of 350 white and 450 purple was then obtained. (a) What is the phenotypic ratio in the F_2? Using the first letter of the alphabet and as many more in sequence as needed, give (b) the genotype of the purple F_2, (c) the genotype of the F_1, (d) the genotypes of the two P individuals.

3–19 The fruit of the weed shepherd's purse (*Capsella bursa-pastoris*) is ordinarily heart-shaped in outline and somewhat flattened, but occasionally individuals with ovoid fruits occur. Crosses between pure-breeding heart and ovoid yield all heart in the F_1. Selfing this F_1 produces an F_2 in which 6 percent of the individuals are ovoid. Starting with the first letter of the alphabet, and using as many more as necessary, give the genotypes of (a) ovoid, (b) F_1 heart, (c) F_2 heart.

3–20 In certain breeds of dog the genotype *C*– produces a pigmented coat, whereas *cc* gives rise to a white coat (*not* albino). Another pair of alleles (*B* and *b*) determines the color of the coat in *C*– dogs such that *C*–*B*– animals are black and *C*–*bb* animals are brown. Assume two animals of genotype *CcBb* are crossed. What phenotypic ratio results in the pups of a large number of such matings?

3–21 Instead of the gene action described in the preceding problem, assume *C*– animals are white, whereas *cc* animals have pigmented coats. Assume also that allele pair *B*, *b* produces color (in the presence of *cc*) as in the preceding problem. If two *CcBb* dogs are crossed, what phenotypic ratio is expected in the F_2?

3–22 Give the F_1 phenotypic ratio resulting from the cross of two *CcBb* dogs if *C*– is a color inhibiting genotype, *cc* a genotype producing a pigmented coat, *B*– animals have brown coats, and *bb* black.

3–23 If, in another breed of dogs, *C*– is responsible for a pigmented coat and *cc* for an unpigmented coat, *B*– produces brown coat and *bb* black, what is the phenotypic ratio in the F_1 of the cross *CcBb* × *CcBb*?

3–24 In some plants cyanidin, a red pigment, is synthesized enzymatically from a colorless precursor; delphinidin, a purple pigment, may be made from cyanidin by the enzymatic addition of one –OH group to the cyanidin molecule. In one cross in which these pigments were involved, purple × purple produced F_1 progeny as follows:81 purple, 27 red, and 36 white. (a) How many genes are involved? (b) What is the genotype of the purple parents? (c) What is the genotype of each of the three F_1 phenotypic classes? (Use as many letters of the alphabet as needed, starting with A.)

3–25 In terms of *enzyme production*, instead of flower color, as a phenotypic character in the data of the preceding problem, what is the F_1 phenotypic ratio? In the enzyme catalyzing in the data of Problem 3–24, what is the F_1 phenotypic ratio? The enzyme catalyzing conversion of precursor to cyanidin may be designated enzyme 1, and that controlling the production of delphinidin from cyanidin may be designated enzyme 2.

3–26 In the plants of problem 3–24, one cross of white × red produced all purple progeny, whereas another white × red cross gave rise to 1 purple:2 white:1 red. What were the parental genotypes in each of these two crosses?

3–27 In addition to the round, oval, and long radishes mentioned earlier, radishes may be red, purple, or white in color. Red × white produces progeny all of which are purple. If purple oval were crossed with purple oval, how many pure-breeding types would occur in the progeny?

3–28 In cattle, "short spine" is lethal shortly after birth; it is caused by the homozygous recessive genotype *ss*. Heterozygotes are normal. A series of matings between roan (see page 16) animals heterozygous for the short spine gene produces what phenotypic ratio (a) at birth and (b) after several weeks?

3–29 In the summer squash, fruits may be white, yellow, or green. In one case, the cross of yellow × white produced an F_1 of all white-fruited plants that, when selfed, gave an F_2 segregating 12 white:3 yellow:1 green. (a) Suggest genotypes for the white, yellow, and green phenotypes. (b) Give genotypes of the P, F_1, and F_2 of this cross. Use genotype symbols starting with the first letter of the alphabet.

3–30 Summer squash fruit shape may be disk, sphere, or elongate. The cross of sphere × sphere produced an F_1 with all disk-shaped fruits. Selfing the F_1 gave 9 disk:6 sphere:1 elongate. (a) Suggest genotypes for disk, sphere, and elongate. (b) Give genotypes of the P, F_1, and F_2 of this cross. Use genotype symbols beginning with the first letter of the alphabet *after* those used for the preceding problem.

3–31 Considering the facts suggested by the two preceding problems, how many different genotypes are responsible for the (a) yellow sphere, (b) elongate green, (c) disk white phenotypes?

3–32 A tetrahybrid white disk plant is selfed. (a) How many phenotypic classes could occur in its F_1? (b) What fraction of the F_1 will be white disk?

3–33 In Duroc Jersey pigs, two pairs of interacting genes, *R* and *S*, are known. (a) The cross of red × red sometimes produces an F_1 phenotypic ratio of 9 red:6 sandy:1 white. What is the genotype of each of these F_1 phenotypes? (b) For each of the crosses that follow, give the parental genotypes:

	P	F_1	F_2
Case 1	Red × red	All red	All red
Case 2	Red × red	3 red:1 sandy	Not reported
Case 3	Red × white	All red	9 red: 6 sandy: 1 white
Case 4	Sandy × sandy	All red	9 red: 6 sandy: 1 white
Case 5	Sandy × sandy	1 red: 2 sandy: 1 white	Not reported

3–34 Determine the genotypic and phenotypic ratios resulting from each of the following dihybrid crosses (assume lethals to exert their effect during early embryo development):

	Gene characteristics		Progeny ratios	
Parental genotypes	*Gene 1*	*Gene 2*	*Genotypic*	*Phenotypic*
(a) $AaBb \times AaBb$	Complete dominance	Complete dominance		
(b) $Aab^1b^2 \times Aab^1b^2$	Complete dominance	Incomplete dominance		
(c) $a^1a^2b^1b^2 \times a^1a^2b^1b^2$	Incomplete dominance	Incomplete dominance		
(d) $AaBb \times AaBb$	Complete dominance	Recessive lethal		
(e) $a^1a^2Bb \times a^1a^2Bb$	Incomplete dominance	Recessive lethal		
(f) $AaBb \times AaBb$	Recessive lethal	Recessive lethal		

3–35 In a hypothetical flowering plant assume petal color to be due to a pair of codominant alleles, a^1 and a^2, heterozygotes being purple, and homozygotes either red (a^1a^1) or blue (a^2a^2). Assume also another pair of alleles, completely dominant B for color, and recessive b for color inhibition. This pair segregates independently from the a^1, a^2 pair. Two doubly heterozygous purple plants, a^1a^2Bb, are crossed. What phenotypic ratio results in the progeny?

3–36 What do the following progeny ratios suggest? (a) $Aa \times Aa \rightarrow$ 259:37, (b) 2:1, (c) 9:7, (d) dominant allele (O) produces brittle bones (osteogenesis imperfecta), yet only 10 percent of persons of genotype (Oo) have brittle bones. Choose from (1) epistasis, (2) expressivity, (3) lethal allele, (4) reduced penetrance, (5) testcross.

CYTOLOGICAL BASIS OF INHERITANCE

CHAPTER

4

In Chapters 2 and 3 genes were assumed to be located in the nucleus. If this assumption is true, some confirming, objective evidence should be found. Such evidence does exist and can be conveniently designated as cytological or chemical. In the history of the science of genetics the former preceded the latter by many years. The cytological evidence is examined here and the chemical evidence is discussed in Chapter 8 at a logical point in the development of our concepts of genetic mechanisms.

THE NONDIVIDING NUCLEUS

As your previous experience in the life sciences has shown you, living cells of most organisms (*eukaryotes*) are characterized by the presence of a discrete, often spherical body, the **nucleus** (Figure 4–1). Notable exceptions are such *prokaryotes* as the Cyanobacteria and the bacteria in which, although "nuclear material" can be shown to be present, the visible, structural organization so typical of the cells of higher organisms is lacking. Prokaryotes lack the conventional mitotic and meiotic nuclear divisions that characterize eukaryotes. In addition to these two types of genetic systems, we can recognize still a third in the viruses. The structure of an important type of virus, the bacteriophages, is described more fully in Chapter 8.

The interphase (nondividing or interkinetic) nucleus is bound by the **nuclear membrane,** which is not clearly visible with the light microscope. Electron micrographs, however, reveal this membrane to be a

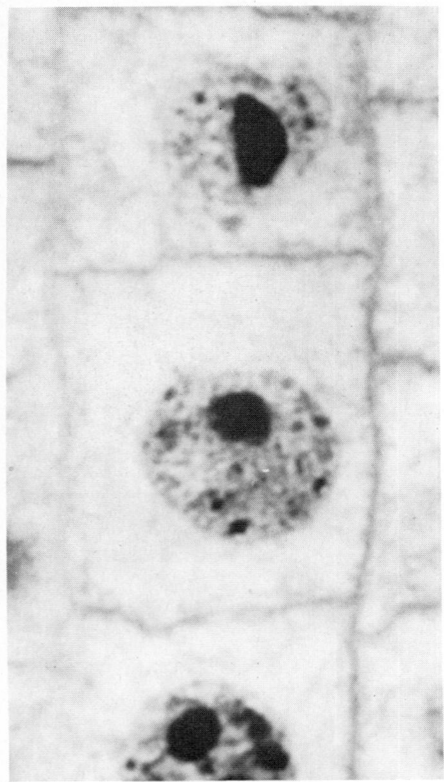

FIGURE 4–1. *Interphase nucleus of onion root tip, as seen with the light microscope. Compare with Figure 4–2. Note the prominent nucleolus.* (Courtesy Carolina Biological Supply Co.)

double layer, provided with numerous "pores" of about 20 to 80 nm in inside diameter. (A nanometer is 1×10^{-9} meters, or 10 angstrom units.) Each pore is surrounded by a thickened, electron-dense ring in the nuclear membrane, giving an outside diameter of up to 200 nm. These pores function in active control of the passage of macromolecules between the nucleus and cytoplasm. The nuclear membrane is continuous with a cytoplasmic double-membrane system, the **endoplasmic reticulum,** to which dense granular structures, the **ribosomes**—some 17 by 22 nm in size—can usually be seen attached in electron micrographs (Figure 4–2). Ribosomes are rich in **ribonucleic acid** (RNA) and play an important part in protein synthesis (see Chapters 9 and 10).

Within the interphase nucleus three major components can be distinguished. The first of these is nuclear sap, or **karyolymph,** a clear, usually nonstaining, largely proteinaceous, colloidal material. The second intranuclear component is a generally spherical, densely staining body, the **nucleolus.** Many nuclei contain two or more nucleoli. The general size range is 2 to 5 μm, but the size varies with a kind of tissue, degree of protein- and RNA-synthesizing activity of the cell (the nucleolus is larger when this activity is high), and nutritional factors. Nucleoli contain RNA, as well as some DNA and protein. Typically each nucleolus is produced by and is physically associated with a nucleolus organizing region (which is the site of synthesis of ribosomal RNA, as described in Chapter 9) of a particular chromosome, the third major nuclear component.

The remainder of the nucleus consists of **chromatin,** fine, threadlike strands of deoxyribonucleic acid (DNA) intimately complexed with various proteins to form *nucleoproteins.* These include (1) five kinds of basic (positively charged) low-molecular-weight proteins called **histones,** and (2) all other proteins of chromatin called **nonhistone proteins,** numbering perhaps 100 different kinds if all eukaryotic species are considered.

THE DIVIDING NUCLEUS

During nuclear division the chromatin strands contract and appear thicker, to become the **chromosomes,** which are visible with an ordinary light microscope. The number and morphology of the chromosomes are specific, distinct, and ordinarily constant for each species, although subspecific taxonomic categories with multiple sets (polyploids) are not infrequent, especially in plants. In some cases, however, morphological differences among different chromosomal races are sufficient to give them species or subspecific rank (Table 4–1 and Chapter 14). Some details of the gross morphology of chromosomes are pointed out later in this chapter, and their molecular structure is described in Chapter 8.

The nature and function of the chromosomes is covered in Chapters 8–11, but it should be noted here that DNA carries the coded information for most of the cellular activities. However, many of these activities, though coded by DNA, take place largely in the cytoplasm.

The interphase cycle may be defined as the entire sequence of events transpiring from the close of one nuclear division to the beginning of the next. The total time varies with the species, maturation, tissue, and temperature, among other factors. Duration of as little as about 3 hours to as much as 174 hours has been reported for various organisms. In a number of kinds of human tissue *in vitro* the interphase cycle typically occupies 18 to 24 hours. For convenience, the following stages are recognized:

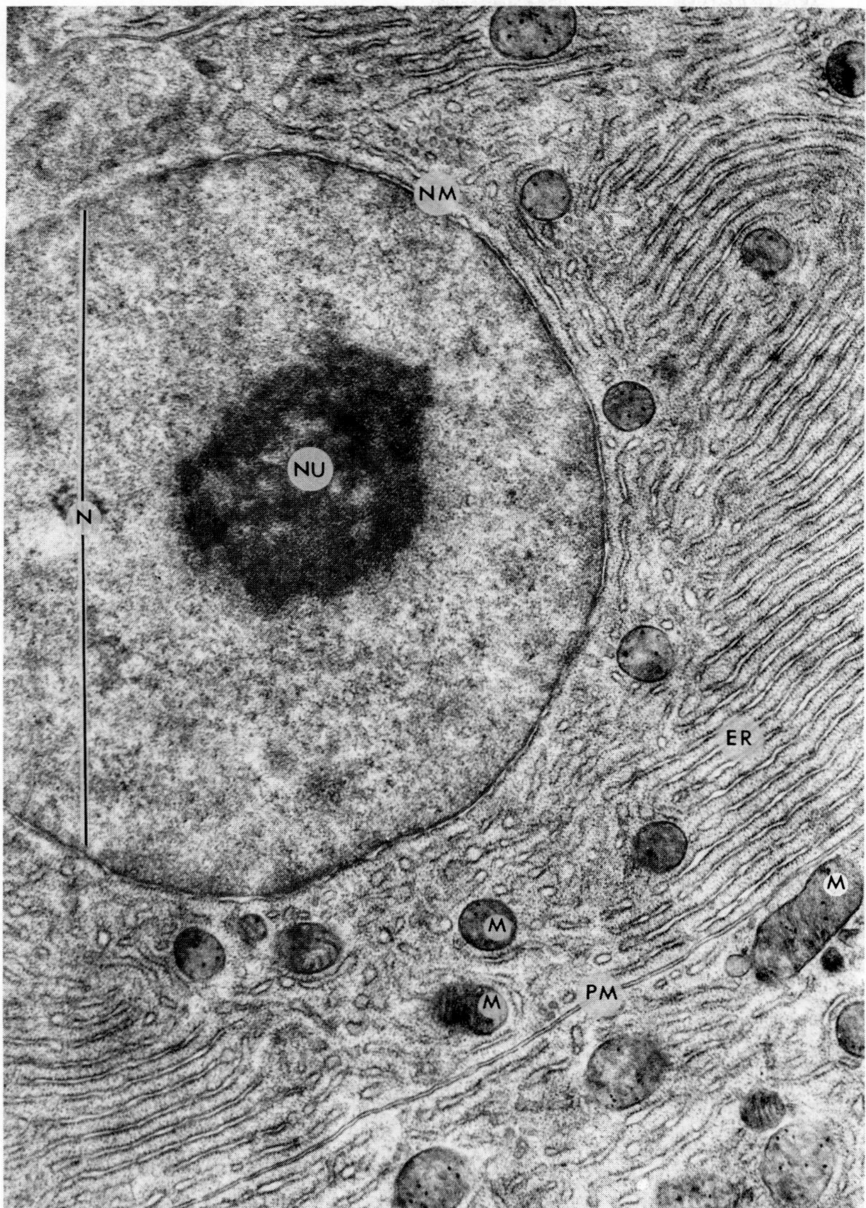

FIGURE 4–2. *Electron micrograph of interphase cell from bat pancreas. N, nucleus; NU, nucleolus; NM, nuclear membrane; ER, endoplasmic reticulum with ribosomes attached; M, mitochondria; PM, plasma membrane delimiting the cell. Note doubleness of the membrane systems.* (From *Cell Ultrastructure* by William Jensen and Roderick Park. Copyright 1967 by Wadsworth Publishing Co., Inc., Belmont California. Reproduced by permission.)

G_1 The first *gap* (or growth) stage of interphase in which nucleus and cytoplasm are enlarging toward mature size begins immediately following cell division. Chromatin is fully extended and not distinguishable as discrete chromosomes with the light microscope. This is a time of active synthesis of RNA and protein, especially (a) enzymes necessary for the DNA replication of the next stage, (2) possibly a protein that acts to trigger nuclear division, and (3) tubulin (the protein component of microtubules) and mitotic apparatus proteins (see page 47) as well as the resuming of normal cell metabolism slowed during division. This is the most variable stage as to duration; it may occupy 30 to 50 percent of the total time of the interphase cycle or be entirely lacking in rapidly dividing cells (for example, those of the early mammalian embryo and those in such lower forms as slime molds and yeast). However, G_1 may last as long as 151 hours in mature cells of corn roots. Differentiated somatic cells that

TABLE 4–1. Chromosome Numbers in Some Plants and Animals

Common name	*Scientific name*	*Haploid chromosome number*
Adder's-tongue	*Ophioglossum reticulatum*	631
Bread mold	*Neurospora crassa*	7
Bread wheat	*Triticum aestivum*	21
Broad bean	*Vicia faba*	6
Cat	*Felis domestica*	19
Cattle	*Bos taurus*	30
Chicken	*Gallus domesticatus*	39
Chimpanzee	*Pan troglodytes*	24
Coleus	*Coleus blumei*	12
Corn	*Zea mays*	10
Cotton	*Gossypium hirsutum*	26
Dog	*Canis familiaris*	39
Donkey	*Equus asinus*	31
Frog	*Rana pipiens*	13
Fruit fly	*Drosophila melanogaster*	4
Grasshopper	*Melanoplus differentialis*	♀ 24, ♂ 23
Green algae	*Chlamydomonas reinhardi*	16
Guinea pig	*Cavia cobaya*	32
Hamster	*Mesocricetus auratus*	22
Honeybee	*Apis mellifera*	16
Horse	*Equus caballus*	32
House fly	*Musca domestica*	6
Human	*Homo sapiens*	23
Kidney bean	*Phaseolus vulgaris*	11
Mosquito	*Culex pipiens*	3
Mouse	*Mus musculus*	20
Oak	*Quercus alba*	12
Oats	*Avena sativa*	21
Onion	*Allium cepa*	8
Pea	*Pisum sativum*	7
Penicillium	*Penicillium* spp.	2,4,5
Pine	*Pinus* spp.	12
Potato	*Solanum tuberosum*	24
Rat	*Rattus norvegicus*	21
Rye	*Secale cereale*	7
Soybean	*Glycine max*	20
Squash	*Cucurbita pepo*	20
Tobacco	*Nicotiana tabacum*	24
Tomato	*Lycopersicon esculentum*	12
Watermelon	*Citrullus vulgaris*	11

no longer divide (for example, neurons) are arrested usually in the G_1 stage. Some authorities prefer to designate this last situation as the $\mathbf{G_0}$ stage. Even though changes in synthesis patterns may occur in such differentiated cells, the condition of their DNA is the same as that of typical G_1 cells. So the question has been raised as to whether G_0 represents any more than a modification of G_1. Watson (1976) aptly refers to G_1 as a sequence of "preprogrammed operations" in preparation for DNA replication. This may include cytoplasmic synthesis of DNA polymerase (an enzyme system catalyzing DNA replication).

S During the *synthesis* stage, replication of DNA and synthesis of histones occur, the former precisely doubling in amount during S. The chromosomes are each composed of two *sister chromatids* (sharing a common centromere, as described later in this chapter) by the end of this stage. This is the most important activity of the S stage. Early

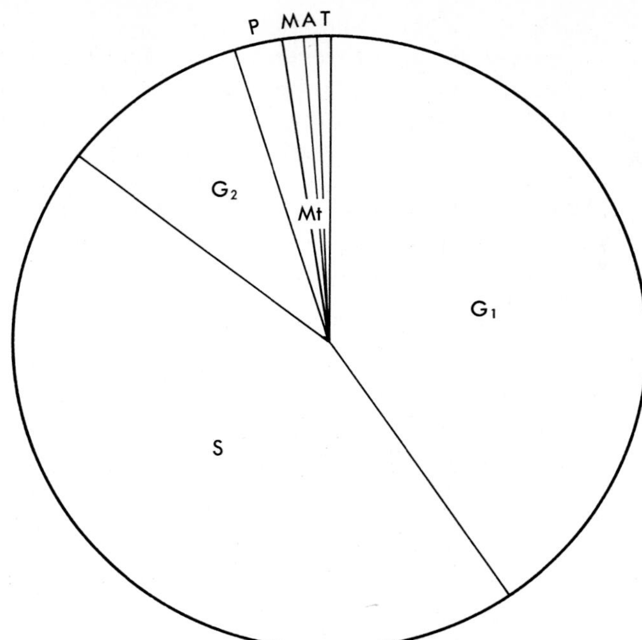

FIGURE 4–3. *Idealized diagram of the relative amounts of time occupied by each stage of interphase and the mitotic cycle. G_1, first growth stage; S, synthesis of DNA and histones; G_2, second growth stage; P, prophase of mitosis; M, metaphase of mitosis; A, anaphase of mitosis; Mt, mitotic period.*

in S there occurs a rapid rise in DNA polymerases (this requires transport from cytoplasm to nucleus) and in an RNA needed for later degeneration of the nuclear membrane, along with other RNAs. This stage may occupy roughly 35 to 45 percent of the interphase cycle; in cultured human cells it extends 6 to 8 hours.

G_2 This is a second *gap* (growth) stage, not as well understood as G_1, in which new DNA is rapidly complexed with chromosomal proteins, and synthesis of RNA (in lesser amounts than in G_1) and proteins continues. It may occupy about 10 to 20 percent of the interphase cycle.

Duration of the several stages of the interphase cycle in cultured HeLa cells (human cancer cells cultured since 1952; see Glossary) is approximately as follows: G_1, 8.2 hours; S, 6.2 hours; G_2, 4.6 hours. An idealized diagram of the relative amounts of time occupied by each stage of the interphase cycle and nuclear division is shown in Figure 4–3. As the G_2 stage draws to a close, the cell gradually enters division, Mt, though governing factors for this transition are not wholly clear.

CELL DIVISION

Growth and development of every organism depend in large part on multiplication, enlargement, and differentiation of its cells, beginning with the zygote. In multicellular individuals attainment of adult form depends on a coordinated sequence of increase in cell number, size, and differentiation from zygote to maturity. In unicellular organisms cell division serves also as a form of reproduction, often the only one. Sexually reproducing forms also depend in most instances directly on cell division for the formation of sex cells, or gametes.

Division of nucleate cells consists of two distinct but integrated activities, nuclear division (**karyokinesis**) and cytoplasmic division (**cytokinesis**). In general, cytokinesis begins after nuclear division is well underway, but in many instances it may be deferred or entirely lacking.

For example, in the development of the female gametophyte of pine (see Appendix B), about eleven nuclear divisions occur before cytokinesis begins, and in some algae and fungi the plant body is a coenocyte, without any walls separating nuclei except for the reproductive cells.

Two types of nuclear division, **mitosis** and **meiosis,** are characteristic of most plant and animal cells. Mitosis is regularly associated with nuclear division of vegetative or *somatic* cells; meiosis occurs in conjunction with formation of reproductive cells (either gametes or meiospores) in sexually reproducing species.

Mitosis

As a process, mitosis is remarkably similar in all but relatively minor details in both plants and animals, from the least specialized to the most highly evolved forms. Mitosis is a smoothly continuous process and is divided arbitrarily into several stages or phases for convenient reference. The following description of the mitotic process in plant cells will adequately serve the purpose of determining whether the mechanism provides a reasonable basis for the genetic assumptions developed in Chapters 2 and 3.

Prophase. As stage G_2 of interphase gives way to prophase of mitosis, chromosomes progressively shorten and thicken to form individually recognizable, elongate, longitudinally double structures (Figure 4–4) arranged randomly in the nucleus. This contraction of the chromosomes is one of the most conspicuous features of prophase. The two **sister chromatids** of each chromosome are closely aligned and somewhat coiled on themselves. The tightening of these coils contributes in large measure to the shortening and thickening of the chromosomes.

The two sister chromatids of each chromosome are held together by strands in a specialized region, the **centromere** (Figure 4–5). The centromere has been shown by electron microscopy to be a complex region of the chromosome. The centromere of each sister chromatid includes a **kinetochore** to which the microtubules of the spindle are attached (these are examined later in this chapter). Kinetochores appear to become organized in prophase only after chromosome contraction has taken place. In the absence of the centromere, chromosome fragments

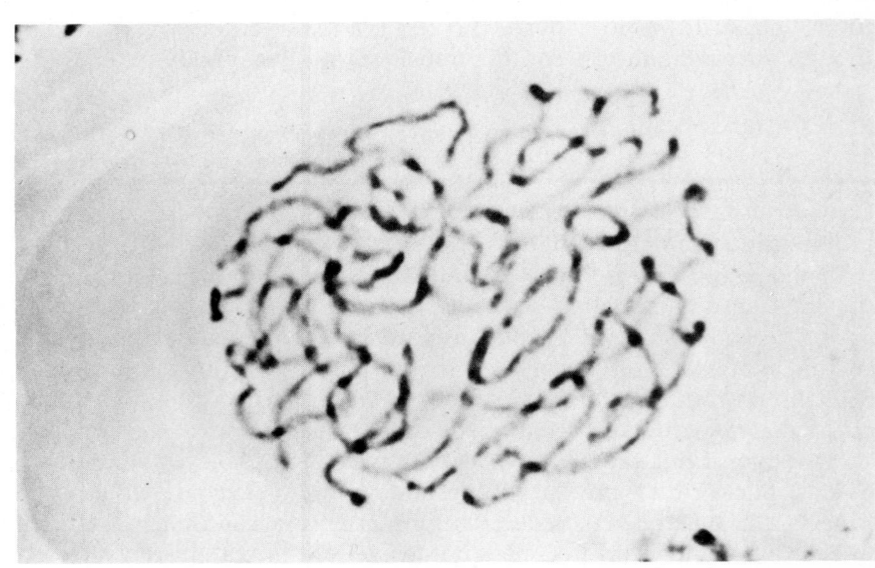

FIGURE 4–4. *Mid-prophase in onion root tip cell. Longitudinal doubleness is evident in some of the chromosomes; the nuclear membrane has become quite indistinct.* (Photo courtesy Dr. W. Tai, University of Manitoba.)

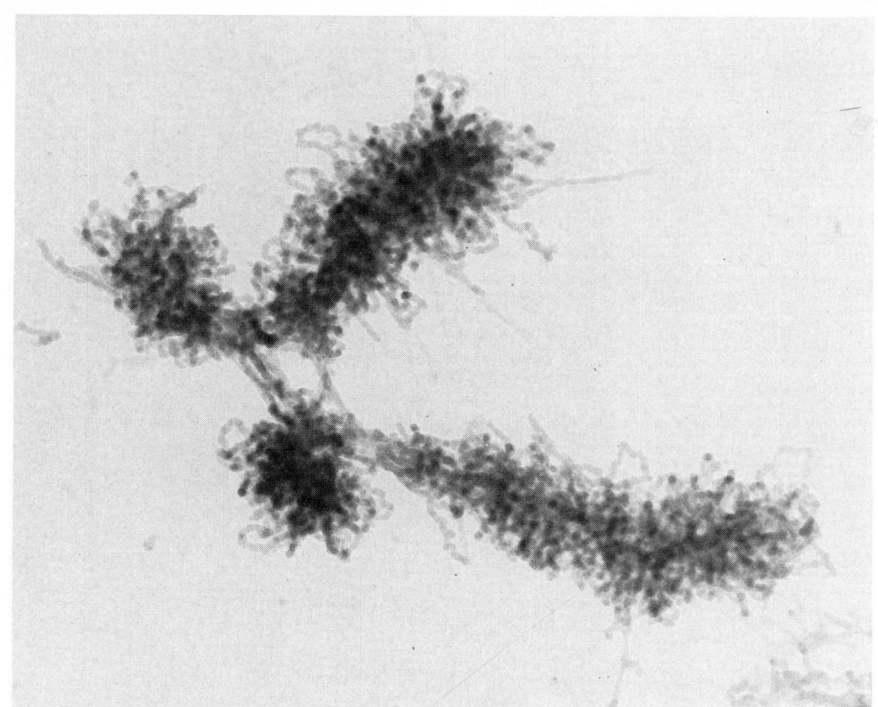

FIGURE 4–5. *Electron micrograph showing two sister chromatids held together by chromatin fibers in the centromere region. This is one of the human chromosomes.* (Photo courtesy Dr. E. J. Du Praw. From E. J. Du Praw, 1970. *DNA and Chromosomes.* Holt, Rinehart and Winston, Inc., New York. Used by permission.)

(acentric fragments) such as those resulting from radiation-induced breakage fail to move normally during nuclear division and are generally lost from one or both of the reorganizing nuclei. This illustrates the essential role of the centromere in chromosome movement during division.

During prophase the nucleolus gradually disappears in most organisms. In some grasses, in the green alga *Spirogyra,* and in *Euglena,* the nucleolus apparently ceases to function during mitosis.

As prophase progresses an important and often conspicuous component, a football-shaped **mitotic apparatus** (spindle and associated structures), begins to form (see Figure 4–6). By the end of prophase this structure occupies a large portion of the cell volume and often extends very nearly from one "end" of the cell to the other. The mitotic apparatus consists of slender spindle fibers resolved by electron microscopy as **microtubules** arranged along the long axis of the spindle. Some of the microtubules are continuous from pole to pole of the mitotic apparatus; these are the **continuous fibers.** Groups of others, the **chromosomal fibers,** extend from one pole or the other to each chromosomal kinetochore to which they are attached (Figure 4–7).

In many organisms prophase comprises the bulk of the time consumed by mitosis (Table 4–2). As prophase draws to a close, the longitudinally double chromosomes move or are moved in the direction of the midplane (equator) of the developing spindle; a period of time often designated as **prometaphase** (Figure 4–12).

Another common feature of prometaphase is the degeneration and disappearance of the nuclear membrane (Figure 4–12). The mechanism is incompletely understood, but the aggregation of mitochondria (especially in animal cells) suggests enzymatic processes. Physical stresses exerted by microtubules, which are attached to the nuclear membrane, may also play a role in membrane degeneration.

FIGURE 4–6. *Spindle formation from prophase to metaphase.*

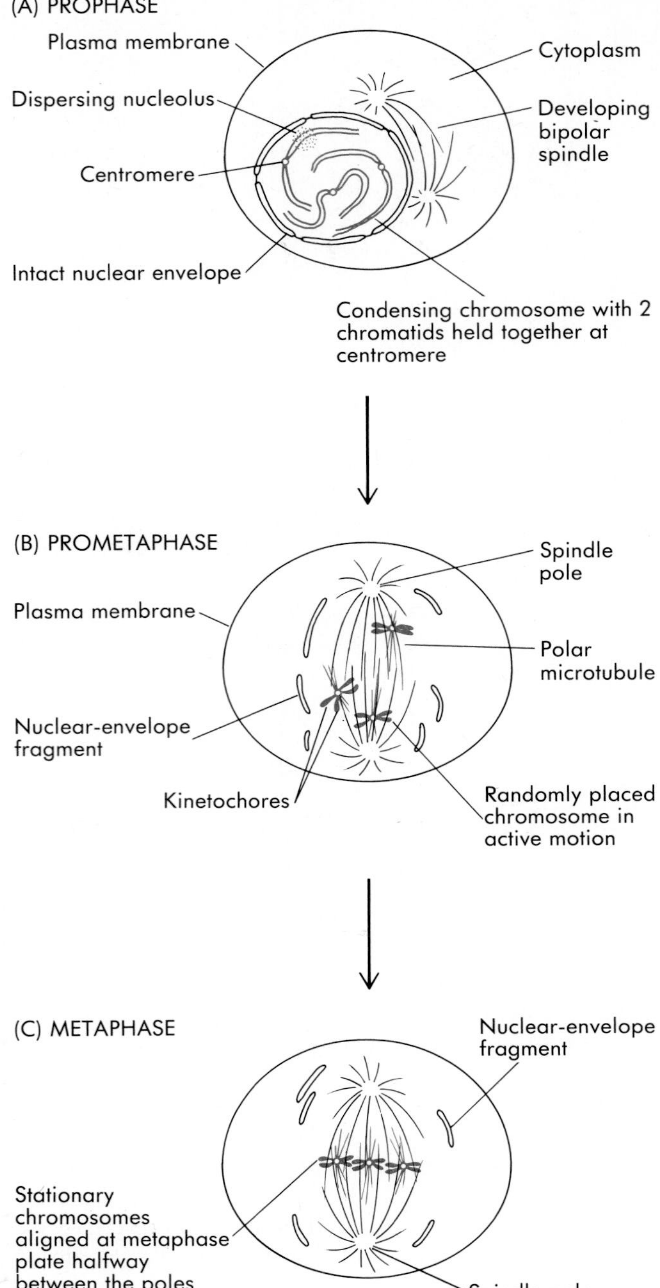

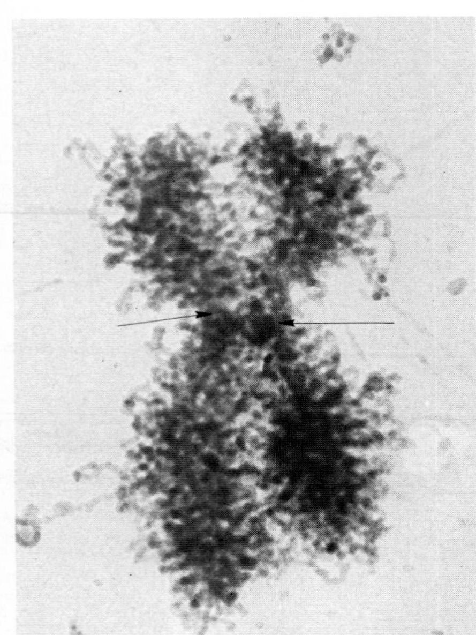

FIGURE 4–7. *Electron micrograph of one of the human chromosomes showing the two kinetochores of the centromere.* (Photo courtesy Dr. E. J. Du Praw. From E. J. Du Praw, 1970, *DNA and Chromosomes*. Holt, Rinehart and Winston, Inc., New York. Used by permission.)

Metaphase. Metaphase (Figures 4–6 and 4–8) is that period of time in which the centromeres of the longitudinally double chromosomes occupy the plane of the equator of the mitotic apparatus, although the chromosomal arms may extend in any direction. At this stage the sister chromatids are still held together by connecting chromatin fibers at the centromere regions. Electron microscopy shows that the kinetochores of the two sister chromatids face opposite poles; this will permit proper separation in the next phase (anaphase).

During metaphase the chromosomes are shortest and thickest. Polar views furnish good material for chromosome counts; both lateral and polar views are useful for studying chromosome morphology.

TABLE 4–2. Duration of Mitosis in Living Cells

Name of organism				*Duration, minutes*				
Common	*Scientific*	*Tissue*	*Temp. °C*	*P*	*M*	*A*	*T*	*Total*
		Plants (angiosperms)						
Onion	*Allium cepa*	Root tip	20	71	6.5	2.4	3.8	83.7
Oatgrass	*Arrhenatherum* sp.	Stigma	19	36–45	7–10	15–20	20–35	78–110
Pea	*Pisum sativum*	Endosperm	—	40	20	12	110	182
Pea	*Pisum sativum*	Root tip	20	78	14.4	4.2	13.2	110
Spiderwort	*Tradescantia* sp.	Stamen hair	20	181	14	15	130	340
Broad bean	*Vicia faba*	Root tip	19	90	31	34	34	155
		Animals						
Fowl	*Gallus* sp.	Fibroblast culture	—	19–25	4–7	3.5–6	7.5–14	34–52
Grasshopper	*Melanoplus differentialis*	Neuroblast	—	102	13	9	57	181
Mouse	*Mus musculus*	Spleen mesenchyme	38	21	13	5	20	59
Salamander	*Salamandra maculosa*	Embryo kidney	20	59	55	6	75	195

Anaphase. Anaphase (Figure 4–9) is characterized by separation of the metaphase **sister chromatids** and their passage as **daughter chromosomes** to the spindle poles. It begins at the moment when the binding of sister centromeres to each other is released and ends with their arrival at the poles. Anaphase therefore accomplishes the *quantitatively* equal distribution of chromosomal material to two developing daughter nuclei. The distribution is also *qualitatively* equal if the replication process of the preceding S stage has been exact. The replication process will be more closely examined at the molecular level in Chapter 8. The mechanism of the anaphasic movement of chromosomes is not understood in spite of intense experimentation and study.

Telophase. The arrival of the (longitudinally single) daughter chromosomes at the spindle poles marks the beginning of telophase; it is, in turn, terminated by the reorganization of two new nuclei and their

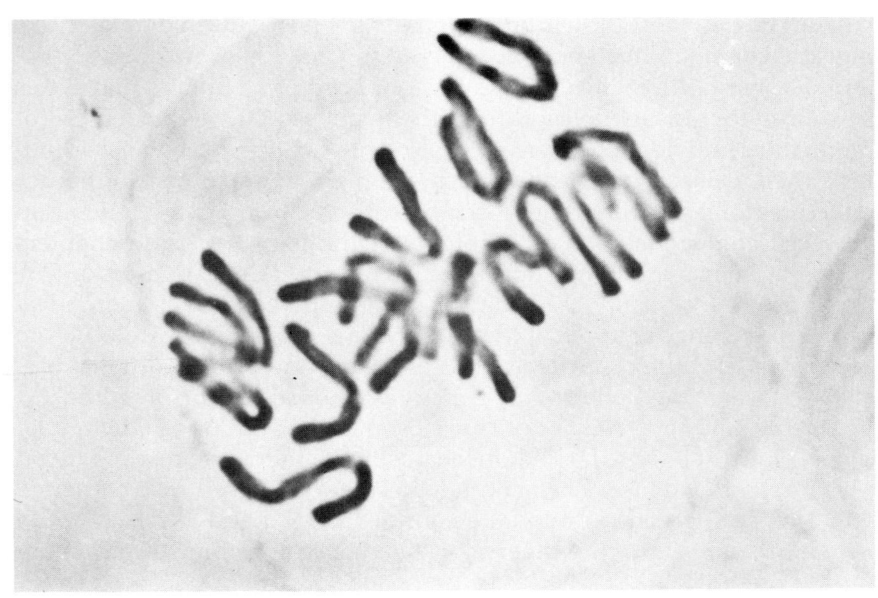

FIGURE 4–8. *Metaphase in onion root tip cell.* (Photo courtesy Dr. W. Tai, University of Manitoba.)

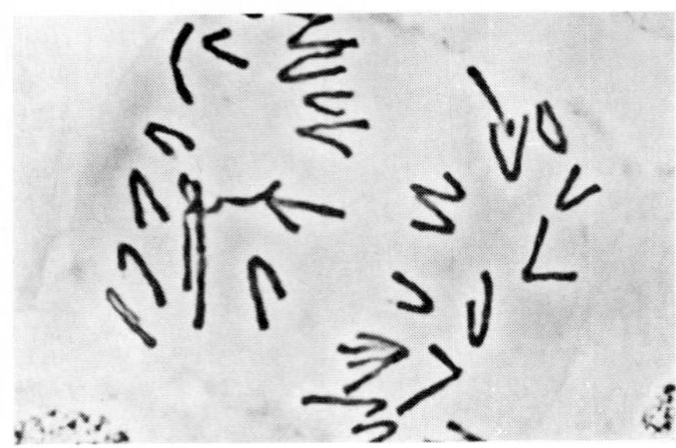

FIGURE 4–9. *Anaphase in onion root tip cell.* (Photo courtesy Dr. W. Tai, University of Manitoba.)

entry into the G_1 stage of interphase (Figures 4–10 and 4–12). In general terms, the events of prophase occur in reverse sequence during this phase. New nuclear membranes are constructed from materials that may be remnants of the original membrane, or derived from the endoplasmic reticulum, or newly synthesized from appropriate cellular components. The mitotic apparatus gradually disappears, the nucleoli are reformed at the nucleolar organizing sites of specific chromosomes, and the chromosomes resume their long, slender, extended form as their coils relax. Replication of chromosomal material, by which each chromosome again consists of two sister chromatids, then occurs in the S stage of the succeeding interphase.

Cytokinesis

If cytokinesis occurs, the process takes place during telophase, though it may be initiated during anaphase. In cells of higher plants cytokinesis is typically accomplished by formation of a **cell plate** (Figures 4–11 and 4–12). Early steps in cytokinesis include formation on vesicles in the midplane of the mitotic apparatus and the coalescence of these vesicles, starting at the center of the spindle, to form a cell plate that is later converted to a **middle lamella,** with new cross walls between the daughter cells deposited on each side of the middle lamella (Figure 4–11). In animal cells cytokinesis is accomplished by the cytoplasm simply pinching the cell in half. This cleavage first begins as a furrowing process around the cell almost always in the plane of the metaphase plate. It continues as a contractile ring, closing in a way similar to purse strings, as it draws tighter around the center of the cell. Filaments of the contractile proteins, actin and myosin, are involved in this process. Figure 4–12 shows the entire sequence of mitosis and cytokinesis in a plant cell.

Significance of mitosis. The process of mitosis and the subsequent replication of chromosomal material in the succeeding interphase has the inevitable result, if the cell "makes no mistakes," of creating from one cell two new ones that are chromosomally identical. Certainly in mitosis the chromosomal material is distributed in equal quantity to two daughter cells in a strikingly precise manner.

If the replication process during the S stage of interphase is qualitatively equal, then the chromosomes will serve quite adequately as physical bearers of the genes considered in Chapters 2 and 3. The pattern of inheritance followed for *Coleus,* for example, requires that genes be transmitted in cell division from the zygote to every somatic cell of the mature individual. The same is also true for human beings. In mitosis there exists a process by which the precise, equal distribution of structures called chromosomes can be carried through cell generation after cell generation. The assumptions of the two preceding chapters regarding genes are borne out if the genes are indeed located on the chromosomes; certainly the behavior of these bodies makes them ideal vehicles for genes in terms of the speculations previously raised. Before accepting the chromosomal locations of genes, however, additional parallels between the behavior of genes, as deduced from breeding experiments, and the behavior of chromosomes, as seen under the microscope, must be noted. The mechanism of mitosis provides good presumptive evidence that many of the theories raised in Chapters 2 and 3 regarding gene location are sound. Physical and chemical proof are still required, but the theory looks sufficiently promising to retain for the present.

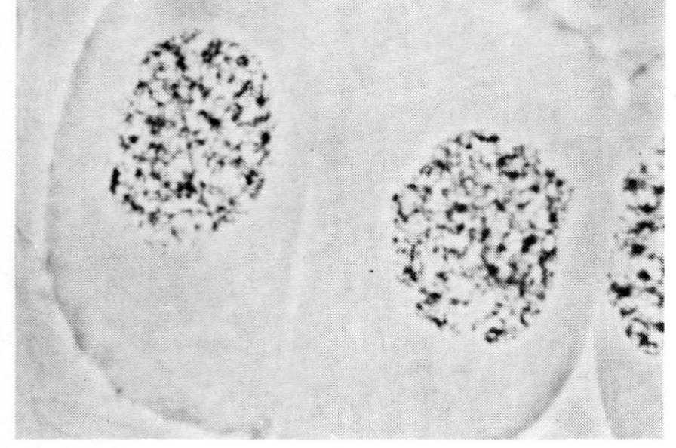

FIGURE 4–10. *Late telophase in onion root tip cell.* (Photo courtesy Dr. W. Tai, University of Manitoba.)

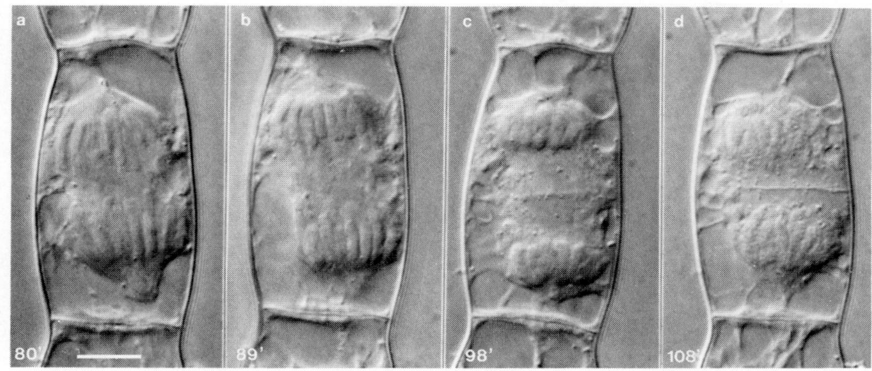

FIGURE 4–11. *Differential interference microscopy of cell plate formation in a stamen hair cell from the spiderwort plant,* Tradescantia virginiana *(cv Zwanenburg Blue). Note time course in minutes for the process.* (Photo courtesy Dr. Steve Wolniak, University of Maryland.)

Although not directly germane to present purposes, it is interesting to note that the rather complicated physical and chemical changes in mitotic nuclei often occur in a surprisingly short time. Table 4–2 summarizes a few studies from the literature.

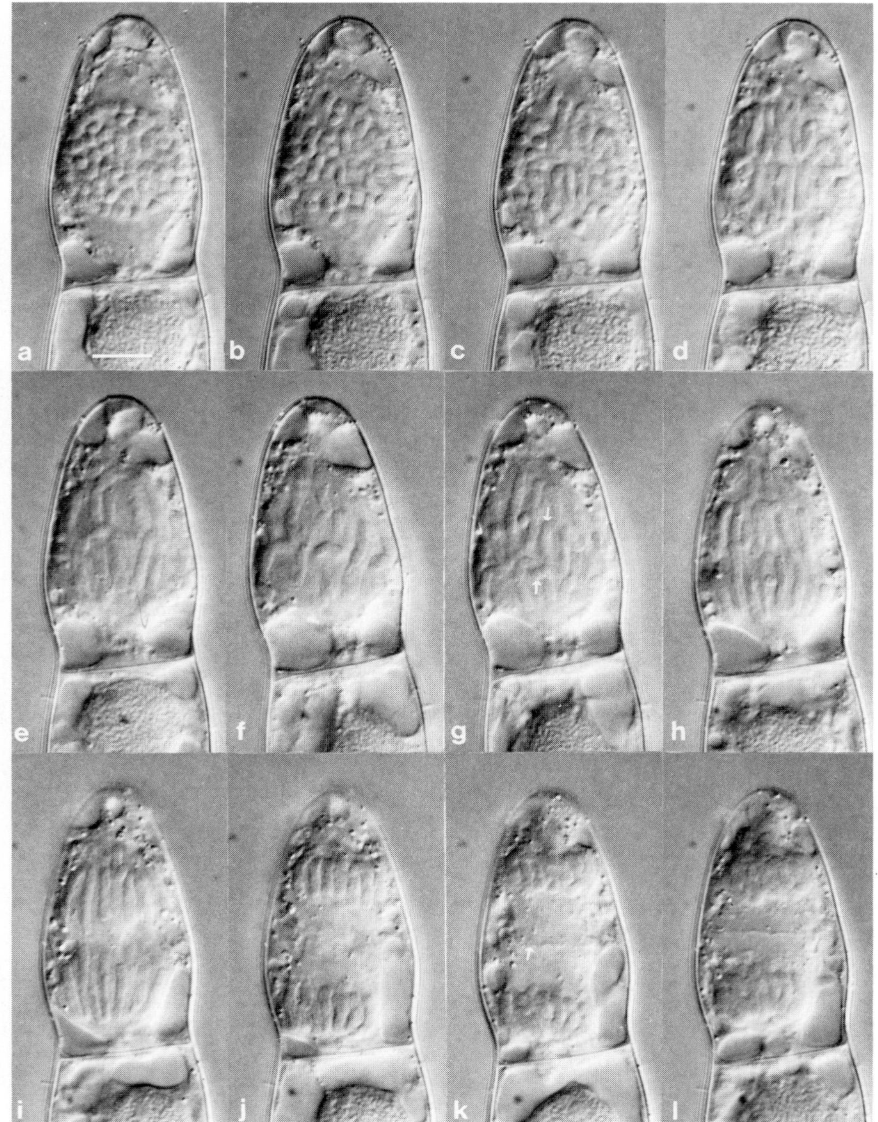

FIGURE 4–12. *Photomicrograph of a stamen hair cell from the spiderwort plant,* Tradescantia virginiana *(cv Zwanenburg Blue), at various stages of mitosis. (a) Prophase. The nuclear envelope has not yet broken down, and the chromatin has condensed into (replicated) chromosomes. The triangular-shaped zones above and below the spherical nucleus are regions where the mitotic spindle apparatus is forming. (b) Prometaphase. The nuclear envelope has broken down and the chromosomes have begun to move toward the center of the spindle. This movement is known as congression. (c, d) Congression continues as the chromosomes continue to become aligned in the center of the spindle. (e, f) Metaphase. The chromosomes are aligned in the central part of the spindle, the metaphase plate. (g) Early anaphase. The chromatids have split at their centromeres and the daughter chromosomes are separating from each other. (h, i) Mid anaphase. The chromosomes continue to separate. (j) Late anaphase. The chromosomes' arms are "contracting" near the spindle pole regions. (k) Telophase. The cell plate vesicles have begun to aggregate in the spindle midzone (arrow). (l) Telophase. The cell plate is becoming increasingly well organized as a wall that will separate the two daughter cells. Soon after this photograph was taken, the nuclear envelopes reformed in each daughter cell and, thereafter, the chromosomes began to condense. All micrographs, differential interference microscopy, × 1200, Bar = 10 μm.* (Photo courtesy Dr. Steve Wolniak, University of Maryland.)

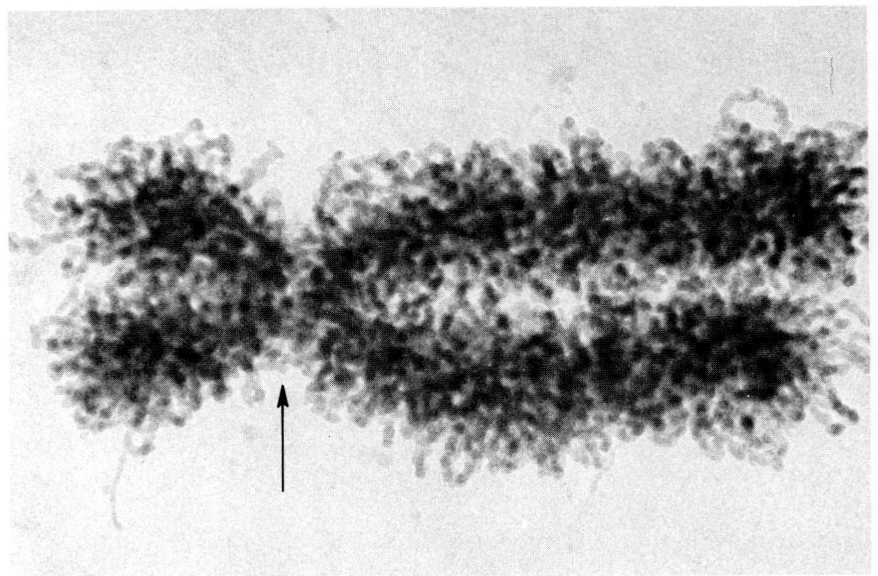

FIGURE 4–13. *Electron micrograph of one of the human chromosomes showing centromere (arrow) and DNA fibrils.* (Courtesy Dr. E. J. Du Praw, from E. J. Du Praw, 1970. *DNA and Chromosomes.* Holt, Rinehart and Winston, Inc., New York. Used by permission.)

Chromosome Morphology

General structure. In stained preparations, as seen in the light microscope, chromosomes appear to possess few definitive morphological characteristics by which individual members of a set may be distinguished from each other. The only useful features in this context are (1) length, (2) centromere location and relative arm lengths, and (3) presence or absence of satellites (distal chromosomal segments separated by a secondary constriction) (Figure 4–16).

During mitotic metaphase, chromosomes are ordinarily at their shortest and thickest (Figure 4–13) as has been pointed out, varying from as short as a fraction of a micrometer (1×10^{-6} meter) to as long as around 400 μm in exceptional cases, and between about 0.2 and 2 μm in diameter. Each chromosome of a complement has its own characteristic length (within relatively narrow limits) and centromere location. However, in some organisms, especially those with large numbers of chromosomes, there is often considerable size similarity among some of the members in a set. This is true for some of the human chromosomes, and until recently, members of different homologous pairs could not be distinguished easily. This difficulty has been especially troublesome in assigning groups of genes to the proper chromosome, as well as in distinguishing various chromosomal aberrations. Before the development of new techniques, which will be described in the next section, it was possible only to photograph smears of metaphase cells (Figure 4–14), cut the chromosomes apart, and arrange them in as nearly matching pairs as could be determined from length, centromere location, and presence or absence of satellites. Photographs so prepared (or drawings made from them) are referred to as **karyotypes** (Figure 4–15).

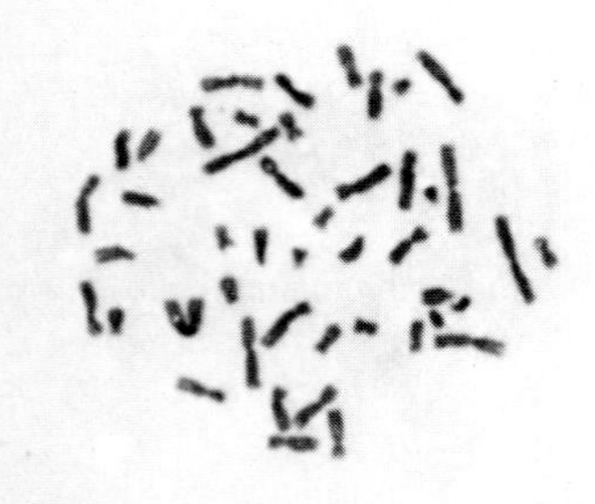

FIGURE 4–14. *Mitotic metaphase of human male chromosomes.* (Photomicrograph by Mr. John Derr.)

Although no such thing as a "typical" chromosome exists, a composite one is diagramed in Figure 4–16. Many chromosomes, however, show a distressing lack of morphological landmarks such as satellites. Every normal one does possess a centromere; its position and therefore the lengths of the arms are relatively constant. Such properties as these do permit to some degree the identification of some of the chromosomal aberrations that are described in Chapters 14 and 15. On the basis of centromere location cytologists recognize four major types of chromosomes (Figure 4–17):

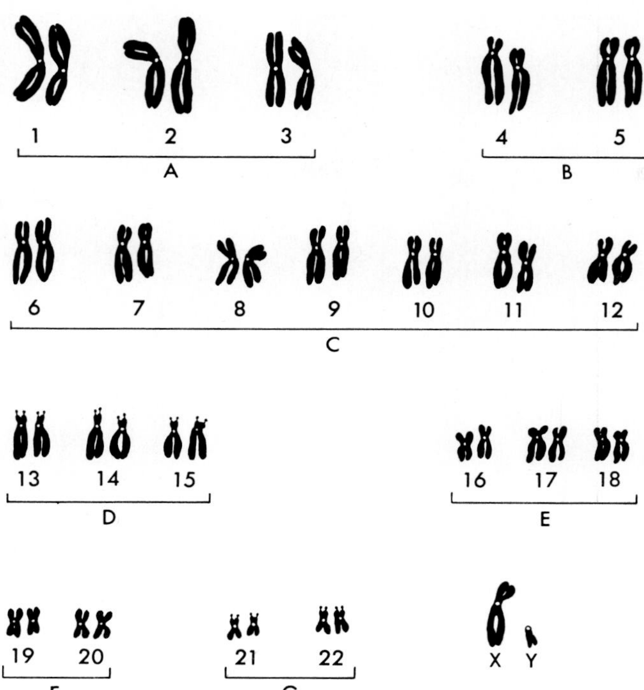

FIGURE 4–15. *The chromosomes of a normal human male arranged as a karyotype. Note the knobs or satellites on several pairs. Because of morphological similarities among several of the human chromosomes, groups are assigned letter designations as shown here.*

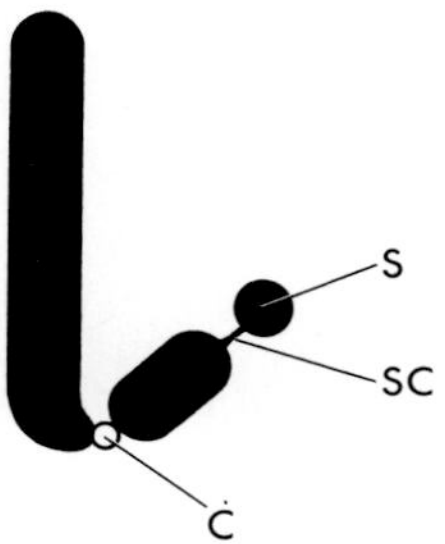

FIGURE 4–16. *Diagram of a composite chromosome as it might appear at high magnification with a light microscope. For simplicity, chromatids are not represented. C, centromere; S, satellite or knob; SC, secondary constriction.*

Metacentric: centromere central; arms of equal or essentially equal length

Submetacentric: centromere submedian, giving one longer and one shorter arm

Acrocentric: centromere very near one end; arms very unequal in length

Telocentric: centromere terminal; only one arm

Heterochromatin and euchromatin. In interphase and early prophase chromatin takes two forms: **heterochromatin,** tightly coiled (condensed) and either more darkly or less darkly staining than **euchromatin,** which is noncondensed. During nuclear division heterochromatin may occur in the vicinity of the centromeres, at the ends of the arms (*telomeric*), or in other positions along the chromosome.

Constitutive heterochromatin may be found in the centromeric region or almost anywhere else along a chromosome, as just noted. However, special locations of intercalary constitutive heterochromatin, though consistent for cytologically "normal" individuals of a given species, can vary among different species. Heterochromatin, once thought to be wholly inert genetically, is now known to include some important genetically active regions. These include nucleolar organizers and genes for some of the RNA molecules as well as more genetically inert regions containing highly repeated nucleotide sequences referred to as *repetitive or satellite DNA.* Centromeric constitutive heterochromatin probably plays an unknown role in anaphase separation of sister chromatids.

With *facultative heterochromatin* one chromosome of a homologous pair becomes partially or wholly heterochromatic and inactive genetically. This is the case with one of the X chromosomes of the human female.

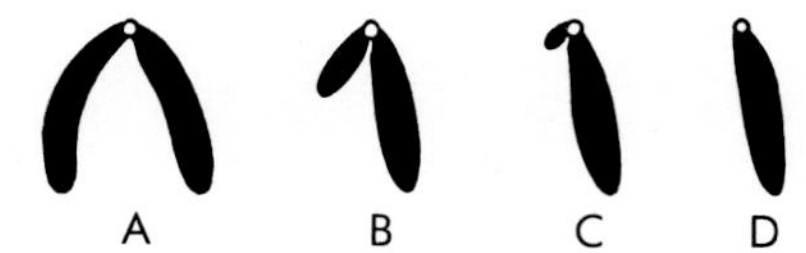

FIGURE 4–17. *Four major morphological chromosome types are recognized on the basis of centromere position and consequent relative arm length. (A) metacentric; (B) submetacentric; (c) acrocentric; (d) telocentric.*

A

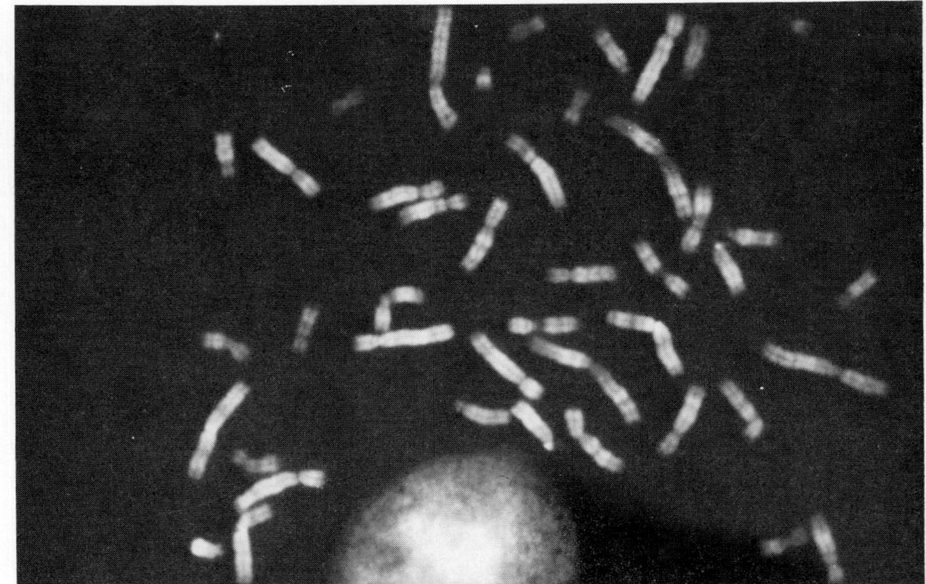

B

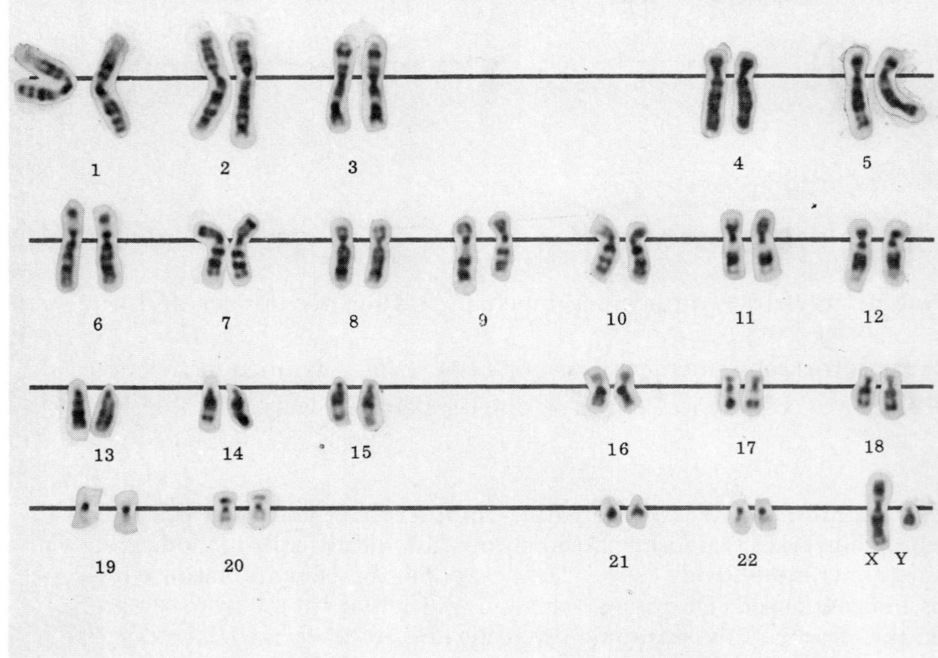

FIGURE 4–18. *(A) Metaphase spread of normal human male chromosomes showing Q-banding. (B) Karyotype of normal human male, Q-banding.* (Photos courtesy Dr. C. C. Lin.)

Chromosome banding. Progress in positive identification of chromosomes came in a fast-breaking series of developments starting in 1968 and 1969 with the work of T. Caspersson and his colleagues in Sweden. Briefly stated, the basis for this breakthrough lies in the fact that many dyes that have an affinity for DNA fluoresce under ultraviolet light. After suitable treatment with such dyes, each chromosome shows bright and dark zones, or bands, which are specific in location and extent for that chromosome. This feature is best seen in metaphase chromosomes. Considerable success has been achieved with *quinacrine mustard,* which produces fluorescent bands of various degrees of brightness. Bands produced by quinacrine mustard are referred to as **Q-bands** (Figure 4–18A).

Banding patterns are unique for each member of the human chromosome complement. Therefore, positive identification and construction of a karyotype are easy (Figure 4–18B). There are, however, some disadvantages to the Q-banding technique, particularly the impermanence of the fluorescence. That difficulty can be overcome by staining with the *Giemsa* dye mixture, which produces a unique set of bands—the **G-bands.** Figure 4–19A presents a metaphase spread from a human male stained with the Giemsa dye. By comparing G-bands (Figure

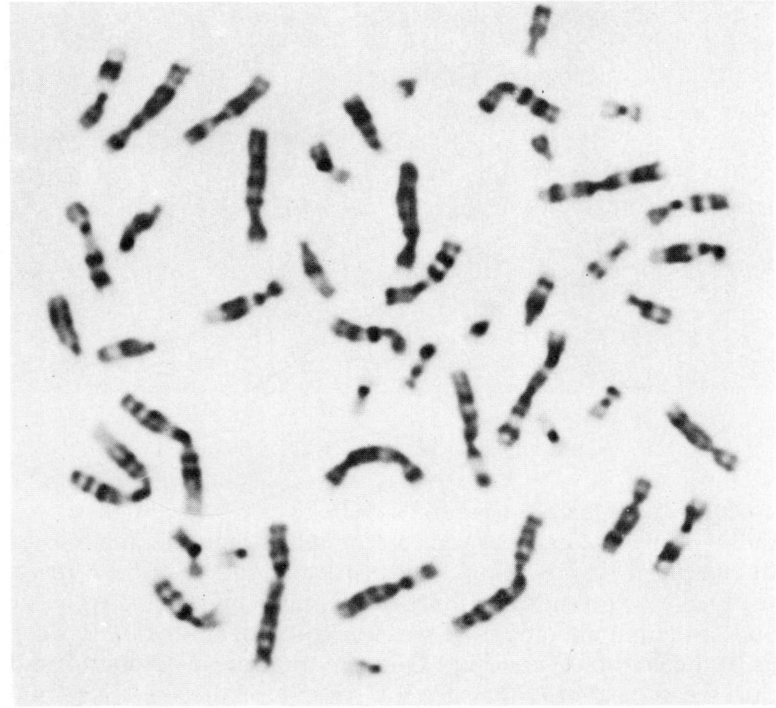

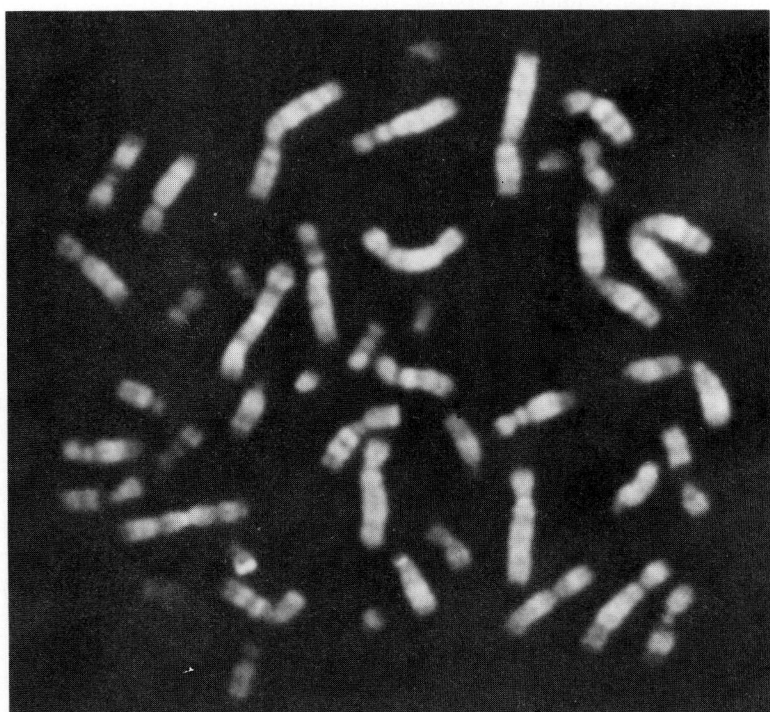

FIGURE 4–19. *(A) Metaphase spread of normal human male chromosomes, G-banding. (B) Metaphase spread of normal human male, Q-banding.* (Photos courtesy Dr. Herbert A. Lubs.)

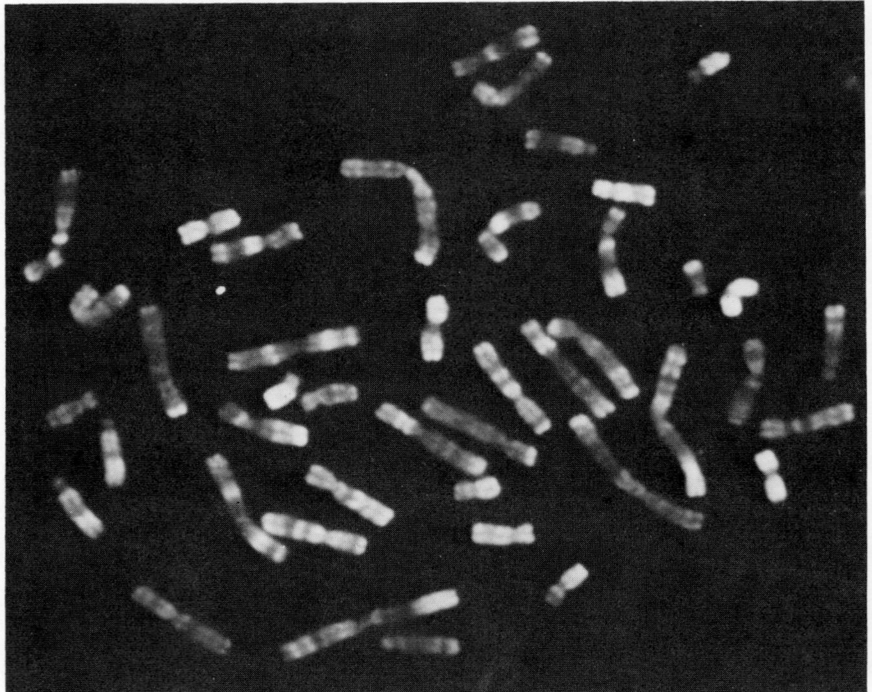

FIGURE 4–20. *Reverse (R−) banding with acridine orange. Metaphase spread of chromosomes of a normal human female.* (Courtesy Dr. C. C. Lin.)

4–19A) with Q-bands (Figure 4–19B) it can be seen that, in general, darkly stained Giemsa bands correspond in large degree to the brightly fluorescing Q-bands, though agreement is not complete. However, virtually every chromosome in a complement can be positively identified, and structural alterations of parts of chromosomes clearly determined, as in the case of Q-banding. One variation of the Giemsa method produces band patterns that are the reverse of the G-bands, that is, the normally dark-stained G-bands are lightly stained R-bands (Figure 4–20) and vice versa. Figure 4–21 shows diagrammatically Q-, G-, and R-band patterns of the human chromosome complement.

In another modification of the technique using Giemsa stain, treatment of cells (fixed to slides) with alkali leads to dense, bright staining in the centromere region. This produces the **C-banding,** which is specific for constitutive heterochromatin. The centromeric region, which includes repetitive DNA, responds only to the C-banding technique, as does any intercalary region that also contains repetitive DNA.

Chromosome banding and taxonomic relationship. Banding techniques have now been applied to a wide range of eukaryotes. As might be suspected, where species have been assigned close taxonomic relationship on the grounds of morphology and physiology, banding similarities have been found. One of the most interesting sets of findings has come from a comparison of banded karyotypes of man (*Homo sapiens*), where 2n = 46, with those of the chimpanzee (*Pan troglodytes*), the gorilla (*Gorilla gorilla*), and the orangutan (*Pongo pygmaeus*). In these latter three species the diploid chromosome number is 48. Some chromosomes of the four karyotypes have been evolutionarily stable; for example, human autosome 1 is closely similar to autosome 1 of the three other species. Others, like human autosome 3, are closely similar in all but the orangutan. Human autosome 2 has been found to have resulted

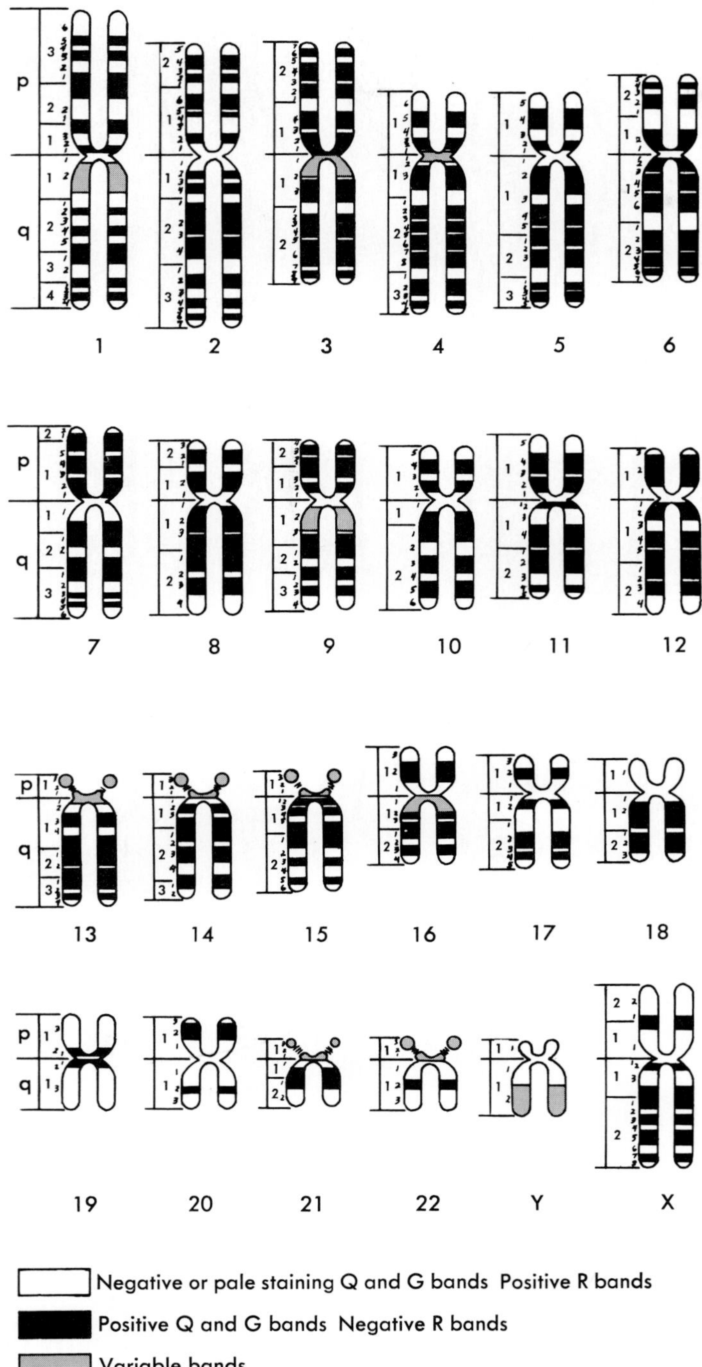

FIGURE 4–21. *Diagrammatic comparison of normal human male karyotype as observed with Q-, G-, and R-banding techniques. The centromere is represented as observed in Q-banding only.* (Redrawn from *Paris Conference (1971): Standardization in Human Cytogenetics.* In Birth defects: Orig. Art. Ser., D. Bergsma, ed. Published by The National Foundation–March of Dimes, White Plains, New York, Vol. III (7), 1972. Used by permission.)

from the fusion of two acrocentric ape chromosomes, numbers 12 and 13, which accounts for the difference in somatic chromosome number between humans and the three ape species. Relationships between the karyotypes are shown in Figure 4–22.

FIGURE 4–22. *(A) G-banded late prophase chromosomes in the 1000-band stage of (left to right) human, chimpanzee, gorilla, and orangutan, showing clearly the extensive chromosomal homology among the four species. (B) Schematic representation of G-banded, late prophase chromosomes from the same four species (left to right) human, chimpanzee, gorilla, and orangutan. The banding homology is even more evident here.* (Photo from Yunis and Prakash, 1982, used with permission.)

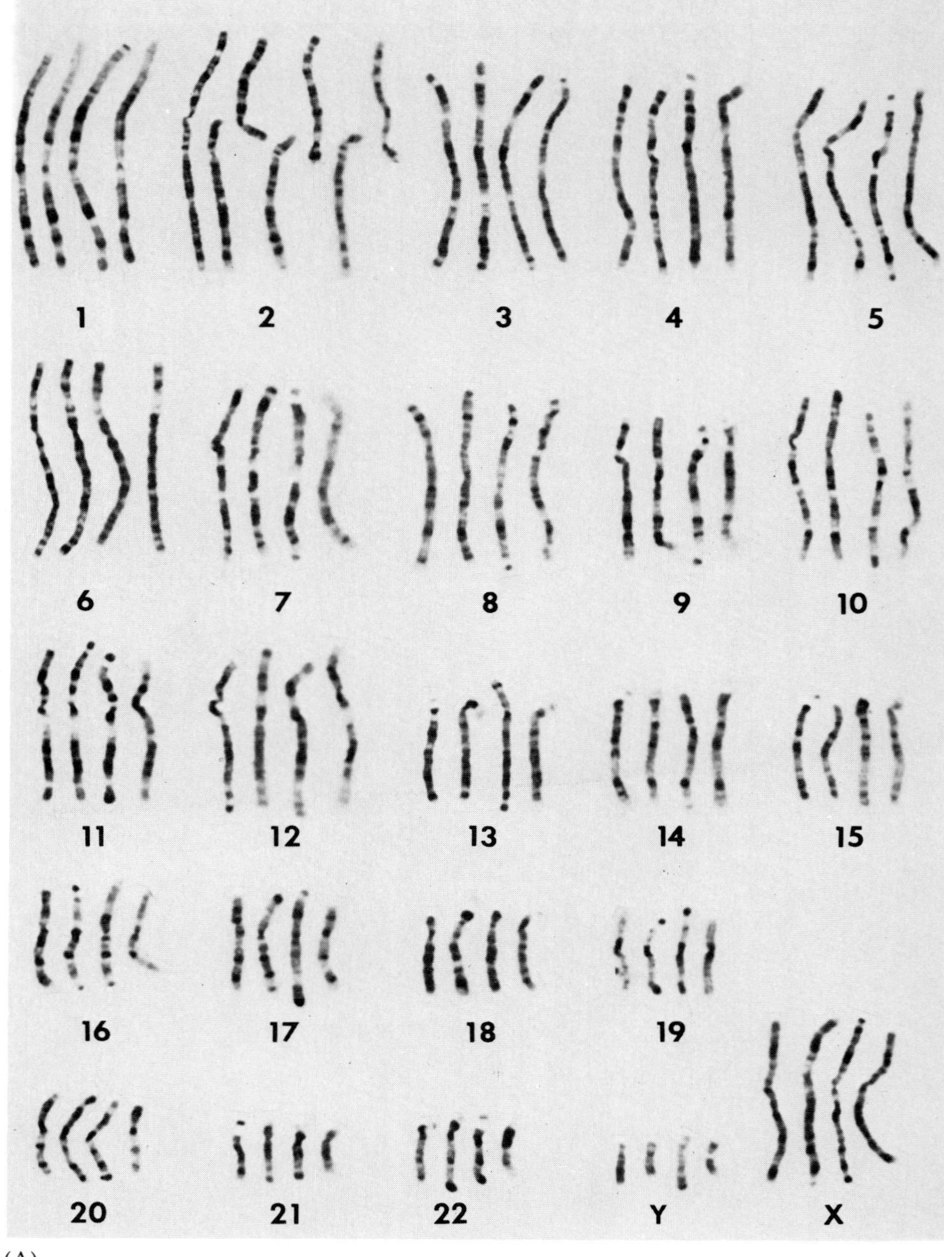

(A)

Clinical importance of chromosome banding. As pointed out in the preceding paragraph, banding patterns are unique and constant for each *normal* chromosome. In the case of a large number of chromosomal abnormalities, such as loss of a very small part, insertion of an additional segment, and addition of whole chromosomes, it has been possible to determine which part of which chromosome is abnormal and in what way. For example, in humans the cat-cry syndrome has been found to be caused by the loss of a small part of chromosome 5, and Down syndrome is seen to be due to an extra chromosome 21 or a part of one. It has even been possible to narrow the chromosomal region responsible to a small part of number 21. These and other disorders associated with abnormalities in chromosome structure or number are discussed in Chapters 14 and 15.

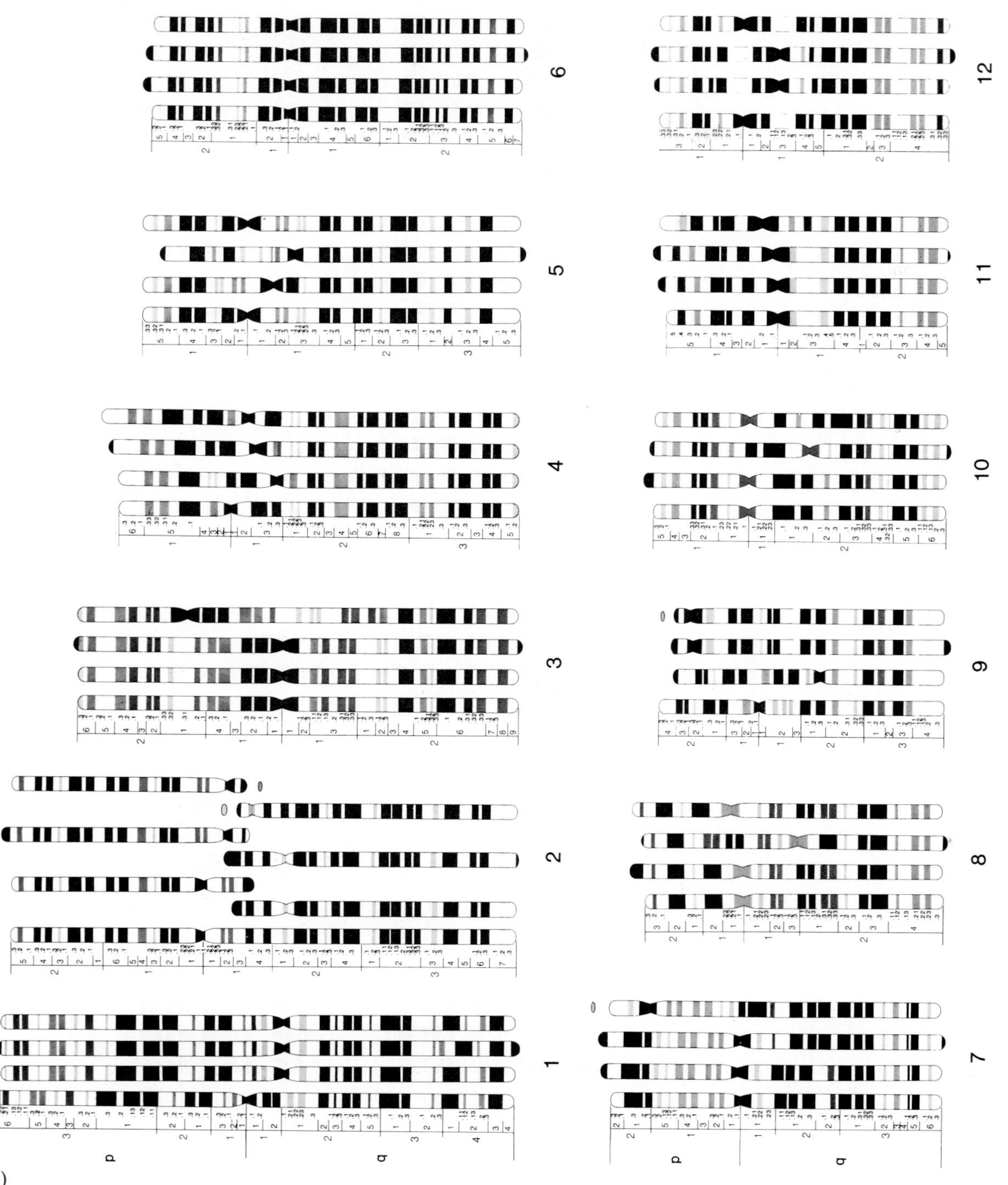

(B)

(Figure 4–22B continues on next page.)

FIGURE 4–22B. *(continued)*

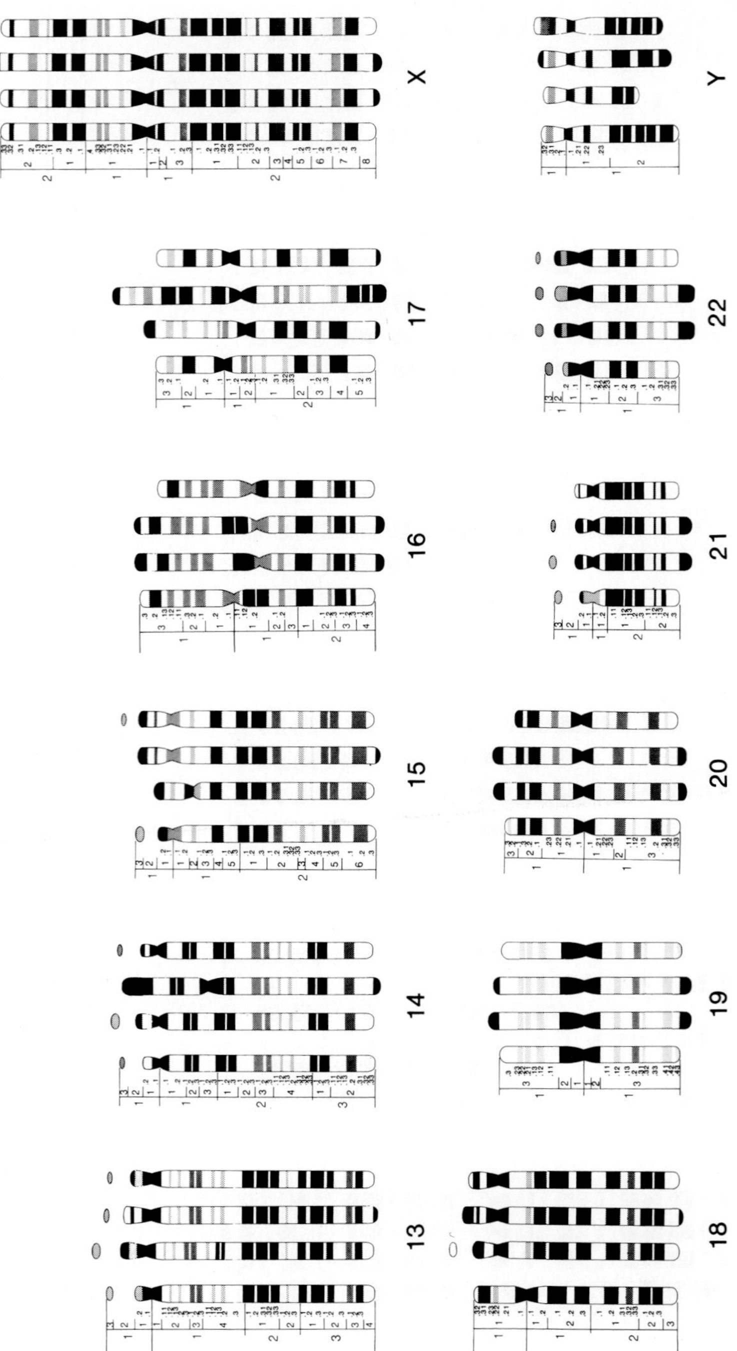

Meiosis

One of two fundamental cytological and genetic events in the life cycle of sexually reproducing plants and animals is syngamy, the union of gametes or sex cells to form a zygote. Studies from as long ago as the nineteenth century clearly indicate that in gametic union the chromosomes contributed by each gamete retain their individual identities in

ARTIFICIAL CHROMOSOMES

A new technique has been developed by which an artificial chromosome can be constructed and then transferred to a yeast cell where it functions almost as well as a real yeast chromosome. The significance of this discovery is that virtually any gene or small block of genes can be added to this artificial chromosome and its expression in the yeast cell subsequently studied. Several laboratories have been able to isolate the various components of yeast chromosomes that are essential for chromosomal activities. These components include the centromere, responsible for chromosome movement; a so-called autonomous replicating sequence (ARC) which is responsible for duplicating the chromosome between cell divisions; and telomeres, the chromosome ends that are required to complete chromosome duplication. When these elements of any chromosome are isolated and fused together and reintroduced back into a yeast cell, they behave during cell divisions almost like a normal chromosome.

This technique has tremendous potential. An exciting aspect of this discovery is that non-yeast genes may be added to the artificial chromosome. Furthermore, the larger the block of non-yeast genes that is added—within certain limits—the better the chromosome seems to function. An important potential use of artificial chromosomes will be to attach human or other foreign genes to these chromosomes and study their expression in the yeast cell. In addition, by culturing the yeast in large quantities, numerous copies of the human gene can be isolated from the genetic material of the cells. Presently this can only be done in bacterial cells, but in bacteria there are limitations on the number of genes that can be added to the yeast chromosome. This limitation is substantially overcome by using artificial yeast chromosomes.

the zygote nucleus. The zygote thus contains twice as many chromosomes as does a gamete or, more accurately, all the chromosomes of both gametes.

This fact is responsible for the presence in diploid or $2n$ cells of matching pairs of chromosomes; each member of a given pair is the **homolog** of the other. In each diploid nucleus, then, there occurs the haploid number of *pairs* of chromosomes, one member of each pair having been contributed by the paternal parent, and the other by the maternal parent. Thus, if among the chromosomes in the sperm, there is a metacentric chromosome (Figure 4–16) of, for example, an average metaphase length of 5 μm, there will be an identical chromosome in the egg. In the zygote and all subsequent cells derived from it by mitosis, *two* metacentric chromosomes of an average metaphase length of 5 μm will be found. This statement applies to all the *autosomes* (those chromosomes not associated with sex of the bearer). As will be seen later, these statements will have to be modified for the *sex chromosomes* in species where these occur.

The result of syngamy is the incorporation into a zygote nucleus of all the chromosomes of each gamete. Such a circumstance would seem to require a counterbalancing event whereby the doubling of chromosome quantity at syngamy is offset by a nuclear division that halves the amount of chromatin per nucleus at some point prior to gamete formation. Such a "reduction division" does take place in all sexually reproducing organisms that have discrete nuclei. This is **meiosis,** the second of the two fundamental cytological and genetic events in the sexual cycle.

Meiosis as a form of nuclear division differs greatly from mitosis. It consists of two successive divisions, each with its own prophase, metaphase, anaphase, and telophase; therefore, it results in four daughter nuclei instead of two as in mitosis (although in many species not all the cellular products are functional). Furthermore, because of fundamental differences between mitosis and meiosis, the nuclear products of a meiotic division have only one set of chromosomes each (haploid) as compared to the two sets (diploid) found in each mitotic product. In addition, if genes and chromosomes have any relationship, the nuclear products of meiosis can be genetically unlike each other compared to the genetically identical products of mitosis.

A **meiocyte** is any diploid cell that is destined to undergo meiosis. In seed plants and other heterosporous forms, for example, meiocytes are represented by the megasporocytes ("spore mother cells"), in homosporous plants (those producing only one kind of spore by meiosis) by sporocytes or "spore mother cells," and in higher animals by primary spermatocytes and primary oocytes. The position of the meiocyte in the life cycle varies greatly from one major group to another. For instance, in the vascular plants the meiocytes give rise directly to megaspores and microspores (that is, the **meiospores**), producing gamete-bearing plants by mitosis. In the seed plants these gamete-bearing plants (the gametophytes) are much reduced in size, number of cells, and complexity; refer to Appendix B. In animals, however, meiocytes give rise directly to gametes. In many algae and fungi the zygote nucleus itself functions as a meiocyte, giving rise to haploid cells that ultimately produce haploid, gamete-bearing plants by mitosis. In summary, meiosis may be sporic (vascular plants and others), or gametic (human and many other animals), or zygotic, and may be performed by (1) sporocytes ("spore mother cells"), (2) primary spermatocytes and oocytes, or even (3) the zygote itself.

First division: prophase-I. Although most cytologists recognize and name at least five stages of prophase-I because of its complexities of chromosome behavior, it will suffice for our purposes merely to emphasize events in sequence. The first meiotic prophase often persists for a very lengthy time, and may be measured in weeks or months or even longer. In the human female it persists in each primary oocyte from a fetal age of 12 to 16 weeks until ovulation, generally once each 28 days, after sexual maturity. Some human oocytes, therefore, persist in arrested prophase-I until the end of the reproductive cycle at menopause, which is about 45 years. In very early prophase-I, the diploid number of chromosomes gradually becomes recognizable under the light microscope as long, slender, threadlike structures (Figure 4–23). In the light microscope the chromosomes *appear* to be longitudinally single in earliest prophase-I (**leptotene**), although chemical and autoradiographic studies show that replication has occurred in the S stage of the preceding interphase. By definition, then, no visible chromatids exist at this time in a meiocyte. Chromosomes shorten and thicken progressively, presumably by the same mechanism as for mitosis.

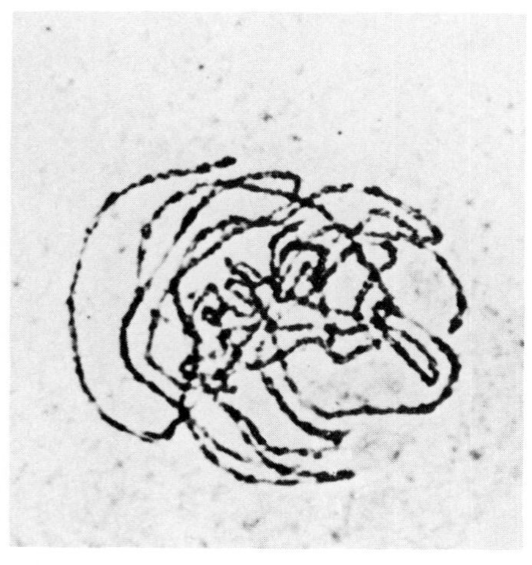

FIGURE 4–23. *Pollen mother cell of crested wheatgrass* (Agropyron cristatum) *in late zygotene of meiosis.* (Photo courtesy Dr. W. Tai, University of Manitoba.)

While this contraction is underway, a second event that characterizes meiosis (as opposed to mitosis) takes place. Recall that each diploid nucleus (including meiocytes) contains *pairs* of homologous chromosomes. In early prophase-I (**zygotene**) the homologs begin to pair, or **synapse.** This **synapsis** is remarkably exact and specific, taking place point by point, with the two homologs usually somewhat twined about each other (Figure 4–24). However, nothing definitive is known about the attractive forces that bring together homologs, as opposed to random chromosomes. Synapsis begins by an unknown mechanism when the ends of the homologs are within about 300 nm of each other.

Electron microscopy shows at synapsis a feature of dimensions too small to be discernible with the light microscope, the **synaptonemal complex** (Figure 4–25). This intricate and complex structure is the physical mechanism by which (1) homologs are held together at synapsis and (2) crossing-over (to be described) is made possible. The synaptonemal complex is composed of three parallel bands, interconnected by fine strands arranged at right angles to them. The two outer bands are axial components of the homologous chromosomes and are some 30 to 60 nm in diameter; the central one may be preformed in the nucleolus. The central component gives rise to the lateral elements and is about 100 nm in width. These elements may be amorphous (both central and lateral in most higher plants and animals, including humans), banded (lateral elements in ascomycete fungi and some insects), or latticelike with fine rods spaced about 10 nm apart (central element in insects).

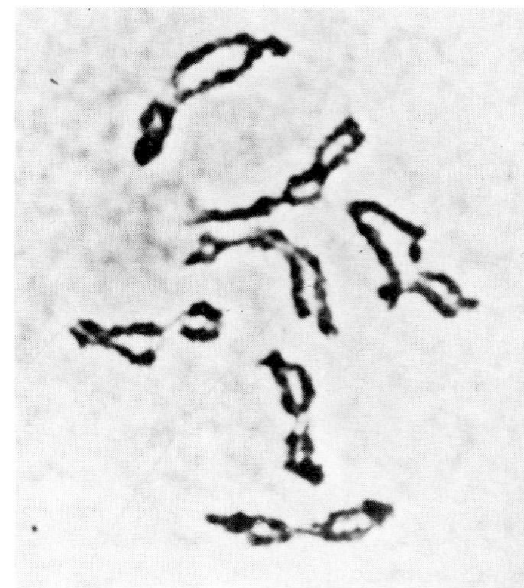

FIGURE 4–24. *Diplotene of meiosis in crested wheatgrass showing synapsis and chiasmata.* (Photo courtesy Dr. W. Tai, University of Manitoba.)

The whole structure is some 140 to 170 nm in width, just below the limit of resolution in light microscopy. Formation of the synaptonemal complex is initiated at several points along the pair of homologs. There is good chemical evidence that breakage and repair of DNA molecules of the chromosomes take place within the synaptonemal complex; this can and often does result in exchange of material between nonsister chromatids (*crossing-over*). Furthermore, synaptonemal complexes are not found in organisms in which crossing-over does not occur (for example, the male fruit fly, *Drosophila*). Synapsed chromosomes continue to shorten and thicken, and in the process, their longitudinal doubleness becomes apparent in light microscopy.

The synaptonemal complex is completed at the next stage of prophase-I (**pachytene**), during which the synapsed homologs are now clearly seen to be composed of two chromatids each. The points at which exchange of material occurs between nonsister chromatids is evidenced by more or less X-shaped configurations, the **chiasmata** (plural for **chiasma**), as seen in Figure 4–25. The longer the chromosome, the greater likelihood of more than one chiasma, although one chiasma appears to interfere with the formation of another in a closely adjacent region of the chromosomes on the same side of the centromere. The basis for this **interference** is not altogether clear.

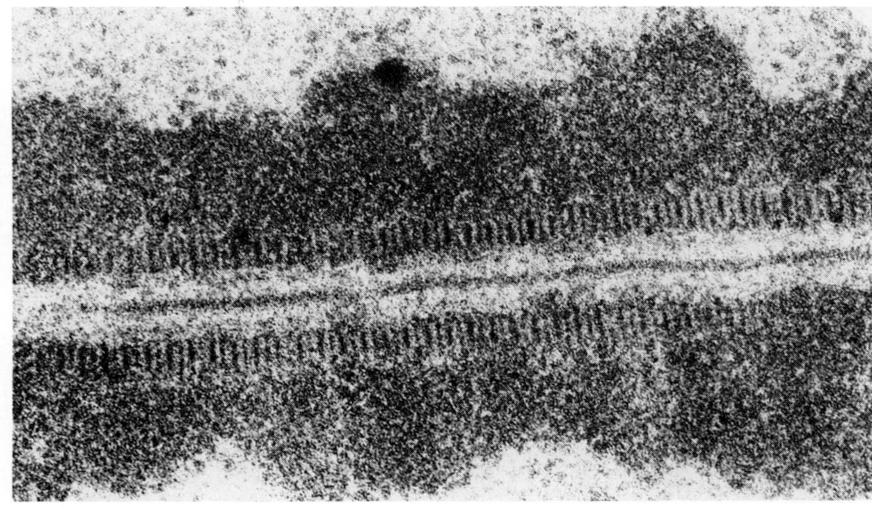

FIGURE 4–25. *Synaptonemal complex in Neottiella.* (Photo courtesy Dr. D. von Wettstein.)

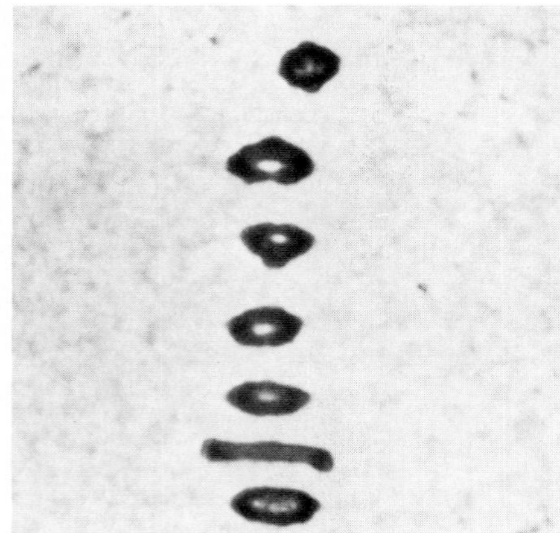

FIGURE 4–26. *Metaphase-I of meiosis in crested wheatgrass.* (Photo courtesy Dr. W. Tai, University of Manitoba.)

Next, in the **diplotene** stage of prophase-I, separation of homologs (except at points where chiasmata occur) is initiated. Chromosomes continue to contract and the nucleolus begins to disappear.

Finally, in the last stage of prophase-I (**diakinesis**), chromosomes reach maximum contraction. At the outset of diakinesis the synapsed homologs become spaced out well in the nucleus, often near the nuclear membrane. Chiasmata gradually *terminalize,* that is, they appear to move toward the ends of the arms and finally "slip off," owing to the continued shortening of the chromosomes. The nucleolus then disappears, the nuclear membrane degenerates, and a spindle is formed.

Metaphase-I. The arrival of synapsed homologous chromosome pairs at the equator of the spindle begins metaphase-I. This phase differs from mitotic metaphase in (1) the arrangement of the haploid number of synapsed homologous chromosome pairs on the equatorial plane and (2) the tendency for the centromere of each homolog to be directed somewhat toward one of the poles (Figure 4–26). An important point to be noted here is the *randomness* of arrangement of the paired homologs; that is, **for a given pair, it is just as likely that the paternal member be directed toward the "north" pole as for the maternal member to be so oriented.** Recall the gamete types produced by a polyhybrid individual (Chapter 3).

Anaphase-I. In anaphase-I actual **disjunction** of synapsed homologs occurs, one longitudinally double chromosome of each pair moving to each pole, thereby completing the process of terminalization. Here is an additional difference from mitosis, mitotic anaphase being marked by separation of sister chromatids, which then move poleward as longitudinally single daughter chromosomes. Thus in mitosis one of each of the chromosomes present travels to each pole, with the result that each new nucleus has just as many chromosomes as the parent nucleus had, whether haploid or diploid. In meiosis, however, whole chromosomes of each homologous pair (as modified by crossing-over that has occurred in prophase-I) separate, so that each pole receives either a paternal or maternal, longitudinally double chromosome of each pair. This ensures a change in chromosome number from diploid to haploid in the resultant reorganized daughter nuclei. In short, whereas mitotic anaphase was marked by the separation of sister *chromatids,* anaphase-

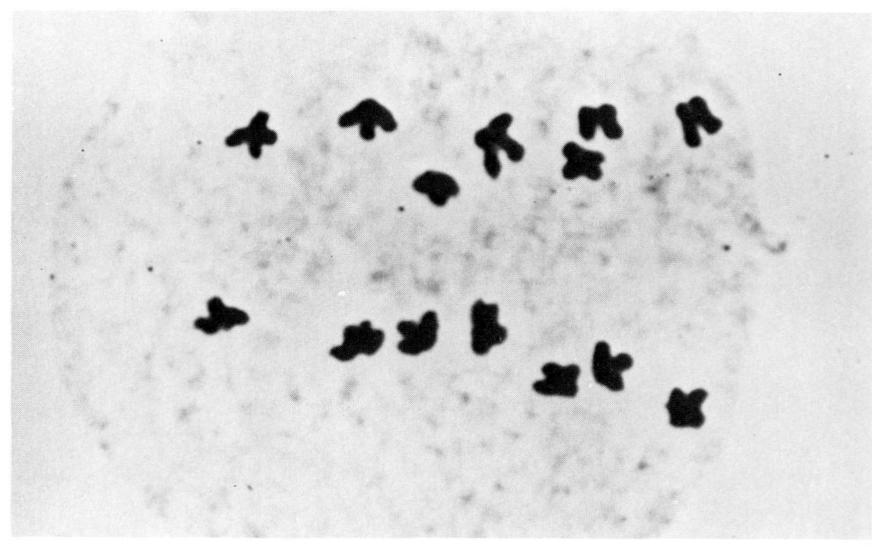

FIGURE 4–27. *Anaphase-I of meiosis in crested wheatgrass.* (Photo courtesy Dr. W. Tai, University of Manitoba.)

I of meiosis is characterized by separation of homologous *chromosomes* (Figure 4–27).

Telophase-I. The arrival of chromosomes at the poles of the spindle signals the end of anaphase-I and the beginning of telophase-I. During this phase the chromosomes may persist for a time in the condensed state, the nucleolus and nuclear membranes may be reconstituted, and cytokinesis may also occur (Figure 4–28). In some cases, as in the flowering plant genus *Trillium,* meiocytes are reported to progress virtually directly from anaphase-I to prophase-II or even metaphase-II; in other organisms there may be either a short or fairly long interphase or interkinesis between the first and second meiotic divisions. In any event, the first division has thus accomplished the separation of the chromosomal complement into two haploid nuclei. There is no replication between meiotic divisions.

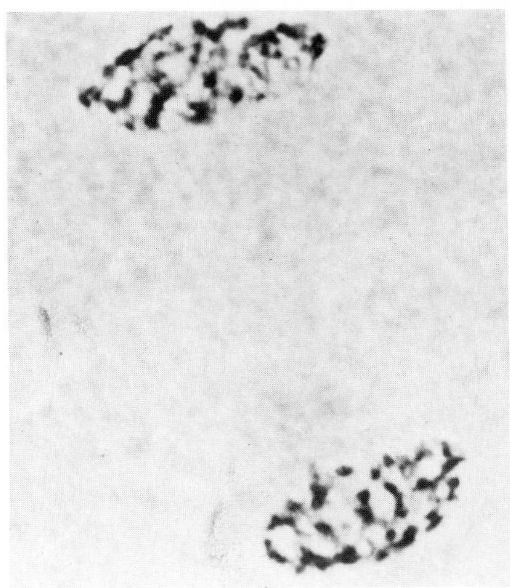

FIGURE 4–28. *Late telophase-I of meiosis in crested wheatgrass.* (Photo courtesy Dr. W. Tai, University of Manitoba).

Second division: prophase-II. Prophase-II is generally short and superficially resembles mitotic prophase, except that sister chromatids of each chromosome are usually widely divergent, exhibiting no relational coiling.

Metaphase-II. On two spindles, generally oriented at right angles to the first division spindle and often separated by a membrane or wall, the haploid numbers of chromosomes, each consisting of two chromatids joined at the centromeres, are arranged in the equatorial plane. This stage is generally brief (Figure 4–29).

Anaphase-II. Centromeres now separate and the *sister chromatids* of metaphase-II move poleward *as daughter chromosomes,* similar to their movement in mitosis. Their arrival at the poles marks the close of this phase.

Telophase-II. Following arrival of the haploid number of daughter chromosomes at the poles, the chromosomes return to a long, attenuate, reticulate conformation, nuclear membranes are reconstituted, nucleoli reappear, and cytokinesis generally separates each nucleus from the others (Figure 4–30).

The process of meiosis is summarized diagrammatically in Figure 4–31.

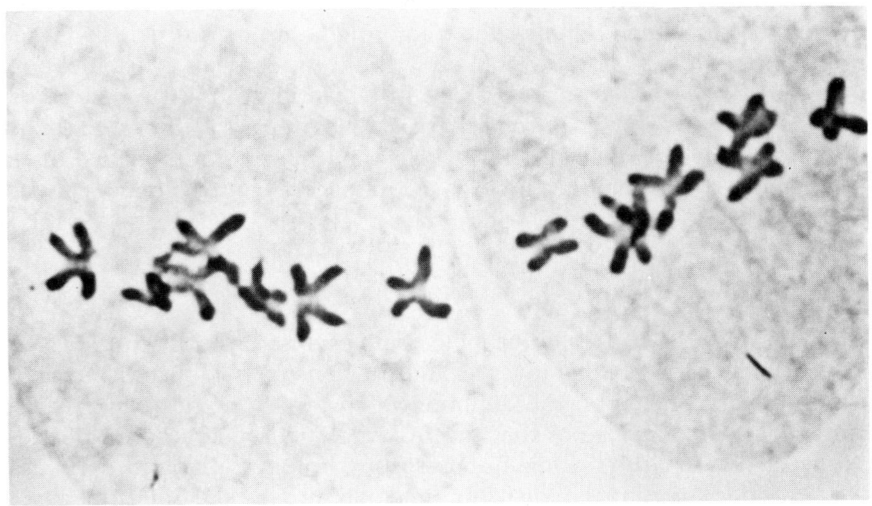

FIGURE 4–29. *Metaphase-II of meiosis in crested wheatgrass.* (Photo courtesy Dr. W. Tai, University of Manitoba.)

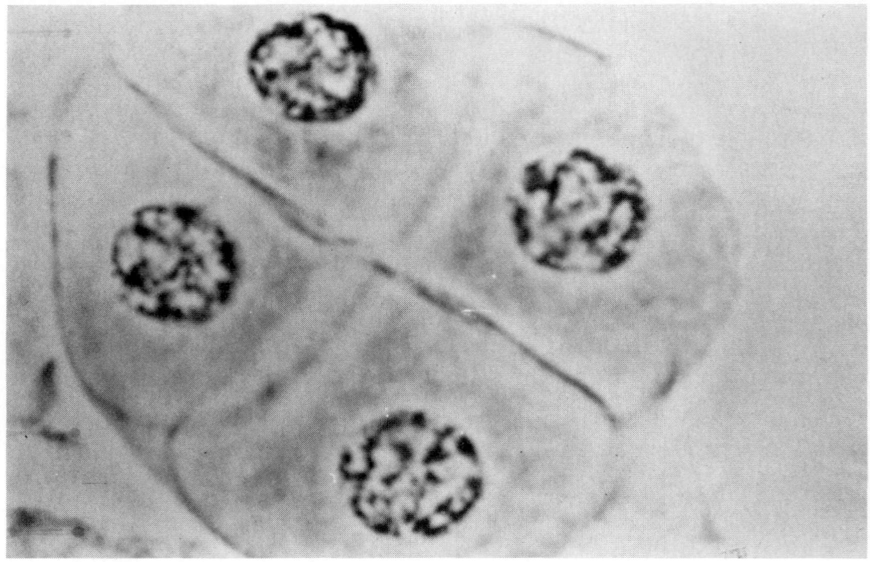

FIGURE 4–30. *Tetrads in crested wheatgrass.* (Photo courtesy Dr. W. Tai, University of Manitoba.)

Concluding view. Electron microscopy has supplied answers to some questions about meiosis, but others remain unanswered. For instance, such tools permit the observation that in anaphase-I *homologs* separate; in mitotic anaphase *sister chromatids* face in opposite directions, one toward each pole; in metaphase-I, the two centromeres of each pair of *synapsed homologs* face different poles. Because of the attachment of microtubules to the kinetochores of the centromeres, then, homologous whole chromosomes are separated at anaphase-I, but sister chromatids separate in mitotic anaphase.

The second meiotic division has been compared to mitosis, but it is quite different. On the spindles in meiosis-II the chromosomes are always present in the haploid number; in mitosis either the haploid or the diploid chromosome number may occur, depending on the tissue in which division is taking place. Chromatids in the second division are widely separated, exhibit no relational coiling, and have often been modified by crossing-over.

Although electron microscope observations of the synaptonemal complex have given some information on the mechanism and time of crossing-over, the precise physical and chemical details of synapsis remain to be elucidated. Finally, the triggering stimulus for meiosis is still unknown, although it appears likely to be hormonal, or at least chemical. Meiocytes excised in interphase often undergo mitosis instead of meiosis *in vitro;* excision of cells successively later in prophase-I results in a progressively larger number of cells that do undergo meiosis. In all probability, commitment to meiosis as opposed to mitosis may occur as early as the S or G_2 stage of interphase.

Significance of Meiosis

Cytologically, the basic significance of meiosis is the formation of four monoploid nuclei from a single diploid nucleus in two successive divisions, thus balancing off the doubling or chromosome number that results from syngamy. Note that the first meiotic division accomplishes the reduction in chromosome number from diploid to haploid, whereas the second is equational in distributing equal numbers of daughter chro-

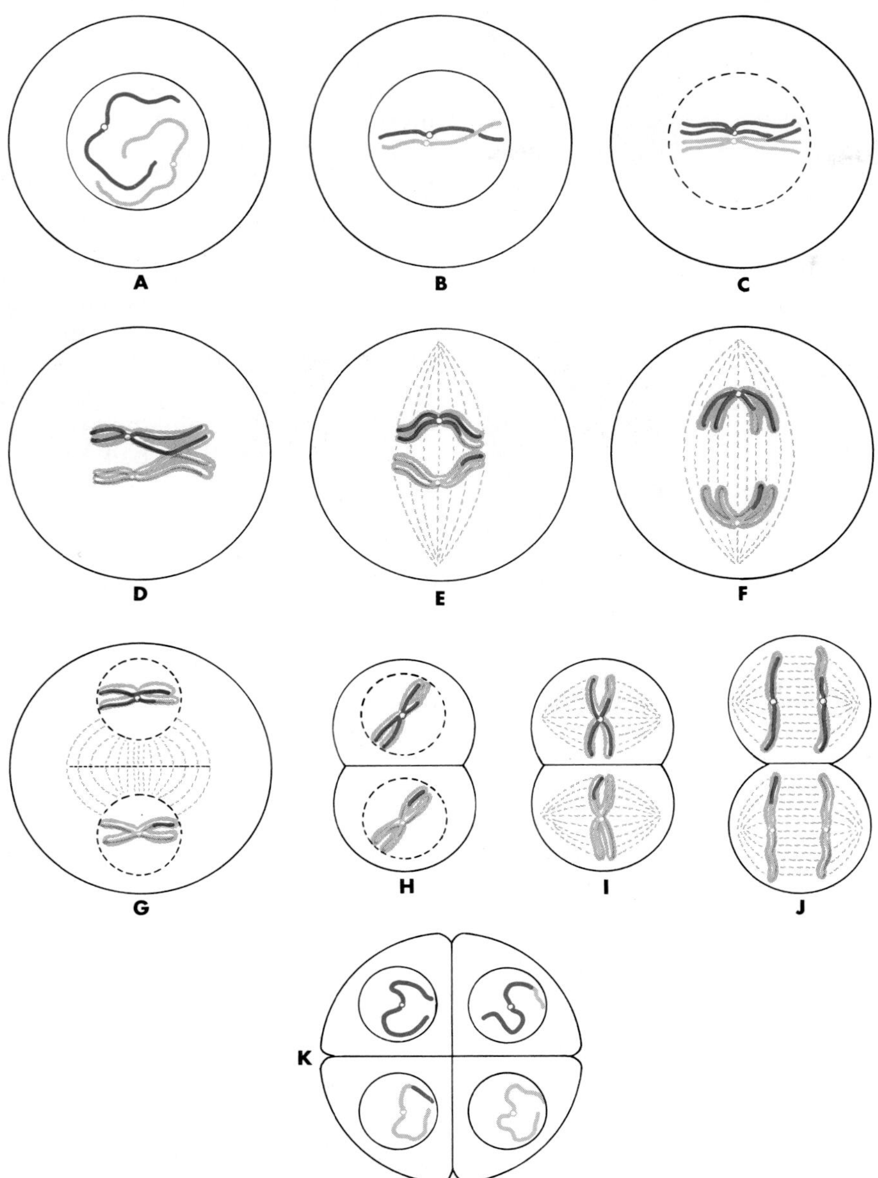

FIGURE 4–31. *Diagrammatic representation of meiosis in a meiocyte showing one pair of homologous chromosomes (2n = 2) based on light microscopy. (A) Prophase-I, chromosomes long and slender, appearing longitudinally single, although sister chromatids had been produced in the preceding interphase; (B) prophase-I, homologous chromosomes synapsing, and still appearing longitudinally single; (C) prophase-I, chromatids now visually evident, with one chiasma; (D) prophase-I disjunction underway, chiasma still evident; (E) metaphase-I chromosomes on midplane of spindle with centromeres divergent; (F) anaphase-I, poleward separation of previously synapsed homologs; note that each chromosome is still composed of two chromatids but some of these have been modified by crossing-over; (G) telophase-I, each reorganizing nucleus now contains the monoploid number of chromosomes that are still composed of two chromatids each; (H) prophase-II; (I) metaphase-II; (J) anaphase-II; (K) a post-meiotic tetrad of monoploid cells. Note that each in this case contains a genetically unique chromosome because of crossing-over.*

mosomes to developing new nuclei. In higher animals the cellular products of meiosis directly become gametes and/or polar bodies (Appendix B–8), but in vascular plants the meiotic products are meiospores that give rise to reduced gamete-bearing plants (Appendix B–6).

But how does meiosis relate to the hypotheses raised concerning probable gene distribution in somatic and gametic cells in the preceding chapter? If genes are located on chromosomes, the process of meiosis generates genetic variability in two important ways: (1) random assortment of paternal and maternal chromosomes and (2) crossing-over. Assume, for example, an organism heterozygous for three pairs of alleles (a trihybrid), *AaBbCc*, in which *ABC* was derived from its paternal parent and *abc* from its maternal parent. If these three pairs of alleles are located on three different homologous chromosome pairs (which might be designated as pair 1, pair 2, and pair 3), then in prophase-I, the paternal and maternal number 1 chromosome (bearing alleles *A* and *a*, respectively) will synapse, as will the number 2s (bearing *B* and *b*) and the number 3s (with alleles *C* and *c*).

Arrangement of each pair of synapsed homologs on the metaphase-I spindle is random; that is, the paternal member of each pair has an equal chance of being oriented toward either pole, as does the maternal member. In anaphase-I each (longitudinally double) chromosome moves toward the nearer pole as it separates from its homolog. Therefore each telophase-I nucleus has an equal chance of receiving a paternal or a maternal homolog of each chromosome. The same chance exists for each remaining chromosome pair. Hence, for three pairs of genes on three different pairs of homologous chromosomes, eight possible telophase-I gametic combinations are possible:

Daughter nucleus no. 1	*Daughter nucleus no. 2*
ABC	abc
ABc	abC
AbC	aBc
Abc	aBC
aBC	Abc
aBc	AbC
abC	ABc
abc	ABC

The second division will simply increase the number of each gametic gene arrangement from one to two. The number of possible gamete types occurring after the second meiotic division is 2^3, or 8. That is, a given gamete in this example has $(\frac{1}{2})^3$ chance of receiving any particular arrangement of paternal and/or maternal chromosomes and genes. Furthermore, in a sample of several hundred gametes from such an individual, each of the eight types would be expected to occur in approximately equal number. For organisms with many chromosomes the number of possible combinations becomes very large. In humans, for instance, with 23 pairs of homologous chromosomes, the probability that any particular gamete will have a specific combination of chromosomes (excluding crossing-over) is $(\frac{1}{2})^{23}$, or about 1 in 8 million. Possible zygotic genotypes resulting from random fusions of these gametes rise to about 64 trillion. A good deal of the variation in natural populations is due to this kind of **independent segregation** of genes normally occurring in the breeding population.

What would be the situation if these three allele pairs are, instead, located on a single pair of homologous chromosomes (that is, **linked**)? Assume, for the purpose of illustration, that one member of the pair of homologs bears genes *A*, *B*, and *C* on a given arm and that the other homolog bears genes *a*, *b*, and *c*. If no chiasmata are formed, indicating no crossing-over, then only two kinds of gametes (*ABC* or *abc*) will be produced, these occurring in equal number. If chiasmata do form between *A* and *B* and between *B* and *C* in at least some meiocytes, as is more often the case, then eight gametic genotypes will again be produced. However, in this case *the fraction of each type produced will depend on the frequency with which crossing-over occurs between A and B and B and C*. This has considerable importance in the mapping of genes, which will be considered in detail in Chapter 6.

PROBLEMS

4–1 In *Coleus* the somatic cells are diploid, having 24 chromosomes. How many of each of the following are present in each cell at the stage of mitosis or meiosis indicated? (Assume cytokinesis to occur in mid-telophase.) (a) Centromeres at anaphase, (b) centromeres at anaphase-I, (c) chromatids at metaphase-I, (d) chromatids at anaphase, (e) chromosomes at anaphase, (f) chromosomes at metaphase-I, (g) chromosomes at the close of telophase-I, (h) chromosomes at telophase-II.

4–2 Corn is a flowering plant whose somatic chromosome number is 20. How many of each of the following will be present in *one* somatic cell at the stage listed: (a) centromeres at prophase, (b) chromatids at prophase, (c) kinetochores at prophase, (d) chromatids in G_1, (e) chromatids in G_2?

4–3 Either from information you already have or based on facts in Appendix B, for a corn plant how many chromosomes are present in each of the following: (a) leaf epidermal cell, (b) antipodal nucleus, (c) endosperm cell, (d) generative nucleus, (e) egg, (f) megaspore, (g) microspore mother cell?

4–4 From your present knowledge, or after consulting Appendix B, how many human eggs will be formed from (a) 40 primary oocytes, (b) 40 secondary oocytes, (c) 40 ootids?

4–5 From your present knowledge, or after consulting Appendix B, how many human sperms will be formed from 40 primary spermatocytes?

4–6 From your present knowledge, or from Appendix B, 20 microsporocytes of a flowering plant would be expected to produce how many (a) microspores, (b) sperms?

4–7 Consult Table 4–1 in answering the following questions. (a) What is the probability in cattle that a particular egg cell will contain only chromosomes derived from the maternal parent of the cow producing the egg? (b) If this cow is mated to its brother, what is the probability that the calf will receive only chromosomes originally contributed by the calf's grandmother? (*Suggestion:* Can the product law of probability be of help here?)

4–8 Triploid watermelons have the advantage of being seedless. (a) What is the somatic chromosome number of such plants? (b) What explanation can you offer for their lack of seeds?

4–9 A cell of genotype *Aa* undergoes mitosis. What will be the genotype(s) of the daughter cells?

4–10 A cell of genotype *Aa* undergoes meiosis. What will be the genotype(s) of the daughter cells if all of them are functional?

4–11 A student examining a number of onion root tips counted 1,000 cells in some phase of mitosis. He noted 692 cells in prophase, 105 in metaphase, 35 in anaphase, and 168 in telophase. From these data what can be concluded about relative duration of the different stages of the process?

4–12 What is the probability that any ascospore of the orange-pink bread mold (*Neurospora crassa*) will have all its chromosomes derived from the + parent? (You may wish to consult Table 4–1 and Appendix B before trying to answer.)

4–13 How many different gamete genotypes will be produced by individuals of the following genotypes if all genes shown are on different chromosome pairs: (a) *AA*, (b) *Aa*, (c) *AaBB*, (d) *AaBb*, (e) *AAbbCc*, (f) *AaBbCcDdEe?*

4–14 A given individual is heterozygous for two pairs of alleles that are located on the *same* homologous chromosome pair, *A* and *B* on one chromosome, with *a* and *b* located on the homologous chromosome. (a) How many gamete types can such an individual produce if there is no crossing-over? (b) What will be those types?

4–15 Use the same individual as in the preceding problem, but assume that crossing-over between the two pairs of alleles occurs with a total frequency of 0.2 (that is, 2 in each 10 gametes will have crossover genotypes).

(a) How many gamete types can this individual produce? (b) What will be those types? (c) Using the product law of probability, with what frequency will *each* of the *crossover* gamete types occur? (d) With what frequency will *each* of the *noncrossover* gamete types occur?

4–16 In corn, recessive gene *hm* determines susceptibility to the fungus *Helminthosporium,* which produces lesions on leaves and in kernels, reducing yield, vigor, and market value of the crop. Its dominant allele determines resistance to the fungus. The recessive allele of a second gene, br_1, produces shortened internodes (segments of stem between successive leaves) and stiff, erect leaves (brachytic plants). Its dominant allele is responsible for normal plant form. These two pairs of genes are linked on chromosome 1. If a resistant, normal plant, heterozygous for both pairs of genes (with dominants of each gene on one homologous chromosome) is pollinated by a susceptible, brachytic plant, (a) would you expect any resistant, brachytic and susceptible, normal individuals in the progeny? (b) Explain the cytological basis of your answer.

4–17 Assume a particular metaphase human chromosome has an average length of 5 μm. Give the equivalent in (a) meters, (b) millimeters, (c) nanometers, (d) angstrom units.

4–18 What possible advantages to a species can you see in crossing-over?

4–19 Can you see possible disadvantages that might result from crossing-over?

4–20 Notice (Table 4–1) that diploid chromosome numbers are, with rare exceptions such as the male grasshopper, even numbers. Explain.

PROBABILITY AND GOODNESS OF FIT

CHAPTER

5

An understanding of the laws of probability is of fundamental importance in (1) appreciating the operation of genetic mechanisms, (2) predicting the likelihood of certain results from a given cross, and (3) assessing how well a progeny phenotypic ratio fits a particular postulated inheritance pattern. One of the fundamental laws of probability (that is, the law of the probability of coincident independent events) has already been applied in studies of dihybrid and polyhybrid ratios based on transmission of two pairs of alleles that are assumed to be different chromosome pairs, as well as allelic pairs that are on a single chromosome pair. This product rule of probability states, in essence, that **the chance (or probability) of the simultaneous occurrence of two or more independent events is equal to the product of the probabilities of each event.** Recall also the *additive* rule of probability from Chapter 2.

TWO INDEPENDENT, NONGENETIC EVENTS

Single-Coin Tosses

The laws of probability can be applied to *any* chance or random event. For example, a coin tossed into the air and allowed to come to rest is likely to land either "heads" or "tails," if the highly improbable chance of its landing on edge is neglected. Therefore, one could predict that, in the total array of possibilities (heads plus tails), the probability of either a head or a tail is 1 in 2, or $\frac{1}{2}$. But if one coin is tossed four successive times, a ratio other than 2 heads:2 tails would not be surprising. Occasionally, such combinations as four heads, or three heads and one tail, or one head and three tails, or even four tails would be expected. If, however, a very large number of tosses were to be made, a result quite close to a 1:1 ratio would be expected. Note that successive tosses are assumed to be independent of each other; that is, the result of one toss has no effect on any succeeding toss.

Two-Coin Tosses

What can be expected if two coins are tossed simultaneously for, say, 50 tosses? Assume two students were each asked to toss two coins simultaneously 50 times by shaking the coins in closed, cupped hands and letting them fall lightly on a table. Ignoring the possibility of a coin landing on edge, only two results can occur: each coin will come to rest flat on the table, with either a head up or the other side, a tail, up. With the two coins, then, the outcome of any single toss will be HH, HT, or TT. The result of the series by student *A* was

HH	12
HT	27
TT	11

Now one has to ask, "Are these the results I could have expected, based on the theory that each coin has an equal chance of landing either heads or tails?" Or, in a genetic experiment, after having made a given cross and obtained certain results, one may ask, "What possible mechanism could be operating in order to produce these results?" In either a coin toss or a genetic breeding experiment, one of the first necessary steps is to formulate certain hypotheses to predict or explain results, and then ask whether, *in terms of those hypotheses, deviation from the predicted result is within limits set by chance alone.* If so, the hypotheses may be used in predicting outcomes of untried cases: if not, the hypotheses will have to be altered. (The chi-square test, to be discussed later in this chapter, will permit a judgment on the validity of hypotheses.)

In order to make a judgment on the question detailed in the preceding paragraph, several **simplifying assumptions** about the coins, as well as the general conditions of the experiment itself, must be made. First, one must ask *whether the coins themselves were unbiased:* that is, were they so constructed physically that each coin had an equal chance of landing either head or tail? Because there is no real evidence to the contrary, it may be assumed that the coins are indeed unbiased. If this assumption should turn out to be untrue, then the ultimate evaluation of the results will have to be changed.

A second question must also be examined: that is, whether the "headness" or "tailness" of the fall of one of the two coins will have any effect on the fall of the second coin. That is to say, does *each* coin have an *equal* chance of coming to rest heads or tails? Certainly, in this experiment there should be no more than a remote chance that the ultimate fall of one coin affects the other. Therefore, a *second assumption* will be that *the behaviors of the coins themselves are independent of each other.*

A *third assumption* will also have to be made, at least at the outset, namely that *successive tosses (events) are also independent of each other.* This would imply, for example, that a toss resulting in two heads will in no way affect the outcome of the next or any other toss.

If it is assumed, therefore, that the **coins are (1) unbiased and (2) independent of each other and also that (3) the several trials or tosses are likewise independent of each other,** what kind of results would be *expected?* Each coin has one chance in two, or a probability of $\frac{1}{2}$, of coming to rest heads and a probability of $\frac{1}{2}$ of landing tails. By the product rule of probability, the chance of *both* coins showing heads in the same toss is $\frac{1}{2} \times \frac{1}{2}$, or $\frac{1}{4}$. Student *A* did, indeed, get two heads in almost exactly one-fourth of his tosses. Likewise, the probability of having two tails simultaneously is also $\frac{1}{2} \times \frac{1}{2} = \frac{1}{4}$. This result was not quite obtained in the example under consideration.

What should be expected with regard to one head and one tail? The chance of coin 1 landing heads is $\frac{1}{2}$; the chance of coin 2 coming to rest tails is also $\frac{1}{2}$. Now $\frac{1}{2} \times \frac{1}{2} = \frac{1}{4}$, but the total array of possibilities thus arrived at ($\frac{1}{4}$ HH + $\frac{1}{4}$ TT + $\frac{1}{4}$ HT) equals only $\frac{3}{4}$, leaving $\frac{1}{4}$ of the possibilities unaccounted for. Note that, by the rule of additive probability, chance of one head plus one tail is really the *sum* of the probability of coin 1 being heads and coin 2 being tails, *plus* the probability of coin 1 being tails and coin 2 being heads; that is, the HT category is really HT + TH, or 2 HT. Therefore the probability of one head and one tail is $2(\frac{1}{2} \times \frac{1}{2})$, or $\frac{1}{2}$.

The general statement for *two independent* events of known probability may be written as

$$a^2 + 2ab + b^2$$

where a represents the probability of a head and b the probability of a tail. If, as in this instance, a and b each equal $\frac{1}{2}$, then the value of this expression upon substitution becomes $(\frac{1}{2})^2 + 2(\frac{1}{2} \times \frac{1}{2}) + (\frac{1}{2})^2$ or $\frac{1}{4} + \frac{2}{4} + \frac{1}{4} = 1$. Notice that the total array of probabilities here is 1, and also that $a^2 + 2ab + b^2$ is the expansion of the binomial $(a + b)^2$, where the exponent represents the number of times the event occurs.

The observations and expectations for this two-coin toss can therefore be summarized as follows:

Class	*Observed*	*Expected*
HH	12	12.5
HT	27	25.0
TT	11	12.5
	50	50.0

Student *B*'s results, however, were a little different:

Class	*Observed*	*Expected*
HH	10	12.5
HT	33	25.0
TT	7	12.5
	50	50.0

Obviously the first set of data is closer to the results expected under the assumptions previously made, yet the second set of results was actually obtained. Each of these sets of data will be examined to see if the departure from the expected results is too great for *chance alone* to have operated within the framework of the three assumptions stated on page 72.

Four-Coin Tosses

Consider an experiment involving four coins tossed together 100 times. The possible combinations of heads and tails, and the results actually obtained in a particular trial, are as follows:

Class	*Observed*
HHHH	9
HHHT	32
HHTT	29
HTTT	25
TTTT	5

Just as a set of ideal results in a two-coin toss has been determined from the expansion of $(a + b)^2$, the binomial $(a + b)^4$ must be expanded here to determine the expected results. The *power* (exponent) of the binomial used is determined by the number of events (coins, children, and so forth) involved. Again, let

$$a = \text{probability of a head for any coin} = \tfrac{1}{2}$$

and

$$b = \text{probability of a tail for any coin} = \tfrac{1}{2}$$

Expand $(a + b)^4$:

$$a^4 + 4a^3b + 6a^2b^2 + 4ab^3 + b^4$$

Substitute the numerical values of a and b:

$$(\tfrac{1}{2})^4 + 4[(\tfrac{1}{2})^3 \cdot \tfrac{1}{2}] + 6[(\tfrac{1}{2})^2 \cdot (\tfrac{1}{2})^2] + 4[(\tfrac{1}{2}) \cdot (\tfrac{1}{2})^3] + (\tfrac{1}{2})^4 = 1$$

or

$$\tfrac{1}{16} + \tfrac{4}{16} + \tfrac{6}{16} + \tfrac{4}{16} + \tfrac{1}{16} = 1$$

The first term of the expression, $(\frac{1}{2})^4$, gives us the probability of all four coins coming up heads simultaneously; the second term, the probability of three heads (a^3) plus one tail (b), and so on. The observed results can now be compared with those calculated for expectancy in a four-coin toss:

Class	*Observed*	*Calculated*	
HHHH	9	6.25	(= $\frac{1}{16}$ of 100)
HHHT	32	25.00	(= $\frac{4}{16}$ of 100)
HHTT	29	37.50	(= $\frac{6}{16}$ of 100)
HTTT	25	25.00	(= $\frac{4}{16}$ of 100)
TTTT	5	6.25	(= $\frac{1}{16}$ of 100)
	100	100.00	

Again, the problem of determining *how well* these observations compare with the expected results must be considered on the basis of the three simplifying assumptions previously discussed—that is, whether the difference between these results and the calculated values should be accepted as *chance deviations* from the results of two- and four-coin tosses where the coins *are* unbiased and independent of each other, as are successive tosses.

THE BINOMIAL EXPRESSION

Answers to many genetic problems involving "either-or" situations, including those with which members of the medical or legal professions must deal from time to time, are easily provided by the **binomial** approach. Therefore, it is necessary to understand the method of binomial expansion before going on. Instead of multiplying out algebraically, one can apply the following simple rules for expansion of $(a + b)^n$.

1. *The power of the binomial chosen, that is, the value of* **n**, *is determined by the number of coins, individuals, and so on.* Thus, for two coins the expression $(a + b)^2$ is used, even if the coins are tossed 50 times; for a four-coin toss, $(a + b)^4$ is used, and so on. The power of the binomial remains constant for a given number of events even though the number of tosses is changed. The same principle applies to families of n children displaying one or the other of two phenotypic expressions.
2. *The number of terms in the expansion is* $n + 1$. Thus $(a + b)^2$ expands to $a^2 + 2ab + b^2$, having three terms, $(a + b)^4$ expanded contains five terms, and so on.
3. *Every term of the expansion contains both* a and b. The power of a in the first term equals n, the power of the binomial, and descends in units of 1 to 0 in the last term (and a^0, which equals 1, is not written in); likewise, b increases from b^0 (not written) in the first term to b^n in the last. Furthermore, the sum of the powers *in each term* equals n. Check this against the expansion of $(a + b)^4$.
4. *The coefficient of the first term* is 1 (not written); the coefficient of the second term is found by multiplying the coefficient of the preceding term by the exponent of a and dividing by the number indicating the position of that preceding term in the series:

$$\text{coefficient of next term} = \frac{\text{coefficient of preceding term} \times \text{exponent of preceding term}}{\text{ordinal number of preceding term}}$$

The same principle may be represented graphically in the form of Pascal's triangle, where each coefficient is shown as the sum of two numbers immediately above. Note that the second coefficient in each horizontal line also represents the power of the binomial. Thus the second number of the last line is 6, so that this line represents the series of coefficients in the expansion of $(a + b)^6$.

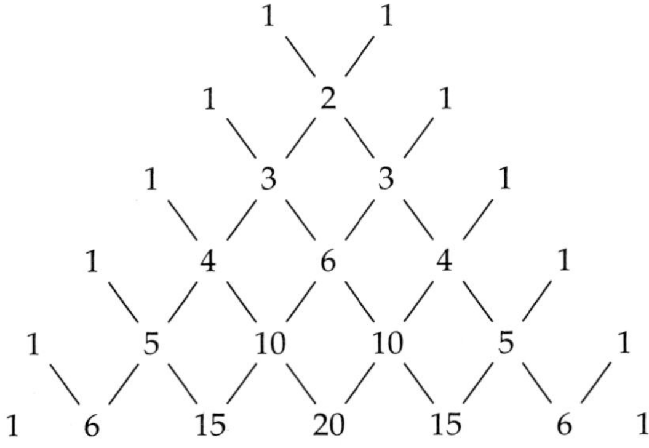

TABLE 5–1. Expansions of Binomials

$(a + b)^1 = a + b$
$(a + b)^2 = a^2 + 2ab + b^2$
$(a + b)^3 = a^3 + 3a^2b + 3ab^2 + b^3$
$(a + b)^4 = a^4 + 4a^3b + 6a^2b^2 + 4ab^3 + b^4$
$(a + b)^5 = a^5 + 5a^4b + 10a^3b^2 + 10a^2b^3 + 5ab^4 + b^5$
$(a + b)^6 = a^6 + 6a^5b + 15a^4b^2 + 20a^3b^3 + 15a^2b^4 + 6ab^5 + b^6$

Thus, in expanding $(a + b)^4$, the first term is a^4 (for $1a^4b^0$). The second is $4a^3b$; that is, from the preceding term, a^4,

$$\frac{4 \times 1}{1} = 4$$

which is the coefficient of the second term. The third term, which from rule 3, we know contains a^2b^2, becomes $6a^2b^2$ in the same way:

$$\frac{4 \times 3}{2} = 6$$

and so on. In this way the binomial expansions listed in Table 5–1 are obtained.

Another useful way of thinking of binomial expansions is to remember that the expression $(a + b)^2$ means, of course, $(a + b) \times (a + b)$:

$$\begin{array}{r} a + b \\ \times\ a + b \\ \hline a^2 + ab + ab + b^2 \end{array}$$

Collecting like terms from this last expression gives us $a^2 + 2ab + b^2$. Similarly, $(a + b)^3$ can be thought of as $(a + b) \times (a + b) \times (a + b)$, or

$$\begin{array}{l} \quad\quad a^2 + 2ab + b^2 \\ \times \quad\quad a + b \\ \hline a^3 + 2a^2b + ab^2 + a^2b + 2ab^2 + b^3 \end{array}$$

which may be rewritten as

$$a^3 + 3a^2b + 3ab^2 + b^3$$

If a and b are each equal to $\frac{1}{2}$, symmetrical *probability curves*, as shown in Figure 5–1, are produced. On the other hand, if $a = \frac{3}{4}$ and $b = \frac{1}{4}$, a skewed curve is obtained, as indicated in Figure 5–2.

GENETIC APPLICATIONS OF THE BINOMIAL

Just as the binomial is used to determine expected results in coin tosses, it may also be used to determine the probability of children showing a given heritable trait in a particular family.

For instance, *ptosis* (drooping eyelids) is an inherited trait in which affected persons are unable to raise the eyelids, so that only a relatively small space between upper and lower lids is available for vision, giving

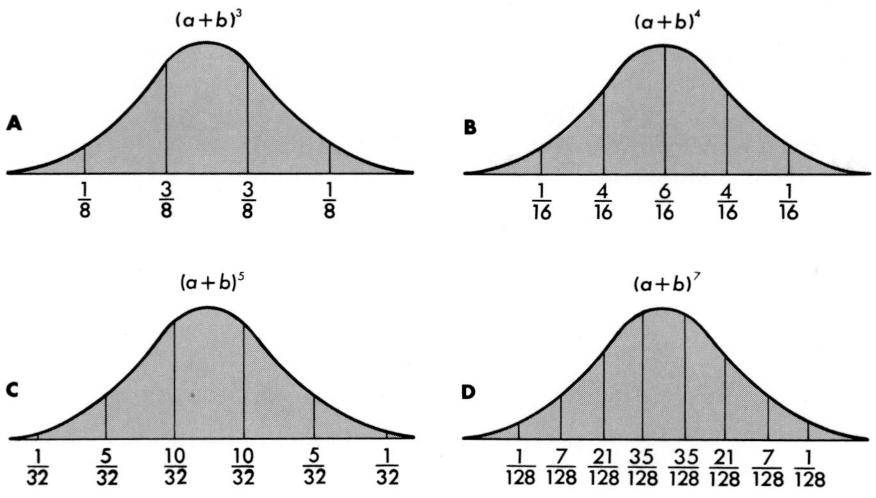

FIGURE 5–1. *Probability curves for binomial expansions where* a = b ½. *(A)* (a + B)³; *(B)* (a + b)⁴; *(C)* (a + b)⁵; *(D)* (a + b)⁷.

them a "sleepy" appearance. Most pedigrees indicate that ptosis is due to an autosomal dominant. Suppose a man with ptosis, whose father also displayed the trait but whose mother did not, wishes to marry a woman with normal eyelids. They consult a physician to determine the likelihood of the occurrence of the defect in their children. If, for example, they plan to have four children, what is the probability of three of those being normal and one having ptosis?

From the known facts, it is evident that the young man is heterozygous because he has ptosis (letting P represent the trait, his phenotype alone tells us he is $P-$), although his mother was normal and, therefore, pp. So having the trait and having necessarily received a recessive gene from his mother, he must be Pp. The woman, on the other hand, is pp, as determined by her normal phenotype. So this is the cross $Pp \times pp$. This is recognized as a testcross, so that the probability of normal children is $\frac{1}{2}$. But what is the chance for a family of *three normal children and one affected child* if they have four children? Given the binomial $(a + b)^4$ where

$$a = \text{probability of a normal child} = \tfrac{1}{2}$$

$$b = \text{probability of a child with ptosis} = \tfrac{1}{2}$$

the expansion of the second term, $4a^3b$ will, upon substitution, yield the information:

$$4a^3b = 4[(\tfrac{1}{2})^3 \times \tfrac{1}{2}] = \tfrac{4}{16} = \tfrac{1}{4}$$

The second term ($4a^3b$) in this expansion is chosen because it contains a^3 (representing three children of the phenotype denoted by a) and b (for one child of the phenotype represented by b). Therefore, there is a probability in this case of 1 in 4, or 0.25, that *if they have four children*, three will be normal and one will have ptosis. Or it can be said that, of all families of four children born of parents with these genotypes, it is expected that one out of four will consist of three children without ptosis and one with the condition. Similarly, there is one chance in sixteen (a^4), or 0.0625, that none of the four will exhibit ptosis or (b^4) that all four will have ptosis.

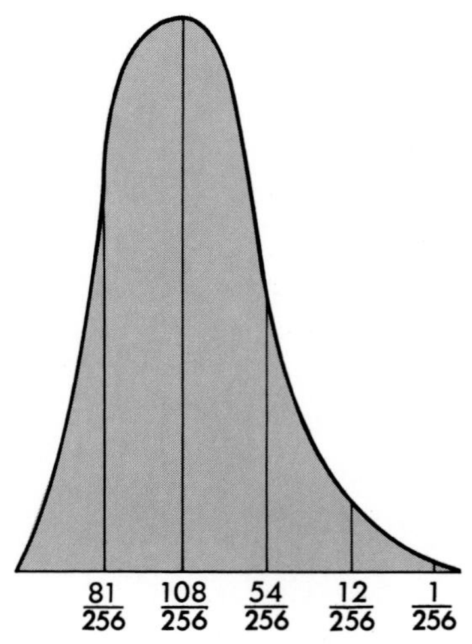

FIGURE 5–2. *Skewed curve for expansion of the binomial* (a + b)⁴ *where* a = ¾ *and* b = ¼.

On the other hand, consider a young man and woman, each of whom has ptosis and is heterozygous. If such a couple should marry and have four children, what is the probability of three being normal and one having ptosis? The potential parents are represented as $Pp \times Pp$. From previous experience with monohybrid crosses of two heterozygotes, it is known that, in case of complete dominance, the expected progeny phenotype ratio is 3:1. That is, the probability here of an affected child should be expected to be $\frac{3}{4}$, and of a normal one to be $\frac{1}{4}$. As in the preceding example, let

$$a = \text{probability of a normal child} = \tfrac{1}{4}$$

$$b = \text{probability of a child with ptosis} = \tfrac{3}{4}$$

Because this case again concerns a family of four children, the binomial $(a + b)^4$ is used once more. From the expansion of this binomial one must again substitute in the second term and solve:

$$4a^3b = 4[(\tfrac{1}{4})^3 \times \tfrac{3}{4}] = \tfrac{12}{256}$$

So with two heterozygous parents the probability of three normal children and one affected child is $\frac{12}{256}$ or 0.047. This is quite different from the 0.25 probability for the preceding case.

Had the question been, say, two normal and two affected children, the third term, $6a^2b^2$, in the expansion of $(a + b)^4$ would have been used. In any such problem it is imperative to select both a binomial to the proper power and the correct term within the expansion. The former is determined by the number of events (children, heads, and so on), the latter by the number of individuals of each of the two alternative types (for example, affected versus normal, heads versus tails). Of course, ptosis, in itself, is not a serious departure from "normal," but this method of calculating probability might be of considerable importance to parents, both of whom are heterozygous for some recessive disabling or lethal allele. It is also imperative to determine the correct values for a and b; they do *not* always equal $\frac{1}{2}$ each.

More than two possibilities can be dealt with by employing a polynomial raised to the appropriate power, for example, $(a + b + c)^2$, $(a + b + c + d)^5$, and so forth. But, as the number of possibilities rises, polynomials become increasingly cumbersome. In such cases, the general formula

$$\frac{n!}{s!\,t!}\,p^s\,q^t$$

may be used. Here n represents the total number of events (for example, 4 children), $n!$ (n factorial) directs the user to multiply $4 \times 3 \times 2 \times 1$, assuming $n = 4$, p and q represent the probabilities of each event (for example, p for the probability of a dominant phenotype, or of a girl, or of a boy, in whom are combined a number of specific genetic traits, etc.), q equals the probability of the alternate event, and s and t are the number of individuals represented by p and q, respectively. This formula will serve in lieu of a binomial or polynomial expansion, but is most useful in cases where the number of events is three or more. An example will serve to illustrate.

Red hair (*rr*) is completely recessive to nonred hair (*R*—) and wavy hair is the heterozygous expression of a pair of incompletely dominant genes. How often in families of three children where both parents are heterozygous for nonred hair (*Rr*) and have wavy hair ($h^1 h^2$) will the children be a red, wavy-haired girl, a nonred, curly-haired boy, and a red, straight-haired girl? Probabilities of each of these three children's phenotypes are:

girl, red, wavy	$\frac{1}{2} \times \frac{1}{4} \times \frac{1}{2} = \frac{1}{16} = a'$
boy, nonred, curly	$\frac{1}{2} \times \frac{3}{4} \times \frac{1}{4} = \frac{3}{32} = b'$
girl, red, straight	$\frac{1}{2} \times \frac{1}{4} \times \frac{1}{4} = \frac{1}{32} = c'$

Substituting in the formula in the above paragraph ($a'b'c'$),

$$= \frac{3!}{1 \times 1 \times 1} (\tfrac{1}{16})^1 (\tfrac{3}{32})^1 (\tfrac{1}{32})^1$$

$$= \frac{3 \times 2 \times 1}{1} \times \tfrac{1}{16} \times \tfrac{3}{32} \times \tfrac{1}{32}$$

$$= 6 \times \tfrac{1}{16} \times \tfrac{3}{32} \times \tfrac{1}{32} = \frac{18}{16{,}384} = 0.0010986$$

or about 0.11 percent. Or, the probability may be expressed as about 11 in 10,000 such families. Had the family consisted of *four* children, say *two* girls with red, wavy hair (instead of one), with the remaining two children as in the preceding example, substituting,

$$\frac{4!}{2 \times 1 \times 1} (\tfrac{1}{16})^2 \times (\tfrac{3}{32})^1 \times (\tfrac{1}{32})^1 = 12 \times \tfrac{1}{256} \times \tfrac{3}{32} \times \tfrac{1}{32} = \frac{36}{262{,}144}$$

or approximately 0.00014; that is, about 14 families in 100,000 such cases.

PROBABILITY OF SEPARATE OCCURRENCE OF INDEPENDENT EVENTS

Just as the probability of the simultaneous occurrence of two independent events is the product of their separate probabilities, **the probability of the separate occurrence of either of two independent events equals the square root of their simultaneous occurrence,** provided, of course, the two events are of equal probability. In the preceding section we saw that the probability of tossing two heads simultaneously is $\frac{1}{4}$. The probability of tossing a head in one toss is $\sqrt{\frac{1}{4}}$ or $\frac{1}{2}$.

The same approach can be applied in genetics. Suppose one wished to determine the frequency of an allele for albinism in a particular population. Studies show, for example, that about 1 in 10,000 babies in Norway and Ireland is an albino, a condition that is the result of the action of a recessive autosomal allele. Each albino child represents the simultaneous occurrence of independent events of equal frequency, namely, the union of two gametes, each of which carries the recessive allele. Inasmuch as the frequency of albinos in this case is 0.0001, the probability that any given gamete in the population carries the allele for albinism is equal to $\sqrt{0.0001}$, or 0.01. Because this allele is one member of a *pair* of alleles whose total frequency, therefore, must be 1, the frequency of the dominant allele for normal pigmentation is $1 - 0.01 = 0.99$.

Having these two values, the frequencies of the three possible genotypes can be calculated readily. Let

a = frequency of the allele for normal pigmentation (A) = 0.99

b = frequency of the allele for albinism (a) = 0.01

Substituting the expansion of $(a + b)^2 = 1$ provides genotypic frequencies:

binomial expansion:	a^2	+	$2ab$	+	b^2
genotype:	AA		Aa		aa
frequency:	$(0.99)^2 = 0.9801$	+	$2(0.99 \times 0.01) = 0.0198$	+	$(0.01)^2 = 0.0001$

This operation has applications in population genetics (Chapter 19).

DETERMINING GOODNESS OF FIT

Nongenetic Events

In the earlier discussion of calculating expected results of coin tosses, one as yet unanswered question was raised: *In terms of the assumptions made concerning the coins and the tosses, is the deviation from the expected results due to chance alone?* This question implies that one does not expect always to achieve results *exactly* equal to one's calculations and raises the problem of how much deviation from calculated probabilities can be accepted as likely being due purely to chance. Reference to Table 2–1 shows that even Mendel's results, involving hundreds of thousands of individuals, did not reflect *exactly* the expected ratios (although they were surprisingly close!). In other words, a mathematical tool is needed to determine "goodness of fit." Such a tool is the chi-square (χ^2) test.

To understand how to use this important statistical test, reexamine the earlier coin tosses. Recall the two sets of two-coin tosses made by students *A* and *B*:

	A's tosses			*B's tosses*	
Class	*Observed*	*Calculated*	*Class*	*Observed*	*Calculated*
HH	12	12.5	HH	10	12.5
HT	27	25.0	HT	33	25.0
TT	11	12.5	TT	7	12.5

Obviously, *A*'s tosses are closer to the expected results, but they do not correspond exactly. Therefore, are the deviations of magnitude in the two sets of results acceptable as chance deviations when the assumptions previously made are correct and complete, and only chance is operating? How large a departure from results expected under a given set of assumptions can be tolerated as representing merely chance, nonsignificant deviations? The chi-square test will make possible a judgment in this question.

The formula for calculating chi-square is

$$\chi^2 = \Sigma\left[\frac{(o - c)^2}{c}\right]$$

where *o* equals observed number of events, *c* equals calculated number of events, and Σ indicates that the bracketed quantity is to be summed for all classes. Both *o* and *c must be calculated in actual numbers and not in percentages.* Moreover, *each class must contain more than five individuals.*

To calculate chi-square for student *A*'s coin tosses (1:2:1 expectation), his data may conveniently be set up as follows:

Class	Observed o	Calculated c	Deviation $o - c$	Squared deviation $(o - c)^2$	$\frac{(o - c)^2}{c}$
HH	12	12.5	−0.5	0.25	0.02
HT	27	25.0	+2.0	4.00	0.16
TT	11	12.5	−1.5	2.25	0.18
TOTALS	50	50.0	0		$\chi^2 = 0.36$

With the value $\chi^2 = 0.36$, our question can now be stated as, "How often, *by chance alone,* will a deviation *this large or larger* be found when a 1:2:1 ratio is expected?" or, "How often, by chance, can one expect a value of $\chi^2 \geq 0.36$?" If this probability is quite high, then one can accept the validity of the assumptions under which the expectancy was determined as well as the thesis that the observed deviation was produced by chance.

The answer to our question may be obtained by consulting a table of chi-square (Table 5–2).

To use the table it is necessary only to know the "degrees of freedom" operating in any particular case. This is one less than the number of classes involved and represents the number of *independent* classes that contribute to the calculated value of χ^2. In *A*'s coin tosses, two of the classes may have any value (between 0 and 50), but once values for these two are set, the third is automatically determined as the difference between the total for *all* classes and the sum of all *other* classes. The values of *P* across the top in the table indicate the *probability* of obtaining a value of chi-square (and, therefore, a deviation) as large or larger, *purely by chance.*

In Table 5–2 then, with two degrees of freedom, the table is read across until either the calculated value of chi-square or two values between which it lies are found. In the case at hand, the value of $\chi^2 =$ 0.36 does not appear in the table, but for two degrees of freedom there are two values, 0.103 and 0.446, between which it lies. Reading up to

TABLE 5–2. Table of Chi-square

Degrees of freedom	*P = 0.99*	*0.95*	*0.80*	*0.70*	*0.50*	*0.30*	*0.20*	*0.05*	*0.01*
1	0.00016	0.004	0.064	0.148	0.455	1.074	1.642	3.841	6.635
2	0.0201	0.103	0.446	0.713	1.386	2.408	3.219	5.991	9.210
3	0.115	0.352	1.005	1.424	2.366	3.665	4.642	7.815	11.341
4	0.297	0.711	1.649	2.195	3.357	4.878	5.989	9.488	13.277
5	0.554	1.145	2.343	3.000	4.351	6.064	7.289	11.070	15.086
6	0.872	1.635	3.070	3.828	5.348	7.231	8.558	12.592	16.812
7	1.239	2.167	3.822	4.671	6.346	8.383	9.803	14.067	18.475
8	1.646	2.733	4.594	5.527	7.344	9.524	11.030	15.507	20.090
9	2.088	3.325	5.380	6.393	8.343	10.656	12.242	16.919	21.666
10	2.558	3.940	6.179	7.267	9.342	11.781	13.442	18.307	23.209

Taken from Table 3 of Fisher, *Statistical Methods for Research Workers,* published by Oliver and Boyd, Ltd., Edinburgh, by permission.

values of P, it is seen that a chi-square value of 0.36 corresponds to a probability value of between 0.95 and 0.80. This means that, for an expectancy of 1:2:1, one can expect a deviation as large or larger than that experienced in between 80 and 95 percent of repeated trials. The observed deviation could, therefore, easily be a result of chance, and both the expectancy and the assumptions on which it was based appear good, that is, there is a *good fit* between observed results and the calculated expectancy. Note that the *larger the value of* P (and, therefore, *the smaller the value of* χ^2), the more closely the data approximate the expected ratio, and the greater the evidence that the original hypotheses are correct. In the same way, the *smaller the value of* P, the "less" the evidence that the hypotheses are correct. Finally, at a level of $P \leq 0.05$, the likelihood of a correct set of hypotheses is so small that the hypotheses are rejected. Two courses of action are open at that point: (1) secure a larger sample, or (2) revise the hypotheses and test the data against the revised hypotheses.

In the same way, χ^2 for *B*'s coin tosses (again a 1:2:1 expectation) is calculated in this way:

Class	*Observed* o	*Calculated* c	*Deviation* $o - c$	*Squared deviation* $(o - c)^2$	$\frac{(o - c)^2}{c}$
HH	10	12.5	−2.5	6.25	0.50
HT	33	25.0	+8.0	64.00	2.56
TT	7	12.5	−5.5	30.25	2.42
TOTALS	50	50.0	0		$\chi^2 = 5.48$

In Table 5–2 it is seen that one may expect, by chance alone, a value of $\chi^2 = 5.48$ in between 5 and 20 percent of such trials. This is obviously not as good a fit between observed and calculated results as was obtained by student *A*, but is it close enough to accept? The answer is "yes," because P is greater than 0.05. At the level of $P = 0.05$ only 1 in 20 such trials will show this large a deviation by chance alone. A chi-square value equal to or greater than that for a probability of 0.05 does not *tell* one that the large deviation necessary to produce such a high value of chi-square could *not* occur purely by chance, only that it is highly unlikely to do so. In the two-coin tosses, then, values of χ^2 up to (but not including) 5.991 indicate a sufficient probability of chance alone to be accepted as the cause of the deviation. Larger values of chi-square for two degrees of freedom, however, would require a careful review of the assumptions under which (in this case) a 1:2:1 expectancy was calculated.

In the same way, χ^2 for the four-coin toss, also reported in this chapter, is calculated to be 5.35. Table 5–2 shows that, for this value of chi-square and four degrees of freedom (remember, five different combinations are possible), $P = 0.30$ to 0.20, which shows this value of chi-square to be well below the level of significance.

Genetic Applications of Chi-Square

Recall the case of hydrocyanic acid in clover discussed in Chapter 3. A series of crosses (pages 32–36) of two parental strains, only one of which produced this substance in its leaves, gave rise to an F_2 of 351 HCN:256 no HCN. Although the 607 F_2 individuals reported by Atwood and

Sullivan resulted from some 23 different crosses, it will suffice to treat them as though they were all sister progeny of the same cross. Chi-square calculation for a 3:1 expectancy gives the value $\chi^2 = 95.49$:

Class	*o*	*c*	*o − c*	$(o - c)^2$	$\frac{(o-c)^2}{c}$
HCN positive	351	455.25	−104.25	10,868	23.87
HCN negative	256	151.75	+104.25	10,868	71.62
TOTALS	607	607.00	0.0		$\chi^2 = 95.49$

Reference to Table 5–2 shows that, for one degree of freedom, such a value is highly significant, since it is likely to occur by chance in far fewer than one in a hundred trials. While the great deviation that produces this extremely high value of chi-square is not impossible when one expects a 3:1 ratio, it is so improbable that the assumptions on which that expected ratio was based must be reexamined. For a 3:1 ratio, these would, of course, include the concept of a single pair of alleles with one allele completely dominant, or (for a 12:4 ratio) two pairs in which, say, *A* and *a* are both epistatic to *B* and *b*, though *A* is dominant to *a*:

9 *A* − *B* −
3*A* − *bb* } 12 if *A* is epistatic to *B* and *b*

3 *aaB* −
1 *aabb* } 4 if *a* is epistatic to *B* and *b* but recessive to *A*

Since 351:256 is considerably closer to a 9:7 expectancy, it is worth determining χ^2 for this ratio:

Class	*o*	*c*	*o − c*	$(o - c)^2$	$\frac{(o-c)^2}{c}$
HCN positive	351	341.4	+9.6	92.16	0.27
HCN negative	256	265.6	−9.6	92.16	0.35
TOTALS	607	607.0	0.0		$\chi^2 = 0.62$

With one degree of freedom, Table 5–2 indicates, for this value of χ^2, a probability of between 0.30 and 0.50. This value of χ^2 is well below the level of significance, and interpolation in Table 5–2 indicates that such a value of χ^2 will occur in slightly more than 44 percent of similar trials when the assumptions underlying a 9:7 ratio are operating. Therefore, until conflicting data are turned up, these assumptions are acceptable. As described in Chapter 3, these include the concept of two genes with "duplicate recessive epistasis" where only the *A* − *B* − individuals are phenotypically distinguishable from other possible genotypes resulting from the crosses studied.

The chi-square test is a very useful one for obtaining an objective approximation of goodness of fit, but, as pointed out earlier, it is reliable only when the observed or calculated value in any class is more than five and numbers of individuals (not percentages) are used. Its proper use in genetic situations can, as seen in this chapter, clarify the operation of mechanisms in particular crosses.

PROBLEMS

5–1 In tossing three coins simultaneously, what is the probability, in one toss, of (a) three heads, (b) two heads and one tail?

5–2 A couple has two girls and is expecting a third child. They hope it will be a boy. What is the probability that their wish will be realized?

5–3 Another couple has eight children, all boys. What is the chance that their ninth child would be another boy?

5–4 How often, in families of eight children, will all eight be girls?

5–5 Five children are born on Christmas in a certain hospital. What is the probability that (a) all five are girls, (b) all five are of the same sex (that is, either all males *or* all females)?

5–6 What is the probability of getting (a) a 5 with a single die, (b) a 5 on each of two dice thrown simultaneously, (c) any combination totaling 7 on two dice thrown simultaneously?

5–7 In crossing two heterozygous deep *Coleus* plants, what is the probability of the occurrence in the F_1 of (a) deep, (b) shallow?

5–8 In crossing two *Coleus* plants of genotype *DdIi*, what is the probability in the F_1 of (a) *DdIi*, (b) *ddII*, (c) deep irregular, (d) shallow irregular?

5–9 (a) Give the third term in the expansion of $(a + b)^8$. (b) If $a = b = \frac{1}{2}$, what is the numerical value of this term? (c) If $a = \frac{1}{4}$ and $b = \frac{3}{4}$, what then is the numerical value of this term?

5–10 For this problem refer back to Problems 3–29 through 3–32. You plant 256 squash seeds from a cross between two tetrahybrid white disk plants (*AaBbCcDd*) and hope for some F_1 plants that bear yellow sphere fruits. If all 256 seeds give rise to fruiting plants, how many of them would be expected to produce yellow sphere fruits?

5–11 Astigmatism is a vision defect produced by unequal curvature of the cornea, causing objects in one plane to be in sharper focus than objects in another. It results from a dominant allele. Wavy hair appears to be the heterozygous expression of a pair of alleles for straight (h^1) or curly hair (h^2). A wavy-haired woman who has astigmatism, but whose mother did not, marries a wavy-haired man who does not have astigmatism. What is the probability that their first child will be (a) curly haired and nonastigmatic, (b) wavy haired and astigmatic? (c) How many different phenotypes could appear in their children with respect to these hair and eye conditions?

5–12 Free ear lobes (*A*—) and clockwise whorl of hair on the back of the head (*C*—) are dominant to attached lobes and counterclockwise whorl, respectively. A husband and wife know their genotypes to be *AaCc* and *aaCc*. They expect to have five children. What is the probability that three will be "free-clockwise" and two "attached-counterclockwise"?

5–13 Red hair (*rr*) and possession of two whorls of hair (*ww*) on the back of the head both appear to be inherited as recessive traits in most pedigrees. (a) How many times in families of three children, in which both parents are *RrWw* (nonred-haired, one whorl), will these consist of one red-haired boy with two whorls and two nonred-haired girls with one whorl? (b) Does $a + b = 1$ in this case? Why?

5–14 For a given recessive allele in a particular population 4 persons in 100 are homozygous. What is the probability that any given gamete in that population carries the recessive allele?

5–15 A man and his wife both have free ear lobes, but are both heterozygous. They have four children. What is the probability that these children include one girl with free ear lobes, one boy with attached ear lobes, and two girls with attached ear lobes? (You might wish to use the formula on page 78.)

5–16 **Multiple telangiectasia** in humans is the heterozygous expression of an allele that is lethal when homozygous. Heterozygotes have enlarged blood vessels of face, tongue, lips, nose, and/or fingers and are subject to unusually frequent, serious nose bleeding. Homozygotes for the trait have many fragile and abnormally dilated capillaries; because of severe multiple hemorrhaging these individuals die within a few months after birth. Two heterozygotes married forty years ago and now have four grown children. What is the probability that two of these are normal and two have multiple telangiectasia?

5–17 A study by Danks et al. (1965) shows that in Australia four in 10,000 live births are individuals who have **cystic fibrosis** of the pancreas, an inherited recessive metabolic defect in digestion of fats, which is ultimately fatal in children homozygous for the allele. This figure is similar to that of other studies in the United States. What is the probability that any given gamete in these populations carries this recessive allele?

5–18 In the Australian and American populations referred to in the preceding problem, what is the calculated frequency of persons heterozygous for cystic fibrosis?

5–19 Chi-square is calculated for a given observed progeny ratio against two expectancies, *A* and *B*. The value of chi-square is lower for *A* than for *B*. (a) Which expectancy, *A* or *B*, better fits the observed data? (b) For which expectancy, *A* or *B*, would the value of *P* be lower? (c) Which expectancy, *A* or *B*, better fits the observed ratio?

5–20 A value of $\chi^2 = 0$ indicates what degree of correspondence between calculated results and those actually observed?

5–21 What is the value of P in the case described in the preceding problem?

5–22 A certain cross yields a progeny ratio of 210:90. Chi-square for a 2:1 expectancy is 1.5, and that for a 3:1 expectancy is 4.0. (a) How many degrees of freedom are there in this case? (b) Is deviation significant in either case? (c) What genetic explanation would you therefore prefer?

5–23 A certain cross produces an F_1 ratio of 157:43. By means of the chi-square test, determine the probability of a chance deviation this large or larger on the basis of a 13:3 expectancy.

5–24 Another cross involving different genes gives rise to an F_1 of 110:90. By means of the chi-square test determine the probability of a chance deviation this large or larger on the basis of (a) a 1:1 expectancy and (b) a 9:7 expectancy. (c) Is the deviation to be considered significant in either case? (d) What do you do with these results?

5–25 Suppose, with the genes involved in Problem 5–24, the F_1 ratio had been 1,100:900. Try a chi-square test with this sample to determine whether there is a significant deviation from (a) 1:1 and (b) 9:7 expectancy. (c) Is the deviation now significant in either case? (d) What is the effect of sample size on the usefulness of the chi-square test?

5–26 Following are listed some of Mendel's reported results with the garden pea. Test each for goodness of fit to the given hypothesis:

Cross	*Progeny*	*Hypothesis*
(a) Yellow × green cotyledons	(F_2) 6,022:2,001	3:1
(b) Green × yellow pods	(F_2) 428:152	3:1
(c) Violet red × white flowers	(F_1) 47:40	1:1
(d) Round yellow × wrinkled green seeds	(F_1) 31:26:27:26	1:1:1:1

5–27 Phenylketonuria (PKU; see Problem 2–22) occurs in 1 in 10,000 Caucasians. This metabolic defect results from a homozygous recessive genotype. What is the calculated frequency of PKUs in the Caucasian population?

LINKAGE, CROSSING-OVER, AND GENETIC MAPPING OF CHROMOSOMES

CHAPTER

6

From the discussion of dihybrid inheritance (Chapter 3) recall that the *Coleus* testcross *DdIi* (deep irregular) × *ddii* (shallow regular) produced four progeny phenotypic classes in a 1:1:1:1 ratio. Also note that this result is to be expected on the basis of the behavior of chromosomes in meiosis, if each of the two pairs of alleles is on a different pair of chromosomes. With this *unlinked* arrangement of alleles, either member of one pair can combine at random with either member of the other pair, resulting in the production of four kinds of gametes, *DI,Di,dI,* and *di, in equal number,* by the doubly heterozygous individual. Random fusion of these four gamete genotypes with the *di* gametes of the completely recessive parent results in the 1:1:1:1 progeny phenotypic ratio. As knowledge of the genetics of various organisms increases, clearly the number of genes per species exceeds the number of chromosome pairs by a considerable margin. For example, it has been estimated on both genetic and molecular grounds that the number of genes in the fruit fly (*Drosophila melanogaster*) is about 5,000. Other estimates have been even higher, so this figure may well be conservative. Because this species has only four pairs of chromosomes, each chromosome must bear many genes. The same expectation holds for all eukaryotic organisms.

All the genes carried on a given pair of autosomes constitute a **linkage group,** and would be expected to be inherited as a block were it not for crossing-over (Chapter 4). Therefore, it should be expected that the number of linkage groups for any organism is equal to the haploid chromosome number. This expectation will have to be amended when the sex chromosomes are considered (Chapter 7). Interestingly enough, linkage was anticipated before it was actually demonstrated. Just three years after the rediscovery of Mendel's pioneer paper, Sutton

(1903) suggested that each chromosome must bear more than a single allele and that alleles "represented by any one chromosome must be inherited together." However, Sutton was unable to support his hypothesis experimentally. Only a few years later, Bateson and Punnett (1905–1908) did have the data with which to do so, but failed to recognize that they were dealing with allelic pairs located on the same chromosome. Bateson and Punnett's method of interpretation is examined in the next section.

LINKAGE AND CROSSING-OVER

Drosophila

In the fruit fly, *Drosophila melanogaster,* a lowercase letter with a superscript plus (+) sign is often used to designate the dominant allele of a pair and the lowercase alone designates the recessive allele. Furthermore, because far more genes are known than there are letters in the alphabet, use of two or more letters to denote a given allele is required. In the fruit fly these genes occur on chromosome II (chromosome number designation in *Drosophila* is by Roman numerals).[1]

h^+	normal bristles	h	hooked bristles
pr^+	red eyes	pr	purple eyes

When the cross $h^+\ pr^+/h\ pr$ ♀ × $h\ pr/h\ pr$ ♂ is made, 99.4 percent of the progeny have either normal bristles *and* red eyes, or hooked bristles *and* purple eyes. The remaining 0.6 percent have either *normal* bristles and *purple* eyes, or *hooked* bristles and *red* eyes. Thus, if this cross yields an F_1 of 1,000 individuals:

normal, red	$h^+\ pr^+/h\ pr$	994
hook, purple	$h\ pr/h\ pr$	
normal, purple	$h^+\ pr/h\ pr$	6
hook, red	$h\ pr^+/h\ pr$	

Because the phenotypic ratio in the F_1 departs so widely from the 1:1:1:1 testcross ratio to be expected with unlinked alleles, it can be safely assumed that the two loci are *linked.*

Bateson and Punnett on Sweet Pea

In the sweet pea (*Lathyrus odoratus*) two pairs of alleles affecting flower color and pollen grain shape occur, each pair exhibiting complete dominance:

R	purple flowers	Ro	long pollen grains
r	red flowers	ro	round pollen grains

Bateson and Punnett crossed a completely homozygous purple long with a red round. The F_1 was, as expected, all purple long, and a 9:3:3:1 phenotypic ratio was expected in the F_2. Results, however, were quite different, as seen in Table 6–1.

[1]Two or more sets of alleles that are on the same chromosome are separated by slash (/) marks.

TABLE 6–1. Bateson and Punnett Sweet Pea Cross

	Observed		Expected (9:3:3:1)	
Phenotype	*Number*	*Frequency*	*Number*	*Frequency*
Purple long	296	0.6932	240	0.5625 (= $\frac{9}{16}$)
Purple round	19	0.0445	80	0.1875 (= $\frac{3}{16}$)
Red long	27	0.0632	80	0.1875 (= $\frac{3}{16}$)
Red round	85	0.1991	27	0.0625 (= $\frac{1}{16}$)
	427	1.0000	427	1.0000

Satisfy yourself that chi-square for a 9:3:3:1 expectancy is 219.28. Inspection of Table 5–2 shows that, for three degrees of freedom, the basis for this expected ratio (independent segregation of the two pairs of alleles) is, therefore, not acceptable. From the arithmetic standpoint, Bateson and Punnett recognized that the observed ratio was "explicable on the assumption that . . . the gametes were produced in a series of 16, viz., 7 purple long, 1 purple round, 1 red long, and 7 red round." For example, if each parent produced *r ro* gametes with a frequency of $\frac{7}{16}$ (= 0.4375), then $(\frac{7}{16})^2$, or $\frac{49}{256}$ (= 0.1914), of the progeny should be red round. The observed frequency of such plants, 0.1991, is quite close to the calculated frequency of 0.1914. Or, to look at it another way, the red round plants represent the fusion of two *r ro* gametes, and, by the product law of probability, the frequency of such gametes must equal the square root of the frequency of the red round plants. The square root of 0.1991, that is, the observed frequency of the red round individuals, is 0.4462, which is acceptably close to 0.4375, the Bateson and Punnett figure of $\frac{7}{16}$. Expressed as decimals, the 7:1:1:7 array of gamete genotypes then would be 0.4375 *R Ro*, 0.0625 *R ro*, 0.0625 *r Ro*, and 0.4375 *r ro*. However, Bateson and Punnett were unable to explain why or how this gamete ratio came about because they did not relate it to the behavior of chromosomes in meiosis. Later genetic studies provided evidence that genes for flower color and shape of pollen grains are linked.

Arrangement of Linked Genes

When two pairs of genes are linked, the linkage may be of either of two types in an individual heterozygous for both pairs: (1) the two dominants, *R* and *Ro*, may be located on one member of the chromosome pair, with the two recessives, *r* and *ro*, on the other, or (2) the dominant of one pair and the recessive of the other may be located on one chromosome of the pair, with the recessive of the first gene pair and the dominant of the second gene pair on the other chromosome. The first arrangement, with two dominants on the same chromosome, is referred to as the *cis* arrangement; the second, having one dominant and one recessive on the same chromosome, is called the *trans* arrangement. Figure 6–1 illustrates these possibilities.

Thus the genotypes of the parental and first filial generations of the Bateson and Punnett cross may be written in standard fashion to reflect linkage:

P *R Ro*/*R Ro* × *r ro*/*r ro*

F_1 *R Ro*/*r ro*

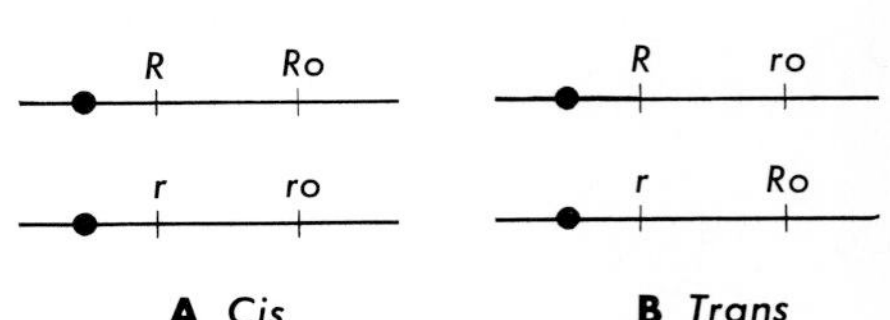

FIGURE 6–1. Cis *and* trans *arrangements for two pairs of linked genes in a diploid cell.*

Without crossing-over, the F_1 would produce but two types of gametes, *R Ro* and *r ro*. Crossing-over, however, produces two additional gamete genotypes, *R ro* and *r Ro*. That these four genotypes are *not* produced in equal frequency could easily be seen from a testcross in which the double heterozygote carries these genes in the *cis* linkage:

P purple long ♀ × red round ♂
R Ro/r ro *r ro/r ro*

If the two pairs of alleles were not linked, a 1:1:1:1 dihybrid testcross ratio would, of course, result. Based, however, on the gamete frequencies that Bateson and Punnett recognized, testcross progeny should occur in these frequencies:

Phenotype	*Genotype*	*Frequency*		*Type*
1. Purple long	*R Ro/r ro*	0.4375	0.875	Noncrossovers or nonrecombinants
2. Red round	*r ro/r ro*	0.4375		
3. Purple round	*R ro/r ro*	0.0625	0.125	Crossovers or recombinants
4. Red long	*r Ro/r ro*	0.0625		

The first two phenotypic classes have received from the pistillate parent an unaltered chromosome carrying either *R Ro* or *r ro*. The latter two classes, however, have received a *crossover* chromosome, either *R ro* or *r Ro*. Gametes carrying each of these altered chromosomes must occur with a frequency of only 0.0625 ($= \frac{1}{16}$), rather than 0.25, as would be the case with unlinked alleles. With genes *R* and *Ro*, the *crossover frequency* is 12.5 percent. This frequency is remarkably constant, regardless of the type of cross or the kind of linkage (*cis* or *trans*).

To understand how this happens, recall the occurrence of chiasmata in prophase-I, as described in Chapter 4. If a chiasma occurs between these two pairs of alleles, *crossing-over* has taken place so that the original linkage, *R Ro/r ro*, becomes *R ro/r Ro*, as indicated in Figure 6–2. For each meiocyte in which this happens, the result is four haploid nuclei of genotypes *R Ro*, *R ro*, *r Ro*, and *r ro*. For each meiocyte in which a crossover fails to occur between these two pairs of genes, the result is four haploid nuclei, two each of genotypes *R Ro* and *r ro*. If the meiocyte carries these genes in the *cis* configuration, as in this case, *R ro* and *r Ro* gametes are referred to as *crossover gametes*, and *R Ro* and *r ro* sex cells as *noncrossover* types. Progeny such as classes 1 and 2, which receive one or the other *intact chromosome* from the heterozygous parent, are referred to as parental types. Progeny making up classes 3 and 4 in this testcross incorporate new combinations of linked genes and are referred to as recombinants.

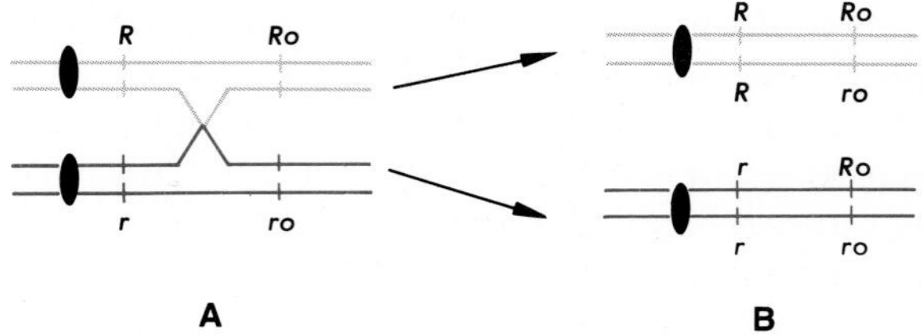

FIGURE 6–2. *Result of crossing-over. A chiasma occurring between linked genes* R *and* Ro, *and involving two nonsister chromatids as shown in (A) may result in "repair" such that the original* cis *linkage is coverted to a* trans *linkage in two of the four chromatids, as depicted in (B).*

LINKAGE, LINKAGE GROUPS, AND MAPPING

The chromosomal basis of heredity became clearly established in the second decade of this century, thus verifying Sutton's earlier hypothesis and supplying experimental data to explain and extend cases that had puzzled Bateson and Punnett. Discovery followed discovery in rapid succession. In 1910, Thomas Hunt Morgan was able to provide evidence for the location of a particular gene of *Drosophila* on a specific chromosome (Morgan 1910a). Within a short time he demonstrated clearly that linkage does exist and that linked genes are often inherited together, but may be separated by crossing-over (Morgan 1910b, 1911a). Even more exciting than these proofs of earlier hypotheses was Morgan's conclusion that a definite relation exists between recombination frequency and the linear distance separating genes within a chromosome. He wrote (Morgan 1911b), "in consequence, we find coupling in certain characters, and little or no evidence at all of coupling in other characters, the difference *depending on the linear distance apart of the chromosomal materials that represent the factors*" (italics added). Morgan used the term *coupling* in the sense developed by Bateson, Saunders, and Punnett (1905), that is, referring to a greater frequency of gametes carrying two dominants (or two recessives) than would occur by random segregation. They called similar association of one dominant and one recessive *repulsion*. Largely through cytological work of Morgan, the concept of coupling and repulsion was replaced by that of linkage and crossing-over. The terms *cis* and *trans* were employed later (Haldane 1942), the former replacing *coupling* and the latter, *repulsion*.

Such techniques made it possible to assign genes to particular chromosomes, as well as to begin construction of *chromosome maps* that show linkage groups and relative distances between successive genes. Linkage groups were developed rapidly for *Drosophila melanogaster* and also for a variety of animals and plants. Perhaps the most remarkable aspect of this burgeoning knowledge was that geneticists could now assign relative positions on chromosomes to genes that could not be seen in the light microscope and whose nature and precise function were not to be clarified for another forty or fifty years.

As this kind of information was accumulated, it became clear that the number of linkage groups in any species is ultimately found to equal its haploid chromosome number, or, in the sex having dissimilar sex chromosomes, one more than the haploid number. Thus, four linkage groups are known in females of *Drosophila melanogaster*, where $n = 4$, five in males of the same species, whose sex chromosomes are nonhomologous (see Chapter 15), and 10 in corn (*Zea mays*), which has 10 pairs of chromosomes. In species whose genetics is less completely worked out, the number of *known* linkage groups is temporarily smaller than its haploid chromosome number. These statements apply only to nuclear genes of eukaryotes; the matter of extranuclear genes will be examined in Chapter 16.

Determination of linkage groups and chromosomal assignments in human beings is proceeding at a rapid pace (see Table 6–2). As early as 1976, geneticists were able for the first time to assign at least one gene to each of the human chromosomes. By 1978, the number of genes assigned to the autosomes (nonsex chromosomes) had risen to 1,300 (McKusick 1983). At the same time 107 were known to be on the X chromosome (plus many "probables"), and one had been assigned to the Y chromosome. (The X and Y chromosomes are the sex chromosomes; they are discussed in Chapter 7.) When one realizes that the estimate for the diploid gene number in *Homo sapiens* may be in the tens of thousands, it is clear that much mapping work remains to be done.

TABLE 6–2. Some Linkage Groups and Chromosome Assignments in Human Beings

Chromosome	*Gene*
1	Succinate dehydrogenase
1	Glucose dehydrogenase
1	Amylase (pancreatic)
1	Amylase (salivary)
1	Actin, skeletal muscle alpha chain
1	Alkaline phosphatase, liver/bone form
1	Duffy blood antigens
1	Xeroderma pigmentosum A
1	Glucose dehydrogenase
1	Fumarate hydrase
1	Peptidase-C
1	Phosphoglucomutase-1
1	Rhesus blood antigens (Rh)
1	HISTONE CLUSTER B: H3, H4
1	Cystic fibrosis antigen
1	Enolase-1
1	Coagulation factor III
1	Radin blood group
1	Guanylate kinase, 1
1	Guanylate kinase, 2
1	Renin
1	Oncogene TRK
1	Oncogene ARG
1	Oncogene MRAS1
1	Oncogene MYC
2	Acid phosphatase-1
2	Elastin
2	IMMUNOGLOBIN KAPPA LIGHT CHAIN GENE CLUSTER
2	Interferon-1
2	Kidd blood group
2	Malate dehydrogenase (soluble)
2	Ribulose 5-phosphate 3-epimerase
2	Collagen III
2	Collagen IV
2	Beta-3-interferon
2	Colton blood group
2	Homeobox-4
2	J region kappa light chain
2	Constant region kappa light chain
2	Oncogene NMYC
2	Oncogene REL
2	CRYSTALLINE, GAMMA POLYPEPTIDE CLUSTER
2	Isocitrate dehydrogenase, soluble
2	Glucagon
2	Tubulin, alpha, testis specific
2	Adenosine deaminase complexing protein-2
2	Alkaline phosphatase, placental
2	Desmin
2	Ornithine decarboxylase-1
3	Herpes virus sensitivity
3	Small-cell cancer of the lung
3	Beta-galactosidase-1
3	Aminoacylase-1
3	Renal cell carcinoma
3	Glutathione peroxidase-1
3	Alpha-2HS-glycoprotein
3	Transferrin
3	Rhodopsin
3	Pseudocholinesterase-1
3	Somatostatin
3	Dihydrofolate reductase pseudogene-4
3	Cellular retinol binding protein
4	Huntington's disease
4	MN blood group
4	Phosphoglucomutase-2
4	Stoltzfus blood group
4	Diabetes insipidus
4	Peptidase S
4	Metallothionein II processed pseudogene
4	Alpha-fetoprotein
4	ALCOHOL DEHYDROGENASE, CLASS I, CLUSTER
4	Formaldehyde dehydrogenase
4	Alcohol dehydrogenase class III
4	Epidermal growth factor
4	FIBRINOGEN GENE CLUSTER
4	Fibrinogen, alpha chain
4	Fibrinogen, beta chain
4	Oncogene KIT
4	Oncogene RAF2
5	Emetine resistance, ribosomal protein S14
5	Oncogene FMA, McDonough feline sarcoma
5	Diptheria toxin sensitivity
5	Hexosaminidase B (beta subunit)
5	Endothelial cell growth factor
5	Chromate resistance, sulfate transport
5	Arginyl-tRNA synthetase
5	Threonyl-tRNA synthetase
5	Leucyl-tRNA synthetase
5	Histidyl-tRNA synthetase
5	Dihydrofolate reductase
5	Histocompatibility: class II antigens, gamma chain
5	Fibroblast growth factor, acidic
5	Oncogene FMA, McDonough feline sarcoma
6	Initiator methionine tRNA
6	Tubulin, beta, M40
6	Insulin-dependent diabetes mellitus
6	Clotting factor XIII, A component
6	Ragweed pollen sensitivity
6	Clotting factor XII; Hageman factor
6	Prolactin
6	MAJOR HISTOCOMPATIBILITY COMPLEX
6	Arginase, liver
6	Phosphoglucomutase-3
6	Malic enzyme, mitochondrial
6	Malic enzyme, cytoplasmic
6	HLA-A tissue type
6	HLA-C tissue type
6	HLA-B tissue type

Chromosome	*Gene*
6	HLA-DZ tissue type
6	HLA-DR tissue type
6	HLA-DQ tissue type
6	HLA-DP tissue type
6	Dihydrofolate reductase pseudogene-2
6	Glyoxalase I
6	Arginase, liver
6	Plasminogen
6	Oncogene, Kirsten rat sarcoma virus
6	Oncogene MCF3
6	Oncogene SYN
6	Oncogene YES-2
6	Oncogene PIM1
6	Oncogene, avian myeloblastosis virus
7	Actin, cytoskeletal beta
7	Collagen I alpha-2
7	Epidermal growth factor receptor
7	HISTONE CLUSTER A: H1, H2A, H2B
7	Phosphoserine phosphatase
7	Homeobox-1
7	Oncogene ERBB
7	Oncogene ARAF2
7	Oncogene PKS1
7	Oncogene MET
7	Malate dehydrogenase, mitochondrial
7	Asparagine synthetase
7	Actin, cytoskeletal beta, pseudogene-5
7	Blue cone pigment
7	Cystic fibrosis
7	Clotting factor XIII, B component
7	Interferon, beta-2
7	Nonhistone chromosomal protein-2
7	Carboxypeptidase A
8	Polymerase, DNA, beta
8	Glutathionine reductase
8	Oncogene MOS
8	Oncogene MYC
8	Oncogene Moloney murine sarcoma virus
8	CARBONIC ANHYDRASE CLUSTER
8	Carbonic anhydrase I
8	Carbonic anhydrase II
8	Carbonic anhydrase III
8	Clotting factor VII
8	Thyroglobin
9	Oncogene ABL
9	ABO blood groups
9	Fibroblast interferon; beta-interferon
9	LEUKOCYTE INTERFERON GENE CLUSTER; ALPHA-INTERFERON
9	Galactosyl transferase-1
9	Nail-patella syndrome
9	Aldehyde dehydrogenase-1
9	Galactose-1-phosphate uridyl transferase
9	Interferon F (fibroblast interferon)
9	Immunoglobin epsilon heavy chain pseudogene
9	Relaxin, H1
9	Relaxin, H2
10	Adenosine kinase
10	Phosphofructokinase, platelet type
10	Hexokinase-1
10	Inorganic pyrophosphatase
10	Aldolase B pseudogene
10	Glutamate dehydrogenase
10	Inorganic pyrophosphatase
10	Terminal deoxynucleotidyl transferase
10	Ornithine amino transferase
10	Phosphoglycerate mutase A
10	Oligomycin resistance; mitochondrial ATPase, ATPM
11	Herpes virus sensitivity
11	Esterase-A-4
11	Acid phosphatase-2
11	Catalase
11	Oncogene EST-1
11	NON-ALPHA GLOBIN CLUSTER; HEMOGLOBIN BETA CLUSTER
11	Hemoglobin beta
11	Hemoglobin delta
11	Hemoglobin epsilon
11	Hemoglobin gamma 136 alanine
11	Hemoglobin gamma 136 glycine
11	Insulin
11	Lactate dehydrogenase A
11	Oncogene HRSAS1 (Harvey rat sarcoma-1)
11	APOLIPOPROTEIN CLUSTER I
11	Apolipoprotein A-I
11	Apolipoprotein C-III
11	Apolipoprotein A-IV
11	PEPSINOGEN A CLUSTER
11	Pepsinogen A3
11	Pepsinogen A4
11	Pepsinogen A5
11	CATHEPSIN GENE CLUSTER
11	Cathepsin D
11	Cathepsin B, H, L
11	Congenital glaucoma
12	Lactate dehydrogenase C
12	Methionyl tRNA
12	Enolase-2
12	Citrate synthetase, mitochondrial
12	Lactate dehydrogenase
12	Oncogene INT1 (murine mammary cancer virus)
12	Oncogene Kras2 (kirsten rat sarcoma virus)
12	SALIVARY PROTEIN COMPLEX
12	Salivary protein Pe
12	Salivary protein Po
12	Elastase-1
12	Major intrinsic protein of lens fiber
12	Homeobox-3
12	Collagen II, alpha-1
12	Peptidase B
12	Interferon, gamma or immune type
12	Phenylalanine hydrolyase
12	Aldehyde dehydrogenase, mitochondrial
13	Retinoblastoma-1
13	Ribosomal RNA

(*continues on next page*)

TABLE 6–2. Some Linkage Groups and Chromosome Assignments in Human Beings (*cont.*)

Chromosome	*Gene*
13	Immunoglobulin E level
13	Clotting factor VII
13	Clotting factor X
13	Collagen IV alpha-1 chain
13	Collagen IV alpha-2 chain
13	Esterase D; S-formylglutathione hydrolase
13	Propionyl CoA carboxylase, alpha subunit
14	T-cell antigen receptor, alpha subunit
14	Ribosomal RNA
14	Tryptophanyl-tRNA synthetase
14	Nucleoside phosphorylase
14	Oncogene FOS, FBJ murine osterosarcoma virus
14	Oncogene AKT1
14	Creatine kinase, brain type
14	T-cell leukemia-1
14	IMMUNOGLOBULIN HEAVY CHAIN GENE CLUSTER
14	Variable region genes
14	D region genes
14	J region genes
14	Constant region of heavy chain of IgM1
14	Constant region of heavy chain of IgM2
14	Constant region of heavy chain of IgD
14	Constant region of heavy chain of IgE2
14	Constant region of heavy chain of IgG4
14	Constant region of heavy chain of IgG3
14	Constant region of heavy chain of IgG1
14	Constant region of heavy chain of IgE
14	Constant region of heavy chain of IgEp1
14	Constant region of heavy chain of IgA1
14	Constant region of heavy chain of IgA2
14	Liver glycogen phosphorlyase
14	Pancreatic ribonuclease
15	Dyslexia-1
15	Hexosaminidase A
15	Oncogene: FES feline sarcoma virus
15	Ribosomal RNA
15	Sorbitol dehydrogenase
15	Factor XI
15	Immunoglobulin heavy chain diversity region-2
15	Actin, cardiac alpha
15	Beta-2-microglobin
15	Isocitrate dehydrogenase, mitochondrial
15	Pyruvate kinase-3
16	Growth rate controlling factor-2
16	Muscular dystrophy, atypical vitelliform
16	ALPHA GLOBIN GENE CLUSTER
16	Phosphoglycolate phosphatase
16	Protein kinase, beta form
16	Congenital cataract
16	Thymidine kinase, mitochondrial
16	Haptoglobin
16	Hemoglobin alpha
16	Hemoglobin zeta pseudogene
16	Hemoglobin zeta
16	METALLOTHIONEIN I CLUSTER
16	METALLOTHIONEIN II CLUSTER
16	Chymotrypsinogen B
16	Esterase-B3
16	Growth rate controlling factor-2
16	Nonhistone chromosomal protein-1
16	Aldolase A
17	Homeobox-2
17	Glactokinase
17	RNA polymerase II, large subunit
17	Oncogene ERBA avian erythroblastotic leukemia virus
17	Thymidine kinase-1
17	MYOSIN, HEAVY CHAIN CLUSTER
17	Myosin, cardiac heavy chain, alpha-adult
17	Myosin heavy chain beta-fetal
17	Myosin heavy chain, adult-1
17	Myosin heavy chain, adult-2
17	Myosin heavy chain, embryonic-2
17	Oncogene ERB-B2
17	Oncogene NGL
17	U2 smRNA GENE CLUSTER
17	Collagen I alpha-1 polypeptide
17	Protein kinase C, alpha form
17	Peptidase E
17	Aldolase C
17	GROWTH HORMONE/PLACENTAL LACTOGEN GENE CLUSTER
17	Growth hormone, normal
17	Growth hormone, variant
18	Oncogene ERV1, endogenous retrovirus-1
18	Gastrin-releasing peptide
18	Oncogene B-cell leukemina/lymphoma-2
18	Oncogene YES-1
18	Thymidylate synthetase
18	Myelin basic protein
18	Peptidase A
18	Asparaginyl-tRNA synthetase
18	Dihydrofolate reductase
19	Bombay phenotype
19	Lewis blood group
19	Lutheran blood group
19	Insulin receptor
19	Lysosomal DNA-ase
19	Polio virus sensitivity
19	Secretor
19	LW; Landsteiner-Weiner blood group
19	Peptidase D; prolidase
19	Green/blue eye color
19	Cytochrome P450, family II, subfamily C
19	APOLIPOPROTEIN CLUSTER II
19	Apolipoprotein E
19	Apolipoprotein C-II
19	Apolipoprotein C-I
19	Glucosephosphate isomerase
19	Creatine kinase, muscle type
19	Protein kinase C, gamma form
19	Ferritin, light chain
19	CHORIONIC GONADOTROPIN, BETA

Chromosome	*Gene*	*Chromosome*	*Gene*
19	Elongation factor-2	X	Ornithine transcarbamylase
19	Coxsackie B3 virus susceptibility	X	Oncogene ARAF1
20	Adenosine deaminase	X	Oncogene PKS2
20	Growth hormone releasing factor, somatocrinin	X	Glucose-6-phosphate dehydrogenase
20	Chromogranin B	X	Gonadal dysgenesis, XY female type
20	Inosine triphosphatase-A	X	H-Y regulator, or repressor
20	Leucine transport, high	X	Hemophilia A; factor VIII
20	Adenosine deaminase	X	Hemophilia B; factor IX
20	Protooncogene SRC, Rous sarcoma	X	Hypoxanthine-guanine phosphoribosyltransferase
21	Cystathionine beta-synthetase	X	Retinitis pigmentosa, X-linked
21	Oncogene ETS-2	X	Duchenne muscular dystrophy
21	Ribosomal RNA	X	Phosphoglycerate kinase-1 pseudogene-1
21	Superoxide dismutase-1, soluble	X	Monoamine oxidase A
21	Antiviral protein; alpha-interferon receptor	X	Phosphoglycerate kinase-1
21	Antiviral protein; beta-interferon receptor	X	Testicular feminization; androgen receptor
21	Phosphofructokinase, liver type	X	Alpha-galactosidase A
21	Crystalline, alpha A	X	X-linked mental retardation
22	Beta-galactosidase-2; GLB protective protein	X	Glutamate dehydrogenase pseudogene-1
22	Myoglobin	X	Becker muscular dystrophy
22	P1 blood group	X	Deutan color blindness; green cone pigment
22	Ribosomal RNA	X	Protan color blindness; red cone pigment
22	IMMUNOGLOBULIN LAMBDA LIGHT CHAIN GENE CLUSTER	X	Ocular albinism, Forsius-Eriksson type
22	Variable region of lambda light chains	X	Ocular albinism, Nettleship-Falls type
22	J region of lambda light chains	X	X chromosome controlling element; X-inactivation center
22	Constant region of lambda light chains	Y	Azoospermia-third factor
22	Aconitase, mitochondrial	Y	Stature
22	Cytochrome P450, family II, subfamily D	Y	H-Y antigen
22	NADH-diaphorase-1	Y	Homolog of X-linked locus for surface antigen MIC2
22	Adenylosuccinase	Y	Maturation rate
22	Aldolase A	Y	Testis determining factor (TDF)
X	Steroid sulfase	Y	Pseudoautosomal segment (PAS)
X	Primary adrenal hypoplasia	Y	Argininosuccinate synthetase pseudogene
X	Agammaglobulinemia	Y	Actin pseudogene
X	Xg blood group		
X	Polymerase DNA, alpha		

Source: V. A. McKusick, *The Human Gene Map,* January 7, 1987.

Cytological Evidence for Crossing-Over

Whenever a particular chromosome pair (bearing certain genes) can be clearly identified because of some structural characteristic, experimental evidence shows that when an interchange of material between two homologs occurs, there is likewise an interchange of genes (that is, genetic crossing-over). In an analysis of such a situation in corn, Creighton and McClintock (1931) furnished such a convincing correlation between cytological evidence and genetic results that their work has rightly been called a landmark in experimental genetics.

In corn, chromosome 9 (the second shortest in the complement of 10 pairs) ordinarily lacks a small knob (satellite). But in one particular strain investigated, a single plant was found to have dissimilar ninth chromosomes. One possessed a satellite at the end of the short arm and also an added segment that had been translocated from chromosome 8 to the long arm. The other member of the pair was normal, lacking both the satellite and the added segment of number 8. Chromosome 9 bears, among other allele pairs, the following:

C	colored aleurone	*Wx*	starchy endosperm
c	colorless aleurone	*wx*	waxy endosperm

Aleurone and endosperm are parts of the triploid food storage tissue in the grain. The plant having these dissimilar ninth chromosomes was heterozygous for both the aleurone and endosperm alleles. From earlier work, Creighton and McClintock knew which chromosome of the pair carried which gene. Thus the two ninth chromosomes of this plant, which they used as the pistillate parent in their crosses, could be identified visually and may be diagramed:

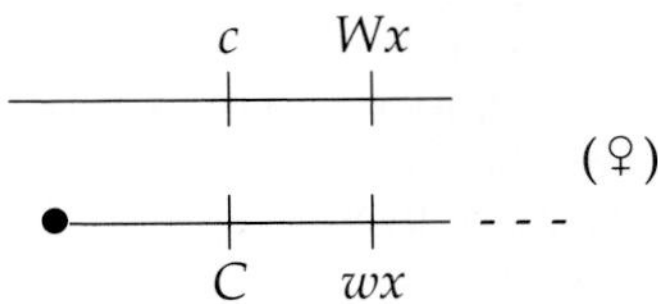

The dashed portion indicates the translocated segment of chromosome 8. An individual possessing such a dissimilar ninth pair of chromosomes was crossed with a *c Wx*/*c wx* plant possessing two knobless ninth chromosomes that also did not have the added segment of chromosome 8 (that is, "normal" ninth chromosomes):

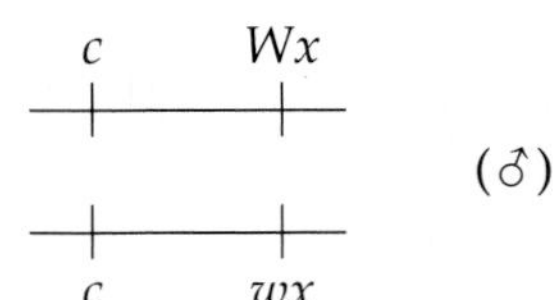

Although only 28 grains resulted from this cross, cytological examination of microsporocytes from adults grown from these grains confirmed the predicted relation between cytology and genetics (except for one class that was not found).

This cross, with the ninth chromosome in each parent and progeny class, is diagramed in Figure 6–3.

The fact that endosperm and aleurone are triploid tissues need not complicate our understanding of this masterful research because Creighton and McClintock used chromosomes of the *diploid* microsporocytes to demonstrate the correlation between cytological and genetic crossing-over.

A similar verification that genetic crossing-over is accompanied by a physical exchange betwen homologous chromosomes was also established in 1931 by Stern for *Drosophila*. He used a strain in which the females had a portion of the Y chromosome attached to one of their X chromosomes. Work by Stern and by Creighton and McClintock indicated a clear relationship between the interchange of material between homologs and genetic crossing-over.

The genetic aspects of localization of genes to the chromosomes covered thus far may be summarized as follows:

1. Certain genes assort *at random;* these are unlinked genes (Chapter 3).
2. *Other genes do not assort randomly, but are linked* (Chapter 6). These *linkage groups* tend to be transmitted in unitary groups.

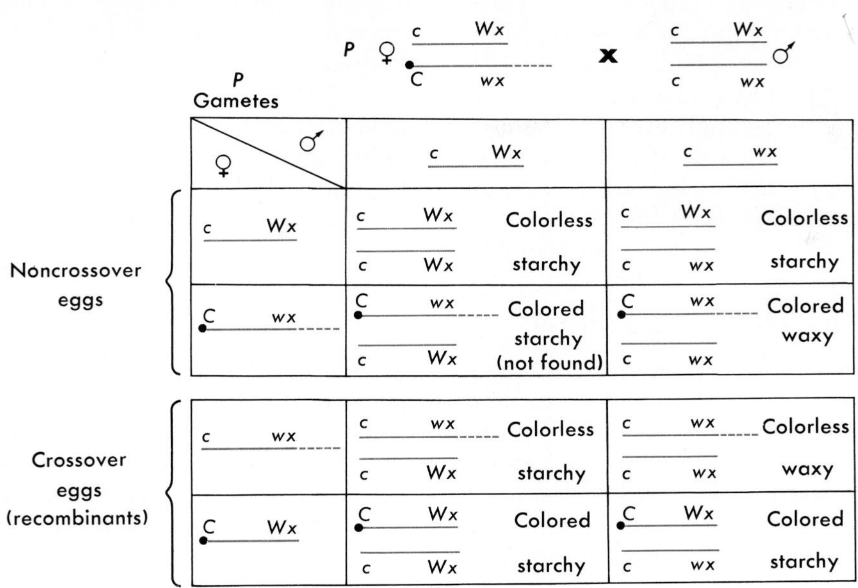

FIGURE 6–3. *A diagrammatic representation of parallelism between cytological and genetic crossing-over, showing the phenotype of aleurone and endosperm of F_1 grains and the chromosome morphology and genotypes for microsporocytes produced by F_1 plants.* (Based on the work of Creighton and McClintock. See text for full explanation.)

3. In diploid cells *chromosomes also occur in pairs* that are normally transmitted as units to daughter nuclei (Chapter 4).
4. Linked genes do not always "stay together" but are often exchanged reciprocally (genetic *crossing-over*) (Chapter 6).
5. Chromosomes may be observed forming *chiasmata* and exchanging parts reciprocally (Chapters 4 and 6). Such exchange is reflected in genetic crossing-over. Furthermore, chiasma formation and genetic crossing-over occur with similar frequencies.

Thus linkage is an exception to the pattern of random segregation of genes, and crossing-over results in an exception to the usual consequences of linkage.

MAPPING NONHUMAN CHROMOSOMES

As Morgan (1911b) predicted, frequency of crossing-over is governed largely by distances between genes. That is, the probability of crossing-over occurring between two *particular* genes increases as the distance between the genes becomes larger, so that crossover frequency appears to be directly proportional to distances between genes. As will be described later, this relationship is quite valid, though not equally so in all parts of a chromosome, because the proximity of one crossover to another decreases the probability of another very close by. The centromere has a similar interference effect; frequency of crossing-over is also reduced near the ends of the chromosome arms. Because of this general relationship between intergene distance and crossover frequency, and because such distances cannot be measured in the customary units employed in light microscopy, geneticists use an arbitrary unit of measure, the **map unit,** to describe distances between linked genes. A *map unit is equal to 1 percent of crossovers* (recombinants); that is, it represents the linear distance along the chromosome for which a recombination frequency of 1 percent is observed. These distances can also be expressed in **morgan units;** one morgan unit represents 100 percent crossing-over. Thus, 1 percent crossing-over can also be expressed as 1 *centimorgan* (1 cM), 10 percent crossing-over as 1 *decimorgan,* and so on. The morgan unit is obviously named in honor of T. H. Morgan (page 91). So for sweet peas, the distance from *R* to *Ro* would be described as 12.5 map units, or 12.5 centimorgans. Most geneticists prefer map units, however.

Interestingly enough, it is now possible to calculate the sizes of many genes, as well as the distances separating them, and to photograph genes in the electron microscope. Such approaches require use of information about the molecular structure of the genetic material, which will be explored in later chapters.

The Three-Point Cross

The most commonly used method in genetic mapping of *Drosophila* chromosomes and those of many other nonhuman eukaryotes is the trihybrid (or "three-point") testcross. Such a cross in *Drosophila melanogaster,* the little fruit fly whose genetics is so well known, will demonstrate this approach. Using a plus sign to denote the so-called wild type, as is customary in mapping problems, a three-point testcross will be examined, using these genes:

Gene symbol	*Phenotype*
+	Normal wing (*dominant*)
cu	Curled wing (*recessive*)
+	Normal thorax (*dominant*)
sr	Striped thorax (*recessive*)
+	Normal bristles (*dominant*)
ss	Spineless bristles (*recessive*)

The numbers of individuals in each progeny class in the following illustration are hypothetical, but the map distances and the genes are real. For the moment, an arbitrary gene sequence will be chosen; the results may either confirm that order or dictate a different one. How to determine the correct gene sequence will be demonstrated after examining the cross:

P normal normal normal ♀ × curled spineless striped ♂
+ + + /*cu ss sr* *cu ss sr*/*cu ss sr*

F_1 *phenotype*	*Maternal chromosome*	*#*	*%*	*Type*
Normal normal normal	+ + +	430	88.2	noncrossovers
Curled spineless striped	*cu ss sr*	452		
Normal spineless striped	+ *ss sr*	45	8.3	*cu-ss* single crossovers
Curled normal normal	*cu* + +	38		
Normal normal striped	+ + *sr*	16	3.3	*ss-sr* single crossovers
Curled spineless normal	*cu ss* +	17		
Normal spineless normal	+ *ss* +	1	0.2	double crossovers
Curled normal striped	*cu* + *sr*	1		
		1,000	100.00	

The first two classes, normal-normal-normal and curled-spineless-striped, in this example are each phenotypically like one parent or the other. For this reason, they are sometimes referred to as parental types. However, in some of the problems at the end of this chapter, it will be noted that the progeny classes that resemble the parents *phenotypically* differ from them *genotypically*. For this reason, it is best to refer to these two classes merely as **noncrossover** types.

By grouping F_1 data as in the table, several important considerations are apparent:

1. **The maternal chromosome received by members of each of the two numerically largest classes** (noncrossover flies), determinable from the phenotypes, **discloses whether** *cis* **or** *trans* **linkage is obtained in the maternal parent.** Here, for example, normal normal normal individuals must have received + + + from the maternal parent and *cu ss sr* from the paternal. Because noncrossover gametes will be more frequent than crossover sex cells, + + +/*cu ss sr* and *cu ss sr*/*cu ss sr* flies will occur in the majority if the linkage is in the *cis* configuration.
2. **Double-crossover individuals** (those that would not be noted in a two-pair cross involving only *cu* and *sr*) **show up.** These are individuals that, as the term implies, result from the occurrence of two crossovers between the first and third genes in order on the maternal chromosome. *Even numbers of crossovers between two successive genes will not be picked up if the same two chromatids are involved in both crossovers:*

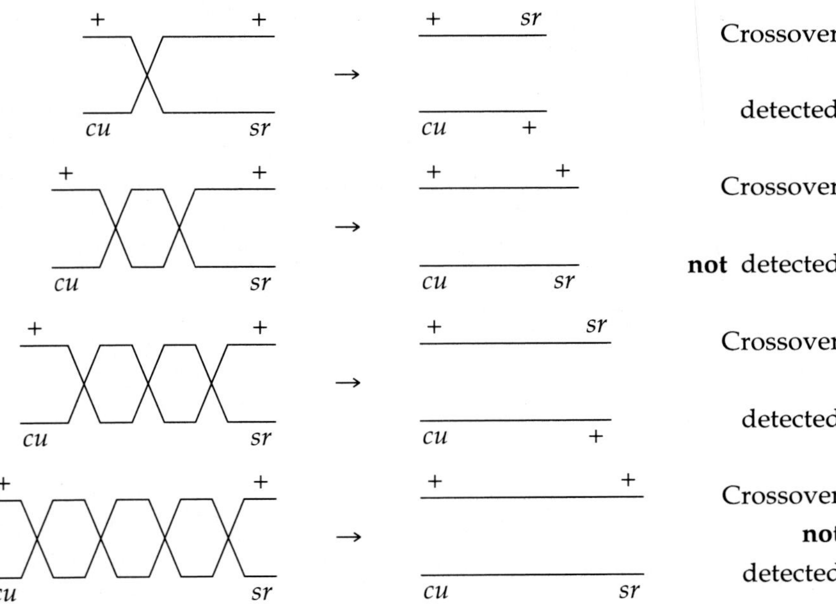

3. **The double crossovers are recognizable as the numerically smallest progeny groups and may be used to determine gene sequence.** Double crossovers are numerically the smallest groups because the product law of probability applies; that is, if the single crossover frequency between genes 1 and 2 is 0.1, and the crossover frequency between genes 2 and 3 is 0.2, then the probability of two crossovers in the *same chromatids,* one between genes 1 and 2 and another between genes 2 and 3 (thus, a double crossover), should be 0.1×0.2, or

0.02—a much smaller frequency than for either single crossover class. Because + *ss* + and *cu* + *sr* are, therefore, here identifiable as double crossovers, there is only one sequence of genes in the *cis* configuration in the maternal parent that would yield these two gene combinations following a double crossover. This becomes clear if we think of the original *cis* arrangement and how the + *ss* + and *cu* + *sr* chromosomes can be derived therefrom. This will be possible only by a crossover between the first and second genes in the sequence, *plus* another between the second and third:

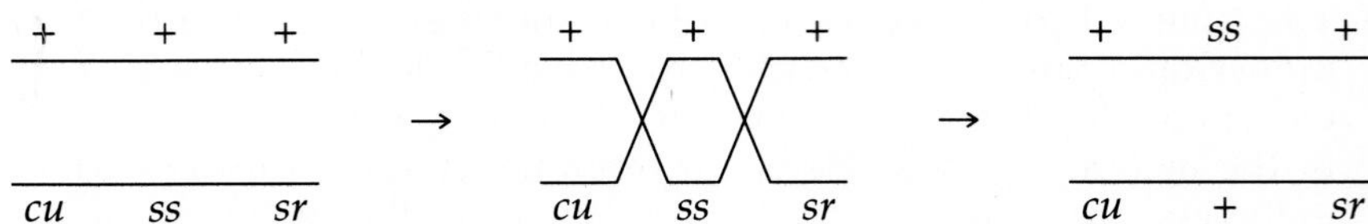

If, on the other hand, these genes were arranged in either of the other two possible sequences (*sr cu ss* or *ss sr cu*), the outcome of double crossing-over would be incompatible with the results observed in this example:

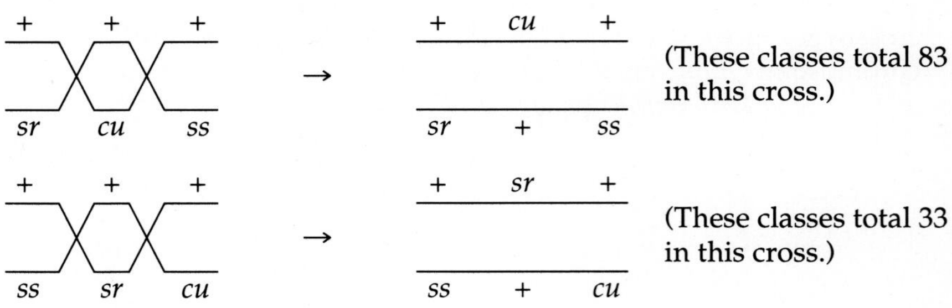

(These classes total 83 in this cross.)

(These classes total 33 in this cross.)

Thus the true sequence here can only be *cu ss sr*.

4. **The true distance** between *cu* and *ss* is, therefore, 8.3 + 0.2 = 8.5 (single crossovers + double crossovers).
5. **The true distance** between *ss* and *sr* is, therefore, 3.3 + 0.2 = 3.5 (single crossovers + double crossovers).
6. **The true distance** between *cu* and *sr* is 8.3 + 0.2 + 3.3 + 0.2 = 12.0 (*cu ss* single crossovers + *ss sr* single crossovers + *twice* the double crossovers). This is true because the double crossovers represent just what their name implies: *two* crossovers, one between *cu* and *ss*, plus a second one between *ss* and *sr*.

The genes here described constitute three members of a linkage group in *Drosophila*. With the cross outlined above, a beginning of genetically mapping one chromosome of *Drosophila* can be made. If these are thought of as the first three to be known in a new linkage group, they can be placed arbitrarily at particular *loci:*

cu *ss* *sr*
0.0 8.5 12.0

Of course, *sr* may be placed at the "left" end and the sequence reversed. If later work shows another gene, *W* (dominant for wrinkled wing), to be located 4.0 map units to the "left" of *cu*, then the map is redrawn and each locus renumbered accordingly:

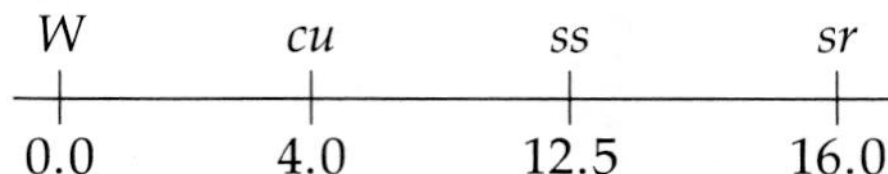

Figure 6–4 shows the locations of these and some of the other genes as presently assigned. Knowledge of which chromosome bears which genes in *Drosophila* (and other dipterans) is aided by a study of giant chromosomes (Chapter 15). A similar linkage map for corn (*Zea mays*) is shown in Figure 6–5. Some of the problems at the end of this chapter further explore the techniques of genetic mapping.

If the reciprocal of the trihybrid testcross given on page 98, namely,

cu ss sr/*cu ss sr* ♀ × + + +/*cu ss sr* ♂

had been made only the two noncrossover phenotypes (normal normal normal and curled spineless striped) would have been recovered in the progeny. This is so because crossing-over does not occur in the male flies. Dipterans are unusual in this respect.

Interference and Coincidence

The discussion of mapping techniques thus far would seem to imply that crossing-over in one part of a chromosome is independent of crossing-over elsewhere in that chromosome. The fact that this is not true was demonstrated as long ago as 1916 by Nobel-Prize-winning geneticist H. J. Muller.

Consider for a moment a chromosome bearing three genes, *a*, *b*, and *c*:

a *b* *c*

region I region II

Let the *a*-*b* portion be designated region I and the *b*-*c* segment as region II. Then, recalling the now familiar product law of probability, the frequency of double crossovers between genes *a* and *c* should equal the crossover frequency of region I times the crossover frequency of region II. Actually this is seldom true unless the distances involved are large.

For example, in the three-point mapping experiment in *Drosophila*, the following crossover frequencies were obtained:

"Region"	*Genes*	*Frequency of crossovers*	*Percentage crossovers*	*Map distance (in map units)*
I	*cu ss*	0.083	8.3	8.3 + 0.2 = 8.5
II	*ss sr*	0.033	3.3	3.3 + 0.2 = 3.5
Double Crossovers	*cu ss sr*	0.002	0.2	

If crossing-over in regions I and II were independent, 0.085 × 0.035, or almost 0.3 percent double crossovers would be predicted, whereas only 0.2 percent was observed. A disparity of this kind, in which the number of actual double crossovers is smaller than the number calculated on the basis of independence, is very common, suggesting that, once a crossover occurs, the probability of another in an adjacent region is reduced. This phenomenon is called **interference.**

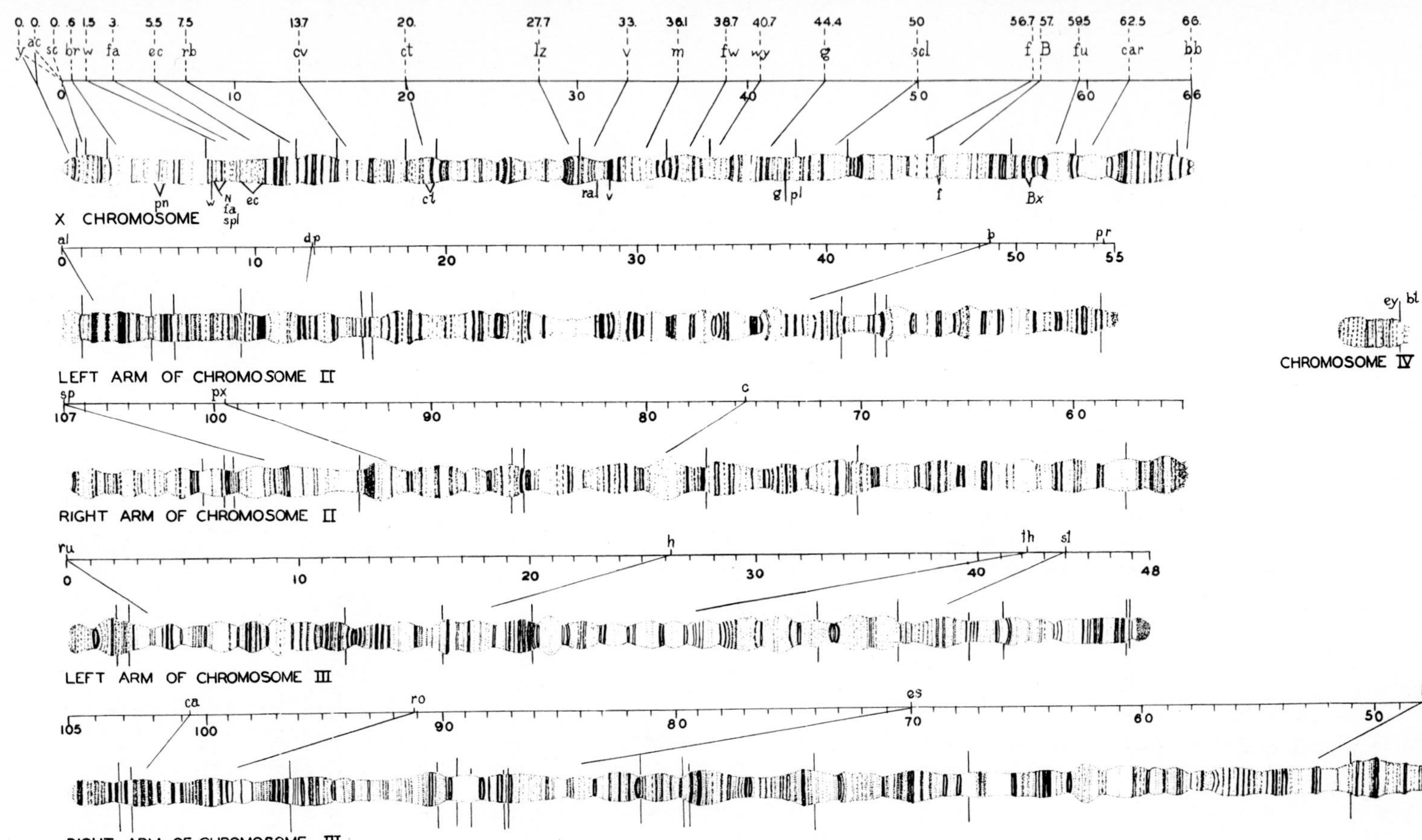

FIGURE 6–4. *Comparison of cytologic and genetic maps of the chromosomes of* Drosophila melanogaster. *For each chromosome, the genetic map is above the cytologic.* (From T. S. Painter, 1934, used with permission.)

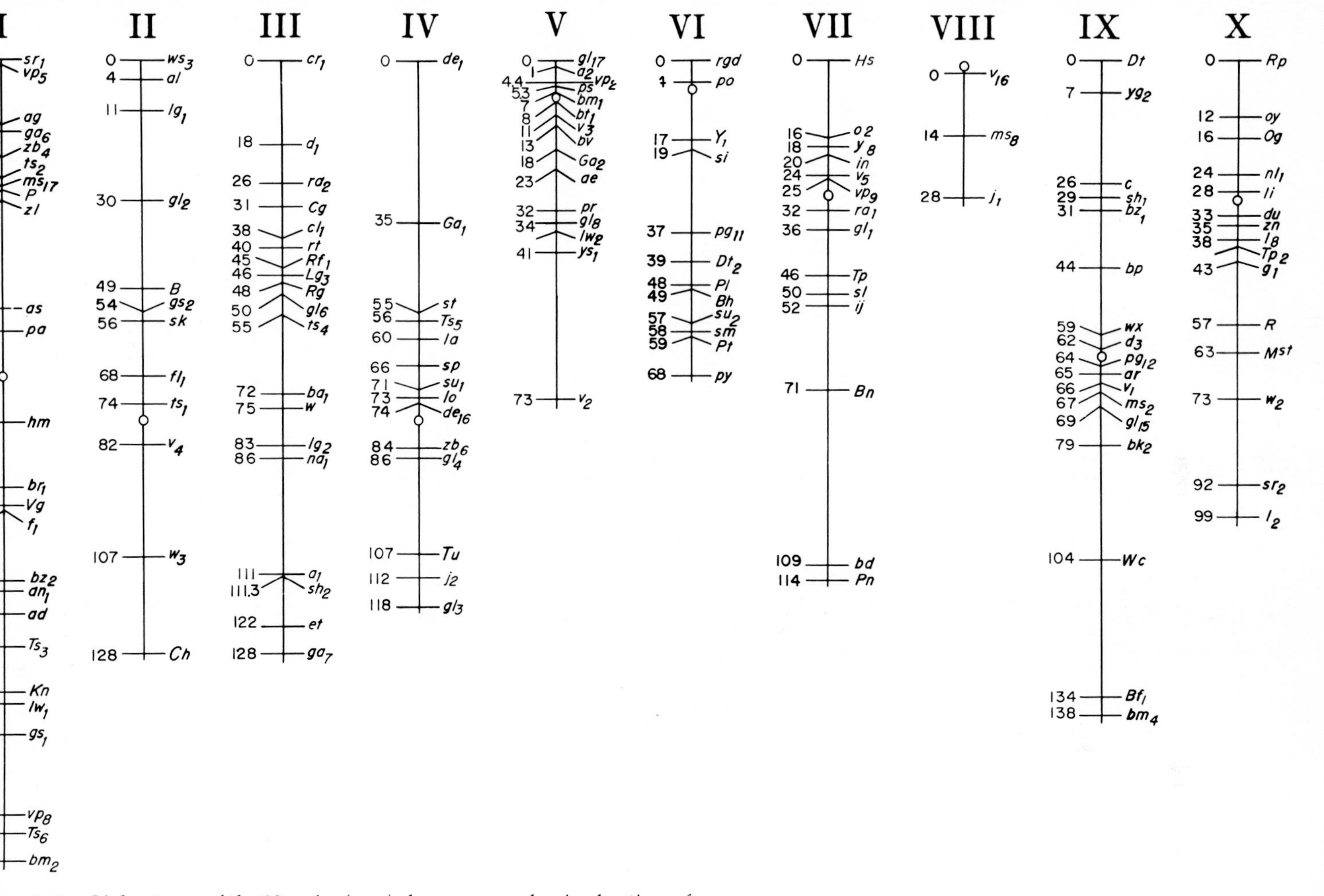

FIGURE 6–5. *Linkage map of the 10 maize (corn) chromosomes, showing locations of various known genes (symbolized by letters). Distances between genes are obtained by breeding experiments described in Chapter 6.* (After Neuffer, 1965, courtesy De Kalb Agricultural Association, Inc.; De Kalb, Illinois, by permission.)

Interference appears to be unequal in different parts of a chromosome, as well as among the several chromosomes of a given complement. In general, interference appears to be greatest near the centromere and at the ends of a chromosome. Degrees of interference are commonly expressed as **coefficients of coincidence** or simply as coincidence:

$$\text{coincidence} = \frac{\text{actual frequency of double crossovers}}{\text{calculated frequency of double crossovers}}$$

In the *Drosophila* example coincidence is

$$\frac{0.002}{0.003} = 0.67$$

As interference decreases, coincidence increases. Coincidence values ordinarily vary between 0 and 1. Absence of interference gives a coincidence value of 1, whereas complete interference results in a coincidence of 0. Coincidence is generally quite small for short map distances. In *Drosophila,* coincidence is zero for distances of less than 10 to 15 map units, but gradually increases to 1 as distances exceed 15 map units. Furthermore, there seems to be no interference across the centromere from one arm of the chromosome to the other.

Similarly, interference is reported from a wide variety of organisms. For example, Hutchison, who discovered the *c-sh* linkage in corn, reported map distances for three genes, *c* (colorless aleurone), *sh* (shrunken grains), and *wx* (waxy endosperm). His data indicated the following crossover frequencies:

"Region"	*Genes*	*Frequency of crossovers*	*Percentage crossovers*	*Map distance (in map units)*
I	*c sh*	0.034	3.4	3.4 + 0.1 = 3.5
II	*sh wx*	0.183	18.3	18.3 + 0.1 = 18.4
Double crossovers	*c sh wx*	0.001	0.1	

Again, if crossing-over in regions I and II were independent, $0.035 \times 0.184 = 0.6$ percent double crossovers would be anticipated, whereas only 0.1 percent was observed, giving a coefficient of coincidence of 0.167. On the other hand, especially in bacterial and bacteriophage genetics, double crossovers may be encountered in excess of random expectation, giving coincidence values > 1. This is referred to as *negative interference.*

MAPPING HUMAN CHROMOSOMES

Because human beings are not bred in laboratories, it is necessary to resort to other procedures in order to map the human genome. A highly successful method, somatic cell hybridization, began around 1960 as an unexpected dividend from a French experiment on cancer cell lines. It was discovered that cells from widely disparate species can and do fuse *in vitro,* even though the evolutionary characteristics of these species are so divergent that mating is not possible. The particular work of concern here involves the production of human-mouse hybrid cells. Mouse (murine) cells are particularly valuable in this kind of experiment because (1) a large number of mouse mutant lines is readily available from commercial laboratories; (2) mouse and human chromosomes can be easily distinguished from each other through their morphology and their different banding patterns; and (3) both human and mouse genes are expressed in the hybrid cells.

When mixed together, a very few (about one in a million) mouse and human cells fuse. Fusion extends even to the nuclei, so that the hybrid cells initially contain all 40 mouse chromosomes and all 46 human chromosomes. In mapping experiments the rate of fusion can be increased by adding either polyethylene glycol or Sendai virus (related to the influenza viruses) that has been inactivated, but not destroyed, by chemical treatment or by ultraviolet irradiation. This virus alters cell surfaces so that the rate of fusion is increased by a factor of as much as 1,000. Mouse and human chromosomes can also be differentiated by the fact that all those of the former species are acrocentric, whereas all human chromosomes except 13, 14, 15, 21, and 22 are metacentric or submetacentric.

As human-mouse hybrid cells are cultured further, human chromosomes are progressively and preferentially lost with each division cycle. (In mouse-rat and mouse-Chinese hamster hybrid cells, however, the *mouse* chromosomes are preferentially lost.) Approximately one to fifteen human chromosomes are preferentially lost during the first several cell generations.

After 100 generations many cells have lost all their human chromosomes. Use of a variety of selective media determines, in many cases, which human chromosomes will more readily be lost or retained, and acts to select for or against growth of strains unable to synthesize certain substances, including enzymes. For example, 5-bromodeoxyuridine (5-BDU) kills cells that produce the enzyme thymidine kinase. Growth of hybrid cells derived from thymidine kinase deficient (TK^-) murine cells and TK^+ human cells occurs in the presence of 5-BDU only if human chromosome 17 is lacking.

Thus, it is concluded that gene *TK* is located somewhere in human chromosome 17. Of course, each human chromosome can be easily identified by its unique banding pattern. Although it has not yet been possible to assign gene *TK* to a specific locus in chromosome 17, this can be achieved in general terms through the study of chromosomal aberrations, whereby part of one chromosome has been transferred to a member of a different (nonhomologous) pair. This is called translocation and is treated, along with other kinds of chromosomal structural aberrations, in Chapter 15. Most of the human chromosome assignments in Table 6–2 have been made through the somatic cell hybridization technique.

If two or more specific human gene products and a given human chromosome are both present in the same hybrid cells, then those genes are located in the same chromosome; that is, they are **syntenic.** The term **synteny** refers to genes that are located on the same chromosome, whether or not they show recombination; **linkage** refers only to genetic loci that have been shown by *recombination studies* to be in the same chromosome. Syntenic genes may be so far apart in their chromosome that they seem to segregate independently; that is, they may show as much as 50 percent recombination—as would be exhibited by nonsyntenic genes.

Further differentiation of mouse versus human enzymes is often possible through electrophoretic patterns of the respective enzymes. *Gel electrophoresis* studies of enzymes and other cell substances involves placing the material in a homogenous gel, such as starch. An electrical

THE MANIC-DEPRESSION GENE

Approximately 1 to 2 million Americans suffer from manic-depression. The disorder is characterized by extreme mood swings. In the mania phase, people exhibit symptoms of either elation or irritability. Thoughts race through their minds and they become increasingly active and talkative. They show generally poor judgment, which can have extreme economic and social consequences. At other times people so afflicted are clinically depressed, with feelings of hopelessness and helplessness. Their sleeping and eating patterns change and they might have suicidal thoughts; some even attempt to kill themselves. In between these periods, manic-depressives are perfectly normal.

A recent study has now located a gene for manic-depression on chromosome 11. The approach of the study is indicative of the degree of difficulty in determining the chromosomal location of human genes. The study was conducted on the Old Order Amish, a group of 12,000 people living in isolation in Lancaster County, Pennsylvania. The advantage of studying this group is that they tend to have large families and keep detailed, well-documented genealogies. Very few Amish enter or leave the community. The entire population is descended from 50 couples who emigrated from Germany between 1720 and 1750. Identifying those with manic-depression was easier because the Amish do not use drugs or alcohol, both of which can mask psychiatric symptoms. Also, the suicide rate was easy to determine because the Amish community is notorious for its absence of crimes or acts of violence.

After identifying individuals with manic-depression, genetic analysis was conducted on samples of white blood cells to determine linkage to an easily identifiable genetic marker located so close to the manic-depression gene that it is inherited with the gene 100 percent of the time. This marker was found to be located on the short arm of chromosome 11 of every individual diagnosed as having manic-depression. As soon as this study was completed, a group of Icelandic families was studied in whom manic-depression was known to be inherited as a single dominant trait. No linkage to chromosome 11 was found, proving that clearly more than one gene may cause manic-depression.

The significance of the findings to date is that they open up a new avenue of investigation to approaches for determining linkage relationships, as well as identifying genes for inherited diseases in humans. Moreover, by showing that manic-depression does have a genetic basis, and therefore, that a diagnosed individual does not have control over his or her behavior, researchers have removed some of the stigma previously associated with the disorder.

field is applied to the gel, and the cell substance moves toward either the negative or positive pole, depending on its own net charge. Enzymes are protein in whole or in part, and proteins are composed of amino acids. Some of these amino acids carry a positive charge, others a negative charge, but most carry none, and thus determine direction of electrophoretic mobility.

Single metaphase chromosomes can be isolated and added to cultured cells, resulting in a product change that can be associated with the particular chromosome transferred. These cells, however, are somewhat unstable, thus limiting the general usefulness of the method.

Still another method of determining gene distances on the X chromosome, one of the sex chromosomes, will be described briefly in the treatment of sex determination (Chapter 7).

LINKAGE STUDIES IN BACTERIA

Present evidence indicates that such bacteria as the common colon bacillus, *Escherichia coli,* contain one to several molecules of deoxyribonucleic acid (Chapter 7), each forming a continuous, closed ("no-end") structure except during reproduction such as conjugation (see next section and Appendix B1). Such a closed circle of DNA is functionally comparable to a chromosome of higher organisms, but is *not* a chromosome in the *structural* sense as found in eukaryotes. As described in

Appendix B1, during conjugation between donor and recipient, one such ring "chromosome" opens at a particular point and passes as a filament in part or in whole into the body of the recipient cell. The length of the transferred DNA molecule depends on the duration of the conjugation process.

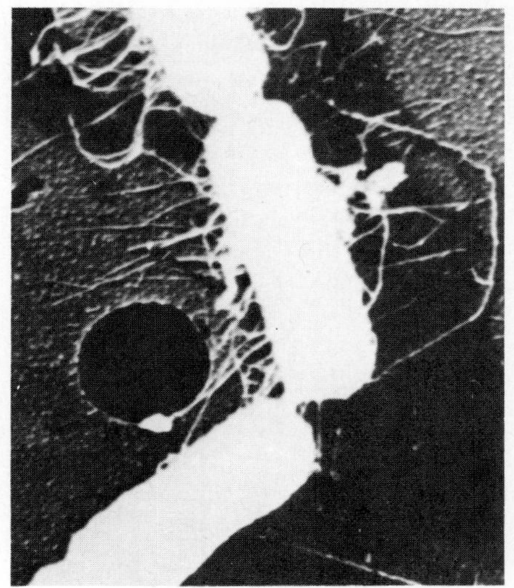

FIGURE 6–6. *Electron micrograph of conjugating* Escherichia coli *cells. The bridge connecting the two conjugants may be seen at the angle between upper and lower partners. It happens that the conjugating cells are both in late stages of fission as well, and one of them is coincidentally being attacked by bacteriophages.* (Photo courtesy Dr. Thomas F. Anderson, Institute for Cancer Research, Philadelphia.)

Evidence from Conjugation

In such haploid organisms, the usual technique of determining linkage distances by means of recombinational frequencies cannot be employed. Instead, donor cells of a known genotype (for several loci) are mixed with a large number of recipients of a different genotype, allowed to **conjugate** (Figure 6–6), then separated at predetermined times by agitating with a Waring blender. After separation, progeny of the recipient cells are tested by inoculation on different *deficiency media* to detect various physiologically deficient strains (*auxotrophs*). Genes are thus found to be located sequentially along the "chromosome" and are transferred in order to the recipient cell. For example, if donor cells of *Escherichia coli* K-12 (see Figure 6–7) are used, with these four known genes:

leu A production of the enzyme α-isopropylmalate (affects synthesis of the amino acid leucine)

lac Y production of the enzyme galactoside permease (affects metabolism of lactose)

tsx resistance or susceptibility to phage (virus; see next section) T6

gal K production of the enzyme galactokinase (affects metabolism of galactose)

By interrupting conjugation after different time intervals, genes are observed to be transferred in this time sequence:

Duration of conjugation (min.)	*Genes transferred*
1.6	*leu A*
7.9	*leu A, lac Y*
9.2	*leu A, lac Y, tsx*
16.7	*leu A, lac Y, tsx, gal K*

Therefore, the order of these genes is as given in the 16.7-minute sequence.

A particular experiment might involve, for instance, these strains:

(donor) *leu* A^-, *lac* Y^-, tsx^+, *gal* K^+ . . .
(recipient) *leu* A^+, *lac* Y^+, tsx^-, *gal* K^- . . .

Here a minus sign superscript indicates inability to synthesize the enzyme listed in the preceding paragraph (that is, **auxotrophic** for that substance) or susceptibility to phage T6 (tsx^-). Such genotypic symbols as *leu* A^+ indicate ability to synthesize the enzyme involved (**prototrophic**). The two strains are mixed in a tube of liquid (complete) medium, then, after conjugation, plated out in an agar plate containing a complete medium. Here a large number of progeny colonies, both auxotrophs and prototrophs, develop. After incubation, colonies are transferred to a series of deficiency media to detect recombinations. This is done by pressing the master plate onto a sheet of velvet whose fibers

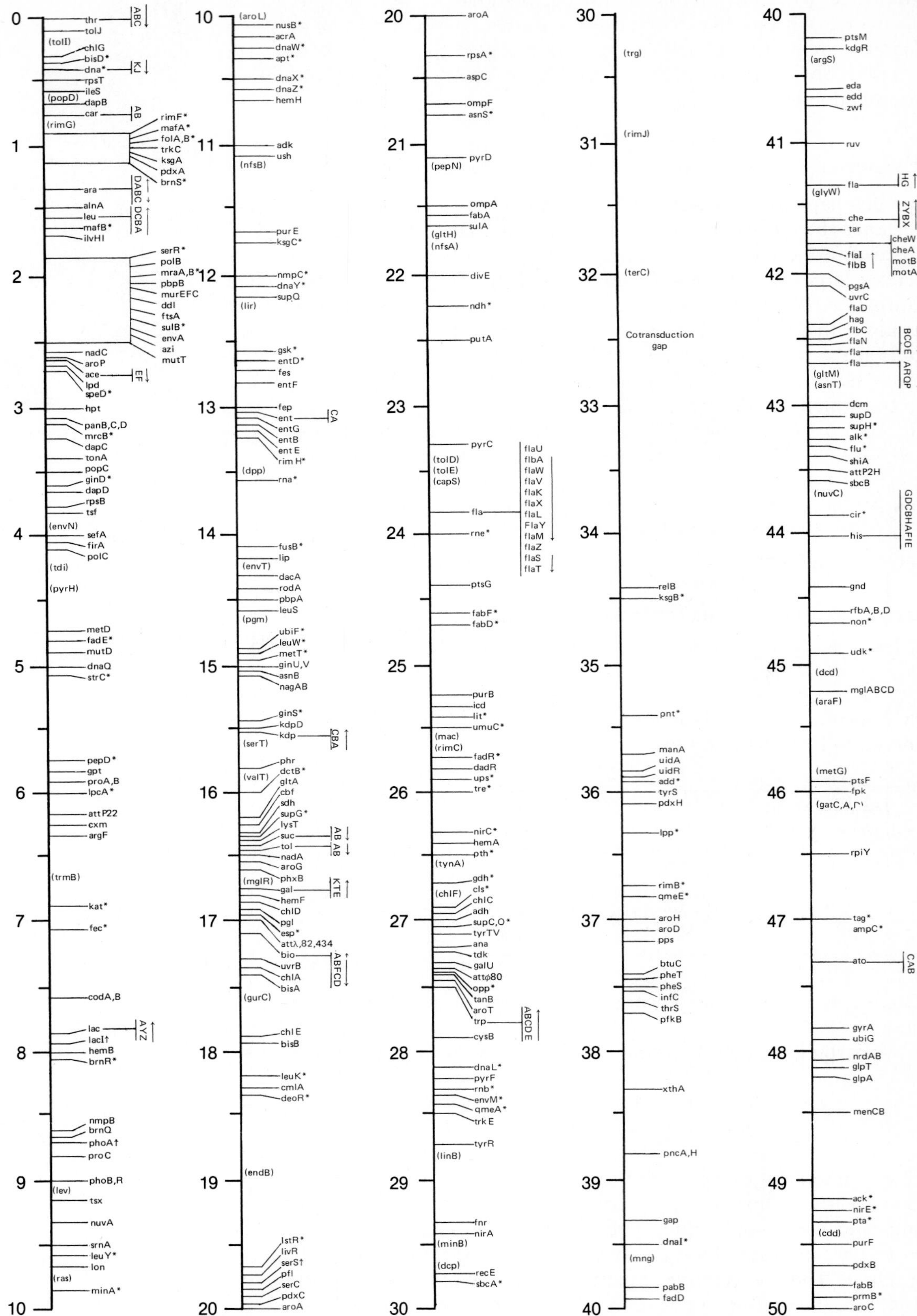

FIGURE 6–7. *Linkage map of* Escherichia coli *K-12. The numbers represent map positions in minutes determined from time-of-entry interruptions during conjugation at 37°C. Because of the large number of genes mapped, the circular chromosome has been represented as linear. More than 1,000 genes are now known in this organism.* (Redrawn from Bachmann and Low, 1980, used by permission.)

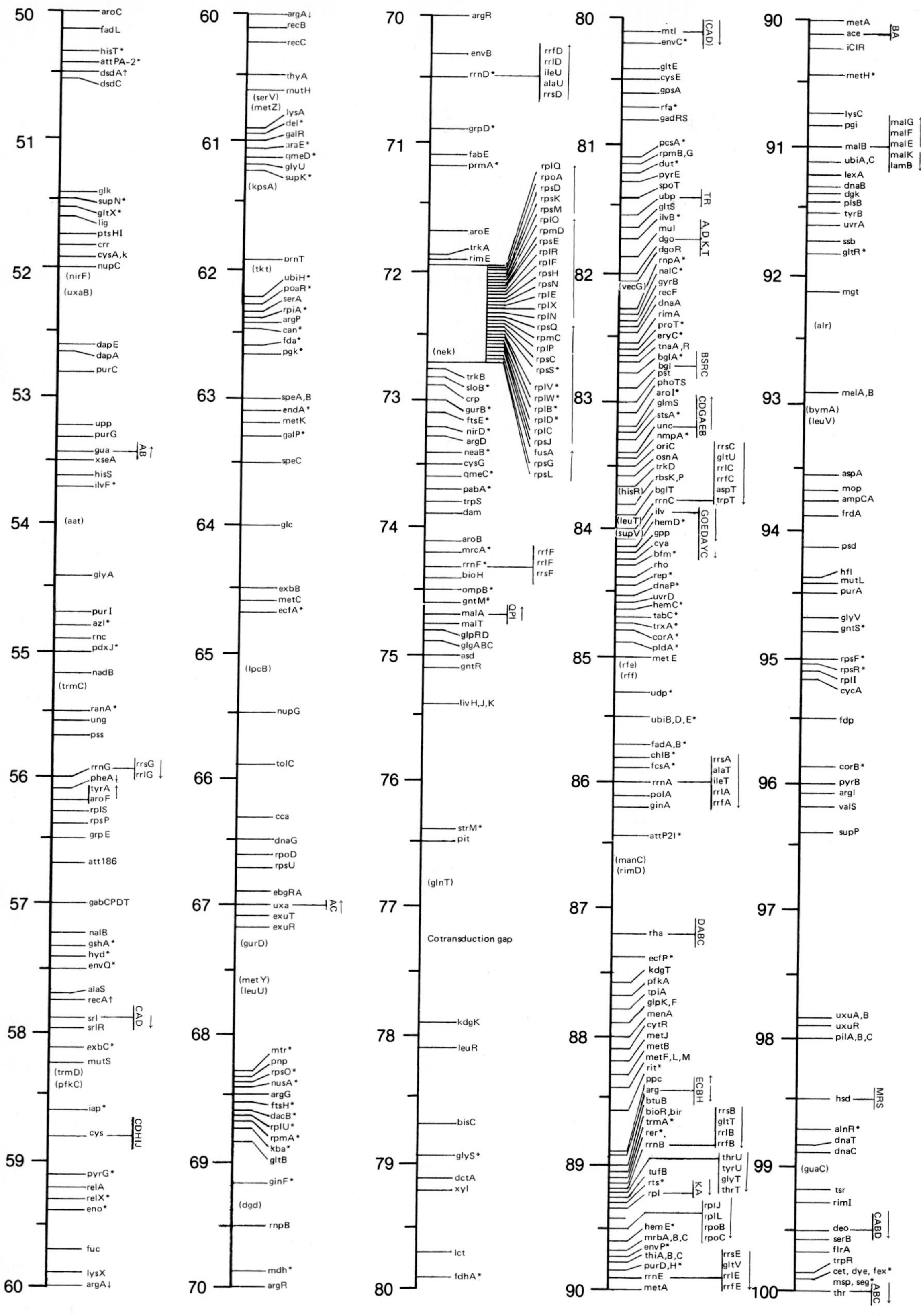

50
aroC
fadL
hisT*
attPA-2*
dsdA↑
dsdC
51
glk
supN*
gltX*
lig
ptsHI
crr
cysA,k
nupC
52
(nirF)
(uxaB)
dapE
dapA
purC
53
upp
purG
gua
xseA
AB
hisS
ilvF*
54
(aat)
glyA
purI
azl*
rnc
pdxJ*
55
nadB
(trmC)
ranA*
ung
pss
56
rrnG
pheA↓
tyrA
aroF
rrsG
rrlG
rplS
rpsP
grpE
att186
57
gabCPDT
nalB
gshA*
hyd*
envQ*
alaS
recA↑
srl
srlR
CAD
58
exbC*
mutS
(trmD)
(pfkC)
iap*
cys
CDHIJ
59
pyrG*
relA
relX*
eno*
fuc
lysX
argA↓
60
argA↓
recB
recC
thyA
mutH
(serV)
(metZ)
lysA
del*
galR
araE*
qmeD*
glyU
supK*
61
(kpsA)
ornT
62
(tkt)
ubiH*
poaR*
serA
rpiA*
argP
can*
fda*
pgk*
63
speA,B
endA*
metK
galP*
speC
64
glc
exbB
metC
ecfA*
65
(lpcB)
nupG
tolC
66
cca
dnaG
rpoD
rpsU
ebgRA
uxa
AC
exuT
exuR
67
(gurD)
(metY)
(leuU)
68
mtr*
pnp
rpsO*
nusA*
argG
ftsH*
dacB*
rplU*
rpmA*
kba*
gltB
69
ginF*
(dgd)
rnpB
mdh*
argR
70
argR
envB
rrnD*
rrfD
rrlD
ileU
alaU
rrsD
grpD*
71
fabE
prmA*
rplQ
rpoA
rpsD
rpsK
rpsM
rplO
rpmD
rpsE
rplR
rplF
rpsH
rpsN
rplE
rplX
rplN
rpsQ
rpmC
rplP
rpsC
rpsS*
rplV*
rplW*
rplB*
rplD*
rplC
rpsJ
fusA
rpsG
rpsL
aroE
trkA
rimE
72
(nek)
trkB
sloB*
crp
gurB*
ftsE*
nirD*
argD
neaB*
cysG
qmeC*
pabA*
trpS
dam
73
74
aroB
mrcA*
rrnF*
rrfF
rrlF
rrsF
bioH
ompB*
gntM*
malA
QPI
malT
glpRD
glgABC
asd
gntR
75
livH,J,K
76
strM*
pit
(glnT)
77
Cotransduction gap
kdgK
78
leuR
bisC
79
glyS*
dctA
xyl
lct
fdhA*
80
mtl
envC*
(CAD)
gltE
cysE
gpsA
rfa*
gadRS
81
pcsA*
rpmB,G
dut*
pyrE
spoT
ubp
TR
gltS
ilvB*
mul
dgo
A,D,K,T
dgoR
rnpA*
nalC*
82
(vecG)
gyrB
recF
dnaA
rimA
proT*
eryC*
tnaA,R
bglA*
bgl
BSRC
pst
phoTS
aroI*
83
glmS
stsA*
unc
CDGAEB
nmpA*
oriC
osnA
trkD
rbsK,P
rrsC
gltU
rrlC
rrfC
aspT
trpT
(hisR)
bglT
rrnC
ilv
hemD*
GOEDAYC
84
(leuT)
(supV)
gpp
cya
bfm*
rho
rep*
dnaP*
uvrD
hemC*
tabC*
trxA*
corA*
pldA*
metE
85
(rfe)
(rff)
udp*
ubiB,D,E*
fadA,B*
chlB*
fcsA*
rrsA
alaT
ileT
rrlA
rrfA
86
rrnA
polA
ginA
attP2I*
(manC)
(rimD)
87
rha
DABC
ecfR*
kdgT
pfkA
tpiA
glpK,F
menA
cytR
88
metJ
metB
metF,L,M
rit*
ppc
arg
ECBH
btuB
bioR,bir
trmA*
rer*
rrnB
rrsB
gltT
rrlB
rrfB
89
thrU
tyrU
glyT
thrT
tufB
rts*
rpl
KA
rplJ
rplL
rpoB
rpoC
hemE*
mrbA,B,C
envP*
thiA,B,C
purD,H*
rrnE
rrsE
gltV
rrlE
rrfE
metA
90
metA
ace
BA
iCIR
metH*
lysC
pgi
malG
malF
malE
malK
lamB
91
malB
ubiA,C
lexA
dnaB
dgk
plsB
tyrB
uvrA
ssb
gltR*
92
mgt
(alr)
melA,B
93
(bymA)
(leuV)
aspA
mop
ampCA
frdA
94
psd
hfl
mutL
purA
glyV
gntS*
95
rpsF*
rpsR*
rplI
cycA
fdp
corB*
96
pyrB
argI
valS
supP
97
uxuA,B
uxuR
pilA,B,C
98
hsd
MRS
alnR*
dnaT
dnaC
99
(guaC)
tsr
rimI
deo
CABD
serB
flrA
trpR
cet, dye, fex*
msp, seg*
thr
ABC
100

pick up individuals of each colony. The velvet is then pressed to a series of replica plates of a deficiency medium.

Many such experiments, usually using triple auxotrophs (for example, $- - - + + + \times + + + - - -$) to reduce to a very low value the probability of mutation as a factor, have resulted in a fairly complete genetic map of *E. coli* K-12, for example, as shown in Figure 6–9. Note that a total time of 100 minutes is indicated for transfer of the entire chromosome during conjugation. Reference to Appendix B1 will show that in the *Hfr* strain (high frequency recombination) the *F* (fertility) factor is integrated into the bacterial nucleoid. At the outset of conjugation the *F* factor is "nicked," for example, part of it then leads the way through the conjugation tube into the recipient (F^-) cell, while the remainder trails at the far end of the donor chromosome. Ordinarily, however, the conjugation process is interrupted by natural stresses far short of transfer of the complete donor chromosome.

In the F^+ strain, the *F* factor exists free in the cytoplasm and is usually the only genetic material transferred in conjugation—thus converting the F^- recipient to an F^+ strain.

If the *F* factor is integrated into the chromosome its replication is synchronous with that of the chromosome so that it also is passed from the parental cell to its fission progeny. In some cells the integrated *F* factor is excised from its chromosome, often with some of the chromosomal genes attached. A fertility factor carrying some attached chromosomal genes is designated F' ("*F* prime"). Conjugation between F' and F^- results in partial diploidy (for the chromosomal genes carried by F'). Occasionally, in such cases, crossing-over between homologous regions occurs, producing recombinants. Partial diploids are called **merozygotes** and the process just described is termed **sexduction.** Sexduction also makes it possible to determine dominance of alleles in these otherwise haploid cells.

LINKAGE STUDIES IN VIRUSES

Bacteriophages, or phages, are viruses that infect bacteria. Phage structure and life cycle are described in Chapter 8. Like all viruses, phages have genetic systems as evidenced by the fact that host and symptom specificities occur. Moreover, changes or mutations of phages can occur in such genetically controlled properties as virulence toward a particular host or in the kind of coat proteins produced. During the 1940s and 1950s, a number of studies showed convincingly that bacteria and viruses indeed do possess genetic material and that it is DNA in some viruses and RNA in other viruses.

Virulent phages destroy, or lyse, their hosts. One mutant in T2 and T4 phages (the T series infects the common colon bacterium, *E. coli*) accomplishes this host destruction more rapidly than does the "wild-type" phage. Rapid lysing strains are designated as T2r, T4r, and so on. If a culture of *E. coli* is infected by a mixture of T2 and T4r phages, *four* progeny types are recovered: the "parental" T2 and T4r and also the recombinants T2r and T4.

Such recombination does not result from a sexual process, but rather by recombination and exchange of genetic material between the phages in conjunction with involvement of the bacterial host's genetic material. The details of this process will be discussed when the molecular nature of the genetic material itself is described, but for present purposes, it is important to note (1) that viruses can be "crossed," and (2) as a result, a map of the viral genome can be constructed. Rather early in phage studies three different linkage groups were proposed for T2, but later work has shown that the phage map, like that of bacteria, is "circular," in that each marker is linked to another on either side. One whole complex of T4r mutants, known as rII (because they were originally assigned to linkage group II of early investigators) has provided important information on the fine structure of the gene.

MECHANISM OF RECOMBINATION

The detailed mechanism whereby the donated chromosome segment becomes integrated into that of the recipient's chromosome in bacterial conjugation is fairly well understood, as are the fine points of crossing-over in eukaryotes. A segment of donor chromosome is not necessarily incorporated as a whole although a mechanism analogous to crossing-over is assumed. A full statement of present understanding must be deferred until the nature and operation of the genetic material has been considered at the molecular level. It does, however, appear that some sort of break-and-exchange mechanism is operative. For immediate purposes it will suffice to point out that the segment of donor chromosome synapses with the homologous segment of the recipient's chromosome and then breaks in both donor and recipient may occur. A segment of the donor chromosome then may replace a corresponding section of the recipient's chromosome. If genes in the two segments differ (for example, *leu* A^+ versus *leu* A^-), recombination will have occurred in the recipient cell. Here *leu* A^+ denotes a gene for ability to synthesize the amino acid leucine; *leu* A^- designates a gene for inability to synthesize that amino acid.

In eukaryotes, the most suggestive evidence concerning the events of recombination is derived from certain plants that have a dominant haploid phase—for example, many algae, most true fungi, and all bryophytes (liverworts and mosses). In bryophytes, for example, meiosis results in a spherical tetrad of four *unordered* meiospores. But in such ascomycete fungi as *Neurospora,* the cells resulting from meiosis are *ordered;* that is, situated in line in the ascus (see life cycle, Appendix B2). Such ordering reflects the pattern of chromosomal arrangement at each meiotic stage. Although meiosis in *Neurospora* produces the usual four meiospores, each divides once by mitosis to produce a total of eight ascospores, sequential pairs of which are genotypically identical. Each ascospore may be removed in order from the ascus and germinated to determine physiological phenotype, or examined visually for such morphological traits as color. Such an analysis is termed **tetrad analysis;** it clearly indicates that recombination via crossing-over must occur at the four-chromatid stage, and thus supports the concept of a break-and-exchange mechanism.

From a cross between an auxotroph (for example, *prolineless*) and the wild-type prototroph, both + and *pro* ascospores are found in equal numbers in each ascus. However, the sequential arrangement of these spores may be either + + + + *pro pro pro pro,* or + + *pro pro* + + *pro pro.* The latter arrangement is possible only if recombination by crossing-over occurs during the four-strand stage (Figure 6–8).

If a double auxotroph for the linked traits *prolineless/serineless* (*pro/ser*) and the wild type (+/+) are crossed, the reciprocal recombinants (+/*ser* and *pro*/+) are produced with equal frequency, along with the nonrecombinant types.

Even more can be learned from studies of segregation of blocks of three linked markers. Tetrad analysis clearly shows that each crossover

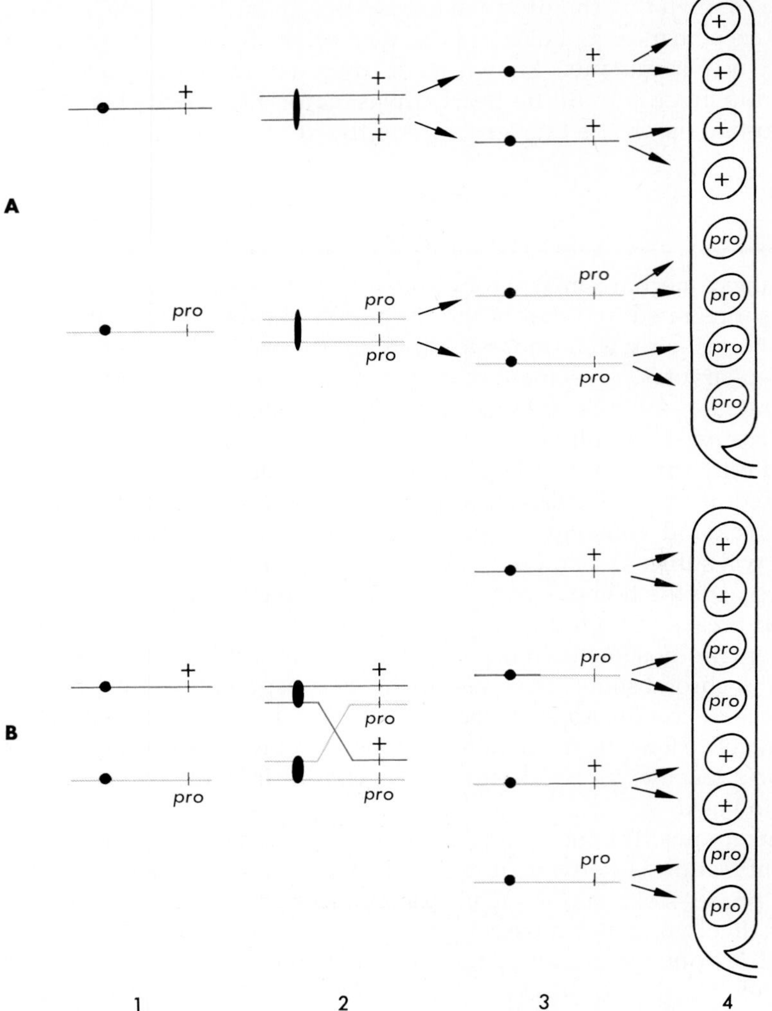

FIGURE 6–8. *Alternative arrangements of ascospores in ascus of* Neurospora. *In (A) crossing-over does not occur or does not involve genes (+) and pro; in (B) crossing-over occurs between the centromere and the (+) pro alleles (B 2). Only if crossing-over occurs in the four-strand stage, involving nonsister chromatids, can the sequence of ascospores shown at (B 4) be attained. In 1 of both (A) and (B), the chromosomes contributing to the zygote by each parental strain are shown; in 2, replication has taken place and it is at this stage that synapsis and crossing-over (if any) occurs; in 3, the chromosomes of each of the meiospores resulting from meiosis are depicted; in 4, a mature ascus and the genotypes of each of its ascospores are shown. Meiosis is taking place between 1 and 3; mitosis occurs between 3 and 4.*

can involve either of the two chromatids of each homologous chromosome so that three different basic types of double crossover tetrads are possible:

1. *Two-strand doubles,* in which the same two chromatids are involved in both crossovers (Figure 6–9A)
2. *Three-strand doubles,* in which three chromatids are involved, one of them participating twice (Figure 6–9B)
3. *Four-strand doubles,* in which each of the crossovers involves a different pair of chromatids (Figure 6–9C)

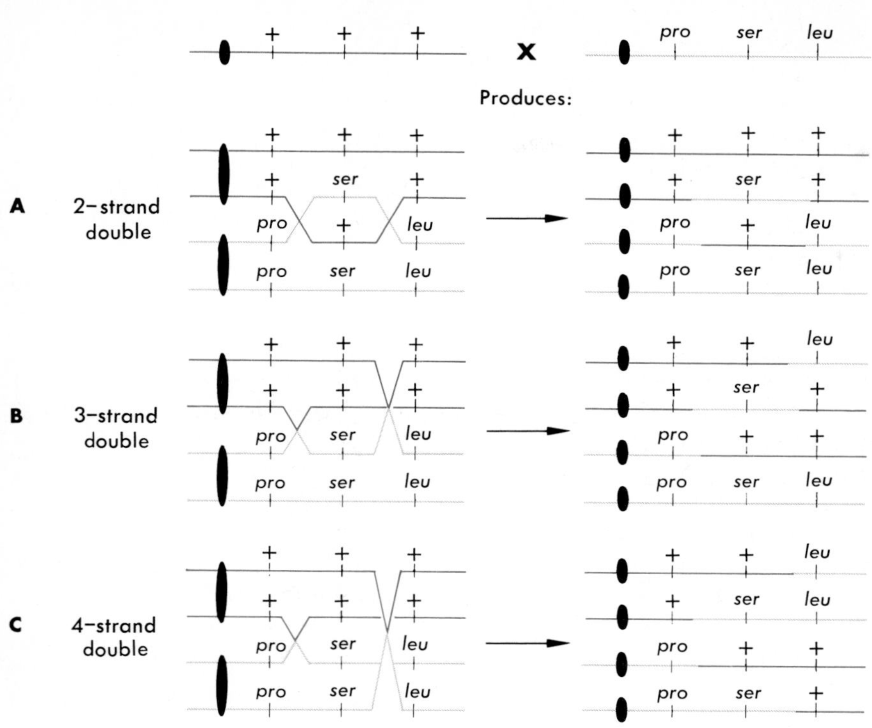

FIGURE 6–9. *Diagram showing possible types of double crossing-over involving (A) two chromatids only, (B) three chromatids, (C) all four chromatids, as inferred from tetrad analysis.*

At any given moment a visible chiasma does not necessarily indicate the point at which crossing-over occurs, because after formation, chiasmata move toward the ends of the chromatids involved. There is, nevertheless, a close correspondence between number of chiasmata and the frequency of crossovers in organisms whose cytology and genetics have been extensively investigated. Three- and four-strand doubles (Figures 6–9B and C) can be explained only if breaks occur before chiasmata are formed and after replication of chromatids has occurred.

SUMMARY

Thus, current evidence demonstrates that

1. The number of genes exceeds the number of pairs of chromosomes as our knowledge of the organism's genetics develops.
2. Therefore certain blocks of genes are linked, and
3. The number of linkage groups is equal to the number of pairs of chromosomes (with appropriate exception for nonhomologous sex chromosomes), but
4. Linkage is not inviolable.
5. A reciprocal exchange of material between homologous chromosomes in heterozygotes is reflected in crossing-over.
6. The frequency of crossing-over appears to be closely related to physical distance between genes on a chromosome and serves as a tool in constructing genetic maps of chromosomes.
7. Crossing-over results basically from an exchange of genetic material between nonsister chromatids by break-and-exchange following replication.

PROBLEMS

6–1 If, in sweet pea, the cross *R Ro/r ro* × *R Ro/r ro* is made, what would be the expected frequencies of (a) parental gametes of each of the possible genotypes, (b) *R Ro/R Ro* progeny, (c) purple long progeny?

6–2 Loci for the human Duffy blood groups, that is, production of either, both, or neither of the antigens Fy^a and Fy^b, and the rare Charcot-Marie-Tooth disease (a severe sensory and motor neuropathy) are both located on autosome 1. Studies suggest a recombination frequency of about 0.15. Give the distance between the two loci in map units.

6–3 **Elliptocytosis,** a rare but harmless condition in which the erythrocytes are ellipsoidal instead of the more common disk shape, is due to the presence of either of two completely dominant genes, El_1 or El_2. Production of rhesus antigen D is also due to a dominant gene, *D;* inability to produce this antigen is associated with genotype *dd*. Persons of genotypes *DD* and *Dd* are referred to as being Rh positive (Rh+). Loci El_1 and *D* are linked. They have been assigned to chromosome 1. In the following case consider only gene pairs El_1, el_1 and *D, d*. An Rh+ man exhibits elliptocytosis as did his Rh− mother. His Rh+ father had normal disk-shaped erythrocytes. (a) Neglecting the small probability of crossing-over in his parents, give this man's genotype. (b) What kind of linkage configuration is this?

6–4 As noted in the preceding problem, gene pairs El_1, el_1, and *D, d*, are linked on human chromosome 1. Some studies suggest a distance of three map units between these loci. If a doubly heterozygous man, known to have these two genes linked in the *cis* configuration, marries an Rh− woman who has normal disk-shaped red blood cells, what is the probability of each of the following phenotypes among their children: (a) Rh+ with elliptocytosis, (b) Rh+ without elliptocytosis?

6–5 The following four pairs of alleles are linked on chromosome 2 of tomato:

Aw, aw purple, green stems
Dil, dil normal green, light green leaves
O, o round, elongate fruits
Wo, wo wooly, smooth leaves

Crossover frequencies in a series of two-pair testcrosses were found to be: *wo-o,* 14 percent; *wo-dil,* 9 percent; *wo-aw,* 20 percent; *dil-o,* 6 percent; *dil-aw,* 12 percent; *o-aw,* 7 percent. (a) What is the sequence of these genes on chromosome 2? (b) Why is the *wo-aw* two-pair crossover frequency not greater?

In the next four questions, the following facts will have to be used. In *Drosophila,* the following alleles occur on chromosome III: *e,* ebony body; *fl,* fluted or creased wings; *jvl* ("javelin"), bristles cylindrical and crooked; *obt* ("obtuse"), wings short and blunt.

6–6 A series of dihybrid testcrosses shows the following crossover frequencies: *jvl-fl,* 3 percent; *jvl-e,* 13 percent; *fl-e,* 11 percent. (a) What is the gene sequence? (b) How do you account for the fact that the sum of the *fl-e* and *fl-jvl* frequencies exceeds the *jvl-e* frequency?

6–7 Another cross discloses a crossover frequency of 19 percent between *jvl* and *obt*. How well can you locate *obt* in the sequence in Problem 6–6?

6–8 If the *e-obt* crossover frequency is next found to be 7 percent, where should *obt* be located in the sequence?

6–9 The *fl-obt* crossover frequency is determined by the cross + + /*ft obt* (♀) × *fl obt/fl obt* (♂) to be 17.5 percent. (a) Does this confirm your answer to Problem 6–7? (b) What should be the frequency of double crossovers in a trihybrid testcross involving genes *fl, e,* and *obt* if there is no interference?

6–10 As pointed out in the explanatory note preceding Problem 6–6, genes *e, fl, jvl,* and *obt* are located on chromosome III, along with many other known genes, thus constituting part of one linkage group. How many linkage groups are there altogether in *Drosophila melanogaster* females?

6–11 Each individual of the Jimson weed (*Datura stramonium*) produces both sperms and eggs. After consulting Table 4–1, give the number of linkage groups in this plant.

6–12 How many linkage groups are there in the (a) female grasshopper, (b) male grasshopper, (c) human female, (d) human male? Check Table 4–1 for help.

6–13 Mendel studied seven pairs of contrasting characters in the garden pea. Why, do you suppose, did he not discover the principle of linkage? (It would be worthwhile to consult Blixt's paper before you reach a final conclusion.)

6–14 (a) Does crossing-over take place in the ♀ *Drosophila* in oogenesis? (b) Does crossing-over take place in the ♂ *Drosophila* during spermatogenesis?

6–15 How many different gamete genotypes are produced by (a) the female in the cross + + /*jvl fl* (♀) × *jvl fl/jvl fl* (♂), (b) the male in the cross + + /*jvl fl* (♀) × + + /*jvl fl* (♂)? (Refer back to Problems 6–6 through 6–9 and your answers to Problem 6–14.)

6–16 How many different gamete genotypes are produced by (a) the female in the cross + + + /*jvl fl e* (♀) × *jvl fl e/jvl fl e* (♂), (b) the male in the cross + + + /*jvl fl e* (♀) × + + + /*jvl fl e* (♂)?

6–17 These two pairs of alleles are linked on chromosome 2 in tomato:

O round fruit	*P* smooth fruit skin
o ovate (elongate) fruit	*p* "peach" (fuzzy fruit skin)

A testcross between a heterozygous "smooth, round"

pistillate plant and a "peach, ovate" staminate one produced the following progeny:

smooth, round	420	peach, round	57
peach, ovate	460	smooth, ovate	63

(a) Which are the crossover progeny? (b) What are the parental genotypes? (c) What kind of linkage does the pistillate parent have? (d) Give the parental gamete genotypes and the frequency of each to be expected from a large number of plants having the same genotype as the pistillate parent in this cross. (e) What is the map distance in map units between loci *o* and *p*?

6–18 Two other pairs of alleles known in tomato are

Cu, "curl" (leaves curled)

cu, normal leaves

Bk, "beakless" fruits

bk, "beaked" fruits, having sharp-pointed protuberance on blossom end of mature fruit.

The cross of two doubly heterozygous "curl beakless" plants yields four phenotypic classes in the offspring, of which 23.04 percent are "normal beaked." Are these two pairs of genes linked? How do you know?

6–19 From the data of the preceding problem can you determine (a) whether, in the parents, *cu* and *bk* were in the *cis* or *trans* configuration; (b) the distance between them in map units (assuming no interference)?

In the next four problems, use this information. Two of the many known pairs of genes in corn are

Pl purple plant

pl green plant

Py tall plant (normal height)

pl pigmy (very dwarf)

These genes are 20 map units apart on chromosome 6. The cross *Pl Py/pl py* × *Pl Py/pl py* is made. Now answer the following four questions.

6–20 What is the gamete genotypic type ratio produced by each parent?

6–21 What percentage of the offspring has the genotype *pl py/pl py?*

6–22 What percentage of the progeny is purple pigmy?

6–23 What percentage of the offspring will be "true-breeding"?

6–24 Look again at the *Drosophila* trihybrid testcross on page 98. If the cross + *ss* + /*cu* + *sr* × *cu ss sr/cu ss sr* had been made instead, what would be the percentage of (a) + *ss* + and *cu* + *sr*, (b) + + + and *cu ss sr* flies in the progeny, assuming the same crossover frequencies and the same interference?

6–25 In *Drosophila,* these genes occur on chromosome III:

+ wild *h* hairy (extra hairs on scutellars and head)
+ wild *fz* frizzled (thoracic hairs turn inward)
+ wild *eg* eagle (wings spread and raised)

The cross + + + /*h fz eg* ♀ × *h fz eg*/*h fz eg* ♂ yielded this F_1:

wild wild wild	393	wild wild eagle	28
hairy frizzled eagle	409	hairy frizzled wild	30
wild frizzled eagle	58	wild frizzled wild	1
hairy wild wild	80	hairy wild eagle	1

(a) Give the sequence of genes and the distances between them. (b) What is the coincidence?

6–26 The cross + + + /*a b c* ♀ × *a b c*/*a b c* ♂ in the fruit fly gives the following crossover results:

a-b single crossovers	5.75 percent
b-c single crossovers	8.08 percent
a-c double crossovers	0.25 percent

What is the coincidence?

6–27 In tomato the following genes are located on chromosome 2:

+ tall plant	*d* dwarf plant
+ uniformly green leaves	*m* mottled green leaves
+ smooth fruit	*p* pubescent (hairy) fruit

Results of the cross + + + /*d m p* × *d m p*/*d m p* were

+ + +	470	+ *m p*	1
+ + *p*	14	*d* + *p*	25
d + +	0	*d m p*	441
+ *m* +	19	*d m* +	30

(a) Which groups in the progeny represent double crossovers? (b) What is the correct gene sequence? (c) What are the distances in map units between the first and second, and between the second and third genes? (d) Is there interference?

The following genes are linked on chromosome II of *Drosophila:*

+ wild	*b*	black body
+ wild	*cn*	cinnabar eyes
+ wild	*vg*	vestigial wings

6–28 A trihybrid cross between a heterozygous ♀ and a homozygous recessive ♂ produced the following 1,000 progeny:

+ + +	39	*b cn* +	1
b + +	416	*b cn vg*	48
+ *cn* +	42	*b* + *vg*	50
+ *cn vg*	402	+ + *vg*	2

(a) Which are the noncrossover progeny classes? (b) Which are the double crossover classes? (c) What kind of linkage occurs in the ♀? (d) What is the middle gene in the sequence? (e) What is the distance in map units between *cn* and *b*; between *vg* and *cn?* (f) Is there interference? (g) What is the coefficient of coincidence to the nearest second decimal place (carry the *expected* frequency to 5 decimal places)?

6–29 From the chromosome map for maize (corn), Figure 6–5, note that genes pg_{12}, gl_{15}, and bk_2 are all on chromosome 9. The testcross + + + /pg_{12} gl_{15} bk_2 × pg_{12} gl_{15} bk_2/pg_{12} gl_{15} bk_2 is made. If there is complete interference, what is the frequency of (a) noncrossovers, (b) pg_{12} gl_{15} single crossovers, (c) gl_{15} bk_2 single crossovers, (d) double crossovers in the progeny? (Assume all genotypes to be equally viable.)

6–30 With the cross of the preceding problem, what would be the frequencies of each of the eight progeny classes if (a) there is no interference, (b) coincidence is 0.5? (Assume all genotypes to be equally viable and crossover probabilities to be equal in all parts of the chromosome.)

SEX DETERMINATION AND INHERITANCE RELATED TO SEX

CHAPTER 7

There are two types of sexually reproducing animals and plants: (1) monoecious (from the Greek *monos,* meaning "only," and *oikos,* meaning "house"), in which each individual produces two kinds of gametes, sperms, and eggs, and (2) dioecious (Greek prefix *di,* meaning "two," and *oikos*), in which a given individual produces only sperms or eggs. In dioecious organisms, the primary sex difference concerns the kind of gametes and the primary sex organs by which these are produced. Each sex also exhibits many secondary sex characters. In humans these include voice, distribution of body fat and hair, and details of musculature and skeletal structure; in *Drosophila* these include the number of abdominal segments, presence (♂) or absence (♀) of sex combs, and so on (Figure 7–1). *Our immediate problem is to examine the mechanisms that determine the sex of an individual. Two basic types,* ***chromosomal*** *and* ***genic,*** *will be discussed, though the distinction is not always an obvious one.*

SEX CHROMOSOMES

Diploid Organisms

XX-XO system. Chromosomal differences between the sexes of several dioecious species were found early in the course of cytological investigations. H. Henking, a German biologist, in 1891 noted that half the sperms of certain insects contained an extra nuclear structure, the X body. The significance of this structure was not immediately understood, but in 1902, Clarance McClung, an American biologist, reported that the somatic cells of the female grasshopper contained 24 chromosomes, whereas those of the male had only 23. Three years later, Ed-

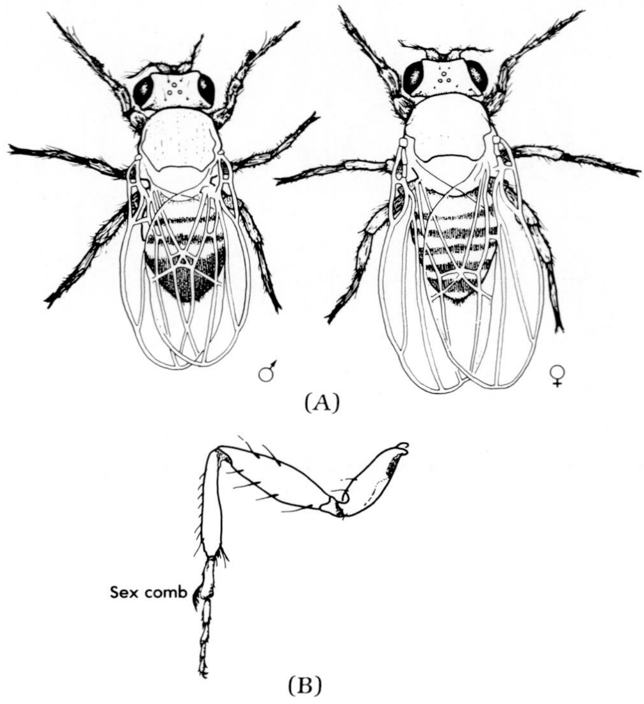

FIGURE 7–1. Drosophila melanogaster: *(A) male, left; female, right. Two distinguishing features are the merging of the posterior bands in the male versus their distinct separation in the female, and the longer more tapering abdomen of the female. (B) Detail of legs of male showing sex comb.*

ward B. Wilson and N. M. Stevens succeeded in following both oogenesis and spermatogenesis in several insects. They realized that the X body was a chromosome, so the X body became known as the X chromosome. Thus in many insects, there is a chromosomal difference between the sexes, that is, females are referred to as XX (having two X chromosomes) and males as XO ("X-oh," having one X chromosome). As a result of meiosis, all the eggs of such species carry an X chromosome. Only half the sperms have one and the other half have none.

XX-XY system. In the same year, 1905, Wilson and Stevens found a different arrangement in other insects. In such cases females were again XX, but males had in addition to one X chromosome, an odd one of a different size, which was called the Y chromosome: thus males were XY. Half the sperms carry an X and half a Y. The so-called XY type of sex differentiation occurs in a wide variety of animals, including *Drosophila,* and mammals, including humans (Figure 7–2), as well as in at least some plants (for example, the angiosperm genus *Lychnis*).

A distinction can thus be made between the X and Y chromosomes associated with sex and those that are alike in both sexes. The X and Y chromosomes are called **sex chromosomes;** the remaining ones of a given complement, which are the same in both sexes, are **autosomes.** In both the XX-XO and XX-XY types described thus far, all the eggs have one X chromosome, whereas the sperms are of two kinds, X and O, or X and Y. Both eggs and sperms have the same number of autosomes. In each case the male is the **heterogametic** sex (producing two kinds of sperms), whereas the female is the **homogametic** sex (producing only one kind of egg).

ZZ-ZW system. A final major type of chromosomal difference between the sexes is one in which the female is heterogametic and the male is homogametic. The sex chromosomes in this case are often designated as Z and W to avoid confusion with instances in which the female is

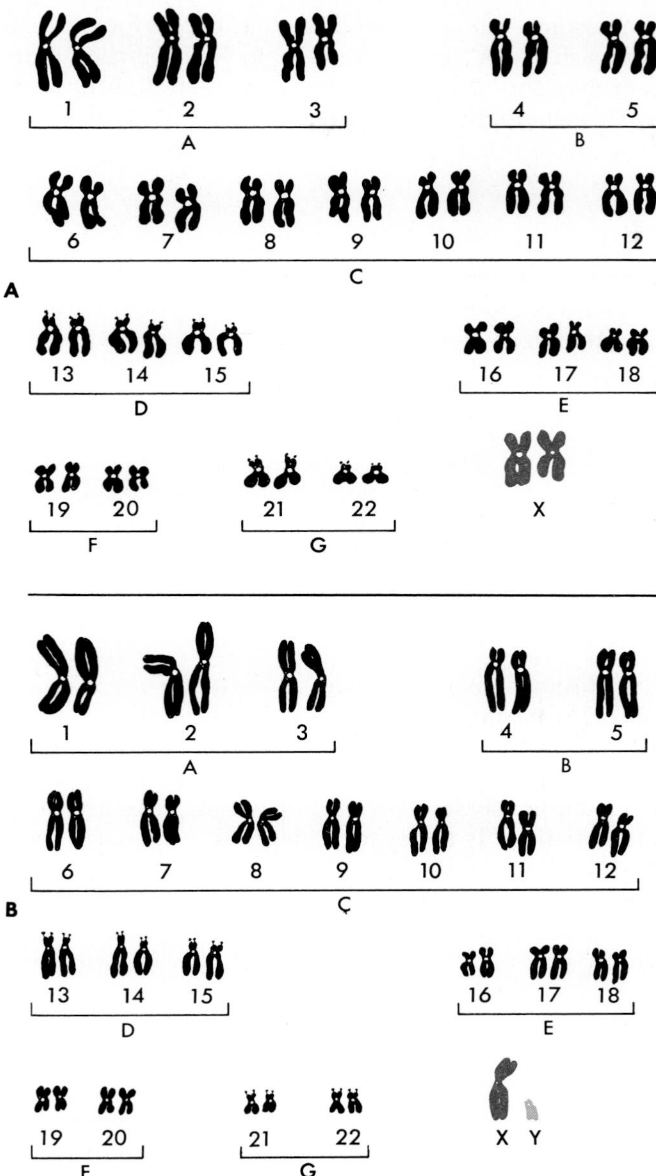

FIGURE 7–2. *Karyotype of human beings showing sex differences. (A) female; (B) male. Chromosome pairs 1–22 are autosomes, customarily arranged from largest to smallest, with groups of similar-size chromosomes designated by letter groups. The X and Y are sex chromosomes; structurally, X is similar to group C, and Y is similar to group G.*

homogametic. Females are thus ZW and males are ZZ. Birds (including the domestic fowl), butterflies, moths, and some fishes belong to this group.

Haploid Organisms

Liverworts. Like all sexually reproducing plants, liverworts (division Bryophyta) are characterized by a well-marked alternation of generations (see Appendix B) in which a haploid sexually reproducing phase or generation (the **gametophyte**) alternates in the life history with a diploid asexually reproducing individual (the **sporophyte**). As early as 1919, the chromosome complement of the sporophyte of the liverwort *Sphaerocarpos* was reported to consist of seven matching pairs, plus an eighth pair in which one of the two chromosomes was much larger than the other. The larger member of this eighth pair has been designated the X chromosome and its smaller partner is the Y chromosome. At meiosis, which terminates the diploid sporophyte generation, X and Y chromosomes are segregated so that of the four meiospores produced from each meiocyte, two receive an X chromosome and two receive a Y chromosome. Meiospores containing an X chromosome develop into female gametophytes; those with a Y develop into males. Thus females are X, males are Y, and asexual sporophytes are XY.

SUMMARY OF SEX CHROMOSOME TYPES

The various types of chromosomal differences between the sexes may be summarized as follows:

♀	♂	*Examples*
XX	XY	*Drosophila*, humans, and other mammals, some dioecious angiosperm plants
XX	XO	Grasshopper; many Orthoptera and Hemiptera
ZW	ZZ	Birds, butterflies, and moths
X	Y	Liverworts

Such chromosomal differences as these raise certain fundamental questions. For example, is a *Drosophila* individual a male because of the presence of the Y or because only one X is present? Is an individual a female because of the absence of the Y or because of the presence of two X chromosomes? Do the autosomes have anything to do with sex determination? Is the system identical in *Drosophila* and humans, both of which have the XX-XY sex difference? What do genes have to do with the situation? What causes sex reversal in which an individual of one sex becomes the other sex? Or what operates to produce individuals that are part male and part female in species where the sexes are ordinarily separate and distinct?

DROSOPHILA

Primary Nondisjunction of X Chromosomes

Work on the genetics of *Drosophila melanogaster* showed that sex determination, at least in that animal (where 2n = 8), was far less simple than the mere XX-XY difference would suggest. From some unusual breeding results in his laboratory, in a masterpiece of inductive reasoning, C. B. Bridges was able to lay the groundwork for a complete understanding of sex determination in *Drosophila*.

The gene for wild type red eyes (+) is carried on the X chromosome; a recessive allele (v) produces vermilion eyes in homozygous females and in all males (which, of course, have only one X chromosome). Ordinarily vermilion-eyed females mated to red-eyed males produce only red-eyed daughters and vermilion-eyed sons:

P	X^vX^v		X	X^+Y
P gametes		♀ X^v		
		♂ $\frac{1}{2}X^+$	$\frac{1}{2}Y$	
F_1	$\frac{1}{2}X^+X^v$ (red ♀)		+	$\frac{1}{2}X^vY$ (vermilion ♂)

However, in rare instances, crosses of this type produce unexpected vermilion-eyed daughters and red-eyed sons with a frequency of one per 2,000 to 3,000 offspring. Bridges surmised that these unusual progeny are due to a failure of the X chromosomes in an XX female to disjoin during oogenesis. Such *primary nondisjunction,* he reasoned, would produce three kinds of eggs, the majority of which would contain the normal single X chromosome, and a small number with either two X chromosomes or no X at all. If each X chromosome is represented by either X^+ (carrying the dominant allele for red eyes) or X^v (bearing the recessive allele for vermilion eyes), and each *set of three autosomes* is represented by an R, Bridges's cross may be represented in this way:

P	$AAX^v \times AAX^+Y$	
	(vermilion ♀) (red ♂)	
P gametes: ♀	AX^v (numerous) + AX^vX^v (rare) + AO (rare)	
♂	$AX^+ + AY$	
F_1	AAX^+X^v	red ♀ (numerous; normal)
	$AAX^+X^vX^v$	red "metafemale" (rare; die)
	AAX^+O	sterile red ♂ (rare)
	AAX^vY	vermilion ♂ (numerous; normal)
	AAX^vX^vY	fertile vermilion ♀ (rare)
	$AAOY$	die (rare)

The metafemales (AAAXXX) are weak and seldom live beyond the pupal stage; AAOY individuals die in the egg stage. Note that the XXY females indicate that the presence of a Y chromosome does not determine maleness itself, though males without it (XO) are sterile.

Secondary Nondisjunction

Bridges next mated the exceptional vermilion-eyed females (AAX^vX^vY) that arose as a result of primary nondisjunction to normal red-eyed males (AAX^+Y), and obtained progeny in these frequencies:

0.46	red ♀	
0.02	vermilion ♀	
0.02	red ♂	
0.46	vermilion ♂	
0.02	metafemales	die
0.02	*OYY*	die

Occurrence of the vermilion-eyed females and red-eyed males is due to *secondary nondisjunction.* Meiosis in XXY females is expected to be somewhat irregular because of pairing problems and, indeed, Bridges's results confirm this. In oogenesis, synapsis may involve either the two X chromosomes (XX type) with the Y chromosome remaining unsynapsed or one X and the Y (XY type) with the other X remaining free. Bridges found XY synapsis to occur in about 16 percent of the cases and XX synapsis in about 84 percent. After XY synapsis, disjunction segregates the X and Y synaptic partners to opposite poles. The unsynapsed X may go to *either* pole so that XY synapsis produces four kinds of eggs with a frequency of 0.04 each: XX, Y, X, and XY (Figure 7–3). On the other hand, XX synapsis is followed by disjunction of the two previously synapsed X chromosomes and their movement to opposite poles. The free Y may, of course, go to either pole, but the result is only two kinds

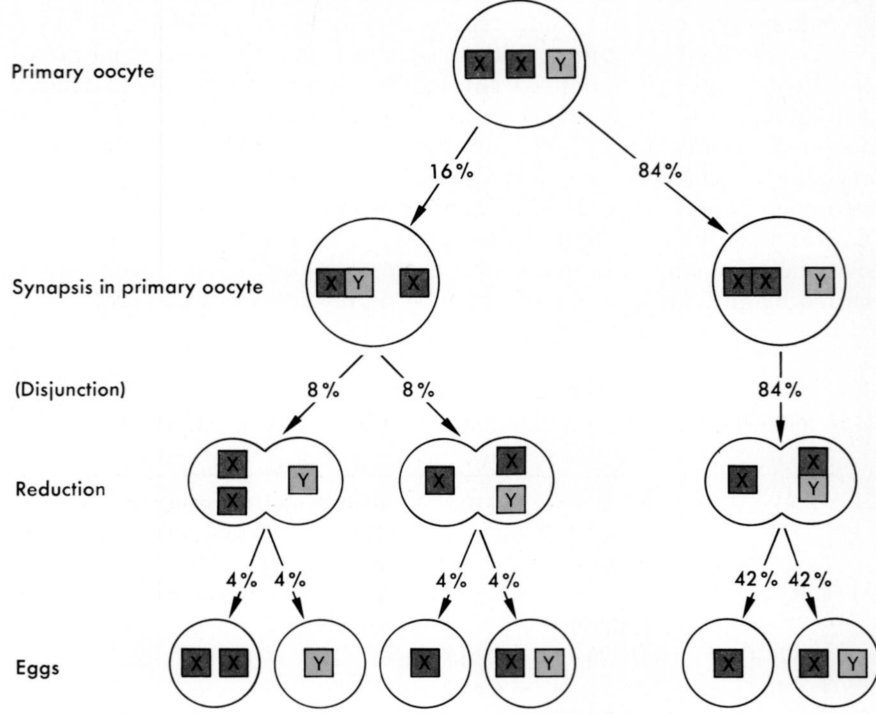

FIGURE 7–3. *Diagram of secondary nondisjunction in XXY female* Drosophila, *resulting in 46 percent X, 46 percent XX, 4 percent Y eggs. For explanation, see text.* (Based on work of Bridges, 1916a.)

of eggs, X and XY, with a frequency of 0.42 each. The overall result of secondary nondisjunction is four kinds of eggs in these frequencies:

$$0.46\ X^vY + 0.46\ X^v + 0.04\ X^vX^v + 0.04Y$$

Fertilization by sperm from a cytologically normal, red-eyed male (X^+Y) produces eight types of zygotes, which may be grouped in six classes:

Frequency	*Zygote*	*Phenotype*
0.23	X^+X^vY	red ♀
0.23	X^+X^v	
0.02	$X^+X^vX^v$	metafemale (die)
0.02	X^+Y	red ♂
0.23	X^vYY	vermilion ♂
0.23	X^vY	
0.02	X^vX^vY	vermilion ♀
0.02	YY	die

Bridges verified the chromosomal constitution of each of the viable genotypes.

As Bridges used the terms, *primary nondisjunction* may occur in either XX females or XY males. In the former it leads to the production of XX and O eggs. Occurrence in the first meiotic division of males produces XY and O sperms. Should it take place during the second division, XX, YY, and O sperms result. *Secondary nondisjunction,* on the other hand, occurs in XXY females, where it gives rise to XX, XY, X, and Y eggs. As the term *nondisjunction* implies, these aberrant gametes are produced only as a result of failure of the sex chromosomes to disjoin after synapsis; they are not physically attached.

Attached-X Flies

Another strain of flies in which nondisjunction of the X chromosomes occurred in *all* females was discovered by L. V. Morgan (wife of T. H. Morgan). When such females were mated to cytologically normal males carrying a recessive gene on the X chromosome, all viable male progeny were phenotypically like the father (but sterile), whereas all viable female offspring were like the mother. In addition, one-fourth of the total progeny were metafemales and another fourth (AAOY) died in the egg stage. Morgan reasoned that in these attached-X females ($\widehat{XX}$) the two X chromosomes were physically attached so that only two kinds of eggs, A$\widehat{XX}$ and AO, were produced. This explanation was soon confirmed cytologically.

Polyploid Flies

Experimentally produced triploid (three whole sets of chromosomes, or 3n) and tetraploid (4n) flies were next incorporated into Bridges's work, so that many kinds of flies with respect to chromosome complements were ultimately produced. As this work continued, it became increasingly clear that, in *Drosophila* at least, the important key to the sex of the individual was provided by the *ratio of X chromosomes to sets of autosomes.* The Y, then, has nothing to do with sex determination but does govern male fertility. These results are summarized in Table 7–1.

Metamales (or supermales) are to the male sex what metafemales (or superfemales) are to the female sex. That is, they are weak, sterile, underdeveloped, and die early. Intersexes are sterile individuals that display secondary sex characteristics intermediate between those of male and female (Figure 7–4).

From all these results the mechanism of sex determination in *Drosophila* may be summarized:

1. Sex is governed by the ratio of the number of X chromosomes to sets of autosomes. Thus, from Table 7–1, females have an X/A ratio = 1.0; males = 0.5. This relationship applies even to polyploid flies as long as the appropriate X/A ratio is maintained.
2. Genes for maleness per se are apparently carried on the autosomes; those for femaleness are carried on the X chromosome.
3. The Y chromosome governs male *fertility,* rather than sex itself, because AAXY and AAXO flies are both male with regard to secondary

TABLE 7–1. Summary of Chromosomal Sex Determination in *Drosophila**

Number of X chromosomes	*Number of sets of autosomes*	*Number of autosomes*	*Total chromosome number*	*X/A ratio*	*Sex designation*
3	2	6	9	1.50	Metafemale
4	3	9	13	1.33	Triploid metafemale
4	4	12	16	1.00	Tetraploid female
2	2	6	8	1.00	Female
3	4	12	15	0.75	Tetraploid intersex
2	3	9	11	0.67	Triploid intersex
1	2	6	7	0.50	Male
1	3	9	10	0.33	Triploid metamale
1	4	12	13	0.25	Tetraploid metamale

*After Bridges, 1916.

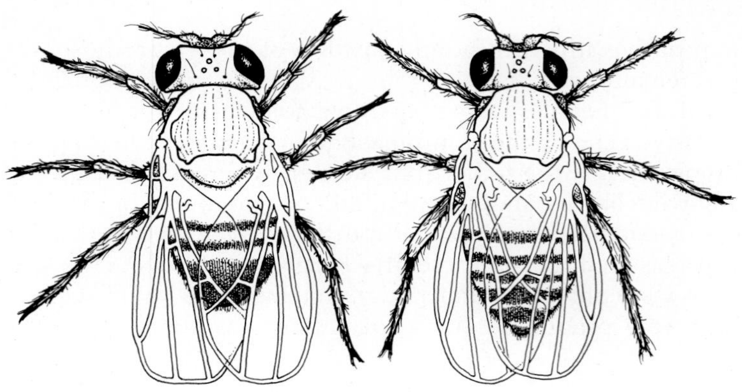

FIGURE 7–4. *Triploid intersexes (X/A ratio 0.67) in* Drosophila.

sex characters, but only the former produce sperms; it has no effect in AAXXY flies, which have an X/A ratio of 1.0 and are female.
4. An X/A ratio greater than 1.0 or less than 0.5 results in certain characteristic malformations (*metafemales and metamales*).
5. An X/A ratio less than 1.0 but greater than 0.5 produces individuals intermediate between females and males (*intersexes*). The degree of femaleness is greater when the X/A ratio is closer to 1.0 and the degree of maleness is greater when that ratio is closer to 0.5.

Other workers have confirmed and extended these observations on intersexes in *Drosophila*. By using X-rays to fragment chromosomes, lines of flies having extra fragments of the X chromosome have been developed. Thus an individual with about 2.67 X chromosomes and 3 sets of autosomes (= 6 autosomes) has an X/A ratio of 2.67/3 = 0.89, and is an intersex. Its secondary sex characteristics, however, are more female in nature than a fly with 2.33 X and 3 sets of autosomes, and therefore an X/A ratio of 2.33/3 = 0.78.

The same system in which the X/A ratio is critical is reported for the flowering plant (angiosperm) *Rumex acetosa*.

The Transformer Gene

One additional complicating factor in sex determination in *Drosophila* is worth examining briefly. A recessive allele, *tra*, on the third chromosome (an autosome), when homozygous, "transforms" normal diploid females (AAXX) into sterile males. The XX *tra tra* flies have many sex characters of males (external genitalia, sex combs, and male-type abdomen) but, as noted, are sterile, as are XXY *tra tra* flies. XY *tra tra* males, however, are normal and fertile.

Gynandromorphs

Concepts of sex determination as developed for *Drosophila* are verified by the occasional occurrence of **gynandromorphs** (or gynanders). These are individuals in which part of the body expresses male characters, whereas other parts express female characters. A bilateral gynandromorph, for example, is male on one side (right or left) and female on the other. The male portions of such flies would be expected to have a male chromosomal composition. Ingenious experiments using known "marker" genes on the X chromosome have provided experimental evidence to confirm this prediction. Lagging of the X chromosome at mitosis can result in daughter cells of the chromosomal complement AAXX and AAXO when the laggard X fails to be incorporated in a daughter nucleus. The portion of the body developing from the former cell will be normal female and the portion developing from the latter (sterile) cell will be male. Gynandromorphs represent one kind of *mosaic,* or an organism made up of tissues of male and female genotypes.

Sex as a Continuum

Instead of an either-or, male or female, XX or XY condition, sex may be viewed as a continuum, which ranges from supermaleness to maleness, to intersexes of varying degree, to femaleness, and on to superfemaleness. Where an individual places in such a continuum is related to the ratio of individual X chromosomes to sets of autosomes. Of course, it is not the chromosomes as gross structures that are the deciding factors, but rather, *genes* on the chromosomes. Thus genes for maleness are associated with the autosomes, and those for femaleness with the X chromosomes. Yet this entire sex-determining arrangement may, in some cases, be upset by a single pair of recessive autosomal alleles (*tra*)!

HUMAN BEINGS

In normal human beings, males are XY and females are XX, just as in *Drosophila*. But is sex here also determined by the X/A ratio? With regard to sex, does the Y chromosome bear genes for male fertility as in *Drosophila* or for male sex per se? To answer these questions it will be helpful to examine (1) the sex chromosomes, (2) the concept of sex differentiation, and (3) the human sex anomalies and their chromosomal makeup.

The Sex Chromosomes

X chromosome. The human X chromosome is medium-length, submetacentric, and intermediate in length between chromosomes 7 and 8 (Table 7–2). The centromere is more nearly central in X than in any of the larger chromosomes of the C group. With the development of fluorescent banding techniques (Chapter 4), visual differentiation of the X chromosomes and morphologically similar autosomes became exact (Figure 7–5). In mitotic metaphase spreads, the X chromosome measures approximately 5.0 to 5.5 μm, depending on the preparation.

Y chromosome. The Y chromosome in most human males averages roughly 2 μm in length, very slightly longer than the members of the shortest or G group of chromosomes (Table 7–2). This chromosome is, however, quite variable in length among different men; in some it is regularly as long or longer than members of the F group (chromosomes 19 and 20), whereas in others it may be less than half the length of the members of the G group. No particular phenotype or syndrome has been consistently associated with either a long or a short Y chromosome. The Y chromosome is acrocentric, even more so than the G chromosomes, and its longer arms generally lie close together. Unlike chromosomes 21 and 22, the Y has no satellites. With fluorescent banding techniques, the longer arm fluoresces brilliantly in good-quality preparations (Figure 7–5). Staining of XY (or XYY) cells with such fluorescent dyes as quinacrine hydrochloride discloses the brightly fluorescing Y

TABLE 7–2. Measurements of Relative Lengths of Human Chromosomes in Percentage of Total Haploid Autosome Length

Chromosome number	*A*	*B*	C
1	9.08	9.11 ± 0.53	8.44 ± 0.433
2	8.45	8.61 ± 0.41	8.02 ± 0.397
3	7.06	6.97 ± 0.36	6.83 ± 0.315
4	6.55	6.49 ± 0.32	6.30 ± 0.284
5	6.13	6.21 ± 0.50	6.08 ± 0.305
6	5.84	6.07 ± 0.44	5.90 ± 0.264
7	5.28	5.43 ± 0.47	5.36 ± 0.271
X	5.80	5.16 ± 0.24	5.12 ± 0.261
8	4.96	4.94 ± 0.28	4.93 ± 0.261
9	4.83	4.78 ± 0.39	4.80 ± 0.244
10	4.68	4.80 ± 0.58	4.59 ± 0.221
11	4.63	4.82 ± 0.30	4.61 ± 0.227
12	4.46	4.50 ± 0.26	4.66 ± 0.212
13	3.64	3.87 ± 0.26	3.74 ± 0.236
14	3.55	3.74 ± 0.23	3.56 ± 0.229
15	3.36	3.30 ± 0.25	3.46 ± 0.214
16	3.23	3.14 ± 0.55	3.36 ± 0.183
17	3.15	2.97 ± 0.30	3.25 ± 0.189
18	2.76	2.78 ± 0.18	2.93 ± 0.164
19	2.52	2.46 ± 0.31	2.67 ± 0.174
20	2.33	2.25 ± 0.24	2.56 ± 0.165
21	1.83	1.70 ± 0.32	1.90 ± 0.170
22	1.68	1.80 ± 0.26	2.04 ± 0.182
Y	1.96	2.21 ± 0.30	2.15 ± 0.137

From *Paris Conference (1971): Standardization in Human Cytogenetics.* In *Birth Defects: Orig. Art. Ser.*, ed. D. Bergsma. Published by The National Foundation-March of Dimes, White Plains, N.Y., Vol. VIII (7), 1972. Used by permission.

Column A: Previous Denver-London Conference data.

Column B: Data from 10 cells by Drs. T. Caspersson, M. Hultén, J. Lindsten, and L. Zech. Cells stained with orcein.

Column C: Data from 95 cells provided by Drs. H. Lubs, T. Hostetter, and L. Ewing from 11 normal subjects (6 to 10 cells per person). Average total length of chromosomes (diploid set) per cell: 176 μm. Cells stained with orcein or Giemsa 9 technique.

Percentages do not add to 100 because they are averages for a given chromosome in several metaphase spreads.

chromosome, not only in mitotic metaphase, but also in interphase (Figure 7–6). The same technique can be used successfully on sperm and on cells in the amniotic fluid.

X- and Y-Chromatin in Interphase Nuclei. A clue to the sex chromosomal complement of an individual may be obtained quite simply by examining squamous epithelial cells from scrapings of the lining of the cheek. When stained, most somatic cells of normal females show a characteristic structure, the *Barr body*, so named after its discoverer, Murray Barr, who first described it in 1949 (Figure 7–7). Females are therefore termed *sex chromatin positive;* normal males, whose cells do not contain a Barr body, are *sex chromatin negative.* Because of various cytological factors, as well as some of the techniques employed, frequency of sex chromatin positive cells in scrapings of oral mucosa from normal females have been reported to range from 36 to 80 percent of cells examined. The Barr body is a small structure (about 1 μm in greatest dimension), hemispherical, disk shaped, rod shaped, or triangular in outline. Ordinarily it lies appressed to the inner surface of the nuclear membrane or, in nerve cells, it may be associated with the nucleolus. In diploid

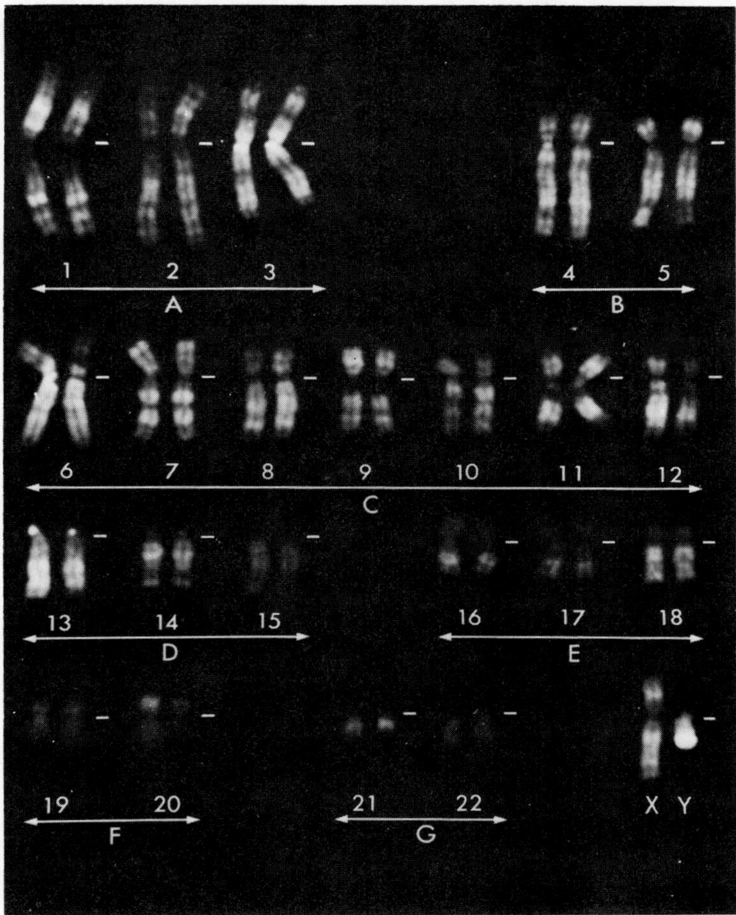

FIGURE 7–5. *Karyotype of normal human male (46, XY) Q-banding. Banding techniques permit much greater accuracy in matching chromosomes and identifying the sex chromosomes. The short bar beside each chromosome pair marks the location of the centromere. Note the brightly fluorescing long arm of the Y chromosome.* (Courtesy Dr. C. C. Lin.)

cells of females the number of Barr bodies is one less than the number of X chromosomes.

In somatic cells having two X chromosomes, one replicates later than the other. Mary Lyon and others have proposed that the late-replicating X chromosome becomes inactivated early in embryonic life and forms the Barr body. Whether the maternal or paternal X chromosome is inactivated in any given cell depends on chance, but once this has occurred in an embryonic cell, indications are that the same chromosome becomes the Barr body in all cells derived therefrom. Thus, females that

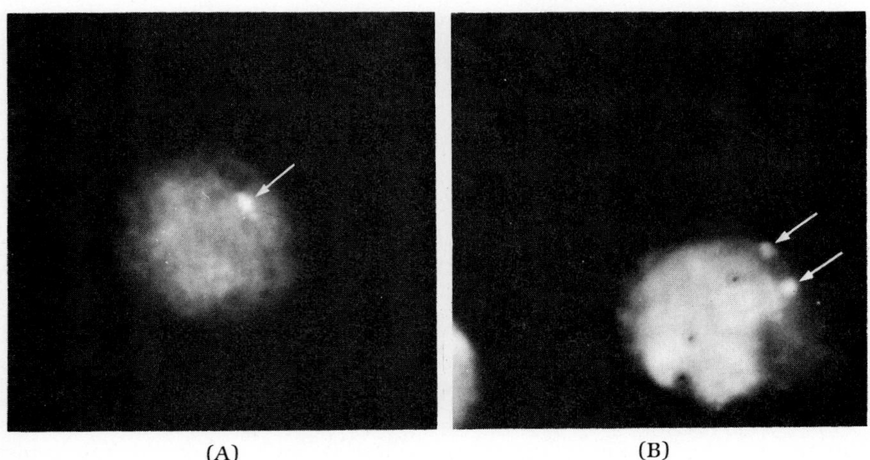

FIGURE 7–6. *The brightly fluorescing Y chromosome (arrows) in quinicrine-stained interphase. (A) 46, XY male, (B) 47, XYY male.* (Courtesy Dr. C. C. Lin.)

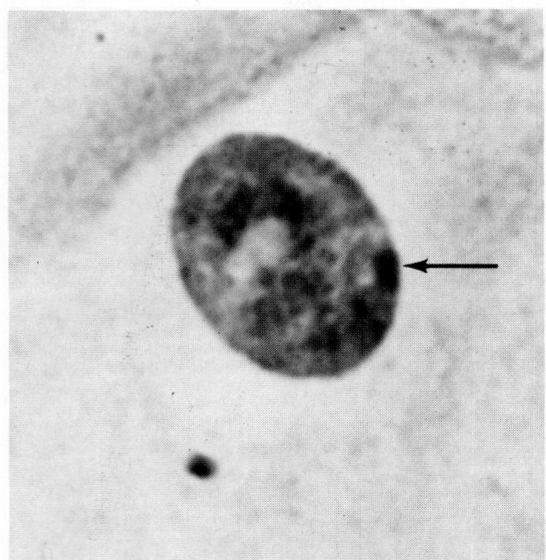

FIGURE 7–7. *Nucleus of normal human female squamous epithelial cell showing prominent, dark Barr body (arrow) against the nuclear membrane.* (Courtesy Carolina Biological Supply Co.)

are heterozygous for genes located on the X chromosome are *mosaics:* Some patches of tissue express the dominant phenotype and others express the recessive phenotype.

Sex Differentiation

Genetic sex. Normal females ordinarily have two X chromosomes; normal males have one X and one Y. As noted earlier, genes on these chromosomes determine femaleness or maleness. Thus one can speak of females as having the *genetic sex* designation XX and males as having the genetic sex designation XY, although exceptional cases do occur.

Gonadal sex. Chemical substances (inductors) produced by embryonic XX cells act on the *cortical* region of undifferentiated gonads to bring about development of ovarian tissue. In XY embryos, however, inductors stimulate production of testes from the *medulla* of the undifferentiated gonads. Hence the XX genetic sex is ordinarily associated with ovarian *gonadal sex,* and XY is associated with testicular gonadal sex.

Genital sex. The embryonic gonads produce hormones that, in turn, determine the morphology of the external genitalia and the genital ducts. XX embryos normally develop ovaries, female external genitalia, and Mullerian ducts. XY embryos, on the other hand, ordinarily develop testes, male external genitalia, and Wolffian ducts. In XX embryos the Wolffian ducts are suppressed; in XY embryos the Müllerian ducts remain undeveloped. Thus there is a distinction between male and female *genital sex.*

Somatic sex. Production of gonadal hormones continues to increase until *secondary sex characters* appear at puberty. These include amount and distribution of hair (for example, facial, body, axillary, pubic), pelvis dimensions, general body proportions, subcutaneous fat over hips and thighs, and breast development in the female, as well as increased larynx size and deepening of the voice in the male.

Sociopsychological sex. In most individuals, genetic sex, gonadal sex, genital sex, and somatic sex are consistent; XX persons, for example, develop ovaries, female genitalia, and female secondary sex characteristics. Ordinarily these persons are raised as females and adopt the feminine gender role under whatever cultural pattern has been established in the society of which they are members. A similar consistency from genetic sex to sociopsychological sex is seen for XY individuals. However, some individuals display an inconsistency of some kind of degree among these levels of sexuality. Intersexuality occurs when the chromosomal contribution is inconsistent with the gonadal or other secondary sexual characters.

Human Sex Anomalies

The Klinefelter syndrome. One in about 500 "male" births produces an individual with a particular set of abnormalities known collectively as the **Klinefelter syndrome** (Figure 7–8). These persons have a general male phenotype; external genitalia are essentially normal in gross morphology. Although there is some variability in other characteristics, testes are typically small, sperms are usually not produced, and intelligence may be normal to slightly retarded. Arms are longer than aver-

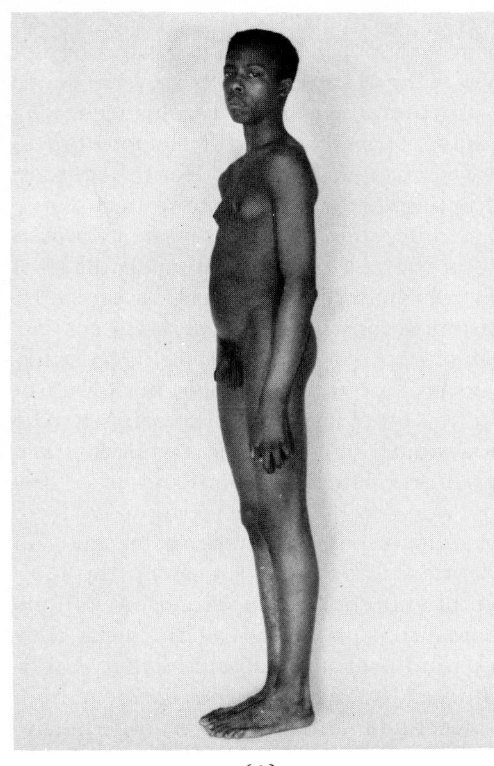

(A)

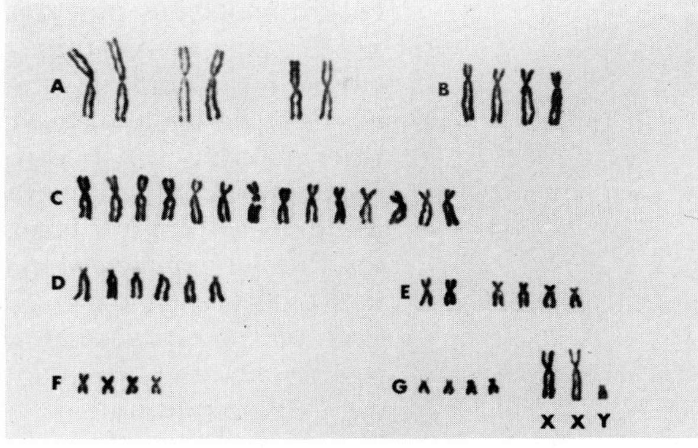

(B)

FIGURE 7–8. *The Klinefelter syndrome in man. Such persons are AAXXY (47, XXY). External genitalia are male type, but there is usually some femalelike breast development. (A) General phenotype.* (Photo courtesy Dr. Victor A. McKusick.) *(B) Karyotype of another Klinefelter individual.* (Photo courtesy Ms. Dawn DeLozier.)

age, some degree of breast development is common, and the voice tends to be higher pitched than in normal males. Klinefelters are sex chromatin positive; the karyotype shows 47 chromosomes, that is, 47,XXY.

The XXY individual may arise through fertilization of an XX egg by a Y sperm or through fertilization of an X egg by an XY sperm. Although the majority of Klinefelters are born to mothers under the age of 30, this is in large part a reflection of the age group in which most births of all types occur (Figure 7–9). After a drop in Klinefelter births from

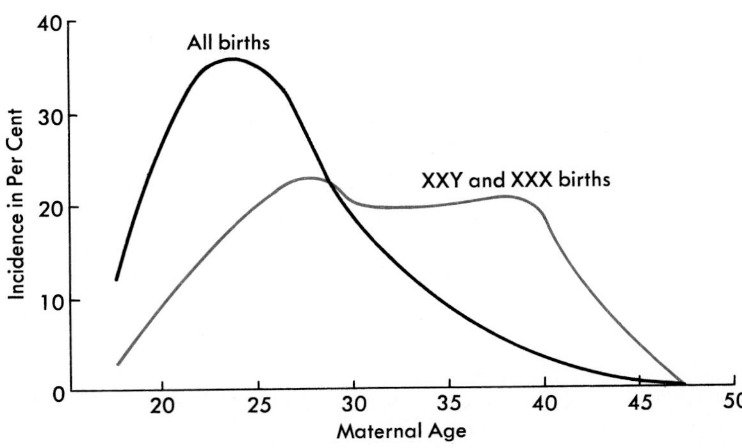

FIGURE 7–9. *The relationship between maternal age and XXY and XXX births.*

ages 27 to 32, there is a small increase again after age 32, whereas total births decrease rapidly in this age group. This fact suggests nondisjunction of the X chromosome in aging oocytes as a somewhat more important factor than XY nondisjunction during spermatogenesis.

Less frequently, Klinefelters have more than two X chromosomes, and even more than one Y (Table 7–3). Generally the greater the number of X chromosomes, the greater the degree of mental retardation.

As a clue to the nature of the sex-determining mechanism in human beings, the important point here is the fact that persons with at least one Y chromosome have the general phenotype of a male, even in the presence of any number of X chromosomes. *Thus, "maleness," however defined, is dictated by the presence of at least one Y chromosome.* One study has disclosed that in 70 percent of the XXY males studied, both X chromosomes were maternal in origin (the other 30 percent, of course, were cases of paternal origin). There is some suggestion of a maternal age effect, but no evidence at this time for a paternal age effect.

The Turner syndrome. A second major sex chromosome anomaly of interest here is the **Turner syndrome** (Figure 7–10), in which the individual presents a general female phenotype, but with certain unique departures. These include short stature, "webbing" of the neck, low hairline on the nape of the neck, and a broadly shield-shaped chest. Slight mental retardation is often noted, but a few Turners achieve high scores on standardized IQ tests. Secondary sex characters do not develop; breast development is absent to very slight, pubic hair is reduced or absent, and axillary hair does not develop. Genitalia remain essentially infantile.

Turners are sex chromatin negative, which suggests the presence of only one X chromosome. This was confirmed cytologically in 1969 by Ford and his colleagues. The karyotype is, therefore, 45,X (that is, AAXO). Studies of X-linked traits, such as the Xg blood group, reveal that in 72 percent to 76 percent of the Turners examined, the *paternal* X chromosome is absent. The reason for this disparity is unknown. Interestingly enough, however, there appears to be a relationship between

TABLE 7–3. **Sex Chromosome Anomalies and Their Estimated Frequencies**

Designation	*Chromosome constitution*	*Sex chromatin*	*Somatic chromosome number*	*General sex phenotype*	*Usual fertility*	*Estimated frequency per 1,000*	*Estimated number in the U.S.**
Klinefelter	AAXXY	+	47	♂	–	2.0	453,000
Turner	AAXO	–	45	♀	(–)†	0.2–0.4	45,300–90,600
Triplo-X	AAXXX	+ +	47	♀	+	0.75	169,880
Tetra-X	AAXXXX	+ + +	48	♀	Unknown	Very low	Very low
Triplo-X, Y	AAXXXY	+ +	48	♂	–	Very low	Very low
Tetra-X, Y	AAXXXXY	+ + +	49	♂	–	Very low	Very low
Penta-X	AAXXXXX	+ + + +	49	♀	–	Very low	Very low
XYY	AAXYY	–	47	♂	±‡	0.7–2.0	186,500–453,000
Klinefelter XXYY	AAXXYY	+	48	♂	Not reported	Very low	Very low
Klinefelter XXXYY	AAXXXYY	+ +	49	♂	Not reported	Very low	Very low

*Based on a 1980 population of 226.5 million with an average birth rate of 2.1 children among women of child-bearing age. (Preliminary 1980 census figures from U.S. Census Bureau.)

†A very few cases of motherhood have been reported for presumed Turners.

‡But generally highly infertile because of low sperm count.

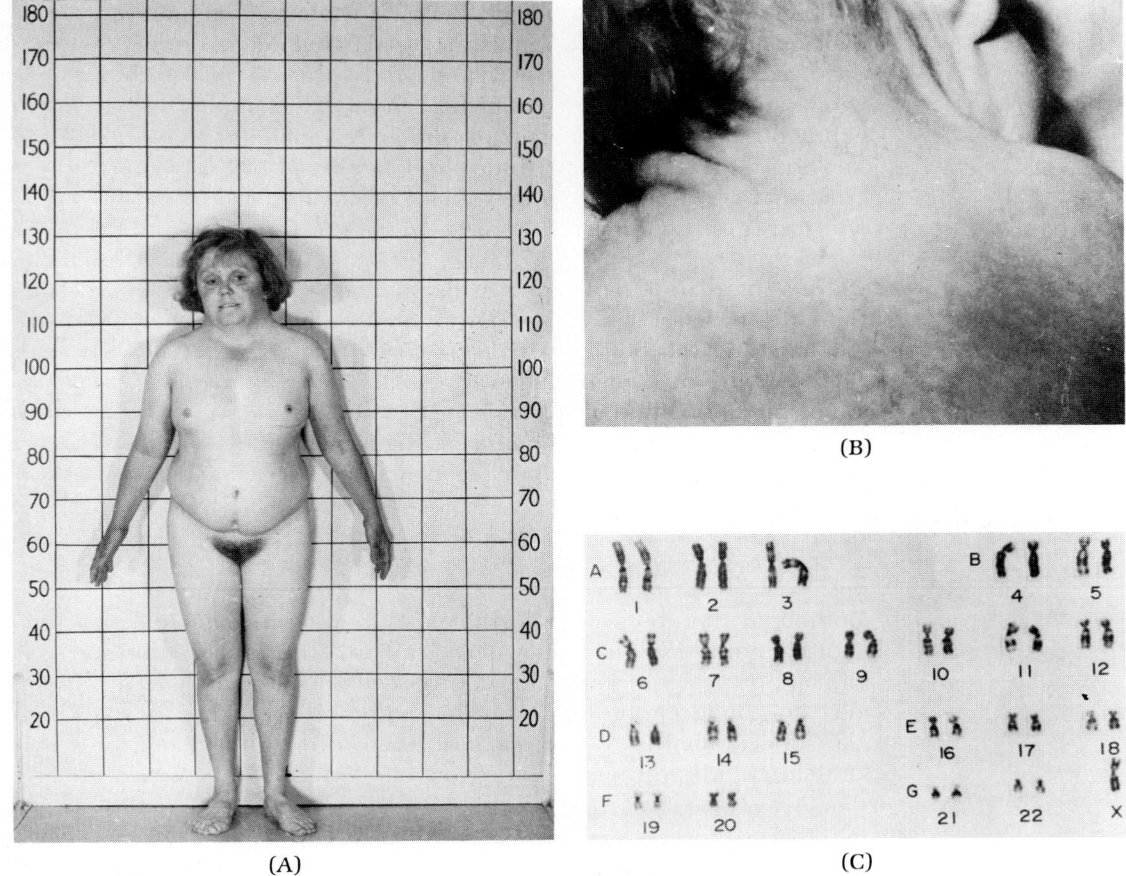

FIGURE 7–10. *The Turner syndrome in the human female. These persons are AAXO (45, X). Note the female genitalia. (A) General phenotype.* (Photo courtesy Dr. Victor A. McKusick.) *(B) Detail of webbed neck. (C) Karyotype of a Turner.* [(B) and (C) courtesy Ms. Dawn DeLozier.]

age of the mother and an increased number of 45,X conceptions, with the incidence being higher in younger mothers.

Ordinarily, Turners are sterile, although at least one normal birth and several pregnancies have been reported for presumed Turners. In these cases, however, mosaicism (45,X/46,XX) has not been ruled out with certainty.

The frequency of live Turner births has been estimated at 2 to 4 per 10,000 live births, although a study of 139 cases of gonadal dysgenesis places the incidence at a mere 1 in 10,000. On the other hand, de Grouchy and Turleau found the frequency to be 4 in 10,000. This low incidence of Turner births is a reflection of a high rate of intrauterine mortality. Reports of spontaneously aborted 45,X fetuses range from 90 to 97.5 of all Turners conceived. Some 20 to 30 percent of all spontaneously aborted fetuses are 45,X. The Turner syndrome is, in effect, often lethal.

The conclusion to be drawn from the Turner syndrome is that the general sex phenotype is female in the absence of a Y chromosome.

Poly-X females. In 1959, the first known case of a triplo-X individual, that is, 47,XXX, was reported. This person was clearly female in general sex phenotype, but at age 22, she had infantile external genitalia and marked underdevelopment of internal genitalia and breasts. She was somewhat retarded mentally. Since that time many more XXX females

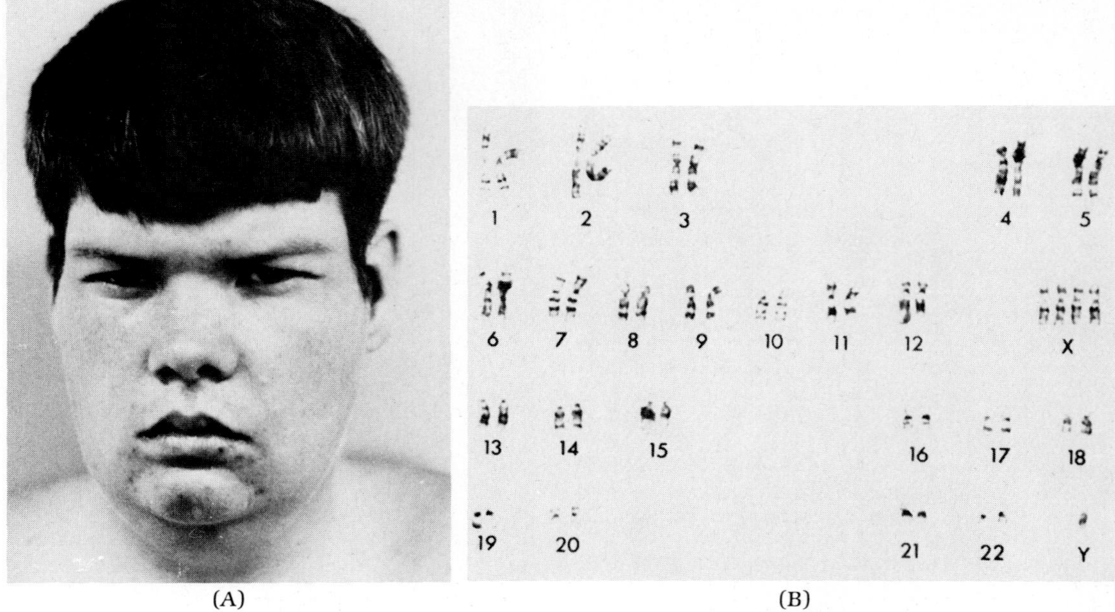

FIGURE 7–11. *(A) Individual with XXXXY syndrome showing male phenotype. This individual had an IQ of 50 at age 24. (B) Karyotype of individual.* (Photographs courtesy Dr. H. E. Carlson.)

have been described, and it is estimated that between 1 in 1,000 and 1 in 2,000 live female births is triplo-X. Some XXX females are essentially normal, but others are retarded and/or show abnormalities of primary and secondary sex characters. Apparently all are fertile, but among more than 30 children of triplo-X mothers, all were XX or XY except for one Klinefelter. A very few tetra-X (48,XXXX) and penta-X (49,XXXXX) persons have been described; manifestations are similar to those of triplo-X individuals but more marked. In general, as the number of X chromosomes increases, intelligence is progressively reduced, as with Klinefelters. Also, it is important to note that multiple-X individuals are female in general phenotype, and show evidence of a maternal age effect (Figure 7–9). Persons with more than two X chromosomes plus a Y are males, but are physically and mentally retarded (Figure 7–11).

The XYY male. Particular interest has recently focused on behavior patterns in XYY individuals. These are all males, above average in height (more than 72 inches), with intelligence quotients ranging from about 80 to 118, and a history of severe (facial) acne during adolescence. Abnormalities of both internal and external genitalia have been noted in some of these persons, but no consistent major anomalies occur. This kind of individual was noticed first in 1965, when a high incidence of XYY males (9 of 315 inmates, or a frequency of about 0.029), housed in the maximum security section of a Scottish criminal institution, was reported. Many similar reports that followed in the 1960s led to the notion that the XYY male is more aggressive and more likely than the XY male to commit crimes of violence. This conclusion is now viewed as having been premature, since after additional studies of the general population, it has become clear that many XYY males do make satisfactory social adjustment; some are indeed of very retiring personality. A 1970 study of 4,366 consecutive births at Yale-New Haven Hospital showed XYY births to occur with a frequency of 0.69 per 1,000. Figure

7–12A shows an individual at 7 years who is XYYYY. This individual has an IQ of 50 and has disturbances in motor development as well as speech disorders. Figure 7–12B shows the G-banded karyotype of an XYYYY individual.

Genetically determined sex reversal. J. German and colleagues (1978) reported an interesting case of a presumed gene-determined sex reversal in humans. This case involved an extensive family pedigree in which three confirmed 46,XY phenotypic females were intensively studied. Each of these three had vestigial gonads or gonadal tumors in place of normal gonads of either sex. Throughout life, to the time of examination, each appeared to be a normal female and was raised as such. All were somewhat tall for females (approximately 67 to 70 inches), with normal female genitalia, normal uterus and fallopian tubes but, as pointed out, with only rudimentary ovarian tissue.

Based on their study of this family, German and his colleagues concluded that this rare type of XY gonadal dysgenesis is due to a recessive allele, probably X-linked, that prevents testicular differentiation in 46,XY embryos. They concluded that the Y chromosome does not act alone in causing undifferentiated embryonic gonadal tissue to develop into testes, but there is an interaction "between a locus near the centromere of the Y and one somewhere on the X." This interaction, they theorize, must be either (1) induction of a structural gene on the X chromosome for testicular development by a controlling element on the Y chromosome, or (2) a structural gene on the Y chromosome being induced by a control on the X. In the family studied, the conclusion was that the X-linked gene is defective and unable to interact with the Y-borne locus.

Inasmuch as initiation of male gonadal development requires the Y chromosome, implications of some Y-linked product in the process is

FIGURE 7–12. *(a) 49 XYYYY individual.* (Photo courtesy Dr. L. Sirota. *(b) 49 XYYYY karyotype.* (Photo courtesy Dr. Juana Pincheira.)

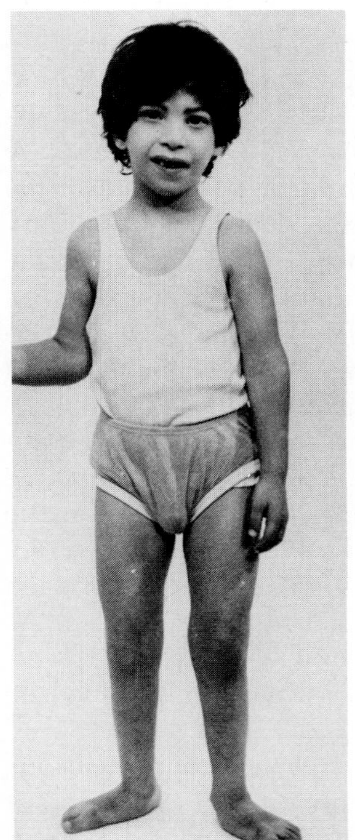

(A)

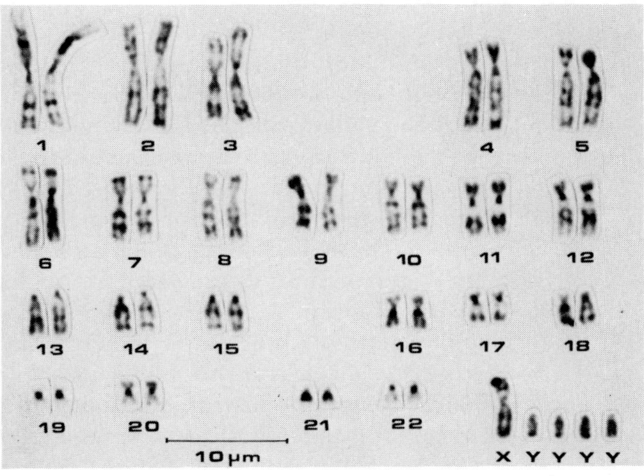

(B)

suggested. Deletions (see Chapter 15) of the short arm of the Y chromosome result in a general female phenotype and rudimentary, nonfunctional gonads. Deletions of the long arm of the Y chromosome, on the other hand, sometimes occur in normal, fertile males. Thus a testis-determining locus would appear to be on or very near the short arm of the Y chromosome.

Other sex chromosome anomalies. A number of other sex chromosome anomalies have been reported, most of which are quite rare (Table 7–3). In every case, though, the presence of even one Y chromosome serves to produce the male sex phenotype, regardless of the number of Xs that may be present. No clear relationship between sex chromosome anomalies in children and any particular parental characteristics, except for some suggestion of a higher risk in mothers with thyroid disorder (hyperthyroid and hypothyroid), has been observed.

Later work has indicated the presence on the short arm of the Y chromosome, near the centromere, of a gene responsible for production of the **H-Y antigen.** This antigen occurs in all male tissue. Furthermore, XYY and XXYY individuals produce twice as much antigen as XY persons. It is not known whether the H-Y antigen is, itself, the direct determinant for testis development; it is possible that it regulates H-Y receptor cells. If these are defective, the undifferentiated gonadal cells fail to respond to the H-Y antigen and testes do not form, leaving the individual to mature as a female. Female breast development occurs at puberty in these persons, but ovaries and parts derived from Müllerian ducts are lacking. External genitalia may be female or ambiguous.

On the other hand, in other anomalous XY individuals, virilization may not occur until puberty, suggesting that the undifferentiated gonads fail to bind the male hormone during embryonic development. In these cases of *testicular feminization,* testes are present, though often located within the body. This syndrome is caused by mutation of a gene that J. de Grouchy and C. Turleau feel is very probably located on the X chromosome. These individuals are invariably 46,XY, but they present a feminine phenotype.

Hermaphroditism. No account of sex-chromosome anomalies would be complete without mention of hermaphroditism and pseudohermaphroditism. *True hermaphrodites,* by generally accepted definition, are individuals that possess both ovarian and testicular tissue. The external genitalia are ambiguous, but often more or less masculinized; secondary sex characters vary from more or less male to more or less female. Some are reared as males and some as females. Ordinarily, true hermaphrodites are sterile because of rudimentary ovatestes. However, at a meeting of The American Society of Human Genetics (1978), J. Brazzel and colleagues reported a pregnancy (which came about in the normal way) that terminated in delivery of a stillborn child after about 30 weeks of gestation to a 25-year-old true hermaphrodite. Even more remarkable was the fact that this individual engaged in male sexual activity in the early years but gradually shifted to a preference for the female role. Surgical procedures disclosed a tumor containing muscle plus ovarian tissue on the left side, with an ovatestis in a sac in the right groin. As these researchers put it, "True hermaphroditism . . . is rare; however, pregnancy and bisexual activity in such an individual must indeed be considered more rare."

The incompletely known, apparently multiple, causes of the various degrees of hermaphroditism require a discussion beyond the scope of an introductory text. However, some cytological and morphological characteristics are interesting in that they demonstrate that sex development can go awry in a variety of ways. In a study of 108 cases of true hermaphroditism, for example, 59 were 46,XX in karyotype, 21 were 46,XY, and 28 were mosaics. All but two of the mosaics were found to possess some Y chromosome cell lines. Apparent 46,XX true hermaphrodites could be accounted for without conflicting with the apparent necessity of a Y chromosome for testicular development by assuming loss of the Y chromosome from a 47,XXY fetus at an appropriate stage of embryonic development. The most suggestive evidence may well come from the 46,XX/46,XY mosaics, a condition that seems to indicate that many true hermaphrodites may originally have had both XX and XY cell lines.

Pseudohermaphrodites have either testicular or ovarian tissue, generally rudimentary, but not both. On the basis of the usual chromosomal constitution (genetic sex), two major classes, *male* and *female*, are distinguished. The former are most often 46,XY or 46,XY/45,X mosaics, and external genitalia are ambiguous. A penislike organ of variable size is present. In adolescence pubic and axillary hair develops and the voice deepens. Often some breast development, as in females, occurs. This condition is referred to as *masculinizing male pseudohermaphroditism*. This is clearly distinguishable from the testicular feminization syndrome discussed earlier.

Feminizing male pseudohermaphrodites are, like the masculinizing variety, usually 46,XY (hence, male) or, less often, a 46,XY/45,X (or other) mosaic. These individuals present a general female sex phenotype, but often (not always) have poorly developed secondary sex characters. Some lead a normal *female* sex life, though they cannot conceive.

Female pseudohermaphrodites are 46,XX (hence, *female*) but present a more or less masculine phenotype. The external genitalia are ambiguous; ovaries are present but are immature or rudimentary. This condition arises most frequently by virilization of a female fetus because of either an inherited proliferation of the adrenal glands or a prenatal hormone imbalance in the mother. Pseudohermaphroditism is summarized in Table 7–4.

Mechanism of sex determination in human beings. The developing mammalian embryo appears to have the potential to become either sex. If the embryo has the female chromosome composition (XX), the gonads differentiate as ovaries; the Müllerian ducts differentiate into fallopian tubes, uterus, and upper vagina; the Wolffian duct system regresses. If

TABLE 7–4. **Summary of Human Pseudohermaphrodites**

Type	*Variety*	*Gonads*	*External genitalia*	*Genetic sex*	*General sex phenotype*	*Sex of rearing*
♂	Masculinizing	Testes ± dystrophic	Ambiguous but ± ♂	XY	♂ or ± ♀	♂ (or ♀)
	Feminizing	Testes ± dystrophic; often inguinal	± ♀	XY	♀	♀
♀	—	Ambiguous; often immature ovaries	Ambiguous but ± ♂	XX	± ♂	♂ (or ♀)

± means more or less.
Parentheses around a sex designation indicate a minority of cases.
For details see text.

HUMAN GENE DETERMINING SEX IS DISCOVERED

The question of what determines the sex of humans has always been an interesting one. The sex of a human fetus at 6 weeks is not yet determined. The fetus can still literally become either sex. Then, during the seventh week, events occur that determine which sex the individual will be. Scientists have recently localized a region on the Y chromosome that seems to contain the genetic information for this genetic "switch." The gene is referred to as the testis-determining factor, or TDF. Its presence or absence determines whether the fetal gonad develops into testes or ovaries.

Surprisingly, so-called "sex-reversed" individuals led to this discovery. Sex-reversed individuals have the chromosome complement that is normally associated with one sex, but are, in fact, the opposite sex. For example, such an individual has the XX chromosomes that determine femaleness, but this person is, in fact, a male. It was found that XX males, who were sterile, carried a very small portion of a Y chromosome on one of their X's. Similarly, XY females lacked the same region on their Y chromosome. One XX male contained only 0.5 percent of the Y chromosome, and one XY female lacked 0.2 percent of the same Y region. So far, in 90 sex-reversed patients, the correlation between the presence and absence of this small Y region and sex has been 100 percent. The identical TDF region has been identified in gorillas, monkeys, dogs, cattle, horses, and goats, further suggesting that this is a common sex-determining mechanism in mammals.

The current theory is that the TDF gene is the master switch that, when turned on, activates an entire series of genes whose function is sex differentiation. One disturbing finding is that a similar region is also found on the X chromosome. Since the Y chromosome is normally associated with maleness, it would have been better if this region was limited only to the Y chromosome. It may just be an inactive form of the gene (pseudogene) on the X chromosome, or the two genes may work together in some way to determine sex.

the embryo has the male chromosome composition (XY), the gonads develop as testes; the Wolffian duct system differentiates into epididymides, vas deferens, and seminal vesicles; the Müllerian duct system regresses. Individuals that have at least one Y chromosome are, with few exceptions, male with regard to external genitalia and general phenotype, though they may be sterile (Table 7–3). In contrast, persons with one or more X chromosomes are *ordinarily* phenotypically female as long as no Y is present, though again infertility sometimes occurs.

Do the autosomes play any part in sex determination in humans? Persons with exceptional numbers of autosomes are still male or female in general phenotype, which is consistent with genetic sex. A model for the situation in human beings and probably mammals in general may be summarized as follows:

1. Autosomes play no part in determining sex.
2. The first gene in the testicular-determination pathway, designated as the testes-determining Y gene, is located on the Y chromosome and starts the developmental sequence toward maleness (provided a controlling—X linked?—gene permits testicular differentiation).
3. The first gene in the ovarian-determining pathway is located on the X chromosome, and initiates the ovarian-determination developmental sequence. This gene on the X chromosome determines femaleness in the absence of any Ys.
4. A "good" male or female phenotype requires the "correct" number of X chromosomes: one for males and two for females in diploid individuals.

The normal genetic controls over sex in mammals function so that only one gonadal pathway is followed in any individual. This control occurs by placing the gene for initiation of testes determination on the Y chromosome, which is unique to males, and also by having the testicular-determination pathway occur earlier in development than the ovarian-determination pathway.

Sex determination in humans, then, although related to the X-Y chromosome makeup, clearly operates differently than in *Drosophila*. Mammals in general appear to follow the mechanism outlined for human beings.

PLANTS

Sex determination in one flowering plant, the wild campion (*Lychnis dioica* formerly the genus *Melandrium*) of the pink family (Caryophyllaceae), has been extensively investigated. This plant illustrates still another variation of the XX-XY system. In angiosperms, the plant recognized by name is the asexual, diploid sporophyte generation. The haploid sexual phase is microscopic and contained largely within the tissues of the sporophyte, on which it is parasitic. Flowers contain either or both of two essential sex organs, stamens and/or pistils. The former produce microspores that develop into male gamete-bearing plants (at one stage these are the well-known pollen grains). Pistils produce and contain the egg-bearing sexual plant. Many plants have so-called perfect flowers, which contain both stamens and one or more pistils. *Lychnis* has imperfect flowers, which bear either stamens or pistils. The species is, moreover, dioecious, so that there are staminate (male-producing) individuals and pistillate (female-producing) ones. Though the nomenclature is not accurate, staminate plants are often referred to as male plants and pistillate plants as female.

In *Lychnis*, staminate plants are XY and pistillate plants are XX. The X/A ratio bears no relation to "sex" but, through studies of plants with multiple sets of chromosomes, the X/Y ratio is found to be critical. X/Y ratios of 0.5, 1.0, and 1.5 are found in plants having only staminate flowers; in plants whose X/Y ratio is 2.0 or 3.0, occasional perfect flowers occur among otherwise all staminate flowers. In plants having four sets of autosomes, four X chromosomes and a Y, flowers are perfect but with an occasional staminate one. From this it could also be concluded that the Y determines staminate plants unless there is an excessive number of X chromosomes. This situation is summarized in Table 7–5 and illustrated in Figure 7–13.

TABLE 7–5. "Sex" and X/Y Ratios in *Lychnis*

Chromosome constitution	*X/Y ratio*	*"Sex"*
2A XYY	0.5	♂
2A XY 3A XY 4A XY	1.0	♂
4A XXXYY	1.5	♂
2A XXY 3A XXY 4A XXY 4A XXXXYY	2.0	♂ with occasional ⚥ flower
3A XXXY 4A XXXY	3.0	♂ with occasional ⚥ flower
4A XXXXY	4.0	⚥ with occasional ♂ flower

♂ = staminate; ♀ = pistillate; ⚥ = perfect.
Each A = one set of 11 autosomes, each X = an X chromosome, each Y = a Y chromosome.

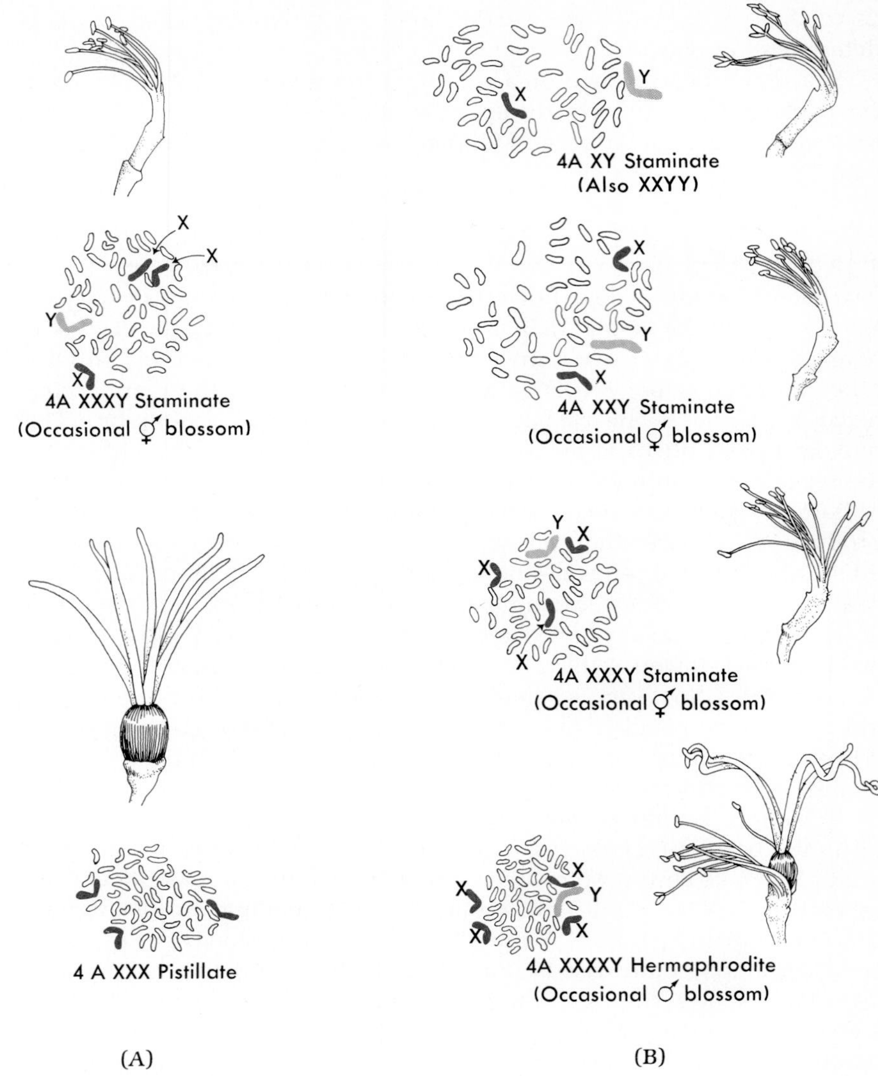

FIGURE 7–13. *(A) Reproductive parts and camera lucida drawings of the somatic chromosomes from 4A XXXY and 4A XXX plants of* Lychnis *(Melandrium). Note that a single Y chromosome produces a staminate plant having occasional perfect flowers; in the absence of the Y chromosome, plants are pistillate. (B) Reproductive parts and camera lucida drawings of the somatic chromosomes of* Lychnis *in a series showing pistillate-determining tendency of the X chromosome. All plants shown are tetraploid with respect to autosomes, but differ in the number of X chromosomes present. Tendency to produce pistillate flowers increases as the number of X chromosomes rises.* (Redrawn from H. W. Warmke, 1946, used by permission.)

Extensive investigation of the cytology of *Lychnis* shows that, where portions of the X or Y chromosome are deleted, the two chromosomes compare as shown in Figure 7–14. The sex chromosomes are quite dissimilar in size, that is, the Y is larger than the X, and each is larger than

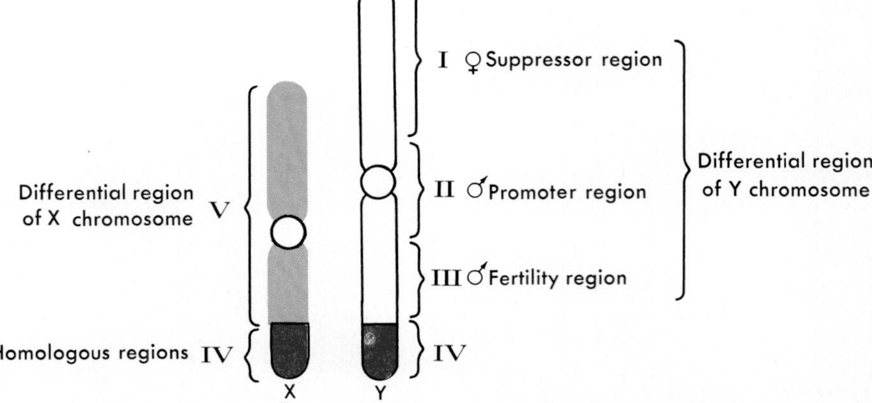

FIGURE 7–14. *Comparison of X and Y chromosomes of the plant* Lychnis. *Regions I, II, and III bear holandric genes, and region V, X-linked genes. Genes in region IV are termed incompletely sex-linked.* Lychnis *is unusual in that its Y chromosome is the larger of the two sex chromosomes.*

any autosome. Only a small portion of the X is homologous with a similar small bit of the Y, as suggested by their synaptic figures in meiosis. When region I is deleted, perfect-flowered plants are produced, whereas loss of region II produces pistillate plants (that is, whereas AAXY plants are staminate, AAXY-II are pistillate), and deletion of region III produces sterile staminate plants with aborted stamens.

GENIC DETERMINATION OF SEX

Sex does not appear to be controlled in all dioecious organisms by numerous genes on two or more chromosomes. Several examples will serve to illustrate some variations on the so-called chromosomal method.

Asparagus

C. M. Rick and G. C. Hanna presented evidence to show that sex in asparagus is determined by a single pair of alleles, with "maleness" (that is, formation of staminate flowers) as the dominant character. Asparagus is normally dioecious with some plants producing only staminate flowers and others producing only pistillate flowers. Rudimentary nonfunctional pistils occur in staminate flowers, and abortive stamens occur in pistillate blossoms. However, sometimes the pistils in staminate flowers function to produce viable seeds. Flower structure in this species is such that these seeds most likely result from self-pollination. In effect, this is a "male × male" cross from the genetic standpoint. The basis for the occasional functional pistils in staminate plants is not known, but both genetic and environmental factors have been suggested.

Rick and Hanna germinated 198 seeds from these uncommon functional pistils produced on staminate plants, and found a progeny ratio of 155 staminate to 43 pistillate. This is a close approximation to a 3:1 ratio, as a chi-square test will indicate. One-third of the staminate progeny, when crossed with normal pistillate plants, yielded only staminate offspring. The rest proved to be heterozygous, and produced staminate and pistillate offspring in a ratio close to 1:1.

If "maleness" is represented as $A-$ and "femaleness" as aa, the original "male × male" cross would be

$$\text{P} \quad Aa \times Aa$$
$$\text{F}_1 \quad \tfrac{3}{4}\,A- \;+\; \tfrac{1}{4}aa$$

On the basis of monohybrid inheritance, one-third of the $A-$ (staminate) plants should be AA and two-thirds should be Aa. Crosses between these genotypes and normal pistillate (aa) individuals would produce the results reported by Rick and Hanna.

Honeybee

Worker and queen bees are diploid females with 32 chromosomes. Drones, on the other hand, are males that have only 16 chromosomes. Through extensive studies on the parasitic wasp *Habrobracon* it was discovered that femaleness in such hymenopterans is determined by heterozygosity at a number of different loci on several chromosomes. Haploid individuals that hatch from unfertilized eggs cannot be heterozygous; hence they are male. Diploid males, however, have been produced experimentally by developing individuals that are homozygous in a sufficient number of chromosome segments.

FIGURE 7–15. *Barren stalk corn,* bs bs. *Note the absence of ears.* (From *The Ten Chromosomes of Maize,* DeKalb Agricultural Association, DeKalb, Illinois. Reproduced by permission.)

Corn

Unlike the organisms so far described, corn (*Zea mays*) is monoecious. The "tassel" consists of staminate flowers and the ear consists of pistillate flowers. Among several controls over "sex" in this plant are two interesting pairs of alleles. The genotype *bs bs* ("barren stalk") results in plants having no ears at all, though a normal tassel is present (Figure 7–15). Such individuals are staminate ("male"). Gene *ts* ("tassel seed"), when homozygous, converts the tassel to pistillate flowers so that ears develop at the top of the plant (Figure 7–16). Therefore *ts ts* individuals are pistillate ("female"). If ♂ represents staminate flowers, and ♀ represents pistillate ones, the various genotypes and phenotypes are as follows.

FIGURE 7–16. *Tassel seed corn,* ts ts. *Note silks of developing ear at top of stem in place of tassel.* (From *The Ten Chromosomes of Maize,* DeKalb Agricultural Association, DeKalb, Illinois. Reproduced by permission.)

Genotype	*Phenotype*	
Bs – Ts –	Normal monoecious	♂ ♀ ♀
bs bs Ts –	Staminate	♂
Bs – ts ts	Pistillate; ears terminal *and* lateral	♀ ♀ ♀
bs bs ts ts	Pistillate; ears terminal only	♀

Plants of genotype *bs bs Ts –* plus either *Bs – ts ts* or *bs bs ts ts* comprise a dioecious race of corn. Such a race may easily be produced by making the cross *bs bs ts ts* × *bs bs Ts ts*, which will segregate pistillate and staminate in a 1:1 ratio. Perhaps a similar sort of development has occurred in the evolution of dioecism. The practical advantage of the dioecious state to the breeder is significant in that it eliminates the need for the time-consuming procedure of emasculation in order to prevent self-pollination.

SUMMARY OF SEX DETERMINATION

Sex represents something of a continuum in many organisms, but it is basically gene-determined. Species differ with respect to (1) the number of genes that appear to play a part and (2) the locations of those genes (X chromosome, Y chromosome, autosomes). The genes an individual receives at the moment of syngamy determine which kind of gonads, and therefore gametes, will be formed. Later processes of sex differentiation are quite distinct from the initial event of sex determination. In insects the critical factors in sex differentiation appear to be intracellular. In humans and other mammals, however, sex differentiation is hormonal. Sex hormones, produced by the gonads, interact with endocrine glands elsewhere in the body to affect the multiplicity of secondary sex characters. These characters may be markedly affected by hormone injections; therefore, the potential of each individual for developing characteristics of either sex is demonstrated.

INHERITANCE RELATED TO SEX

The fact that males in *Drosophila* and in humans have an X and Y chromosome, whereas females have two Xs and no Y, raises some interesting genetic possibilities. This is especially true in view of the fact that the sex chromosomes are not entirely homologous, and therefore inheritance patterns related to the sex of the individual should be expected to be somewhat different from those examined previously for autosomes. For example, genes that are only on the X chromosome will be represented twice in females and once in males; recessives of this type might be expected to show up phenotypically more often in males. Genes located exclusively on the X chromosome are called **sex-linked genes** or **X-linked genes.** On the other hand, genes that occur only on the Y chromosome can produce their effects only in males; these are **holandric genes** (Greek, meaning *holos*, "whole," and *andros*, "man"). Still other mechanisms are known whereby a given trait is limited to one sex (**sex-limited genes**), or even in which dominance of a given allele depends on the sex of the bearer (**sex-influenced genes**). Genes that occur on homologous portions of the X and Y chromosomes are called **incompletely sex-linked.** The first four of these types of sex-related inheritance will be examined in detail.

SEX LINKAGE (X-LINKAGE)

Drosophila

From 1904 to 1923, T. H. Morgan taught at Columbia University and attracted many students who were later to become brilliant geneticists. Most of these people worked in what became known fondly as "the fly room." This must have been a remarkable and stimulating group; as one of them (R. H. Sturtevant) described it, "There was an atmosphere of excitement in the laboratory, and a great deal of discussion and argument about each new result as the work rapidly developed."

For example, in a long line of wild-type, red-eyed flies, an exceptional white-eyed male was discovered. To the Columbia group, this was a new character, apparently produced through mutation or change of the gene for red eyes (Figure 7–17). Morgan and his students crossed this new male to his wild-type sisters; all the offspring had red eyes, which indicated that white was recessive. An F_2 of 4,252 individuals was obtained; 3,470 were red-eyed and 782 were white-eyed. This is not a good representation of the expected 3:1 ratio, but later evidence indicated that white-eyed flies do not survive as well as their wild-type sibs and are therefore less likely to be counted. However, the important point here is that all 782 white-eyed F_2 flies were males! About an equal number of males had red eyes.

You will recall that the vermilion eye color alleles discussed earlier, and the white versus red alleles, are on the X chromosome. Again using Y to represent the Y chromosome (which does not carry a gene for eye color), + for red eyes, and *w* for white eyes (alleles on the X chromosome), Morgan's crosses may be represented as follows:

P	X^+X^+ red ♀	×	X^wY white ♂	
F_1	$\frac{1}{2}X^+X^w$ red ♀	and	$\frac{1}{2}X^+Y$ red ♂	
F_2	$\frac{1}{4}X^+X^+$, red ♀	$\frac{1}{4}X^+X^w$, red ♀	$\frac{1}{4}X^+Y$, red ♂	$\frac{1}{4}X^wY$ white ♂

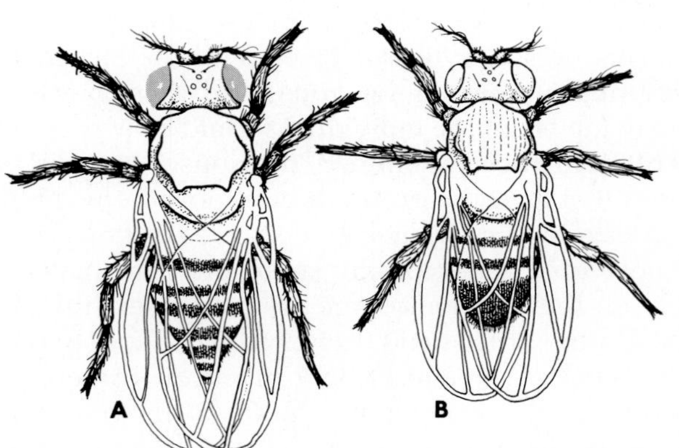

FIGURE 7–17. *Eye color mutation in* Drosophila melanogaster. *(A) Red-eyed female; (B) white-eyed male.*

Morgan correctly predicted that white-eyed females would be produced by the cross $X^+X^w \times X^wY$. A perpetual stock of white-eyed flies of both sexes was then established by mating white-eyed males and females.

An important characteristic of sex-linked inheritance emerges from an examination of the original Morgan cross. Note that the F_2 white-eyed males have received their recessive gene from their F_1 *mothers*, and these, in turn, have received allele *w* from their own white-eyed *fathers*. This "crisscross" pattern from father to heterozygous daughter (often termed *carrier*) to son is typical for a recessive sex-linked allele. Moreover, in this particular pedigree, about half of the sons in the F_2 show the trait, whereas none of the daughters do. Of course, in the cross $X^+X^w \times X^wY$ half the daughters as well as half the sons have the recessive phenotype. Notice that normal females (AAXX) carry two of these alleles and thus may be either homozygous (X^+X^+ or X^wX^w) or heterozygous (X^+X^w), but normal males (AAXY) can only be **hemizygous** (X^+Y or X^wY) with each gene present in only a single dose.

Sex Linkage in Humans

More than 150 confirmed or highly probable X-linked traits are known in humans; most of these are recessive (see McKusick and Table 6–2). The first to be described in the literature, a type of red-green color blindness in which the green-sensitive cones are defective (*deutan* color blindness), is due to an X-linked recessive allele. It affects about 8 percent of human males, but only about 0.7 percent of females. Females, of course, may be homozygous normal, heterozygous, or (rarely) homozygous for the defective allele, inasmuch as females have two X chromosomes and, therefore, a greater chance of receiving an allele for normal vision from at least one parent. Heterozygous women vary in the degree to which their color vision is affected, depending on the proportion of their retinal cells expressing the normal allele. Males, on the other hand, receive either a dominant allele for normal or a recessive allele for defective red-green color vision (from the *mother*, who contributes their X chromosome). Deutan color blindness appears to be the most commonly encountered sex-linked trait in human beings. A different kind of red-green color blindness, the *protan* type, produces a defect in the red-sensitive cones, but is much less common than the deutan type, occurring in only some 2 percent of males, and in only 4 women out of 10,000. Still other forms of color blindness, some X-linked and some autosomal, are also known in humans.

Hemophilia, a well-known disorder in which blood clotting is deficient because of a lack of the necessary substrate thromboplastin, is likewise a sex-linked recessive condition. Two types of sex-linked hemophilia are recognized:

1. *Hemophilia A,* characterized by lack of antihemophilic globulin (Factor VIII). About four-fifths of the cases of hemophilia are of this type.
2. *Hemophilia B,* or "Christmas disease" (after the family in which it was first described in detail), which results from a defect in plasma thromboplastic component (PTC, or Factor IX). This is a milder form of the condition.

From an unusual family in which both types of hemophilia were segregating, H. J. Woodliff and J. M. Jackson in 1966 concluded that the two loci involved were far apart on the X chromosome. Incomplete evidence suggests a distance of more than 40 map units. Hemophilia A is well known in the royal families of Europe, where it is traceable to Queen Victoria, who must have been heterozygous (Figure 7–18). No hemophilia is known in her ancestry; hence it is surmised that her hemophilia allele arose from a mutant gamete.

Male hemophiliacs occur with a frequency of about 1 in 10,000 male births (0.0001) and heterozygous females may be expected in about twice that frequency. (In Chapter 19 methods of calculating such probabilities will be explored.) Under a system of random mating, hemophilic females would be expected to occur once in $10{,}000^2$, or 100 million, births. But this probability is reduced by the likelihood that male hemophiliacs will die before reaching reproductive age (unless medically treated in order to extend life expectancy). Moreover, a hemophilic girl would be likely to die by adolescence. Consequently, few cases of female hemophiliacs are known, though they have been reported in some pedigrees involving first-cousin marriages. Because clotting time in different hemophiliacs varies somewhat, it has been suggested that the condition is affected by a number of modifying genes. Heterozygous women can be detected by a small increase in clotting time as well as lower levels of Factor VIII. D. Y. Wissell and co-workers found that Factor VIII levels in heterozygous women are distributed on a normal curve whose mean is lower than that for homozygous normal females.

Deficiency of the enzyme glucose-6-phosphate dehydrogenase (G6PD) is another important trait that results from action of a recessive X-linked allele, which is estimated to affect up to 100 million persons. This enzyme is directly involved in a minor glycolytic pathway in red blood cells. G6PD deficiency is common in blacks (about 10 percent of American black males) and in persons in Mediterranean areas (10 to 40 percent of Sardinian males in malarial areas, but less than 1 percent in Sardinian males of nonmalarial areas). The disorder is rare in whites in areas where malaria is not indigenous. Aside from the apparent advantage of G6PD deficiency in affording some protection against malaria,

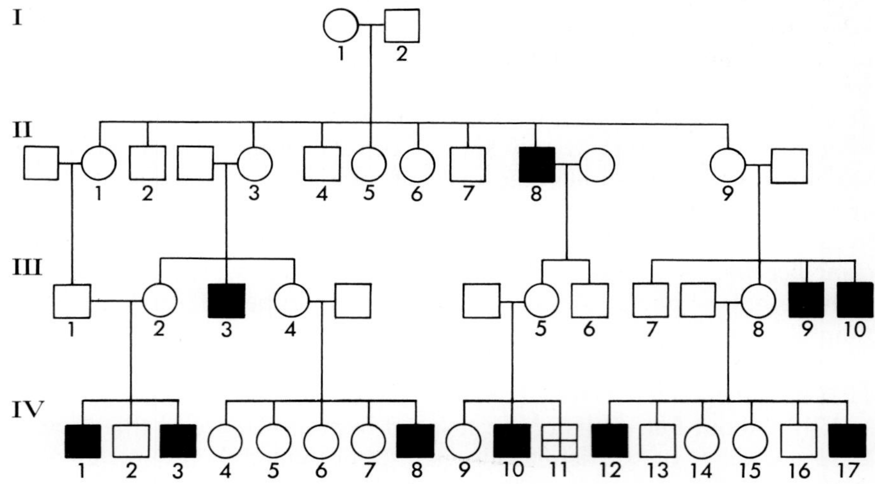

FIGURE 7–18. *Pedigree of some of the descendants of Queen Victoria showing incidence of hemophilia (shaded symbols). (I-1) Queen Victoria; (II-8) Leopold, Duke of Albany; (II-3) Fredrick of Hesse; (III-8) Victoria Eugene, who married Alfonso XIII of Spain; (III-9) Lord Leopold of Battenberg; (III-10) Prince Maurice of Battenberg; (IV-1) Waldenmar of Prussia; (IV-3) Henry of Prussia; (IV-8) Tsarevitch Alexis of Russia; (IV-10) Rupert, Viscount Trematon; (IV-12) Alfonso of Spain; (IV-17) Gonzolo of Spain; (IV-11) died in childhood; (II-2) represents Edward VII of England, great grandfather of Elizabeth II.*

the condition is noteworthy for the destruction of erythrocytes, with consequent severe hemolytic anemia, when certain drugs are administered. These drugs include para-amino salicylic acid, the sulfonamides, napthalene, phenacetin, and primaquine (an antimalarial agent). Inhalation of the pollen or ingestion of seeds of the broad bean (*Vicia faba*) produces the same result. Anemia caused by the broad bean is known as **favism.** In the absence of these drugs or plant parts, neither hemizygous recessive males nor heterozygous and homozygous recessive females suffer ill effects.

Many variants of this recessive allele are known, each of which affects a different portion of the polypeptide chain of the enzyme. These variants range in effect from severe through mild to no enzyme deficiency; some even produce increased enzyme activity.

Some of the other sex-linked traits in humans include two forms of diabetes insipidus, one form of anhidrotic ectodermal dysplasia (absence of sweat glands and teeth), absence of central incisors, certain forms of deafness, spastic paraplegia, uncontrollable rolling of the eyeballs (nystagmus), a form of cataract, night blindness, optic atrophy, juvenile glaucoma, and juvenile muscular dystrophy. Most of these are fairly clearly due to a recessive allele. On the other hand, hereditary enamel hypoplasia (*hypoplastic amelogenesis imperfecta*), in which tooth enamel is abnormally thin so that teeth appear small and wear rapidly down to the gums, is due to a dominant sex-linked allele.

Mapping of X-Linked Genes

Once two or more genes are clearly shown to be X-linked (by a consistent crisscross pattern of inheritance), some pedigree data extending over at least three generations can be used to determine preliminary map distances. Necessary information includes: (1) appearance of the traits in grandfather and grandson, but not in the mother of the latter, (2) identification of grandson having a doubly heterozygous mother, and (3) determination of *cis* or *trans* linkage in the mother.

An illustration will clarify this. Assume the following symbols and phenotypes for two known X-linked genes:

Xm+	production of blood antigen Xm	*R*	normal deutan color vision
Xm−	nonproduction of antigen Xm	*r*	deutan color blindness

A hypothetical family pedigree might look like this:

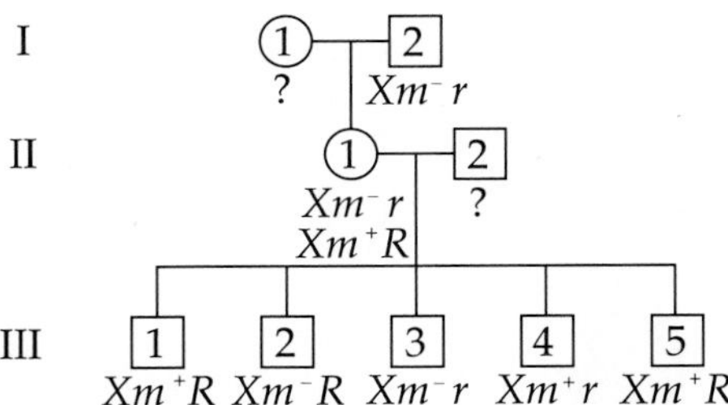

In this pedigree III-2 and III-4 are recombinants, which suggests (on the basis of this very small sample) a crossover frequency of $\frac{2}{5}$, or 40 percent. If enough data are accumulated to form a larger sample, the map distances between these two loci can be more accurately determined. Actual data suggest a much smaller map distance than the 40 map units arrived at here. This method of X-chromosome mapping is often referred to as the *grandfather method*. Note that it is not necessary to have even the phenotypes of the father (II-2) or of the maternal grandmother (I-1).

Sex Linkage in Other Organisms

Sex linkage in XX-XO species is, of course, just as it is in *Drosophila* and humans, because sex-linked genes are, by definition, those on the X chromosome. Here again only females can be heterozygous.

In birds, where the female is the heterogametic sex, the situation is reversed, although the mechanism is unchanged. Sex-linked genes follow the crisscross pattern, but from mother through heterozygous sons to granddaughters.

Although the barred feather pattern, which results from action of a dominant sex-linked allele in Plymouth Rock chickens (Figure 7–19), is a frequently cited illustration, a pair of alleles governing speed of feather growth is a more interesting one. Both traits have been used for identifying chick sex, but feather color patterns are sometimes modified by other genes. Gene *k*, for slow feather growth, eliminates such complications and can be detected within hours after hatching. For example, since males are homogametic, the cross Z^+W (♀) × Z^kZ^k (♂) results in two kinds of progeny, Z^+Z^+ (males with normal feather growth) and Z^kW (females with slower feather growth). Gene *k* has no effect on other characters of commercial value.

Sex-Linked Lethals

The gene for hemophilia is actually a recessive, **sex-linked lethal,** since it may often cause death. Slight scratches, accidental injuries, or even bruises, which would not be serious in normal persons, may result in fatal bleeding for the hemophiliac, although internal bleeding (from bruises, internal lesions, and so forth) is often more important. By bringing about death, sex-linked lethals will alter the sex ratio in a progeny.

Duchenne (or progressive pseudohypertrophic) muscular dystrophy is a disorder in which the affected individual, though apparently normal in early childhood, exhibits progressive wasting away of the muscles, resulting in confinement to a wheelchair by about age 12, and death in the teen years. Figure 7–20 shows the progression of the disease in a series of boys of increasing age. Like hemophilia, this disorder is due to a recessive sex-linked allele. At present, no means of arresting or preventing this condition is known; a given genotype dooms the bearer at conception to death in adolescence. The allele responsible is a lethal,

FIGURE 7–19. *Barred feather patterns, caused by a dominant sex-linked gene, in Plymouth Rock female (left) and male (right).*

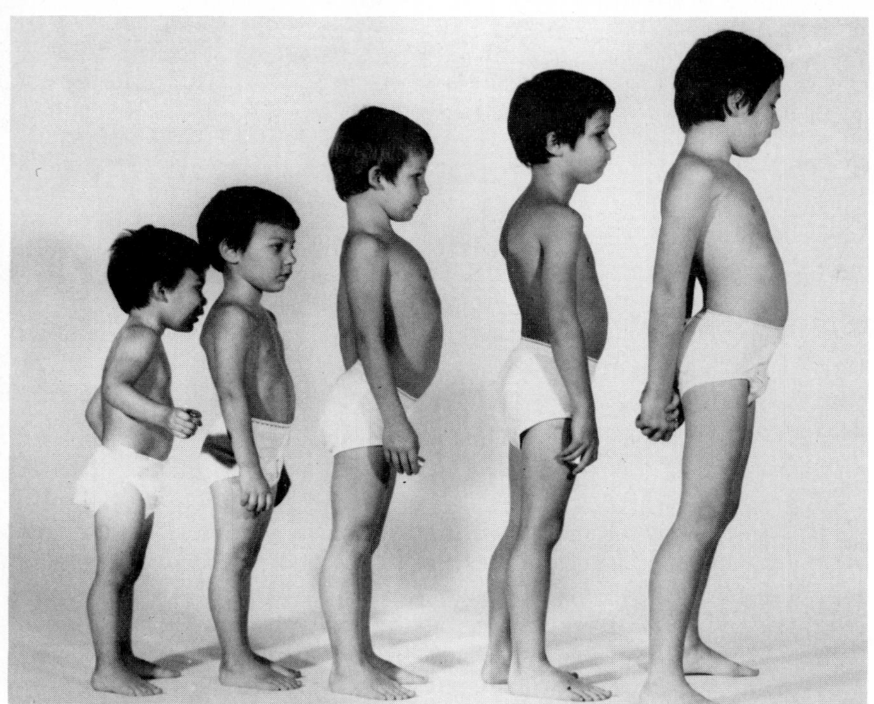

FIGURE 7–20. *Progressive development of Duchenne (pseudohypertropic) muscular dystrophy, caused by a recessive X-linked allele. This case will terminate in death.* (Courtesy Muscular Dystrophy Association of America, Inc., New York.)

and will change the sex ratio in a given group of offspring over time. If we let + represent the normal (dominant) allele and *d* represent the allele for muscular dystrophy, and consider children of many marriages between heterozygous women and normal men, $+d \times +Y$, the offspring would be expected in a ratio of $\frac{1}{4}$ $++$, $\frac{1}{4}$ $+d$, $\frac{1}{4}$ dY, and $\frac{1}{4}$ +Y at birth. Individuals of the last genotype, however, die before age 20. So an initial approximately 1:1 female-to-male ratio will later become 2:1 female-to-male. If a recessive sex-linked lethal kills before birth, the ratio of female to male live births is changed from nearly 1:1 to 2:1. A 2:1 female-to-male ratio is always a strong indication of a sex-linked recessive lethal allele.

HOLANDRIC GENES

Holandric genes are those that occur normally on the Y chromosome only and, therefore, are not expressed in females. As noted earlier, the human Y chromosome does bear a gene responsible for the H-Y antigen, on the short arm (Table 6–2). Histocompatibility antigen genes have been located on the Y chromosome in humans, mouse, rat, and guinea pig. It has been tentatively suggested that gene(s) controlling spermatogenesis are on the long arm of Y. Some of the genes that contribute to height are thought also to be on this chromosome, as well as genes that determine slower maturation of the individual. These and other loci on the human Y chromosome are listed in Table 6–2. Also, as noted earlier, the male fertility factor is on the Y chromosome of *Drosophila*.

SEX-LIMITED GENES

Sex-limited genes are autosomal genes whose phenotypic expression is determined by the presence or absence of one of the sex hormones. Their phenotypic effect is thus *limited* to one sex or the other.

FIGURE 7–21. *Hen-feathering (left) and cock-feathering (right) in domestic fowl. Cock-feathering is characterized by long, pointed, curving neck and tail feathers.*

Perhaps the most familiar example occurs in the domestic fowl where, as in many species of birds, males and females may exhibit pronounced differences in plumage. In the Leghorn breed, males have long, pointed, curved, fringed feathers on tail and neck, but feathers of females are shorter, rounded, straighter, and without the fringe (Figure 7–21). Thus males are cock-feathered and females are hen-feathered. In such breeds as the Sebright bantam, birds of both sexes are hen-feathered. However, in others (Hamburg or Wyandotte) both hen- and cock-feathered males are seen, but all females are hen-feathered.

It has been shown that feathering type depends on a single pair of alleles, *H* and *h*, in the following manner:

Genotype	♀	♂
HH	hen-feathered	hen-feathered
Hh	hen-feathered	hen-feathered
hh	hen-feathered	cock-feathered

Thus Sebright bantams are all *HH*, Hamburgs and Wyandottes may be $H-$ or *hh*, and Leghorns are all *hh*. Cockfeathering, where it occurs, is *limited* to the male sex.

A trait in humans of interest here is dizygotic (two egg) twinning. Obviously a trait limited in expression to females, it is transmitted by members of both sexes. Occurrence of multiple births after administration of pituitary gonadotropins suggests the physiological basis, but whether this is a monogenic or polygenic trait is difficult to determine. Family pedigrees indicate recessiveness of the trait. Insufficient data on monozygotic twinning precludes a conclusion on its genetic basis.

Sex-limited inheritance patterns are quite different from those of sex-linked genes. The latter may be expressed in either sex, though with differential frequency. Sex-limited genes express their effects in only one sex or the other, and their action is clearly related to sex hormones. They are principally responsible for secondary sex characters. Beard development in human beings is such a sex-limited character as men normally have beards, whereas women normally do not. Yet studies indicate no significant difference between the sexes in number of hairs per unit area of skin surface except in their development. This appears to depend on sex hormone production, changes in which may result in a bearded lady.

SEX-INFLUENCED GENES

In contrast to sex-limited genes, where expression of a trait is limited to one sex, **sex-influenced genes** are those whose dominance is *influenced* by the sex of the bearer.

Although baldness may arise through any of several causes (for example, disease, radiation, thyroid defects), "pattern" baldness exhibits a definite genetic pattern. In this condition, hair gradually thins on top, ultimately leaving a fringe of hair low on the head (Figure 7–22). Pattern baldness is more prevalent in males than in females, where it is rare and usually involves marked thinning rather than total loss of hair on top of the head. Numerous pedigrees clearly show that pattern baldness is due to a pair of autosomal alleles operating in this fashion:

Genotype	♂	♀
b_1b_1	Bald	Bald
b_1b_2	Bald	Not bald
b_2b_2	Not bald	Not bald

Gene *b* behaves as a dominant in males and as a recessive in females; it appears to exert its effect in the heterozygous state only in the presence of male hormone. A number of reports in the literature dealing with abnormalities that lead to hormone imbalance or with administration of hormones support this view (Hamilton). Some authors prefer to classify pattern baldness as sex-limited because the trait is generally less completely manifested in females, but it seems best to restrict the concept of sex-limited genes to those cases where one of the phenotypic expressions is *limited entirely to one sex* or the other because of anatomy.

A few well-known cases of sex-influenced genes occur in lower animals—for example, horns in sheep (dominant in males) and spotting in cattle (mahogany and white dominant in males, red and white dominant in females; Figure 7–23). Some hypothetical crosses involving these characters are explored in problems at the end of this chapter.

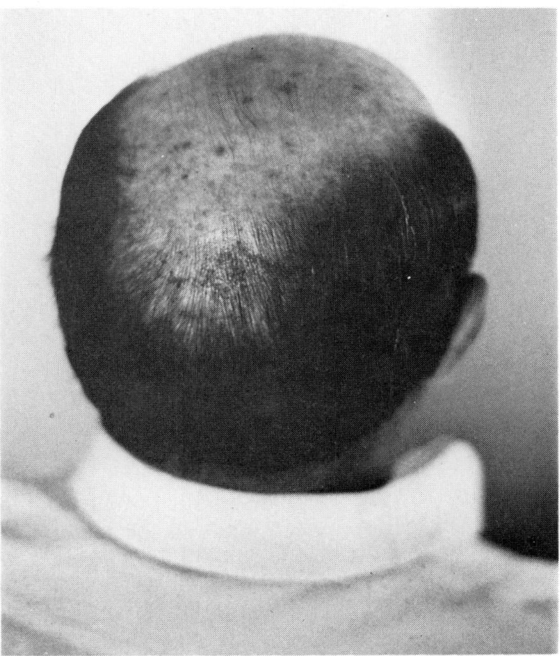

FIGURE 7–22. *Pattern baldness in a man, a sex-influenced trait dominant in males and recessive in females.*

FIGURE 7–23. *A sex-influenced trait in cattle. (A) Mahogany and white, dominant in males, recessive in females; (B) red and white, dominant in females, recessive in males.* (Courtesy Ayshire Breeders' Association, Brandon, Vermont.)

PROBLEMS

7–1 What is the sex designation for each of the following fruit flies (each A = one set of autosomes, each X = one X chromosome): (a) AAXXXX, (b) AAAAAXX, (c) AAXXXXXX, (d) AAAAAXXX, (e) AAAAXXXX, (f) AAAXY?

7–2 What is the sex designation for the following human beings: (a) AAXXX, (b) AAXXXYYY, (c) AAXO, (d) AAXXXXXY, (e) AAXYYY?

7–3 What phenotypic ratio results in corn from selfing *Bs bs Ts ts* plants?

7–4 In *Drosophila*, what fraction of the progeny of the cross *Tra tra* XX × *tra tra* XY is "transformed"?

7–5 What is the sex ratio in the progeny of the cross given in Problem 7–4?

7–6 In poultry, the dominant gene, *B*, for barred feather pattern is located on the Z chromosome. Its recessive allele, *b*, produces nonbarred feathers. What is the genotype of a (a) nonbarred female, (b) barred male, (c) barred female, (d) nonbarred male?

7–7 Give the phenotypic and sex ratios in the progeny of the following crosses in poultry: (a) nonbarred ♀ × heterozygous barred ♂, (b) barred ♀ × heterozygous barred ♂.

7–8 In poultry, removal of the ovary results in the development of the testes. Thus a female can become converted to a male, producing sperm and developing male secondary sex characters. If such a "male" is mated to a normal female, what sex ratio occurs in the progeny?

7–9 An XXY *Drosophila* female is mated to a normal male. The progeny include 5 percent metafemales. If secondary nondisjunction occurred, what was the frequency of XX eggs?

7–10 Based on your answer to 7–9, and the assumption that Y eggs occur with a frequency of 0.1, what percentage of the progeny of the cross of Problem 7–9 will be phenotypically normal females?

7–11 An attached-X female *Drosophila* is mated to a normal male. (a) What fraction of the zygotes become viable females? (b) What fraction of the zygotes are metafemales in chromosomal constitution? (c) What fraction of the viable, mature progeny are sterile males?

7–12 If the female parent of problem 7–11 is $AA\widehat{X^{+}X^{w}}$ and the male $AAX^{w}Y$, what will be the eye color of (a) viable, mature male progeny and (b) viable, mature female progeny?

7–13 How can the human Y chromosome be distinguished in suitably stained preparations?

7–14 If you assume parents to be AAXX and AAXY, how could you account in humans for children of each of the following types: (a) AAXYY, (b) AAXXY, (c) AAXO?

7–15 Numerous cases of human mosaics are reported in the literature. How could you account cytologically for an AAXX-AAXO mosaic?

7–16 No individuals, either live births or spontaneously aborted fetuses (abortuses), completely lacking any X chromosomes (for example, AAOY) have ever been discovered. Why is this to be expected?

7–17 From the standpoint of survival of the species, which system of sex determination, that of *Asparagus* or of humans, seems to offer the greater advantage? Why?

7–18 For humans, give the genetic sex of (a) most true hermaphrodites, (b) masculinizing male pseudohermaphrodites, (c) feminizing male pseudohermaphrodites, (d) female pseudohermaphrodites.

7–19 Can you suggest any reason why such persons as those having the karyotype 48,XXXY are highly sterile?

7–20 A human blood antigen, Xg^{a}, is due to a dominant allele (symbolized also as Xg^{a}) on the short arm of the X chromosome. Its recessive allele may be symbolized *Xg*. Phenotypes are designated as either XG^{a} + or Xg^{a} −. Examine the following pedigree of a Turner individual and determine the source of the single X chromosome in the Turner child (was it maternal or paternal?).

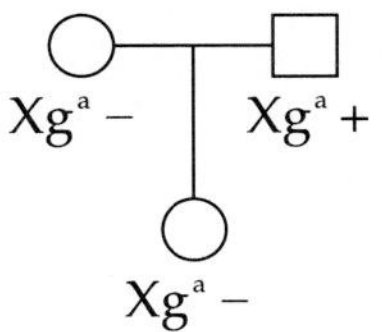

7–21 In terms of the gene symbols given in the preceding problem, what is the genotype of each of the three persons in 7–20?

7–22 Examine the following pedigrees of a Klinefelter individual. Which parent was the source of the extra X chromosome?

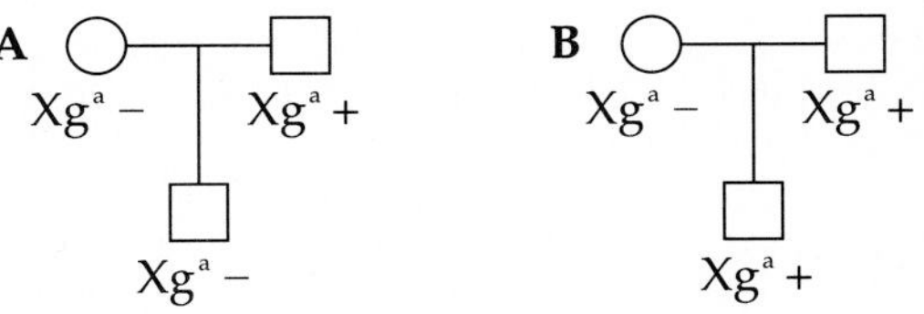

7–23 In terms of the gene symbols given in Problem 7–20, what is the genotype of each of the three persons in pedigree 7–22?

7–24 "Bent," a dominant sex-linked allele, *B*, in the mouse, results in a short, crooked tail; its recessive allele, *b*, produces normal tails. If a normal-tailed female is mated to a bent-tailed male, what phenotypic ratio should occur in the F_1?

7–25 Nystagmus is a condition in humans characterized by involuntary rolling of the eyeballs. The allele for this condition is incompletely dominant and sex-linked. Three phenotypes are possible: normal, slight rolling, severe rolling. A woman who exhibits slight nystagmus and a normal man are considering marriage and ask a geneticist what the chance is that their children will be affected. What will he tell them?

7–26 "Deranged" is a phenotype in *Drosophila* in which the thoracic bristles characteristically disarranged and the wings are vertically upheld. Crosses between deranged females and normal males result in a 1:1 ratio of normal females to deranged males in the progeny. What is the mode of inheritance and how would you describe the dominance of this allele?

7–27 In poultry, sex-linked allele *B*, which produces barred feather pattern, is completely dominant to its allele, *b*, for nonbarred pattern. Autosomal allele *R* produces rose-comb; its recessive allele, *r*, produces single comb in the homozygous state. A barred female, homozygous for rose-comb, is mated to a nonbarred, single-comb male. What is the F_1 phenotypic ratio?

7–28 Members of the F_1 from Problem 7–4 are then crossed with each other. What fraction of the F_2 is barred rose, and are these male or female?

7–29 In what ratio does (a) barred, nonbarred and (b) rose, single segregate in the F_2 of problem 7–28?

7–30 A family has five children, three girls and two boys. One of the latter died of muscular dystrophy at age 15. The others have graduated from college and are concerned over the probability that their children may develop the disease. What would you tell them?

7–31 "Jimpy" is a trait in the mouse characterized by muscular incoordination which results in death at an age of three to four weeks. Crosses between heterozygous females and normal males produce litters in which half the males are jimpy. What type of gene is jimpy?

7–32 At locus 0.3 on the X chromosome of *Drosophila* there occurs a recessive allele, *l*, which is lethal in the larval stage. A heterozygous female is crossed to a normal male; what F_1 adult sex phenotypic ratio results?

7–33 A woman with defective tooth enamel and normal red-green color vision, who had a red-green blind father with normal tooth enamel and a mother who had defective tooth enamel and normal red-green vision, marries a red-green blind first cousin with normal tooth enamel. What is the probability of their having a child with normal tooth enamel and red-green color blindness if crossing-over does not occur?

7–34 What is the probability, if they have three children, that these will be two red-green blind girls with normal teeth and one boy with defective teeth but normal red-green color vision?

7–35 The bald phenotype can sometimes be distinguished at a relatively early age. A nonbald, red-green blind man marries a nonbald, normal-visioned woman whose mother was bald and whose father was red-green blind. What is the probability that they will have each of the following children: (a) bald girls, (b) bald normal-visioned boys, (c) nonbald boys, (d) red-green blind children of either sex?

7–36 In addition to the allelic pair determining pattern baldness (*B*, *b*) described in this chapter, consider early baldness to be due to another autosomal allele (*E*) on a different pair of chromosomes and also dominant in males. The phenotype for *ee* may be either late baldness or nonbaldness, depending on sex and the genotype for the *B*, *b* alleles. Two doubly heterozygous persons (*BbEe*) marry. (a) What is or will be the phenotype of the male parent? (b) What is or will be the phenotype of the female parent? (c) What is the phenotypic ratio among *male* children of couples such as this one? (d) What is the phenotypic ratio among *female* children of couples such as this one?

IDENTIFICATION OF THE GENETIC MATERIAL

CHAPTER

8

Up to this point, a variety of genetic observations have been made in a wide assortment of organisms. It has been seen that all these observations can be explained by theorizing that genes occur at specific loci and in linear order on chromosomes. The following questions, however, must now be considered:

1. What is the genetic material? Is it the same in all organisms?
2. How does this genetic material operate to produce detectable phenotypic traits, and can its function explain such phenomena as dominance and recessiveness?
3. In terms of the genetic material, what actually is a gene? Can its molecular configuration be determined?
4. In terms of the nature of the genetic material and of the gene, what is mutation? That is, when a gene changes and produces a different effect, what happens to it at the level of its molecular structure?
5. Differentiation in multicellular organisms, as well as many intermittent biochemical reactions in both eukaryotes and prokaryotes, suggest that not all genes function all the time. If this is so, how is gene expression regulated?
6. Is this genetic material (are these genes) all located in the chromosomes of eukaryotes, or does the cytoplasm play any part in inheritance? How does the answer to this question relate to prokaryotes?
7. Can answers to such questions as these be used in any way to mitigate or offset the effect of disadvantageous or lethal genes, or even to replace them?

These questions will be pursued in the next chapters; in this one the problem of identifying the genetic material itself will be examined. Basically, the successful approaches have used microorganisms. It is much easier to find answers to these questions in bacteria and viruses than it is in rather complex individuals such as humans. Three means of genetic exchange between bacteria, **transformation, transduction,** and **conjugation** (conjugation is covered in Appendix B-1), were crucial to the answering of these important questions. Interestingly enough, knowledge gained from studies on these simpler forms is many times found to be applicable to all the more highly evolved ones.

BACTERIAL TRANSFORMATION

The Griffith Effect

In 1928, Fredrick Griffith published a paper in which he cited a number of remarkable results for which he had no explanation. His observations involved a particular bacterium, *Diplococcus pneumoniae,* which is associated with certain types of pneumonia. This organism occurs in two major forms. The first is *smooth* (S), whose cells secrete a covering capsule of polysaccharide materials, causing its colonies on agar to be smooth and rather shiny. The smooth (S) form is virulent in that it produces septicemia in mice. The cells of a second form of *Diplococcus pneumoniae, rough* (R), lack a capsule and their agar colonies have a rough, rather dull surface. This form is nonvirulent. The above evidence suggests that virulence is related to the presence of a capsule.

The smooth (S) types can be distinguished by their possession of different capsular polysaccharides (designated as I, II, III, IV); the specific polysaccharide is antigenic and genetically controlled. Mutations from smooth to rough occur spontaneously with a frequency of about one cell in 10^7, though the reverse is much less frequent. Mutation of, say, a S-II to rough may occur, and may rarely revert back to smooth. When that happens the smooth revertants are again S-II. In the course of his work, Griffith injected laboratory mice with living R-II pneumococci; the mice suffered no ill effects. Injection of mice with a living S-III or any other smooth culture was fatal. However, when cells of either R or S types were heat-killed before injection into a mouse, no disease was produced. Simultaneous inoculation of the animals with a mixture of live R-II bacteria and heat-killed S-III bacteria resulted in high mortality. This surely was an unexpected turn of events. No living S-III cells were injected into the mice, yet both living R-II and S-III organisms could be isolated following autopsy of the dead mice.

An explanation was not immediately forthcoming. It was as though the killed S-III individuals were somehow restored to life, but this was patently absurd. The recently understood phenomenon of mutation might be implicated, but the frequency of occurrence in the Griffith experiments was far too high to be compatible with known mutation rates. Scientists were left only with the idea that in some way the heat-killed cells conferred virulence on the previously nonvirulent strain; in short, the living R-II cells were somehow *transformed* by the heat-killed S-III cells. So the Griffith effect gradually became known as *transformation* and turned out to be the first major step in the identification of the genetic material.

Identification of the Transforming Principle

Sixteen years after Griffith's work, Oswald Avery, Colin MacLeod, and MacLyn McCarty reported successful repetition of the earlier work, but *in vitro*, and were able to identify the transforming principle or the chemical substance that brings about transformation. They tested fractions of heat-killed cells for transforming ability. Completely negative results were obtained with fractions containing only the polysaccharide capsule, various cell proteins, or ribonucleic acid (RNA); only extracts containing deoxyribonucleic acid (DNA) were effective. Even highly purified fractions containing DNA and less than 2 parts protein per 10,000, for example, retained the transforming ability. DNA, plus even minute amounts of protein, plus proteolytic enzymes, was fully effective. But DNA, plus DNase, an enzyme that destroys DNA, lost its transforming capability.

Therefore, it began to appear *beyond any reasonable doubt that DNA must be the genetic material.* Moreover, this landmark of genetic research indicated that the *genetic material, DNA, differs from the end product (polysaccharide) it determines.* The process of transformation involves the incorporation of a single strand of a foreign DNA molecule into the recipient molecule (see Figure 8–5). Cotransformation of closely linked genes is a standard method of gene mapping in bacteria. Transformation has also been successful for cells in culture of both animals and plants.

TRANSDUCTION

The clear implication of DNA as the genetic material that was furnished by transformation experiments was extended and confirmed by a series of experiments begun by Norton Zinder and Joshua Lederberg in 1952 on the mouse typhoid bacterium (*Salmonella typhimurium*). Their experiments involved the process of **transduction** in which a bacterium-infecting virus (**phage**) serves as the vector transferring DNA from one bacterial cell to another. To understand this process fully we must first examine the phage structure and life cycle.

Structure of T-even Phages

There is no typical virus structure, but there are several modifications of a basic structure. In general terms, viruses consist of an outer, inert, nongenetic protein "shell" and an inner "core" of genetic material. In many cases this genetic material is DNA, but in some cases it is RNA. Perhaps the best-known viruses from a structural standpoint are the so-called T-even phages (for example, T2, T4) that infect the colon bacillus *Escherichia coli.*

The T phages are of a general "tadpole" shape, differentiated into a *head* and a *tail* region. The former is an elongate, bipyramidal six-sided structure composed of several proteins (Figures 8–1 and 8–2). Its diameter is approximately 0.065 μm and its length is about 0.1 μm. Within the head is a DNA molecule about 68 μm in length. The dimensions of the head are such that the DNA molecule must be packed tightly within it.

The tail is a hollow cylinder, most of which can contract lengthwise. The approximate uncontracted dimensions are 0.08 μm in length and 0.0165 μm in diameter. Contraction changes these dimensions to 0.035 μm in length and 0.025 μm in diameter. The tail can be visualized by electron microscopy as bearing 24 helical striations. Contraction involves compression of this striated sheath region. The core diameter is approximately 0.007 μm and encloses a longitudinal canal 0.0025 μm in diameter. Six spikes and six tail fibers, the latter bent at an angle about midway in their length, arise from a hexagonal plate at the distal end

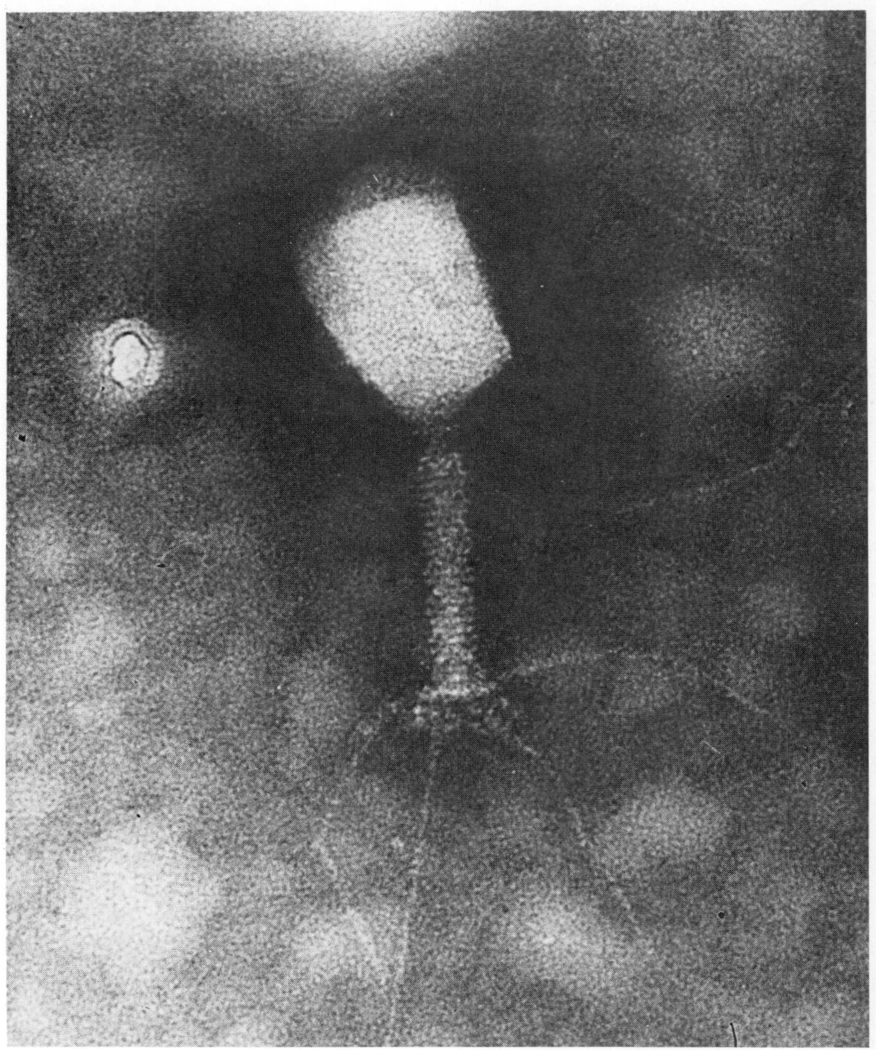

FIGURE 8–1. *Electron micrograph of a T4 bacteriophage, 630,000 X.* (Courtesy Dr. Thomas F. Anderson, Institute for Cancer Research, Philadelphia.)

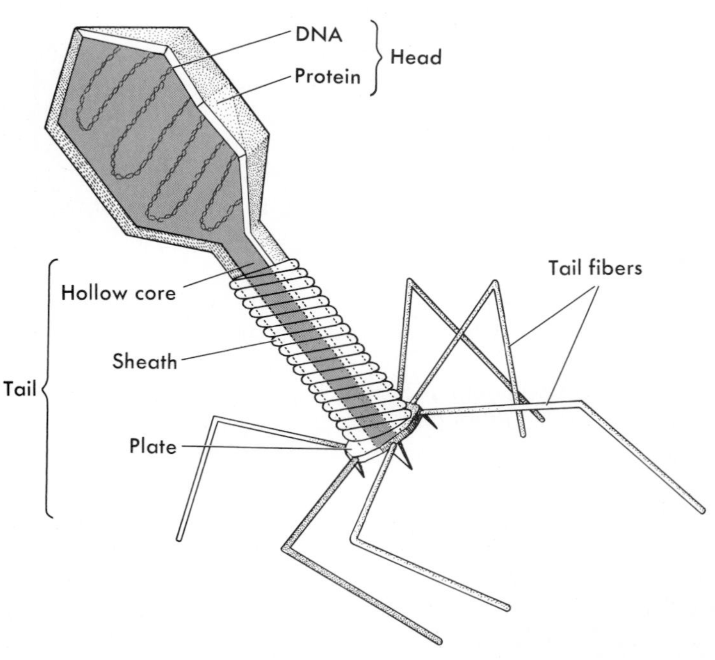

FIGURE 8–2. *Diagrammatic longitudinal section of a T4 bacteriophage. Only a small portion of the total DNA content is shown.*

of the tail. Tail fibers are not visible in the mature phage until it is absorbed to the surface of a bacterial cell. Phage T4 is shown in electron micrograph view in Figure 8–1 and in diagrammatic sectional view in Figure 8–2. Other phages are quite different in morphology. ϕX174, for example, appears in electron micrographs as a cluster of 12 identical, adherent spherical subparticles but morphologically distinct.

Two general types of phage replication cycles can be recognized: (1) *virulent* phages, in which infection is followed by *lysis* (bursting) of the host cell and the release of new, infective phages, and (2) *temperate* phages, in which infection only rarely causes lysis.

Life Cycle of a Virulent Phage

Infection begins with a chance collision between phage and bacterial cell, followed by attachment of phage to one of the numerous receptor sites on the bacterial cell (Figure 8–3); the sheath then contracts, driving the core through the host cell wall. The phage DNA then enters the bacterial cell.

Determination of immediately succeeding events is facilitated by the fact that the protein outer shell contains sulfur but no phosphorous, whereas DNA contains phosphorous but no sulfur. Using two samples of phage, one of which contains radioactive ^{35}S and the other radioactive ^{32}P, Alfred Hershey and Martha Chase were able to show in 1952 that all the phage DNA enters the host cell following attachment; most of the protein remains outside. Details of the Hershey and Chase experiment are diagramed in Figure 8–4. An *eclipse period* ensues, during which the phage DNA replicates numerous times within the bacterial cell. Toward the end of the eclipse period the phage DNA directs the production of protein coats and assembly of some 50 to 200 new, infective viral particles. Within a short time (for example, 13 minutes for phage T1, and 22 minutes for phage T2) a phage-produced enzyme (lysozyme) brings about lysis of the host cell and release of the mature phages (Figure 8–3). The following steps in the process can be recognized:

1. *Attachment* of phage tail to specific receptor sites on the bacterial cell wall.

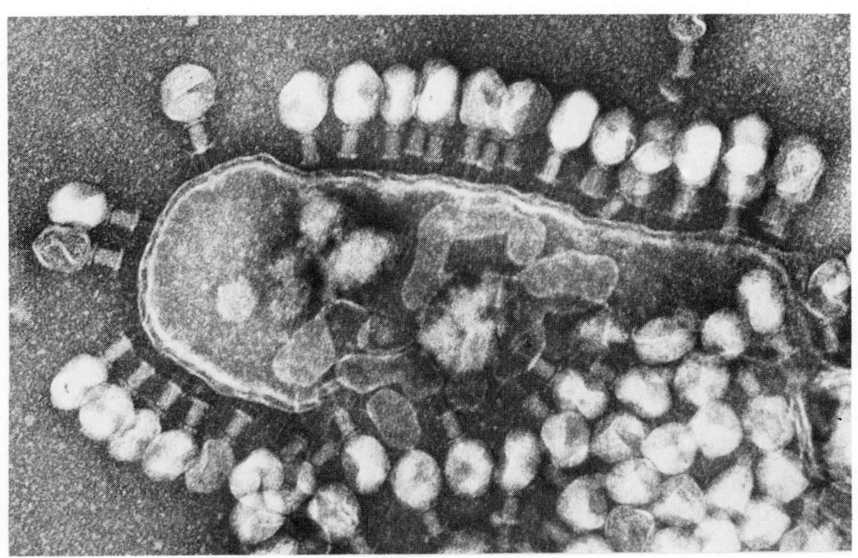

FIGURE 8–3. *Electron micrograph of T4 phage attacking* Escherichia coli. *Not only can phage particles be seen attached by their tail fibers to the cell wall of the bacterium (top and right), but new phage particles are shown being released from the lysed bacterial cell (left).* (Photo by Dr. L. D. Simon, courtesy Dr. Thomas F. Anderson, Institute for Cancer Research, Philadelphia.)

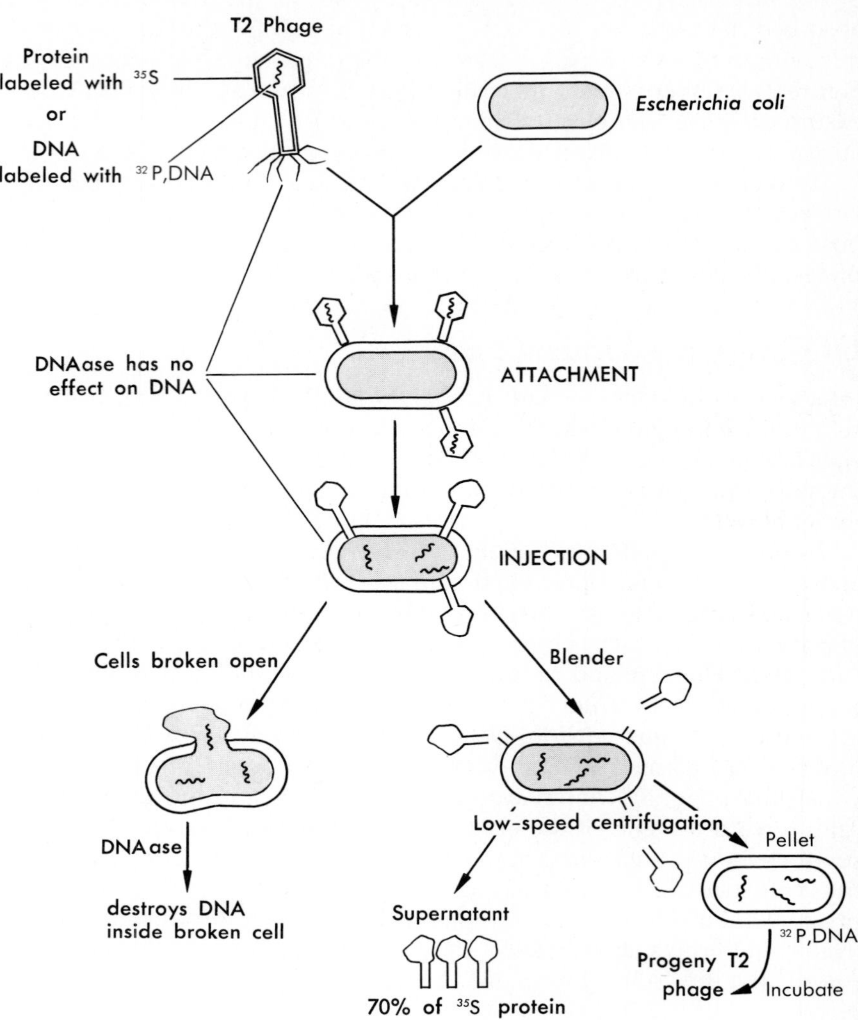

FIGURE 8–4. *Diagram of life cycle of a virulent phage showing details of the Hershey–Chase experiment. Phage particles and bacterial cells are not to scale.* (From Fraser, *Viruses and Molecular Biology*, Macmillan, 1967. By permission.)

2. *Injection* of phage DNA.
3. *Eclipse* period in which no infective phage is recoverable if the bacterial cell is artificially lysed and during which synthesis of new phage DNA and protein coats is taking place.
4. *Assembly* of phage DNA into new protein shells.
5. *Lysis* of host cell and release of infective phage particles.

Life Cycle of a Temperate Phage

Temperate phages do not ordinarily lyse their host; in such cases the phage DNA becomes integrated into the bacterial DNA as a *prophage*. Prophages replicate synchronously with host DNA; progeny bacterial cells then contain this bacterial-plus-phage DNA. Under these conditions both infection by a virulent phage and maturation of infective virus particles are prevented. Infrequently, however, the association between host and phage DNA may be terminated (excision); the lytic cycle then follows as in the case of a virulent phage. Because the bacterial hosts in this case are potentially subject to lysis, they are termed *lysogenic*. Both phages lambda (λ) and P22, for examples, behave as temperate phages.

Generalized Transduction

In the early 1950s, Zinder and Lederberg looked for indications that the recently described process of conjugation (Appendix B–1) might occur in the mouse typhoid bacterium *Salmonella typhimurium*. In one experiment they cultured two different strains of this bacterium, met^- his^- and phe^- trp^- tyr^- (which signify defects in the ability to synthesize the amino acids methionine, histidine, phenylalanine, tryptophan, and tyrosine, respectively). Neither strain would grow on a medium that lacked the amino acids that the strain could not synthesize. In one of their experiments, the two strains were grown in liquid medium in a U-tube but separated by a membrane whose pores were too small to permit passage of the bacterial cells. Thus, cell-to-cell contact (and, therefore, conjugation) could not occur. When, however, filters of a pore size that permitted passage of the temperate phage P22 (but too small for the bacteria) were used, wild-type cells were recovered.

Additional work demonstrated clearly that P22 was, in fact, the agent of recombination. Transformation was excluded as a possibility by the simple test of treating *cell-free* extracts of one strain with DNase to destroy any naked DNA. Again recombination occurred. So, rather than demonstrating either conjugation or transformation, Zinder and Lederberg had discovered a new type of recombination, one mediated by a virus. To this process they gave the name **transduction.** Its distinguishing characteristic is that the vector of recombination is a phage.

A temperate phage usually exists as a *prophage*, that is, integrated into the bacterial DNA. In an occasional cell the prophage becomes detached and, like a virulent phage, lyses the host cell. For example, with phage P22, instead of phage DNA, a segment of host DNA, which happens to be about the same length as the normal phage genome, becomes incorporated in a phage coat. *Any* portion of host DNA of the appropriate length may be so incorporated. In the new host, the transducing DNA may be integrated into the newly infected cell's genome. If transduced genes and those of the new host are alleles, the transduced gene(s), upon inclusion in the host DNA by a process of crossing-over, produce recombinant progeny. This type of transduction, in which the transducing DNA may involve any of the bacterial genes, is referred to as **generalized transduction** (Figure 8–5).

Generalized transducing phages carry *only* bacterial genes, and their DNA may replace a corresponding segment of DNA in the newly infected cell by crossing-over. For example, if the donor DNA includes genes tyr^+ trp^+ (which are close together on the *E. coli* chromosome), and the recipient is tyr^- trp^-, any tyr^+ trp^- and tyr^- trp^+ progeny must be recombinants. The distance between the two loci may be calculated from the ratio of recombinants to total transduced cells. Recombinant frequency varies inversely with distance between loci just as in eukaryotes.

Specialized Transduction

Some phages are able to integrate into the host cell's DNA only at certain positions. Phage lambda (λ), for example, infects *E. coli* and can occupy only a site between the *gal* (galactose) and *bio* (biotin) genes. Occasionally the phage DNA is excised incorrectly with the result that the excised DNA carries some bacterial and some viral genes. This hybrid DNA is then incorporated into phage shells as usual. Upon infecting a new host cell, the phage may integrate (at its normal site), and create a bacterial cell that is partially diploid. Such *merozygotes* can be used in (1) mapping some of the bacterial genes on either side of the phage insertion point, and (2) determining which host allele, normal or defective, is dominant. The pieces of DNA that may exist either free in the bacterial cytoplasm or integrated into its DNA are designated *episomes*. These are discussed further in Chapter 16.

Note the distinction between generalized and **specialized transduction.** Generalized transducing phages carry *only bacterial genes*, and these may *replace* those of the newly infected host; specialized transducing phages carry *both bacterial and viral genes*. Whereas genes carried by generalized transducing phages may replace those of the bacterial cell, genes born by specialized transducing phages may be *added* to the genome of the newly infected cell.

Behavior of DNA in Transformation and Transduction

In both transformation and transduction, DNA from another bacterium enters the cell that is about to be "converted." In transformation, naked DNA comes from dead cells, and quite a bit less than its DNA complement penetrates the living cell. Cells can be transformed only when they are in a state of "competence." In transduction a small portion of the bacterial genome is "injected" from a virus. The transducing DNA may be either all bacterial or bacterial-plus-viral DNA. A summary of the two processes is found in Figure 8–5.

If foreign DNA is to bring about transformation or transduction, it must be integrated into the host DNA molecule. This may occur in two different ways. First, in transformation, the uptake of double-stranded DNA into a competent recipient cell occurs. In the process of entering, one of the strands of DNA is digested by nucleases from the recipient cell, leaving a single strand. This single strand aligns with the appropriate homologous region on the bacterial chromosome, and through an enzymatic process the strand replaces its homologous strand in the recipient DNA molecule. The displaced strand is degraded. The resultant DNA molecule contains one strand of original DNA and one strand of transforming DNA, and is called a **heteroduplex.** After one round of DNA replication and cell division, one cell contains the original chromosome, and one cell contains the transformed chromosome (Figure 8–5A).

In transduction, when a donor cell is lysed, the chromosome is broken up into small pieces. Then the newly forming phage particles mistakenly incorporate a length of pure bacterial DNA into a phage head. When the contents of this phage are injected into another bacterial cell, the transduced DNA is released into the cell, and some genes may be incorporated by a recombination process (Figure 8–5B).

When it is recognized that the material transferred during conjugation (Chapter 6 and Appendix B) is also DNA, it is clear that the three processes—conjugation, transformation, and transduction—all implicate DNA as the genetic material. The essential feature of heredity, then, is the unchanged transmission of this "information tape" from one generation to the next. If we are to identify genes, we must understand the structure of this all-important molecule that has been appropriately called the "thread of life."

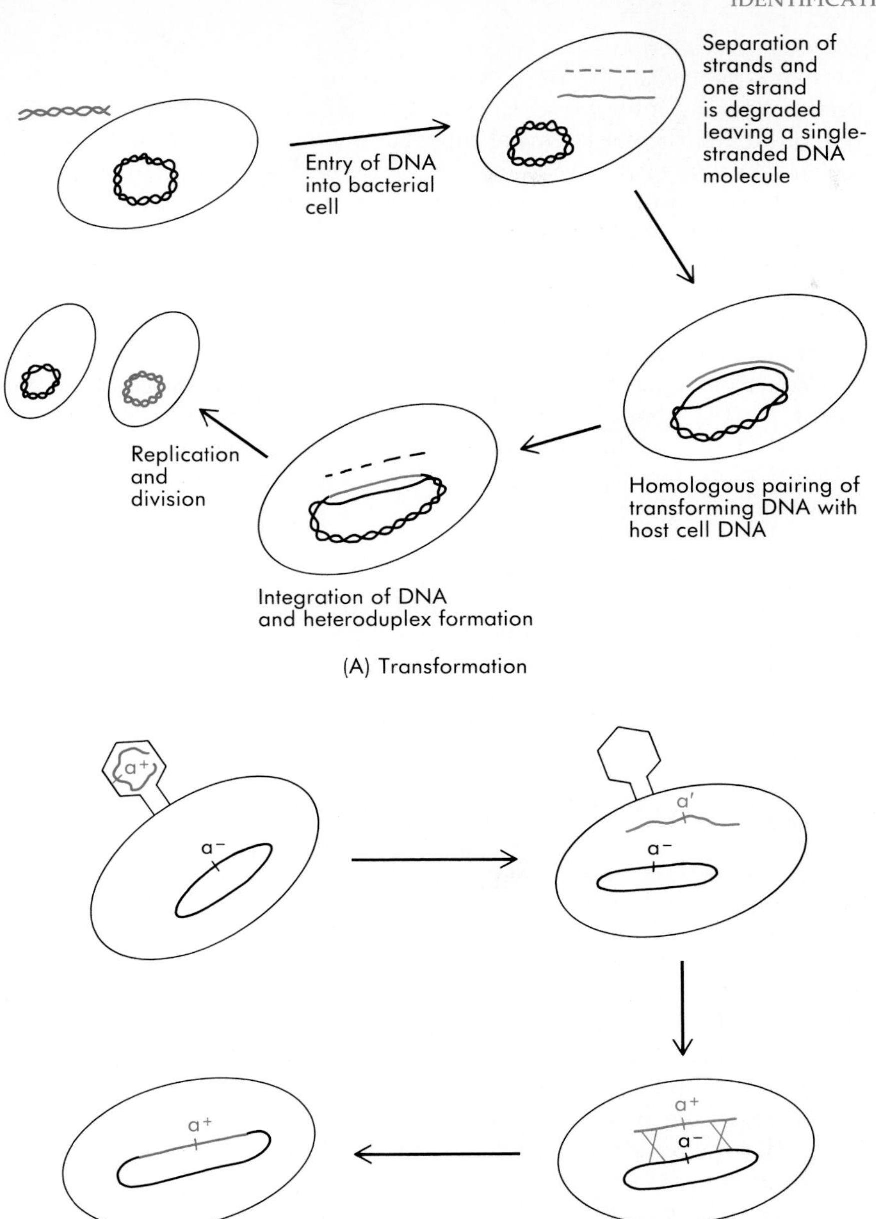

FIGURE 8–5. *Diagrammatic representation of (A) transformation and (B) generalized transduction.*

DEOXYRIBONUCLEIC ACID

History

Interestingly enough, recognition of DNA as the genetic material was slow in coming. By 1869, Friedrich Miescher, a 22-year-old Swiss physician, had isolated from pus cells obtained from discarded bandages in the Franko-Prussian War, and from salmon sperm, a previously unidentified macromolecular substance, to which he gave the name *nuclein*. Although he was unaware of the structure and function of nuclein, he submitted his findings for publication. The editor who received the paper was dubious about some aspects of the report and delayed publication for two years while he tried repeating some of the more questionable aspects of Miescher's work. Finally, in 1871, Miescher's report was published, but it made little immediate impact. He continued his careful work up to his death in 1895, recognizing (with the help of his student, Altmann, in 1889) that nuclein was of high molecular weight and was associated in some way with a basic protein, to which he gave the name *protamine*. By 1895, the pioneering cytologist E. B. Wilson speculated that "inheritance . . . may be affected by the physical transmission of a particular chemical compound from parent to offspring."

Nuclein was later renamed *nucleic acid,* and work on it continued slowly in several laboratories. In the early twentieth century the biochemist Kossel identified the constituent nitrogenous bases of nucleic acid, as well as its 5-carbon sugar, and phosphoric acid. Kossel's work and the later investigations of Ascoli, Levine, and Jones during the first quarter of the 1900s disclosed the two kinds of nucleic acid, deoxyribonucleic acid and ribonucleic acid. Development of DNA-specific staining techniques by Feulgen and Rossenbeck in 1924 enabled Feulgen to demonstrate in 1937 that most of the DNA content of a cell is located in the nucleus. H. F. Judson has written a highly readable account of the discoveries in genetics and molecular biology revolving around this DNA story.

Structure

In 1953, James Watson and Francis Crick proposed a molecular model for DNA after about a year and a half of joint work at Cambridge University. So completely was their model substantiated by subsequent investigations that this team shared a Nobel Prize in 1962 with Maurice Wilkins.

Publication of their proposal was the outcome of intensive work that involved Maurice Wilkins, Rosalind Franklin, and Linus Pauling. Both Wilkins and Franklin prepared X-ray diffraction pictures on which the ultimate model was based. They and Pauling considerably sharpened Watson and Crick's notion of the structure of the molecule in discussions and by correspondence. Watson, in his engrossing account of those years (*The Double Helix,* 1968), emphasized chiefly his and Crick's contributions (which were considerable); however, it appears that not only did Franklin come to some of the correct conclusions earlier than did Watson and Crick but she might have shared in the Nobel Prize if not for her untimely death at age 37.

The basic unit of structure of the DNA molecule is the **nucleotide.** This is also the basic unit incorporated into DNA during synthesis. The nucleotide consists of three parts: a phosphate, a sugar, and a nitrogen base. The structure of the phosphate and sugar are shown in Figure 8–6. The sugar is a form of ribose called **deoxyribose** because it is missing the hydroxyl (OH) group at the number 2 carbon similar to the OH present at the number 3 carbon. The structures of the nitrogen bases are found in Figure 8–7. They include two purines—**adenine** and **guanine**—and two pyrimidines—**thymine** and **cytosine.** The structure of the nucleotides, showing the relationship of the three components, is seen in Figure 8–8. A nucleic acid molecule (**polynucleotide**) consists of a large number of nucleotides joined together between the sugars and phosphates by phosphodiester bonds in the manner shown in Figure

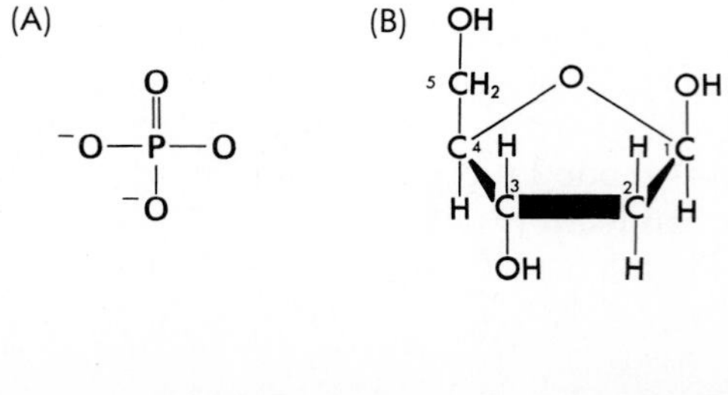

FIGURE 8–6. *Molecular structure of (A) phosphate and (B) deoxyribose, the sugar of DNA.*

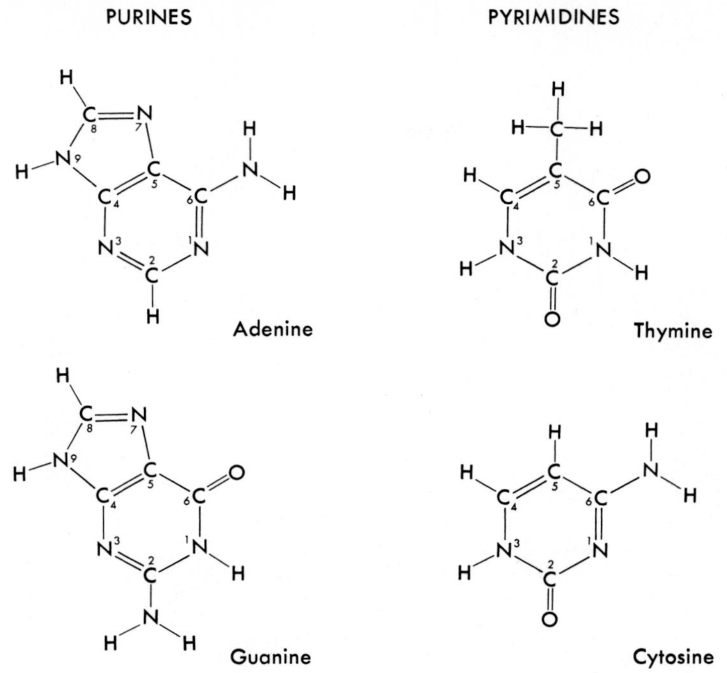

FIGURE 8–7. *Molecular structure of the four nitrogenous bases of DNA.*

8–9. The sugar–phosphate linked structure forms the so-called *backbone* of the molecule. The sequence of attached bases gives each molecule of nucleic acid its uniqueness.

The three-dimensional structure of DNA for which Watson, Crick, and Wilkins received the Nobel Prize was based on X-ray diffraction and biochemical evidence. When crystals of a molecule are exposed to X-rays, the rays are scattered according to the structure of the molecule. The pattern of scatter is unique to each type of molecule. Figure 8–10 is a picture of DNA produced through X-ray crystallography. This kind of data on DNA indicated that it is a helical molecule with a regularly repeating pattern.

FIGURE 8–8. *The four deoxyribonucleotides: (A) deoxyadenylic acid, (B) deoxythymidylic acid, (C) deoxycytidylic acid, (D) deoxyguanylic acid.*

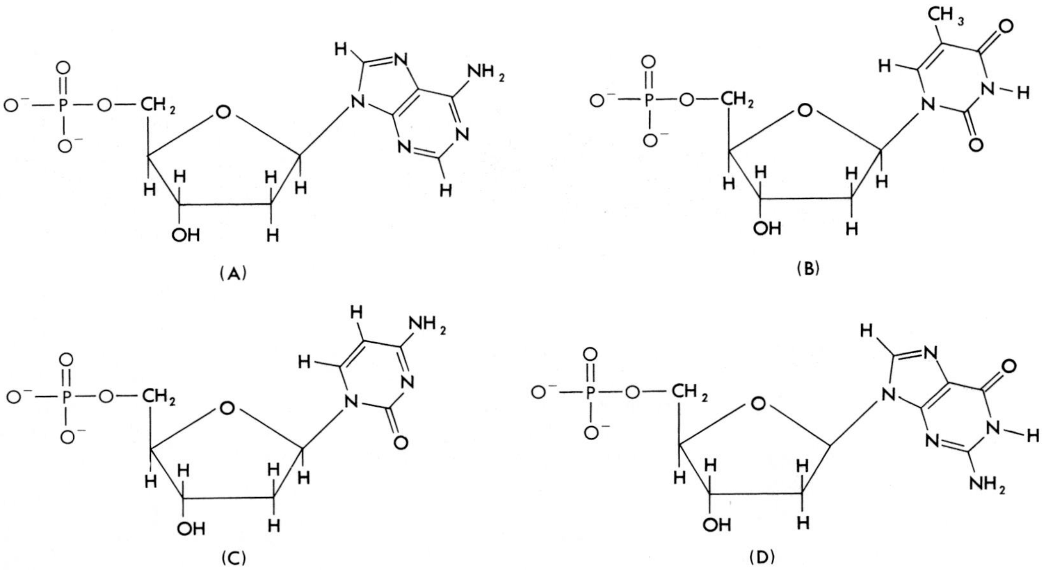

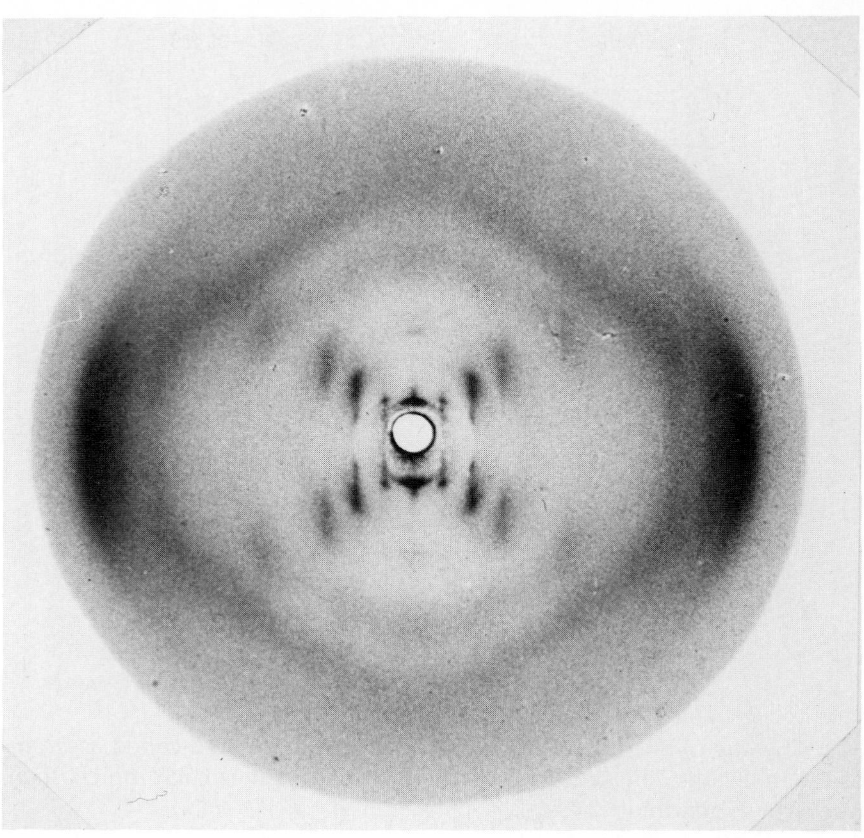

FIGURE 8–10. *X-ray diffraction pattern of DNA. It was determined from this data that DNA was a helical molecule with a regularly repeating pattern.* (With permission of the Biophysics Department, King's College, London University.)

FIGURE 8–9. *Polynucleotide chain showing the phosphodiester bonding between adjacent sugars of the backbone of the molecule. This is called a phosphodiester bond because phosphoric acid has been joined to two alcohols (OH groups of sugars) by ester linkages on both sides.*

The biochemical evidence came from the work of Erwin Chargaff in the late 1940s. He found that regardless of the source of the DNA, the concentration of adenine always equaled the concentration of thymine, and the concentration of guanine always equaled the concentration of cytosine. Also, the A/T ratio or the G/C ratio always approached unity. A list of these values for DNA from several sources is found in Table 8–1.

TABLE 8–1. **Comparison of Nucleotide Composition of DNA**

Source	*A*	*T*	*G*	*C*	*A*/*T*	*G*/*C*	(*A* + *T*)/(*G* + *C*)
Human sperm	31.0	31.5	19.1	18.4	0.98	1.03	1.67
Salmon sperm	29.7	29.1	20.8	20.4	1.02	1.02	1.43
Euglena nucleus	22.6	24.4	27.7	25.8	0.93	1.07	0.88
Euglena chloroplast	38.2	38.1	12.3	11.3	1.00	1.09	3.23
Escherichia coli	26.1	23.9	24.9	25.1	1.09	0.99	1.00
Mycobacterium tuberculosis	15.1	14.6	34.9	35.4	1.03	0.98	0.42
Phage T2	32.6	32.6	18.2	16.6*	1.00	1.09	1.87
Phage ϕX174	24.7	32.7	24.1	18.5	0.75	1.30	1.35
Drosophila	27.3	27.6	22.5	22.5	0.99	1.00	1.22
Corn (*Zea*)	25.6	25.3	24.5	24.6	1.01	1.00	1.04

*5-hydroxymethyl cytosine.

The DNA of phage ϕX174 is single-stranded (except during its replicative phase), hence its values for A/T and G/C depart considerably from unity. In fact, whenever this kind of deviation is found it is indicative of single-strandedness. Note also that values for (A + T)/(G + C) vary widely from well below to well above 1; that is, although the relationships A = T and C = G are valid, it is also true that A + T ≠ C + G in most cases. Certainly the structure of DNA does not require equality in that relationship. In fact, *DNA of different species is distinguished in large measure by the relative numbers of AT and CG pairs, their sequence, whether these occur as AT or TA and as CG or GC, and the number of such base pairs* (hence the length of the DNA molecules).

The biochemical evidence suggested to Watson and Crick that the bases were paired to the inside of the molecule, and that the structure of DNA best fit a model made up of *two* polynucleotide strands joined together by the paired bases in the form of a helix. The basic structure is therefore a long molecule of high molecular weight composed of two polynucleotide strands being held together by the paired bases in the form of a helix. The molecule may be compared to a twisted ladder in the form of a helix as shown in Figure 8–11.

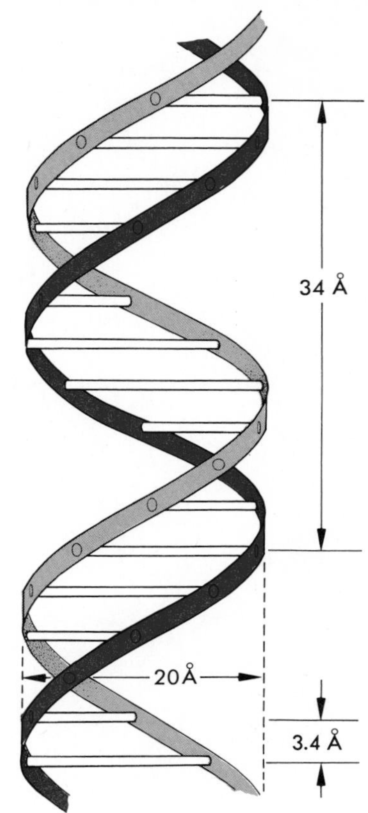

Figure 8–11. *The general structure of double-stranded deoxyribonucleic acid. (1 A = 0.0001 μm or 0.10 nm)*

The purine and pyrimidine bases are spaced 3.4 angstrom units (0.34 nm or 3.4 × 10^{-4} μm) apart, which gives 10 base pairs per complete turn of the sugar–phosphate backbone. This results in 34 angstroms per each complete turn of the helix. Each pair of bases is oriented 36 degrees clockwise from the preceding pair. The diameter of the molecule is about 20 angstrom units as measured from the X-ray diffraction studies. This corresponds exactly to the distance occupied by one base pair inside the two sugar–phosphate backbones.

The double helix is able to assume a number of forms. The most biologically important form is the **B form.** This was the form used in the early X-ray diffraction studies. It is more hydrated than the **A form.** The A form is more compact, with 11 base pairs per turn of the helix, and it is 23 angstroms in diameter. The bases are tilted more in relation to the axis of the helix than in the B form. The A form may only occur under experimental conditions. Possibly in equilibrium with the B form in cells, however, is **Z-DNA** (Figure 8–12).

In solutions of high ionic strength, such as 2 M NaCl, DNA is in a left-handed helical form. This left-handed structure is called Z-DNA because of its zigzag structure. In this form, the molecule still consists of two antiparallel chains, but otherwise it is quite different from the A or B forms. The helix is 18 angstroms in diameter, and there are 12 base pairs per turn. If the Z form exists in cells in equilibrium with the B form, one must wonder about its biological role. The binding of proteins at localized areas on the molecule could produce an ionic environment high enough to cause the transition. Regulatory proteins bound to DNA might be able to do this. Therefore, at least localized areas of Z-DNA are believed to occur within the cellular DNA, and there is evidence that Z-DNA is present in *Drosophila* chromosomes.

Typically, only the four different nitrogenous bases occur in DNA. In the double-stranded molecule adenine is connected to thymine by two hydrogen bonds, and cytosine is connected to guanine by three hydrogen bonds. This explains the data of Chargaff, and is called **complementary base pairing.** As a result, if the sequence of bases along one strand of the molecule is known, the complementary sequence on the opposite strand will be automatically known.

Techniques involving enzymatic splitting of the DNA molecule have made it possible to determine the exact arrangement of its parts. The nitrogenous bases are linked to deoxyribose at the 1′ carbon (forming

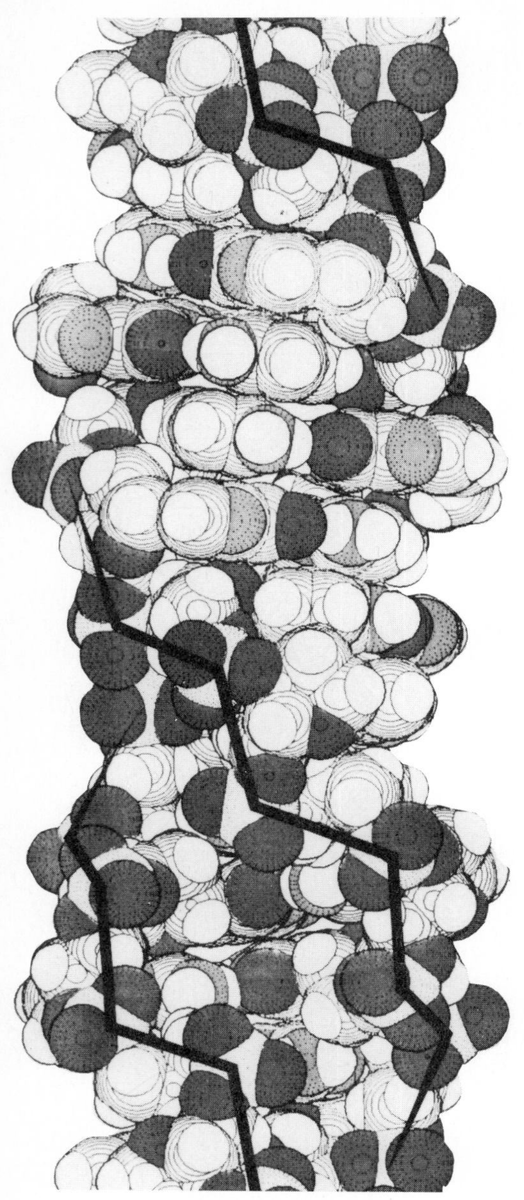

(*a*) Z DNA

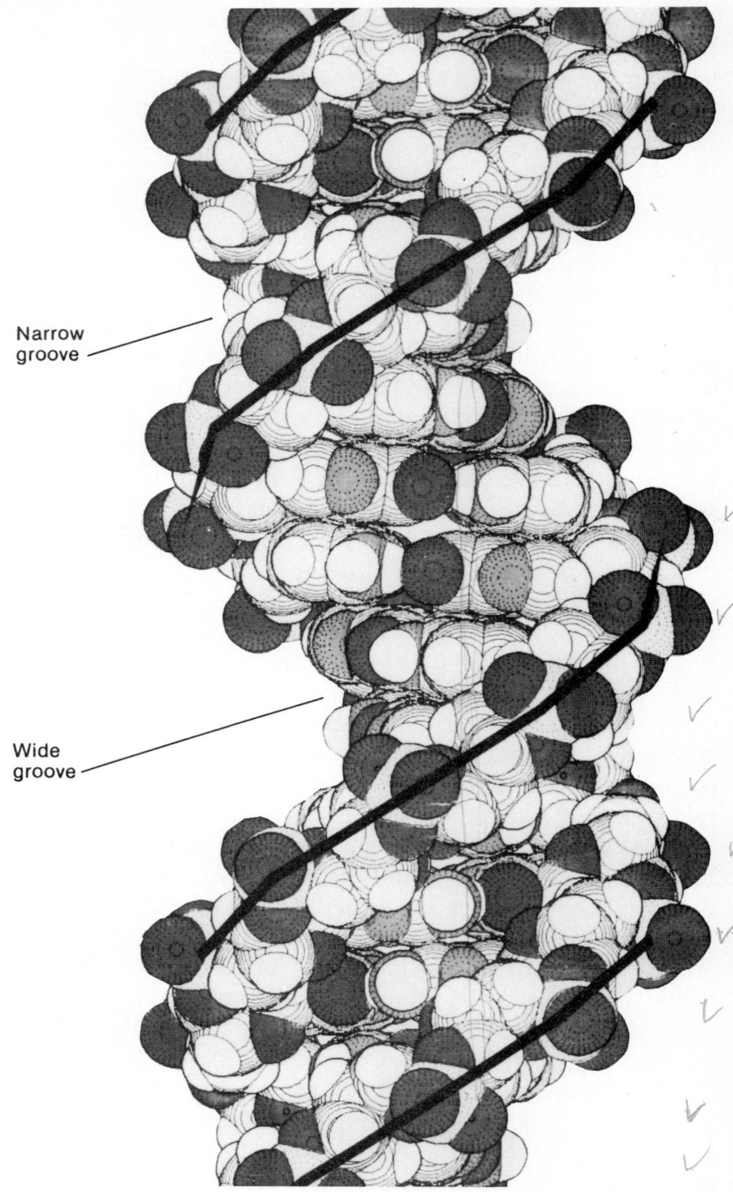

(*b*) B DNA

FIGURE 8–12. *Left-handed helix of Z form of DNA, compared to right-handed B form.*

nucleosides), and the phosphate is attached to the 5′ carbon (forming nucleotides), as shown in Figure 8–9. Successive nucleotides are linked by 3′, 5′ phosphodiester bonds; the structure of a segment of DNA consisting of four deoxyribonucleotide pairs is shown in Figure 8–13. The 3′ and 5′ hydroxyl groups of two different deoxyribose molecules form a double ester with the PO_4 groups as seen in Figure 8–13.

The two sugar–phosphate strands are oriented in opposite directions. In reading the *left* strand from top to bottom the sugar–phosphate linkages are 5′ → 3′, whereas in the *right* strand they are 3′ → 5′. The two "backbone" strands are thus described as being *antiparallel*. The double helical nature of the molecule would not be possible without this antiparallel orientation of the strands. It was determined by Watson that hydrogen bonding between nucleotide pairs gives measurements very similar for AT and GC pairs (Figure 8–14). Even in a molecule such as this, with only the four "usual" bases, the possible kind and number of sequences are virtually infinite. The DNA of different species is, in

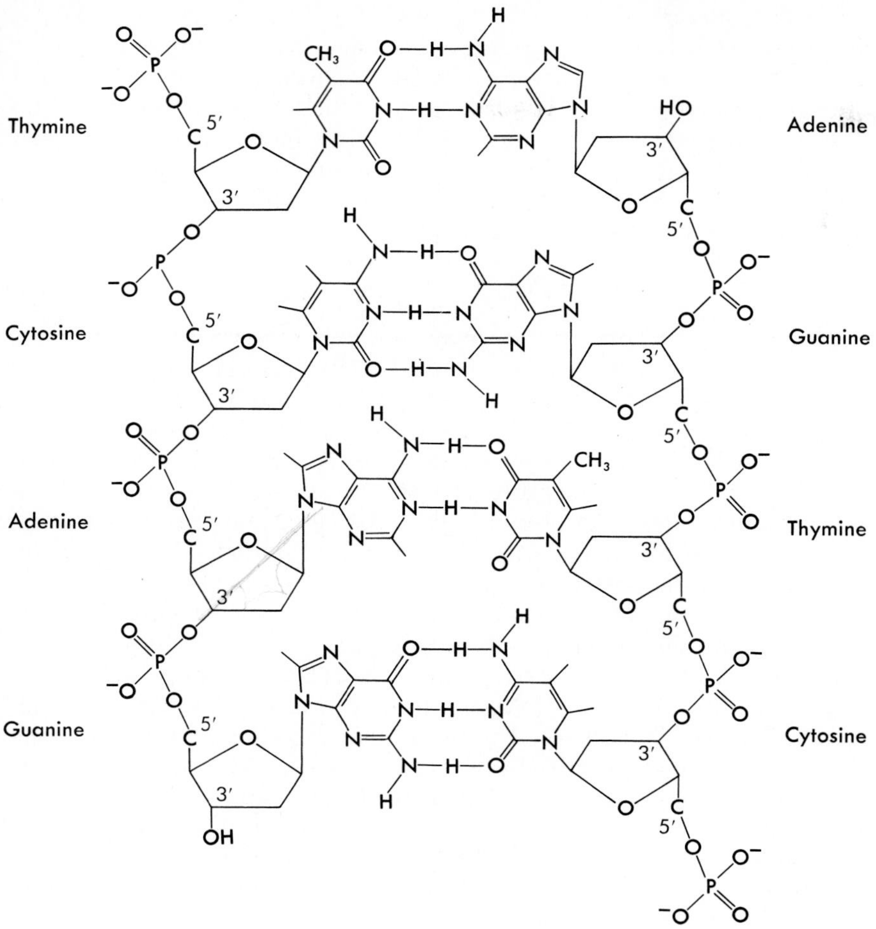

Thymine
Adenine
Cytosine
Guanine
Adenine
Thymine
Guanine
Cytosine
5′
3′
HO
OH
CH₃

FIGURE 8–13. *Molecular model of a four nucleotide pair segment of DNA. For simplicity the helical nature of the molecule is not shown. Note antiparallel orientation of the sugar–phosphate strands.*

Adenine
Thymine
0.285 nm
0.29 nm
51°
1′—C
of sugar
1.11 nm
C—1′
of sugar
Cytosine
Guanine
0.283 nm
0.286 nm
0.284 nm
1.08 nm

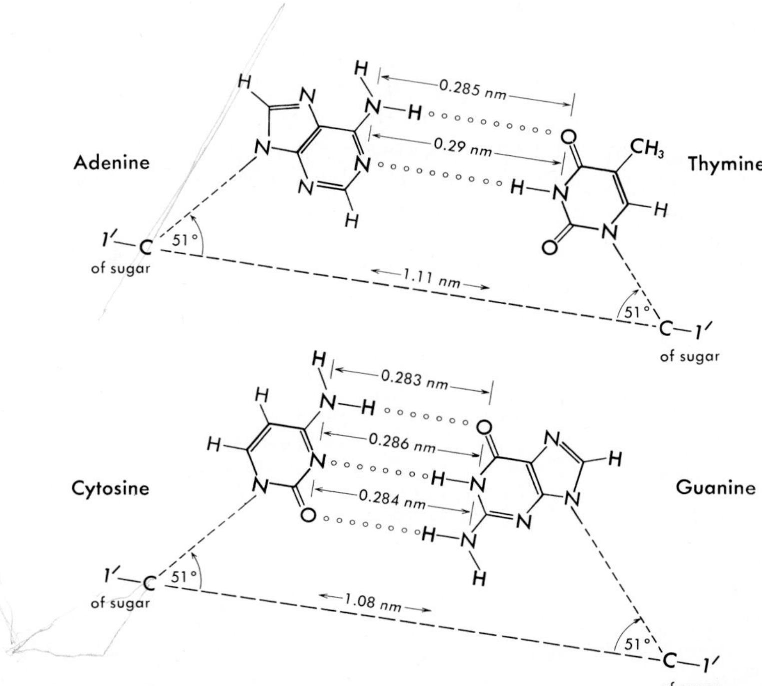

FIGURE 8–14. *Details of hydrogen bonding between deoxyribonucleotide pairs. Note the close similarity of measurements of AT and GC pairs. (0.1 nm = 1 A).*

fact, distinctive through such sequence differences in nucleotide pairs. The DNA of some phages (for example, φX174) is single-stranded. It does, however, become temporarily double-stranded after infection of a host cell as a preliminary to replication. In some DNA molecules, other, rarer bases regularly replace some of the four common ones. The T-even phages, for instance, contain 5-hydroxymethyl cytosine instead of cytosine, but in an amount equal to the guanine content, thus indicating its pairing qualities. Other similar substitutions are known, but the exact role of these rare bases is only partly understood.

LOCATION OF DNA IN CELLS

As was pointed out earlier, the Feulgen stain technique is specific for DNA; any cellular structure containing DNA retains a purple color. Not only is this process specific for DNA but the intensity of staining is directly proportional to the amount of deoxyribonucleic acid present. Measurements of light absorption by structures colored by the Feulgen process can be used to determine the relative amounts of this nucleic acid present. Such microspectrophotometric techniques also employ ultraviolet light, the peak absorption near 260 nm, which is very near its most effective mutagenic wavelength (254 nm). Such techniques, applied to many different kinds of eukaryotic cells, show DNA to be almost entirely restricted to the chromosomes. (The only exception is that small amounts of cellular DNA are located in the chloroplasts and mitochondria; see Chapter 16). The physical arrangements of DNA in the chromosomes of eukaryotes have received intensive attention in recent years, and our knowledge has been considerably enhanced. It has been known for some time that the amount of DNA per cell nucleus, and per bacterial cell, is many times the volume of the containing structures so that when "spilled out" the DNA occupies a larger volume of space (Table 8–2 and Figure 8–15).

Obviously, DNA, which behaves as a linear molecule in crossing-over and bears genes in apparent linear sequence, cannot simply extend in a straight line from one end of a eukaryotic chromosome or bacterial cell to the other. However, it has been demonstrated in a few cases and is generally accepted that each eukaryotic chromatid consists of a single continuous DNA molecule. From a physical standpoint, the great length of DNA molecules relative to that of the structures containing them imposes stringent requirements on "packaging." When chromatin is spread in low-salt buffers for electron microscopy, at regular intervals of about 100 angstroms, nucleoprotein units occur that appear as "beads" on a string. These units are **nucleosomes,** and are shown in

TABLE 8–2. Approximate Quantitative Characteristics of the DNA of Some Organisms and a Phage

Organism	*Approximate molecular weight, DNA*	*Number of deoxyribonucleotide pairs*	*Length, μm*	*Length, cm*
Phage T2	1.3×10^8	2×10^5	68	0.008
Escherichia coli	2.8×10^9	4.3×10^6	1,465	0.15
Drosophila melanogaster (diploid cell)	2.2×10^{11}	3.38×10^8	1.15×10^5	11.43
Human (2n)	3.9×10^{12}	6×10^9	2.04×10^6	204

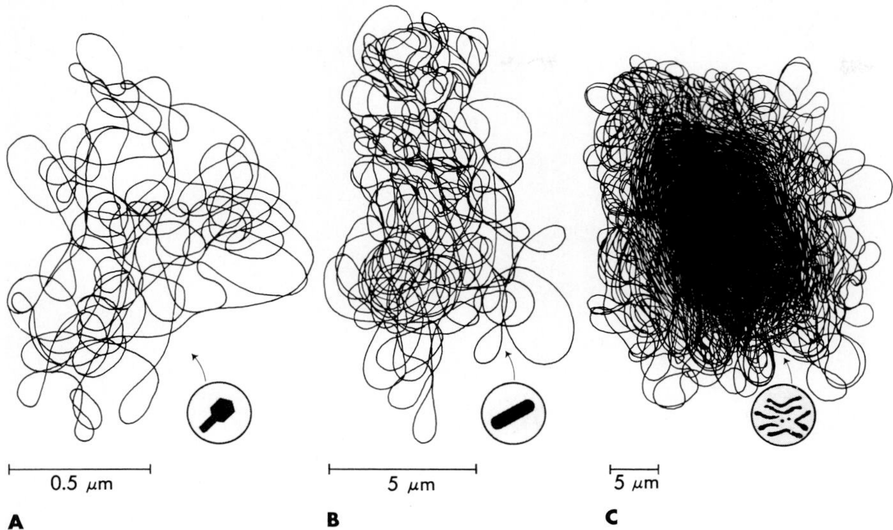

FIGURE 8–15. *Total length of DNA of (A) bacteriophage T-4 (about 68 μm); (B) the bacterium* Escherichia coli *(about 1,100 μm); (C) the fruit fly* Drosophila melanogaster *(about 16,000 μm). Compare these lengths to the size of the structures into which the DNA is packed.* (Redrawn from F. W. Stahl, *The Mechanics of Inheritance.* Copyright 1964, Prentice-Hall, Inc. Englewood Cliffs, NJ, by permission of author and publisher.)

Figure 8–16A. This organization of eukaryotic DNA into nucleosomes constitutes the first level in a stepwise packaging of the continuous double helix of DNA into chromatin and ultimately into metaphase chromosomes.

The three-dimensional shape of the current nucleosome model (Figure 8–16B) is an ellipsoid, with a length of 110 angstroms and a diameter of 65 to 70 angstroms. The shape is somewhat similar to that of a rugby ball. Nucleosome cores consist of an octamer of two molecules each of four histones, designated as H2A, H2B, H3, and H4. The surface of the core particle is surrounded by a superhelical strand of DNA of about 165 base pairs wrapped in two turns.

Internucleosomal DNA, or **linker DNA,** can have a linear measurement of about 50 to 340 angstroms, depending on the cell type, tissue, and species. A fifth kind of histone, H1, is located on the linker DNA between the nucleosome particles. H1 is larger than any of the core histones, and much more variable in component amino acids. It appears to have been less rigorously conserved in evolution. It is believed that the H1 histones play a role in the next level of packaging by having the nucleosome-packaged chromatin fiber twisted into a helix held together by the H1 histones. Even though the nucleosome plays a major role in the first level of packaging of DNA, it allows each DNA molecule to behave genetically as if it were continuous within the chromosomal structure. The structure as described herein is the working model at present, and there is not uniform agreement as yet as to the exact three-dimensional structure of the nucleosome particle.

In addition, **supercoiling** of DNA is probably the natural form of DNA in living cells and must also contribute to the packaging of the DNA in the cell. When a helix forms a circle and the ends fuse, it can lie flat on a plane. However, if the double helix is untwisted several turns before the ends join, it tries to resume the normal twist. Since the ends are joined, it cannot do that without the double helix twisting

FIGURE 8–16. *(A) Electron micrograph of chicken erythrocyte nucleus spread showing nucleosomes (arrows) and spacer DNA connecting them. (B) Current model of nucleosome particle. The DNA wraps 1¾ turns around the histone octomer core similar to a spring.*

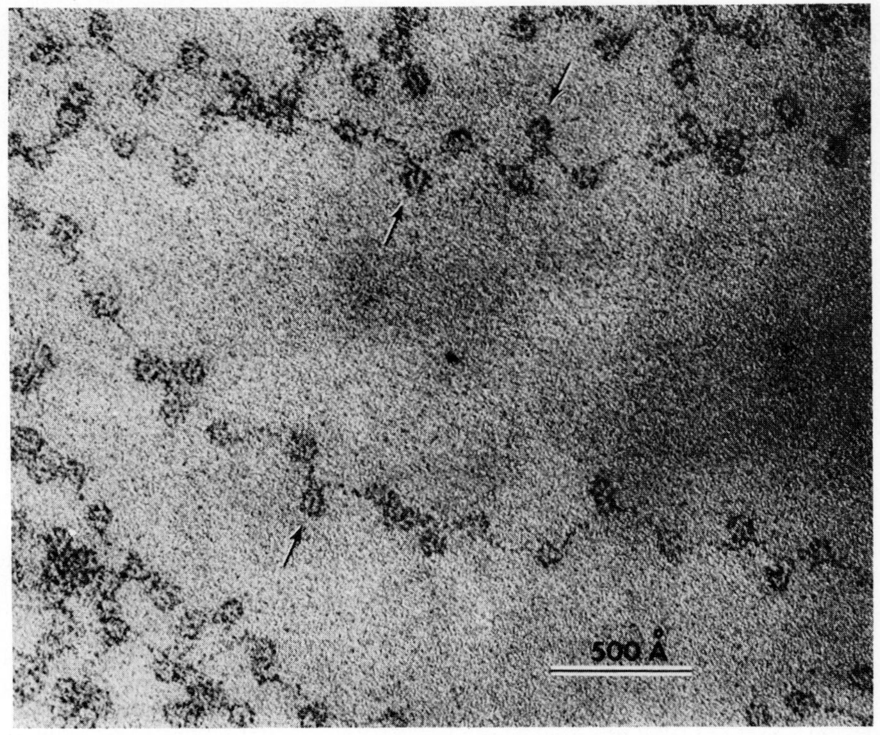

A

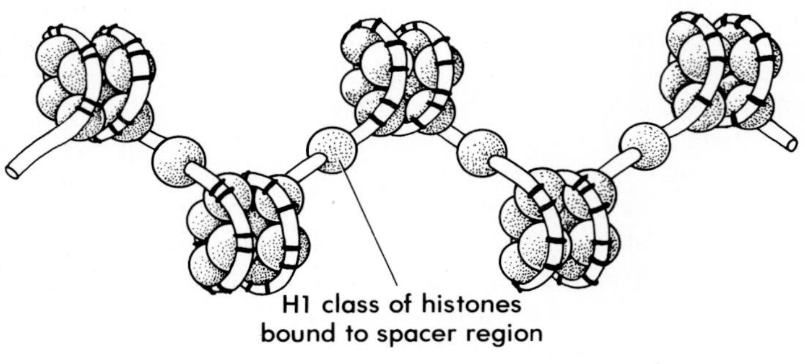

B

around itself and forming a superhelix with loops at the two ends. Figure 8–17 illustrates this process. Supercoiling also occurs with linear molecules when loops of DNA are twisted around each other. In addition to nucleosome packaging, supercoiling provides a powerful method of regulating the activity of genes by preventing regulatory elements from reacting with the DNA in specific regions.

Electron microscopy, particularly scanning electron microscopy, does not clearly show the physical arrangement of DNA in eukaryotic chromosomes. All that can be seen in Figure 8–18 in the micrograph of a metaphase chromosome of the Chinese hamster is the highly coiled chromatin (nucleoprotein) fibers or *microconvules*. In general, these fibers range from about 50 to 500 angstroms, depending somewhat on the species, but largely on the technique used in preparation. Interestingly enough, the centromeric region appears as a constriction in each chromatid but displays no obvious structural element that might be correlated with the kinetochore (compare with Figure 4–9). Figure 8–19 is a

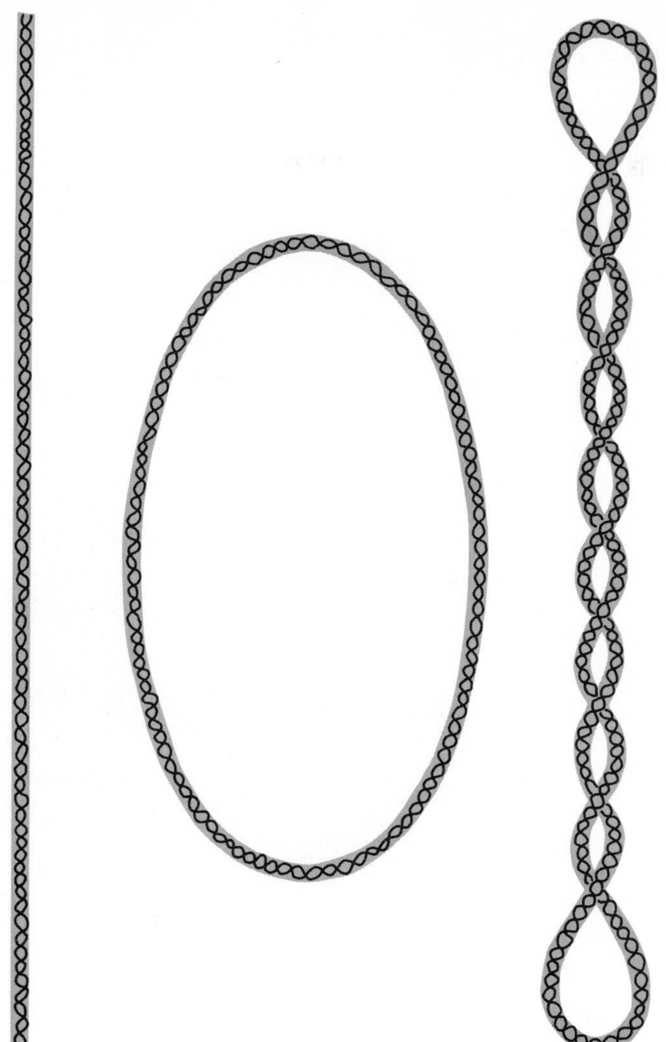

FIGURE 8–17. *DNA molecules in a linear form (left), relaxed circular form (middle), and supercoiled form (right).*

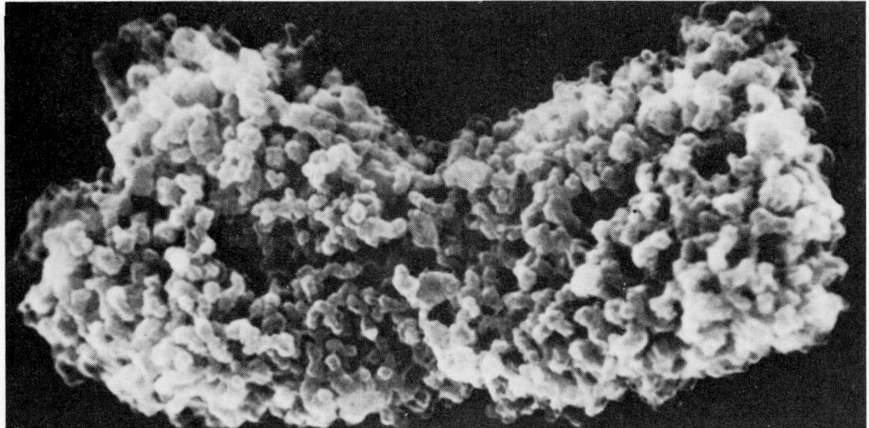

FIGURE 8–18. *Scanning electron micrograph of an isolated metaphase chromosome showing the highly coiled chromatin fibers (microconvules) that comprise the chromatids, 28,000X. The bar (upper right) represents 0.5 μm.* (Micrograph kindly supplied by Dr. Wayne Wray. From M. L. Mace, Jr. et al. 1977, used by permission.)

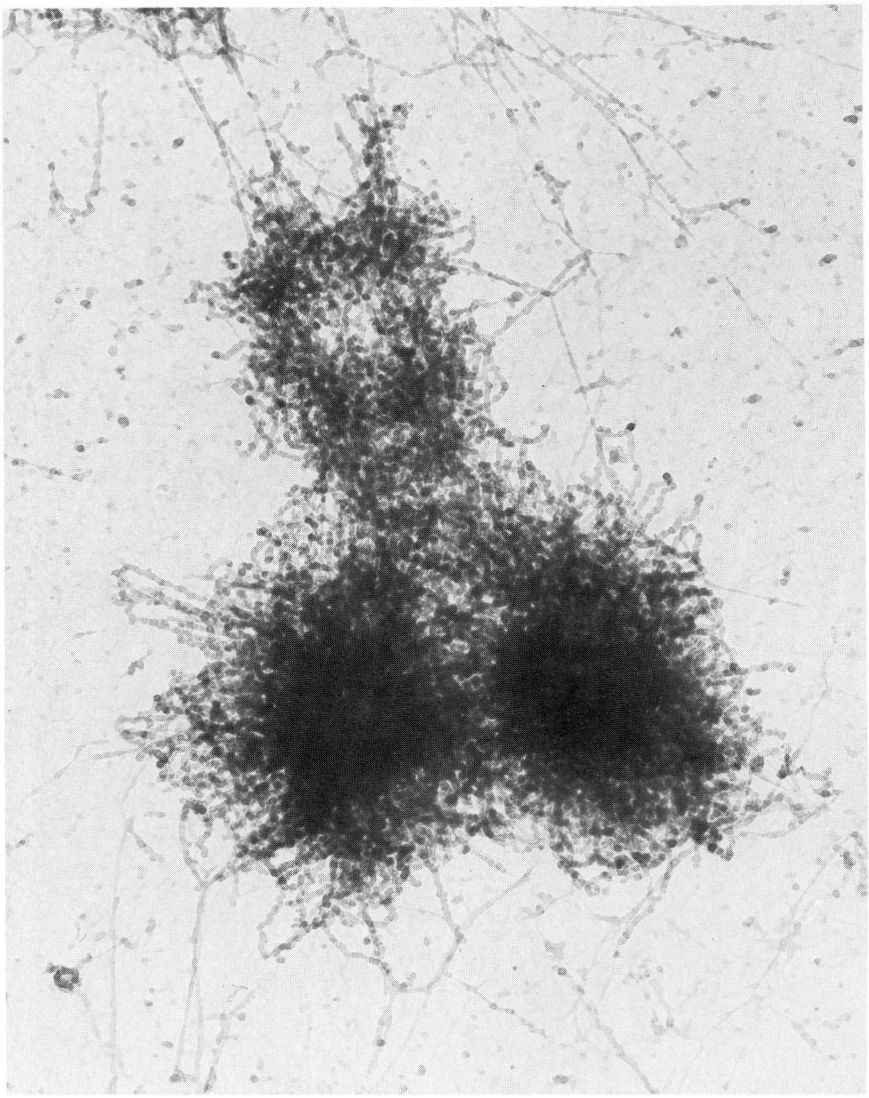

FIGURE 8–19. *One of the G group of human chromosomes. Satellites and their extensions, appearing like horns, are at the upper portion of the chromosome, just above the centromeric region. Transcription takes place with great precision in this tangled mass of nucleoprotein fibers.* (Scanning electron micrograph courtesy Dr. F. Bahr. From G. F. Bahr, 1977. "Chromosomes and Chromatin Structure." In J. J. Yunis, *Molecular Structure of Human Chromosomes,* published by Academic Press, New York. Used by permission of author and publisher.)

transmission electron micrograph of one of the small G-group human chromosomes. Again the seemingly tangled network of nucleoprotein fibers is evident.

On the other hand, the bacterial "chromosome" is unequivocally a single, long DNA molecule in the form of a "naked" (that is, not histone-complexed), continuous, closed, no-end structure. That it is highly folded and coiled can be determined from electron microscopy, as well as from the quantitative relationship of the bacterial DNA molecule and the cell that contains it. For example, an *E. coli* cell is about 2 μm in length, whereas its DNA molecule is 700 times longer. Moreover, a bacterial cell may contain more than one of these DNA molecules, depending on its stage in the life cycle.

Cyclic Quantitative Changes in DNA Content

By appropriate quantitative measurements of DNA content of cells, a number of striking parallels with chromosome behavior and our earlier postulates regarding genes can be seen:

SEQUENCING THE HUMAN GENOME

In the summer of 1986, a controversial scientific plan was proposed to embark on a massive project to determine the complete nucleotide sequence of the human genome. Proponents of the plan feel that knowing the complete nucleotide sequence of the human genome would provide a "tool for the investigation of every aspect of human function." Very simply, knowing the entire human genome sequence would allow study on almost any human gene and would allow increased speed in the search for human disease genes.

Many have questioned the scientific value of knowing the complete human nucleotide sequence. Most estimates predict that with present technologies the 3 billion nucleotide sequence could be determined for about $1 per nucleotide or $3 billion. Support for this massive project is decreasing because of the cost, and the project seems doomed until technological advances can make it cheaper and faster. Also, the U.S. Congress would have to be convinced that this is a worthwhile endeavor so that the money would not be taken from other areas of science. And, finally, no one can agree on which government agency should supervise this massive project.

Presently, the idea of complete sequencing of the human genome has given way to the need for a detailed physical map of the genome. A physical map would consist of a series of overlapping DNA fragments of approximately 40,000 base pairs. With such a map, the location of a specific gene could be associated with a particular fragment, and then the nucleotide sequence of the fragment would be determined in order to study that gene. The Japanese have developed a highly automated sequencing effort that, by 1990, will be able to determine a sequence 1 million bases long in a single day. Their objective is to sequence the smallest human chromosome (number 21) that has 48 million base pairs in a period of five months.

Work is also proceeding on the standard genetic map. Over 1,000 genes have already been mapped (see Table 6–2); however, it is estimated that the number of mapped genes would have to be increased at least to 3,500 in order for the resolution to be useful. The cost of this effort has been estimated to be at least $50 million. The objective is to have genetic and physical maps of increasing quality, with a detailed map by the late 1990s. A report from the National Academy of Science has recommended that the first goals should be determining the genetic and physical maps before undertaking extensive sequencing of the entire genome. The report also suggested that another priority should be the development of faster, more automated methods of sequencing.

1. The quantity of DNA detected in gametes is, within limits of experimental error, half that of diploid meiocytes.
2. Zygotes contain twice the amount of DNA in gametes.
3. The amount of DNA increases during interphase by a factor of 2. Telophase nuclei have half the DNA content of late-interphase (G_2) or early-prophase nuclei.
4. DNA content of polyploid nuclei is proportional to the number of sets of chromosomes present; that is, tetraploid nuclei can be shown to have twice as much DNA as diploid cells.

Sequencing of DNA Molecules

It should be emphasized that in the discussion so far the most important feature about DNA is the sequence of base pairs in a given molecule. This determines its genetic function as well as its uniqueness. It is therefore very important to be able to determine the sequence of bases in specific regions or even entire genes. Initially, this was very difficult, and it was possible to determine the base sequence of only a very short piece of DNA (6–10 base pairs long). More recently, however, technical developments have allowed DNA sequencing to become a standard laboratory procedure. Sequencing can even be an automated process. In 1980, F. Sanger and W. Gilbert received a share of the Nobel Prize

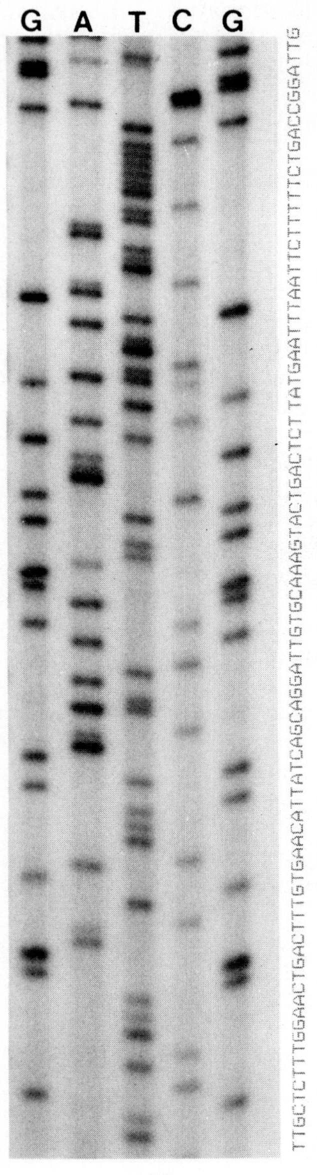

(B)

3′ ——— AACAGTCAT ——— 5′

DNA Polymerase

3²P-ATP GTP CTP TTP + ddttp	3²P-ATP GTP CTP + ddCTP TTP	3²P-ATP GTP + ddGTP CTP TTP	3²P-ATP + ddATP GTP CTP TTP
T TT TTGT TTGTCAGT	TTGTC	TTG TTGTCAG	TTGTCA TTGTCAGTA

	T	C	G	A	Sequence
9				—	A
8	—				T
7			—		G
6				—	A
5		—			C
4	—				T
3			—		G
2	—				T
1	—				T

(A)

FIGURE 8–20. *(A) Steps in sequencing DNA molecules by the Sanger dideoxy method (see text for description). (B) Polyacrylamide sequencing gel showing sequence of a region downstream from the soybean seed lectin gene. The lanes correspond to each of the four bases, with G used as a marker on both end lanes. The correct sequence is read up from the bottom of the gel and is given on the right.* (Photo courtesy Dr. Lila Vodkin, Department of Agronomy, University of Illinois.)

in chemistry for their pioneering work on the development of the techniques for sequencing nucleic acids. DNA sequences are determined in short single-stranded blocks of 300 or more nucleotides with longer sequences being built up from the shorter overlapping sequences. As an example, the dideoxy method developed by Sanger will be discussed. The steps in this method are illustrated in Figure 8–20A.

Fragments of DNA are produced by restriction enzymes (see Chapter 20) and the ends labeled with ^{32}P ATP. The complementary strands are separated from each other and a short primer is added to the single-stranded molecule. The procedure is then to incubate the primed single-stranded molecule with DNA polymerase (the DNA synthetic enzyme; see next section), the four nucleoside triphosphates—ATP, GTP, TTP, and CTP—plus an additional form of one of the bases, a dideoxynucleotide triphosphate. The dideoxynucleotide triphosphate has H atoms at both the 2′ and 3′ carbons of the sugar:

PPPO $-CH_2$ — [sugar ring: O, Base, H, H]

The H atoms are incorporated into DNA normally, but since a 3′ OH group is required to complete the bond with the phosphate of the next nucleotide, and none is present in a dideoxynucleoside triphosphate, the synthesis of the chain is terminated at that point. The result is that at the end of the incubation there is a mixture of different-size fragments, each terminating with a dideoxynucleotide triphosphate. The experiment is repeated four times, each with a different dideoxynucleoside triphosphate. Within each experiment synthesis is always terminated at the same base, the one present in the dideoxy form. The synthesized fragments from the four experiments are then separated by parallel electrophoresis on an acrylamide gel under denaturing conditions. This separates the various fragments by size, with the smallest migrating the farthest in the gel. Fragments differing by only one nucleotide are separable by this technique. The actual sequence is then read from the bottom to the top of the gel. A typical sequencing gel is shown in Figure 8–20B. Sequences of 300 base pairs or more can be read from a single gel. Longer sequences are established by building up overlapping shorter sequences. Presently this sequence information is entered into a computer for actual matching of long overlapping sequences. Sequencing techniques are an essential step in modern approaches to genetic manipulation as well as a requirement for understanding gene regulation and expression.

REPLICATION OF DNA

The structure of DNA as established by Watson and Crick led them to propose a model for replication of the molecule. They suggested that the hydrogen bonds between the base pairs would be broken, and the two polynucleotide chains could unwind. Each chain would then act as a template for the formation of another complementary chain. When the hydrogen bonds reformed between the old and new strands, the result would be two identical molecules of DNA. This pattern of replication is described as **semiconservative replication.** Each new molecule consists of one strand from the old parent molecule and one strand that is newly synthesized. This pattern of replication has been experimentally verified by several methods.

Semiconservative Nature of DNA Replication

Under the strand-separation hypothesis, each newly replicated double helix must consist of one "old" strand and one "new" strand. By a series of ingenious experiments, Matthew Meselson and Franklin Stahl showed that this does indeed occur. Bacteria were cultured for some time in a medium containing only the heavy isotope of nitrogen, ^{15}N, until all the DNA were labeled with heavy nitrogen. These cells were next removed from the ^{15}N medium, washed, and transferred to a medium whose nitrogen was the usual isotope, ^{14}N. After one bacterial generation, a sample of cells was removed and the DNA was extracted. All the DNA that replicated once in the ^{14}N was "hybrid"; that is, it consisted of one strand whose nitrogen was ^{15}N ("old") and one strand whose nitrogen was ^{14}N ("new"), as suggested in Figure 8–21.

Differentiation between ^{14}N-DNA and ^{15}N-DNA is readily made by *density gradient centrifugation*. In this technique, particles of different densities (such as ^{14}N-DNA and ^{15}N-DNA) are suspended in a concentrated solution of the salt of the highly soluble heavy metal, cesium chloride, and subjected to intense centrifugation in an ultracentrifuge.

FIGURE 8–21. *The Meselson and Stahl experiment demonstrating the semiconservative replication of DNA. See text for details. Strands labeled with ^{15}N are darkly shaded; ^{14}N-labeled strands are lightly shaded.*

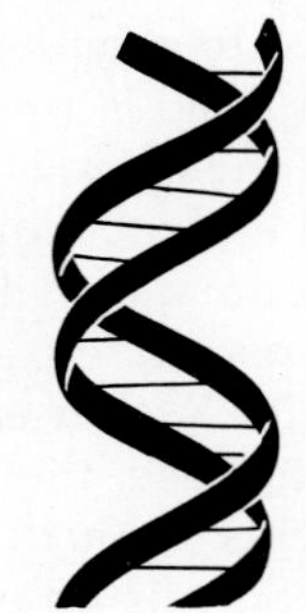

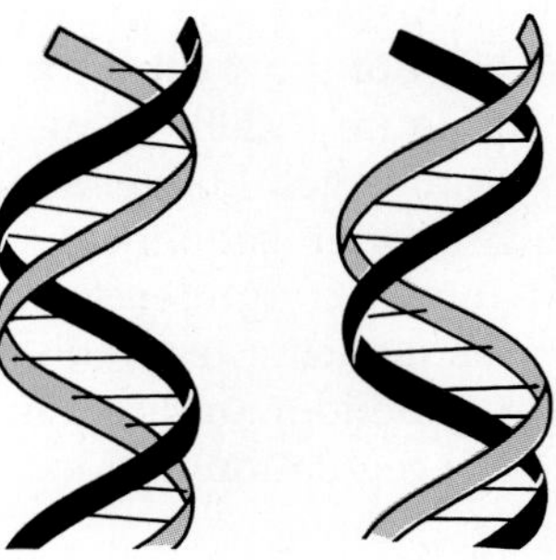

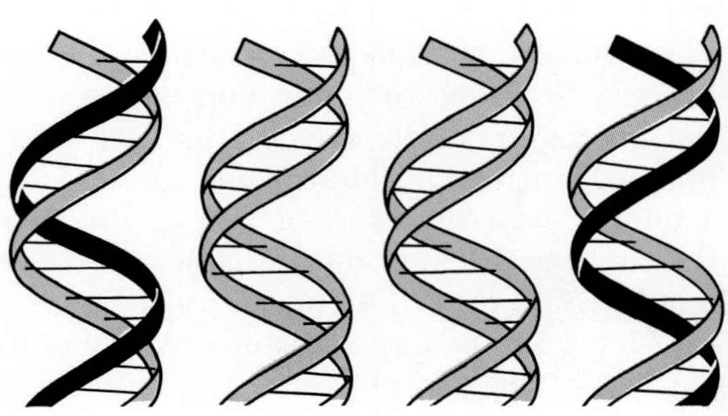

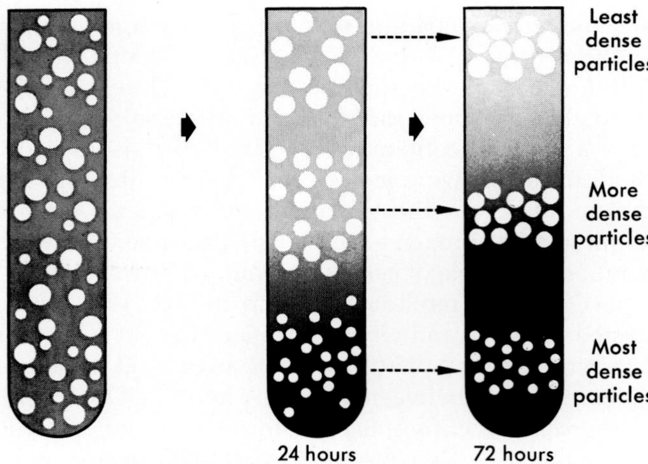

FIGURE 8–22. *The mechanics of density gradient centrifugation in a cesium chloride solution.* (Adapted from D. Fraser, *Viruses and Molecular Biology*, Macmillan, 1967. By permission.)

The centrifugal field produces a density gradient in the tube of both the salt and the suspended particles. After several hours the gradient reaches an equilibrium in which densities of the solution and suspended material match at some level in the tube and all the particles of like density collect in a narrow band (Figure 8–22). Here the particles will remain, subject only to their diffusibility. It is possible to separate particles differing in density by extremely small amounts with this technique.

When the hybrid DNA of Meselson and Stahl was ultracentrifuged in a cesium chloride solution, the DNA was found by ultraviolet absorption (260 nm) to have taken a position exactly intermediate between that previously established for ^{15}N-DNA and ^{14}N-DNA. After two cell generations, in which DNA has replicated a second time, half would be expected to contain only ^{14}N and half would be hybrid. Therefore, there should be two bands of DNA in the ultracentrifuge tube, one at the intermediate position, and one at the "all" ^{14}N position. This is precisely what was observed. This pattern of replication, in which each of the old strands serves as one of the two strands in each new DNA molecule, is referred to as *semiconservative replication*.

The results of these experiments are entirely consistent with the strand-separation theory (Figure 8–23). Meselson and Stahl eliminated any lengthwise extension of double-stranded DNA molecules by sepa-

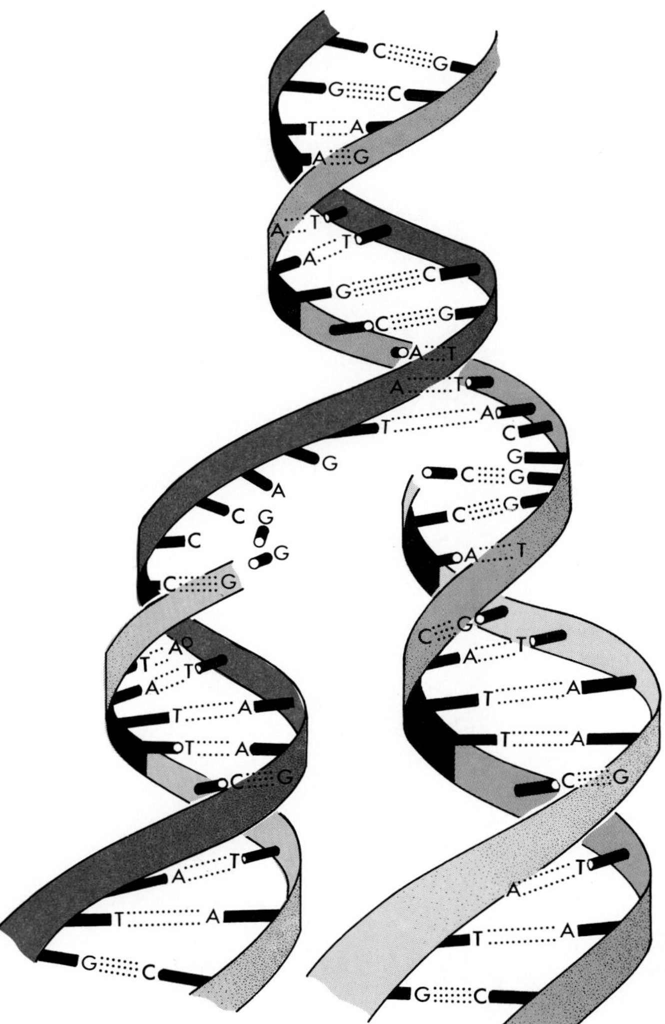

FIGURE 8–23. *DNA replication according to the strand separation theory where "old" and "new" duplexes are assumed to unwind and wind simultaneously as nucleotides are assimilated into newly developing strands.* (Redrawn from D. Fraser, *Viruses and Molecular Biology*, Macmillan, 1967. By permission.)

rating the newly synthesized strands and showing them to be pure ^{14}N. In addition, autoradiographs of DNA from *E. coli* that had been allowed to replicate in radioactive thymidine show that replicating DNA has a Y-shaped configuration even in the circular DNA molecule. Although the question of how the strands unwind to permit this type of replication is only partially answered, the nature of the triggering mechanism seems to relate to certain initiator nucleotide sequences. Considerable progress has been made in the last few years in elucidating the process of replication, and a virtually complete understanding may be imminent. The possibility of errors in the replication process (including "proofreading" mistakes) suggests a mechanism for mutation. This will be explored in Chapter 12.

The currently accepted model of DNA replication is not too unlike the original proposal of Watson and Crick, and conforms to the experimental results of Meselson and Stahl. The molecule unwinds in a localized area, thus forming replication "forks." The replication process is generally referred to as a *semidiscontinuous* process. Because the actual polymerizing enzyme can add nucleotides only in the 5′→3′ direction, synthesis in one strand (*leading strand*) is continuous in the 5′→3′ direction toward the fork from the point of initiation. In the other strand (*lagging strand*), as the fork opens, multiple sites of initiation are exposed. Synthesis then proceeds in short segments in the 5′→3′ direction, away from the fork. This process is shown in Figure 8–24, and discussed in more detail in the following sections.

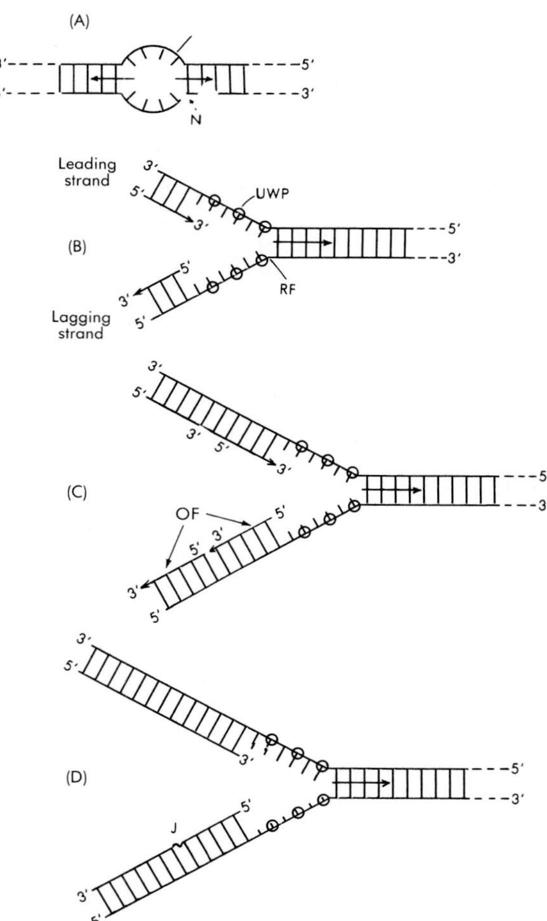

FIGURE 8–24. *Replication of DNA. (A) Replication bubble(s) (RB) forms at unique site(s) determined by a nick (N) produced in one sugar–phosphate strand by endonuclease. In many organisms replication bubbles "open up" replication forks (RF) bidirectionally (horizontal arrows). (B) Unwinding of template strands through action of unwinding proteins (UWP). Only one of the replication forks is shown here. (C) Semidiscontinuous synthesis of the daughter strands in their 5′ → 3′ direction through addition of new nucleotides at their 3′ ends, producing Okazaki fragments (OF). (D) Continued discontinuous synthesis of daughter strands; the Okazaki fragments are joined (at J) through action of DNA ligase. For greater simplicity, the RNA primers (see text) are not shown here.*

Initiation of replication in *E. coli*. In *E. coli* the continuous, no-end DNA molecule, or "chromosome," retains its circularity during replication. Replication involves an initial unwinding of the double helix to form two single-strand templates (Figure 8–24A). This is accomplished by an **endonuclease,** an enzyme that cuts ("nicks") one of the sugar–phosphate backbone strands at specific points. Next the weak hydrogen bonds that link purine–pyrimidine bases are broken, a process that is often referred to as "unzipping" of double-stranded DNA. The result is the formation of a *replication bubble,* which subsequently extends as a Y-shaped **replication fork** (Figure 8–24B). If this unzipping occurs anywhere except at the end of the molecule, replication becomes *bidirectional.*

Unwinding of template DNA. Before the template strands of DNA can be replicated, an unwinding must occur in order to produce greater lengths of single strands. This is accomplished by **unwinding** or **helix-destabilizing proteins,** 200 of which bind to single-stranded DNA in the vicinity of the replication fork, destabilizing the double helical portion of the molecule (Figure 8–24B). Each molecule of unwinding protein complexes with about 10 deoxyribonucleotides. About 2,000 unpaired nucleotides are thus available on each single-stranded arm of the replication fork for complementary pairing of the bases of the newly synthesized strand.

Priming. The DNA polymerases, which are responsible for adding new complementary deoxyribonucleotides to the template strand, cannot function without available 3′OH groups. Generation of 3′OH groups to which the DNA polymerases add new nucleotides is provided by synthesis of a short segment of primer ribonucleic acid (RNA), catalyzed by an RNA polymerase. The primer is synthesized as a copy of a short sequence of one strand of DNA and hydrogen bonded to the template strand. One of the DNA polymerases then adds DNA monomers to the RNA primer, which is later removed enzymatically, leaving a gap that is closed.

The DNA Polymerases

Three principal DNA polymerases, I, II, and III, are responsible for the template-directed condensation of new deoxyribonucleoside triphosphates in bacteria. Two inorganic phosphates are then split off (Figure 8–25), leaving deoxyribonucleoside monophosphates to make up the growing strand. Synthesis of the complementary strands proceeds from the 3′OH terminus of the RNA primer, and extends the newly synthesized DNA strand in its 5′→3′ direction.

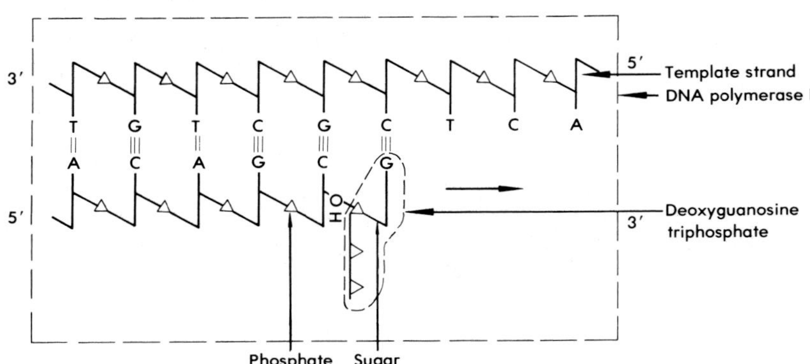

FIGURE 8–25. *Template-directed condensation of new deoxyribonucleoside triphosphates, as catalyzed by DNA polymerase I in* Escherichia coli. *See text for details.*

DNA polymerase I is a single polypeptide chain of a molecular weight of 109,000; about 400 molecules are present in each *E. coli* cell. This enzyme binds to single-stranded DNA and is largely reponsible for filling in gaps between small precursor oligonucleotides (a short polymer of 2–10 nucleotides). It also exerts a "proofreading" function in that although its polymerizing activity is in the 5′→3′ direction (by adding deoxyribonucleotide monomers in antiparallel fashion on the daughter strand), it also "edits" the daughter strand by removing incorrect, nonpairing nucleotides, due to a 3′ exonuclease capability. In addition, DNA polymerase I excises thymine dimers (damaged pairs of bases caused by ultraviolet radiation) in the 5′→3′ direction. In the same fashion, this exonuclease activity is able to digest the RNA primer in the 5′→3′ direction and, at the same time, fill in the gap with deoxyribonucleotides. Recent work, however, indicates that DNA polymerase I, while necessary for cell viability, is not indispensable for DNA *replication.*

DNA polymerase II is composed of a single polypeptide chain of a molecular weight of 120,000. It has the same polymerization capability as DNA polymerase I, as well as 3′→5′ exonuclease activity, but lacks a 5′→3′ exonuclease function. Moreover, it cannot utilize a nicked primer or one that is extensively single-stranded. Its functions are not completely known, although it does appear to bring about repair of ultraviolet lesions. There are about 100 molecules of this enzyme per *E. coli* cell.

DNA polymerase III holoenzyme (the functional enzyme complex) consists of seven individual subunits and has a total molecular weight of 379,000. About 10 molecules occur in each *E. coli* cell. This is the principal DNA-polymerizing or DNA-replicating enzyme in bacteria in that it catalyzes addition of deoxyribonucleosides to the RNA primer. It is distinguished from the other two DNA polymerases not only by its molecular structure but also by (1) inability to utilize single-stranded DNA, (2) lack of a 5′→3′ exonuclease activity, (3) greater instability, and (4) lower affinity for deoxyribonucleoside triphosphates. Some characteristics of the DNA polymerases are summarized in Table 8–3.

TABLE 8–3. Some Characteristics of the DNA Polymerases of *E. coli**

	DNA polymerases		
	I	*II*	*III*
Molecular weight	109,000	120,000	379,000
Structure	single chain	single chain	seven subunits
Molecules/cell	400	100	10
Product of gene†	pol A	pol B	dna E
Polymerization 5′→3′	+	+	+
Exonuclease 5′→3′	+	−	−
Exonuclease 3′→5′	+	+	+
Lability at 37°C	−	−	+
Nucleotides polymerized per min at 37°C	~1,000	~50	~15,000
Template-primer:			
Primed single strands‡	+	−	+
Nicked duplex	+	−	−
Polymer synthesis *de novo*	+	−	−
Affinity for triphosphate precursors	low	low	high

*Data from Kornberg (1980).
†See Figure 6–7 for map locations in *E. coli,* strain K-12.
‡A long single strand with a short complementary strand annealed with it.

Discontinuous synthesis and Okazaki fragments. The DNA polymerases add new deoxyribonucleotides in the 5′→3′ direction by addition at the 3′ end of the nascent complementary strand; that is, both daughter strands are extended only in the 5′→3′ direction (Figure 8–24C and D). This means that the lagging strand, and to a lesser extent the leading strand, is synthesized in discontinuous segments. The major evidence for discontinuous synthesis was obtained by Okazaki and his colleagues who showed that if replication was interrupted in progress, small fragments of 1,000 to 2,000 nucleotides could be isolated, compared to the result at completion when the isolated DNA was 20 to 50 times larger and contained no fragments. These short fragments of DNA are now called *Okazaki fragments.* They represent an intermediate stage of DNA replication and establish the principle of *discontinuous synthesis.* Because none of the polymerases can join a 3′ OH group to a 5′ phosphate group, as would be the case between these fragments, another enzyme called **DNA ligase** is required to join these groups of the adjacent base-paired deoxyribonucleotides together so that the daughter strand becomes continuous (Figure 8–24D).

In summary, DNA polymerase III is the major DNA-replicating enzyme in *E. coli,* catalyzing nucleotide addition in the 5′→3′ direction on the leading strand, as well as adding nucleotides to spaced primers on the lagging strand, again in the 5′→3′ direction. It also acts as a 3′→5′ exonuclease, hydrolyzing single-stranded DNA into 5′ mononucleotides. DNA polymerase I replaces polymerase III when replication encounters an RNA primer and replaces the RNA primer with DNA again in the 5′→3′ direction. DNA polymerase II can elongate primed DNA that is complexed with DNA-binding protein. When accompanied by certain other prerequisite conditions, DNA polymerase II or III will elongate primed DNA that is combined with elongation factors I and II.

Strand separation and unwinding of helical DNA molecules. Our model of DNA replication requires the separation of the two complementary strands. This cannot be accomplished by DNA polymerase III and only slightly by DNA polymerase I. Other classes of proteins have been identified, which generally function to unwind the strands and stabilize the replication fork. These proteins are called **helicases.** Additional proteins serve to stabilize the separated single strands and prevent reannealing so that replication can proceed. These proteins are called **helix-destabilizing proteins.**

One additional problem is caused by strand separation that requires either rotation of the single strands at the fork or rotation of the unreplicated double-stranded portion of the molecule. The structure of DNA makes this rotation difficult or impossible. Additional proteins have been identified that function to release this rotational tension in the molecule. A general class of enzymes called *DNA topoisomerases* perform this function by making and then repairing single- or double-stranded breaks in the molecule some distance from the replication fork, allowing

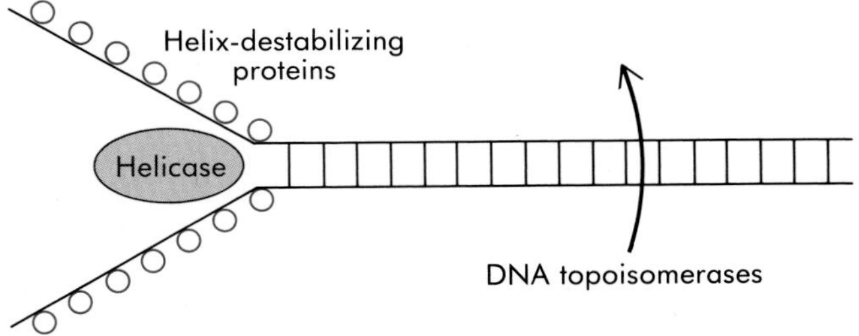

FIGURE 8–26. *Action of proteins that separate the double strands of DNA (helicases), stabilize the single strands (helix destabilizing proteins), and release rotational tension away from the replication fork (DNA topoisomerases).*

FIGURE 8–27. *Thetalike configuration assumed by replicating, circular, no-end, double helical DNA molecule of* Escherichia coli. *Arrows indicate the two replication forks. These have progressed bidirectionally from a single replication bubble.*

the tension caused by rotation to be released. This does not disrupt the continuity of the molecule. The overall action of these proteins is summarized in Figure 8–26.

Visualization of replicating bacterial DNA. Countless electron micrographs of replicating bacterial DNA have been published, giving objective evidence for the series of steps just outlined. In both *E. coli* and phage lambda (λ), a single replication bubble is formed, near gene O in lambda and near the *ilv* locus on the *E. coli* K-12 map (Figure 6–7). From the replication bubble the replication fork proceeds bidirectionally in both of these cases. In *E. coli,* for instance, these events result in the formation of an θ-like (thetalike) configuration as replication advances (Figure 8–27).

In bacterial conjugation and in intracellular multiplication of many viruses, a *rolling circle* model has been invoked. Here again, replication begins with a nick in one of the template strands, which thereby has a 5′ end and a 3′ end. Synthesis of a daughter strand begins by addition of complementary deoxyribonucleotides on the nicked template strand. As replication proceeds, the "unraveling" of the single linear strand is extended by the rotation of the closed (unnicked) strand about an imaginary axis; hence the term *rolling circle model* (Figure 8–28).

DNA Replication in Eukaryotes

In general terms, replication of eukaryotic DNA proceeds somewhat along the lines of the process in *E. coli.* At least three DNA polymerases are involved, designated as α, β, and γ. The first of these, DNA polymerase α, reportedly has a molecular weight of about 175,000 in humans; that of other sources (for example, calf thymus) has somewhat different characteristics. Its action is the main enzyme of DNA replication.

DNA polymerase β has a molecular weight of 30,000 to 45,000, depending on the animal source. It is widely distributed in multicellular animal species but has not been found in bacteria, plants, or protozoa. Even though its function remains unidentified, its conservation through 500 million years of biological evolution (from sponges to human beings) suggests an indispensable role.

DNA polymerase γ is found only in mitochondria. The eukaryotic polymerases seem to function just as the bacterial polymerases and require the same divalent cations (Mg^{++} or Mn^{++}).

The replicon. Since the DNA in each eukaryotic chromosome is substantially longer than the DNA in a virus or bacterium, replication does not simply start at one origin (either at one end or internally) and proceed outward. Instead, replication is initiated at several origins along the DNA molecule and proceeds bidirectionally away from the origin. The sequentially replicating segments thus formed each constitute a *replicon* or replication unit. Figure 8–29A depicts diagrammatically this mode of bidirectional replication in two replicons, and Figure 8–29B is an electron micrograph and interpretation of a *Drosophila* chromosomal DNA molecule in the process of replication, showing several replicons.

Accuracy and speed of the replicative process. Whatever the precise details of the operation of the replicative mechanism, its accuracy and speed are quite high. Kornberg and his associates were able to synthesize biologically active ϕX174 DNA of approximately 5,400 nucleotides

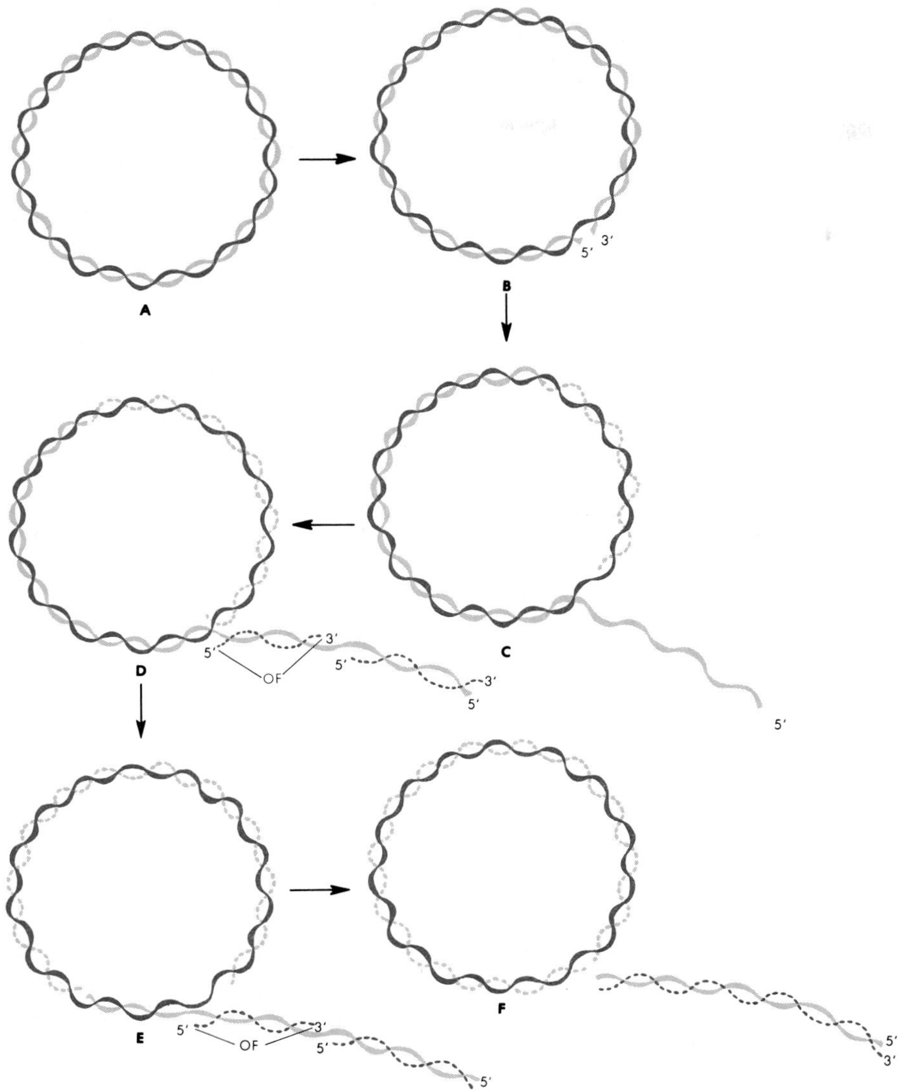

FIGURE 8–28. *The rolling circle model of DNA replication. (A) Closed circle of DNA before replication. (B) Nicking of one strand by an endonuclease, providing free 5′ and 3′ ends. (C-F) Successive stages in rolling of "old" continuous strand and "peeling off" of the complementary old strand with synthesis of new complementary strands. In (E) Okazaki fragments (OF) are seen on the template old strand. These have been joined by action of DNA ligase in (F). Circularization follows after the stage shown at (F).*

(recall that the DNA of this phage is single-stranded except in its replicative form). Infection of susceptible *E. coli* cells with this synthetic DNA resulted in replication of φX174 particles identical with natural phage, as well as lysis of host cells in the usual fashion. Speed of synthesis *in vitro* has been reported as 500 to 1,000 nucleotides per minute, but *in vivo* speeds as high as 100,000 per minute have been calculated.

Single-stranded viruses such as φX174 cannot, of course, follow exactly the procedure outlined for double-stranded bacterial DNA. The single strand present in the infective phage serves as a template in the host cell for synthesis of a complementary strand, and forms a double-stranded molecule (the replicative form) that replicates many times. Finally, however, the replicative form begins to produce only single (infective) strands, which are then assembled within protein coats as mature phage and released.

However, certain RNA tumor viruses (for example, Rous sarcoma virus and Rauscher mouse leukemia virus) either produce or induce an enzyme, *RNA-dependent DNA polymerase (reverse transcriptase)*, which catalyzes synthesis of DNA from an RNA template. Replication of these and some other RNA viruses takes place through a DNA intermediate

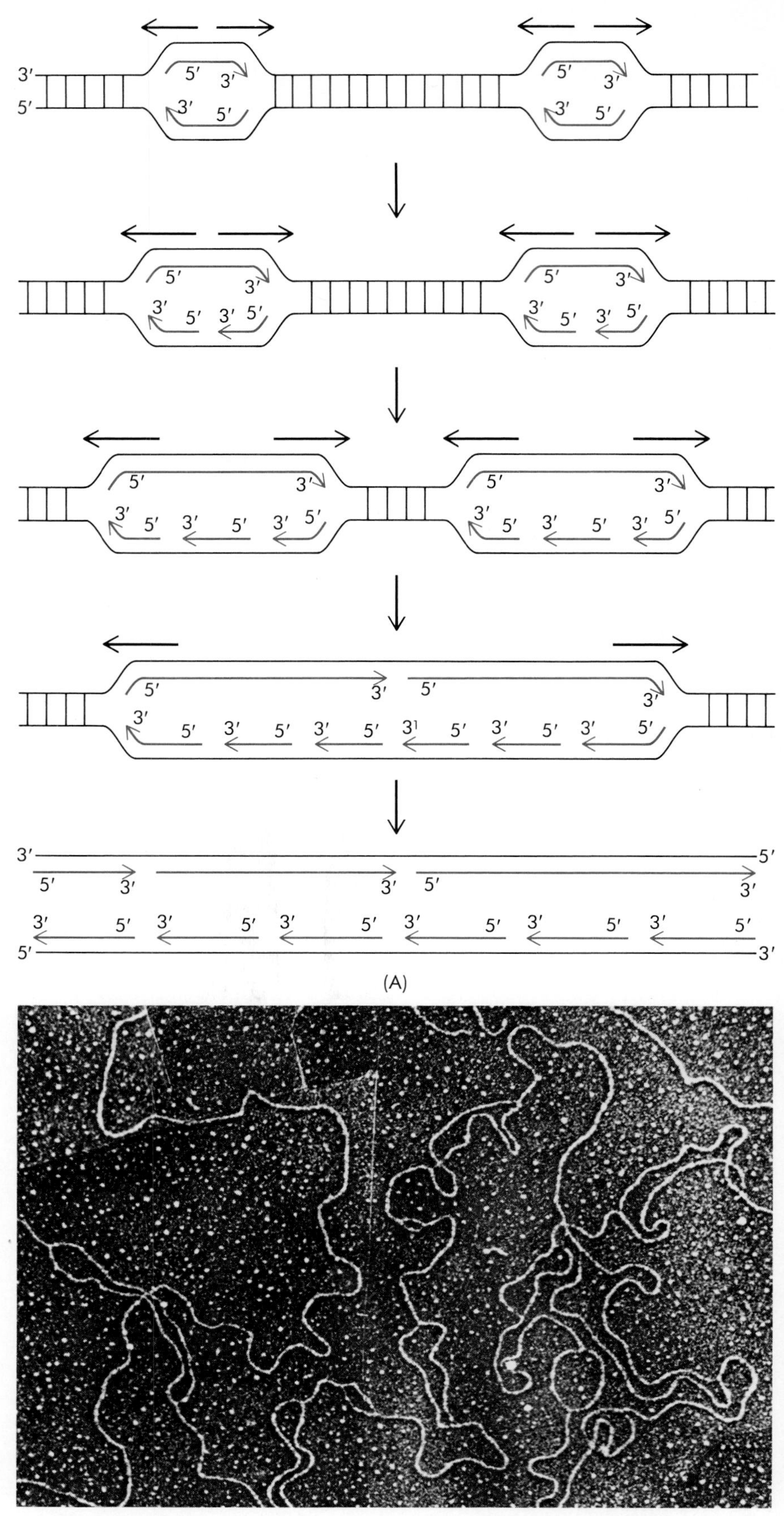

FIGURE 8–29. *(A) Pattern of replication of long DNA molecules with more than one origin of replication. Replication proceeds out bidirectionally from each origin. (B) Picture of a replicating* Drosophila *DNA showing several regions of replication.* (Photo courtesy of Dr. David Hogness. From H. J. Kriegstein and D. S. Hogness, 1974, used by permission.)

rather than through an RNA intermediate as described for other RNA viruses. Furthermore, this discovery is providing more information on carcinogenesis by RNA viruses.

SUMMARY

In summary, the essential features of DNA replication are as follows:

1. The overall pattern of replication is semiconservative; that is, each new molecule produced is made up of one intact strand of the old parent molecule and one newly synthesized strand.
2. Replication begins at specific points called origins. There are one or more origins per DNA molecule.
3. Initiation requires a short piece of RNA onto which the nucleotides are added.
4. Replication proceeds bidirectionally from the origin unless the origin is at the end of the molecule.
5. Nucleotides are added only in the 5′→3′ direction.
6. The replication process is semidiscontinuous, occurring fairly continuously on one strand and discontinuously on the other strand. The short fragments produced are joined together to complete synthesis.

PROBLEMS

8–1 What are some of the genetic phenomena that can be investigated in corn, fruit flies, peas, and human beings that cannot readily be studied in *Neurospora*, bacteria, and viruses?

8–2 If one strand of DNA is found to have the sequence 5′ AACGTACTGC 3′, what is the sequence of nucleotides on the 3′, 5′ strand?

8–3 For a molecule of *n*-deoxyribonucleotide pairs, give a mathematical expression that can be used to calculate the number of possible sequences of those nucleotide pairs if only the "usual" bases are present.

8–4 Develop a formula for determining the length in micrometers of a DNA molecule whose number of deoxyribonucleotide pairs is known. (Note: 1Å = 1×10^{-4} μm.)

8–5 Phage T2 DNA is estimated to consist of about 200,000 deoxyribonucleotide pairs. What is the length in micrometers of its DNA complement?

8–6 The molecular weight of the DNA of one somatic (that is, diploid) cell of corn is given as 9×10^{12}. Calculate (a) the number of deoxyribonucleotide pairs, (b) its length in micrometers, and (c) its length in inches. Express answers to (a) and (b) as powers of 10.

8–7 One calculation for the molecular weight of the DNA in human autosome 1 is 0.5×10^{12}. (a) Of how many deoxyribonucleotide pairs is this DNA composed? (b) What is its length in micrometers? (c) What is its length in inches? Answer (a) and (b) as powers of 10. Carry calculations to the nearest first decimal place.

8–8 Determine the molecular weight of (a) adenine, (b) thymine, (c) cytosine, (d) guanine.

8–9 What is the molecular weight of each of these deoxyribonucleosides: (a) deoxyadenosine, (b) thymidine, (c) deoxycytidine, (d) deoxyguanosine?

8–10 Calculate the molecular weight of each of the following: (a) deoxyadenylic acid, (b) thymidylic acid, (c) deoxycytidylic acid, (d) deoxyguanylic acid.

8–11 From your answers to the preceding problem, give an average molecular weight for any single deoxyribonucleotide pair. (Consider only the "usual" bases.)

8–12 From your answer to the preceding question, and the information given in Problem 8–5, what is the molecular weight of phage T2 DNA?

8–13 If the molecular weight of *Escherichia coli* DNA were taken as 2.7×10^9, (a) of how many deoxyribonucleotide pairs would *E. coli* DNA consist, and (b) what would be its length in μm?

8–14 In this chapter the generation time for virulent T2 phage was given as 22 minutes. It appears that there is no replication of DNA for about the first six minutes and that, thereafter, DNA replicates at an exponential rate for five minutes, after which the rate of replication declines to the end of the 22-minute period. In that five-minute period, about 32 phage DNA molecules appear to be formed for each infecting phage particle. At what rate are phage DNA molecules being replicated during this exponential period?

8–15 From your answer to the preceding problem and information given in Problem 8–5, at what rate are deoxyribonucleotide pairs being synthesized per minute in T2?

8–16 Assume that you have just determined the adenine content of the DNA of *Bacillus hypotheticus* to be 20 percent. What is the percentage of each of the other bases?

8–17 If you determined the adenine–guanine content of another species to total 25 percent, should you believe it?

8–18 Note the deoxyribonucleotide composition listed in Table 8–1 for φX174. Another report in the literature for this phage gives the following deoxyribonucleotide percentages: A, 27.6; T, 27.8; G, 22.3; C, 22.3. (a) What are the A/T and G/C ratios? (b) What does your answer to part (a) suggest regarding the structure of the phage DNA whose composition is given in this problem? (c) How would you account for the considerable difference in these ratios from those given in Table 8–1?

8–19 What significance might there be to the fact that *Euglena* chloroplasts contain DNA and that this chloroplast DNA has a markedly different nucleotide composition from nuclear DNA reported for the same organism as given in Table 8–1?

8–20 DNA of a species of yeast, a microscopic fungus, is reported to have a thymidylic acid content of 32.6 percent. The A/T ratio was reported as 0.97. On this basis, what is the percentage composition of deoxyadenylic acid?

8–21 A culture of *Escherichia coli* was grown in a ^{15}N medium until all of the cells' DNA was so labeled. The culture was then transferred to a ^{14}N medium and allowed to grow for exactly two generations, so that all DNA replicated twice in the ^{14}N medium. If the DNA of the latter culture were then subjected to density gradient centrifugation, how much of the DNA should be (a) ^{15}N only, (b) "hybrid," (c) ^{14}N only?

8–22 Differentiate between transformation and transduction.

8–23 Knowing that 25,400 μm = 1 inch, develop a formula for determining the length of a DNA molecule in inches (L_{in}) if (a) the length in μm is known, and (b) if only the number of deoxyribonucleotide pairs is known.

GENES AND PROTEINS

CHAPTER

9

With its precise replication, deoxyribonucleic acid (DNA) serves to carry genetic information from cell to cell and from generation to generation. The next task will be to determine the way in which this information is translated into proteins that determine the phenotype. According to R. F. Doolittle, "If DNA is the blueprint of life, then proteins are the bricks and mortar. . . . Your genes supply the information, but you are your proteins." Virtually all the phenotypes examined thus far are the result of biochemical reactions that occur in the cell. All of these reactions require enzymes, and enzymes are proteins, either wholly or in part. More than 2,000 enzymes have been identified. Each is a unique molecule catalyzing a specific chemical reaction. Other phenotypes are due primarily to the kinds and amounts of nonenzymatic proteins present, for example, hemoglobin, myoglobin, gama globulin, insulin, or cytochrome C. Proteins are composed of one or more long linear polymers of amino acid residues (**polypeptide chains**) that are synthesized almost exclusively in the cytoplasm. The topic of this chapter is how DNA, located primarily in the nucleus, mediates the synthesis of proteins in the cytoplasm, and more specifically, how the information encoded in the sequence of bases in DNA is translated into a sequence of amino acids in proteins. In Chapter 3, several cases appearing to suggest a connection between genes and proteins were cited. A few additional examples will furnish clear proof of this gene–enzyme relationship.

GENES AND ENZYMES

Phenylalanine Metabolism

The groundwork for a functional relationship between genes and enzymes was laid in 1902, when William Bateson reported that a rare human defect, *alkaptonuria,* was inherited as a recessive trait. Then in 1909 the English physician Archibald Garrod published a book, *Inborn Errors of Metabolism,* that was far ahead of its time in suggesting a relationship between genes and specific biochemical reactions. Alkaptonuria was among the heritable disorders that he discussed at length. This condition, manifested by a darkening of cartilaginous regions and a proneness to arthritis, results from a failure to break down alkapton (2,5-dihydroxyphenylacetic acid, or homogentisic acid). Alkapton accumulates and is excreted in the urine, which turns black upon exposure to the air. By the 1920s, it was discovered that the blood of alkaptonurics is deficient in an enzyme, *homogentisic acid oxidase,* that catalyzes the oxidation of alkapton.

This defect is but one of a group, all related to the body's metabolism of the essential amino acid phenylalanine. As diagrammed in Figure 9–1, human beings receive their supply of this substance from ingested protein. Once in the body, phenylalanine may follow any of three paths. It may be (1) incorporated into cellular proteins, (2) converted to phenylpyruvic acid, or (3) converted to tyrosine, another amino acid.

Persons of genotype *pp* (see Figure 9–1) fail to produce the enzyme phenylalanine hydroxylase, with the result that phenylalanine accumulates in the blood, with excesses up to a gram a day being excreted in the urine. Such persons are said to have *phenylketonuria,* or PKU, which is accompanied by serious mental and physical retardation. Un-

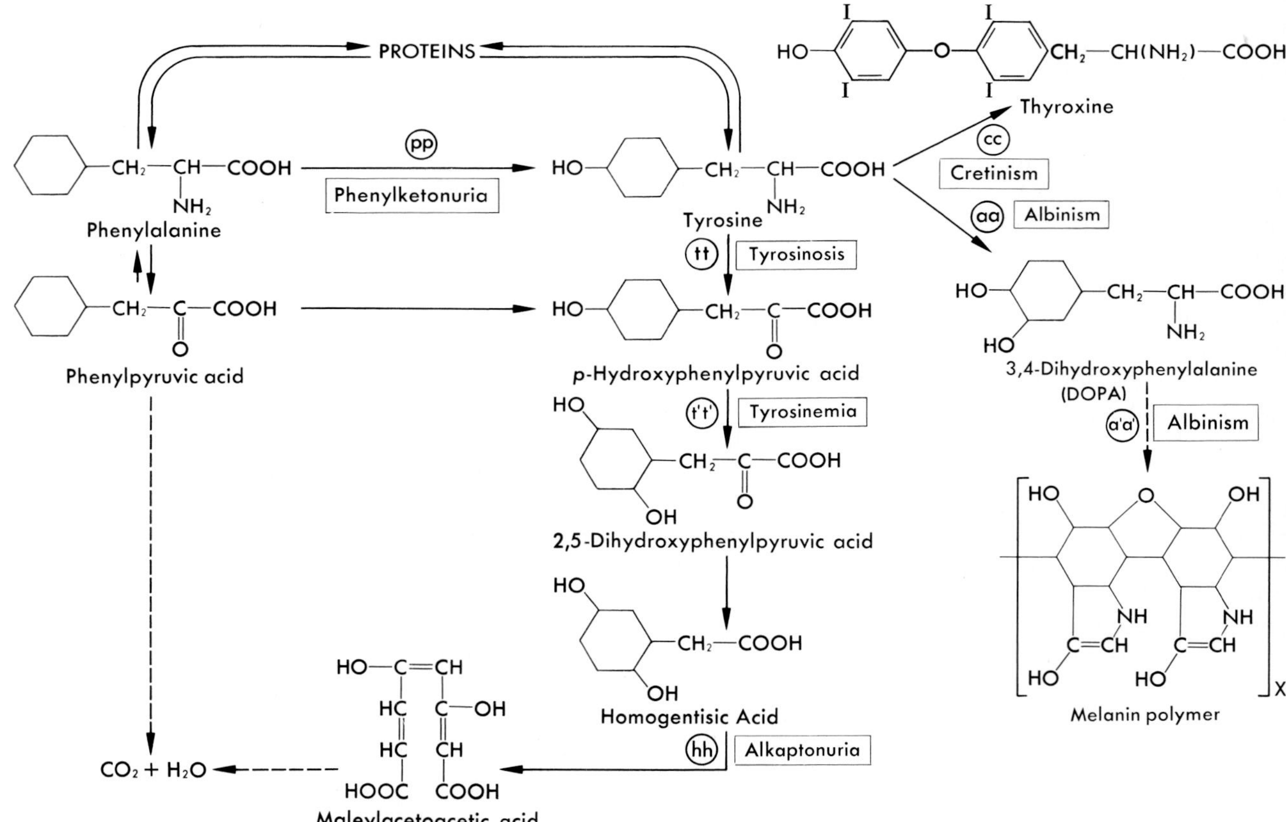

FIGURE 9–1. *Metabolic pathways of the amino acid phenylalanine in humans. Mutations in different allelic pairs;* pp, cc, aa, a'a', tt, t't', *and* hh *cause blocks at various points because of the failure to produce a given enzyme, resulting in such inherited metabolic disorders as phenylketonuria (PKU), tyrosinosis, tyrosinemia, cretinism, or albinism.*

treated PKU individuals have a mean IQ of 65. However, if symptoms are diagnosed very early and the infant is put on a diet low in phenylalanine (normal development requires some) for at least the first five years of life, brain development is quite normal. Screening of newborn infants for PKU is now carried out in virtually all states.

Another recessive allele (t in Figure 9–1) blocks the conversion of tyrosine to parahydroxyphenylpyruvic acid, probably through failure to produce liver tyrosine transaminase. This leads to accumulation of tyrosine, excesses of which are excreted in the urine. This condition is called *tyrosinosis; no serious symptoms appear to be associated, although the disorder has been reported for only one human subject. A somewhat less uncommon but rare disorder, tyrosinemia,* is due to inability of persons who are homozygous recessive for another allele (t′ in Figure 9–1) to produce the enzyme parahydroxyphenylpyruvate oxidase. In this condition large quantities of parahydroxyphenylpyruvic acid, as well as lactic and acetic acid derivatives, are excreted in the urine. If not treated with low-tyrosine diets, this condition leads to serious liver problems, culminating in early death.

Other metabolic blocks associated with faulty utilization of phenylalanine, and all caused by recessive alleles, lead to *albinism* or to *cretinism*. Two kinds of albinism have been recognized for some time. The majority of albinos are unable to produce tyrosinase (these are *aa* persons in Figure 9–1), which catalyzes the conversion of tyrosine to DOPA; hence they do not produce melanin. Some albino persons (*a′a′* in Figure 9–1), however, can produce DOPA but cannot utilize tyrosinase in oxidizing DOPA to DOPA quinone, thus breaking the melanin production process at a later point. Goitrous cretinism is also of several types, each involving a particular block in the conversion of tyrosine to thyroxine. Cretinism is accompanied by a considerable degree of physical and mental retardation in addition to thyroid defects.

In each case described for phenylalanine metabolism defects, a particular (recessive) allele appears to be associated with nonproduction of a specific enzyme that catalyzes a particular biochemical reaction. Occurrence or nonoccurrence of the reaction then determines a related phenotypic effect. Heterozygotes do not display the disorder; one "dose" of the normal allele results in enough enzyme to permit the reaction to proceed. Because serious inherited metabolic defects such as PKU are (in the absence of highly uncommon mutation) the result of a marriage between heterozygotes, it is very desirable to be able to detect heterozygosity with certainty in phenotypically normal persons. Considerable progress is being made in this direction, and persons heterozygous for PKU (carriers) can now be detected by the phenylalanine test.

Levels of phenylalanine rise higher and are maintained longer in *Pp* individuals than in homozygotes after administration of a standard dose of this amino acid. Detection of heterozygosity is now possible in more than 100 inherited disorders, although not all tests are completely reliable. Some of the disorders for which carrier tests are available include Tay-Sachs, sickle-cell trait, PKU, galactosemia, cystic fibrosis, glucose-6-phosphate dehydrogenase (G6PD) deficiency, hemophilia A, Huntington's disease, and Duchenne muscular dystrophy.

Depending on the nature of the defect, heterozygosity may be determined even prenatally by amniocentesis (see Chapter 20). O'Brien and co-workers, for example, reported success in diagnosing the Tay-Sachs disorder by amniocentesis during the sixteenth to twenty-eighth week of pregnancy by testing the sloughed fetal cells present in the amniotic fluid for enzyme level. In fifteen different pregnancies they were able to detect normal homozygous embryos in seven cases, heterozygotes in two cases, and Tay-Sachs condition in six cases. Five of the latter were therapeutically aborted; tests of various organs of these five showed hexosaminidase A to be absent. The remaining one was born and was developing marked symptoms of the Tay-Sachs disorder by the age of nine months.

Sickle-Cell Anemia

In addition to specifying enzymes, genes specify many proteins that play other important roles in cells. One of the most important proteins in humans is the oxygen-carrying pigment *hemoglobin*. It is now known that a single amino acid change in this molecule is the cause of sickle-cell anemia in humans. The genetics of this disease was discussed in Chapter 2. The disease is inherited as a single-gene trait. Human adult hemoglobin HbA is made up of four protein chains: two identical α chains and two identical β chains. In 1957, Vernon Ingram showed that in one of the β chains of HbS (hemoglobin from a person with sickle-cell anemia), a single amino acid change had occurred, with the substitution of one amino acid for another. Such a simple amino acid change in hemoglobin severely alters its oxygen-carrying capacity and results in this very serious inherited disorder. This became the first known inherited change in the molecular structure of a protein.

Neurospora

Studies of metabolic defects in humans, important as they are, do not provide a set of experimental conditions for adequate testing of the gene–enzyme relationship. In fact, most studies before the 1940s had concerned themselves largely with morphological characters, many based on such complex biochemical reactions that impeded analysis and used genetically inefficient subjects such as humans. The real breakthrough was furnished by George Beadle and Edward Tatum in 1941 and in a series of later reports by them and their colleagues.

Beadle found that an ascomycete fungus, the orange-pink bread mold *Neurospora crassa*, was an ideal organism for investigation of metabolic defects. *Neurospora*, easily grown in the laboratory, has a simple life history (Appendix B) in which only the zygote is diploid (hence dominance is not a problem and even recessive genes are expressed), and it has many easily detected physiological variants. It can make everything necessary for normal growth from a medium containing sugar, a nitrogen source, biotin, and inorganic salts. This is called a "minimal" medium. Some mutants would not usually grow on the minimal medium without the addition to the medium of a single complex substance, such as a vitamin or an amino acid. This indicated that there was a biochemical block in the synthesis of the compound, which would normally be lethal, but could be overcome easily by adding the required compound to the culture medium. The aim of Beadle and Tatum was to induce heritable metabolic deficiencies by X-irradiation, to identify them by observing ability to grow on minimal medium and various supplemental media, and thus to determine the specific metabolic block.

The next step was to cross nutritional mutants with normal strains and observe segregation, thus showing that the defect is an inherited trait. Furthermore, the phenotype of the mutants could be shown to be due to loss of activity of the enzyme responsible for the synthesis of the specific compound. These studies led to the **one gene–one enzyme–one phenotype** hypothesis. Now, we know that not all proteins are enzymes, for example, hemoglobin, and that many enzymes are not composed of a single polypeptide chain, but instead are aggregates of several peptide chains, each being the product of a different gene. A change in any of the polypeptides may change the phenotype. The present concept then is a **one gene–one polypeptide–one phenotype** relationship.

PROTEIN STRUCTURE

As indicated earlier, almost all enzymes are protein, at least in part, and the catalytic property of an enzyme is conferred by its protein makeup, with or without added cofactors. In addition, proteins serve structural and gene regulatory roles in cells. Proteins are large, heavy, generally complex molecules of great variety and biological significance. The basic structural units of all proteins are the amino acids, numbers of which are linked together to form long chains. All the physical and chemical properties of protein molecules depend on this amino acid sequence and the three-dimensional form into which the chain folds. Only 20 amino acids are important as constituents of proteins, although others are found metabolically in other pathways. Since the number of amino acids in a protein may vary from as few as 8 or 9 to over 1,000, the number of protein molecules that are possible with just these 20 amino acids is virtually unlimited.

Each amino acid may be represented by the general structural formula

```
   H   H   O
   |   |   ||
H—N—C—C—OH
       |
      (R)
```

where $-NH_2$ represents the amino group, $-COOH$ represents the carboxyl group, and R is the side chain. The amino group, the carboxyl group, and the side chain are all linked to the same carbon. Differences in the side chain determine the individuality of amino acids. Thus, in glycine, R is merely an H atom

```
   H   H   O
   |   |   ||
H—N—C—C—OH
       |
       H
```

and in alanine, R is a methyl group

```
   H   H   O
   |   |   ||
H—N—C—C—OH
       |
      CH3
```

whereas the addition of a benzene ring to the side chain of alanine produces phenylalanine

$$
\begin{array}{c}
\quad H \quad H \quad O \\
\quad | \quad\;\; | \quad\;\; \| \\
H-N-C-C-OH \\
\qquad\;\; | \\
\qquad\;\; CH_2 \\
\qquad\;\; | \\
\qquad\;\; C_6H_5
\end{array}
$$

and so on. The structures of the 20 amino acids used in proteins are found in Figure 9–2.

Amino acids are linked together when the amino group of one amino acid joins the carboxyl group of the next. This C--N bond is formed with the loss of water and is termed a **peptide bond.** The formation of a dipeptide by condensation of two molecules of alanine may be represented by

$$
H_2N-CH(CH_3)-CO-OH + H-NH-CH(CH_3)-COOH \rightarrow H_2N-CH(CH_3)-CO-NH-CH(CH_3)-COOH + H_2O
$$

in which CONH is the peptide bond. A **polypeptide** is thus a series of amino acids joined by peptide bonds. Each polypeptide or protein molecule has a free $-NH_2$ at one end (the amino terminal end) and a free $-COOH$ group on the other end (the carboxyl end). Through these residues, the side chains confer a number of properties. Some are large and complex; others are small and simple. Some carry a positive charge (lysine, arginine); others carry a negative charge (glutamic acid, aspartic acid); and many are electrically neutral. Differences in bulk and electrical charge along a polypeptide, plus an intricate folding of the protein molecule into a three-dimensional structure, impart much of the specificity of enzyme–substrate and antigen–antibody relations.

The characteristic of a protein on which all other characteristics are based is the sequence of amino acids, called the **primary structure.** As we shall see, the sequence of amino acids in a protein is determined by the sequence of nucleotides in DNA. This primary structure of a protein molecule determines its final conformation (shape) as well as its biological activity. Conformation occurs through folding of the molecule in a way that minimizes the free energy, that is, the chain assumes the most energetically stable shape. This folding brings certain side chains into proximity so that they may interact with each other as well as with other molecules.

In addition, certain conformations are used over and over in many proteins. One of the best known is the **alpha helix** (Figure 9–3A) in which a part of the molecule is twisted into a helical shape with a turn every 3.6 amino acids, and stabilized by hydrogen bonds between amino acids. Another stable shape is the **beta pleated sheet** conformation (Figure 9–3B) in which polypeptide chains lie side by side with

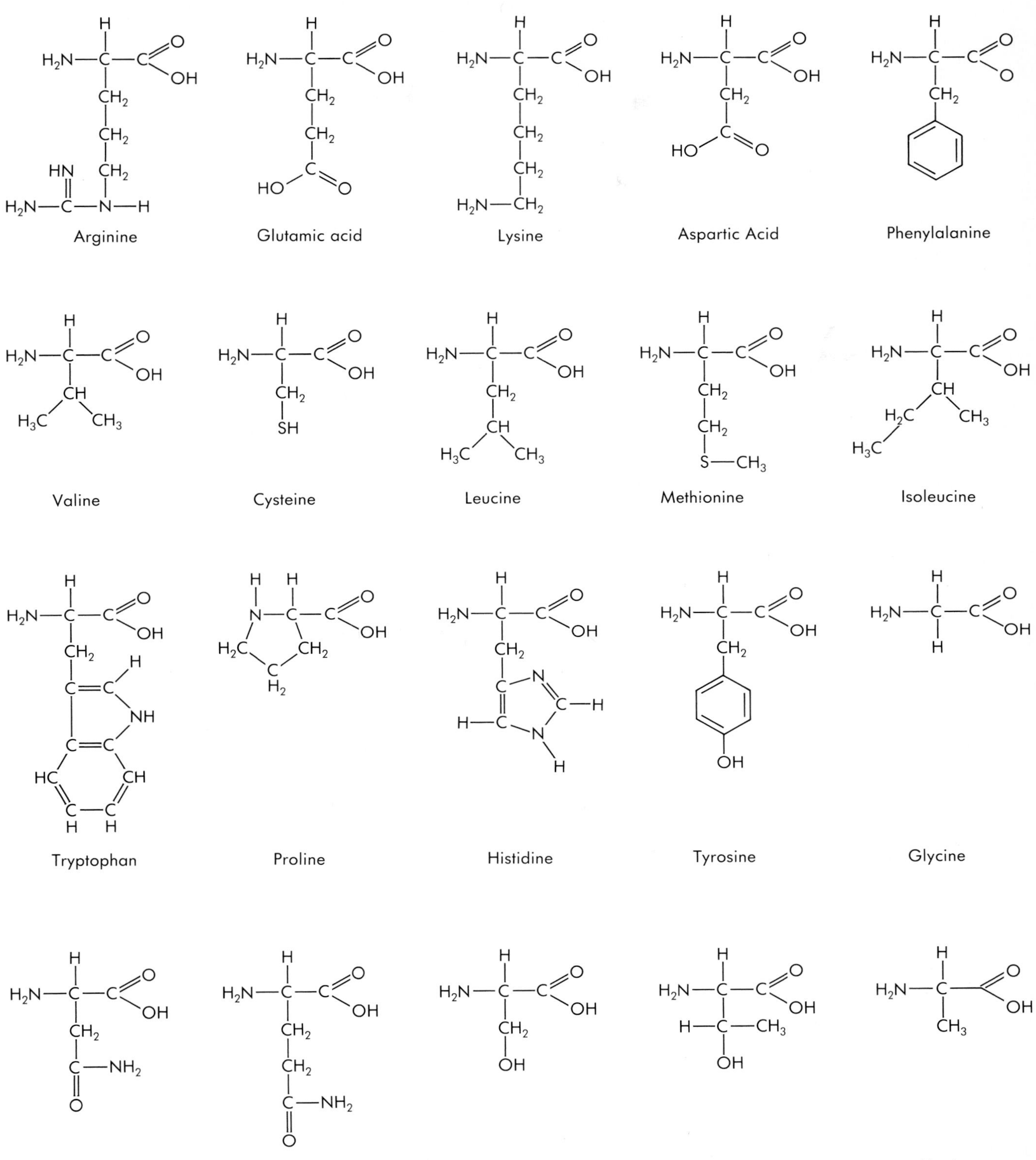

FIGURE 9–2. *Structures of 20 biologically important amino acids.*

hydrogen bonds holding them together. Many proteins are heterogeneous mixtures of these two types of structures. Certain combinations of these structures form **domains** made up of 30 to 150 amino acids,

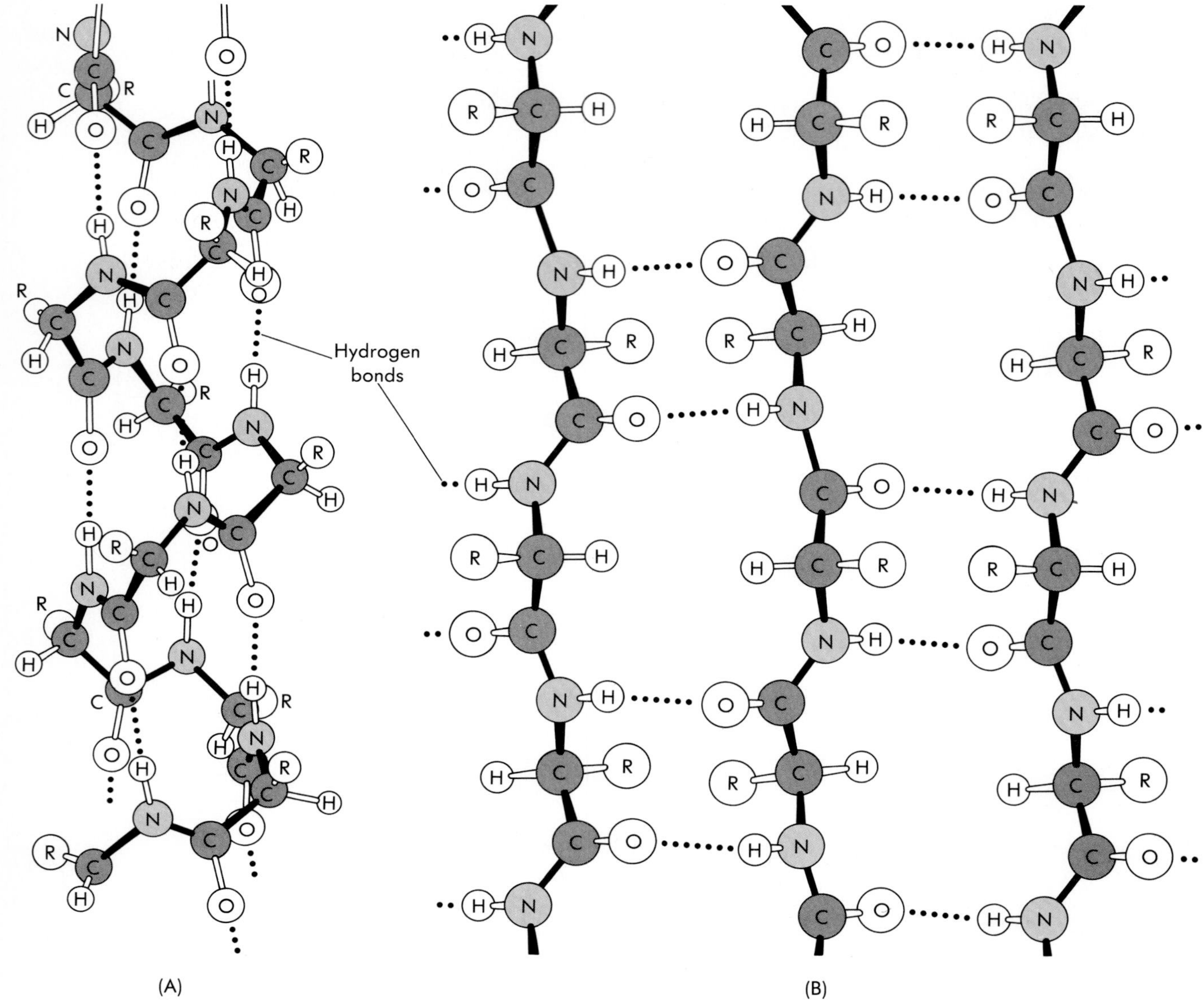

FIGURE 9–3. *Protein conformations: (A) alpha helix, (B) beta sheet.*

and serve as basic structural units of proteins. A very common domain consists of two beta sheets connected to an alpha helix. These shapes are only possible with certain amino acid sequences, and since they are found in many proteins, they must be considered to be of high evolutionary value.

POLYPEPTIDE CHAIN SYNTHESIS— THE COMPONENTS

Because of its infinite variety of possible group arrangements of the four nucleotides, DNA is admirably suited for carrying the information needed to direct the synthesis of an almost unlimited number of different proteins. But DNA, except for small amounts in chloroplasts and mitochondria, is located in the eukaryotic nucleus, whereas protein synthesis occurs almost entirely in the cytoplasm. Moreover, DNA is not degraded or "used up" in performing its function, and it differs from the polypeptides for which it codes.

The spatial separation of the genetic information from the site of protein synthesis in eukaryotic cells makes necessary the presence of an intermediate class of molecules, carrying information between genes in the nucleus and the cytoplasm, where this information is used to direct the synthesis of proteins. This flow of information is illustrated as

DNA ⟶ RNA ⟶ Protein

where the arrows indicate the flow of information. DNA through replication produces more DNA. The information then flows from DNA through the intermediate RNA, and is used to direct the synthesis of specific proteins.

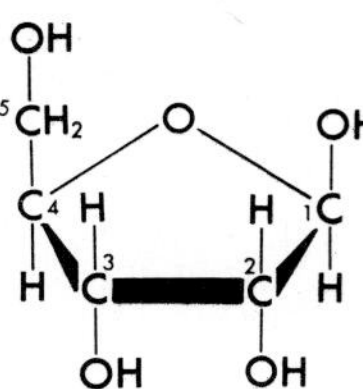

FIGURE 9–4. *Molecular structure of ribose, the sugar of ribonucleic acid. Compare the number 2 carbon of ribose with the number 2 carbon of deoxyribose (Figure 8–6.)*

RNA Structure

Ribonucleic acid (RNA) differs from DNA in several important ways:

1. The sugar of the sugar–phosphate backbone is ribose, not deoxyribose (Figure 9–4).
2. Nitrogenous bases are the same as in DNA except that the pyrimidine uracil (U) (Figure 9–5) replaces the pyrimidine thymine.
3. Successive *ribonucleotides* are joined by 5′-3′ phosphodiester bonds as in DNA.
4. Molecules of RNA are generally *single-stranded* polymers of ribonucleotides, although most RNA molecules assume a secondary structure by folding back on themselves, and form hydrogen bonds in a double-stranded structure.

The ribonucleosides and ribonucleotides are listed in Table 9–1.

TABLE 9–1 **Ribonucleosides and Ribonucleotides**

Base	*Ribonucleoside*	*Ribonucleotide*
Adenine	Adenosine	Adenylic acid
Uracil	Uridine	Uridylic acid
Cytosine	Cytidine	Cytidylic acid
Guanine	Guanosine	Guanylic acid

Uracil

Thymine

FIGURE 9–5. *Uracil found in RNA compared to thymine in DNA.*

RNA Synthesis

RNA of all types is *transcribed* from a template strand of DNA, catalyzed by several enzymes collectively referred to as **DNA-dependent RNA polymerase** or simply **RNA polymerase.** The nucleotide sequence of RNA is complementary to that of the template DNA strand, except that uracil replaces thymine. The process of synthesizing an RNA strand complementary to one DNA strand is called **transcription.** The strand of DNA used as the template for RNA synthesis is called the *sense* strand; the other strand is the *antisense* strand. The same strand is not always the sense strand throughout the entire DNA molecule or chromosome.

An outline of the essential details of transcription is found in Figure 9–6. During the process of transcription, there is a localized unwinding

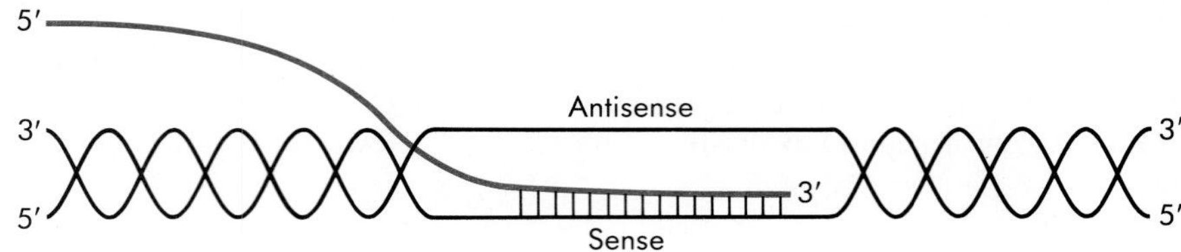

FIGURE 9–6. *Transcription of RNA from the DNA sense strand. The DNA double helix unwinds and RNA is synthesized complementary to the sense strand.*

of the double helix, with transcription proceeding in the 5′→3′ direction from the single-stranded region of the DNA template. This localized unwinding moves along the molecule followed by recoiling of the helix behind the newly synthesized RNA. The region of DNA actually transcribed into RNA is called the **coding region.** However, there are sequences preceding (upstream from) the actual coding region that are required for transcription. These sites, called **promotors,** are the sites of recognition and interaction between RNA polymerase and DNA that signal transcription to begin.

Two special promotor regions have been identified that appear to be present in all organisms. In a region five to ten bases preceding the coding region is a sequence of seven bases that reads

TATAATG

with minor variations. A sequence such as this is called a **consensus sequence** because it is a sequence observed to occur with very little variation in many different organisms. In bacteria this region is called the **Pribnow box** and in eukaryotes the same region has the sequence

TATAAAT

and is called the **Hogness box,** each after the person who originally described the region. This region is generally referred to as the **TATA box** and is believed to orient the enzyme RNA polymerase in preparation for transcription.

Another important region, further upstream from the TATA box, is located approximately thirty-five bases upstream from the coding region. It consists of a nine-base consensus sequence and is thought to be the actual site of binding of the RNA polymerase. The enzyme, once bound to these promotor sequences, is then capable of beginning transcription at the start of the coding region.

Prokaryotic RNA polymerases. The RNA polymerase from prokaryotes is composed of five subunits with a total molecular weight of about 500,000. This is clearly one of the largest enzymes known. One of the subunits, *sigma,* can be disassociated from the core enzyme without loss of catalytic activity. The suggestion has been made that the function of this subunit is mainly in recognition of promotors and is not involved directly with transcription. This is based on the fact that in the absence of sigma, transcription is random and may start anywhere. The core enzyme can synthesize RNA in the presence of a DNA template, the four ribonucleoside triphosphates, and magnesium ions. There appears to be only one RNA polymerase used to make the several types of RNA in an *Escherichia coli* cell.

Eukaryotic RNA polymerases. Unlike prokaryotes, eukaryotes produce three different RNA polymerases, each with a distinctive function.

Polymerase I, located in the nucleolus, synthesizes only ribosomal RNAs. **Polymerase II,** found in the nucleoplasm, synthesizes a pre-messenger RNA. **Polymerase III,** also found in the nucleoplasm, synthesizes only transfer RNA and 5*s* RNA. These polymerases have been distinguished from one another on the basis of their sensitivity to *alpha amanatin,* the toxin produced by the poisonous mushroom from the genus *Amanita.* Polymerase I, from wheat embryos, is not sensitive to amanitin. Polymerase II is inhibited by low (0.05 μg/ml) concentrations of the toxin. Polymerase III is inhibited only by high (5 μg/ml) concentrations of the toxin.

RNA is synthesized in the nucleus of eukaryotes, using one of the strands of DNA as a template (the one containing the correct promotor sequence). Therefore the sequence of ribonucleotides is complementary to one of the DNA strands. This is the case except for the modification of some of the usual bases after RNA synthesis. The basic function of RNA is the carrying of the genetic message of DNA into the cytoplasm, where it is responsible for directing the synthesis of polypeptide chains.

Classes of RNA

There are three major classes of RNA molecules: messenger RNA (mRNA), ribosomal RNA (rRNA), and transfer RNA (tRNA). Each is transcribed from different genes, and each has a unique function in cells.

When one studies large molecules and molecular aggregates, a standard method of characterization is their sedimentation properties during ultracentrifugation. In an ultracentrifuge, molecules are subjected to forces up to 700,000 times gravity. The velocity of sedimentation through a solution is a function of a molecule's shape and molecular weight. The unit of measurement of sedimentation used to compare molecules is the sedimentation coefficient, *s*. This is equal to the velocity of a molecule in a given solution divided by the centrifugal force applied to it. The units of *s* are **Svedbergs.** One Svedberg is equal to 10^{-13} sec. It is therefore common to refer to molecules or aggregates of molecules by the number of *s* units, such as 30*s* or 50*s*. The larger the molecule or particle, the larger the *s* value. However, because shape has a direct influence on *s*, the values are not additive. These units are very commonly used to compare RNA molecules and aggregates of RNA and proteins, such as those found in ribosomes.

Messenger RNA (mRNA). Specificity for a particular amino acid sequence is determined by **messenger RNA.** This is shown, for example, by the fact that phage mRNA utilizes bacterial ribosomes made before infection to bring about synthesis of phage proteins. This ribonucleic acid directs incorporation of various amino acids from the intracellular pool into proteins at the ribosome surface where enzymes effect the peptide linkage and form polypeptide chains.

Transcription of *eukaryotic* DNA to produce mRNA begins with the synthesis of long precursor molecules by RNA polymerase II from the template strand of DNA. This enzyme functions by catalyzing formation of 5′→3′ phosphodiester bonds of the RNA "backbone" by "reading" the DNA template in the 3′→5′ direction. The developing mRNA molecule is antiparallel and its nucleotides are complementary to those of the DNA template strand (Figure 9–7). Messenger RNA chain growth is rapid—from 15 to 100 nucleotides per second *in vitro.*

The immediate product of transcription in *eukaryotes* is a molecule of many more ribonucleotides than that comprising the ultimate, func-

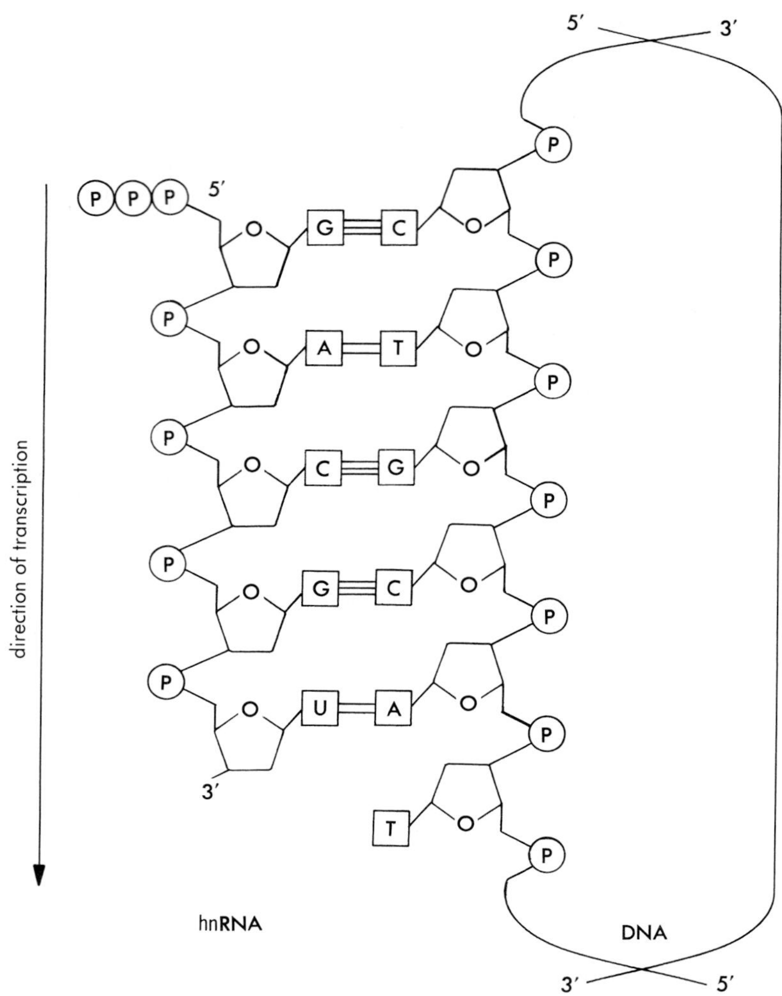

FIGURE 9–7. *Transcription of hnRNA from the template strand of DNA showing that it is antiparallel and complementary to the template DNA strand. hnRNA is synthesized in the 5′ → 3′ direction by addition of successive ribonucleotides at the 3′ end of the growing mRNA molecule.*

tional mRNA. This primary transcript in eukaryotes is generally referred to as **heterogeneous nuclear RNA (hnRNA).** Molecules of hnRNA are estimated to range from 500 to 50,000 nucleotides long. They are very rapidly processed into much smaller molecules. Some estimates predict that as little as 10 percent of most original transcripts eventually reach the cytoplasm as mRNA. Following its transcription in eukaryotes, hnRNA is processed into functional mRNA in three principal steps (Figure 9–8).

First, soon after initiation of transcription, there is a modification of the 5′ end of the hnRNA by the formation of a **cap.** This consists of addition of a methylated guanosine in a rare 5′-5′ linkage (Figure 9–8). This cap sometimes also includes methylation of additional sugars of both the 5′ nucleotides. It appears that this cap is required in eukaryotes for formation of an mRNA-ribosome complex and its subsequent operation in polypeptide synthesis.

Second is the **RNA splicing** step. This is the controlled excision of large **intervening sequences** or **introns** from the transcript and rejoining of the remaining fragments, called **coding sequences** or **exons,** together to produce the finished mRNA. The actual mechanism of cutting and splicing is not completely understood; however, it is known that the border regions of each intron usually contain similar sequences, usually started with a GU and ending with an AG, and that a small RNA-protein particle called the **snRNP particle** is involved. The exons comprise the

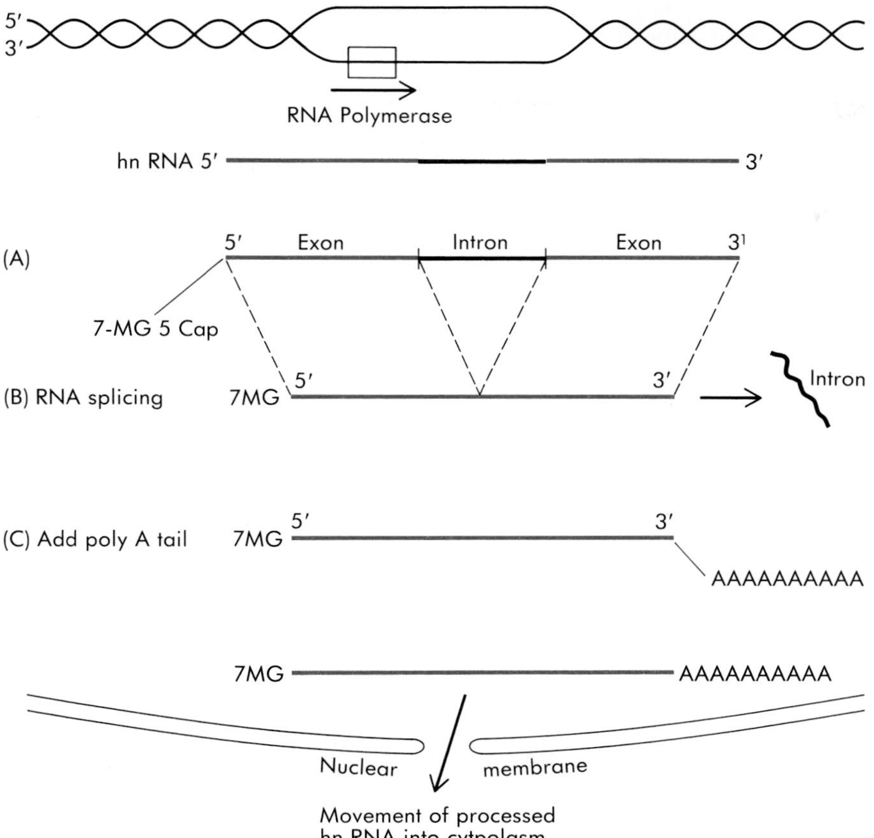

FIGURE 9–8. *Transcription and processing of heterogeneous nuclear RNA (hnRNA). (A) Adding of 7-methyl guanosine. (B) Removal of introns in splicing step. (C) Addition of poly A tail on 3′ end. After the processing is completed, the mRNA moves through pores in the nuclear membrane into the cytoplasm.*

actual sequences that are ultimately translated into proteins. A few introns of excised hnRNA have been sequenced and do not seem to contain any special or unusual sequences such as repeats or symmetrical sequences.

Usefulness of the discarded portions of hnRNA is not understood. In at least one case, such as the yeast cytochrome b gene located in the mitochondrial DNA, sequences of an intron are actually translated into a protein. This was originally discovered by mapping mutations in the first intron of the gene. These mutations block production of the cytochrome b mRNA. The suggestion is that these mutations block the production of some diffusible product of an early intron responsible for removal of a later intron. It was eventually found that before excision of the first intron between exon 1 and 2, the first exon and intron are translated into a protein called an **RNA maturase.** This maturase is necessary for excision of the first intron from the hnRNA (Figure 9–9). Mutations in sequences of the first intron prevent production of a functional maturase, resulting in blocking of the production of the functional mRNA.

The third step in RNA processing involves enzymatic addition of a segment of up to 200 adenylic acid residues called the **poly-A tail** to the 3′ end of the molecule. A poly-A tail seems to be present on all eukaryotic mRNAs except those responsible for histone synthesis. Poly-A addition appears to be an essential part of posttranscriptional processing inasmuch as chemicals that inhibit poly-A synthesis also prevent appearance of mRNA in the cytoplasm. Although the poly-A tail is not required for actual polypeptide synthesis, progressive loss of nucleotides from the poly-A segment ultimately reaches a critical minimum

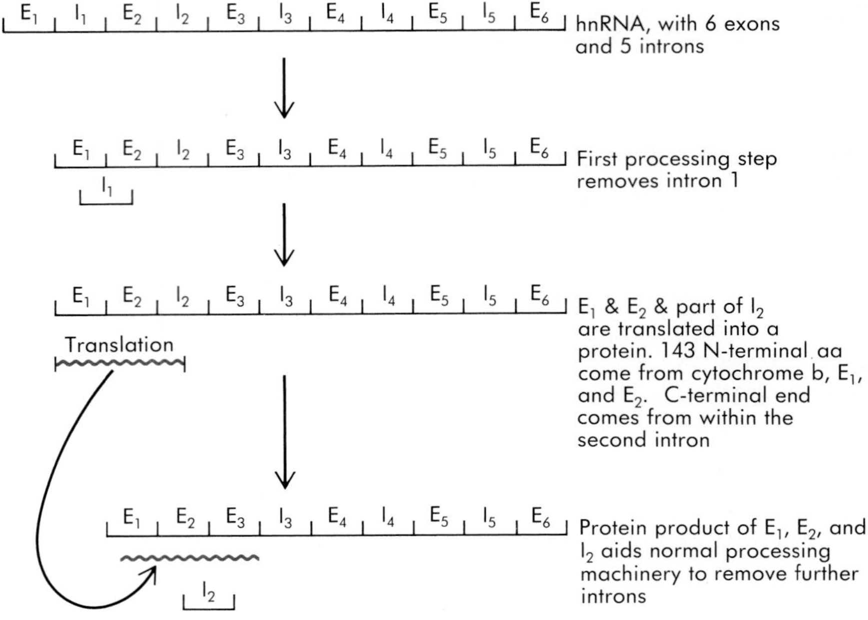
E1 I1 E2 I2 E3 I3 E4 I4 E5 I5 E6
hnRNA, with 6 exons and 5 introns
E1 E2 I2 E3 I3 E4 I4 E5 I5 E6
I1
First processing step removes intron 1
E1 E2 I2 E3 I3 E4 I4 E5 I5 E6
Translation
E1 & E2 & part of I2 are translated into a protein. 143 N-terminal aa come from cytochrome b, E1, and E2. C-terminal end comes from within the second intron
E1 E2 E3 I3 E4 I4 E5 I5 E6
I2
Protein product of E1, E2, and I2 aids normal processing machinery to remove further introns

FIGURE 9–9. *Cytochrome b gene processing in yeast mitochondrial DNA in which a translation product of the first two exons and part of the second intron is used for further processing of the messenger RNA.*

for mRNA stability. Furthermore, length of the poly-A tail differs for different types of messengers.

The number of ribonucleotides and, therefore, the length of any mRNA molecule, depends in large part on the particular polypeptide for whose synthesis mRNA is responsible. In general, lengths of several hundred to several thousand ribonucleotides have been reported. In bacteria, mRNA is produced near, or even at, its functional length. In *E. coli* the average mRNA length is about 1,000 ribonucleotides. Messenger RNA molecules responsible for the α and β polypeptide chains of human hemoglobin consist of more than 400 ribonucleotides each.

Messenger RNA of *E. coli* is quite short-lived, functioning for only a few minutes. It appears to retain its stability only as long as it is attached to polysomes (a multiple group of ribosomes on the same m-RNA) with the result that bacterial cells do not become "cluttered" with large amounts of mRNA. For example, the antibiotic actinomycin D blocks the synthesis of new mRNA in bacteria; studies using actinomycin-D-treated bacteria indicate that any particular bacterial mRNA is used only ten to twenty times. Therefore if different "species" of mRNA can be selectively produced (see Chapter 11), the cell is able to synthesize a variety of proteins at different times and under different conditions.

Recent studies on stability of mRNA in eukaryotes indicate somewhat longer functional periods than were suspected earlier. Half-lives of several hours to two or three days were found, a time span quite a bit greater than was determined in bacteria. The mRNA associated with hemoglobin production in reticulocytes and mature erythrocytes persists and functions for several days after degeneration of the nucleus. In most cases, mRNA is continuously produced, utilized, and degraded.

In those viruses whose genetic material is RNA instead of DNA—for example, tobacco mosaic virus (TMV), Rous sarcoma virus (RSV), polio virus, and phage Qβ—there is no transcription of mRNA from a DNA template. One of the best understood is the phage Qβ (Figure 9–10). The genome of this virus is single-stranded RNA, which has a molecular weight of about 1×10^6. Shortly after infection of susceptible bacterial cells, the viral RNA directs the production of (1) a coat protein, (2) a maturation protein, and (3) an RNA replicase subunit. The latter combines with host proteins to form functional *RNA replicase*. This enzyme attaches to the 3′ end of the virus RNA molecule (referred to as the plus strand), which it transcribes in the 3′→5′ direction to produce antiparallel polyribonucleotide molecules (the minus strands). Some experimental data suggest that this **replicative form (RF)** of RNA is a double helix through extensive base pairing, but other work has been interpreted as indicating that extensive base pairing does not occur. In any event, the minus strands thus produced then serve as templates for synthesis of several new plus RNA strands (Figure 9–10C). The complex of developing viral plus strands and their complementary RNA templates is called **replicative intermediates (RI).** In this type of virus, then, the single-stranded RNA genome (+) serves as a template for transcription of antiparallel, complementary minus strands, which, in turn, serve as templates for production of numerous new plus strands. Transcription is in the 3′→5′ direction of the template, and the new strands are synthesized in their 5′→3′ direction.

A different pattern of replication is demonstrated by the majority of plant viruses that are single-stranded RNA. An outline of this process is shown in Figure 9–11. The genome of the virus released into the plant cell is also called the (+) strand. This (+) strand acts as a messenger RNA and directs the synthesis of the viral **replicase,** the enzyme that will replicate the viral chromosome. The replicase first makes a (−) strand complementary to the (+) strand. This (−) strand is then used by replicase to make more (+) strands. Several (+) strands can be made from the (−) strand at once (similar to the Qβ example), and can serve as additional messengers for more replicase synthesis, as well as the synthesis of other viral proteins. Eventually the new (+) strands and proteins are assembled into new virus particles.

Perhaps the most interesting variation of RNA transcription occurs in a group of tumor-producing agents in most vertebrates including humans. Their genome consists of a single-stranded molecule of RNA and each particle contains two identical copies of the RNA. The replicating enzyme, contained in the virus, first makes a complementary

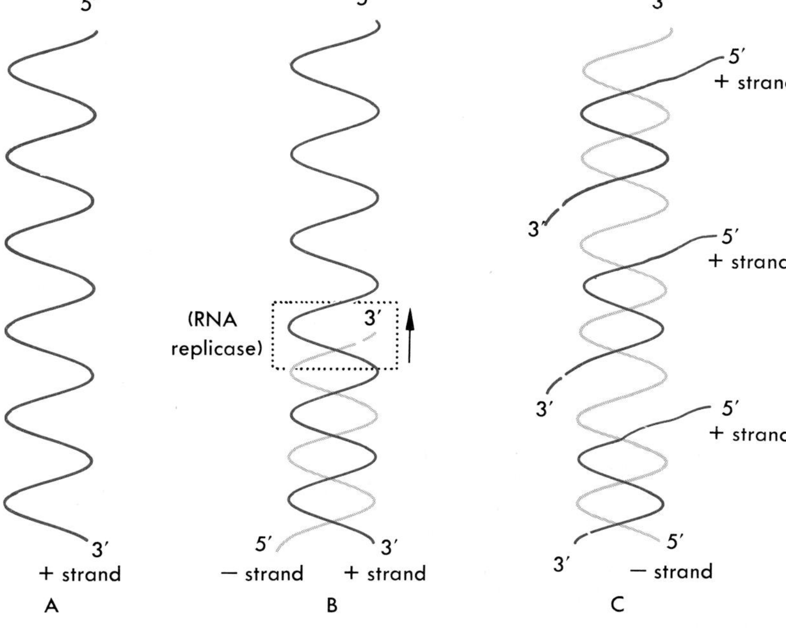

FIGURE 9–10. *Diagrammatic representation of formation of replicative form and replicative intermediates in the phage QB. Any base pairing is ignored. See text for details. (A) Single-stranded plus RNA strand of the phage. (B) Beginning of the formation of the replicative form (RF); synthesis of the (−) strand by RNA replicase occurs in the 5′ → 3′ direction. (C) Synthesis of new plus strands from the minus RNA template, forming a replicative intermediate. Several (+) strands are synthesized from a single (−) strand. Enzyme attachment is not shown.*

ACQUIRED IMMUNE DEFICIENCY SYNDROME (AIDS)

One of the most currently prevalent and publicized retroviruses is the one linked with acquired immune deficiency syndrome, or AIDS. According to the Centers for Disease Control, "AIDS is characterized by the presence of Kaposi's sarcoma and/or life-threatening opportunistic infections in a previously healthy individual less than 60 years of age who has no underlying immunosuppressive disease and has not received immunosuppressive therapy." More current definitions also include brain dysfunction, dramatic weight loss, and a broader list of AIDS-indicator diseases.

A diagram of the AIDS virus is found below. Each virus particle is covered by two lipid layers from the host cell. Two glycoproteins (proteins attached to sugars) are associated with these layers. **gp41** is within the lipid membrane, and **gp120** extends out beyond the lipid membrane and creates the second layer. This double-layered envelope covers a double-layered protein core made up of **p24** and **p18** proteins. The viral RNA is carried inside the core, along with molecules of reverse transcriptase.

The first cases of AIDS were reported in 1981 and three years later its cause was shown to be due to a human retrovirus: human immunodeficiency virus or HIV. The host cell for the virus is the T4 cell, noted for its central role in regulating the various processes of the immune system. Once the virus kills enough of these T4 cells, the victim is susceptible to opportunistic infections by a variety of microorganisms not dangerous to a healthy person. In addition to affecting the immune system, the AIDS virus also affects the brain and spinal chord. People infected with HIV are also at risk for developing three different types of cancers: Kaposi's sarcoma, which is characterized by skin tumors that spread to internal organs and cause internal bleeding and death; cancerous growths of the skin and mucous membranes; and tumors originating in the B lymphocytes.

AIDS is the final stage of HIV infection. The virus seems to move in a progression, that is, it seems to follow a regular course. The HIV infection and incubation period varies from three to ten years. During this time an infected person may show no symptoms. After this initial incubation period, however, infected people will develop the AIDS-related complex (ARC). People with ARC have a depletion of T-helper cells and continuously swollen lymph glands. ARC is usually followed by the appearance of AIDS.

The AIDS virus is transmitted four ways: by sexual contact; by contaminated blood or blood products; by the sharing of drug needles contaminated with the virus; and by an infected woman passing the virus to her child before, during, or shortly after birth. There is no evidence that the virus can be transmitted by casual contact such as shaking hands, hugging, or through food or water. The virus is easily killed by common disinfectants, including household bleach. The Public Health Service estimates that 270,000 people in the United States will have AIDS or will have died from the disease by 1991. Antiviral vaccines and drugs are being developed but will probably not be tested until 1990.

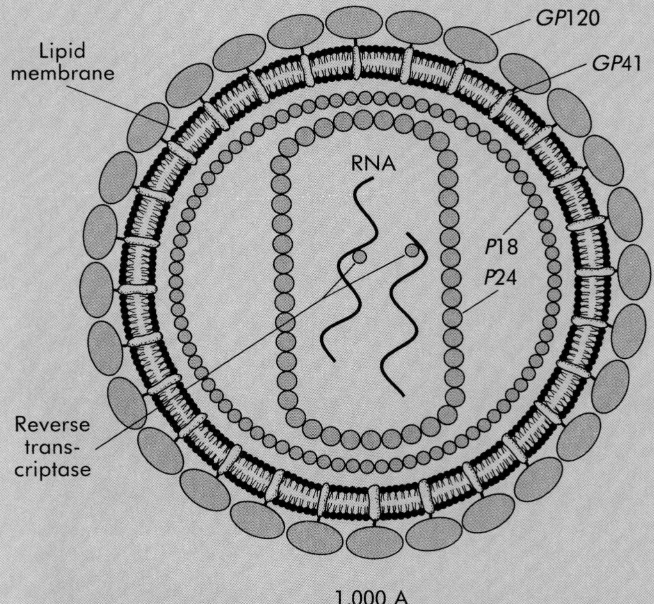

(Reprinted by permission of Macmillan Publishing Company from Microbiology, *5th edition, by George A. Wistreich and Max D. Lechtman. Copyright © 1988 by Macmillan Publishing Company.)*

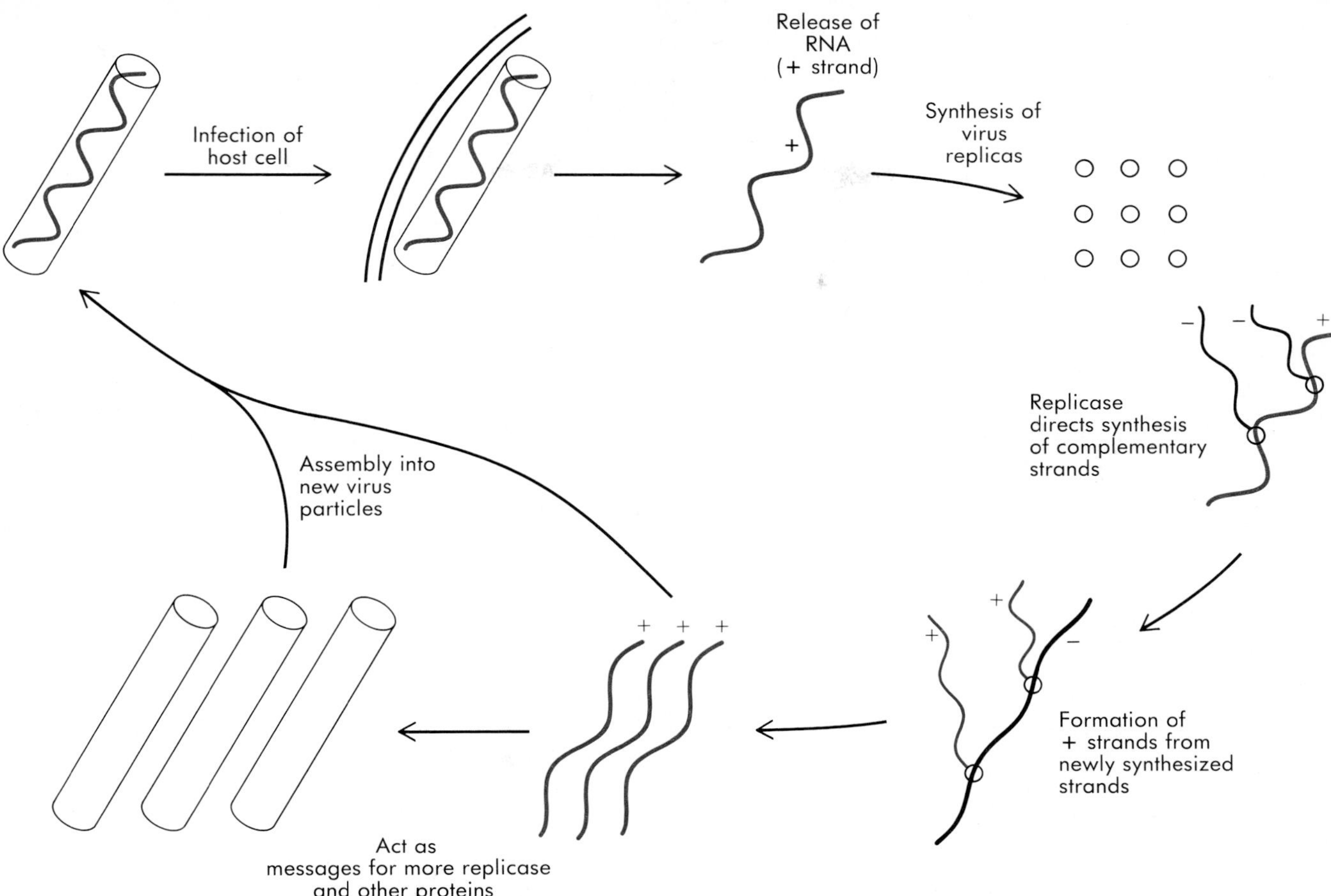

FIGURE 9–11. *Schematic representation of replication of single-stranded RNA plant viruses.*

double-stranded DNA copy of the viral RNA. This DNA integrates into the host chromosomal DNA, and through normal transcription of this DNA, new virus chromosomes are made. Since this process is the reverse of transcription, the enzyme is called **reverse transcriptase.** This backward copying of RNA into DNA has earned these viruses the name **retroviruses** as a group. The integration of the DNA copy into the host cell's chromosome undoubtedly has a role in transforming the cell into a tumor cell.

Transfer RNA (tRNA). Transfer RNA molecules serve as adapter molecules to bring amino acids to the site of protein synthesis. There is at least one tRNA that is specific for each of the twenty amino acids. As will be seen, there is more than one tRNA molecule for each amino acid.

The structure of **transfer RNA** is now well understood. This line of research was opened in 1965 when, after seven years of intensive work, Robert Holley and his colleagues reported the complete nucleotide sequence of alanine tRNA of yeast (Holley received a Nobel Prize in 1968 for this work). Nucleotide sequences are now known for more than 100 different "species" of tRNA. Literature on this molecule has now become quite voluminous and new papers are appearing frequently.

Transfer RNA has several unique characteristics: (1) It is a relatively small molecule of 75 to 90 ribonucleotides and is thus smaller than either mRNA or any of the rRNAs, and has a sedimentation coefficient of 4*s*; (2) a number of "unusual" nucleotides are found in tRNA, many of

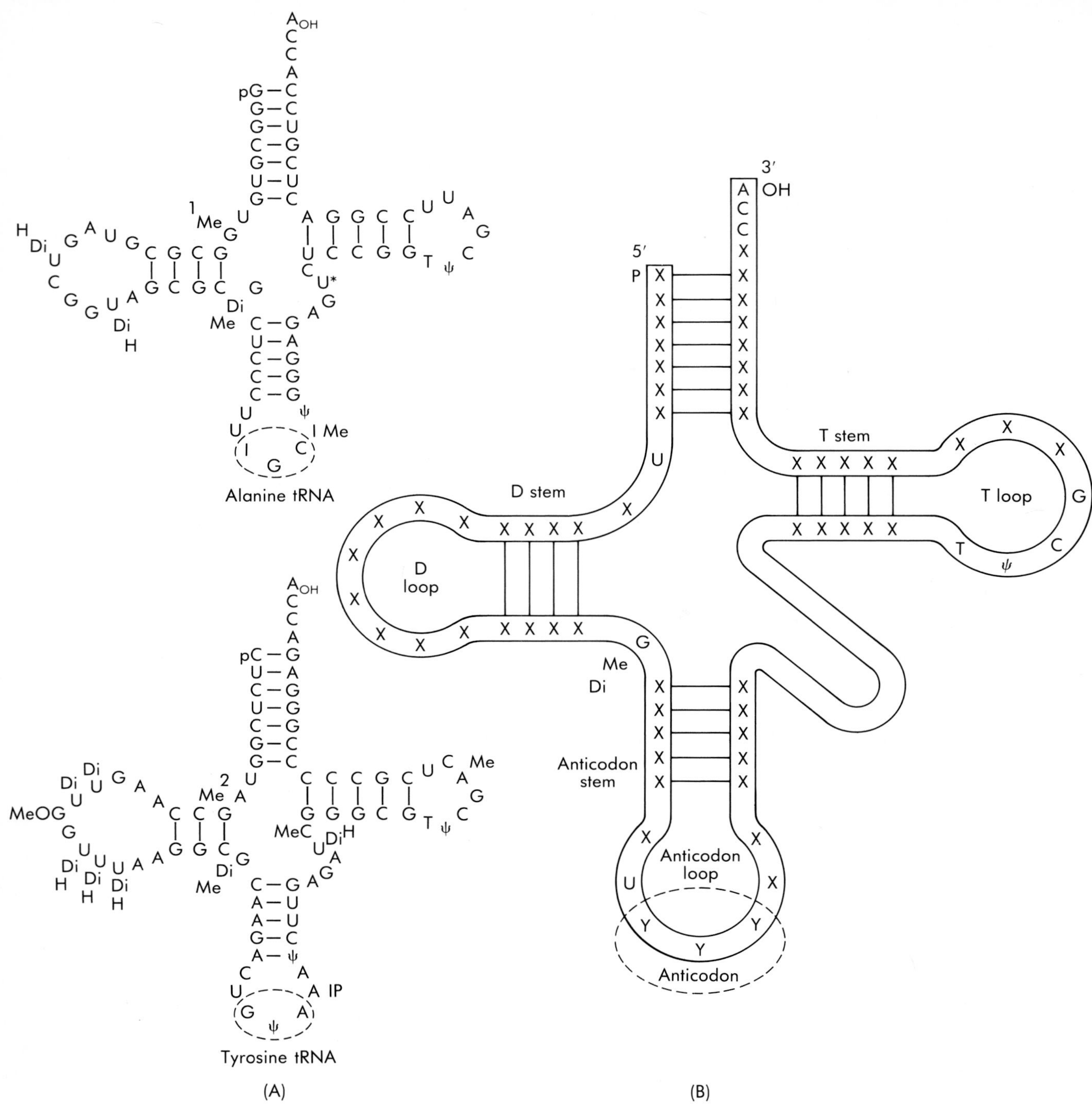

FIGURE 9–12. *(A) Cloverleaf model structure for yeast alanine tRNA and tyrosine tRNA. Note that internal pairing is far from complete, and a terminal -A-C-C-A occurs in each molecule. A large number of "unusual" bases, such as pseudouridine, occurs. The presumed anticodon (see text) is enclosed by the dashed oval.* (Courtesy Dr. J. T. Madison, 1966, used by permission.) *(B) Generalized two-dimensional cloverleaf model of tRNA, based on analyses of several yeast tRNA molecules by various investigators. Note the common 3′ terminal -CCA, the TCG in the T loop, the anticodon-U (here YYYU) in the anticodon loop, the DiMeG between the anticodon loop and D loop, and the U as the first unpaired base on the 5′ strand. As diagramed here, the anticodon is read from right to left (3′ → 5′). The number of paired bases in each lobe shown here appears to be constant for all tRNA species so far reported. (A = adenosine, C = cytidine, G = guanosine, T = ribothymidine, U = uridine, ψ = pseudouridine, DiMeG = dimethylguanosine, Y = any base of the anticodon.)*

them methylated derivatives of the common ones (for example, l-methylguanylic acid, 2N-methylguanylic acid, 1-methyladenylic acid, ribothymidylic acid, 5-methylcytosine), whose importance lies in the fact that the presence of methyl groups prevents formation of complementary base pairs and thus affects the three-dimensional form of the molecule; (3) hydrogen bonding between usual bases, and its lack between many of the unusual ones, also affects the shape of the tRNA molecule. Research from the late 1960s and early 1970s makes it clear that tRNA molecules may have either three or four large single-stranded lobes, depending on the kind, number, and sequence of nucleotides. This gives them a so-called cloverleaf shape. Two-dimensional models are shown in Figure 9–12A, and a generalized diagram is in Figure 9–12B.

There are some common characteristics of the cloverleaf configuration: (1) the 3′ end carries a terminal C-C-A sequence to which an amino acid is covalently attached during polypeptide synthesis; (2) the "stem" or amino acid helix consists of seven paired bases; (3) the T-stem is composed of five base pairs—the last (that is, nearest the T-loop) is C-G; (4) the anticodon stem includes five paired bases; (5) the anticodon loop consists of seven bases, the third, fourth, and fifth of which (from the 3′ end of the molecule) constitute the **anticodon,** which permits temporary complementary pairing with three bases on mRNA; (6) the base on the 3′ side of the anticodon is a purine; (7) immediately adjacent to the 5′ side of the anticodon, uracil and another pyrimidine occurs; (8) a purine, often dimethylguanylic acid, is located in the "corner" between the anticodon stem and the D-stem; and (9) the D-stem is composed of three or four base pairs (depending on the "species" of tRNA). The D-loop is also variable in size. The extra arm (shown extending toward the lower right in Figure 9–12B) is variable in nucleotide composition and is lacking entirely in some tRNAs. X-ray diffraction studies are now providing insight into the three-dimensional structure of tRNA (Figure 9–13).

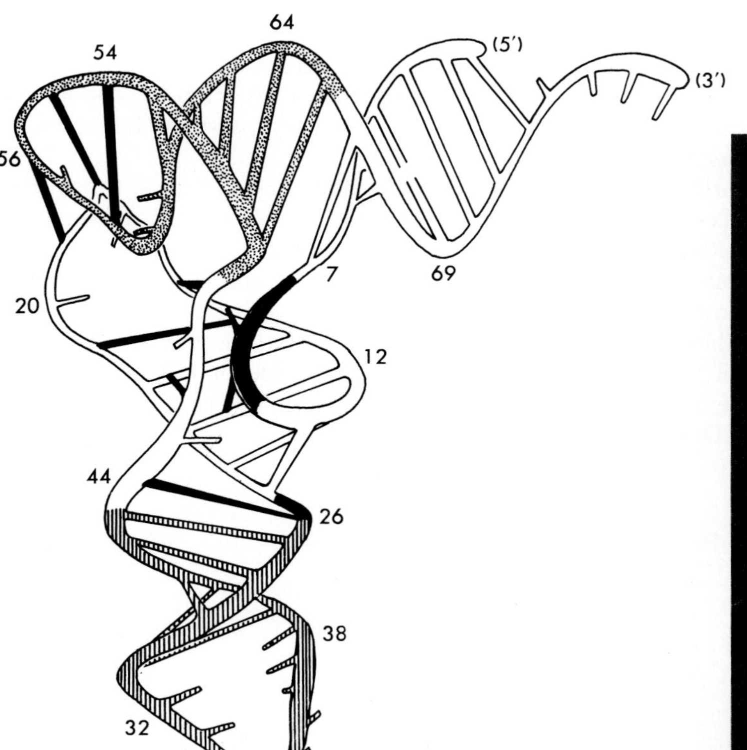

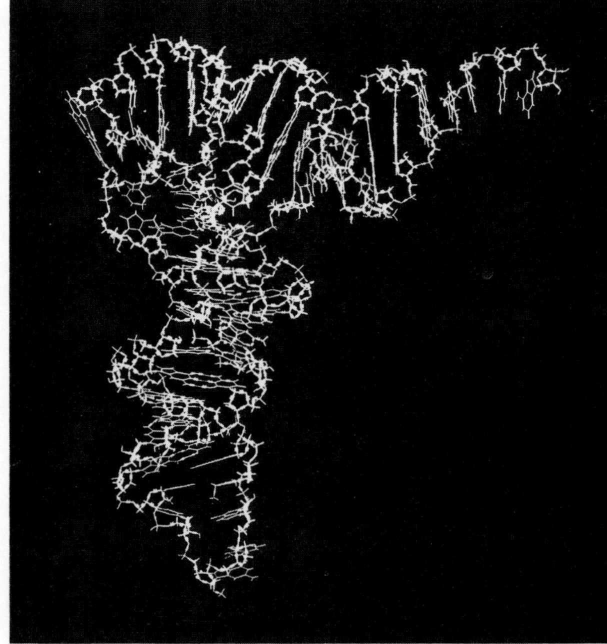

FIGURE 9–13. *(A) Schematic three-dimensional model of yeast phenylalanine tRNA molecule. The ribose–phosphate "backbone" is shown as a continuous cylinder with bars indicating hydrogen-bonded base pairs. Unpaired bases are indicated by shorter "branches." The T C arm is shown in heavy stippling; the anticodon arm is indicated by vertical lines. Black portions represent tertiary interactions.* (Redrawn from S. H. Kim et al. 1974 with permission.) *(B) Actual three-dimensional model of yeast phenylalanine tRNA molecule.* (Courtesy S. H. Kim, used by permission.)

Transfer RNA is transcribed from several particular sites on template DNA and comprises about 15 percent of the RNA present at any one time in an *E. coli* cell. Transfer RNAs are also processed from larger precursors. In prokaryotes, processing includes removal of nucleotides from the precursor and modification of some internal nucleotides. For example, in *E. coli* the precursor for the amino acid tyrosine (pre-$tRNA^{tyr}$) consists of 128 bases; in processing to $tRNA^{tyr}$ forty-one nucleotides are removed from the 5′ end and two are removed from the 3′ end.

In addition to processing, methylation and inclusion of other "unusual" intercalary bases takes place after transcription, as does the addition of the 3′ terminal -C-C-A. The final, functional molecule has 85 bases. In general terms, processing of tRNA precursors involves several major steps: (1) processing out of "extra" nucleotides, (2) methylation and other modifications of many intercalary bases, (3) addition of the 3′ terminal -C-C-A if it was not part of the original transcript, and (4) assumption of the final three-dimensional shape through hydrogen bonding. Also in eukaryotes, tRNAs are processed from larger precursors, many of which contain intervening sequences processed out in a manner similar to mRNA processing. The remaining bases are then joined together, and some of the intercalary bases are modified. Processing of eukaryotic tRNA is often much more complex than suggested here.

Ribosomes and ribosomal RNA (rRNA). The site of polypeptide synthesis is the **ribosome,** a small particle that averages approximately 175 × 225 Å, composed of protein and **ribosomal RNA.**

Ribosome structure and composition are becoming better understood, although many details remain to be elucidated. Ribosomal characteristics are best known in *E. coli*. The intact bacterial ribosome has a sedimentation coefficient of 70*s* and can be divided into two subunits: 30*s* and 50*s*. The 30*s* subunit contains 21 different proteins, each represented by one molecule. The amino acid sequence has been determined for some of these proteins. They range from 74 to 203 amino acid residues. A single 16*s* molecule of ribosomal RNA (rRNA), consisting of 1,600 ribonucleotides whose sequence has been partially determined, is also present.

The 50*s* ribosomal subunit of *E. coli* contains 34 proteins; those that have been sequenced are made up of 46 to 178 amino acid residues. A single 23*s* rRNA molecule supplies about one-half the mass of the subunit. Sequencing of this 23*s* rRNA molecule of about 3,200 ribonucleotides has not yet been completed. There is also a 5*s* rRNA molecule of some 120 ribonucleotides that has been completely sequenced. In *Bacillus subtilis* it is cleaved from a precursor molecule of 179 nucleotides. Some of the main characteristics of *E. coli* ribosomes are summarized in Table 9–2. The eukaryotic ribosome is quite similar to the prokaryotic

TABLE 9–2. 70*s* Ribosome Composition in *Escherichia coli*

		rRNA		
Ribosomal subunit	*Proteins*	*Sedimentation coefficient*	*Molecular weight*	*Ribonucleotides*
30*s*	21	16*s*	5.5×10^5	1,600
50*s*	34	5*s*	4×10^4	120
		23*s*	1.1×10^6	~3,200

TABLE 9–3. 80s Ribosome Composition in Rat Liver

Ribosomal subunit	*Proteins*	*RNA* *Sedimentation coefficient*	*Molecular weight*	*Ribonucleotides*
40*s*	30	18*s*	7×10^5	2,100
60*s*	50	5*s*	4×10^4	120
		5.8*s*	—	—
		28*s*	1.8×10^6	5,300

ribosome except that it is somewhat larger (see Table 9–3). Prokaryotic and eukaryotic ribosomes also differ in their sensitivity to antibiotics, which inhibit protein synthesis.

Understanding of the structure of the bacterial ribosome has been greatly increased by the reconstruction experiments of Nomura and his colleagues. By mixing the various components (RNA and proteins) of ribosomes and assaying each mixture for function they were able to determine such things as how ribosomes are assembled, the degree of relatedness between prokaryotic and eukaryotic ribosomes, and the importance of each component when it is left out of the reconstitution mixture. The important findings are: (1) the 30*s* particle requires no nonribosomal component for assembly. In other words, it is capable of self-assembly if all the components are present. The basic small subunit particle will assemble when the 16*s* rRNA reacts with about fifteen proteins. After some rearrangement, the remaining six proteins associate to form the small subunit. (2) Functional reconstitution will not occur if eukaryotic proteins are mixed with bacterial RNA or bacterial proteins are mixed with eukaryotic RNAs. (3) When reconstitution experiments are conducted with single proteins omitted from the mixture, sometimes no particle forms, or if one does, it is not functional. We know much less about structural relationships between proteins and subunits of ribosomes in eukaryotes.

In eukaryotes, the genes for the 18*s*, 5.8*s*, and 28*s* rRNA molecules exist next to each other in the order mentioned. Together they comprise a **transcriptional unit.** That is, a primary transcript is made of all three genes, including spacer regions between the genes. The transcript is then processed into the functional rRNA molecules in a similar manner to other RNA processing. In addition, the three genes are tandemly repeated many times. The 5*s* gene is at another location. The primary transcript is about 720 bases long and is processed into the functional 120-base 5*s* rRNA molecule. These 5*s* rRNA genes are also tandemly duplicated many times at their chromosome location.

Ribosomes that function in polypeptide synthesis occur in linear groups connected by messenger RNA; such groups of ribosomes are called **polysomes.** In *E. coli* and other prokaryotes the ribosomes are scattered through the cytoplasm, but in immature erythrocytes (reticulocytes) and in other eukaryotic cells they are situated on the endoplasmic reticulum. There may be 10,000 ribosomes in a single *E. coli* cell, and an even greater number in a eukaryotic cell.

Minimum Necessary Materials

The stage is now set for an examination of the role of each of these components in the synthesis of polypeptides. Success in such synthesis in *in vitro* cell-free systems shows that minimal necessary equipment and supplies are as follows:

1. Amino acids.
2. Ribosomes containing proteins and rRNA.
3. mRNA
4. tRNA of several kinds
5. Enzymes
   a. Amino-acid-activating system
   b. Peptide polymerase system
6. Adenosine triphosphate (ATP) as an energy source
7. Guanosine triphosphate (GTP) for synthesis of peptide bonds
8. Soluble protein initiation and transfer factors
9. Various inorganic cations (for example, K^+ or NH_4^+, and Mg^{2+})

POLYPEPTIDE CHAIN SYNTHESIS—THE PROCESS

Translation

It is now time in the flow of information between genes and proteins to change the language being used. In going from DNA to RNA the language (nucleotide sequences) remained the same. In going from RNA to protein the language is changed from a nucleotide sequence to an amino acid sequence. Just as in the process of translating from one language to another, this process of using information in RNA to make a protein is called **translation.** Amino acids alone do not come to the ribosome to be incorporated into protein. Instead, they are brought to the ribosome by their appropriate tRNA. The first step in incorporating an amino acid into a protein involves the amino acid's attachment to its correct tRNA. This is a two-step process and proceeds as follows.

Amino acid activation. Each of the twenty kinds of amino acids must be *activated* before they can attach to their tRNAs. In activation, an enzyme (aminoacyl-tRNA-synthetase) catalyzes the reaction of a specific amino acid with adenosine triphosphate (ATP) to form aminoacyl-adenosine monophosphate (aminoacyl adenylate) and pyrophosphate:

$$\text{AA} + \text{ATP} \xrightarrow[]{\text{amino acyl synthetase}} \text{AA} \sim \text{AMP} + \text{pyrophosphate}$$

The amino acyl adenylate (AA-AMP) is referred to as an **activated amino acid,** which is linked to adenosine by a phosphate ester bond, and is now raised to an energy level whereby it can react with its tRNA.

By its architecture, each kind of enzyme molecule must "recognize" and be able to fit with a particular amino acid. Each of the various "species" of aminoacyl-tRNA-synthetase possesses two binding sites. Each of these "recognizes" one, and only one, amino acid of the twenty. As many as twenty different amino acids can be involved; hence each cell must have at least twenty different kinds of aminoacyl synthetase.

The activated amino acid, however, is strongly attached to its enzyme and cannot, in this condition, be joined with another to begin formation of a polypeptide chain. This process requires transfer of the activated amino acid to the tRNA.

Transfer of activated amino acid to tRNA. The same amino acyl synthetase that catalyzed the activation of the amino acid next attaches to a receptor site at the terminal adenine ribonucleotide of a particular tRNA molecule when the two come into contact as a result of a chance collision. Then tRNA is said to be charged:

$$AA \sim AMP + tRNA \rightarrow AA \sim tRNA + AMP$$

The aminoacyl adenylate attaches to the free 3′ – OH of the ribose of the 3′ terminal adenylic acid of tRNA by forming an acyl bond with the alpha-carboxyl group of the amino acid. Note that the enzymes must be able to bind specifically to both a particular amino acid and also to a particular tRNA molecule. Therefore, in addition to the minimum of twenty different kinds of synthetases, each cell must have at least twenty different tRNA species. In fact, there is good experimental evidence for the existence of two or more different species of tRNA having the same degree of specificity for the same amino acid.

The specificity of the binding of the synthetase to its amino acid and the transfer to the correct tRNA is dependent, in part at least, on the structure of the amino acid's R group and specific regions probably in the T- or D-loop of the tRNA molecule. Since the differences between the R groups of some amino acids are small (Figure 8–2), occasional errors are made in this process and the wrong amino acid is attached to a tRNA. Recent evidence suggests that the aminoacyl synthetase checks for correct binding before release of the tRNA-AA. If an error has been made, the wrong amino acid is removed and the correct one is attached.

Assembly of Polypeptides

Following amino acid activation and binding to transfer RNA, the charged tRNA diffuses to the ribosome, where actual assembly into polypeptide chains takes place. The following description of peptide bond formation applies to *E. coli* in particular and to bacteria in general, organisms in which the process is best known. It apparently differs only in minor detail from the following series of events in eukaryotes.

When mRNA binds to the smaller (that is, 30*s*) subunit of the bacterial ribosome, it displays at least two groups there of three ribonucleotides each, which constitute sites capable of accepting charged tRNA molecules. (Evidence for three ribonucleotides, rather than some other number, is explored in the next chapter.) These sites are designated the **aminoacyl (A) site** and the **peptidyl (P) site,** respectively. At or near the 5′ end of the associated mRNA molecule a chain-initiating group of (three) ribonucleotides must be present. The first ribosome attaches to this 3-base initiating site, often as the mRNA is still being transcribed (Figure 9–14). In *E. coli,* the first amino acid to be incorporated is N-formylmethionine. This is methionine with a formyl group,

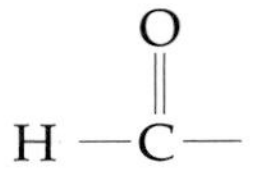

$$\begin{array}{c} \text{O} \\ \| \\ \text{H}-\text{C}- \end{array}$$

attached to the amino group:

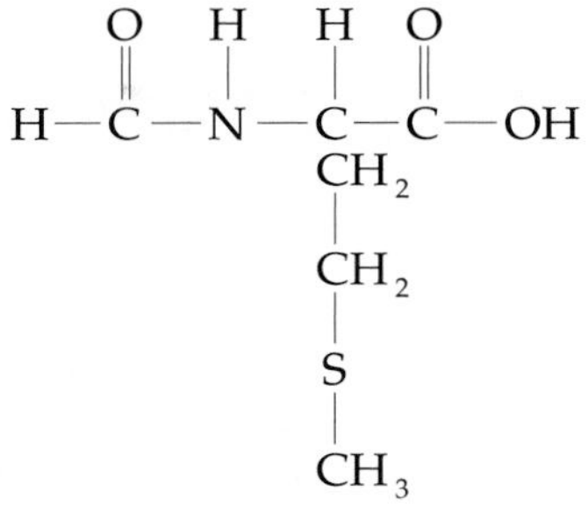

$$\begin{array}{ccccccccc} & \text{O} & & \text{H} & & \text{H} & & \text{O} & \\ & \| & & | & & | & & \| & \\ \text{H}- & \text{C} & - & \text{N} & - & \text{C} & - & \text{C} & -\text{OH} \\ & & & & & \text{CH}_2 & & & \\ & & & & & | & & & \\ & & & & & \text{CH}_2 & & & \\ & & & & & | & & & \\ & & & & & \text{S} & & & \\ & & & & & | & & & \\ & & & & & \text{CH}_3 & & & \end{array}$$

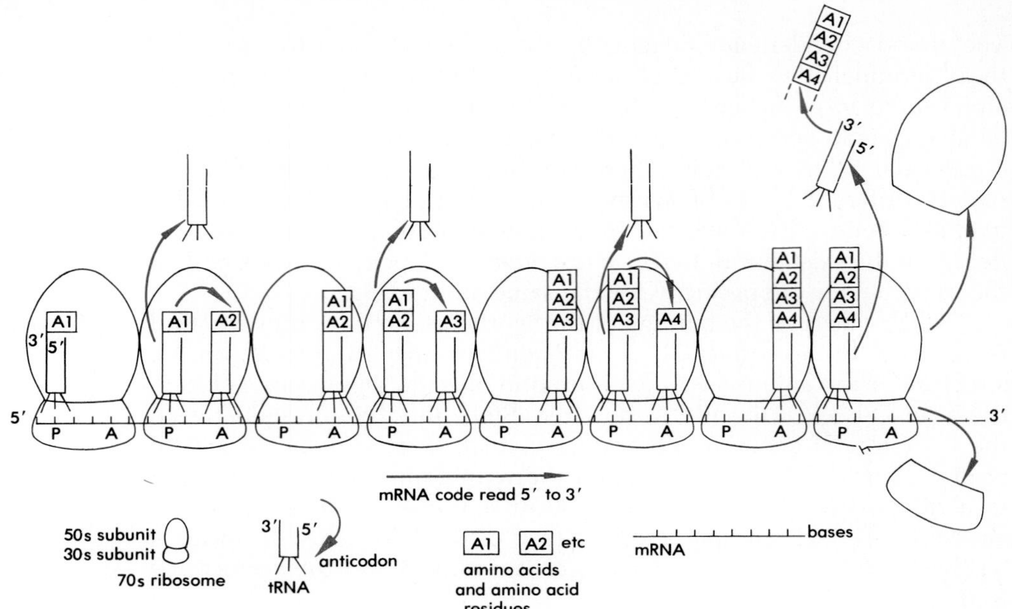

FIGURE 9–14. *Diagrammatic representation of polypeptide synthesis in* Escherichia coli. *Ribosomes first engage mRNA toward its 5′ end at a "start translation" codon, then progress toward the 3′ end. At that point a chain-termination codon is encountered and recognized by protein-release factors; the polypeptide is then separated from the ribosome, and the latter dissociates into its component subunits. In this diagram the ribosome shown as having just been separated into its subunits (extreme right) represents the first to have engaged the mRNA molecule. See text for detailed outline of the steps involved.*

Methionine is formylated enzymatically after attachment to its tRNA. Not all methionine-tRNA complexes (met-tRNA) can be formylated and there are two kinds of met-tRNA, only one of which permits formylation. The formyl group blocks formation of a peptide bond with the carboxyl group of another amino acid. Depending on the particular polypeptide being synthesized, the N-formylmethionine may or may not later be removed enzymatically before the polypeptide chain becomes functional, though in most cases it is removed.

The process of the sequential incorporation of amino acids into proteins by peptide bond formation can be divided into three stages: Initiation, Elongation and Termination (see also Figure 9–14).

Initiation

1. The 30*s* subunit of the first ribosome attaches to mRNA at an *initiator* site at a fixed distance before the initiation codon, which is the nucleotide sequence AGGAGGU, and called the Shine-Delgarno sequence after its discoverers.
2. $tRNA^{fmet}$, charged with the first amino acid (N-formylmethionine), binds to the 30*s* subunit at the P site at three ribonucleotides. Three protein initiation factors (IF) plus guanosine triphosphate (GTP) are required.
3. Immediately upon completion of step 2, the larger 50*s* subunit attaches to the 30*s* subunit and completes assembly of the first ribosome, which still carries the tRNA charged with N-formylmethionine at the P site.

Elongation

4. The second charged tRNA binds to this first ribosome at the latter's A site with help from the **elongation factor** protein, so that both sites are occupied; that is, the P site with the first tRNA, the A site with the second.
5. A peptide bond is formed enzymatically (by peptidyl transferase, which is a protein component of the 50s subunit) between the amino group of the second amino acid and the carboxyl group of the first amino acid.
6. Translocation, which consists of three steps, then occurs:
   a. ejection of the tRNA^{fmet} from the P site
   b. movement of the tRNA-dipeptide from the A site to the P site
   c. movement of the mRNA such that the effect is the apparent movement of the ribosome toward the 3′ end of the mRNA (by three nucleotides)
7. Step 6 brings the previous A site (with its dipeptide) into location at the P site with the aid of **elongation factor G.**
8. A third charged tRNA moves into position at the new A site with the aid of the same elongation factor previously used in Step 4.
9. The process outlined in Steps 4–8 is then repeated again and again, with new ribosomes successively engaging the initiation site and moving or being moved along mRNA in the 5′→ 3′ direction. The details of this elongation phase are illustrated in Figure 9–15.

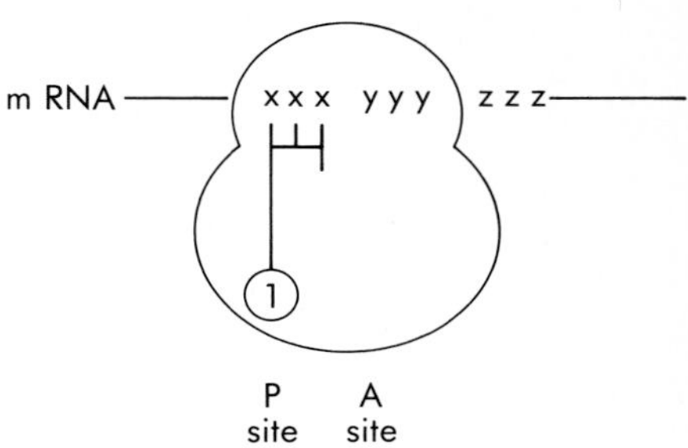

(A) Binding of charged t-TNA at P site

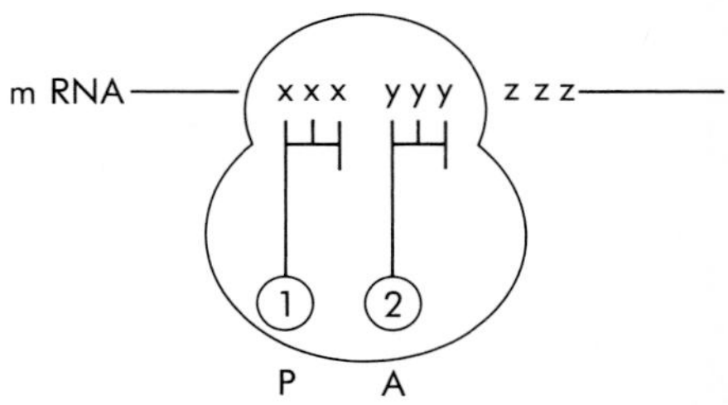

(B) Binding of charged t-RNA at A site

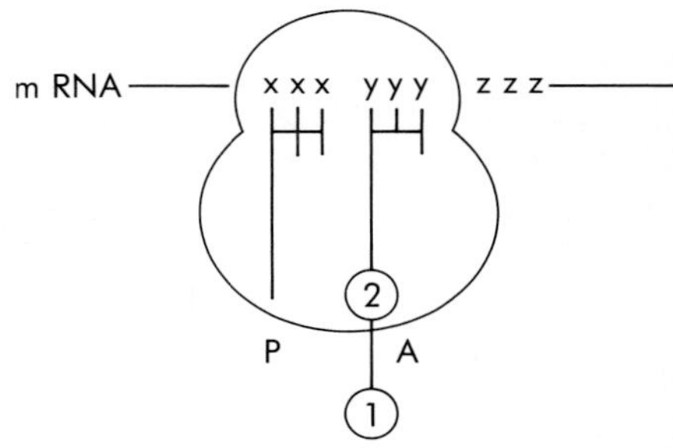

(C) Peptide bond formation by peptidyl synthetase

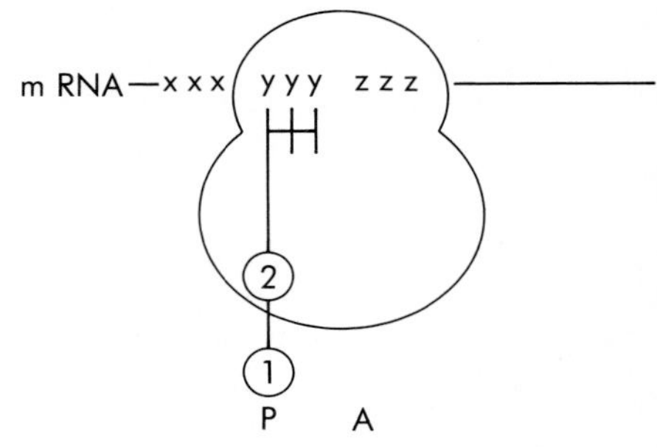

(D) Translocation movement of mRNA–t-RNA complex, so tRNA is at P site. A site is open for the next charged t-RNA

FIGURE 9–15. *Details of elongation phase of protein synthesis.*

Termination

10. These steps are continued until the first, and then successive, ribosomes reach a *termination* group of three ribonucleotides, at which point the ribosomes (each bearing a completed polypeptide) are successively ejected from the polysome with the aid of three protein release factors. The ribosomal subunits dissociate and are able to repeat the entire process.

Extent and Speed of Translation

As stated earlier, the life of any mRNA molecule is finite. It continues to function in the manner described as long as ribosomes attach to its 5′-end, but this terminus is vulnerable to ribonucleases. These are enzymes that hydrolyze RNA, and degrade it into its constituent nucleotides, which may then be used to construct wholly new and different mRNAs. Half-life for much bacterial mRNA appears to be of the order of one to three minutes, but longer in eukaryotes—from a few hours in mouse liver to several or many days in reticulocytes. Relatively little quantitative data are available on the stability of eukaryotic mRNA.

Translation occurs with considerable speed. In hemoglobin production, for example, one polypeptide chain of nearly 150 amino acid residues requires about 80 seconds for synthesis. This means addition of one amino acid in a little more than 0.5 second!

In Vitro Protein Synthesis

A superb confirmation of the general outline of protein synthesis was furnished by Von Ehrenstein and Lipmann, who succeeded in synthesizing hemoglobin in a cell-free system. Such a system included:

1. Ribosomes in the form of polysomes from rabbit reticulocytes (immature red blood cells in which hemoglobin is almost the only protein synthesized) that included mRNA for rabbit hemoglobin.

2. An energy-yielding triphosphate.
3. tRNA from the bacterium *E. coli,* previously charged with amino acids using *E. coli* enzymes.

The most exciting aspect of this work is that Von Ehrenstein and Lipmann were able to synthesize a protein of one species using its mRNA and the tRNA of a vastly different organism. Not only is universality of the general mechanism of protein synthesis strongly suggested but the fundamental "kinship" of all living organisms through their DNA seems inescapable. It is abundantly clear that DNA is the genetic material of all living organisms (and many viruses, too) and that it is an information vocabulary of four "letters" (adenine, thymine, guanine, and cytosine). Differences among species thus reside in the sequences of the nucleotide "letters" to be formed into certain code "words" (amino acids), which will be combined sequentially into protein "sentences" and phenotypic "paragraphs."

But there is still another problem in translation that must be explored, namely, just how the sequence of nucleotides in mRNA is responsible for the particular sequence of amino acid residues of a given polypeptide and just how synthesis of such chains starts and stops. The nature of the *genetic code* will be examined in the next chapter.

PROBLEMS

9–1 Could PKUs be helped by administration of phenylalanine hydroxylase? (You may have to consult sources outside your textbook to answer this question.)

9–2 If alkaptonurics were given large quantities of parahydroxyphenylpyruvic acid, would they excrete larger quantities of alkapton?

9–3 Would increased intake of maleylacetoacetic acid increase the excretion of alkapton?

9–4 In 1963, P. D. Trevor-Roper reported a family of four normally pigmented children born to a husband and wife, both of whom were albinos. In terms of the gene symbols used in Figure 9–1, give the genotypes of these albino parents and their children.

Use the following information to answer the next four problems.

All microorganisms utilize thiamine (vitamin B_1) in their metabolism. Final steps in its synthesis involve enzymatic synthesis of a thiazole and of a pyrimidine, followed by the enzymatic combination of these two substances into thiamine. Consider the following mutant strains of *Neurospora*: strain 1 requires only simple inorganic raw materials in order to synthesize thiamine; strain 2 grows only if thiamine or thiazole is supplied; strain 3 requires thiamine or pyrimidine; strain 4 grows if thiamine or both thiazole and pyrimidine are supplied. Call the enzyme responsible for synthesis of thiazole from its precursor *enzyme "a,"* the one that catalyzes formation of pyrimidine *enzyme "b"* and the enzyme that catalyzes the combining of thiazole and pyrimidine into thiamine *enzyme "c."*

9–5 Which of the strains described above is prototrophic?

9–6 Which enzyme or enzymes is strain 2 incapable of producing?

9–7 Which enzyme or enzymes cannot be produced by strain 4?

9–8 If we assign gene symbol *a* to any strain incapable of producing enzyme "a," symbol *b* to strains not making enzyme "b," and symbol *c* to those not producing enzyme "c," with + signs denoting the ability to produce a given enzyme, each strain can be represented by three gene symbols, plus signs, and/or letters in the appropriate grouping. Give the genotype, according to this plan, for each of the four strains listed above.

9–9 Beadle and Tatum studied a group of arginine auxotrophs in *Neurospora.* Prototrophs are able to synthesize the required amino acid arginine in a three-step process as follows: precursor → ornithine → citrulline → arginine. Auxotrophic strain 1 grows only on media containing arginine; strain 2 grows provided ornithine, or citrulline, or arginine is supplied; strain 3 grows if either arginine or citrulline is supplied. Call the enzyme that catalyzes the first metabolic step *"a,"* the one that catalyzes the second step *"b,"* and the enzyme that catalyzes the third step *"c."* Which strain is incapable of producing (a) enzyme *a,* (b) enzyme *b,* (c) enzyme *c?*

9–10 A short segment from a long DNA molecule has this sequence of nucleotide pairs:

3′ATCTTTACGCTA5′
5′TAGAAATGCGAT3′

(a) If the 5′, 3′ ("lower") DNA strand serves as the template for mRNA synthesis, what will be the sequence of ribonucleotides on the latter?
(b) Give the ribonucleotide at the 5′-end of the mRNA molecule thus transcribed.
(c) What would your answer to part (b) be if the 3′, 5′ DNA strand served as the template?

9–11 A sedimentation coefficient is calculated as 2×10^{-11} second. Express this value in Svedberg units.

9–12 Assume 101 deoxyribonucleotide pairs are responsible for a particular portion of a certain mRNA molecule. What is the length of that stretch of mRNA in (a) angstrom units; (b) micrometers?

9–13 Assume a certain RNA molecule consists of only the "usual" ribonucleotides cytidylic, uridylic, adenylic, and guanylic acids, which occur in equal numbers. You determine the molecular weight of that RNA molecule to be about 27,000. (a) Of how many ribonucleotides does it consist? (b) How long would this molecule be in micrometers if it were laid out in a straight line? (c) Identify the kind of RNA molecule.

9–14 Recall that the genome of phage Qβ is single-stranded RNA that has a molecular weight of 1×10^6. Of about how many ribonucleotides does the genome of this phage consist, if only the four "usual" bases are assumed to occur and in equal number?

9–15 If a portion of the + strand of phage Qβ has the ribonucleotide sequence 3′ A U C G G U U A G . . . 5′, give the ribonucleotide sequence of the − strand.

9–16 A certain codon is determined to be AUG. (a) Of what nucleic acid molecule is this codon a part? (b) What is the corresponding anticodon? (c) Of what nucleic acid molecule is this anticodon a part? (d) What is the deoxyribonucleotide sequence responsible for this codon?

9–17 Differentiate between transcription and translation.

9–18 How do the functions of rRNA, mRNA, and tRNA differ?

9–19 Human hemoglobin polypeptide chain alpha (α) consists of 141 amino acid residues. Would you expect the DNA segment responsible for the ultimate synthesis of this chain to be shorter than, longer than, or about the same length as the functional mRNA molecule for this chain?

THE GENETIC CODE

CHAPTER

10

In preceding chapters it has been established that

1. DNA is the genetic material in most viruses.
2. DNA is responsible for phenotypic expression through transcription of DNA templates to produce mRNA.
3. Specific nucleotide sequences on tRNA interact with complementary base groups of mRNA to translate mRNA base sequences into polypeptides on ribosomal surfaces.

Studies of phenylalanine metabolism, Tay-Sachs syndrome, nutritionally deficient strains of the fungus *Neurospora*, and a host of others make it clear that occurrence of a given biochemical reaction depends on the presence of a specific enzyme (protein) that, in turn, is due to action of a particular genetic locus. From this emerges the concept that the classical particulate gene of Mendel is really a series of deoxyribonucleotides (or ribonucleotides in the RNA viruses) with the relationship

$$\text{gene} \rightarrow \text{RNA} \rightarrow \text{polypeptide} \rightarrow \text{phenotype}$$

This concept of information transfer thus raises the question: What is the specific relationship between sequences of the four bases in DNA and sequences of the twenty amino acids in proteins? Since a one-to-one relationship cannot exist, this implies the existence of a **genetic code** in which some number of nucleotides specify the incorporation of a single amino acid into a protein. It is this genetic code that relates the information found in the sequence of nucleotides in DNA to the information found in the sequence of amino acids in proteins.

COLINEARITY BETWEEN GENES AND PROTEINS

Direct evidence for the specific relationship between genes and proteins came from studies that compared the genetic location of a mutation within a gene to the location of the corresponding amino acid change in the protein coded for by the gene. The classic example of this is the work of Charles Yanofsky on the *tryptophan synthetase* gene in *E. coli*.

The complete enzyme molecule is a complex of two different subunits, A and B. The A component is a single polypeptide chain of 267 amino acid residues whose exact sequence is now known; the B component is a dimer of two identical polypeptide chains. The enzyme catalyzes three of the steps in tryptophan synthesis (Figure 10–1). Both the A and B subunits are required for reaction AB, whereas only the A component is necessary for indole production (reaction A), and the B component alone is needed for reaction B. Both the A and B components are required, then, for synthesis of tryptophan.

Some mutants that are incapable of forming functional tryptophan synthetase do produce a protein that, although enzymatically inactive, gives an immunological reaction with tryptophan synthetase immune serum prepared from rabbit blood. This protein has been named *cross-reacting material* (CRM); both CRM^+ (producing cross-reacting material) and CRM^- (not producing cross-reacting material) mutants have been detected. It is believed that the difference between CRM^+ and CRM^- mutants is that the latter produce either small fragments of the enzyme molecule or grossly abnormal molecules as the result of nonsense and missense mutations (proteins in which missing or the wrong amino acids occur).

Charles Yanofsky and his colleagues were able to map various mutations and identify their specific site within the A gene. By sequencing

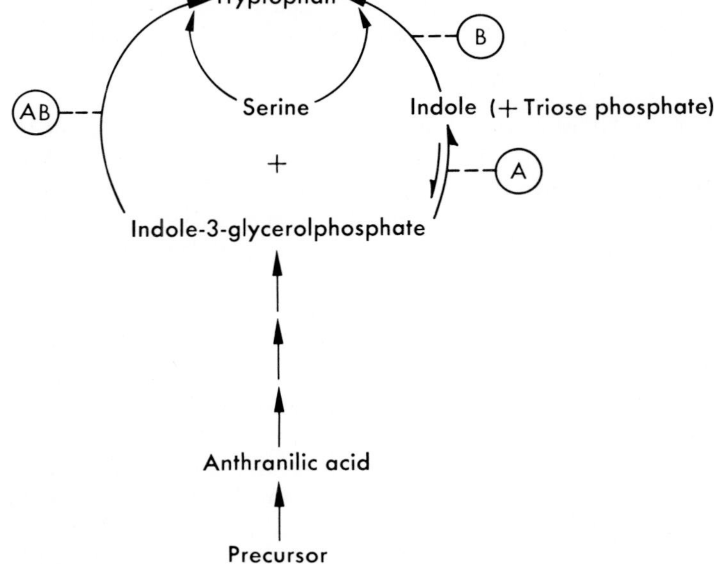

FIGURE 10–1. *Pathways of tryptophan synthesis involving tryptophan synthetase. The enzyme consists of two subunits, A and B, both required for the reaction indole-3-glycerol-phosphate + serine → tryptophan (reaction AB). The A subunit is required for production of indole and triose phosphate (reaction A), and the B subunit is required for the reaction indole + serine → tryptophan (reaction B). Thus, A-deficient mutants require indole or tryptophan for growth, and B-deficient mutants must be supplied with tryptophan.*

TABLE 10–1. Some A Mutants in the Tryptophan Synthetase System of *Escherichia coli*

Mutant	*Amino acid position**	*Wild type amino acid residue*	*Mutant amino acid residue*
A3	48	Glutamic acid	Valine
A33	48	Glutamic acid	Methionine
A446	174	Tyrosine	Cysteine
A487	176	Leucine	Arginine
A223	182	Threonine	Isoleucine
A23	210	Glycine	Arginine
A46	210	Glycine	Glutamic acid
A187	212	Glycine	Valine
A78	233	Glycine	Cysteine
A58	233	Glycine	Aspartic acid
A169	234	Serine	Leucine

*Amino acid residues are numbered consecutively from 1 (amino end) to 267 (carboxyl end). Based on work of Yanofsky et al.

the A polypeptide, they could determine where each amino acid change had occurred in the protein. Their finding was that the location of each mutant site within the A gene corresponded exactly to the position of the altered amino acid in the A polypeptide. A summary of these results is found in Table 10–1, which lists some of the mutations with the corresponding amino acid change. This direct relationship between base sequences and amino acid sequences is called **colinearity** between genes and proteins. It implies a direct spatial relationship between base sequences in genes and amino acid sequences in protein. We will now examine the specific nature of this relationship.

PROBLEMS OF THE NATURE OF THE CODE

Investigators of the genetic code faced many questions concerning its nature. Those to be examined in this chapter include:

1. How many ribonucleotides code for a given amino acid? That is, does each code word or **codon** consist of one, two, three, or more ribonucleotides?
2. Are codons, whatever their nucleotide number, contiguous, or is there some sort of spacer "punctuation" that separates the codons? In other words, is the code **commaless**?
3. If a codon is composed of two or more nucleotides, is the code **overlapping** or **nonoverlapping**?
4. What is the **"coding dictionary"**? That is, precisely which codons code for which amino acids?
5. Is a given amino acid coded for by more than one codon? That is, is the code **degenerate**?
6. Does one codon code for more than one amino acid, leading to the concept of **ambiguity**?
7. Are there any codons that serve as "start" or "stop" signals for translation? In other words, are there **chain-initiating** and **chain-terminating** codons?
8. Is the code **universal** for all organisms? Does the same codon specify the same amino acid in phages, bacteria, corn, fruit flies, and human beings?

The basic problem with the genetic code is that there are 20 amino acids that must be coded by some sequence of four nucleotides in DNA or their complements in mRNA. The mRNA nucleotide or nucleotide sequence that codes for a particular amino acid is called a **codon**. If a codon were to consist of only a single base, the genetic code would have to be very *ambiguous*; that is, the same codon would have to code for different amino acids under different conditions. This would certainly lead to much confusion in the synthesis of proteins.

Codons of two bases present the same kind of problem, providing only $4 \times 4 = 16$ codons for the 20 amino acids. But three-base groups provide 64 codons, *more* than enough for the number of amino acids involved. However, the system will work if the code is *degenerate* in that several codons code for the same amino acid. From the theoretical viewpoint, such a *triplet code* could include both *sense codons* (those that specify particular amino acids) and *nonsense codons* that do not specify any amino acid. Nonsense codons might have some other function, such as signaling "start" or "stop" for polypeptide chain synthesis. Crick and colleagues provided a significant clue not only to the length of the codon but also to the question of "punctuation" and overlap. To understand the relevance of their experiments one of the important classes of substances that they employed must first be examined.

Frame Shifts

The acridine dyes (proflavin, acridine orange, and acridine yellow, among others) are substrates that bind to DNA and, at least in phages, act as mutagens by causing additions or deletions in the nucleotide sequence during DNA replication. An acridine molecule may become intercalated between the stacks of base pairs and double the normal distance between adjacent nucleotide pairs. This may either allow insertion of a new nucleotide during replication or result in the deletion of a base, and therefore alter the sequence.

Crick and his colleagues studied a number of acridine-induced mutations in the rII region of the DNA of phage T4. (The RII mutants are described more fully in the next chapter.) Briefly, they found that mutant types arose through additions or deletions at any of a large number of sites within the rII region, but these may revert to the wild type or a very similar one (pseudo-wild type) through deletions or additions elsewhere in the rII region. That is, in many cases, one mutation may be suppressed by another, particularly if the second change is located relatively close to the first. To illustrate, assume the code is triplet, nonoverlapping, and commaless, and that a repeated deoxyribonucleotide sequence is involved. The order of reading, or **reading frame,** is indicated by the solid lines over the codons:

$$\overline{\text{TAG}}\ \overline{\text{TAG}}\ \overline{\text{TAG}}\ \overline{\text{TAG}}\ \overline{\text{TAG}}\ \ldots \qquad (1)$$

Now if the second T (the fourth base) in this hypothetical series is deleted, the reading frame is shifted by one base starting right after the deletion and becomes

$$\overline{\text{TAG}}\ \overline{\text{AGT}}\ \overline{\text{AGT}}\ \overline{\text{AGT}}\ \overline{\text{AG—}}\ \ldots \qquad (2)$$

A deletion thus alters reading of all following groups; this alteration is termed a *frame shift*. Transcription from the original DNA sequence

(1) produces mRNA of repeating AUC codons, and determines a peptide consisting only of the amino acid sequence coded by AUC; and from (2) produces the codon sequence AUC UCA UCA UCA UC—. This could well be expected to result in a change of amino acid composition of the peptide chain beyond the deletion, if it is assumed that codons AUC and UCA code for different amino acids. We could represent the amino acid of the wild type (1) as aaX aaX aaX aaX . . . , but after deletion, and the frame shift of the mutant type (2) might produce the amino acid sequence aaX aaY aaY aaY . . . (a *missense* mutant). On the other hand, an *a priori* possibility might be that UCA codes for no amino acid (a *nonsense* mutant).

Now, still carrying this deletion, assume a thymidylic acid (T) to be inserted at a different position in sequence (2), between the sixth and seventh nucleotides. The DNA sequence then becomes

$$\overline{TAG}\ \overline{AGT}\ \overline{TAG}\ \overline{TAG}\ \overline{TAG}\ \ldots \qquad (3)$$

and the wild-type reading frame is restored with the third triplet; only the second produces a misreading. Insertion of any of the other three nucleotides at the same point produces only a slightly longer faulty segment. Suppose deoxycytidylic acid (C) is inserted instead of thymidylic acid at the same point; the sequence then becomes

$$\overline{TAG}\ \overline{AGT}\ \overline{CAG}\ \overline{TAG}\ \overline{TAG}\ \ldots \qquad (4)$$

Wild-type sequence in (4) is restored after two "wrong" triplets instead of one. Thus, an insertion farther down the reading sequence corrects for an earlier deletion, regardless of whether the inserted base is the same as the deleted one or not. A single *frame shift*, therefore, may be expected to result in a protein so altered in amino acid sequence that it is nonfunctional. On the other hand, if the altered reading between two opposing events (for example, a deletion followed by an insertion) is of small enough magnitude, and if the missense mutation is of such a nature as to alter slightly or not at all the function of the ultimately produced protein, then the second event suppresses the first. The closer the points at which these opposing alterations occur, the higher the probability that this suppression will take place. In essense, this is precisely what the work of Crick and his group showed.

The greatest significance in this work came from the finding that when three addition (+) mutations or three deletion (−) mutations

		Phenotype
Normal reading frame	GCA ATG CTG CAG TAC TGA TGC GAT	Normal
(+)	GCA ATG CTG CAG TAC TGA TGC GA	Mutant
(++)	GCA ATG CT GCA GTA CTG ATG CG	Mutant
(+++)	GCA ATG CT GCA GT ACT GAT GC	Normal
(−)	GCA - TGC TGC AGT ACT GAT GC	Mutant
(−−)	GCA - TGC TGC A - TAC TGA TGC	Mutant
(−−−)	GCA - TGC TGC A - TAC TGA T - CTG	Normal

FIGURE 10–2. *Diagram showing how three (+ + +) or (− − −) mutations in DNA restore correct reading frame outside limits of the changes, but (+), (+ +), (−), or (− −) mutations result in reading frame being out of sequence.*

occurred, the phenotype often was not mutant. This is because when there are 3 (+) or three (−) mutations, the reading frame is altered only between the first and third change. The region of the message before and after the change is read in the correct order of codons. If the three changes are not too far apart, a reasonably normal protein can be produced. This finding is illustrated in Figure 10–2.

A Triplet, Commaless, Nonoverlapping Code

Note that several aspects of the genetic code are made clear by these experiments. First, the code is likely to be *triplet*, because a single frame shift results in missense, as do two, four, or five frame shifts, but three close additions or deletions will frequently restore sense. Pairs of opposite kinds of frame shifts also restore sense. Still other experimental data make it clear that the code is, indeed, triplet.

In addition, this kind of restoration of sense sequences clearly suggests that (1) there is no "punctuation" between the codons; that is, each codon is immediately adjacent to the next with no intervening "spacer" bases, and (2) the codons are *nonoverlapping*. Another line of evidence for nonoverlapping arises in the fact that amino acid residues appear to be arranged in completely random sequence when different polypeptide chains are analyzed; no one amino acid always, or even usually, has the same adjacent neighbors. This could be the case if the code is *nonoverlapping*. If the code did overlap, a given amino acid would always have the same nearest neighbors.

An mRNA sequence beginning, for example, A-A-C-C-G-A-G-G-A-. . . consists of three triplets, AAC, CGA, and GCA, which code for asparagine, arginine, and alanine, respectively (Table 10–2). If the code

TABLE 10–2. The mRNA Code as Determined *in vitro* for *E. coli.* (Degeneracies are shown; ambiguities are omitted.)

FIRST BASE	SECOND BASE				THIRD BASE
	G	A	C	U	
G	GGG Glycine	GAG Glutamic Acid	GCG Alanine	GUG Valine	G
	GGA Glycine	GAA Glutamic Acid	GCA Alanine	GUA Valine	A
	GGC Glycine	GAC Aspartic Acid	GCC Alanine	GUC Valine	C
	GGU Glycine	GAU Aspartic Acid	GCU Alanine	GUU Valine	U
A	AGG Arginine	AAG Lysine	ACG Threonine	AUG Start Chain Methionine	G
	AGA Arginine	AAA Lysine	ACA Threonine	AUA Isoleucine	A
	AGC Serine	AAC Asparagine	ACC Threonine	AUC Isoleucine	C
	AGU Serine	AAU Asparagine	ACU Threonine	AUU Isoleucine	U
C	CGG Arginine	CAG Glutamine	CCG Proline	CUG Leucine	G
	CGA Arginine	CAA Glutamine	CCA Proline	CUA Leucine	A
	CGC Arginine	CAC Histidine	CCC Proline	CUC Leucine	C
	CGU Arginine	CAU Histidine	CCU Proline	CUU Leucine	U
U	UGG Tryptophan	UAG End Chain *	UCG Serine	UUG Leucine	G
	UGA End Chain	UAA End Chain ‡	UCA Serine	UUA Leucine	A
	UGC Cysteine	UAC Tyrosine	UCC Serine	UUC Phenylalanine	C
	UGU Cysteine	UAU Tyrosine	UCU Serine	UUU Phenylalanine	U

*Originally called *amber*.
‡Originally called *ochre*.

Acidic | Aromatic | Basic | Neutral | Contains sulfur

overlapped by two bases, this sequence of nine bases would consist of the triplets AAC, ACC, CCG, CGA, GAG, AGC, and GCA, coding for asparagine, threonine, proline, arginine, glutamic acid, serine, and alanine. Thus, in this particular sequence of ribonucleotides, arginine would always occur between proline and glutamic acid. Furthermore, an overlapping code would restrict adjacent amino acids to only four different possibilities on either side, and this is clearly not the case.

Although an overlapping code might seem more efficient in that more "messages" can be coded by fewer nucleotides, there is also a strong disadvantage to overlapping. In the nucleotide sequence previously described, consider the result of a substitution (that is, a mutation) in the third base (the first deoxycytidylic acid) to deoxyguanylic acid, so that the sequence becomes

A-A-G-C-G-A-G-C-A- . . .

The mRNA codons, if the code were overlapping by two bases, would then become UUC, UCG, CGC, . . . , which codes for phenylalanine, serine, and arginine, respectively, that is, an entirely different sequence than in the original sequence. Thus, a single base pair substitution in this case would change *three* amino acids rather than one in a triplet, nonoverlapping code. This does not occur!

THE CODING DICTIONARY

Once the genetic code was established as a *commaless, nonoverlapping, triplet code,* the question of which triplets code for which amino acids was pursued vigorously. Marshall Nirenberg and J. H. Matthaei pioneered in efforts to crack the code. They used a discovery by Marianne Grunberg-Manago and Severo Ochoa, who had isolated a bacterial enzyme that catalyzes the breakdown of RNA in bacterial cells. This enzyme is called **polynucleotide phosphorylase**. They found that outside of the cell, with high concentrations of ribonucleotides, the reaction could be driven in reverse, and an RNA molecule could be made. Incorporation of bases into the molecule is random and does not require a DNA template. Nirenberg and Matthaei used this enzyme to construct synthetic polyribonucleotides and tested them in cell-free (*in vitro*) systems for their ability to direct incorporation of specific amino acids into polypeptide chains.

Using a mixture of amino acids, with a different amino acid radioactively labeled in each run, in a cell-free suspension derived from *E. coli* (tRNA, ribosomes, ATP, GTP, necessary enzymes), they were able to bring about polypeptide synthesis *in vitro*. They found that when their synthetic message contained only uracil (poly U), only one amino acid, phenylalanine, was incorporated into a polypeptide. This suggested that the triplet UUU specifies only phenylalanine. They likewise found that a synthetic poly A message directed only the incorporation of lysine and a poly C message coded only for proline. A poly G message is nonfunctional *in vitro*. These experiments allowed identification of three codons. The next approach, developed by G. Khorana, was to give the enzyme mixtures of two bases and construct random **copolymers,** synthetic messages comprised of random sequences of only two bases. These determinations reinforce the concept of a commaless code because, in a code that includes punctuation, U and A would each have to serve both in coding for their respective amino acids as well as to

PROTEIN FOLDING: THE SECOND HALF OF THE GENETIC CODE

It was more than twenty years ago that the genetic code was identified, leading to the discovery of the three bases in mRNA that specifies each amino acid in a protein. It was thought that everything possible could be known about any protein by its amino acid sequence. Unfortunately, this prediction has not yet come to pass. While the rules governing the amino acid sequence of a protein can be known, the rules governing the three-dimensional folding of that sequence into its active form in the cell have not yet been discovered. It is not known why a particular sequence folds into the alpha helix or a pleated sheet structure, for example. This folding has two important implications. First, in biotechnology, where bacteria are being used to produce large quantities of animal proteins, sometimes the protein does not turn out like it should because it does not fold properly. Second, there are certain genetic disorders where the protein does not fold properly.

An example of a protein in which the "folding rules" are somewhat clear is collagen. The protein is a rigid rod made from three polypeptide strands twisted together. The structure consists of repeating units of three amino acids. At every first position of the repeating unit there is a glycine; at the third position there is a proline or hydroxyproline. A great deal of variation can occur at the second position. When collagen is made, the three polypeptide chains zip together. Anything that disrupts this zipping action destroys the collagen structure. Since collagen is the single most abundant protein in the body, defects in its folding could prove fatal. In fact, patients with the connective tissue disorders osteogenesis imperfecta, Marfan's syndrome, and Ehlers-Danlos syndrome have mutations that prevent the correct folding of the molecule.

Solving the folding rules for collagen has been difficult. Knowing the amino acid sequence is clearly not enough. Some researchers feel that there are intermediate stages in the folding process, and that their discovery and examination will reveal some answers. Other researchers are specifically modifying amino acid sequences in proteins by mutations and observing the consequence on folding. While this research is encouraging, it has not yet produced any fundamental folding rules.

Finally, computers are being used to predict folding patterns given certain amino acid sequences. One idea being explored its that proteins with similar sequences should fold in a similar fashion. The information from a million possible structures is analyzed by a computer in the hope that one or two distinct folding patterns will be uncovered. In addition, computers are being used to analyze all the possible ways a protein could be put together from its amino acid sequence. Even the use of computers has thus far not yielded any fundamental rules. Thus, the "second half" of the coding problem is not close to being solved.

function as commas. It is hard to conceive how they could do this in a U/A copolymer. From this time on, the notion of a commaless code won general acceptance.

Testing random copolymers provides some additional information on the code. When synthetic mRNA is constructed, it is found that the sequence of nucleotides is random, so that the relative frequency of incorporation of particular nucleotides is mathematically determined. Thus a mixture containing *two parts uracil to one part guanine* is found to produce triplets in these proportions:

UUU	$\frac{2}{3} \times \frac{2}{3} \times \frac{2}{3} = \frac{8}{27}$	GUU	$\frac{1}{3} \times \frac{2}{3} \times \frac{2}{3} = \frac{4}{27}$
GGG	$\frac{1}{3} \times \frac{1}{3} \times \frac{1}{3} = \frac{1}{27}$	GGU	$\frac{1}{3} \times \frac{1}{3} \times \frac{2}{3} = \frac{2}{27}$
UGU	$\frac{2}{3} \times \frac{1}{3} \times \frac{2}{3} = \frac{4}{27}$	UGG	$\frac{2}{3} \times \frac{1}{3} \times \frac{1}{3} = \frac{2}{27}$
UUG	$\frac{2}{3} \times \frac{2}{3} \times \frac{1}{3} = \frac{4}{27}$	GUG	$\frac{1}{3} \times \frac{2}{3} \times \frac{1}{3} = \frac{2}{27}$

If each of the possible triplets were to code for a different amino acid, then various ones would be incorporated in the proportions shown. Except for the fact that *each* such three-base group does not code for a *different* amino acid, this is essentially what occurs. Use of a synthetic poly-UG, in the relative proportions just given, produces a mixture of polypeptides of which only $\frac{8}{27}$ are polyphenylalanine.

In this way, it was found that poly-UG that contains 2U:1G codes for valine, but the sequence of bases and therefore the exact codon(s), cannot be determined from such random copolymers. Nirenberg and Leder devised a method by which short-chain polyribonucleotides of known sequence could be obtained. They, and also Khorana and his group, introduced these short-chain polyribonucleotides of known sequence into cell-free systems that included ribosomes and a variety of tRNA molecules charged with their amino acids. As in earlier work, one amino acid in each experimental run was labeled with ^{14}C. The synthetic polyribonucleotides bind to the ribosome and direct the binding of only one aminoacyl tRNA. By replicating the experiment with a different amino acid radioactively labeled in each reaction vessel, it is possible to identify the correct trinucleotide codon for each aminoacyl tRNA. The collective results of all these experiments led to the **coding dictionary** shown in Table 10–2. In addition to cracking the genetic code, these results also help to establish that it is a triplet, commaless, nonoverlapping code. With the ability to sequence actual genes that code for proteins, the code has been confirmed in every case studied so far. Elucidation of the genetic code has been rightly called "one of the principal triumphs of molecular biology."

DEGENERACY

Although the code is extensively degenerate, as seen in Table 10–2, with several codons specifying the same amino acid (the only exceptions are AUG for methionine and UGG for tryptophan), a certain order to this degeneracy can be discerned. In many instances it is the first two bases that are the characteristic and critical parts of the codon for a given amino acid, whereas the third may be read either as a purine only (for example, glutamine, CAA, and CAG), or as a pyrimidine only (for example, histidine, CAU, and CAC). In other cases, however, the third position is read as any base, for examples, AC—for threonine, CC—for proline, and so forth.

In 1966, Crick proposed the **wobble hypothesis** to account for the lack of specificity in the 5′ base of the **anticodon,** which may pair with any of several bases at the 3′ end of the codon. Combination possibilities are, however, not limitless, but are restricted to those listed in Table 10–3. The reason for this is due to the conformation of the anticodon loop. The base at the first position of the anticodon influences the number of possible bases recognized at the third position of the codon. For example, if the first base of the anticodon is U, then either an A or G can be recognized at the third position of the codon. If the first anticodon base is G, the third codon base can be either C or U. This accounts for the "wobble." First anticodon bases A or C only allow recognition of U or G, respectively, in the codon. Because it decreases the fidelity of hydrogen bonding, inosine in the anticodon may pair with uracil, cytosine, or adenine ribonucleotides of the codon. Inosine is derived from adenine by the deamination of the 6-carbon to produce a 6-keto group. It is shown paired with cytosine in Figure 10–3.

In addition to the wobble of the first, or 5′-base, of the anticodon, some amino acids have as many as six codons (for example, leucine, CU—UUA, and UUG). Different leucine tRNAs exist for some of the six leucine codons, all of which obey the pairing relationships shown in Table 10–3. In addition, pseudouridine (ψ), which occurs as the mid-

TABLE 10–3. **Pairing Between Codon and Anticodon at the Third Position**

Anticodon	*Codon*
A	U
C	G
G	U, C
I (inosine)	U, C, A
U	A, G

Based on the work of Crick, 1966.

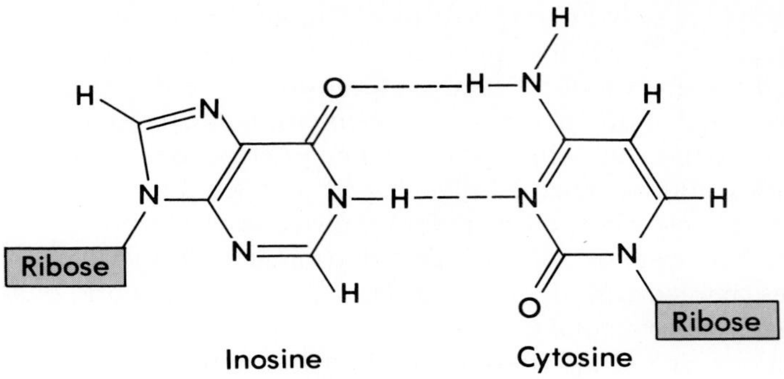

FIGURE 10–3. *An inosine-cytosine base pair. Inosine also pairs with uracil and adenine by two hydrogen bonds.*

dle base of the anticodon of yeast tyrosine tRNA, pairs with adenine. Methionine and tryptophan, which have only one codon each (AUG and UGG, respectively), are exceptions to the general rule of degeneracy.

AMBIGUITY

Essentially, the code is nonambiguous *in vivo* under natural conditions. This means that each codon specifies the same amino acid all the time. Ambiguity is encountered chiefly in cell-free systems under certain conditions. In such a system prepared from a streptomycin-sensitive strain of *E. coli,* UUU, which ordinarily codes for phenylalanine, may also code for isoleucine, leucine, or serine on the presence of streptomycin. This ambiguity is enhanced at high magnesium-ion concentrations. Poly-U, in a cell-free system from thermophilic (heat-tolerant) bacterial species, has been found to bind leucine at temperatures well below the optimum for growth of living cells of the same species. Changes in pH or the addition of such substances as ethyl alcohol also result in ambiguity. However, ambiguities are not ordinarily encountered *in vivo* under normal growing conditions for a given species.

CHAIN INITIATION

It appears clear that polypeptide chains in *E. coli* are initiated with N-formylmethionine, which, in many instances, is enzymatically removed after synthesis of the protein is completed. This is also true for many phages. The coat protein of the RNA phage R17 begins with the N-terminal sequence alanine-serine-asparagine-phenylalanine-threonine. When synthesized in *in vitro* systems this sequence is preceded by N-formylmethionine, suggesting that the enzyme necessary to remove the formylated methionine is absent.

Only one codon, AUG, exists for methionine (Table 10–2), so the question naturally arises of how chain-initiating N-formylmethionine and internally located methionine are distinguished in the biosynthesis of polypeptides. The answer in *E. coli,* at least, lies in part in the occurrence of two different tRNAs for methionine. One of these, symbolized as $tRNA^{fmet}$, is only for initiation of polypeptide synthesis. The methionine is formylated and serves only this purpose. The other, methionyl-tRNA ($tRNA^{met}$), does not allow methionine to serve as a substrate for the formylating enzyme and inserts its methionine only

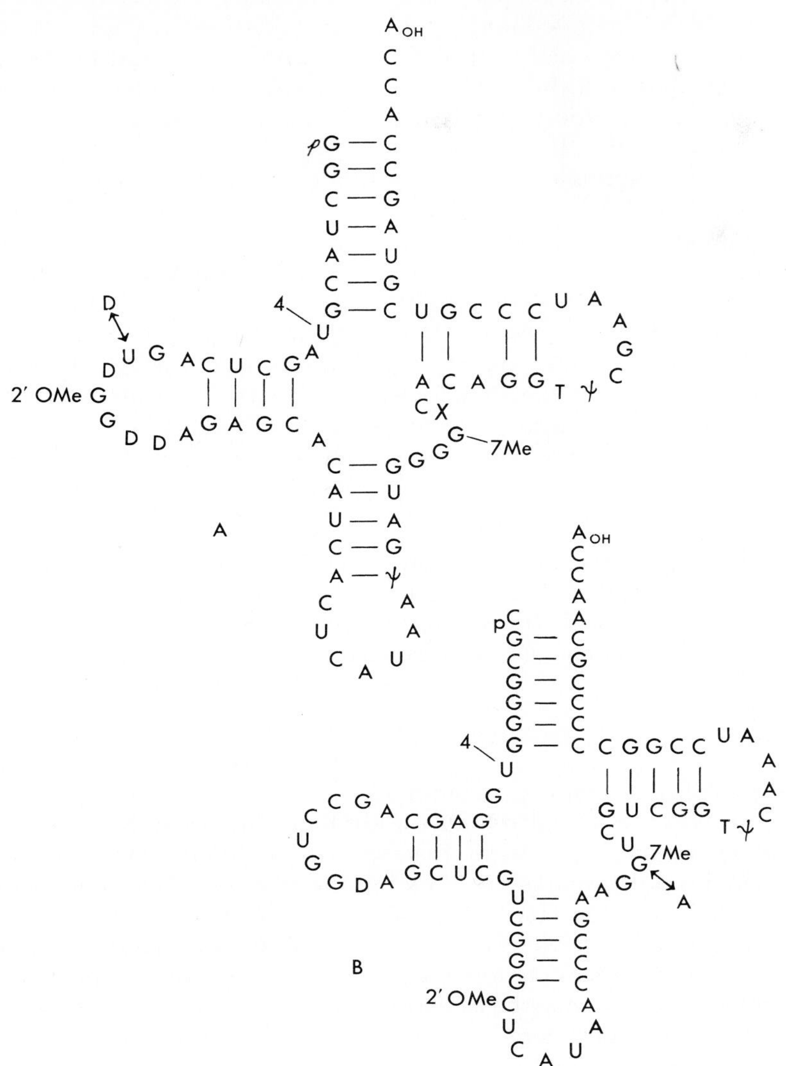

FIGURE 10–4. *Ribonucleotide sequences of (A) $tRNA_m$, and (B) $tRNA_f$. Unusual ribonucleotides are symbolized as follows: D, dihydrouridylic acid; ψ, pseudouridylic acid; T, thymidylic acid; 7MeG, 7-methylguanylic acid; 2MeC, 2 O-methylcytidylic acid; 2 OMeG, 2 O-methylguanylic acid; 4U, 4-thiouridylic acid.* (From S. K. Dube, et al. 1968, and S. Corey, et al. 1968, used by permission.)

into internal positions; thus it functions in elongation rather than in initiation. The anticodon for both kinds of tRNA is 3′URC5′. Therefore, the same AUG codon codes for both formylated and nonformylated methionine; $tRNA^{fmet}$ and $tRNA^{met}$ do, however, differ in many of their internal nucleotides (Figure 10–4).

It is reported by some investigators that $tRNA^{fmet}$ is required for the initiation of all bacterial proteins, although the formyl group, and in some cases the entire methionine residue, is enzymatically removed from the polypeptide chain after completion of synthesis. There is some evidence that methionine, but not N-formylmethionine, functions in initiation in mammalian cells. A special met-tRNA is responsible for insertion of methionine into the N-terminal position, although it is removed in some cases, for example, rabbit globin, before the polypeptide chain is incorporated into hemoglobin. Initiation involves more than an initiation codon alone; at least three protein initiation factors (IF) are involved as well. In some rare cases GUG can also be read as the initiation codon specifying N-formylmethionine. Internally, GUG only specifies valine tRNA.

AUG determines the reading frames of mRNA. The synthetic ribonucleotide AUGGUUUUUUUU . . . is translated only as N-formylmethionine-valine-phenylalanine-phenylalanine . . . ; that is, the reading is AUG-GUU-UUU-UUU. Translation, once initiated, continues until a "stop" codon (see next section) is encountered.

There is some evidence that additional sequences upstream of the initiation codon also are important in the overall initiation process. In prokaryotes approximately four to seven bases preceding the initiation codon have a sequence that reads

A G G A G G

and is called the **Shine-Dalgarno sequence** (see also Chapter 9). It is believed that this sequence interacts with the 3′ end of the 16*s* rRNA of the 30*s* ribosomal subunit during the initiation process. In phage T7, mutations that change the sequence to

A G G A A G

destabilize the initiation complex.

Ribosomes in eukaryotes do not bind directly to a translation initiation site on the message. Instead, the 40*s* subunit binds to the methylated cap that marks the 5′ end of the message. That this is necessary is shown by the fact that, *in vitro,* messengers lacking caps are not translated effectively. An additional group of initiation factors called *cap-binding proteins* are involved. The distance between the cap and the initiation codon may be as short as 40 bases or as long as 300. The 40*s* subunit migrates along the *leader* sequence until it encounters the first AUG codon where it is joined by the 60*s* subunit and is stabilized when binding to the messenger. There are only two other differences in initiation in eukaryotes compared to prokaryotes. First, only AUG is used, and second, the methionine is not formylated. There are, however, differences between the methionyl tRNAs used in initiation and those responding to AUG internally.

CHAIN TERMINATION

Three codons, UAA, UAG, and UGA, do not code for any amino acids and hence are termed *nonsense codons.* Before their base sequences were determined, UAA was known as "ochre," UAG as "amber," and UGA as "opal"—terms that are still employed for these codons as a matter of convenience. Evidence from both *in vitro* and *in vivo* experiments demonstrates that all three nonsense codons are involved in both chain termination and release from the ribosome. Apparently the appearance of a nonsense codon at the A site stimulates binding of release factors, which disengage the polypeptide bound to a tRNA at the P site.

UNIVERSALITY

The current evidence is that for the most part the code is universal for all living organisms and for viruses. This means the same triplets code for the same amino acids in a wide variety of organisms. This concept is clearly supported when eukaryotic genes are inserted into bacterial cells (see Chapter 20) and are correctly transcribed and translated. In the last few years, as studies on the coding properties of mitochondrial DNA have been carried out, a few exceptions to the concept of universality have been discovered. These few exceptions are listed in Table 10–4.

TABLE 10–4. **Deviations from the Universal Code by Mitochondrial DNA**

Codon	*Universal code*	*Mitochondrial DNA code*
UGA	term	tryptophan (yeast/ human) termination (plants)
AUA	ileu	methionine (human)
CUA	leu	threonine (yeast)
CGG	arg	tryptophan (plants)

EVOLUTIONARY RELATIONSHIPS

Evolution of the Genetic Code

The actual origin and evolution of the genetic code is still an unsolved problem. Until recently, with the discovery of the variations in the code in mitochondrial DNA, the code was thought to be completely universal in all organisms. These deviations in the code have actually aided the considerations regarding the origin and development of the present code.

To review what has already been discussed in this chapter, but with the viewpoint of evolutionary process, we must consider the genetic code actually to be a tRNA code, not an mRNA code as it is usually presented (Table 10–2). Amino acids are attached to tRNA molecules having anticodons rather than codons. They in turn find complementary codons to translate. There are 64 codons of mRNA of which 61 are translated. Some anticodons pair with 2 or 3 codons, resulting in fewer anticodons than codons.

The mitochondria are thought to have evolved from the purple photosynthetic bacteria (see Chapter 16), and we will assume that the mammalian mitochondrial DNA arose as an evolutionary simplification of the original bacterial DNA. This simplification resulted in the minimum number of 22 anticodons, which would translate the 60 codons for the 20 amino acids. Table 10–5 shows the minimal anticodon–amino acid assignments for the mammalian mtDNA.

The code would then evolve from this minimal state to its present state in the following way originally suggested by Thomas Jukes. The process is illustrated for the valine anticodon URC and is outlined in Figure 10–5. The first step in code evolution is duplication of the existing anticodon. This duplication is followed by nucleotide substitution of G for U in one of the duplicates. The new anticodon, GAC, pairs with the codons GUU and GUC. The other anticodon, UAC, pairs with GUA

TABLE 10–5. **Anticodons and Amino Acid Assignments in Mammalian Mitochondrial Code (from Jukes, 1983)**

GAA Phe UAA Leu	UGA Ser	GUA Tyr	GAC Cys UCA Trp
UAG Leu	UGG Pro	GUG His UUG Glu	UCG Arg
GAU Ile UAU Met	UGU Thr	GUU Asn UUU Lys	GCU Ser
UAC Val	UGC Ala	GUC Asp UUC Glu	UCC Gly

and GUG. The original anticodon can undergo a second duplication and nucleotide substitution in one of the duplicates to produce the anticodons UAC and CAC. The latter pairs only with a GUG codon. A third duplication and substitution in the GAC anticodon results in the formation of an AAC anticodon. Eventually, following a change in the aminoacylation site in one of the tRNA duplicates, a completely different amino acid is specified by the changed anticodon.

This pattern of duplication and substitution in anticodons, followed by changes in aminoacylation specificities, has resulted in increasing the number of anticodons for amino acids as well as contributing to degeneracy in the code. Note that A and C were not used in the first anticodon position in the primitive codes. Since they are presently, we can only assume that as evolution proceeded, A and C were added to the system. The process continued until living systems became too complex to tolerate further changes in the codon assignments and the code became frozen for twenty amino acids in its present state.

As to the origin of the code, there are two theories. One theory states that the code is a "frozen accident" and that it could not be changed without severe consequences to all living organisms. The other theory states that the code is simply a product of evolution and it is the best possible system that could have evolved. The frozen accident theory has more support than the evolutionary theory due to the observation that the present code is probably not the best for presently living organisms. For example, arginine has six codons, but probably needs only two or three, based on the occurrence of this amino acid in proteins.

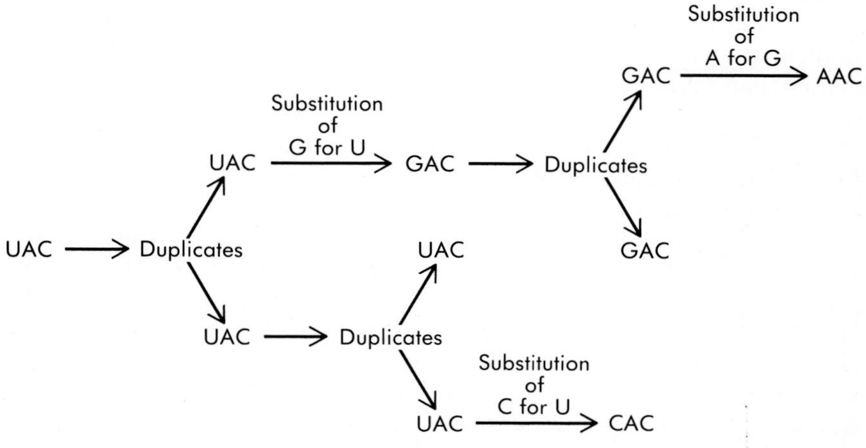

FIGURE 10–5. *Pattern of code evolution by duplication of anticodons followed by substitution of bases at the first position.*

Similarly, lysine has only two codons, but based on use of lysine in proteins, it probably could use at least two more. Whatever actually happened, the freezing of the code for twenty amino acids was a very crucial event, as these amino acids are reponsible for the properties and activities of proteins in all the diverse living organisms on the earth today.

Evolution of DNA Sequences and Species Interrelationships

DNA from different species is expected to exhibit degrees of similarity in nucleotide sequences in proportion to the closeness of evolutionary relationship of those species. Zoologists have long agreed that humans and monkeys are more closely related than either are to fish or bacteria, for example. Significant chemical proof was demonstrated in 1964 by Hoyer, McCarthy, and Bolton.

Their method was simple in concept but delicate in operation. DNA to be tested was first made to undergo strand separation by heating (denaturation), then cooled quickly to prevent recombining, and immobilized in agar. To this was added from another species short strands of denatured DNA that previously had been made radioactive by incorporation of carbon-14 or phosphorus-32. Pairing of homologous nucleotides from the two species occurred during several hours of incubation. The "hybrid" DNA was then recovered and assayed for radioactivity; this resulted in a measure of base-pair homology between the two species. The more base pairing that occurred implied greater relationship.

This technique has been improved over the years, and the latest approach is to carry out the hybridizations, and then slowly raise the temperature again and measure the denaturation of the hybrid molecules. This was done in a study of evolutionary relationships between the hominoid primates. Reciprocal hybridizations are made using single-stranded DNA containing single-copy sequences between two different sources. After the hybrids are made, the temperature is slowly raised in 2.5°C increments, and the temperature is noted when 50 percent of the DNA is denatured and 50 percent is still hybridized. This single parameter is called the $T_{50}H$ value. The more closely related two species are, the lower the $T_{50}H$ value, and the less closely related, the higher the value. Based on these experiments, C. G. Sibley and J. E. Ahlquist were able to draw a phylogenetic tree for this group (Figure 10–6). Note the basic similarity between this tree and the one based on mitochondrial DNA shown in Figure 16–6. Recall also that, in Chapter 4, it was shown that humans and the great apes have many karyotypic homologies.

Also of significance is the fact that *in vitro* protein synthesis, which uses such disparate ingredients as rabbit reticulocyte mRNA and *E. coli* tRNA (plus other necessary components), is successful. The result is production of normal rabbit hemoglobin, which shows that bacterial tRNAs can "recognize" mRNA and polysomes from a taxonomically very different organism. It is true that taxonomically divergent organisms do differ in the degree to which a given codon responds to a particular species of tRNA. For example, AAG codes readily for lysine in vertebrates, but does so more weakly in *E. coli.* But this is only a difference in *degree,* not in kind.

Furthermore, such proteins as cytochrome C show uniformity in several sequences of amino acid residues in very different groups, such

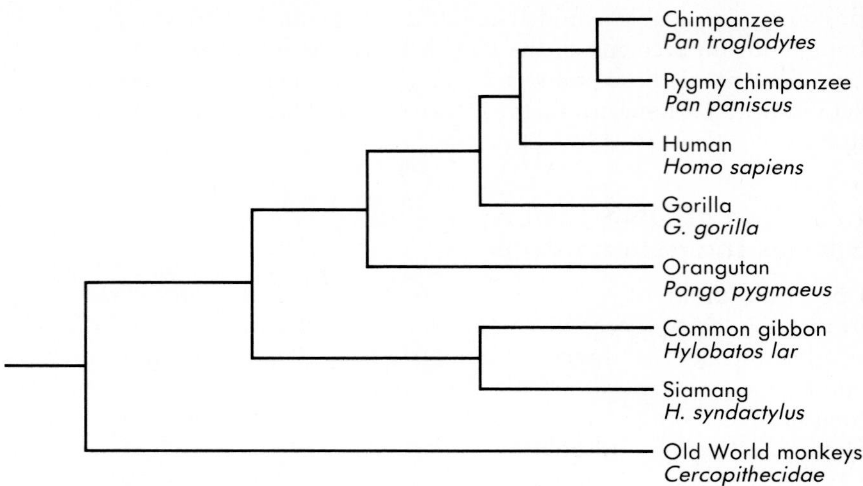

FIGURE 10–6. *Phylogenetic tree of hominoid primates based on DNA-DNA hybridization.* (Redrawn from Sibley and Ahlquist, 1984.)

as mammals, fishes, yeasts, and bacteria, even though their evolutionary divergence must have taken place many hundreds of millions of years ago. In comparing structures of proteins, consideration has been given to how the number of amino acid replacements are related to the time that has elapsed since any two species had a common ancestor. By just looking at amino acid replacements and ignoring the location of the replacement, it was found that proteins behave as biological clocks. Amino acid replacements appear to take place at some constant rate. This, of course, is related to base changes and therefore evolution in DNA.

What has been found, however, is that the clock for changes in the actual function of the protein does not run at a constant rate. Those changes that affect the functioning of the protein seem to be occurring primarily at a slower rate. Also, the rate of change in the third position of a codon can be more rapid than rates of change at the first or second position. When all this is considered, the final picture of molecular evolution appears to be long periods of inactivity, due to the production of mutations that do not affect the functioning of a protein, punctuated by bursts of change, due to mutations occurring that directly affect protein function.

The subject of evolution of the genetic code is, as might be surmised, replete with problems. Its very universality complicates the question, as does the paradoxical situation wherein operation of the code requires precise functioning of many enzymes that apparently could not be produced without the translation mechanism for which they are required.

Summary of Code Characteristics

In summary, the evidence shows that the genetic code is *triplet, commaless, nonoverlapping, degenerate, essentially nonambiguous under natural conditions, colinear, and universal.* In addition, polypeptide chain *initiation in prokaryotes is signaled by certain codons (notably AUG) that bind tRNAs carrying blocked amino acids, although specific protein initiation factors are also required. Chain termination* is governed by three nonsense codons (UAA, UAG, and UGA); chain *release* requires protein release factors as well.

PROBLEMS

10–1 Assume a length of template DNA with the deoxyribonucleotide sequence 3′ T A C C G G A A T T G C 5′. (a) If the code is triplet, nonoverlapping, and commaless, of which amino acid residues (in sequence) will the polypeptide chain for which this stretch of DNA is responsible consist? (b) If the code is triplet, overlapping by two bases, and commaless, how would you answer the preceding question? (c) Of what significance is the TAC triplet in the DNA template?

10–2 For the DNA template of the preceding problem assume the second C to be deleted. What now is the sequence of amino acid residues coded for if the code is assumed to be triplet, nonoverlapping, and commaless?

10–3 In the DNA length of Problem 10–1, assume the second C to be deleted and a T to be inserted after the GG sequence so that the DNA strand now reads 3′ T A C G G T A A T T G C 5′. (a) How does the amino acid sequence now coded for compare with your answer to part (a) of Problem 10–1? (Assume the code to be triplet, nonoverlapping, and commaless.) (b) Does your answer to the preceding part of this problem illustrate a sense, a missense, a nonsense mutation, or none of these?

10–4 Synthetic mRNA is constructed from a mixture of ribonucleotides supplied to a cell-free system in this relative proportion: 3 uracil:2 guanine:1 adenine. What fraction of the resulting triplets would be (a) UGA; (b) UUU?

10–5 The human hemoglobin molecule includes four polypeptide chains, two α chains of 141 amino acid residues each, and two β chains of 146 amino acid residues each. Neglecting chain-initiating and chain-terminating codons, as well as introns, (a) of how many ribonucleotides does the mRNA molecule responsible for the α chain consist? (b) What is the length in micrometers of that molecule? (c) Is your answer to part (b) likely to be too high, too low, or about right?

10–6 Assume an alanine tRNA charged with labeled alanine is isolated, and the amino acid is chemically treated to change it to labeled glycine. The treated amino acid–enzyme-tRNA complex is then introduced into a cell-free peptide-synthesizing system. At which of two mRNA triplets, say GCU or GGU, would this tRNA now become bound? Why?

10–7 If single-base changes occur in DNA (and therefore in mRNA), which amino acid, tryptophan or arginine, is most likely to be replaced by another in protein synthesis?

10–8 Wittmann-Liebold and Wittman studied a number of mutants in the coat protein of the RNA-containing tobacco mosaic virus. Two of their mutants were

Mutant	*Amino acid position*	*Replacement*
A-14	129	isoleucine → threonine
Ni-1055	21	isoleucine → methionine

By reference to Table 10–2, explain what has happened in each of these cases.

10–9 Yanofsky et al. studied a large number of mutants for the tryptophan synthetase A polypeptide chain of *Escherichia coli*. This polypeptide chain consists of 267 amino acid residues. In the wild-type enzyme, a part of the amino acid sequence is: -tyrosine-leucine-threonine-glycine-glycine-glycine-glycine-glycine-serine-. In their mutant A446, cysteine replaces tyrosine; in mutant A187, the third glycine is replaced by valine. By reference to the genetic code, suggest a mechanism for each of these amino acid replacements.

10–10 How many different mRNA nucleotide codon combinations can exist for the internal pentapeptide threonine-proline-tryptophan-leucine-isoleucine?

10–11 From Table 10–1, note that (a) UUU and UUC code for phenylalanine; on the other hand, (b) So and Davie report that, with relatively high concentrations of ethyl alcohol, the incorporation of leucine, and isoleucine to a lesser degree, is sharply increased, whereas incorporation of phenylalanine is decreased, for these same codons. Which of the described situations, (a) or (b), represents ambiguity, and which degeneracy?

10–12 Assume a series of different one-base changes in the codon GGA, which produces these several new codons: (a) UGA, (b) GAA, (c) GGC, (d) CGA. Which of these represent(s) degeneracy, which missense, and which nonsense?

10–13 Assume the average molecular weight of an amino acid residue to be 125 daltons, and a eukaryotic polypeptide of 50,000 daltons. Assume further the mRNA transcript specifying that polypeptide to include one initiation and one termination codon. (a) Of how many ribonucleotides does that messenger RNA molecule consist? (b) Do introns need to be taken into account in answering part (a)? Why?

GENE STRUCTURE AND ORGANIZATION

CHAPTER

11

In the preceding chapter on the genetic code, an important question was raised: In terms of the operation of the genetic code, just how much of the nucleotide sequence comprises a gene? Even by raising this question, scientists have come a long way from the early, understandably vague, concept of the gene as a "bead on a string," separated from adjacent genes by nongenetic material. In this chapter we attempt to develop a concept of the gene from an informational standpoint.

FINE STRUCTURE OF THE GENE

In a classical sense we know that genes occupy distinct chromosomal regions. They can be defined as such by analysis of recombination data. Since a gene is responsible for some phenotype, it must perform a function in each cell in which it is active. Finally, since genes can undergo mutation, a permanent change in the organization of the genetic material is also possible. However, these criteria of recombination, function, and mutation do not necessarily define the same unit of structure. For example, as we have seen, a single base pair change in the hemoglobin gene alters the protein and results in a major change in the phenotype. With the identification of the structure and functioning of the DNA as the genetic material, scientists began to look more closely at the gene in terms of function, recombination, and mutation.

Starting in the mid-1950s, Seymour Benzer carried out genetic analysis on a gene of the bacteriophage T4, giving us the first real picture of the fine structure of any gene. When normal T4 lyses *Escherichia coli* it produces a small circular clear area in the medium where all the

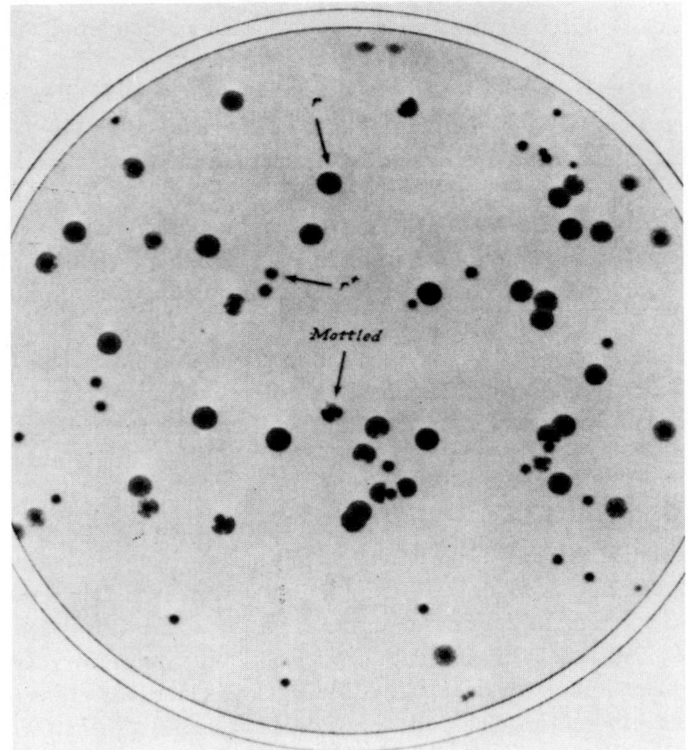

FIGURE 11–1. *Rapid lysis (r) mutants in bacteriophage T2. r^+ = wild type; r = rapid lysis mutant, which forms a larger plaque in the same interval. The mottled plaques owe their appearance to the growth of both r and r^+ phages in the same site.* (From *Molecular Biology of Bacterial Viruses* by Gunther S. Stent. W. H. Freeman and Company. Copyright © 1963.)

bacteria have been killed. This clear area is called a **plaque.** Normally the process of producing a visible plaque in a bacterial culture takes several hours. Benzer studied a class of T4 mutants that were able to accomplish the task of lysis in about twenty minutes. As a result, in a few hours, when normal plaques had begun to appear, the mutants had already produced very large plaques easily distinguishable from plaques produced by normal T4. The phenotype of these mutants has been termed **rapid lysis**. Figure 11–1 shows phenotypes of both normal and rapid lysis viruses in T2, a closely related virus.

Benzer used a specific rapid lysis mutant. This mutant is a **conditional lethal,** meaning that it can grow under some conditions and cannot grow under others. Specifically, this type of mutant, termed **rII,** can enter, replicate, and lyse *E. coli* B cells, but cannot lyse *E. coli* K12 cells (B and K12 are different strains of the bacterium). This differential growth property of rII mutants allowed Benzer to observe very rare genetic events.

The Cistron

Benzer was able to isolate over 2,000 individual mutants of T4, all with the same rapid lysis phenotype. His next question was: Do all of these mutations involve the same function? He used the inability of individual rII mutants to lyse *E. coli* K12 to answer the question. The test he applied is called the **complementation test.** There are two requirements to carry out this test. The cell in which complementation occurs must be both diploid and heterozygous. The approach in the complementation test is outlined in Figure 11–2. This is easily satisfied in most eukaryotes. The test is conducted by making a cross between two mutants. The F_1 of interest is the heterozygote with the trans arrangement (m_1/m_1^+ and m_2^+/m_2). If the mutants are alleles of the same gene, each parent con-

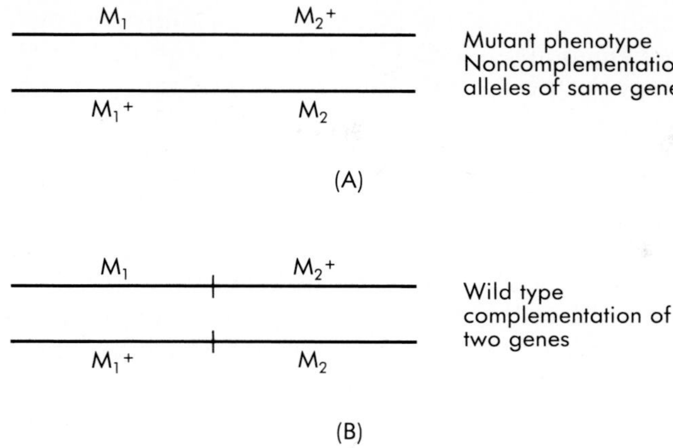

FIGURE 11–2. *Test for complementation with heterozygous mutants in trans arrangement. Two mutations that are alleles (A) will not complement each other in the trans arrangement. Two mutations that are not alleles (B) will complement each other in the trans arrangement.*

tributes a different mutant allele, so the F_1 heterozygote is still phenotypically mutant (Figure 11–2A).

On the other hand, if the mutations are not alleles, the phenotype of the F_1 individual is normal (Figure 11–2B). Under this condition the two mutants are said to complement each other. This means that the normal alleles contribute the normal function to the heterozygote. The fact that the normal phenotype results (complementation occurs) indicates that the mutants are in different functional regions or are not alleles even though they produce the same phenotype. To summarize this point, two mutations that are alleles will not complement each other in the trans arrangement. Two mutations that do complement each other in the trans arrangement are nonallelic, or are members of different functional groups.

Benzer devised a way to carry out the complementation test using rII mutants and *E. coli* K12. The solution is called **mixed infection.** This is the simultaneous infection of K12 cells with two rII mutants. This seems to satisfy, inside the K12 cell at least, the requirements for the complementation test. The results were that although rII mutants could not individually infect K12 cells, some combinations of mutants could complement each other, leading to lysis of the cells (Figure 11–3).

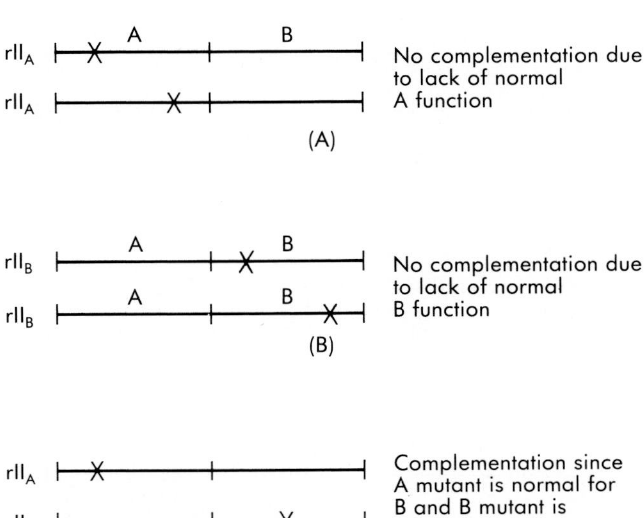

FIGURE 11–3. *Explanation of rII complementation. (A) Mixed infection with two rIIA mutants. (B) Mixed infection with two rIIB mutants. (C) Mixed infection with rIIA and rIIB mutants.*

After subjecting his mutants to the complementation test, Benzer found he could divide them into two groups on the basis of their complementation properties. Group A mutants would not complement each other, but would complement mutants from group B. Similarly, group B mutants would not complement each other, but would complement mutants from group A. The conclusion from these complementation studies is that the genetic region responsible for the rapid lysis phenotype can be divided into two units of function. A mutants are defective in the function that is normal in B mutants, and B mutants are normal in the function that is defective in A mutants. Alone they cannot lyse a K12 cell, but together, A mutants supply the normal B function, and B mutants supply the normal A function, so that they can lyse a K12 cell (Figure 11–3). Benzer coined the term **cistron** for the unit of function. This is simply the genetic region within which there is no complementation between mutants. Presently a cistron is equivalent to a gene, and corresponds to that amount of DNA coding for one functional polypeptide chain. Today the term gene is used more commonly than cistron, but both refer to the same entity.

The Recon

Benzer also carried out crosses between rIIA or rIIB mutants. The way this was done was to infect *E. coli* B with a large number of rIIA or rIIB mutants. Mutants of either class can infect and lyse strain B and show the rapid lysis phenotype. However, the rapid lysis viral progeny still cannot infect K12. Some normal plaques were recovered from strain B and these can infect K12. This suggests that these "normal" recombinants were produced by crossing-over between mutant sites within either the A cistron or the B cistron (Figure 11–4). Analysis of these recombination data can be used to map mutant sites within each cistron, and leads to the conclusion that the cistron is made up of a linear array of mutable sites that can be separated by recombination.

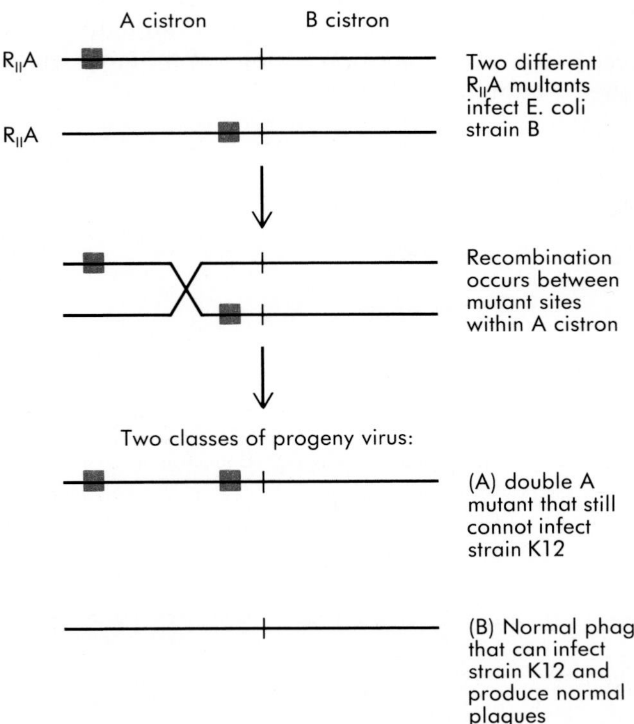

FIGURE 11–4. *Recombination within a cistron. The frequency of this event is determined by the distance between mutant sites. This is used to construct a map of mutant sites within the cistron.*

A measure of the resolving power of this recombinational analysis can be made. The smallest map distance found in the rII region is 0.02 map units, and the entire T4 chromosome is 1,500 map units long, consisting of 1.8×10^5 base pairs. Therefore, the 0.02 map units correspond to approximately two base pairs ($0.02/1{,}500 \times 1.8 \times 10^5 = 2.4$). Since some base pair changes within a codon would not change the amino acid in a protein due to degeneracy in the code, and since nothing regularly occurs in DNA every two base pairs, it is reasonable to conclude that recombination *is* in some cases occurring between adjacent base pairs. Benzer coined the term **recon** for the unit of recombination and this is generally accepted to mean one base pair.

The Muton

Because on what is known about the mode of action of the base-altering and base-replacing chemical mutagens discussed in Chapter 12, and with the mapping experiments being able to resolve distances between adjacent base pairs, one would predict that these sites of mutation being mapped would be equivalent to one base pair. This led Benzer to coin the term **muton** as the smallest mutational unit within a gene. Since that time, additional experimental evidence, such as the sickle-cell hemoglobin and the tryptophan synthetase examples already discussed, confirm the base pair as the mutational unit.

Definition of a gene. As a result of these studies we can add to our already existing information and more accurately define a gene. A definition given by Watson seems to include the several types of criteria. A gene is a distinct chromosomal region responsible for a single cellular function, consisting of a linear array of potentially mutable sites between which recombination can occur.

STRUCTURE OF THE EUKARYOTIC GENE

The complementation test is easily carried out with microorganisms where very large numbers of individuals make possible detection of rare genetic events. This is not so easy with eukaryotes. Since it is more difficult to maintain and examine very large numbers of individuals, many systems have not been subjected to the complementation test. However, there are two examples, the *rosy* locus in *Drosophila* and the *waxy* locus in barley, in which complementation analysis has been carried out. The rosy locus that controls pigmentation in the eye was found to contain two complementation groups or cistrons. The waxy locus that controls the type of starch deposited in the seed contains six complementation groups or cistrons. It would thus seem that in principle the definition of a gene based on complementation would also be generally applicable in eukaryotes.

However, eukaryotic genes have one structural aspect not found in the prokaryotes. It was discovered in the late 1970s that many eukaryotic genes contain nucleotide sequences that are not translated into the amino acid sequences of a protein. This discovery led to the concept of **split genes.** That is, the eukaryotic gene is split into two kinds of sequences: **coding sequences,** which are ultimately translated into proteins, and **intervening sequences,** which are not translated into proteins. These terms correspond to the previously used terms of **exon,** signifying the DNA sequences that are expressed (translated into proteins), and **introns,** signifying the DNA sequences that are not translated into a protein.

The basis for this finding lies in the formation of **heterogeneous nuclear RNA** and RNA processing discussed in Chapter 9. In short, a large initial RNA transcript is processed in the nucleus into the functional messenger. During this processing the intervening sequences are removed, leaving a shorter transcript made up of only coding sequences. The amount of RNA removed can be more than 50 percent of the primary transcript. A generalized scheme of RNA processing of intervening sequences was given in Figure 8–8. The presence of intervening sequences is limited to eukaryotic genes, with the exception of a few genes in viruses, such as adenovirus, which can only replicate in the eukaryotic nucleus.

Thus far, with two exceptions, histone and interferon genes, all eukaryotic genes that code for proteins contain intervening sequences. The range can be as low as two introns in the globin gene, seven in the chick ovalbumin gene, and around fifty in the alpha-collagen gene. In the four yeast tRNA genes that have been sequenced, a single intron is found at the start of the anticodon loop. In lower eukaryotes, all ribosomal RNA genes are interrupted, while in higher eukaryotes, most of the rRNA genes are interrupted.

The occurrence of split genes in eukaryotes seems to be a violation of colinearity. We still may regard the gene (cistron) as a sequence of DNA responsible for the synthesis of a single protein. However, we must recognize that there are sequences within that linear array of base pairs we call a gene that are not reflected in an amino acid sequence of a protein. Nevertheless, this DNA sequence is still made up of mutable sites, which can be separated by recombination. The occurrence of introns and exons does not in any way alter our concept of a gene.

A final question might be: Why have split genes been preserved? One answer originally proposed by Gilbert is **exon shuffling.** The shuffling or exchange of exons within or between genes through either normal recombination or chromosome breakage would allow for the trying out of new "meaningful" combinations of exons. This would be especially appropriate if an exon coded for a particular functional domain of a protein. New combinations, deletions, or duplications of exons could take place to create new proteins with new or improved functions in the cell.

Two types of examples support this theory. In rats there are two insulin genes, and in other mammals there is only one. The single insulin gene in these other mammals has two introns. In rats, one gene has two introns, while the other gene has only one. The implication is that one ancestral gene in rats had two introns, and its duplication allowed the precise removal of one intron and combined two exons in the one gene.

A similar situation occurs in the oxygen-binding proteins, myoglobin in animals, and leghemoglobin in leguminous plants. This example is summarized in Figure 11–5. Leghemoglobin has three introns, two of which occur at identical points in the globin genes of animals. The middle intron of leghemoglobin separates two exons that together code for a protein corresponding to that produced by a central myoglobin exon. It is easy to speculate that the central myoglobin exon originated from some ancestral gene by exon fusion with the primitive sequence remaining in the leghemoglobin gene.

The second type of evidence seems to be two examples of actual exon shuffling. The first is in the gene for low-density lipoprotein (LDL) receptor. This gene appears to be comprised of exons from another gene. One stretch of 400 amino acids in the LDL gene shows 33 percent homology with the precursor of epidermal growth factor (EGF), a com-

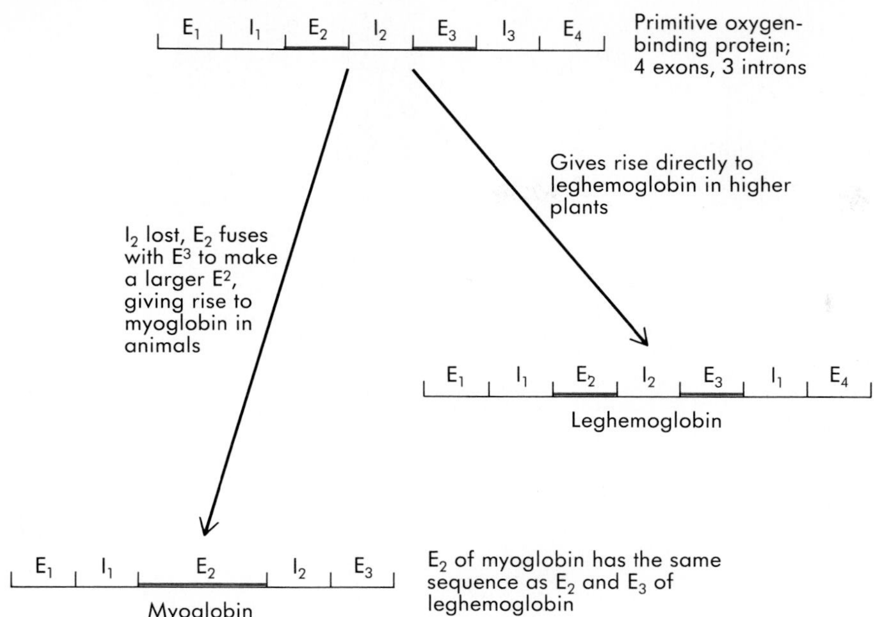

FIGURE 11–5. *Schematic representation of the evolution of two independent oxygen binding proteins, Leghemoglobin in plants and myoglobin in animals from a common primitive oxygen binding protein. The process involves fusion of exons and loss of one intron to form the animal myoglobin molecule.*

pletely different protein. This region of homology is encoded by eight contiguous exons. Five of the nine introns in these two genes are at exactly the same place. Also, a 40-amino-acid sequence in this region is repeated three times. This is encoded by one exon repeated twice in the LDL gene and the EGF gene, and also found in a single exon in a different gene for factor IX, a blood-clotting protein. Furthermore, this same exon is probably also present in two other blood-clotting-factor genes.

A second example is found in a group of glycolytic enzymes: glyceraldehyde phosphate dehydrogenase (GAPDH), pyruvate kinase (PK), and triose phosphate isomerase (TIM). All three genes in chicken share several exons in common, the only variation being the number and location of the introns. They all seem to have been assembled from several common exons. When the TIM gene from maize was sequenced and compared with the chicken gene, three of five introns were at exactly the same location. This also suggests that the ancestral gene was probably already broken up into exons before the divergence of plants and animals. Gilbert now believes that this information should be telling us something about the "rules for creating proteins."

ORGANIZATION OF GENES

Even though observations based on mapping and expression of genes would indicate that each gene is present singly in the genome of an organism, it is becoming more evident that this is not always the case. A gene for a particular function may undergo duplication and variation to the point that the genetic material responsible for a particular phenotype may comprise a **gene family.** These genes appear to be clustered together or scattered around the genome. The members of a family of genes have identical function. Their timing of expression may be coordinated together or in a sequential pattern during development. A classic example is the genes for ribosomal RNA. They are clustered together in large numbers at the nucleolar-organizing region and are expressed

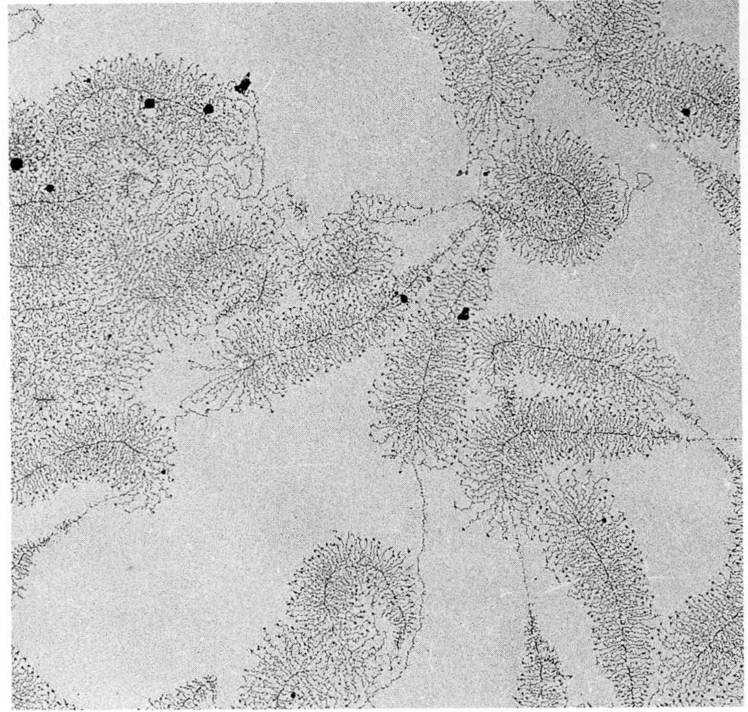

FIGURE 11–6. *Family of ribosomal RNA genes in Xenopus. Each is actively synthesizing ribosomal RNA.* (Photo courtesy of Dr. O. Miller, University of Virginia, used by permission.)

together. Figure 11–6 shows this family of genes actually transcribing ribosomal RNA. This organization is very helpful where the gene product is needed in large amounts, such as rRNA.

The α- and β-globin genes in humans are also clustered into the families. The β-gene cluster is found on chromosome number 11, and extends over 60,000 base pairs. The α-gene cluster is on chromosome number 16 (see Table 6–2). The expression of these genes is coordinately regulated during development. The globin genes are structurally very similar in all animals, with three exons and two introns all located at constant positions. This has led many to conclude that this globin gene family originated from some ancestral gene through duplication and mutation.

The histone genes also represent an interesting family. First of all, none of the five genes contain introns. The five genes exist in clusters around the genome. In *Drosophila* about 100 copies of these five genes are present. In other organisms, such as humans, while being linked on the same chromosome, the histone genes do not exist in tandemly repeated clusters. In many cases, the tRNA genes are tandemly duplicated in large families. Furthermore, they are transcribed into one great long transcript that is processed into the individual tRNA molecules.

Gene families are presently classified into three groups: simple multigene families, complex multigene families, and developmentally controlled complex multigene families. A **simple multigene family** is one in which only one or very few genes are repeated in a tandem array. (An example would be the genes for the 18s, 5.8s and 28s ribosomal RNA discussed in Chapter 9.) A **complex multigene family** consists of related genes in a cluster, but transcribed independently. (The five histone genes are an example.) A **developmentally controlled complex multigene family** codes for a number of different forms of the same protein and are expressed at different times in development. The different forms of α and β hemoglobin expressed at the embryonic, fetal, and adult stages of development are excellent examples.

The Immunoglobulin Gene Family

Genes that code for the immunoglobulins show a most interesting case of organization. The proteins responsible for the immune response in vertebrates are called the **antibodies,** which are synthesized in response to **antigens.** These molecules are characterized by their tremendous diversity. They may occur in a million or billion slightly different forms in an organism with the result that they can respond to the millions and billions of foreign antigens with which the organism comes into contact throughout its lifetime. The question is, how does an organism produce this tremendous variety of proteins? Is there a gene for each slightly different antibody?

First, it is necessary to discuss antibody structure. Each antibody molecule is made up of four **immunoglobulin** chains. There are two identical heavy chains, of 330 or 440 amino acids, and two identical light chains, of 220 amino acids each. The four immunoglobulin chains are held together in such a way as to form a Y-shaped molecule (Figure 11–7). Virtually all antibody molecules are identical except on the arms of the Y. Here, at the amino terminal ends of the four polypeptide molecules, each antibody differs slightly from every other antibody. This is also where the antigenic specificity of each antibody exists. Based on the amino acid sequence, it would appear that each polypeptide chain of an antibody is constructed of smaller subunit proteins called **domains.** These are discrete amino acid sequences within a polypeptide that are associated with a particular function. The constant portion of each polypeptide is made of the same repeating 110 amino acid domain. The variable part of each polypeptide contains a different amino acid domain depending on the specificity of the antibody.

To summarize, each antibody is made up of four polypeptide chains, containing large constant regions of amino acid sequences and small variable regions of amino acid sequences. The uniqueness and specificity of each antibody lies in the diversity of variable regions that can be present.

In the 1980s, we have gained much insight into the organization of the genes that code for these proteins, which is necessary to explain how so many different proteins can be produced in a single organism.

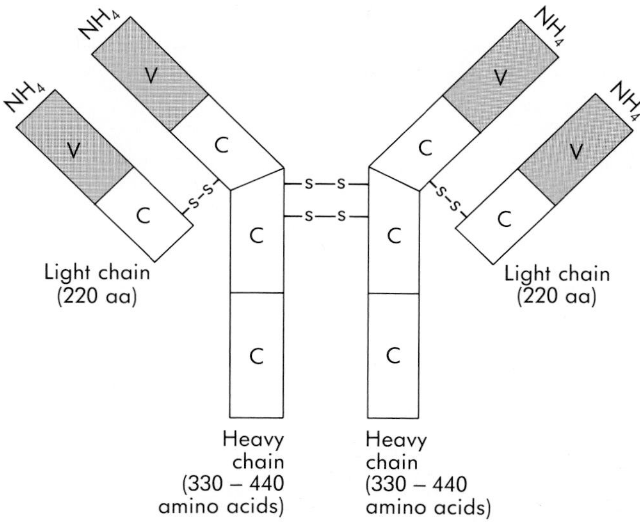

FIGURE 11–7. *Association of four immunoglobin chains into a functional antibody.*

The most accurate theory thus far is based on **somatic recombination** of gene segments for the various protein domains making up a functional antibody molecule. In the germ line of each organism are DNA segments coding for all the possible domains for a mature antibody. There are at least four different types of segments: approximately ten C segments coding for the constant regions of both chains, several hundred V segments coding for the variable portion of the protein, four distinct J segments used for joining V and C segments, and about twenty different D segments for diversity for the heavy chain.

A functional antibody gene is made up in B lymphocytes (somatic cells derived from the germ line) by somatically recombining V, J, D, and C segments into a functional antibody gene (Figure 11–8). The gene is assembled enzymatically by bringing together V, J, and D segments and deleting all the DNA that separates them, which in some cases can be considerable. An RNA transcript is then made of the DNA, including all the V-J or V-D-J segments, the C region, and a large intron separating the C region from the other segments. The intron is processed out, and the functional mRNA is exported from the nucleus and translated into a protein. Combining a few hundred V segments with twenty D segments and four J segments generates approximately 10,000 combinations. When heavy- and light-chain combinations are included, there are over 10 million unique antibody molecules possible.

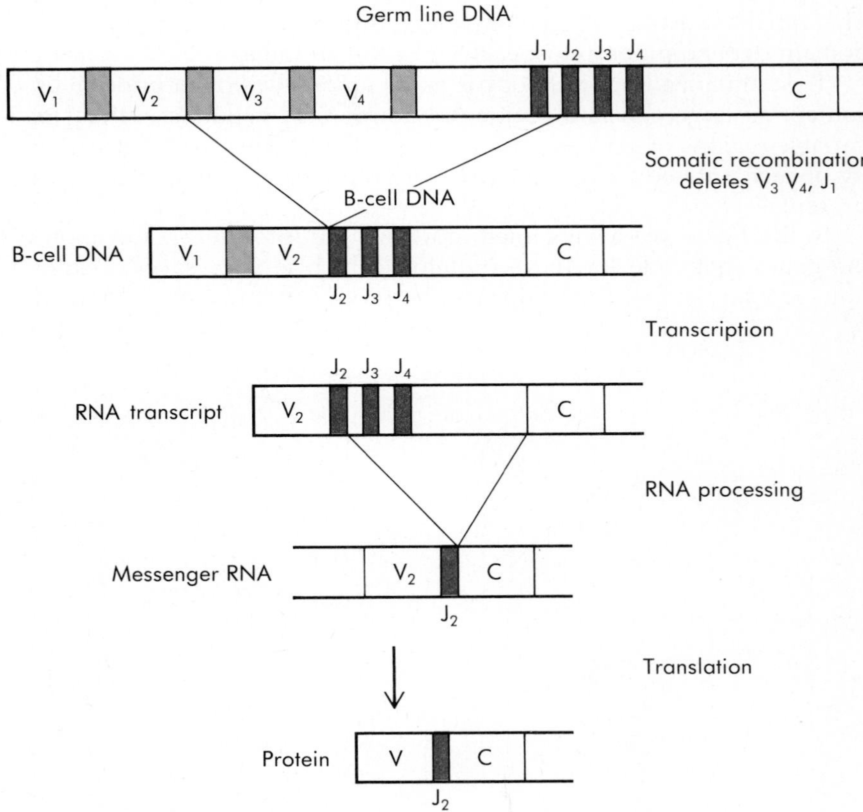

FIGURE 11–8. *Assembly process for a functional immunoglobin gene. Somatic recombination of germline DNA produces a line of B cells containing one unique combination of U, J, and C segments. Selective transcription and processing of the RNA transcript further generates a unique combination of segments used to make a specific antibody.*

MONOCLONAL ANTIBODIES

Antibodies to a particular antigen are usually obtained by preparing the antigen in as pure a form as possible, injecting it into an animal, bleeding the animal, and then separating the antiserum from the blood. The problem is that the immune system responds to even tiny amounts of contaminating antigens as well as to the one of interest. Therefore, the antisera obtained contains a mixture of antibodies to several antigens. In studies that earned them a share of the 1984 Nobel Prize, C. Mielstein and G. J. F. Kohler discovered a way to produce only a single antibody at a time in pure form.

Antibody-forming cells have a very short lifetime in culture. Mielstein and Kohler's objective was simply to immortalize some antibody-forming cells by fusing them with plasma-cell tumors, called **myelomas,** from mice; myelomas divide indefinitely in culture. The hope was that the tumor cell line would cause the hybrid cell line (now called a **hybridoma)** to divide indefinitely and produce the antibody indefinitely. When cultured on a selective medium that only allows the hybridoma cells to grow, a cell line was obtained that grew indefinitely, and not only continually produced antibodies, but produced antibodies of a single type. Since these antibodies are a single type of molecule, they represent a chemically pure reagent. These antibodies are therefore referred to as **monoclonal antibodies,** and they can be obtained in unlimited amounts.

The applications of monoclonal antibodies seem almost unlimited. Since they can be made to almost any antigen, molecular studies on the immunoglobulin molecule itself become easier. Human antigens associated with susceptibility to several diseases such as rheumatoid arthritis, juvenile diabetes, and multiple sclerosis could be much more clearly identified with monoclonal antibodies. Antibodies complexed with toxins and drugs could be used in treating tumors. In molecular biology, monoclonal antibodies could be used to obtain a specific protein as a gene product from a mixture of molecules in a cell. The only limitation is in being able to isolate the antigen to be used. The discovery of monoclonal antibodies demonstrates how an experiment with a very simple objective can lead to a major discovery with a multitude of applications.

Two additional mechanisms are used to generate even more antibody diversity. It seems that the DNA-splicing machinery that fuses V, J, and D segments lacks precision, resulting in slight variations at the junction of the segments. In addition, the rate of spontaneous mutation in the segments for the V region is several orders of magnitude higher than the average rate for eukaryotes. Therefore, even if the B lymphocyte (cells that produce a specific antibody) tried to make the same antibody twice, it could not.

Finally, it is important to complete this story by describing the fate and functioning of these antibody genes and cells in the immune response. The best explanation of the overall phenomenon lies in the **clonal selection theory.** The B lymphocytes originate from bone marrow cells. Each cell produces antibodies against only specific antigens by the process of somatic recombination described above, and therefore carries only specific gene constructs containing single combinations of V, J, D, and C segments. As this specific cell type proliferates, it produces a small clone of identical cells, all producing the same antibodies. The antibodies remain bound to the cell membrane. Whenever they bind to their specific antigen, the cell is stimulated to proliferate rapidly to greatly increase the clone and respond quickly to the specific antigen. This is clonal selection. This also explains how once immunity is developed against a foreign antigen, it remains throughout the lifetime of the individual. The individual always has a small population of B cells circulating in the bloodstream. When the antigen appears, the cells that contain the gene for the antibody specific for the antigen selectively proliferate, producing a rapid antigenic response.

Here we see clusters of gene fragments, recombining to produce functional genes in a cell line. The tremendous protein diversity comes about due to the various combinations of gene segments.

Pseudogenes

The final result of gene development beyond gene families is development to a point that the gene is no longer functionally expressed. This is called a **pseudogene,** a gene that has sequences related to functional genes but through base additions, deletions, and rearrangements, including loss of a promotor, no longer can be expressed. A number of these have been mapped in the human genome (see Table 6–2). In terms of gene evolution, pseudogenes are a dead end.

CONCLUSIONS

The last four chapters have given the following important conclusions regarding the gene:

1. The genetic material is DNA in all organisms except a few viruses, where it is RNA.
2. A gene is a specific linear sequence of nucleotides and may be considered at more than one level of organization.
   a. The functional unit, the *cistron,* is responsible for specifying a particular polypeptide chain and consists of three adjacent deoxyribonucleotides for each amino acid residue in the chain, plus a minimum of one chain-initiating triplet and one or more chain-terminating triplets.
   b. Eukaryotic cistrons often consist of transcribed and translated sequences (exons), interrupted by untranslated sequences (introns).
   c. Each cistron is divisible into (1) *mutons,* the smallest number of nucleotides independently capable of producing a mutant phenotype. In many cases this is a single nucleotide or nucleotide pair; (2) *recons,* the smallest number of nucleotides capable of recombination. This likewise is as small as one nucleotide. Recons and mutons are structurally identical.
3. Specification of the amino acid sequence of a polypeptide chain by a cistron is made possible through the latter's particular deoxyribonucleotide sequence.
   a. Mediation of polypeptide synthesis operates through an intermediate, messenger RNA, which is ordinarily a single-stranded, base-for-base, complementary transcript of the nucleotide sequence on one DNA strand, plus a 5′ "cap" and a poly-A "tail."
   b. Messenger RNA, in conjunction with ribosomes, tRNA, energy sources, and a battery of enzymes, is directly responsible through its sequence of ribonucleotides (the genetic code) for polypeptide biosynthesis at ribosomal surfaces, chiefly in the cytoplasm (exceptions will be taken up in Chapter 16). Functional proteins are then assembled from the polypeptide chains thus formed.

PROBLEMS

11–1 Suppose a certain cistron is found to consist of 1,500 deoxyribonucleotides in sequence. (a) What is the maximum number of mutons of which this cistron could consist? (b) Is this number likely to be too high, too low, or about right for an actual organism? Why?

11–2 For how many codons is the tryptophan synthetase A cistron responsible, if initiating, terminating, and possible introns are neglected?

11–3 If initiating, terminating, and possible introns are neglected, how many nucleotides are there in the tryptophan synthetase A cistron?

11–4 One calculation for the molecular weight of the DNA in a single (*haploid*) set of human chromosomes is 1.625×10^{12}. (a) If 650 is assumed as the average molecular weight of a pair of deoxyribonucleotides, how many nucleotide pairs comprise the DNA of one *diploid* somatic nucleus? (b) What, then, is the maximum number of mutons in each of your somatic cells? (c) What is the total length in micrometers (microns) of the DNA in the nucleus of one of your *diploid* somatic cells? (d) At these values, for how many mRNA codons is the DNA of *one* of your (*haploid*) sets of chromosomes responsible? (e) If you assume an average of 300 amino acid residues per polypeptide chain, and neglect introns and start-stop signals, how many cistrons do you have in *one* of your sets of 23 chromosomes?

11–5 Explain why one mutation in a given cistron may be lethal, whereas another in the same cistron may produce no adverse phenotypic effect and be detectable only by laboratory tests.

11–6 How might a gene family arise at a given locus during the evolution of a particular species?

11–7 Distinguish among these three concepts: cistron, muton, and recon.

11–8 How do complementation and recombination differ?

11–9 In examining gene families one often finds embryonic, fetal, and adult forms of the protein such as in hemoglobin. To what phenomenon do these terms refer? What single term might be used to describe this type of gene family?

11–10 Outline the mechanism for generating antibody diversity?

11–11 What is meant by the variable and constant regions of an antibody molecule?

REGULATION OF GENE ACTION

CHAPTER

12

Even though all the cells of an organism contain the same basic set of genetic information, the cellular phenotypes of multicellular organisms vary greatly. Even a bacterial cell does not produce all the enzymes it is capable of producing all the time. Instead, these cells produce only those enzymes necessary when a specific substrate is present. The existence of such adaptive enzymes or gene activity as a result of stage of development or in response to the environment suggests that there is, in all cells, selective regulation of gene activity governing the synthesis of specific proteins in the cell. This chapter examines the methods by which genes are regulated in order to produce specific proteins.

GENE REGULATION IN PROKARYOTES

Inasmuch as the ultimate product of a gene is a polypeptide chain, which frequently functions as an enzyme, a simple system of control could operate at the level of gene regulation. Evidence accumulated in recent years indicates that there are, in fact, several levels of gene regulation. One regulates the *synthesis* of enzymes by controlling the transcription of mRNA for these enzymes (transcriptional control). Second, control occurs at translation (discussed in Chapter 9). And finally, there is control over the activity of the enzymes whose synthesis is not regulated (posttranscriptional control). One of these is through the **operon,** a group of adjacent genes whose transcription is regulated by other sites, and another is by the process of **end-product inhibition.**

Transcriptional Regulation Through the Operon

The *lac* operon in *Escherichia coli*. Among the many genes of the colon bacillus, two in particular have shed considerable light on the system of transcriptional regulation of prokaryotic genes. Two enzymes are required for lactose metabolism in this organism: *galactoside permease*, responsible for transport of lactose into the cell and its concentration there, and *beta-galactosidase*, which catalyzes the hydrolysis of lactose to galactose and glucose. The wild-type *E. coli* produces these enzymes and a third enzyme, *thiogalactoside transacetylase*, in very small amounts—about ten molecules per cell, in the absence of lactose. When and if lactose is supplied, the rate of synthesis of these enzymes increases as much as 1,000 times. One might say the wild-type cells are *induced* to synthesize the enzymes that function in lactose metabolism when the enzymes' substrate is present.

After some initial uncertainty over whether enzyme synthesis or enzyme activity was affected by lactose, it has become clear that the presence of lactose does indeed *induce* synthesis of enzymes in the inductive strain. Lactose, the substrate, is acting here as an enzyme inducer. Enzymes whose production is increased when a substrate is present are designated as **inducible enzymes.** Although inducers are quite specific, structural analogs may often function in the same way.

The *lac* segment in the *E. coli* chromosome is known to include three genes, *z* for beta-galactosidase, *y* for galactoside permease, and *a* for thiogalactoside transacetylase. The three genes are closely linked and regulated together. Francois Jacob and Jaques Monod and their associates developed a model of a genetic system that regulates lactose metabolism, to which they gave the name **operon.** With minor refinements, this model explains many observations of control of gene expression in bacteria (Jacob and Monod shared the Nobel prize in 1964 for this work). Several other similar regulatory systems have subsequently been described for this and other bacteria. Virtually all bacterial genes are regulated through operons.

Three genes, *z*, *y*, and *a*, are each responsible for one enzyme (Figure 12–1); they occupy adjoining positions on the DNA molecule and map in that sequence. The genetic information of these three genes is transcribed into a single mRNA molecule that is subsequently translated into the polypeptide chains of the three enzymes. RNA polymerase attaches to the *promoter site (p)*. Then transcription proceeds in the 5′→3′ direction, beginning with the *operator site (o)* that lies to the "right" or "downstream" of *p*, that is, in the direction of the three structural genes (Figure 12–1), exercising a control over transcription. Mapping experiments show the sequence of these genetic elements of DNA to be

$$p\ o\ z\ y\ a$$

The three genes, together with the promotor and operator sites, comprise the lac operon. **An operon, then, consists of a system of genes, operator, and promotor sites.** The basis for the regulation of the synthesis of the three enzymes lies in still another site, the **regulator** (*i*), and the interaction of its protein product with lactose and with the operator site. The regulator site is directly adjacent to the promotor in the *lac* operon. The complete map sequence involved, then, may be represented as

$$i\ p\ o\ z\ y\ a$$

and is composed of about 4,700 nucleotides.

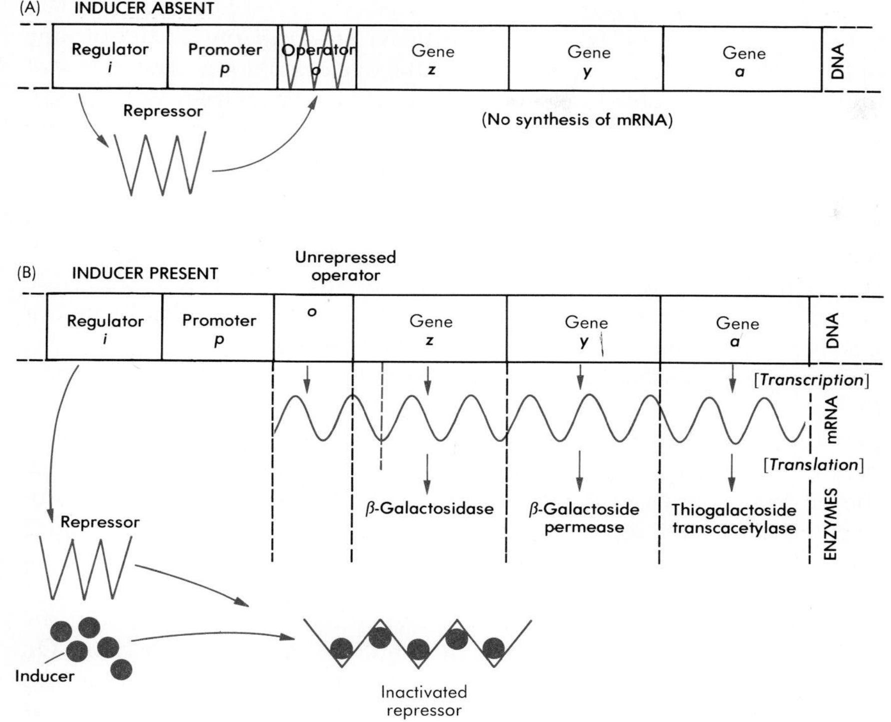

FIGURE 12–1. *Diagrammatic representation of the lac operon of* Escherichia coli. *Transcription is initiated at the operator site (o). The enzyme DNA-dependent RNA polymerase attaches to the promotor site (p) in the presence of cyclic AMP receptor, which must bind first to the promotor at or near the 5′ end of the latter region. Transcription proceeds in the 5′ → 3′ direction beginning as shown within the operator. If the repressor protein is present, in the absence of lactose it will bind to the operator, preventing transcription of the three genes. If lactose is present, it combines with the repressor to inactivate it, allowing transcription to proceed. The regulator* (i) *is directly adjacent to the promotor. For further details, see text.*

The regulator site produces a repressor protein of 360 amino acid residues. In the absence of lactose, the repressor protein binds to the wild-type operator (o^+), and thereby prevents the attachment of RNA polymerase to the promotor so that transcription of the three genes is prevented. But lactose, if present, binds to the repressor and thereby changes its shape, preventing it from binding to the operator. The result is that the three genes are transcribed. Lactose, the substrate for beta-galactosidase and galactoside permease, thus acts as an *inducer* of the operon. Transcription is initiated in the operator site, which is transcribed as part of the overall mRNA product. *Translation,* however, begins in the *z* gene with the *eleventh* nucleotide rather than with the first (Figure 12–2).

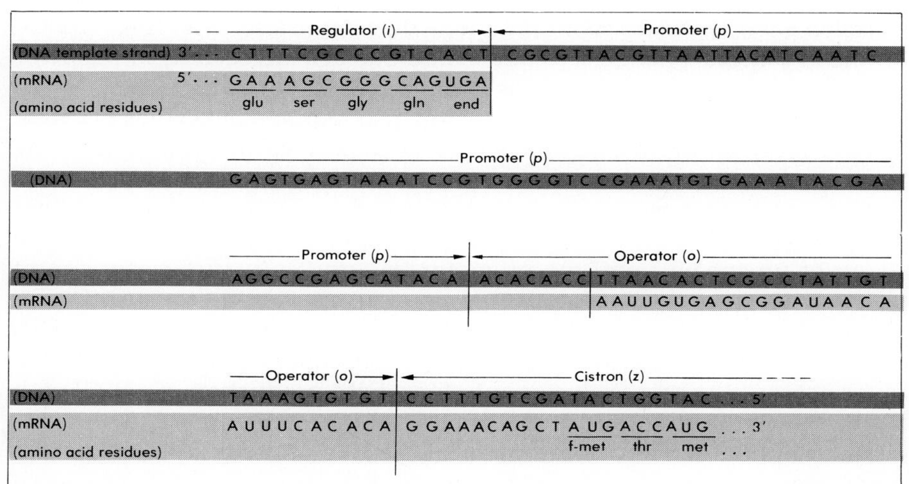

FIGURE 12–2. *Nucleotide sequences of the template DNA strand and the transcribed RNA molecule for a portion of the regulator* (i), *all of the promotor* (p) *and operator* (o), *and the beginning of the* z *gene of the lac operon. Note the translation begins only at the eleventh ribonucleotide of the* z *gene's messenger. Amino acids are abbreviated as follows: f-met, N-formyl-methionine; gln, glutamine; glu, glutamic acid; gly, glycine; met, methionine; ser, serine; thr, threonine. (Read all necleotide sequences from left to right.)*

Other strains of the colon bacillus, derived by mutation from the inductive wild type, produce these enzymes continuously, whether lactose is present or not. These are **constitutive** strains. Constitutive strains of *E. coli* are characterized by either a defective regulator (i^-) or a mutant operator that is unable to bind the repressor. For example, i^+ p^+ o^+ z^+ y^+ a^+ is an *inductive* strain; i^+ produces a repressor that binds to o^+ and prevents transcription of z^+, y^+, and a^+ (Figure 12–1). Presence of lactose, of course, inactivates the repressor so that o^+ permits transcription. On the other hand, i^- p^+ o^+ z^+ y^+ a^+ is constitutive because a defective repressor, which will not bind to o^+, is produced, so transcription of the three genes is continuous. An i^+ p^+ o^c z^+ y^+ a^+ (an *operator-constitutive* strain) is also constitutive because the repressor binding site, rather than the repressor, is defective.

Interestingly enough, studies of merozygotes (partial diploids) produced through conjugation show i^+ to be dominant to i^-. In a cell of genotype i^+ p^+ o^+ z^+ . . ./i^- p^+ o^+ z^- . . . one might expect normal beta-galactosidase to be produced inductively and a modified form of the enzyme (inactive or only weakly active) to be produced constitutively. This is not so; both types of beta-galactosidase are produced inductively. The repressor is able to diffuse from the i^+ p^+ o^+ z^+ . . . "chromosome" to the other and repress z^-, even though the latter is linked to i^-. The regulator gene thus functions in either the *cis* (on the same chromosome) or *trans* (on a different chromosome) position. On the other hand, the operator (*o*) functions only in the *cis* position; for example, the partial diploid i^+ p^+ o^c z^+ . . ./i^- p^+ o^+ z^- . . . produces normal beta-galactosidase constitutively, but no modified enzyme. This is to be expected, inasmuch as the operator is the place where transcription begins.

This aspect of the *lac* operon illustrates *negative control,* the basis for which is production of a *repressor.* In this case, the repressor blocks the expression of the lac genes unless the inducer (lactose) is present to inactivate the repressor.

Another group of mutants involves the repressor gene, *i*. These mutations in *i* act on the *z-y-a* structural genes in either the *cis* or *trans* configuration in merozygotes. The i^s mutant produces a *superrepressor* that prevents production of the enzyme products of all three genes. From study of partial diploids, it is seen to be dominant to i^+ so the dominance relationship among these three alleles is $i^s > i^+ > i^-$.

Because the direction of transcription is from promotor to operator to genes *z, y,* and *a,* in that order, mutations in one gene will affect the expression of genes further along the operon. This type of mutation is called a *polar mutation.* For example, a defect in *z* affects all three genes, one in *y* affects only genes *y* and *a,* and one in *a* affects only *a.* Behavior of several of these genotypes is summarized in Table 12–1.

"Turning on" of the lactose operon also requires molecules of catabolite receptor protein (CRP) bound to molecules of cyclic adenosine monophosphate (cAMP). CRP is the product of gene *crp;* cAMP is made from ATP (adenosine triphosphate) with the aid of the enzyme adenylate cyclase produced by gene *cya* (Figure 12–3). Both genes are to the left of (upstream from) the promotor (Figure 12–3). If glucose, one of the products of the breakdown of lactose, is *not* present, the cAMP-CRP complex is produced and binds to the *lac* promoter (Figure 12–3). This *enhances* RNA polymerase binding and consequent transcription of the *lac* operon. *Presence* of glucose interferes with production of cAMP with the result that the cAMP-CRP complex cannot be formed. This halts transcription of the lactose operon. Defective mutants of both *crp* and *cya* have been produced; either crp^- or cya^- individuals produce such

TABLE 12–1. **Production of Beta-Galactosidase in Several lac Genotypes in *E. coli****

Genotype	*Enzyme production* Constitutive Normal	Constitutive Modified	Inductive Normal	Inductive Modified	Basis
$i^+ o^+ z^+ \ldots$	–	–	+	–	Wild-type
$i^- o^+ z^+ \ldots$	+	–	–	–	Defective regulator
$i^+ o^c z^+ \ldots$	+	–	–	–	Operator constitutive
$i^s o^+ z^+ \ldots$	–	–	–	–	Superrepressor
$i^+ o^+ z^+ \ldots / i^- o^+ z^+ \ldots$			+	+	Wild-type/defective regulator
$i^+ o^+ z^- \ldots / i^- o^+ z^+ \ldots$	–	–	+	–	i^+ dominant to i^-
$i^+ o^+ z^+ \ldots / i^s o^+ z^+ \ldots$	–	–	–	–	i^s dominant to i^+

*A plus sign indicates beta-galactosidase production; a minus sign indicates lack of production of the enzyme.

low levels of their respective products that their effect is minimal or absent. Because the *z*, *y*, and *a* genes are turned on in the presence of the cAMP-CRP complex, this complex is a *positive* regulator of the *lac* operon. This means that the *lac* operon is regulated negatively (turned off) by the repressor and positively (turned on) by the cAMP-CRP complex.

The *his* operon. The *lac* operon involves an active repressor; however, not all operons function in this manner. The *his* operon in *Salmonella typhimurium* (closely related to *E. coli*) is responsible for the synthesis of the amino acid histidine. It includes 13,000 deoxyribonucleotide pairs. Synthesis of histidine in this species occurs in ten enzymatically controlled steps, involving nine enzymes (one functions in two different steps) produced by nine genes. The regulator is responsible for an *inactive* repressor protein that is unable to repress the operator in the *absence* of histidine. The result, then, is that RNA polymerase proceeds through the nine genes and transcribes one long messenger that codes for all the proteins necessary for histidine synthesis. If excess histidine is present, histidine combines with the repressor and changes the repressor into an active form. As such, histidine is called the **corepressor.** This active form binds to the operator and blocks transcription of the operon, and consequently histidine synthesis occurs. Although the operation of the *his* operon differs somewhat from that of the *lac* operon,

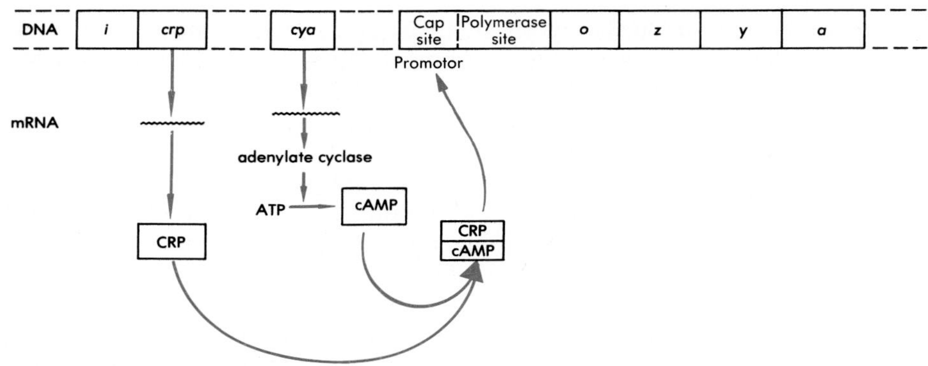

FIGURE 12–3. *The cAMP-CRP positive control of the lac operon in* Escherichia coli. *See text for details.*

it is, nevertheless, a *negative* control because a *repressor* protein is involved in turning the operon off.

The *ara* operon. Regulation of the three genes responsible for the production of the three enzymes that catalyze the three-step breakdown of the pentose sugar arabinose to xylose-5-phosphate (another pentose sugar, involved in cellular oxidations) in *E. coli* represents an interesting variation in operon operation. The segment of the DNA molecule involved (*E. coli* k-12 map position is about 1.3 minutes; see Figure 6–7) consists of three genes, each responsible for one of the three enzymes, an initiator site, an operator, and a regulator. The regulator produces a protein that serves both as an activator and a repressor of the three genes. In the absence of arabinose, this regulatory protein serves as an active repressor by binding to the operator. When present, arabinose will bind to the regulatory protein and remove it from the operator. In addition, this complex becomes an activator by stimulating RNA polymerase. So, in the *absence* of the inducer (here arabinose, the metabolite), the regulatory protein functions as a *repressor* by binding to the operator and preventing transcription; this represents, of course, negative control. The same protein is converted to an *activator* in the presence of the inducer (arabinose) by binding to it and allowing transcription to take place; this represents positive control.

Summary of Types of Control of Gene Expression in Operons

These systems demonstrate both positive and negative methods of control of gene activity. The *lac* operon demonstrates negative control in which the active repressor blocks gene expression by binding to the operator. In order for the genes to be expressed, the repressor must be actively removed from the operator by the inducer (lactose). Catabolite repression of the *lac* operon by glucose is a form of positive control. Here gene expression is dependent on the presence of a small molecule (glucose) serving as the activating signal. The histidine operon is an example of negative control since the histidine repressor binds to the operator only after binding to histidine. Finally, the *ara* operon is an example of both negative and positive control because in the absence of the inducer the regulatory protein represses the operon, and in the presence of the inducer it actually stimulates the operon.

Posttranscriptional Control by End-Product Inhibition

In studies on isoleucine synthesis in *E. coli*, it was demonstrated that addition of isoleucine (the end product of a five-step conversion of threonine; Figure 12–4) to a culture of the bacterium resulted in immediate blocking of the threonine → isoleucine pathway. In the presence of added isoleucine, the cells preferentially use this *exogenous* end product and cease their own isoleucine synthesis. Moreover, it has been shown that *production* of each of the five enzymes is not interfered with, but action of the enzyme responsible for the deamination of threonine to alpha-ketobutyrate (Figure 12–4) is *inhibited* by the end product, isoleucine.

Interestingly, this inhibition results from a binding of end product to the enzyme so that it competes with the substrate for a site on the enzyme molecule. But it is known that this competition is not for the same site; the enzyme apparently has two specific recognition sites, one

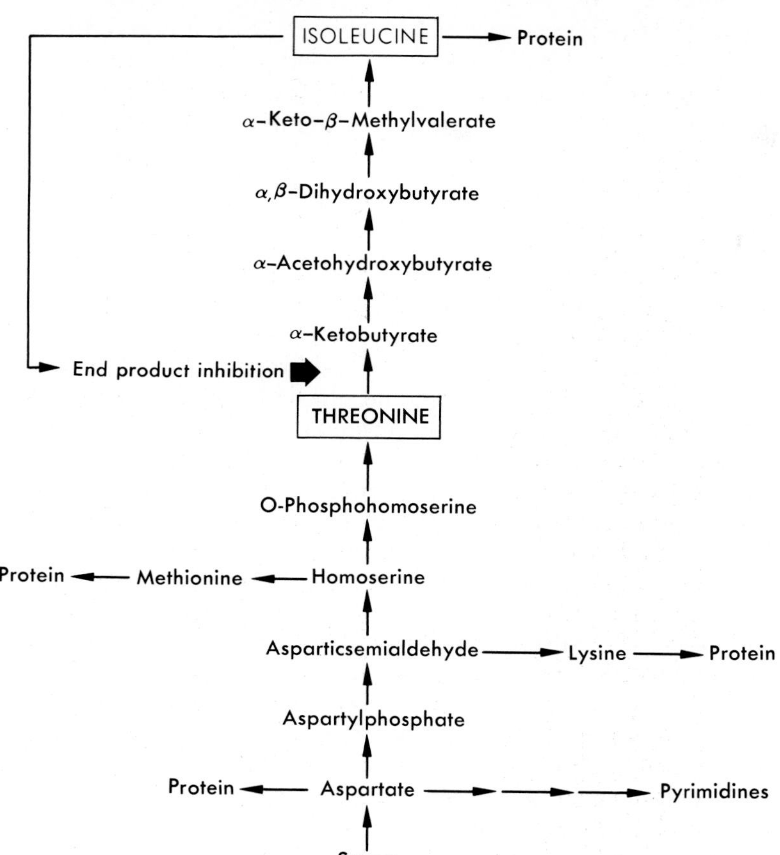

FIGURE 12–4. *Some steps in isoleucine synthesis. In end-product inhibition in* Escherichia coli *enzyme inhibition occurs in the deamination of threonine to alpha-ketobutyrate at the point indicated.*

for its substrate and another for the inhibiting end product. However, attachment at the inhibitor site affects the substrate site. This comes about by a change in shape of the enzyme molecule that subsequently alters the interaction between the substrate site and the substrate. The term *allosteric interaction* is applied to such changes in enzyme activity due to a change in the shape of the enzyme molecule produced by binding of a second substance at a different and nonoverlapping site.

GENE REGULATION IN EUKARYOTES

Multicellular organisms normally develop by an orderly process of differentiation from a single cell, the zygote, to an adult form of many different kinds of cells, tissues, and organs. But because of the behavior of chromosomes in mitosis, all the cells of a complex, multicellular soma may be presumed to have the same genotype. In short, between zygote and adult stages, cells of the organism *differentiate,* both physiologically and physically, yet the genome of all cells should be identical under normal conditions. The problem may be quite simply stated: *How do genetically identical cells become functionally different?*

The key to orderly structural and functional differentiation of multicellular organisms lies in the fact that their cells do not produce the same proteins all the time. Apparently some mechanism exists in the cell to turn genes "on" or "off" at different times and/or in different environments. In fact, it is now clear from considerable experimental evidence that differentiation is the consequence of orderly, temporal

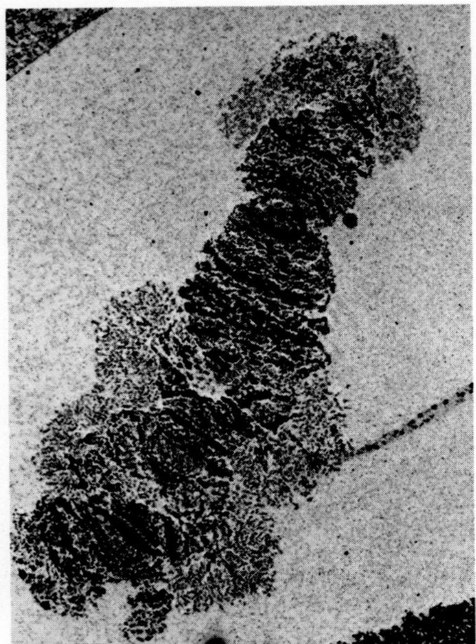

FIGURE 12–5. *Chromosome puffs in polytene chromosome IV of* Chironomus tentans. (Photo courtesy B. Daneholt, used by permission.)

activation and repression of genetic material. Gene regulatory mechanisms may operate at any of four levels: transcriptional; posttranscriptional, involving RNA processing; translational; and posttranslational, involving modification of the protein after it is synthesized. As might be expected, regulatory systems are dissimilar in eukaryotes and prokaryotes.

Giant Chromosomes

In the insect order *Diptera*, the chromosomes of larval tissues such as the salivary glands regularly exhibit unusual behavior that has shed some light on gene regulation. Although the cells of these tissues do not divide, their chromosomes replicate repeatedly while permanently synapsed in homologous pairs, and produce polytene or many-stranded giant chromosomes. If these chromosomes are examined at several stages of the individual's development, specific areas (sets of bands) are seen to enlarge into prominent **puffs** or Balbiani rings (Figure 12–5). The puff is described as a loosening of the tightly coiled DNA into long, looped structures. These puffs appear and disappear in a given tissue at certain chromosomal locations as development proceeds; those at particular locations are correlated with specific developmental stages of the insect. The pattern of puffing varies in a regular and characteristic way with the tissue and its stage of maturation.

By means of differential staining techniques, biochemical tests, and use of radioactive isotopes, it has been demonstrated that each puff is an active site of transcription. Since genes are associated with particular bands, this temporal puffing clearly indicates changes in gene activity over time. Tests show that the mRNA synthesized in one puff is characteristic and differs from that produced by other puffs.

Injection of very small amounts of the hormone ecdysone, produced by the prothoracic glands and inducing molting, causes formation of the same puffs that occur normally prior to molting in untreated larvae. The prothoracic glands are activated by the flow of brain hormone from neurosecretory cells. This has been demonstrated in an ingenious experiment reported by Amabis and Cabral. They found that tying off the anterior part of the larva of the dipteran *Rhynchosciara* just behind the brain resulted both in failure of normal puffing and a decrease in size of puffs already initiated at the time of ligation. In other experiments, injection of the antibiotic actinomycin D (which inhibits mRNA synthesis) prevented puff formation for several hours, even when ecdysone was used simultaneously. Radioactive uridine, injected into larvae, accumulates only in the puffs and nucleoli, but fails to do so if it is preceded by injection of actinomycin D.

All these results clearly point to the puffs as sites of RNA synthesis, according to a pattern closely associated with the development of the individual. Evidently then, genes undergo reversible changes in activity (transcription) that are related to the developmental stage of the organism.

Lampbrush Chromosomes

A similar phenomenon of "turning on and turning off" genes is evident in the large chromosomes of the amphibian oocyte, the **lampbrush chromosomes.** They were given the name because they resemble the brush used in cleaning the lamp chimneys used before the development of electricity. A more accurate name in modern terms might be "test tube brush" chromosomes. They are isolated from oocytes in meiotic pro-

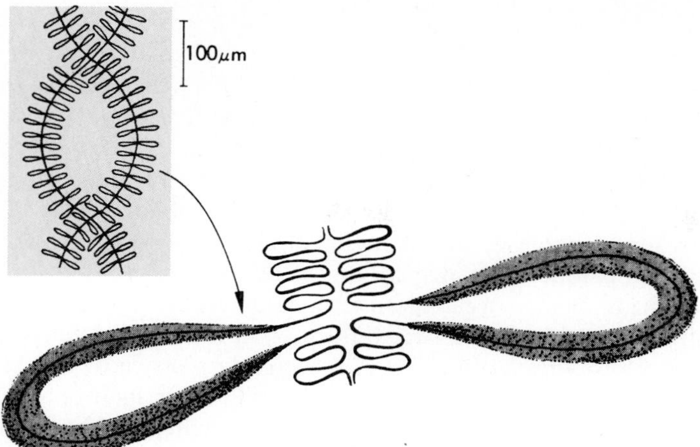

FIGURE 12–6. *Diagram of portion of a lampbrush chromosome as seen in an amphibian oocyte.*

phase. The homologues are paired and held together by chiasmata, but are not condensed as usual chromosomes would be. Instead, they are very long and stretched out (Figure 12–6). Each sister chromatid pair forms a series of pairs of loops, connected by a fine central axis. The central axis contains two duplexes of DNA that are continuous throughout the chromatid (Figure 12–6). Each loop is made up of one duplex of DNA that is continuous with the DNA of the central axis. Some excellent scanning electron microscope pictures of these chromosomes are found in Figure 12–7.

The lampbrush loops are actively involved in RNA synthesis. As such, they are structurally and functionally similar to the puffing observed in polytene chromosomes. In both cases, the DNA packaging of the chromosome has been relaxed, thus permitting transcription to occur. The RNA is produced in a cyclic fashion, suggesting the transcriptional activity of specific genes is being regulated.

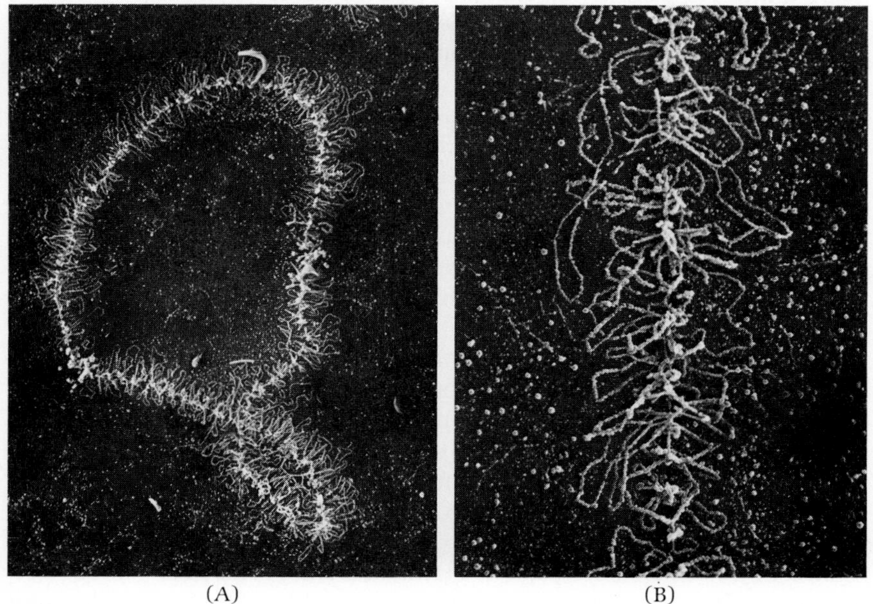

FIGURE 12–7. *Scanning electron micrograph of lampbrush chromosome. (A) Entire chromosome with chromatids; (B) close-up of small segment of one chromatid.* (Photo courtesy Dr. N. Angielier, used by permission.)

Transcriptional Control by Chromosomal Proteins in Eukaryotes

As was noted in Chapter 4, DNA of eukaryotic cells is complexed with low molecular weight, basic proteins, and histones (except sperm nuclei, where protamines are involved). These positively charged molecules owe their charge to high levels of arginine and lysine, and complex with the negatively charged PO_4 groups of the DNA molecule. Their presence increases the diameter of DNA (20 Å) to the 350 Å chromatin fiber. Histones enhance coiling of DNA and provide for the first level of chromosome organization, the nucleosome (see Chapter 8). In addition, when histones are tightly complexed with DNA, the transcriptional properties of the DNA are greatly restricted. However, only five kinds of histones occur in a given cell; their lack of structural variety suggests a general, rather than gene-specific, regulatory function.

On the other hand, the more acidic nonhistone proteins (rich in glutamine and aspartic acid) occur in great variety. They differ markedly in composition among different eukaryotic tissues and species, and thus suggest a more specific role in gene regulation than would be possible for the histones. It is thought that nonhistone proteins may combine with histones to bring about the decomplexing of the latter with DNA, which permits transcription. Which genes are actively transcribing at any given time, then, appears to be closely tied to the histone and nonhistone proteins that complex with DNA. The former class of proteins probably serves as general regulators, the latter as more specific regulators of particular genes.

Indirect evidence for this role of nonhistone proteins is extensive; they (1) possess considerable structural diversity; (2) exhibit tissue and nucleotide sequence binding specificity; (3) occur in much higher levels in transcriptionally active cells, whereas histone levels remain essentially constant; and (4) are closely associated with initiation and continuation of transcription both *in vivo* and *in vitro*. Additional experimental proof of the specificity of the nonhistone proteins was obtained through reconstitution experiments. When DNA and histones from thymus cells were reconstituted with nonhistone proteins from bone marrow cells, the proteins made in the greatest abundance were those characteristic of bone marrow cells.

The degree of histone–nucleic acid complexing appears, in at least some organisms, to vary with the body part and with time. In developing pea embryos, for example, the cotyledons (food storage organs of the embryo) develop rather rapidly in embryogeny, then cease to grow further, even in germination. Tests indicate that almost all the DNA of mature cotyledons is histone-complexed, whereas the terminal meristems, active growth regions that produce all the root and shoot of the seedling at germination, have more noncomplexed DNA.

This suggests an inverse relationship between histone complexing and protein synthesis. James Bonner and his colleagues have presented excellent confirmation of this hypothesis. Pea cotyledons produce a protein, seed reserve globulin, that is not produced in vegetative tissues. DNA from cotyledons can be made to produce the mRNA necessary for *in vitro* synthesis of this protein. DNA from vegetative parts of the plant does not yield mRNA for globulin unless the histone is chemically removed! It has also been shown that maize embryos begin growth and development into seedlings only when most of the DNA is not histone complexed. Histone removal may therefore be involved in the physiological mechanism that initiates germination of the grain.

Chromatin Structure and Regulation

The question of regulation of eukaryotic genes is still not completely solved. Chromatin in these organisms consists of DNA literally covered with proteins. The histones clearly play a role in the packaging of the DNA into chromatin fibers as well as functioning in the expression of genes. The nucleosome, as discussed in Chapter 8, clearly provides this first order of packaging. Coiling due to nucleosome packaging of transcriptionally active regions of DNA appears to be more relaxed than in nontranscribing regions. These more open coils of transcribing DNA are sensitive to **DNase I,** an enzyme that attacks and digests active euchromatin. Weintraub and his colleagues have found that active globin genes in chicken red blood cells are ten times more sensitive to this enzyme than are inactive globin genes. The DNase I-sensitive regions in these erythrocytes is very long (10,000 bases), whereas the globin genes themselves are only 1,000 to 2,000 bases in length.

"Hot spots" for action of DNase I have been found; these sequences are only 100 to 200 nucleotides long, but are a hundred times more sensitive to DNase I than other stretches of chromatin. Such hypersensitive zones may occur "before," within, and "after" genes. Those located upstream of genes are detectable whenever a gene is transcriptionally active, and have been found in a variety of eukaryotes—rat, chicken, yeast, and *Drosophila.* Interestingly enough, these hypersensitive regions upstream in regions *before* transcription initiation suggest exposure of promoter regions to RNA polymerase.

These "hot spots" also can be cut by the enzyme **S1 nuclease,** which usually cuts single-stranded DNA and not chromatin, thus causing deletions. This suggests an even more relaxed coiling in the hypersensitive regions and has been verified in a number of species. Because of their increased accessibility to enzymes, it is also likely that these regions are more accessible to protein-initiation factors. Deletions outside the hypersensitive regions still permit accurate and efficient transcription. It is quite possible that gene expression depends on relaxing of the coils of the nucleosomal unit. Although other factors are undoubtedly involved in eukaryotic gene regulation, it is clear that chromatin structure is an important part of the story.

Heterochromatin

As was noted in Chapter 4, chromatin exists in two forms, based on its degree of packaging throughout the cell cycle. **Heterochromatin** remains tightly condensed and thus deeply staining through the cell cycle. **Euchromatin** is only condensed during cell division and otherwise exists in a noncondensed state. **Constitutive heterochromatin** is permanently inactive and is usually found near the centromere and on the ends of each chromosome. This DNA is called **satellite DNA,** because upon ultracentrifugation, it separates from the main component of DNA. This is because this DNA consists of very highly repetitive sequences (short sequences repeated many times).

Facultative heterochromatin contains genes that are permanently inactivated in some cell lines but not necessarily in others. Such is the case of the mammalian X chromosome. In the male the X chromosome is primarily euchromatic. In the female, one X is randomly inactivated through becoming heterochromatic. This allows for an even balance of X-linked genes between males and females (see Chapter 7). The randomness of this X inactivation is shown in heterozygous female cats that develop variegated or patchy coat color patterns depending on which X-linked coat color allele is inactivated. Such is the case with the calico cat where large patches of black and orange coat color develop due to random X chromosome inactivation. The nature of this X chromosome inactivation producing the Barr body was discussed in Chapter 7.

Thus it seems that the transcriptional activity of an entire chromosome may be related to its structural organization. In this case, permanent tight condensation of the genetic material results in genetic inactivity. Whatever the means, the process appears permanent.

HOMEOTIC MUTANTS IN 100,000 *DROSOPHILA:* A UNIQUE EXAMPLE

Since the genome of a higher organism may contain as many as 100,000 genes, it seems unlikely that in all cases each gene would be regulated individually. This has led to the **master gene** concept whereby a single master gene controls the expression of several other genes for some common process or developmental sequence. In the 1980s, such master genes controlling development have been identified through **homeotic mutations** that transform one body part in *Drosophila* into another normally found in a different segment. For example, wings will grow where eyes should be, or legs become antennae, or legs grow in place of antennae. Since these mutations involve very complex developmental sequences, it is unlikely that they are due to a single gene, but rather many genes controlling the location and development of the structure, whose expression might be under the control of some single master gene.

Most of these genes seem to be located in two clusters: *antennapedia,* which are genes that determine adult structures of the head and the anterior thoracic segments; and *bithorax,* which governs the posterior thoracic and abdominal segments (Figure 12–8). These complexes of genes are very large. Antennapedia consists of 100,000 base pairs, for example. However, a common short sequence has been found within

FIGURE 12–8. *Examples of the homeotic mutations (A) antennapedia, and (B) bithorax.* [(A) Photo courtesy Dr. Turner, Indiana University, used by permission, and (B) photo courtesy Dr. E. B. Lewis, Cal Tech, used by permission.]

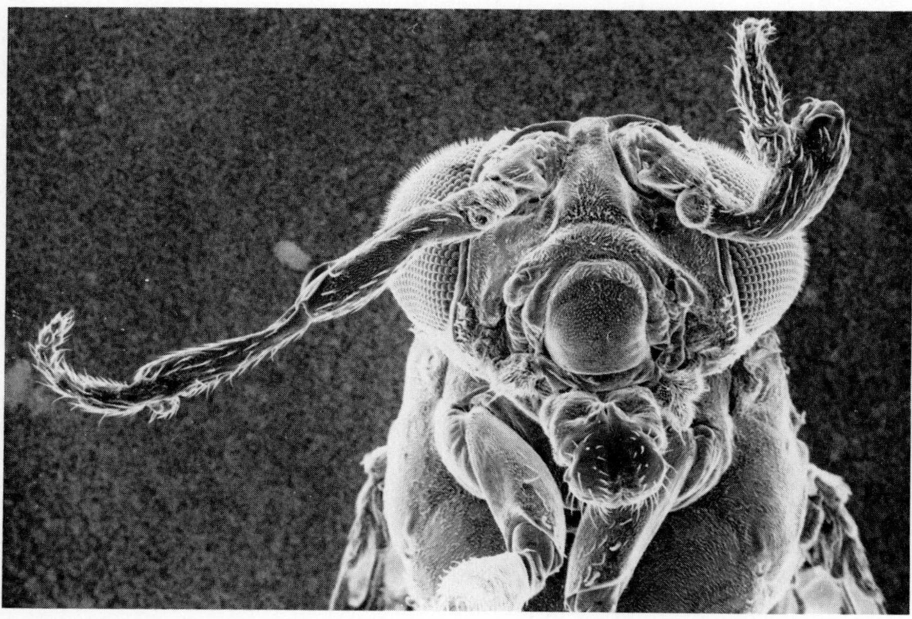

(A)

(B)

each complex. This sequence is called the **homeobox.** The homeobox sequence is approximately 180 base pairs long, codes for 60 amino acids, and is found in organisms from *Drosophila* to humans. The base sequences of the homeoboxes from a diverse group of organisms show 60 to 80 percent homology, but due to degeneracy in the code, the protein coded for by this homeobox could be as much as 90 percent identical. For example, the amino acid sequence of the *Dosophila* homeobox and the same one found in *Xenopus,* a salamander, differs by only one amino acid out of 60.

The homeobox appears then to code for a protein domain, or a functional segment of a protein. It is referred to as a *homeodomain.* There is now experimental evidence that this protein can bind to DNA, making it at least biochemically compatible with the master gene model. The problem now is to understand exactly where these homeodomains bind to DNA and how they themselves are regulated. Some evidence with the *Drosophila* homeoboxes is that they actually regulate each other.

SUMMARY OF GENE REGULATION

Genetic material has turned out to be fairly complicated in its regulation as well as in its action and structure. Yet it is a relatively simple system for the production of an endless variety of phenotypes. Gene action appears to be related to certain proteins and/or to other genes or even to certain metabolites. So to our summary of gene function and fine structure we need to add the following points regarding the regulation of gene action.

1. The activity of many genes is regulated so that their polypeptide products are formed only under certain chemical conditions and/or at certain developmental stages.
2. Although the function of histones and nonhistone proteins is not yet fully known, it has been suggested on credible grounds that they regulate gene activity in those cells in which they occur by affecting DNA coiling and transcription.
3. In bacteria, there is clear evidence for a regulator–operon complex in which the regulator produces a protein that:
   a. Represses an operator with the result that transcription does not occur, but can be inactivated by an inducer, which allows transcription to take place (negative control, as in the *lac* operon), or
   b. The operon may be under positive control, as in the effect of cAMP-CRP in the *lac* operon, or
   c. The repressor fails to repress the operator (inactive repressor) unless activated by combining with the inducer (negative control, as in the *his* operon), or
   d. Represses the operator site in the absence of the inducer (negative control), or serves as an activator after combining with the inducer (positive control), as in the *ara* operon.

PROBLEMS

12–1 In what way or ways are operator and regulator sites similar? Dissimilar?

12–2 Differentiate between repressors and inducers.

12–3 Distinguish between positive and negative control.

12–4 Synthesis of mRNA by the *lac* operon of *Escherichia coli* increases with addition of the inducer. Is this evidence for action of the inducer at the transcriptional or at the translational level?

12–5 Will production of *normal* beta-galactosidase be constitutive, inductive, or absent for each of the following genotypes:
(a) $i^+ p^+ o^+ z^+ y^+ a^+$, (b) $i^- p^+ o^+ z^+ y^+ a^+$, (c) $i^+ p^+ o^c z^+ y^+ a^+$, (d) $i^+ p^+ o^+ z^- y^+ a^+$, (e) $i^+ p^+ o^+ z^- y^+ a^+ / i^+ p^+ o^c z^+ y^+ a^+$?

12–6 Will each of the three enzymes coded by the *lac* operon be produced constitutively or inductively, or not produced in each of the following merozygote genotypes: (a) $i^s p^+ o^+ z^+ y^+ a^+ / i^+ p^+ o^+ z^+ y^+ a^+$; (b) $i^+ p^+ o^c z^+ y^- a^+ / i^+ p^+ o^+ z^- y^+ a^-$?

12–7 Assume an *Escherichia coli* culture in which all the individuals are of genotype $crp^+ \ cya^+ \ldots i^+ p^+ o^+ z^+ y^+ a^+$; will the mRNA for the three *lac* enzymes be transcribed or not? (a) Glucose and lactose both present; (b) glucose and lactose both absent; (c) glucose absent, lactose present.

12–8 Assume that the fourth transcribed group of three deoxyribonucleotide pairs in the wild type *z* gene of the *lac* operon of *Escherichia coli* is $\begin{matrix} 5' \ldots \text{CAA} \ldots 3' \\ 3' \ldots \text{GTT} \ldots 5' \end{matrix}$. Through tautomerization this group is changed to $\begin{matrix} 5' \ldots \text{TAA} \ldots 3' \\ 3' \ldots \text{ATT} \ldots 5' \end{matrix}$. The 3′, 5′ strand is the template for transcription. (a) Will β-galactosidase be produced following this mutation? (b) Will β-galactoside permease and/or thiogalactoside transacetylase be produced after this mutation has occurred?

12–9 An experimental strain of *Escherichia coli* is developed in which the 35 deoxyribonucleotide pairs of the operator are deleted. No other changes occur. Will production of β-galactosidase, β-galactoside permease, and/or thiogalactoside transacetylase be inductive, constitutive, or lacking?

12–10 Would you expect a nonsense mutation in one of the structural genes of the *his* operon to affect transcription or translation?

12–11 The tryptophan *(trp)* operon of *Escherichia coli* is responsible for coordinate production of the enzymes involved in the synthesis of the amino acid tryptophan. The operon includes five structural genes, an operator, and a promoter. A regulator site produces a repressor that is activated by tryptophan. (a) Is this an illustration of positive or of negative control? (b) In the absence of tryptophan in the cell, are the enzymes of the *trp* operon produced or not?

12–12 In what way is the regulation of a gene that is active in a differentiated cell inherently different from regulation of a bacterial gene?

12–13 Why are related eukaryotic genes rarely regulated by controlling the synthesis of a single polycistronic mRNA, as is common in bacteria?

12–14 Two morphologically distinct classes of heterochromatin are identified, euchromatin and heterochromatin. Even without knowing anything about the base sequence of each type, what evidence suggests that they are different from each other at the molecular level?

12–15 What kinds of mutations would completely eliminate translation of the entire sequence encoding a polypeptide?

12–16 In a eukaryotic system, what barriers would an extracellular repressor have to pass through before ultimately binding to DNA?

12–17 When ^{3}[H] uracil is injected into the developing larvae of the diptera *Chironomus tentans*, only certain regions of the chromosome (puffs) in certain tissues contain labeled uracil. Explain this observation.

12–18 *Drosophila* embryos that are homozygous for the homeotic mutation ophthalmoptera (wing develops in place of the eye) are raised at 17°C to produce adults with normal eyes. The same embryos raised at 29°C have wing tissue replacing most of the eye. Eggs were collected at 17°C and shifted to 29°C at various stages of development, or collected at 29°C and shifted to 17°C at various stages of development. The results show that whatever the temperature change, second instar larval stage determined the phenotype. For example, if the temperature was 17°C at the second instar, eyes developed normally irregardless of the shift to 29°C. Conversely, if the temperature was 29°C at the second instar, wings developed in place of the eye even though the larva had been shifted back to 17°C. What do these results suggest concerning expression of this homeotic mutation?

GENE MUTATION

CHAPTER

13

Even though one of the most important requirements for the genetic material is its stability, the capacity for change is also necessary. When this does occur, a **mutation** is said to have taken place. A mutation then is a sudden heritable change in the structure of the genetic material. This change may lead to a corresponding change in the phenotype. As such, mutations are an extremely important source of genetic variability in living populations. In fact, mutations are the only source of new genetic information. Recombination, which is the other major source of genetic variation, simply rearranges already existing genetic information.

Traditionally, mutations involve changes within the DNA molecule. However, changes in chromosome structure and number can lead to heritable phenotypic changes as well. These events are discussed in the next two chapters. In fact, it was chromosomal structural changes that led DeVries in 1901 to develop the first "modern" statement on mutations to explain the variation he had observed in the evening primrose, *Oenothera lamarkiana*.

In addition to mutations contributing to the genetic variation of an organism, they are equally important to the geneticist in an attempt to understand basic processes. Without mutations, no variation would be available for the understanding of gene behavior. Mutations serve as the window, providing insight into basic biochemical processes that involve such things as gene expression and development. We have already studied several examples of the use of mutations to identify basic processes and concepts: the work of Beadle and Tatum on biochemical genetics, Zinder and Lederburg on transduction, Yanofsky on colinearity, Jacob and Monod on the nature of gene regulation in *E. coli*, and Crick on the triplet nature of the code. Clearly without mutations our basic understanding of the science of genetics would be severely limited.

BIOCHEMICAL BASIS OF MUTATION

Very simply, a gene mutation occurs as a result of base pair substitutions. These substitutions lead to what is generally referred to as a **point mutation.** As a consequence of substitution of a base pair, the amino acid sequence of a protein can be altered. If the change in amino acid sequence alters the biological activity of the protein, a phenotypic change can occur. Such is the case with hemoglobin in sickle-cell anemia and insulin in diabetes, where one amino acid in the protein is changed due to a base pair substitution in a gene. Alternatively, the base pair substitution may change the codon into a termination codon. This could result in early termination of the synthesis of a protein, and possibly result in a lethal condition.

Wherever base pairs are added or deleted, **frameshift** mutations occur, altering the amino acid composition of the entire protein. On the other hand, because of redundancy in the genetic code, not all base pair changes may lead to an altered amino acid change in a protein. Such **silent mutations** do occur but have no phenotypic effects and can only be identified by comparing base pair sequences between normal and mutant genes.

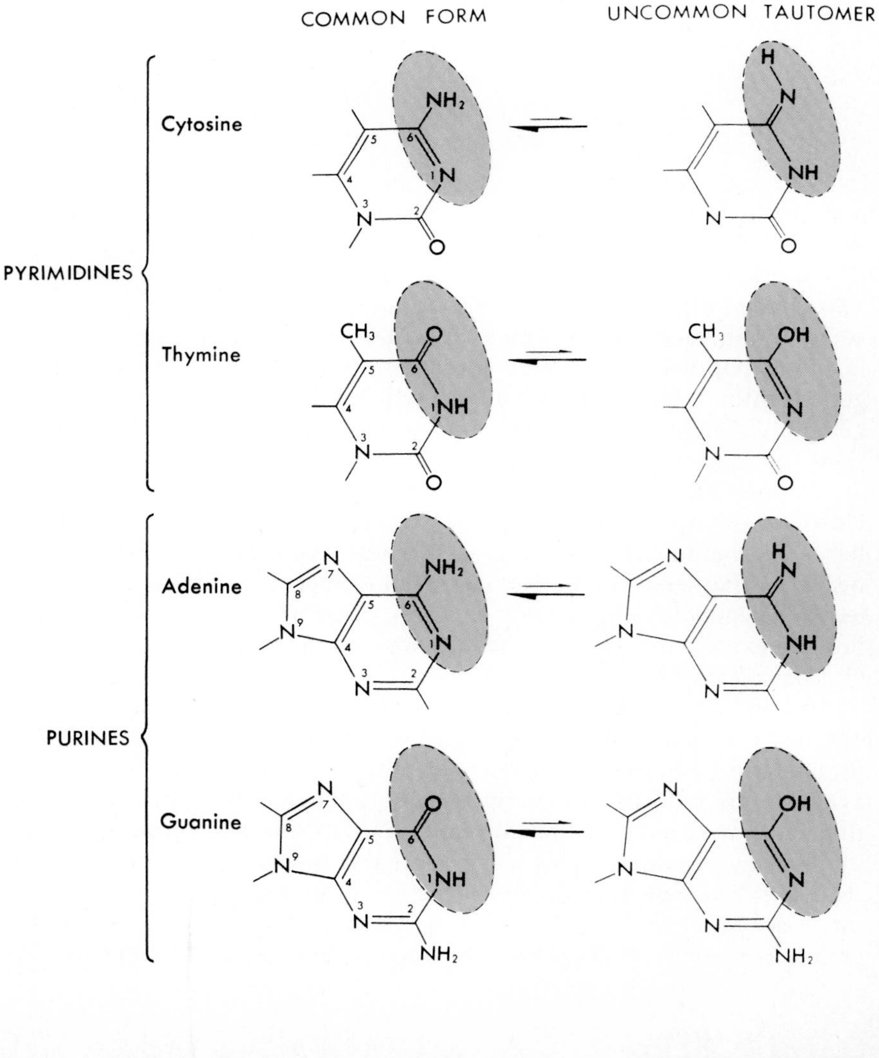

FIGURE 13–1. *(A) Comparison of common forms of DNA bases with their rare tautomers. Tautomers, which may occur in several forms, differ from each other by rearrangements of protons and electrons. This is symbolized in the structural formula by changes in positions of H atoms and double bonds within the dashed ovals. Only the tautomer that is considered important in each case as a mutation is shown here; its occurrence is quite rare.*

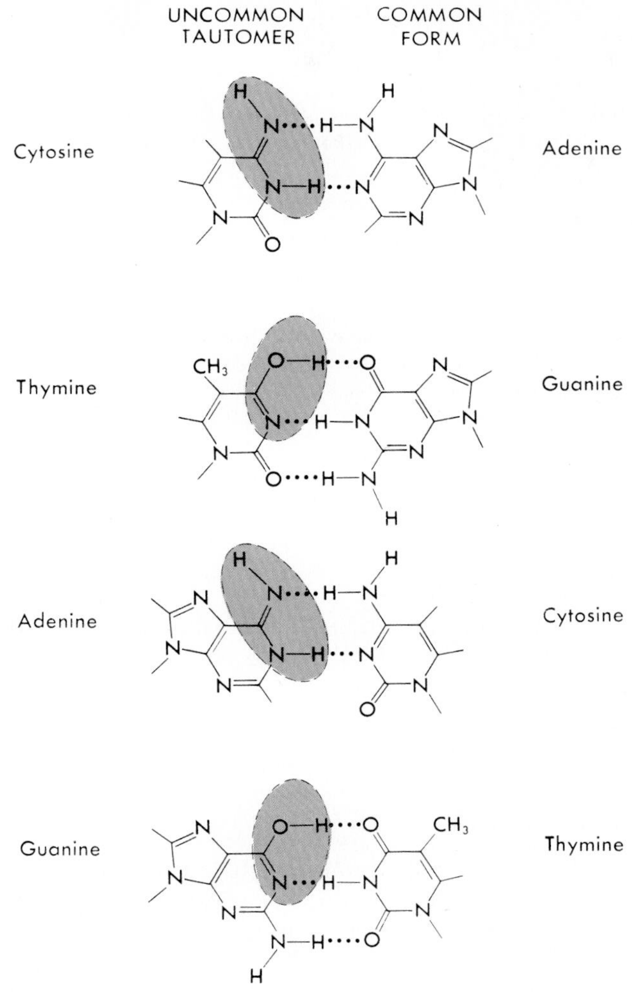

UNCOMMON TAUTOMER — COMMON FORM

Cytosine — Adenine

Thymine — Guanine

Adenine — Cytosine

Guanine — Thymine

FIGURE 13–1. (CONTINUED)
(B) Pairing qualities of the rare tautomers of the four bases. Consequences of this "erroneous pairing" are discussed in the text.

The pathways by which these single base-pair changes take place are important to consider.

Tautomerization

The purines and pyrimidines in DNA and RNA may exist in several alternate forms, or **tautomers.** Tautomerism occurs through rearrangements of electrons and protons in the molecule. Uncommon tautomers of adenine, cytosine, guanine, and thymine, shown in Figure 13–1A, differ from the common forms in the position at which one H atom is attached. As a result, some single bonds become double bonds, and vice versa.

Transitions

The significance of these tautomeric shifts lies in the changed pairing qualities they impart. The normal tautomer of adenine pairs with thymine in DNA; the rare (imino) form pairs with the normal tautomer of cytosine, as depicted in Figure 13–1B. The rare tautomer is unstable and usually reverts to its common form by the next replication. If tautomerism occurs in an already incorporated adenine, the result is a conversion of an **A:T** pair to **G:C** (Figure 13–2A). On the other hand, if

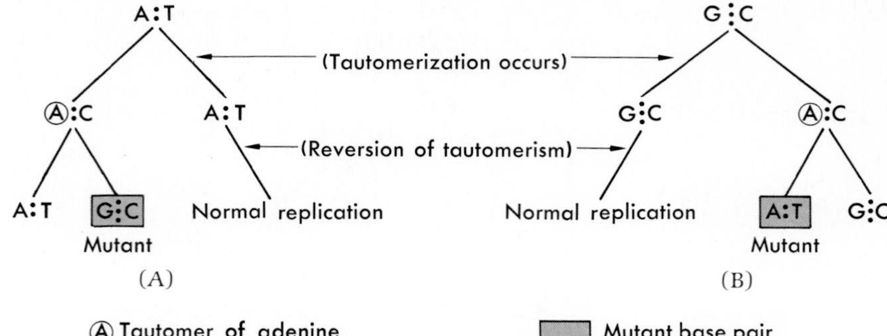

FIGURE 13–2. *(A) Conversion of an A:T pair to a G:C pair by tautomerization of an already incorporated adenine during the first replication of DNA. Reversion of the tautomerization in the second replication leads to the production of one mutant G:C strand. (B) Conversion of a G:C pair to an A:T pair by incorporation of a tautomer of adenine in the first replication of DNA. Assuming reversion of the tautomerism in the second replication, a mutant A:T pair is produced in place of the normal G:C pair. In both figures, the tautomer of adenine is indicated by the circled A; the mutant base pair is boxed.*

tautomerism occurs in an adenine about to be incorporated, the result is a conversion of a **G:C** pair to **A:T** (Figure 13–2B). Such substitution of one purine for another, or of one pyrimidine for another, is termed a **transition.** Transitions may come about in a number of other ways, as described in the following sections.

Deamination

It is well known that various chemicals produce mutations; these are especially well documented in bacteria, yeasts, and phages. Their effects are summarized again later in this chapter. Nitrous acid (HNO_2) is one such *mutagenic* substance. It brings about changes in DNA bases by replacing the amino group ($-NH_2$) with an $-OH$ (hydroxyl) group. Thus adenine, which has an $-NH_2$ at the number 6 carbon (Figure 13–1A), is deaminated by nitrous acid to hypoxanthine:

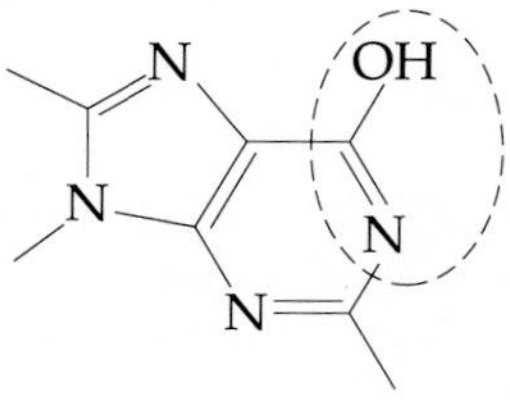

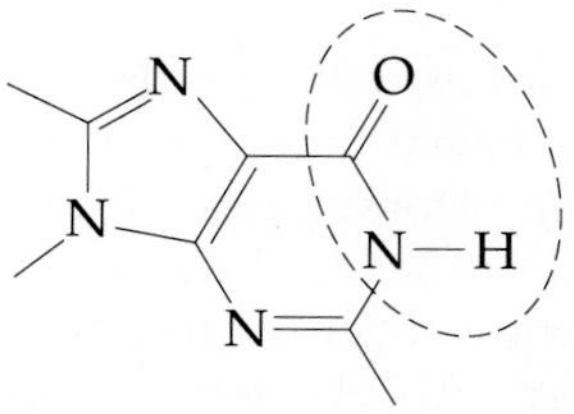

By a tautomeric shift, a more common (keto) tautomer is formed, which pairs with cytosine. Thus, an A:T pair can be converted to a **G:C** pair:

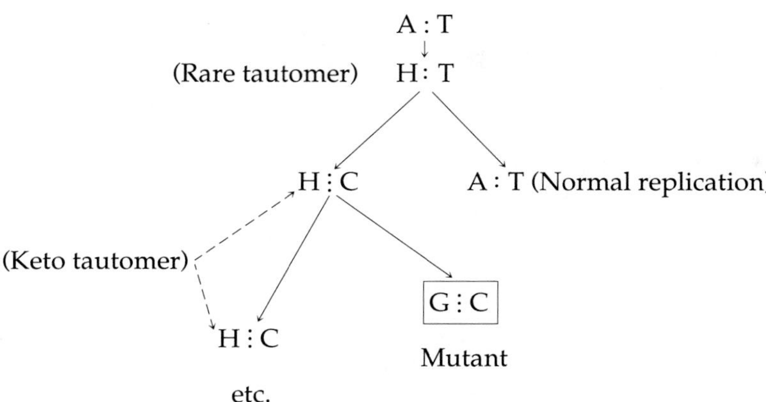

Similarly, deamination converts cytosine to uracil, which pairs with adenine (thus **C:G** becomes **T:A**), and guanine to xanthine, which pairs by two H bonds with cytosine.

Base Analogs

Certain substances have molecular structures so similar to the usual bases that such **analogs** may be incorporated if they are present into a replicating DNA strand. One example will suffice to indicate the process and consequence. For instance, 5-bromouracil in its usual (keto) form

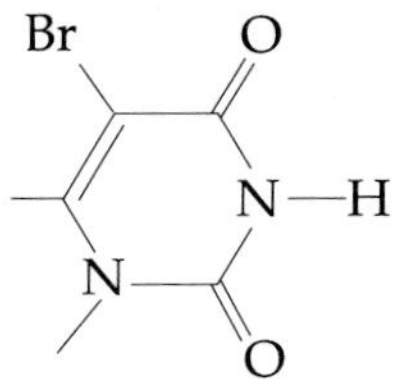

will substitute for thymine, which it closely resembles structurally. Thus an **A:T** pair becomes and remains **A:Bu** (Figure 13–3). There is some *in vitro* evidence to indicate that Bu immediately adjacent to an adenine in one of the DNA strands causes the latter to pair with guanine. But in its rarer (enol) state

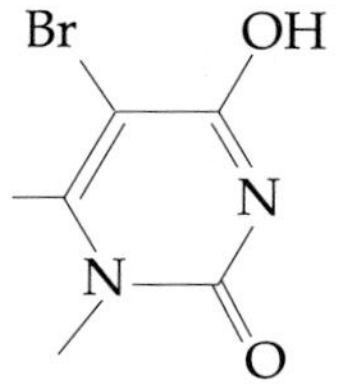

5-Bu behaves similarly to the tautomer of thymine (Figure 13–1) and pairs with guanine. This converts **A:T** to **G:C,** as shown in Figure 13–3). Studies show that 5-Bu increases the mutation rate by a factor of 10^4 in bacteria.

Nitrous acid and base analogs like 5-Bu can produce transitions as well as cause reversion of transition mutants in phages and bacteria to

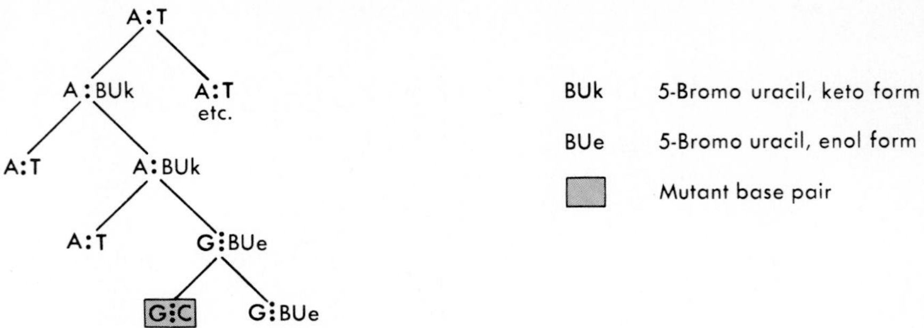

FIGURE 13–3. *Substitution of the common keto tautomer of 5-bromouracil for thymine; its subsequent tautomerization to the rarer enol form, if it occurs converts an A:T pair to a G:C pair. Moreover, the presence of 5-Bu itself in the DNA sequence may cause imperfections in RNA synthesis.*

their original state, regardless of what the initial mutagen may have been. Some rII mutants of phage T4 are transitions and can be reverted in this fashion.

GENETICS OF HEMOGLOBIN

Structure of the Hemoglobin Molecule

That mutation may involve as little as a single deoxyribonucleotide pair has been clearly shown by the pioneering work of Vernon Ingram and J. Hunt on the chemical differences between normal and variant hemoglobins. Human hemoglobin is a protein with a molecular weight of about 67,000. The globin consists of four polypeptide chains, two alpha chains, and two beta chains, each with iron-containing heme groups. As pointed out in Chapter 10, the alpha chain includes 141 amino acid residues, and the beta contains 146. In one molecule there are thus $(2 \times 141) + (2 \times 146)$, or 574 amino acid residues. Nineteen of the twenty biologically important amino acids are included, and their exact sequence in both the α and β chains has been determined.

Tryptic digestion. Before biochemists had analyzed the complete sequence of amino acids in the alpha and beta chains of hemoglobin, Ingram was able to report on chemical differences between hemoglobin of normal persons (hemoglobin A, or Hb-A) and that of individuals suffering from sickle-cell anemia (hemoglobin S, or Hb-S). Because the molecule was too large and complex for total analysis at that time, Ingram digested it with trypsin. This enzyme breaks the peptide bonds between the carboxyl group of either argenine or lysine and the amino group of the next amino acid. Because there are about 60 of these amino acids in the hemoglobin molecule, the molecule was broken up into approximately 60 shorter polypeptides. Each of these was then analyzed by Ingram for amino acid content.

"Fingerprinting" peptides. Ingram placed small samples of the trypsin-digested hemoglobin (A or S) on one edge of a large square of filter paper, then subjected the peptide mixture to an electrical field in the process called **electrophoresis.** Under these conditions, differently charged portions will migrate characteristically. Next, the filter paper with its invisible, spread-out "peptide spots" was dried, turned 90°, and placed with one edge in a solvent (normal butyl-alcohol, acetic acid, and water). Because of differences in solubility of the peptides in this solvent, additional migration and spreading ensued. The paper was finally sprayed with ninhydrin, which produces a blue color in reaction

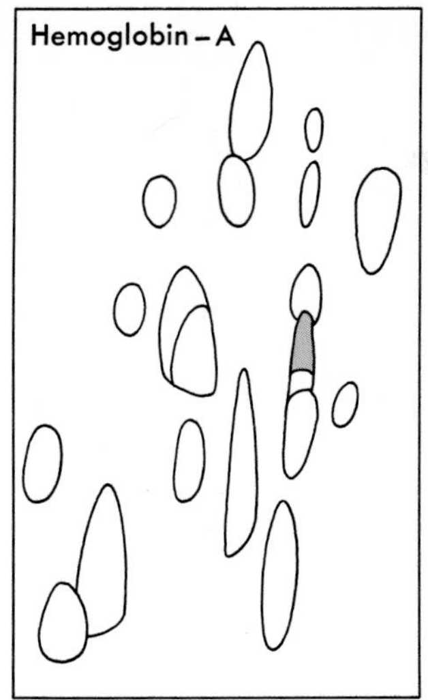

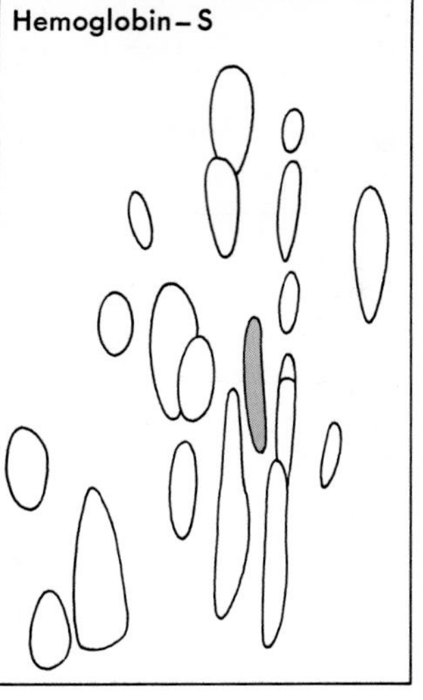

FIGURE 13–4. *Drawings of "fingerprint" chromatogram of the hemoglobins A (normal) and S (sickle cell). Each enclosed area represents the final location of a specific peptide; peptide 4 of Ingram is indicated by shading.*

with amino acids. The resulting chromatogram was called a "fingerprint" by Ingram, an apt description because different proteins treated in this way give unique results (Figure 13–4).

When Ingram fingerprinted the peptides produced from hemoglobin A and S by tryptic digestion, he discovered all "peptide spots" (except one, which he called "peptide 4") of each to be identical in their locations on filter paper. This means that the long α and β chains of hemoglobins A and S are identical *except for one peptide* out of thirty kinds. This one carries a positive charge in Hb-S and no charge in Hb-A, a difference that by 1960, Hunt and Ingram were able to relate directly to a *single amino acid change* in the sequence of 146 amino acids composing the beta chain (see Table 13–1).

Analysis of Several Hemoglobins

With the development of the electrophoretic–chromatographic technique for peptide analysis, studies of many different hemoglobins progressed rapidly. In addition, the complete amino acid sequence for both polypeptide chains is now known. The amino acid sequence for the β chain of hemoglobin A begins with these eight amino acids, starting with the NH_2 end: valine, histidine, leucine, threonine, proline, glutamic acid, glutamic acid, and lysine (Table 12–1). Glutamic acid carries a negative charge, valine and glycine carry no charge, and lysine carries a positive charge.

TABLE 13–1. **Comparison of the First Eight Amino Acids of the β Chain of Four Hemoglobins**

Hb-A	*Hb-S*	*Hb-C*	*Hb-G*
Valine	Valine	Valine	Valine
Histidine	Histidine	Histidine	Histidine
Leucine	Leucine	Leucine	Leucine
Threonine	Threonine	Threonine	Threonine
Proline	Proline	Proline	Proline
Glutamic acid	VALINE	LYSINE	Glutamic acid
Glutamic acid	Glutamic acid	Glutamic acid	GLYCINE
Lysine	Lysine	Lysine	Lysine

		GENE →								
DNA	Hemoglobin A	GTA CAT	CAT GTA	CTT GAA	ACT TGA	CCT GGA	GAA CTT	GAA CTT	AAA TTT	. . . . . .
mRNA Codons		GUA	CAU	CUU	ACU	CCU	GAA	GAA	AAA	
Amino Acids		val	his	leu	thr	pro	glu	glu	lys	
DNA	Hemoglobin S						GTA CAT			
mRNA Codons							GUA			
Amino Acid							val			
DNA	Hemoglobin C						AAA TTT			
mRNA Codons							AAA			
Amino Acid							lys			
DNA	Hemoglobin G							GGA CCT		
mRNA Codons								GGA		
Amino Acid								gly		

FIGURE 13–5. *Suggested derivation of three mutant B-chain hemoglobins from hemoglobin A by single nucleotide changes. The codon for particular amino acids is arbitrarily chosen when several may code for the same amino acid. Altered nucleotides are enclosed by solid lines.*

Many other types of hemoglobinopathies have been similarly analyzed. Each differs from Hb-A in the change of at least one amino acid; some of these are in the α chain, and others are in the β chain. An extensive list of specific amino acid substitutions is given in McKusick.

Transversions. Hemoglobin S differs from hemoglobin A (Table 13–1) by the substitution of valine for glutamic acid as the sixth amino acid in the β chain. Now the codons for glutamic acid (Table 10–2) are GAA and GAG, whereas valine is coded by GUA, GUG, GUC, and GUU. If, say, a GAA (glutamic acid) codon undergoes a replacement to become GUA, valine will, of course, be coded (Figure 13–5). Such a replacement in mRNA will occur if the DNA trinucleotide CTT suffers a change to CAT; that is, the purine adenine replaces the pyrimidine thymine. The substitution of a purine for a pyrimidine (or vice versa) is called a **transversion.**

So the far-reaching and important phenotypic difference between hemoglobin A and that of persons with either sickle-cell trait or sickle-cell anemia is the result of transversion, the change of one nucleotide pair in DNA from **A:T** to **T:A.** A single nucleotide in mRNA thus may spell the difference between life and death.

Transitions versus transversions. By definition, transitions and transversions represent different kinds of base substitutions. A comparison of the two may be represented by:

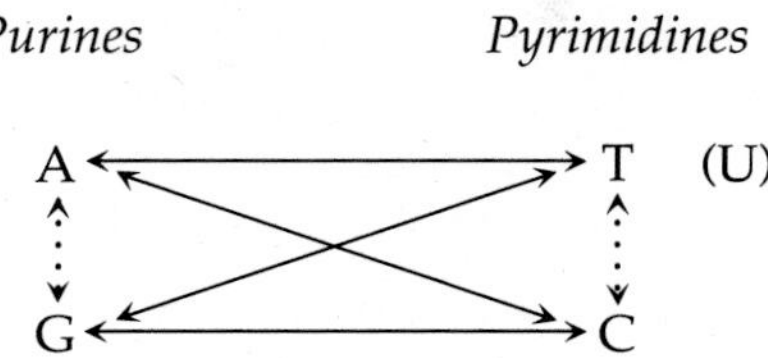

The dotted arrows represent transitions, and the solid ones are transversions.

Inheritance of hemoglobin variants. As described in Chapter 2, production of hemoglobin S is caused by a single allele that is codominant with the allele for hemoglobin A. It is now clear that alleles for hemoglobins A, S, C, and G, all of which affect the β chain, are multiple alleles. The α and β loci are not linked. Mutations that produce changes in the α chain, therefore, do not cause changes in the β chain, and vice versa.

CLASSIFICATION OF MUTATIONS

A number of ways have been used to classify or describe mutations. Each is unique and none are exclusive of all the others. In addition, any mutation may fit into more than one category.

Phenotype

An obvious way to classify a mutation is by the type of phenotype it exhibits. The easiest phenotypic change to see is a **morphological** change. In this case, the mutation results in a phenotypic change that alters the morphology, location, or perhaps color of a structure. **Biochemical** mutations, on the other hand, may not always be observed visually, but can be identified only by biochemical analysis. Sickle-cell hemoglobin is an example. Change in the nutritional state from prototrophic to auxotrophic condition is another example. Some mutations are **conditional.** Their phenotype is only expressed under certain conditions. Temperature-sensitive or drug-resistant mutants fit into this category also. Mutations may be **regulatory** in nature. This would mean that their expression is detected by the inability to control another gene. Constitutive mutants in the lactose operon are such phenotypes (see Chapter 11). Finally, a mutation may be **lethal** in that the organism carrying the mutation cannot survive. Albino mutations in plants are examples of these.

Somatic–germinal mutations. If the mutational event is not found in cells of the germ line, and is limited to somatic tissue, it is a **somatic** mutation. These types of mutations frequently arise in plants as mutant branches or *sports,* due to the mutation being limited to some but not all of the meristematic layers of the apical meristem. This means that the mutation will not be sexually transmitted through the gametes. In plants, these mutations can be propagated vegetatively and retain a

FIGURE 13–6. *Mutant flower heads (inflorescences) of* Zinnia. *The somatic mutation was from "peppermint" (splotched color) to solid color.*

constant phenotype. Figure 13–6 shows an example of a somatic mutation in Zinnia. Useful somatic mutations in plants include the navel orange and the Delicious apple. **Germinal** mutations are those that occur in cells that will ultimately form gametes and therefore are sexually transmitted.

Spontaneous mutations. Spontaneous mutations are those mutations that arise randomly in nature without any readily apparent cause. They may be due to errors in DNA replication or to mutagenic chemicals or radiations present in the environment. Each organism has a characteristic spontaneous mutation rate. Table 13–2 shows spontaneous mutation rates for several organisms. It is clear that spontaneous mutation rates are relatively low, with most occurring one in a million or more. The spontaneous rate for different genes of the same organism may vary greatly.

Induced mutations. Any mutation that occurs in response to an obvious externally applied agent is an induced mutation. This was first shown for X-ray-induced mutations in *Drosophila* by H. J. Muller in 1927. This finding was extended to plants in 1928 by L. J. Stadler, who showed that X-rays increased the mutation frequency in barley. Muller later received a Nobel prize for his important contribution. Charlotte Auerbach was the first to report the induction of mutations by chemical compounds in 1943. She showed that nitrogen and sulfur mustard gas, which had been used in World War I, induced mutations in *Drosophila*.

TABLE 13–2. **Representative Spontaneous Mutation Rates for Various Organisms**

Organism	*Rate*
E. coli	2×10^{-6} to 1×10^{9} per cell division
Chlamydomonas reinhardi	1×10^{-6} per cell division
Neurospora crassa	6×10^{-8} per asexual spore
Zea mays	1×10^{-6} per gamete/generation
Drosophila melanogaster	4×10^{-5} per gamete/generation
Homo sapiens	3×10^{-5} per gamete/generation

Since these original discoveries, a great deal of research has been conducted on identification and use of agents that induce mutations.

Mutagenic Agents

Basically, agents that induce mutations can be divided into two major groups: **physical** mutagens and **chemical** mutagens. The physical mutagens include high-energy ionizing radiations, such as X-rays, gamma rays, neutrons, and beta and alpha particles, and the lower-energy nonionizing radiation, ultraviolet light. Each of these induces mutations through their reaction with DNA.

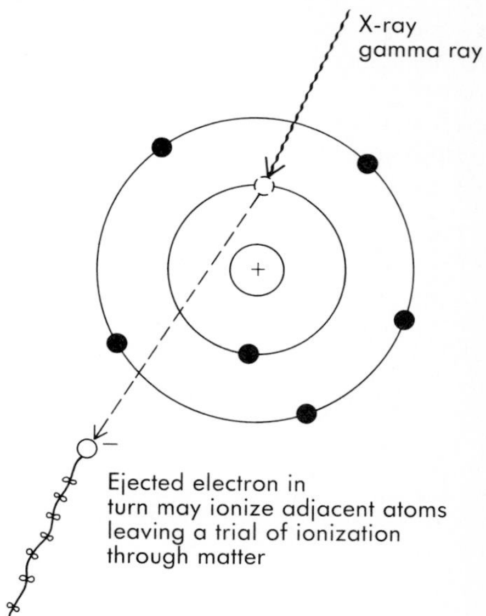
X-ray
gamma ray
+
Ejected electron in
turn may ionize adjacent atoms
leaving a trial of ionization
through matter

FIGURE 13–7. *Ionization by electromagnetic radiation. Removal of a negatively charged electron from an atom leaves a positively charged ion. The positively charged ion and the negative electron create an ion pair. High-energy X-rays or gamma rays produce trails of ion pairs in biological systems.*

Ionizing radiation. The spectrum of electromagnetic radiation is a broad one; it extends from the long radio waves, which may have wavelengths as great as several kilometers, down through cosmic rays, which may be as short as 10^{-14} cm. As wavelength becomes progressively shorter, the energy that the electromagnetic photons contain becomes progressively greater, with the result that they penetrate cells and tissues. In this process photons of sufficient energy content may collide with one of the orbiting electrons of an atom and knock it out of orbit. In this way a previously neutral atom becomes positively charged (an ion), because there is then a greater positive charge in the atomic nucleus than there are negative charges in the orbiting electrons. This process of ion pair formation is outlined in Figure 13–7.

Ionized atoms, and the molecules in which they occur, are chemically much more reactive than neutral ones. In addition, the electrons lost from an atom in this process move off at high speed and cause other atoms to become ionized. Each electron lost from an atom is ultimately gained by another, which causes it to become a negatively charged ion. The overall result of this repeated ionization process is a trail of *ion pairs* along the path of the high-energy photons. Mutagenic effects result from the chemical reactions undergone by ions as their charges are neutralized.

One outcome of the ionization process is breakage of the sugar–phosphate strand(s) of DNA. This, of course, can lead to chromosomal structural aberrations discussed in Chapter 15. But should strand breakage occur at two or more closely spaced points, one or more nucleotide pairs may be lost, and as a result, the reading frame and the transcribed mRNA become altered (see Chapter 10). The ultimate outcome can then be a protein or an enzyme so altered as to be somewhat reduced in capacity to function, or even inoperative. The latter event may well be lethal, particularly in homozygotes. If this occurs in a reproductive cell and is dominant, a phenotype, new for that particular line of descent, may be created; in short, a *mutation* occurs.

The dosage (quantity) of an ionizing radiation is based on the amount of ion pairs produced or the amount of energy deposited in the tissue. One unit of measurement is the roentgen (r), which is equal to roughly 2×10^9 ion pairs per cubic cm of air. The dose of radiation based on the amount of energy absorbed is expressed in rads; one rad is the amount of radiation that deposits 100 ergs of energy in a gram of matter. The rad is only slightly larger than the roentgen, inasmuch as a gram of tissue exposed to 1 r of gamma rays absorbs about 93 ergs. The fact that this ionization causes mutations is demonstrated by the observation that mutation frequency is directly proportional to dosage. Figure 13–8 shows such a relationship for somatic mutations in *Tradescantia* stamen hairs. The same type of response can be shown for virtually any ionizing radiation.

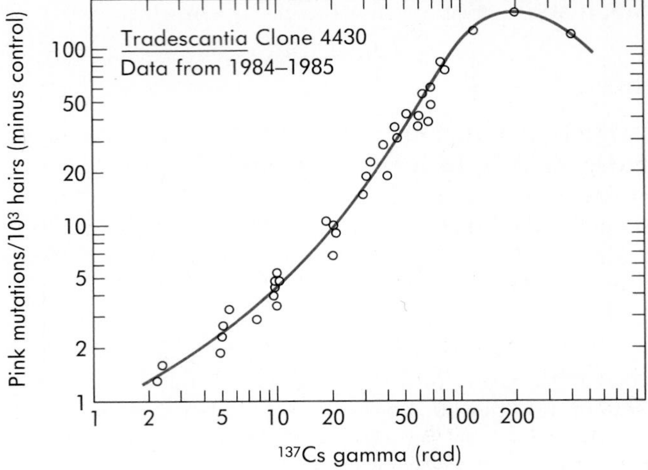

FIGURE 13–8. *Radiation dose response for mutations in* Tradescantia. (Photo courtesy L. Schairer, Brookhaven National Laboratory, used by permission.)

However, there is a difference whether the exposure is received as a "one-time" (acute) exposure or gradually (chronic) over a period of time. For example, for equal dosages of radiation applied to mouse spermatogonia, chronic low-intensity exposure produces a significantly smaller number of detectable mutations than the same total dose applied acutely. Exposure of mouse spermatogonia to 90 r per minute produces about four times the number of mutations as does the same amount of radiation administered at the rate of 90 r per week. It is believed that enzymatically controlled repair processes take place during low-intensity exposures, whereas at the higher rate of exposure, these processes either cannot cope with the larger number of points of simultaneous damage or are themselves impaired.

All persons receive some radiation during their lifetime from cosmic rays and also from radioactive materials on the earth's surface. In addition, medical X-ray examinations add to the accumulation of ionizing radiation exposure, although only that received by the gonads contributes to the genetic load of future generations. The average person in developed countries probably receives less than 50 r of radiation to the gonads during a thirty-year reproductive period as a result of medical examinations. For example, a single X-ray examination of the teeth is reported to provide about 0.0008 r to the gonads; a chest X-ray, about 0.0006 r in males, but 0.002 r in females; an abdominal X-ray, 0.13 r in males, and 0.24 r in females; and a fluoroscopic examination of the pelvic region, about 4 r to 6 r.

A number of calculations for the amount of radiation required to double the rate of mutation have been accumulated for mouse and for *Drosophila*. Applying such figures to human beings requires making several assumptions, the bases for which are not yet clear. Nevertheless, many authorities presently use the estimate of about 150 r as the **doubling dose** (the dose necessary to double the mutation frequency from the spontaneous rate) for humans. Although the amount of radiation received by the gonads for a given type of examination will vary among different institutions, it is helpful for each person to keep a record of the number and kind of X-ray and fluoroscopic examinations received.

Nonionizing radiation. Ultraviolet light (UV), while not possessing enough energy to ionize DNA, imparts enough energy to the molecule to cause mutations. The specific wavelength that is absorbed by DNA

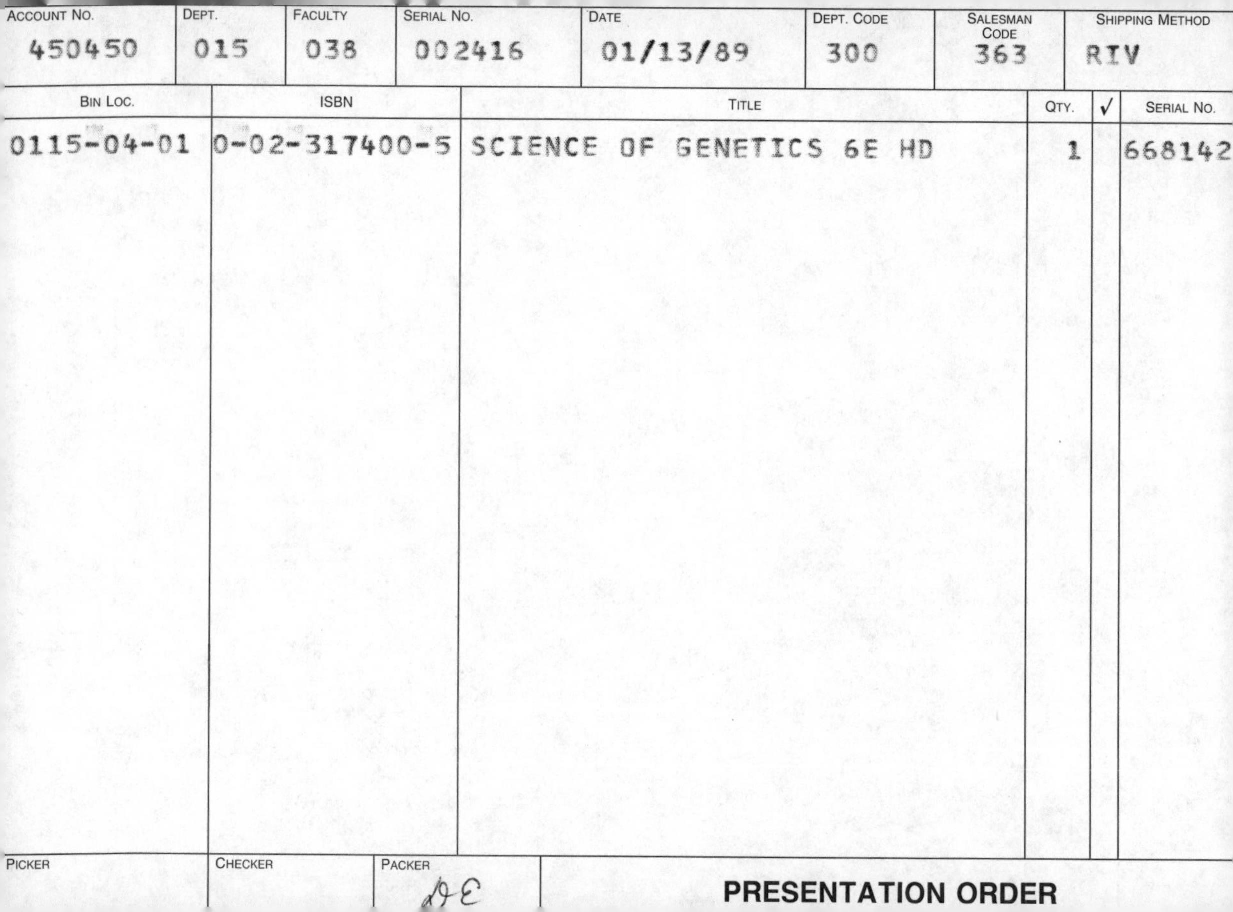

ACCOUNT NO.	DEPT.	FACULTY	SERIAL NO.	DATE	DEPT. CODE	SALESMAN CODE	SHIPPING METHOD
450450	015	038	002416	01/13/89	300	363	RIV

BIN LOC.	ISBN	TITLE	QTY.	√	SERIAL NO.
0115-04-01	0-02-317400-5	SCIENCE OF GENETICS 6E HD	1		668142

PICKER	CHECKER	PACKER
		dE

PRESENTATION ORDER

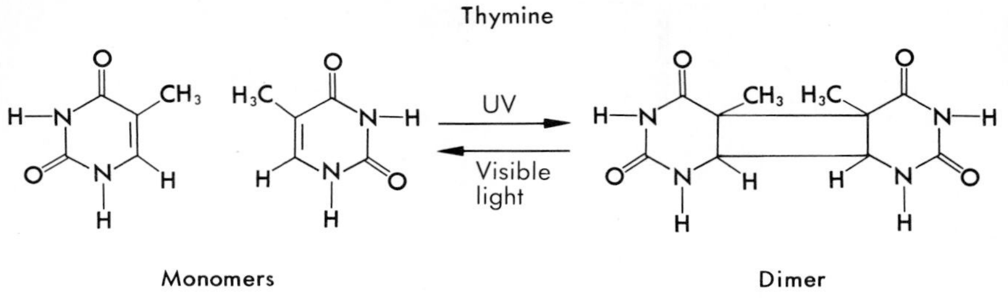

FIGURE 13–9. *The monomer ↔ dimer conversion in thymine under ultraviolet light.*

is 254 nm, and although it is not deeply penetrating in tissues, it is effective in killing bacteria, fungi, and increasing the incidence of skin cancer in humans. The best-known action of UV on DNA is the induction of **pyrimidine dimers.** This is the induction of direct carbon–carbon bonding between adjacent pyrimidine residues with thymine being the most common. This results in a distortion of the molecule or cross-linking between adjacent molecules, which temporarily stops DNA replication. Figure 13–9 shows the formation of a thymine dimer. As a result of attempts to repair the dimer, (see later in this chapter), gaps in the daughter strand are produced upon replication. When these gaps are not repaired correctly a mutation results.

Chemical mutagens. Chemical mutagens are classified in four major groups on the basis of their specific reaction with DNA. A **base analog** is a molecule whose structure is close enough to the natural base that it is sometimes incorporated into DNA in place of the natural base. The analog then mispairs with the wrong complementary base, leading to base-pair substitution mutations. As previously discussed, 5-bromouracil is such a mutagen.

Alkylating agents react with DNA by adding ethyl or methyl groups to the bases. This results either in mispairing of the affected base or its loss entirely, creating a gap. The primary base affected by alkylating agents is guanine, although other bases can also be alkylated. The nitrogen and sulfur mustards, identified as mutagens by Auerbach, are alkylating agents as are ethyl methane sulfonate (EMS) and diethyl sulfate (DES).

Acridine dyes are a class of chemicals that intercalate between the bases of DNA. This distorts the molecule and disrupts the alignment and pairing of the bases. This distortion results in deletion or addition of base pairs during replication. Puromycin, used by Crick to show the triplet nature of the genetic code, is an example of an acridine dye. Finally, there are chemicals that react with DNA and directly alter the base structure rather than incorporating into the molecule. Such **direct-acting** chemicals like nitrous acid deaminate adenine, cytosine, and guanine. This results in mispairing and base-pair substitution mutations.

REPAIR OF DAMAGED DNA

When DNA is physically damaged by a mutagen, an elaborate system of repair processes is activated in the cell to attempt to repair the damage. Much of what is known comes mainly from the repair of UV-induced pyrimidine dimers. The first repair system was discovered when it was found that if *E. coli* cells were exposed to blue light following UV irradiation, virtually no mutations occurred. This repair process

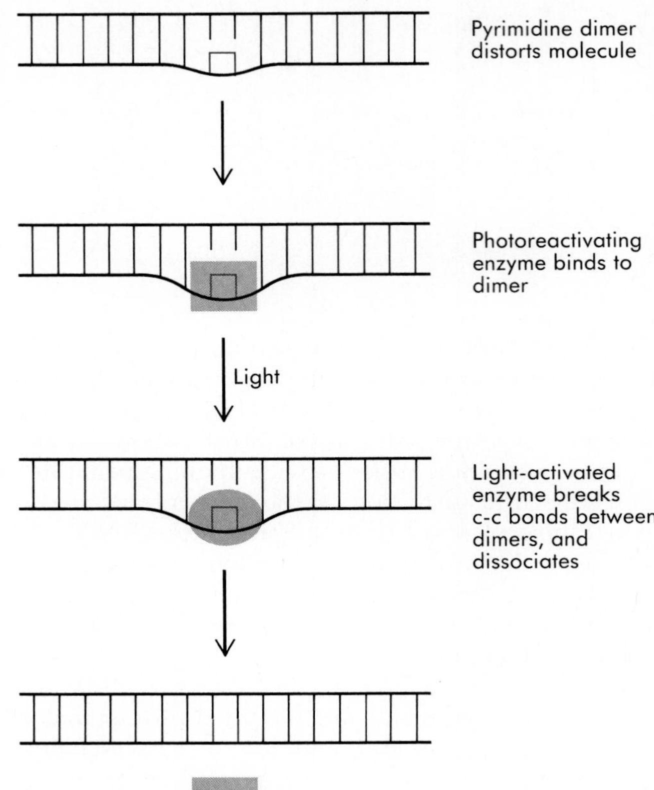
Pyrimidine dimer
distorts molecule
Photoreactivating
enzyme binds to
dimer
Light
Light-activated
enzyme breaks
c-c bonds between
dimers, and
dissociates

FIGURE 13–10. *Action of the photoreactivating enzyme on pyrimidine dimers. (See text for more details.)*

is called **photoreactivation.** It is due to an enzyme that binds to the pyrimidine dimers, using the energy from the light to break the bonds forming the dimer and restoring the original structure of the molecule. The action of the photoreactivating enzyme is shown in Figure 13–10. If the cells were kept in the dark for a few days following UV treatment, the frequency of induced mutations increased greatly.

In the dark, there is still evidence that some repair of UV-induced pyrimidine dimers takes place. This has been found to be due to a multienzyme repair process termed **excision repair.** The process is diagramed in Figure 13–11. It consists of four steps. First, an enzyme made up of gene products of three different genes recognizes the distortion caused by the pyrimidine dimer. The enzyme makes a single-stranded cut near the dimer. This is called the *incision step.* DNA polymerase I then synthesizes a new strand, displacing the old strand containing the dimer. This is called *repair synthesis.* A second cut is then made beyond the dimer, removing it from the DNA molecule, a step called *excision.* Finally, the newly synthesized strand is joined to the original strand by DNA ligase.

In *E. coli,* mutations in any of the three genes coding for the enzyme making the original incision in the DNA molecule completely block this repair process. In humans, an inherited disease called **xeroderma pigmentosum** is caused by a recessive mutation that blocks this excision repair process. People with this disease are very sensitive to sunlight and particularly the UV wavelengths in sunlight. They demonstrate a very high incidence of skin cancer, including malignant melanomas on areas of the skin exposed to sunlight. Cell cultures from humans with

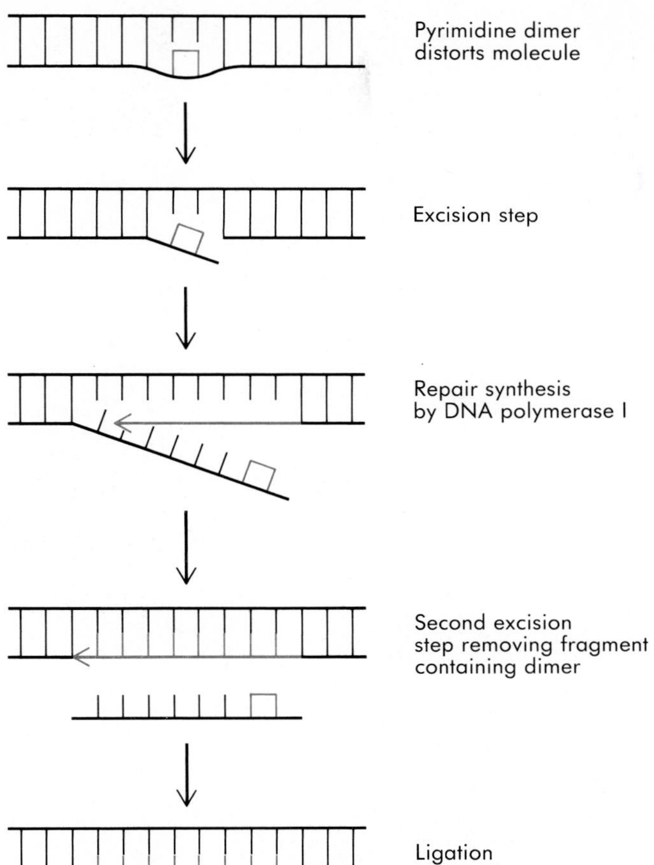

FIGURE 13–11. *Procedure for excision repair of a pyrimidine dimer. (See text for more details.)*

xeroderma pigmentosum are defective in excision repair and are killed at very low doses of UV light.

If a pyrimidine dimer survives the photoreactivation and excision repair systems, another repair system acts during replication. This process is called *postreplication repair*. Whenever replication comes to a dimer it is stopped due to the distortion in the molecule. Eventually replication resumes, but it skips the region around the dimer. This results in "gaps" in the daughter strand corresponding to the dimer on the parent strand. The daughter strand gaps can eventually be closed by repair synthesis. This process is illustrated in Figure 13–12. It is important to note that in postreplication repair, the dimer is not excised or broken, but is retained in the DNA molecule.

It should be clear from this discussion that most cells have developed several sophisticated means to repair damage to the genetic material. The seriousness of any defects in the repair process is illustrated by the xeroderma pigmentosum example. On our planet, we are constantly being exposed to UV radiation from the sun, and several systems to repair UV damage to the genetic material are essential in order to ensure survival. If this damage was not repaired in such an efficient, error-free manner, the mutational burden would have been too great for life to continue as long as it has. This is the reason for the recent concern about excessive exposure to UV radiation from sun bathing and tanning booths adding to the incidence of skin cancer.

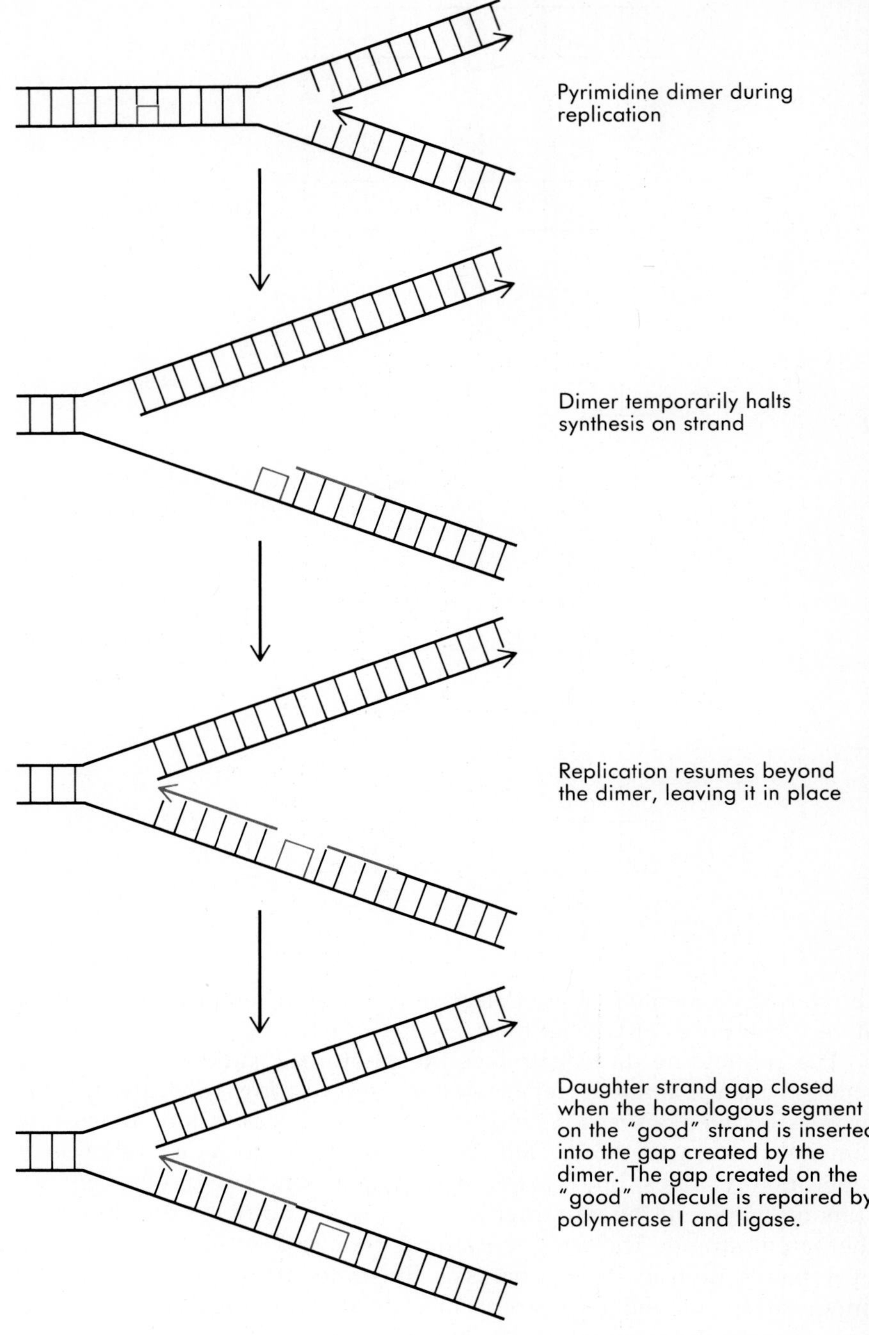
Pyrimidine dimer during replication
Dimer temporarily halts synthesis on strand
Replication resumes beyond the dimer, leaving it in place
Daughter strand gap closed when the homologous segment on the "good" strand is inserted into the gap created by the dimer. The gap created on the "good" molecule is repaired by polymerase I and ligase.

FIGURE 13–12. *Pattern of postreplication repair. Note that the dimer is retained within the molecule. (See text for more details.)*

TRANSPOSABLE ELEMENTS

In the mid-1940s, Barbara McClintock began making observations on some interesting unstable mutations in corn. The mutations involved variegation in the coloration of the kernels. Figure 13–13 illustrates the type of variability in seed color patterns shown by these mutations. Through a very complicated genetic analysis, McClintock attributed these unstable mutations to the action of **controlling elements.** These are mobile genetic elements that cause mutations and chromosomal rearrangements as well as influence the expression of other genes. The main feature of controlling element-induced mutations is that they revert back to the normal state at a relatively high frequency. As a result, the organism becomes a mixture of mutant and revertant tissues. This gives the variegated appearance. Much of the variation in kernel color in Indian corn used for decorations is due to such controlling elements.

The first discoveries in this area go back to the early 1900s and the work of R. A. Emerson, who first observed these highly unstable mutations in seed color in corn. Later in the 1930s, Marcus M. Rhoades showed the genetic instability to be conditional. That is, this type of mutation becomes unstable only in the presence of another gene. Rhoades studied a mutation in a gene responsible for the synthesis of purple pigment (anthocyanin) in the aleurone, the outer layer of the endosperm just under the seed coat. The mutation changed the aleurone color from purple to colorless. However, when another gene was present, the mutation reverted back to purple, but since it occurred late in development, the purple color appeared as spots of purple derived from cells in which the reversion occurred. Rhoades called the spot-generating allele *dotted*.

It was Barbara McClintock who showed that dotted was a type of genetic element—termed controlling element—that was mobile. This discovery of mobile genetic elements by McClintock was recognized as one of the most significant discoveries of the century. McClintock was awarded the Nobel prize in 1983 for this discovery. Mobile-controlling elements or **transposable elements** have now been found in virtually every organism that has been examined for their presence.

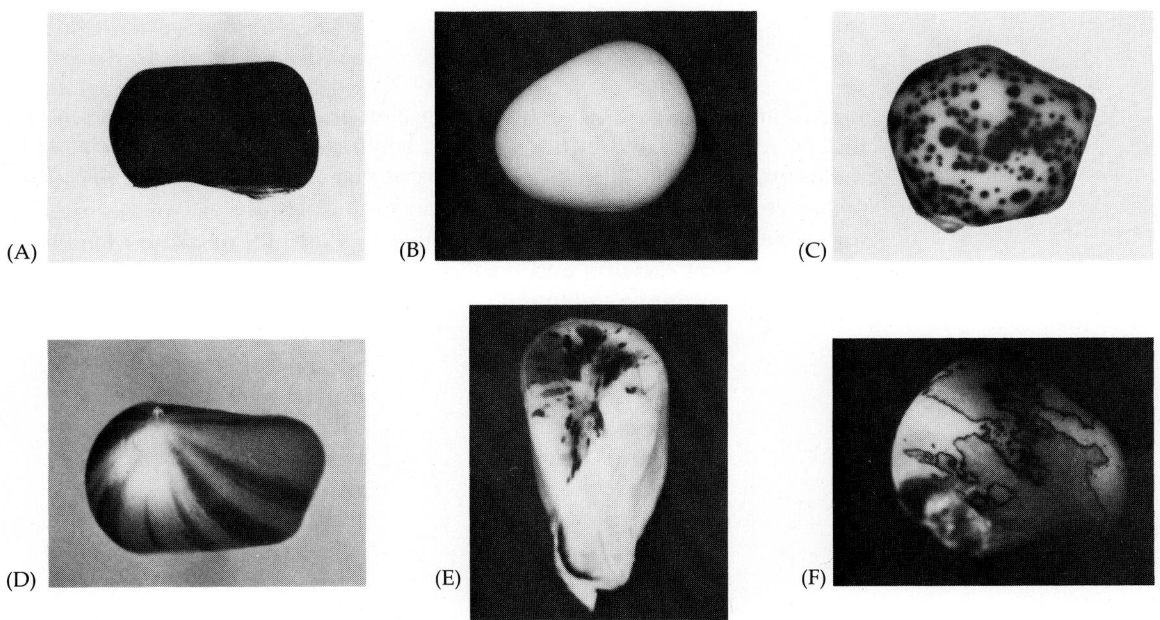

FIGURE 13–13. *Transposable element-induced mutations in corn. (A) Wild-type deeply pigmented kernel; (B) stable recessive null allele of a gene in anthocyanin pigment biosynthesis; (C) unstable mutations caused by suppressor-mutator transposable element, at A_1 locus in anthocyanin biosynthesis (expressed in aleurone); (D) unstable mutations in P locus caused by insertion of Activator (Ac) element (expressed in pericarp); (E) unstable mutation in* Waxy *locus caused by insertion of Activator element; (F) Pattern of marker loss associated with chromosome breakage at site of insertion of Dissociation (Ds) element, promoted by* trans-*acting Ac element.* (Photo courtesy Dr. Nina Fedoroff, Carnegie Institution of Washington, used with permission.)

BARBARA McCLINTOCK AND THE CONSTANCY OF THE GENOME

Although Dr. Barbara McClintock was awarded the 1983 Nobel Prize in Physiology and Medicine "for her discovery of mobile genetic elements" (these are DNA elements that can move from one chromosomal site to another), her contributions to genetics span a period of 60 years. She virtually established the field of maize cytogenetics as a science by developing the cytological techniques to identify and characterize maize chromosomes. Several scientists have suggested that she deserved a Nobel Prize for her maize cytogenetics work alone. Although she did not receive the prize until much later, she was highly recognized for her work by being elected to the prestigious National Academy of Science in 1945. At that time, she was only the third woman given this honor.

Working alone at the Carnegie laboratory at Cold Spring Harbor, she focused her attention on genetic instability in kernel and leaf pigmentation in maize. She first found that chromosome rearrangements appeared to be associated with this genetic instability. She also found that the pattern of mutation was not random, but could be explained by the movement of certain genetic elements from one place to another in the genome. From this work came the idea that there were two kinds of genes, one directly responsible for some phenotype, and another that could control the activity of the first gene.

McClintock's work made very little impact because of "the dogma of the constancy of the genome," a theory that did not allow for the behavior of such mobile genetic elements. She predicted in 1956 that these genetic elements represent systems in the nucleus that operate to control gene action. It was only after the discovery of bacterial transposons (mobile genetic elements in bacteria) in the mid-1970s that McClintock began to receive the recognition she deserved. Today, these mobile genetic elements are found in virtually every genome that is examined. Barbara McClintock's 1983 Nobel Prize has given her the recognition and credit she earned for her discoveries 30 years ago.

Mutations caused by a controlling element occur when the element moves and is inserted into the coding sequence of a gene. This disrupts the expression of that gene in heterozygotes, and causes it to show the recessive mutant phenotype. Insertion mutations are of two types. The first type is inherently unstable because it is due to an element capable of independent excision and transposition (movement). This element is called an **autonomous element.**

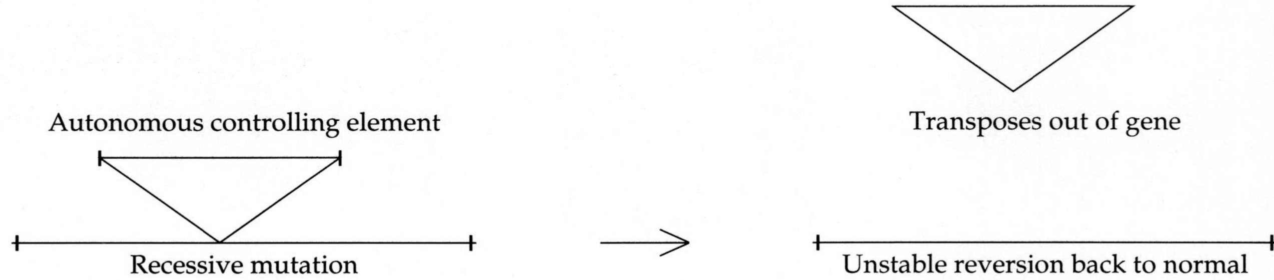

The second type of mutation is not inherently unstable, but becomes so only in the presence of the autonomous element elsewhere in the genome. The element causing this mutation is not capable of independent transposition and is referred to as a **nonautonomous element.**

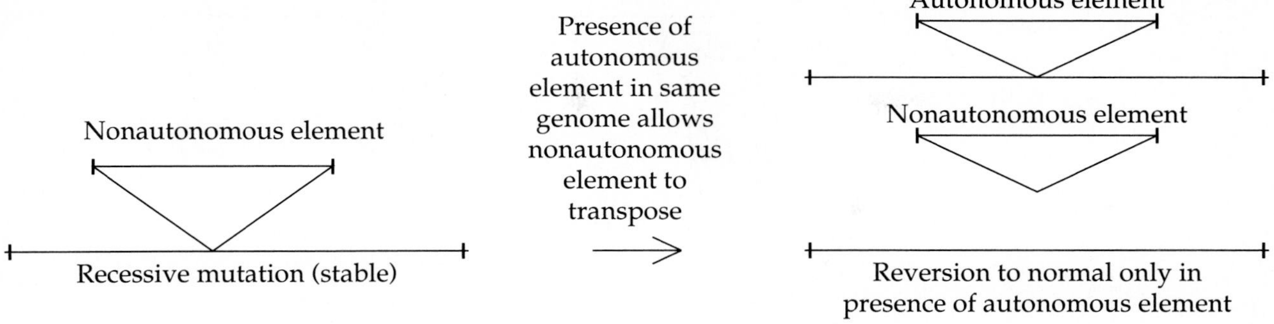

The Ac-Ds family of controlling elements described by McClintock illustrates the operation of these two types of elements. **Ds** or **dissociation** is an element that provides a site for chromosome breakage or gene mutation. Ds, however, is capable of transposition to a new location only in the presence of the second element **Ac** or **activator.** This makes Ds the nonautonomous element. Ac, capable of independent transposition, is the autonomous element in the system. Integration of Ds into or near a new locus causes a recessive mutation in that locus. In the continued presence of Ac, Ds may transpose again, and in doing so results in a reversion of the mutation back to normal. Therefore, in the presence of Ac, Ds-induced mutations are unstable (revertible). In the absence of Ac, Ds cannot transpose, thus making the mutations stable (nonrevertible). Ac as the autonomous element can transpose from one locus to another, producing unstable mutations.

The study of transposable elements in the 1960s and 1970s shifted to bacteria. In these systems it was found that certain mutations were not due to base pair changes, but instead to insertion of large pieces of DNA into genes. In addition, it was discovered that antibiotic resistance could spread rapidly through a bacterial population because the resistance genes were located on transposable elements. The bacterial **transposons** as they are called were characterized at the molecular level, and one of the best known is **Tn 3.** It is approximately 5,000 nucleotides long and has three genes. The genetic map of Tn 3 is found in Figure 13–14. Two genes are for enzymes used in transposition, and one for resistance to the antibiotic ampicillin. The two transposition enzymes are called **transposase,** which initiates transposition, and **resolvase,** which finishes the process. The ends of the transposon are made up of noncoding sequences called **inverted repeats;** the same sequence appears at both ends of the element, but in the reverse order. The inverted

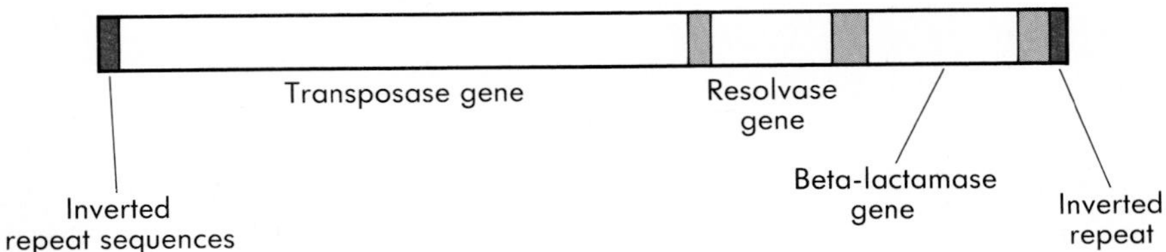

FIGURE 13–14. *Map of bacterial transposon Tn3. This transposon is about 5,000 nucleotide pairs and contains three genes. The transposase and resolvase proteins are involved in transposition, and the beta-lactamase enzyme inactivates the antibiotic ampicillin. The inverted repeats on the ends are recognition signals for the transposition enzymes.*

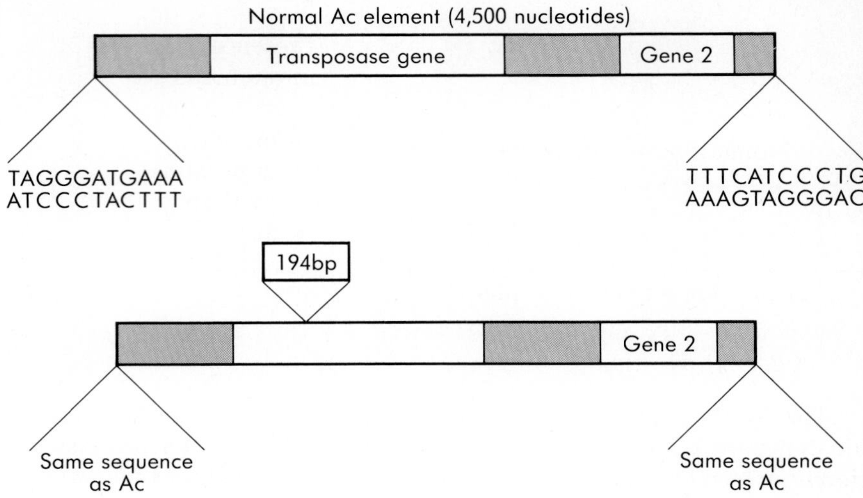

FIGURE 13–15. *Map of normal Ac transposable element and Ds element containing 194 base pair deletion.*

repeats serve to mark the ends of the sequence to be transposed, and also function in the transposition process.

The molecular genetic analysis of the Ac-controlling element from corn has recently been carried out by Nina Fedoroff. A map, including the inverted repeat sequences, is found in Figure 13–15. The element is approximately 4,500 base pairs long and has two coding sequences. It is not as completely defined as Tn 3, but shows remarkable similarity. One of the most exciting discoveries about the Ac-Ds elements is that the Ds element is almost identical to the Ac element, except 194 base pairs of the transposase gene appear to be deleted. It is apparently able to transpose into a locus and cause a mutation, but it must use the normal gene product from the transposase gene of Ac in order to exit from a locus. Thus we have molecular evidence why Ds is a nonautonomous element. Fedoroff feels that contolling elements can be divided into families on the basis of the recognition of the specific transposition signals by the transposition enzymes.

One might contemplate the importance of transposable elements. McClintock believes that since they control gene expression, they represent normal control mechanisms gone awry. They may also play a role in control of gene expression during development. They could have evolutionary significance in restructuring the genome through their role in chromosome breakage. Finally, as they become better known, they may be experimentally useful as gene vectors for genetic engineering approaches (see Chapter 20).

ENVIRONMENTAL MUTAGENS

Thus far we have considered only those chemical or physical agents that are known to be mutagens. In addition to these compounds, there is suspected to be a vast array of chemicals the human population is exposed to that are potentially mutagenic. These include air and water pollutants, food additives and preservatives (over 2,000 are now being used), and agricultural chemicals. In addition, many potentially mutagenic compounds may be **carcinogens,** or capable of inducing cancer in humans. Many cancer-producing compounds are known. These include benzo(a)pyrene, a hydrocarbon produced in the cooking of meat; aflatoxin B1, a fungal toxin produced on seeds of crop plants such as peanuts and corn; and several chemical components of commercially available hair dyes.

Since it has been estimated that 80 percent of human cancer is due to environmental causes, a major concern is the relationship between the carcinogenicity and the mutagenicity of these compounds in humans. Because it is not acceptable to extrapolate mutagenicity data from bacteria to humans, this relationship is very difficult to know. Many carcinogens that would be identified as mutagens in bacteria normally would be detoxified by mammalian metabolism. Furthermore, compounds that would not be mutagenic in bacteria could be converted to a mutagen by the mammalian metabolism in an attempt to detoxify it.

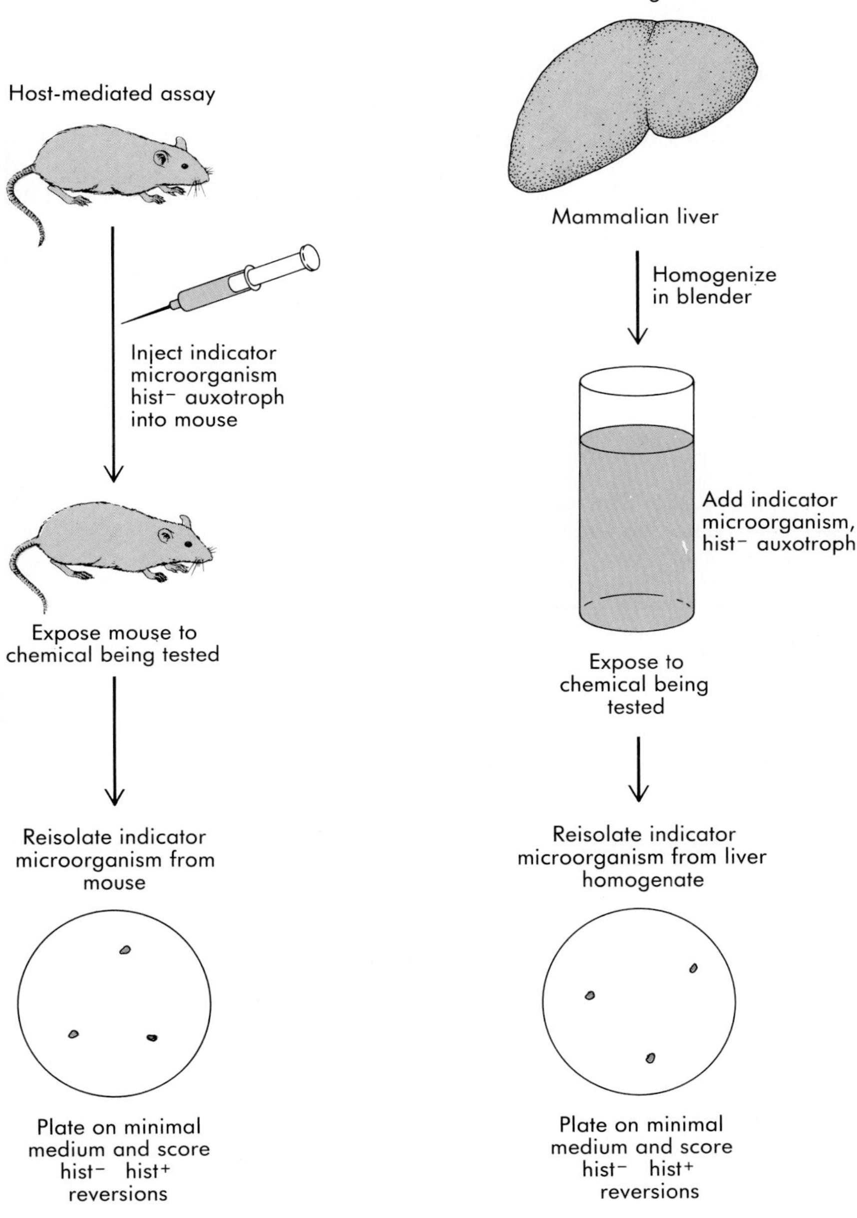
Host-mediated assay
Inject indicator microorganism hist⁻ auxotroph into mouse
Expose mouse to chemical being tested
Reisolate indicator microorganism from mouse
Plate on minimal medium and score hist⁻ hist⁺ reversions
(A)
Liver homgenate
Mammalian liver
Homogenize in blender
Add indicator microorganism, hist⁻ auxotroph
Expose to chemical being tested
Reisolate indicator microorganism from liver homogenate
Plate on minimal medium and score hist⁻ hist⁺ reversions
(B)

FIGURE 13–16. *Schematic diagram of the host-mediated assay (A) and Ames liver homogenate test (B) used for screening potential chemical mutagens.*

The concern here is not only for the present generation but for future generations.

Since many carcinogens are also mutagens, it is very important to be able to identify them as such and any other possible mutagens in the environment. Because of the expense and time to conduct mutagenesis experiments directly on animal populations, two tests have been developed to screen chemicals for mutagenicity in a mammalian system. They are summarized in Figure 13–16. The first test system developed by Marvin Legator is called the host-mediated assay. In this test, an indicator microorganism is exposed to the chemical in question indirectly through a mammalian system. The indicator microorganism is injected into laboratory animals. The animal is then exposed to the chemical in question either through ingestion on inhalation. The microorganism is then reisolated from the mammal and screened for mutations. The typical indicator microorganism is usually an amino acid auxotrophic strain of either *Salmonella typhimurium* or *Neurospora crassa*. The mutation screened is the reversion back to the prototrophic state.

Since the mammalian liver is the main tissue involved in detoxification of foreign compounds in the body, Bruce Ames developed a similar *in vitro* approach to testing for mutagenicity. This test (now called the **Ames test**) uses homogenized liver tissue from either a laboratory animal or a human. The indicator microorganism is introduced into the liver extract, and then exposure to the chemical is carried out. Again, as in the host-mediated assay, the microorganism is scored for mutations from auxotrophy to prototrophy. The Ames test has been refined further to be able to detect mutagens that cause either base pair substitutions or frame shift mutations. A summary of some of the known carcinogens identified as being mutagenic in these two tests is found in Table 13–3.

TABLE 13–3. **List of Chemicals Known as Carcinogens, and Shown to be Mutagenic in the Host-mediated Assay or Liver Homogenate Systems**

2-acetylaminofluorene
aflatoxin B1
1′-acetoxysafrole
2-aminoanthracene
4-aminobiphenyl
2-aminofluorene
1-aminopyrene
2-amino 5-nitrophenol (hair dye component)*
Benzo(a)pyrene
Benzidine
Benzyl chloride
7,12-dimethylbenz(a)anthracene
2,4-diaminoanisole (hair dye component)*
2,5-diaminoanisole (hair dye component)*
2,4-diaminotoluene (hair dye component)*
Furylfuramide (food additive)
2-naphthylamine
p-phenylenediamine (hair dye component)*
Sterigmatocystin

*Mutagenicity of hair dye components is increased by treatment with hydrogen peroxide.

The result from all these tests thus far is that over 70 percent of the carcinogens tested have also been shown to be mutagenic. Any compound that gives a positive test for mutation in these microbial tests should be considered a hazard to humans. As such, they should be carefully approached with regard to benefits and risks to the human population. The answer to the problems of birth defects and cancer may be more in prevention than cure.

MUTATIONS AND CANCER

One of the most exciting discoveries in recent years is that of **oncogenes,** or genes that cause cancer. These genes were originally discovered in the class of viruses called **retroviruses.** It was discovered that when a virus carrying an oncogene invades a cell, the growth pattern of the cell can be converted into one identified as a cancerous (**neoplastic**) type of growth. As long as the virus remains, the cell shows the tumorous growth pattern. Originally it was thought that the oncogene originated with the virus and was the carcinogenic agent. However, it was later found that DNA sequences, related to viral oncogene sequences, were already present in many vertebrate cells. These cellular sequences that have similar structure to eukaryotic genes with introns and exons have survived long periods of evolution and perform normal functions in the cell. This has led to the idea that the viral oncogenes are activated by the virus's removing them from the cell, and in the genetic background of the virus they become oncogenes. Support for this idea comes from the observation that a viral promotor, added to a cellular gene, can transform a normal cell into a cancerous cell.

Two different possibilities are now being explored to explain how these oncogenes of viruses are pathogenic. First, when transferred to the host cell, these genes are controlled by viral promotors and not cellular promotors. This may allow the genes to be expressed constantly and at high levels. Second, oncogenes acquire mutations in the process of becoming oncogenes from protooncogenes. This genetic damage prevents the gene from being controlled in the normal manner, and thus it becomes pathogenic. The mutation theory has much support. In one case there is direct evidence for a single base change: a G into a T, being responsible for the conversion of a protooncogene into an active oncogene.

A common feature of most types of cancer is that of uncontrolled cellular proliferation. Thus when investigators went looking for the molecular mechanism by which the change in state from normal to neoplastic growth (**transformation**) occurs, there was focus first on gene products (proteins), and second on how they relate to cell growth. The gene product of several oncogenes has been found to be an enzyme called a *protein kinase.* This is an enzyme that phosphorylates amino acids in proteins. Furthermore, the kinase from oncogenes phosphorylates only tyrosine. In the process of being transformed into a tumor cell the amount of tyrosine phosphorylation in a cell increases at least tenfold. However, the phosphorylated proteins are not in the nucleus where one might expect, but instead are localized bound to the plasma membrane that is believed to play a role in regulation of cell growth.

It is known that growth factors, such as epidermal growth factor (EGF) and platelet-derived growth factor (PDGF), bind to the plasma

ONCOGENE COPY NUMBER AND BREAST CANCER

In the United States, 1 out of every 14 women will develop breast cancer. The evidence linking proto-oncogenes to the cause or maintenance of human malignancies is becoming increasingly strong. In a very interesting study the number of copies of the *HER-2/new* oncogene was found to vary from 2 to more than 20 per cell in human breast cancer cell lines (cell cultures of breast tumor cells). Furthermore, the results indicate a relationship between the number of copies of the oncogene and the number of lymph nodes that test positive for the cancer. In fact, any patient with more than 3 axillary lymph nodes involved with the disease showed a significant increase in the number of copies of the gene. Thus far, the number of lymph nodes that test positive is the best factor for predicting whether the disease will recur and what the patient's chances are for survival.

The relationship between nodal status and oncogene copy number indicates that oncogene copy number would also be a good predictor of whether the cancer will recur and the chance for survival. A very strong and statistically significant correlation was found between the oncogene copy number and both the elapsed time before a patient had a relapse and the patient's survival rate. In fact, the copy number of *HER-2/new* was superior to all other prognostic factors except for lymph nodes that tested positive for cancer. Patients with more than 5 copies of the oncogene had a much shorter cancer-free survival time and, overall, a much shorter survival time than those with lower oncogene copy number. These new results correlate very well with the data on N-myc oncogene copy number as a cause of human neuroblastoma.

This finding is important for those women whose cancer has not spread to the lymph nodes. These women are usually given a good prognosis for complete recovery and no radiation or chemotherapy treatments. Still, 25 to 30 percent of these women have a recurrence of the cancer. If the oncogene copy number results hold up, those patients whose lymph nodes tested negative for cancer could be further screened and, on the basis of the oncogene copy number revealed by this screening, could be given radiation and chemotherapy treatments, thereby increasing their survival chances.

Oncogene copy number could also provide a clue to what actually causes cancer. It is known that the gene codes for a protein kinase, which is closely related to the epidermal growth factor (EGF) receptor. No one knows what binds to this receptor protein, but blocking it may possibly arrest or cure the disease.

membrane, causing the cell to begin division. Additionally, a tyrosine-specific kinase is stimulated when the EGF molecule binds to the plasma membrane. One hypothesis on cancer induction is that an oncogene present in the cell produces large amounts of a particular gene product, thus overloading the cell's normal processes of growth control and resulting in the transformation of the cell into a tumor cell. Additional support for this idea of overloading the system comes from a fast-growing malignancy of B cells of the immune system called **Burkitt's lymphoma.** This disease is caused by a chromosomal rearrangement resulting in an oncogene being translocated into a region adjacent to the immunoglobulin genes. This leads to a very high level of expression of the oncogene along with the immunoglobulin genes. This enhanced activity of the oncogene is the basic cause of the lymphoma.

As the survey of known oncogenes continues, one additional finding is important. That is, cellular oncogene sequences from several sources are somewhat related. This may mean that they all belong to the same gene family. If this is true, it could mean that cancer is not a series of diseases characterized by the type of tumor, but rather a small number of molecular events (some mutational) bringing about a number of different tumor phenotypes.

PROBLEMS

13–1 From this chapter and any other sources available to you, explain why most mutations are deleterious.

13–2 From this chapter and any other sources available to you, explain why most mutations are recessive.

13–3 From this chapter and any other sources available to you, evaluate the short-term and long-term effects of mass irradiation of the human population.

13–4 Look again at Figure 13–8. Would you say that there is a threshold dose of irradiation below which no mutation is induced?

13–5 X-linked recessive mutations are more easily studied in appropriate organisms than are autosomal ones. Why?

13–6 Which of the following would be likely to suffer the greatest, and which the least, genetic damage from radiation exposure: (a) a haploid, (b) a diploid, (c) a polyploid?

13–7 Three species of oats (*Avena*) vary in chromosome number. Irradiating many samples of these three species with the same amount of X-irradiation resulted in the following rates of induced detectable mutations (given as $\bar{x} \pm s$ per sample):

Species	*Chromosome no.*	*Irradiated*	*Control*
A. brevis	14	4.1 ± 1.1	0.2 ± 0.03
A. barbata	28	2.4 ± 0.6	0.05 ± 0.01
A. sativa	42	0.0	0.0

Give a reason for this kind of result.

13–8 Suppose samples of the species of wheat (*Triticum*) 2n = 14, 28 and 42, were each given the same dose of X-irradiation. Assuming that some radiation-induced mutation occurred in at least some of those species, which species would you expect to show the highest frequency of mutation?

13–9 In an imaginary flowering plant, petal color may be either red, white, or blue. White is due to the presence of a colorless precursor from which red and blue pigments may be synthesized in two consecutive, enzyme-controlled processes. The cross of two blue-flowered plants yields an F_1 ratio of nine blue, three red, and four white. (a) How many pairs of genes are involved? (b) Suggest the correct sequence of enzyme-controlled steps and products. (c) Starting with the first letter of the alphabet and using as many more in sequence as needed, give the genotype of (1) the P individuals; (2) the F_1 blue plants. (d) Give genotypes for (1) red and (2) white F_1 plants. (e) According to the system you have worked out, which pair of genes controls the first step of the process you have worked out in part (b) of this problem?

13–10 By deamination a $\begin{matrix}5' \text{ T T T } 3' \\ 3' \text{ A A A } 5'\end{matrix}$ segment of a DNA molecule is changed to a $\begin{matrix}5' \text{ T T G } 3' \\ 3' \text{ A A C } 5'\end{matrix}$ segment. What change in amino acid incorporation does this produce, assuming the 3′, 5′ ("lower") strand serves as the template for transcription?

13–11 Look again at Figure 13–3. Does the enol form of 5-bromouracil cause a transition or a transversion?

13–12 Table 10–1 shows two tryptophan synthetase mutants, A78 and A58, that affect amino acid residue 233. In A78 wild-type glycine at this position is replaced by cysteine; in A58 glycine is replaced by aspartic acid. Among the mRNA codons for glycine is GGC; one of the two for cysteine is UGC; and one of the two for aspartic acid is GAC. Which of the mutant strains, A78 or A58, involves a transition and which a transversion in the DNA template?

13–13 If a piece of DNA containing a transposable element was isolated, the complementary strands separated and allowed to renature separately, what would the renatured single-stranded molecules look like in the electron microscope?

13–14 Some individuals have a patch of blond hair in a head of brown hair. What type of mutation would this be?

13–15 How many base pairs would have to be deleted in a mutational event to eliminate a single amino acid from a protein and not change the rest of the protein?

CHROMOSOMAL ABERRATIONS: CHANGES IN NUMBER

CHAPTER 14

In earlier chapters it was pointed out that each species of plant and animal is characterized by a particular chromosome complement or set, represented once in haploid cells (for example, gametes and spores) and twice in diploid cells. Possession of such sets of chromosomes, or **genomes,** gives to each species a specific chromosome number (see Table 4–1). But irregularities sometimes occur in nuclear division, or "accidents" (as from radiation) may befall interphase chromosomes so that cells or entire organisms with aberrant genomes may be formed. Such chromosomal aberrations may include whole genomes, entire single chromosomes, or just parts of chromosomes. Thus cytologists recognize (1) **changes in number of whole chromosomes (heteroploidy)** and (2) **structural modifications.** Heteroploidy may involve either entire sets of chromosomes (**euploidy**), or loss or addition of single whole chromosomes (**aneuploidy**). Each may produce phenotypic changes, modifications of phenotypic ratio, or alteration of linkage groups. Many are of some evolutionary significance.

EUPLOIDY

Haploidy

Euploids are characterized by possession of entire sets of chromosomes; haploids carry one genome (n), diploids carry two ($2n$), and so on. Haploidy is rare in animals (the male honeybee is an outstanding exception), but common in plants. In most sexually reproducing algae and fungi and in all bryophytes (liverworts and mosses), the haploid phase represents the dominant part of the life cycle and is the plant recognized in natural populations. In vascular plants, this stage is short-lived and microscopic. Occasionally, adult haploids may be recognized in natural populations but they are ordinarily weak, small, and highly sterile. The first such case in a higher plant appears to have been reported in 1924 in the Jimson weed (*Datura stramonium*); there is good evidence of development from unfertilized eggs.

Sterility in haploids is due to extreme irregularity of meiosis because of the impossibility of chromosomal pairing and the very low probability of chromosome distribution in complete sets to daughter nuclei. For example, consider a hypothetical haploid corn plant ($n = 10$). Only one set of ten chromosomes is present in the sporocytes of such a plant. Each of these unsynapsed chromosomes has one chance in two of going to a given pole of the spindle (if normal and essentially simultaneous anaphase-I movement is assumed); that is, $(\frac{1}{2})^1$ chance of going to, say, the "upper" pole. Thus, in this kind of plant the probability of a normal monoploid nucleus at the end of the first meiotic division is $(\frac{1}{2})^{10} = \frac{1}{1{,}024}$. The odds against a normal haploid gamete increase dramatically with increase in number of chromosomes in a set.

Polyploidy

Euploids with three or more complete sets of chromosomes are called **polyploids.** This condition is rather common in the plant kingdom but rare in animals. Whereas one $4n$ plant, for example, would produce $2n$ gametes and could, in many species, be self-fertilized to produce more $4n$ progeny, the probability of *two* such *animals* (one of each sex) occurring and mating is extremely low. Furthermore, the imbalance of sex-determining mechanisms that would result from polyploidy would be expected to result in sterility because of aberrant meiosis, or to produce individuals that, for many morphological reasons, might be at a considerable disadvantage in mating.

But this is not true in many *plants*. Many plant genera include species whose chromosome numbers constitute a euploid series. The rose genus, *Rosa*, includes species with the somatic numbers 14, 21, 28, 35, 42, and 56. Notice that each of these numbers is a multiple of 7. Therefore, this is a euploid series of the basic haploid number 7, which gives diploid, triploid, tetraploid, pentaploid, hexaploid, and octoploid species. All but the diploids may be collectively referred to as polyploids. One authority has estimated that at least two-thirds of all grass species are or include polyploids.

In many instances in plants, morphological differences between diploids and their related polyploids are not great enough for taxonomists to give the latter forms species rank, although the polyploids can usually be distinguished visually. It was discovered that a large amount of morphological variation within a single species of swamp saxifrage (*Saxifraga pensylvanica*) is associated with polyploidy and that diploids, triploids, and tetraploids differ consistently in several respects. The greater chromosome number of the polyploids was found to be reflected in larger cell size (Figure 14–1).

Measurement of lower leaf epidermal cell size, for example, of randomly selected diploids and tetraploids showed mean cell area of diploids to be 1,606 μm^2 and tetraploids to be 2,739 μm^2. The greater cell size of tetraploids was associated with larger size of plant and plant parts, and a lower length-to-width ratio of leaves, which gives $2n$ and $4n$ plants distinctly different appearances in the field. Leaves of tetraploids are noticeably wider for their length than those of the $2n$ plants (Figure 14–2). Such differences between diploids and tetraploids were found to be significant; statistical tests provided a very high degree of

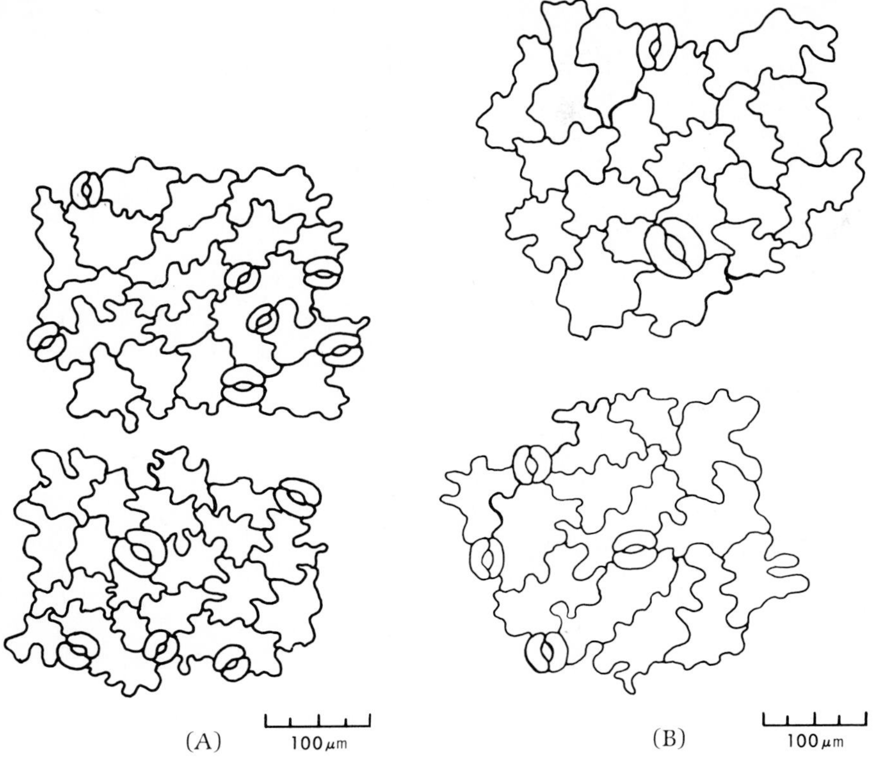

FIGURE 14–1. *Leaf epidermal cells of saxifrage* (Saxifraga pensylvanica). *(A) Diploids; mean cell area ranges from 1,147 μm² to 1,897 μm². (B) Tetraploids; mean cell area ranges from 2,378 μm² to 3,408 μm².*

FIGURE 14–2. *Herbarium specimens of* Saxifraga pensylvanica. *(A) Diploid; (B) tetraploid. Note differences in leaf length–width ratio. Other characteristic dissimilarities occur in shapes and sizes of flower parts, fruits, and seeds.*

FIGURE 14–4. *(A) Diploid and (B) tetraploid snapdragons. Note larger and more numerous flowers in the tetraploid (the somewhat open character of the diploid flowers is not related to chromosome number).* (Courtesy Burpee Seeds.)

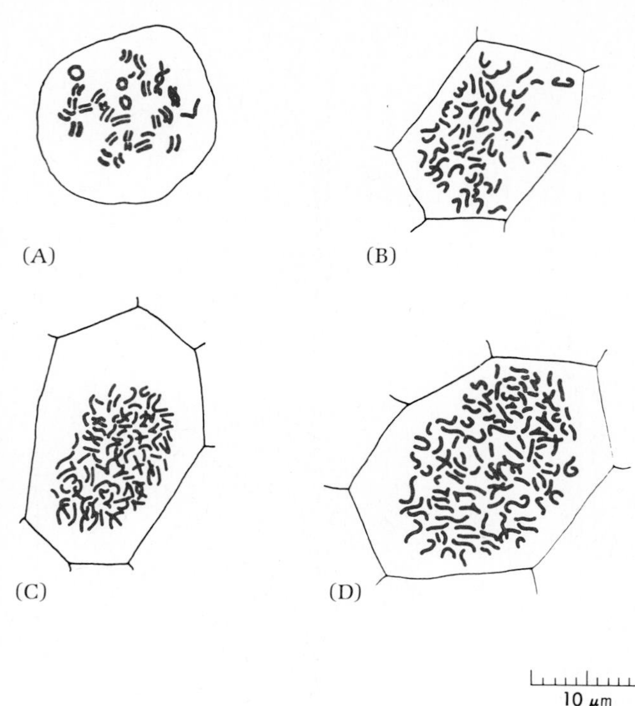

FIGURE 14–3. *Camera lucida drawings of the chromosomes of* Saxifraga pensylvanica. *(A) Microsporocyte at synapsis; (B) root cell, diploid; (C) root cell, triploid; (D) root cell, tetraploid.*

confidence in the premise that two different populations exist in length-to-width ratios of leaves, for example.

The difference in sample means is $4.67 - 3.44 = 1.23$, which is 6.34 times greater than the standard error of the difference in sample means. Chromosomes of the various members of the series are illustrated in Figure 14–3.

In general, tetraploids are often hardier, are more vigorous in growth, have larger flowers and fruits, and are able to occupy less favorable habitats. The tetraploid swamp saxifrage extends a little farther west into drier habitats in Minnesota than the diploid. The same habitat difference is true for tetraploids and diploids for *Tradescantia ohioensis* in Ohio.

In many cultivated plants, tetraploid varieties are commercially more desirable than their diploid counterparts and are commonly available from seed or plant suppliers (Figure 14–4). Many important crop plants are polyploids or include polyploid (usually tetraploid) forms: alfalfa, apple, banana, coffee, cotton, peanut, potato, strawberry, sugar cane, tobacco, and wheat are a few examples. Some scientists have even suggested that the polyploids have a better flavor than their diploid relatives. As might be expected, polyploid varieties with an even number of genomes (for example, tetraploids) are often fully fertile, whereas those with an odd number (for example, triploids) are highly sterile. This latter fact is applied in marketing seeds that contain triploid embryos. Triploid watermelons, for example, are nearly seedless and are listed in a number of seed catalogs (Figure 14–5). Triploids are commonly created by crossing the normal diploid, whose gametes are n, with a tetraploid, whose gametes are $2n$.

(A)

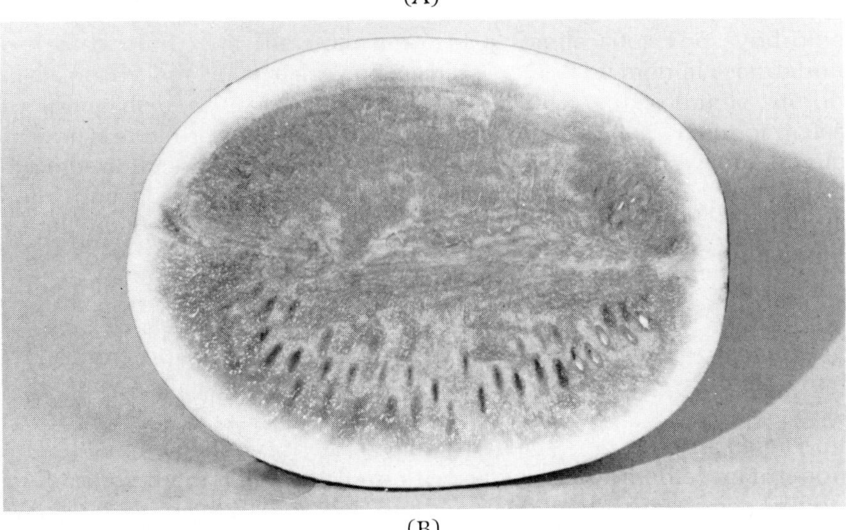

(B)

FIGURE 14–5. *(A) Diploid watermelon with numerous seeds; (B) triploid variety with few and imperfect seeds. (Courtesy Burpee Seeds.)*

Production of Polyploids. Polyploids may arise naturally or may be artificially induced. In plants it appears that diploidy is more primitive and that polyploids have evolved from diploid ancestors. In natural populations this may arise as the result of interference with cytokinesis once chromosome replication has occurred and may occur either (1) in somatic tissue, which gives rise to tetraploid branches, or (2) during meiosis, which produces unreduced gametes. It has been found that chilling may accomplish this in natural populations.

Application of the alkaloid colchicine, derived from the autumn crocus (*Colchicum autumnale*), either as a liquid or in a lanolin paste, induces polyploidy. Although chromosome replication is not interfered with, normal spindle formation is prevented and the double number of chromosomes becomes incorporated within a common nuclear membrane. Subsequent nuclear divisions are normal, so that the polyploid cell line, once initiated, is maintained. Polyploidy may also be induced by other chemicals (acenaphthene and veratrine) or by exposure to heat or cold.

Autopolyploidy and allopolyploidy. If a tetraploid is developed by the colchicine treatment, for example, its cells contain four genomes, all of the same species. Such a polyploid is an **autotetraploid.** Autopolyploids may, of course, exist with any number of genomes. The same situation results if an individual is formed from the fusion of two diploid gametes from the same species; the four sets of chromosomes all belong to the same species.

Synapsis in autotetraploids usually involves synapsis in *pairs* of homologs (that is, two pairs of bivalents; Figure 14–6A); this is the case in the previously mentioned autotetraploid of *Saxifraga pensylvanica.* However, quadrivalents, or a trivalent and a univalent, sometimes occur (Figure 14–6B). Particularly where bivalents are formed, meiosis and the subsequent gamete formation are essentially normal. Random pairing and disjunction in an autotetraploid of genotype *AAaa*, where bivalents are formed, will produce a gamete genotypic ratio of 1 *AA*:4 *Aa*:1 *aa*, as may be seen in the following diagram. Lines connecting alleles show possible combinations:

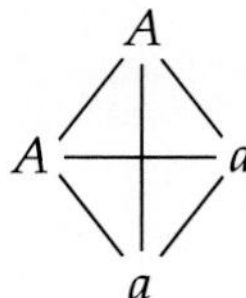

Selfing such an autotetraploid produces some unusual progeny phenotypic ratios as will be seen shortly.

Crossing-over between linked genes in an autotetraploid complicates gamete genotypic ratios. Progeny ratios produced by, say, an *AABB/aabb* individual depend on (1) distance between the *A* and *B* loci, and (2) whether synapsis involves two bivalents, one quadrivalent, or a univalent plus a trivalent. Fertility is generally not significantly reduced if bivalents regularly form, but the other two possibilities (quadrivalent or trivalent plus a univalent) lead to a degree of sterility commensurate with the frequencies of each type of configuration.

On the other hand, polyploids may develop (and be developed) from hybrids between different species. These are **allopolyploids;** the most commonly encountered type is the allotetraploid, which has two genomes from each of the two ancestral species. The Russian cytologist,

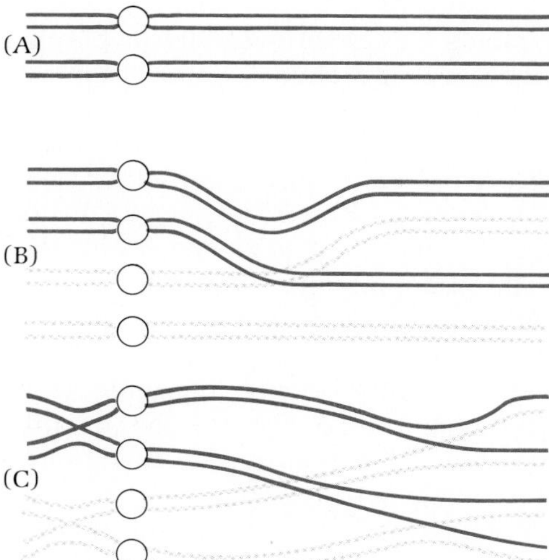

FIGURE 14–6. *Diagram of synapsis in autotetraploids. (A) Bivalents; (B) one trivalent and one univalent; (C) tetravalent. (B) usually leads to high frequencies of chromosomally abnormal gametes and a high degree of sterility.*

Karpechenko, synthesized a new genus from crosses between vegetables belonging to different genera, the radish (*Raphanus*) and the cabbage (*Brassica*). These plants are fairly closely related and belong to the mustard family (*Cruciferae*). Each has a somatic chromosome number of 18, but those of radish have many genes that do not occur in cabbage chromosomes, and vice versa.

Karpechenko's hybrid had in each of its cells 18 chromosomes, 9 from radish and 9 from cabbage. Members of the very different genomes failed to pair in meiosis, and the hybrid was largely sterile. A few 18-chromosome gametes were formed, however, and a few allotetraploids were thereby produced as an F_2. These were completely fertile, because two sets each of radish and cabbage chromosomes were present and pairing between homologs occurred normally. The allotetraploid, or **amphidiploid,** was named *Raphanobrassica*. Unfortunately, the resulting plant has the root of cabbage and leaves of radish and is of no direct economic importance. The method does, however, offer a means of producing fertile interspecific or intergeneric hybrids.

Polyploidy in humans. Complete polyploid human beings are, as might be expected, quite rare and the few cases known are either spontaneously aborted fetuses (**abortuses**) or stillborn. A few live for a matter of hours. Gross and multiple malformations occur in all cases. Table 14–1 presents a sampling of human polyploids reported in the literature. In one study of 227 abortuses, some chromosome anomaly was found in 50 of these; 2 were triploids and 1 was tetraploid. As far as could be determined, none of the latter were mosaics made up of cell lines of different chromosomal constitution. It has been estimated that about 15 percent of all spontaneously aborted fetuses are either triploids or tetraploids. A large number of abortuses and live-born children that die very early are mosaics for diploid–polyploid cell lines.

Origin of polyploid human embryos is difficult to explain satisfactorily. Although fusion of a normal haploid gamete with a diploid one would produce a triploid zygote that might even progress to some stage of embryo development, little concrete evidence exists for occurrence of unreduced gametes in mammals. These could be produced in theory, however, by incorporation of the chromosomes of the first or second polar body within the egg nucleus, or by an analogous abnormality during spermatogenesis, or by meiosis of an exceptional tetraploid oocyte or spermatocyte. Fertilization of a normal haploid egg by more than one sperm has also been suggested, because polyspermy has been found to occur occasionally in rabbits and rats, but there is no clear evidence for this in humans. In summary, polyploidy in humans, whether complete or as a mosaic, leads to gross abnormalities and death.

TABLE 14–1. Polyploidy in Human Abortuses and Live Births

Reference	*69,XXY*	*69,XXX*	*69,XYY*	*92,XXXX*	*92,XXYY*
Penrose and Delhanty (1961)	2				
Makino et al. (1964)	3				
Szulman (1965)	4	1			
Carr (1965)	6	2	1	1	1
Patau et al. (1963)		2			
Schindler and Mikamo (1970)	1*				
Butler et al. (1969)		1*			

*Born alive; died within hours.

FIGURE 14–7. Spartina alterniflora. *(A) View of pure stand in bloom bordering salt water. (B) Single flowering spike.* (Cape Cod National Seashore Park, Eastham, Mass.)

(A)

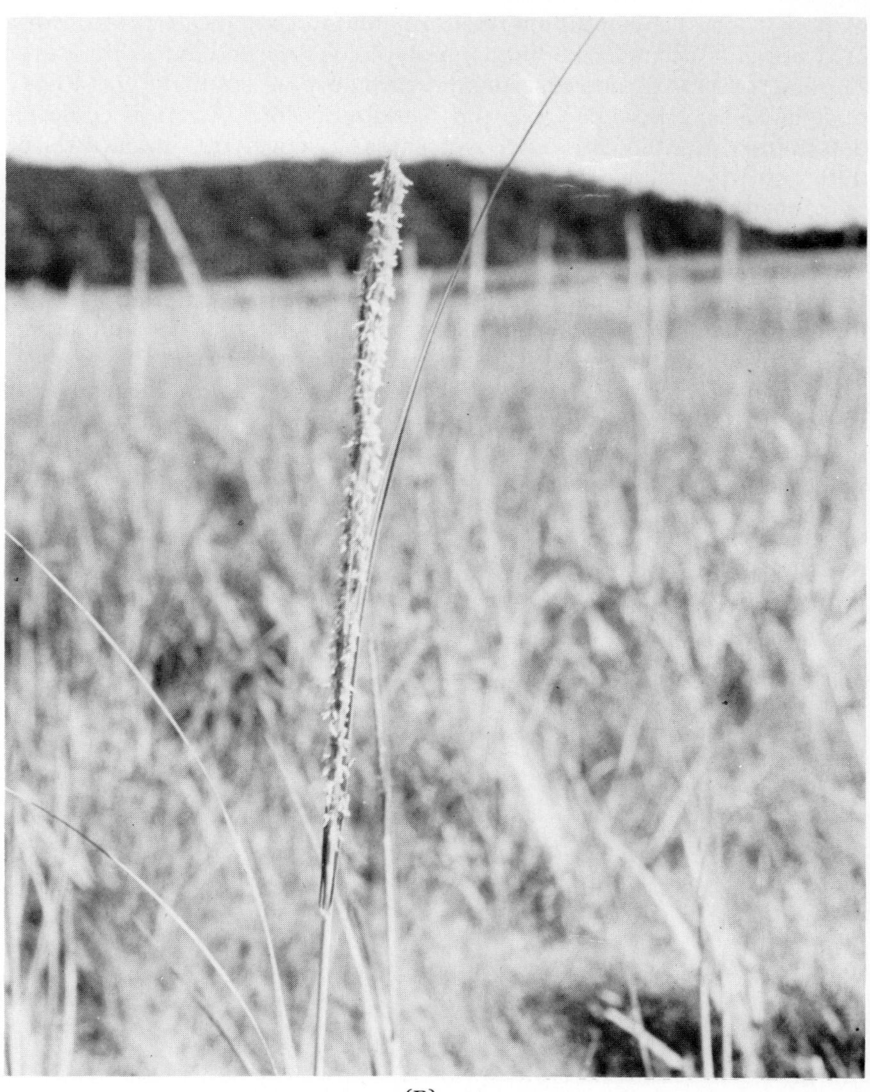

(B)

Evolution through polyploidy. Interspecific hybridization combined with polyploidy offers a mechanism whereby new species may arise suddenly in natural populations. Cytogenetic investigations of such instances of speciation in many cases have involved real "detective work" and even culminated in the artificial production of the new species.

An excellent example is furnished by the work of Huskins (1930) and of Marchant (1963) on the saltmarsh grass *Spartina*. A European species, *S. maritima*, occurs along the Atlantic coast of Europe and adjacent Africa. *S. alterniflora*, an eastern North American species (Figure 14–7), was introduced accidentally into Great Britain in the eighteenth century. The two species are morphologically quite different from each other and have different chromosome numbers, $2n = 60$ for *S. maritima* and 62 for *S. alterniflora*.

The American species gradually spread after introduction in western Europe, and the two species grew intermixed. In the 1870s, a sterile putative F_1 hybrid (now called *S.* × *Townsendii*) was collected. This form has a somatic chromosome number of 62, rather than the expected 61 ($= \frac{60}{2} + \frac{62}{2}$), apparently because of meiotic abnormality in one of its parents. Chromosomes in this genus are small and not easily distinguished from each other on the basis of morphology. Thus it is nearly impossible to determine the species origin of the chromosome complement of *S.* × *Townsendii* with reference to *S. alterniflora* and *S. maritima*. In any event, the hybrid is sterile because of the lack of meiotic pairing in its mixture of *alterniflora* and *maritima* chromosomes.

In the 1890s, a new, vigorous, fertile *Spartina*, which quickly spread over the British coasts and into France, was discovered. This new form, named *S. anglica*, has a somatic chromosome number of 124 in most cases, although some with 122 and 120 have been reported. From its morphology and cytology, *S. anglica* appeared to be an amphidiploid of the cross *S. alterniflora* × *S. maritima*. This did prove to be the case, for *S. anglica* has been created (or recreated) in experimental plots, a triumph of cytological research (Figure 14–8).

Similarly, other workers investigated the origin of New World cotton (*Gossypium*). Briefly, Old World cotton has 26 rather large chromosomes, whereas a Central and South American species has 26 much smaller ones. The cultivated cotton has 52, of which 26 are large and 26 are

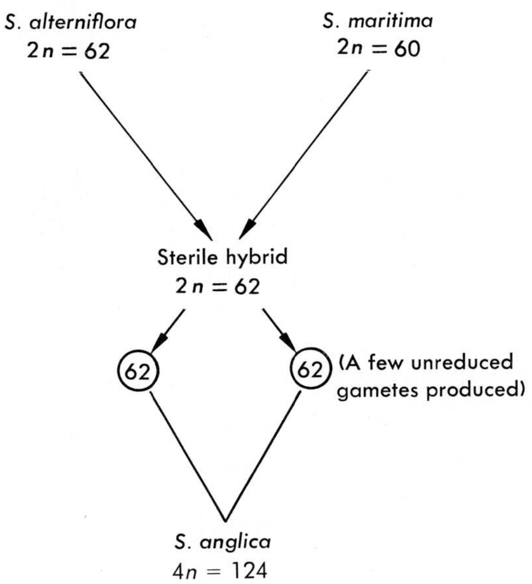

FIGURE 14–8. *Development of* Spartina anglica, *an allotetraploid (amphidiploid). The cross* S. alterniflora × S. maritima *produced a sterile hybrid (named* S. × townsendii *by Marchant, 1963) having 62 chromosomes instead of the expected 61. Production of infrequent gametes having all 62 chromosomes, followed by their fusion, results in the fertile hybrid* S. anglica. *Some amphidiploid plants with 120 and 122 chromosomes have also been reported.*

smaller, and was suspected of being an allotetraploid of a cross between Old and New World species. Eventually the cultivated species was reconstructed by crossing the two putative parents and using colchicine to double its chromosome number.

Caution must be exercised in considering polyploidy as a mechanism of evolution. Basically, polyploidy adds no new genes to the gene pool but, rather, results in new combinations, especially in allopolyploids. Phenotypic effects of autopolyploidy generally represent merely exaggerations of existing characters of the species. Possession of multiple genomes reduces the likelihood that a recessive mutation will be expressed until and unless its frequency becomes quite high in the population. Polyploidy, therefore, has the potential for actually decreasing genetic variation. So, although entities recognized as new species do arise through allopolyploidy, and although vigor and often geographic range may be increased by autopolyploidy, polyploidy must be viewed in its proper perspective as both a positive and negative force of evolution. Simmonds (1976) and Stebbins (1966 and 1971) include extensive discussions of the evolutionary aspects of polyploidy.

ANEUPLOIDY

Aneuploidy is the situation where the chromosome number is not an exact multiple of the characteristic haploid number for the species.

The Jimson weed (*Datura stramonium*) shows a considerable amount of morphological variation in many traits, particularly in fruit characters.

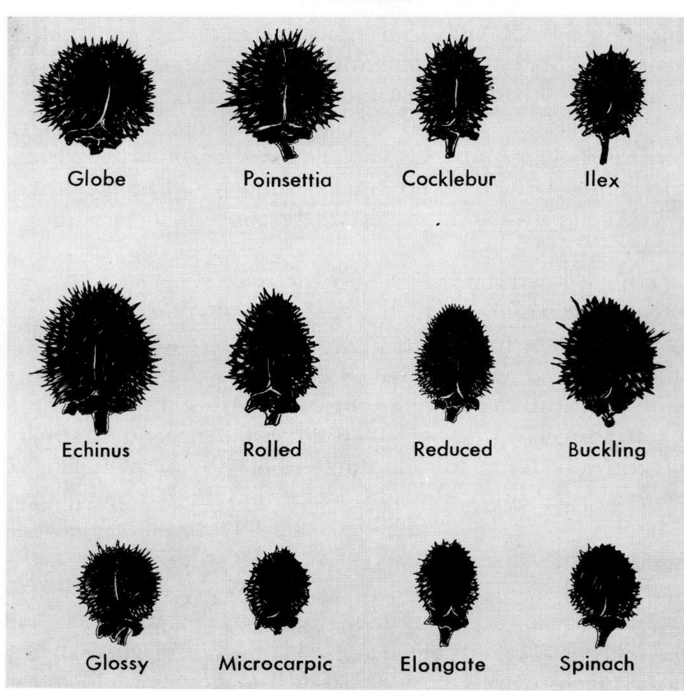

FIGURE 14–9. *Fruits of normal diploid Jimson weed (top) and its 12 possible trisomics (below). Each of the latter was produced experimentally by Blakeslee and Belling.* (Redrawn from A. B. Blakeslee and J. Belling.)

The normal chromosome number for this plant is $2n = 24$, but in a classical study, A. F. Blakeslee and J. Belling (1924) showed that each of several particular morphological variants had 25 chromosomes. One of the 12 kinds of chromosomes was found to be present in triplicate; that is, the somatic cells were $2n + 1$. Such a **trisomic** plant has 3 of each of the genes carried by the extra chromosome. Because the Jimson weed has 12 pairs of chromosomes, 12 recognizable trisomics should be possible, and Blakeslee and his colleagues succeeded in producing and describing all of them (Figure 14–9). Trisomics usually arise through nondisjunction so that some gametes contain two of a given chromosome.

When trisomics are crossed, ordinary Mendelian ratios do not result. For example, the trisomic "poinsettia" (Figure 14–9) has an extra ninth chromosome (that is, triplo-9), which carries gene *P* (purple flowers) or its allele *p* (white flowers). One possible genotype, therefore, for a purple-flowering "poinsettia" is *PPp*. Crossing two such plants produces a 17:1 phenotypic ratio. Pollen (and hence sperms) in *Datura* carrying either more or less than 12 chromosomes is nonfunctional. Megaspores (and subsequently developed eggs), however, are not so affected.

Meiosis in trisomics ordinarily results in two of the three homologs going to one pole, and one to the other. As a result, some gametes carry various combinations of two homologs and others carry one. With this in mind, the cross *PPp* × *PPp* may be represented in this way:

	P	*PPp* ♀ "purple poinsettia"	×	*PPp* ♂ "purple poinsettia"
P gametes:				
	♀	$\frac{1}{6}$ ea.: $P + P + Pp + Pp + PP + p$		
	♂	$\frac{1}{3}$ ea.: $P + P + p$		

F^1	$\frac{4}{18}PP$	homozygous purple diploid
	$\frac{4}{18}Pp$	heterozygous purple diploid
	$\frac{5}{18}PPp$	heterozygous purple trisomic ("poinsettia")
	$\frac{2}{18}Ppp$	heterozygous purple trisomic ("poinsettia")
	$\frac{2}{18}PPP$	homozygous purple trisomic ("poinsettia")
	$\frac{1}{18}pp$	homozygous white diploid

Purple and white segregate in a 17:1 ratio. Other ratios are possible with different parental genotypes.

Aneuploids other than trisomics are reported in the literature, but those in the Jimson weed are best known because of the extensive work of Blakeslee and his associates. Types of aneuploids are summarized in Table 14–2 along with the other modifications of chromosomes considered thus far.

Trisomy in Humans

In humans it has been estimated that approximately 4 percent of all clinically recognized pregnancies carry some form of trisomy. Furthermore, among aborted human fetuses, every chromosome has been found in trisomic condition. The autosomal trisomics comprise 47.8 percent of all abnormal fetuses. The most common are found in groups D, E, and G. Those involving the A, B, C (exclusive of the X), and chromosome 16 never come to full term, suggesting such a drastic genetic imbalance that survival is not possible. Among the most common are trisomy-21, 18, and 13, which are discussed in detail here.

TABLE 14–2. Summary of Variations in Chromosome Number (Heteroploidy)

Type	Designation	Chromosome complement (where one set consists of four chromosomes, numbered 1, 2, 3, and 4)
Euploids		
Haploid	n	1-2-3-4
Diploid	$2n$	1-2-3-4 1-2-3-4
Triploid	$3n$	1-2-3-4 1-2-3-4 1-2-3-4
Autotetraploid	$4n$	1-2-3-4 1-2-3-4 1-2-3-4 1-2-3-4
Allotetraploid	$4n$	1-2-3-4 1-2-3-4 1′-2′-3′-4′ 1′-2′-3′-4′
etc.		
Aneuploids		
Trisomic	$2n + 1$	1-2-3-4 1-2-3-4 1
Triploid tetrasome	$3n + 1$	1-2-3-4 1-2-3-4 1-2-3-4 1
Tetrasomic	$2n + 2$	1-2-3-4 1-2-3-4 1-1
Double trisomic	$2n + 1 + 1$	1-2-3-4 1-2-3-4 1-2
Monosomic	$2n - 1$	1-2-3-4 2-3-4
Nullisomic	$2n - 2$	2-3-4 2-3-4

Down syndrome. The most significant instance of trisomy in humans involves Down syndrome (DS), named for the nineteenth-century British physician, J. Langdon Down, who first described the syndrome in 1866 (Figure 14–10). The incidence in the general population is 1 in 650 to 700 births. DS is the most frequently observed chromosome abnormality in humans. About 4,000 children with DS are born each year in the United States. There are about 50 physical characteristics shown by infants with DS soon after birth. These include eyes that slant up and

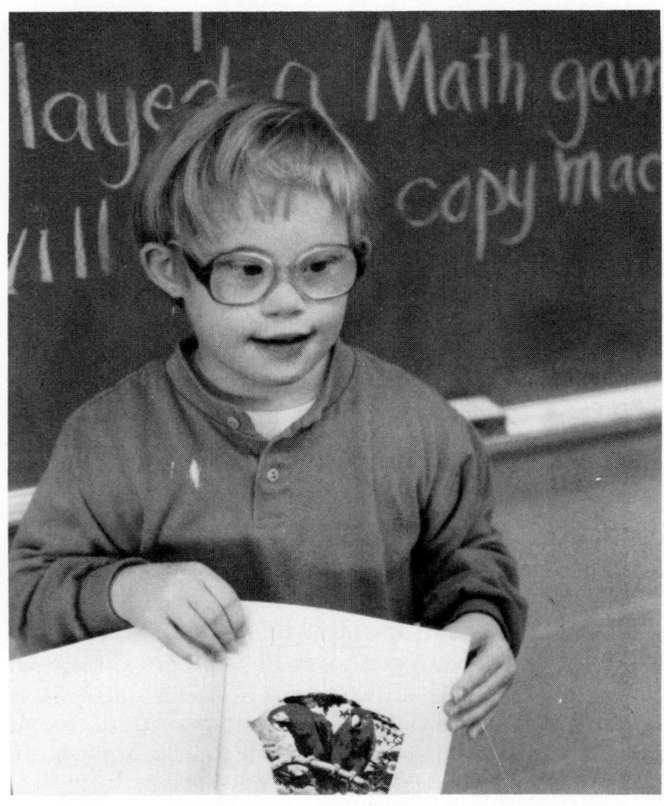

FIGURE 14–10. *Down's syndrome in a child.* (Photo courtesy Mary Lou Wall, Duckworth Education Center, Beltsville, MD.)

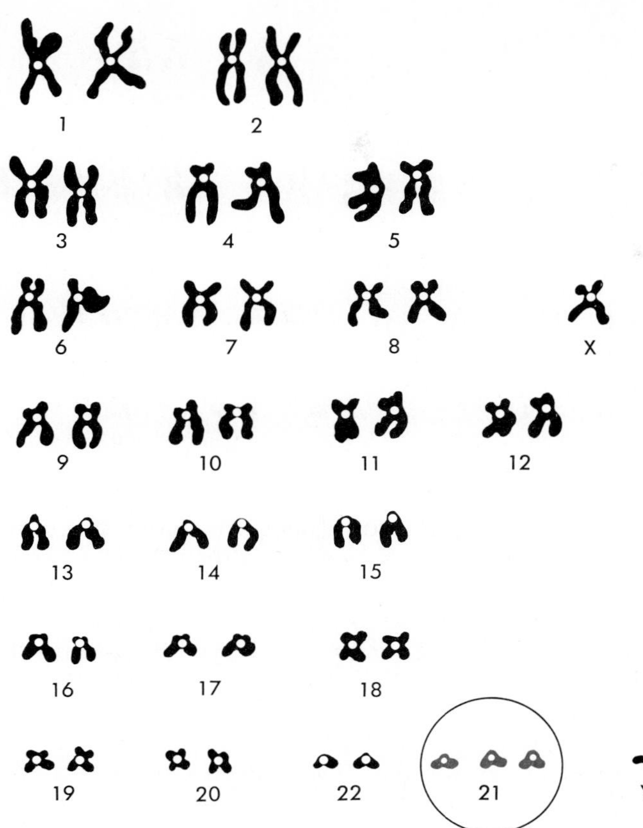

FIGURE 14–11. *Karyotype of trisomic Down's syndrome male having 47 chromosomes (triplo-21).* (Courtesy National Foundation, New York.)

out with internal epicanthal folds, a tongue that is large and protruding, small and underdeveloped ears, an enlarged liver and spleen, and a single palmar crease. Some complications that are easily corrected with surgery include congenital heart defects and intestinal obstructions. DS children are mildly to moderately retarded, but with recent medical, educational, and social advances they can develop further than earlier generations thought possible. They can learn self-care, reading, and job skills. Most are important loving and contributing family members.

Individuals with DS were shown in 1959 to be trisomic for one of the G group of autosomes (Figure 14–11), originally identified as number 21, the second smallest of the autosomes. Recent work with fluorescence microscopy, however, shows that in actuality, the smallest (formerly designated 22) chromosome is present in triplicate. Because the notation triplo-21 has become so entrenched in the literature, it has been agreed to renumber the two members of the G group, with 22 becoming 21 and vice versa. Although this is contrary to the otherwise standard system of assigning successively higher numbers to successively shorter autosomes, confusion between older and newer literature on DS is prevented. The important fact is the presence of an extra one of the shortest of the autosomes. The determinants of the Down phenotype have been localized to trisomy of a segment of the longer arm of chromosome 21 (the proximal part of band q21A).

From the cytological viewpoint, two types of DS may be recognized: (1) *triplo-21*, in which the affected individual is trisomic for autosome 21, with a total somatic complement of 47 chromosomes (about 92.5 percent of persons with Down syndrome are triplo-21), and (2) *translocation*, in which the extra twenty-first chromosome has become at-

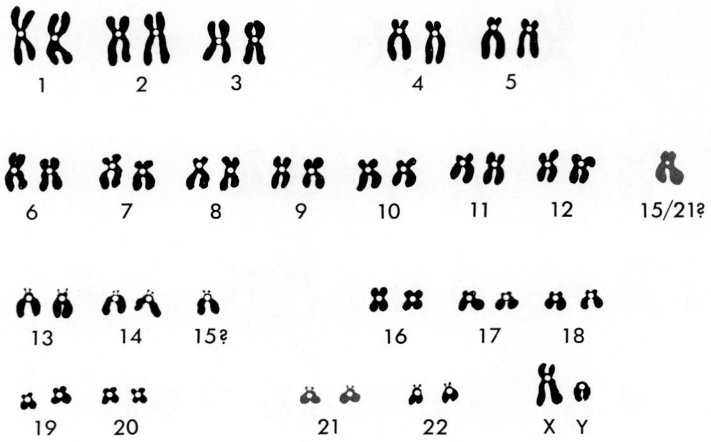

FIGURE 14–12. *Karyotype of Down's syndrome male. This individual has 46 chromosomes, but only one normal 15 and a large one formed by union of a 21 and the other 15. Two normal 21s are also present, giving the individual three "doses" of chromosome 21. More commonly the translocation is 14-21.*

tached to another autosome, most frequently one of the D group, now generally agreed to be number 14. More rarely the translocation involves 15 and 21, or even one of the G group, probably another 21. A karyotype of a translocation Down syndrome, therefore, shows 46 chromosomes, one of which is 15–21 (Figure 14–12). Both types involve an extra number 21: In one case it is a separate entity; in the other it has become attached to another chromosome. Phenotypes of both triplo-21 and translocation syndromes are identical.

Incidence of DS has always been closely related to maternal age (Figure 14–13). Women under 35 years of age give birth to more than 80 percent of the children with DS. Triplo-21s arise through nondisjunction, chiefly in oogenesis. A female is born with all the oocytes she will ever produce—nearly 7 million. These remain in an arrested prophase-I of meiosis from before birth until ovulation, generally at the rate of one per menstrual cycle (about 28 days) after puberty. A given oocyte may remain in this state of suspended development for about 12 to 15 years. On the other hand, the testes produce about 200 million sperms a day, and meiosis in a spermatocyte requires 48 hours or less to complete. Nevertheless, paternal origin of the extra chromosome has now been verified in about 20 percent of the cases.

Nondisjunction during oogenesis is thus a function of senescence of

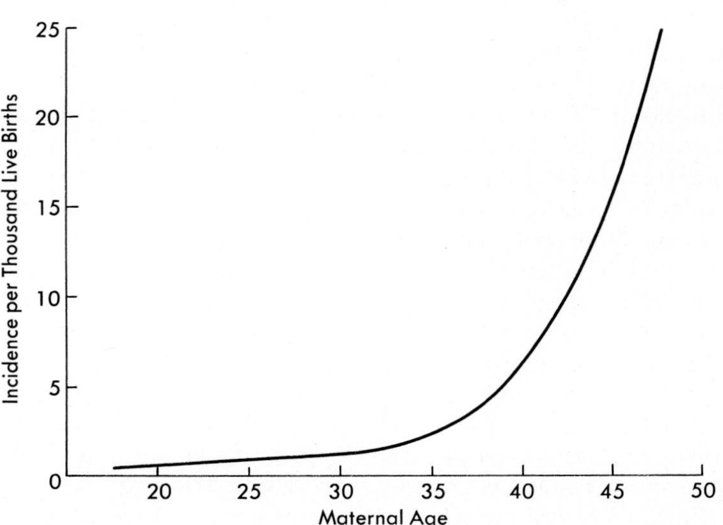

FIGURE 14–13. *Relationship between maternal age and birth of children with Down's syndrome.*

ALZHEIMER'S DISEASE AND DOWN SYNDROME

Alzheimer's disease (AD), a degenerative disorder of the human central nervous system resulting in progressive impairment of memory, seems to have some common neuropathologic changes in common with older patients who have Down syndrome (DS). Since some examples of AD occur in families and are caused by a genetic defect that is transmitted in an autosomal dominant fashion, it was speculated that the gene for AD might be located on chromosome 21. Genetic markers for the disease have now confirmed this location. Trisomy of chromosome 21 is the known cause of Down syndrome.

One of the identifying features of AD is the presence of numerous plaques in the cells of people who have died with AD. The degree of intellectual impairment of a person with AD seems to be correlated with the frequency of these plaques in the cortical cells. A small protein of 42 amino acids, called the brain amyloid protein, seems to be the main component of the plaques. The brains of aged individuals with DS also have the same protein with the identical amino acid sequence. The chromosomal location of the gene for the brain amyloid protein has been located on chromosome 21. Furthermore, other researchers have found the same protein accumulation in the brains of several other species of aged mammals.

Amyloid protein deposition is a normal process linked to age. However, in Alzheimer and Down individuals some yet unknown mechanism leads to increased deposition. Collectively, all of this new information suggests very strongly that one or more genes causing familial Alzheimer's disease are located together on chromosome 21. In addition, it suggests that a gene in the same region codes for the brain amyloid protein that accumulates in both the brains of Alzheimer victims and aged people with Down syndrome. Therefore, both Alzheimer's disease and Down syndrome seem to be related by the fact that the genetic region for both disorders is located on chromosome 21.

the oocytes, whereby separation of homologs is interfered with in some way. This may be through the breakdown of chromosomal fibers or through deterioration of the centromere. Presence of a virus as well as radiation damage (whose cumulative effect would be greatest in older individuals) has been suggested as among the causative factors. Cells are especially susceptible to damage by viruses and by radiation during division. Of course, the older the woman, the longer these agents will be able to act on her oocytes. Up to three times greater incidence of DS has been reported in children born to mothers (age group for age group) who have had infectious hepatitis prior to pregnancy.

Some family studies suggest a specific gene that interferes with disjunction; such genes are known in some organisms. The probability that autoimmune diseases may play a part in certain cases is evidenced by the fact that there is about a threefold increase in the risk of a Down syndrome child at any given maternal age where high thyroid antibody levels are present. In one study, 30 of 177 mothers of Down children had some form of clinical thyroid disorder, compared with only 11 of 177 control mothers (that is, those without Down syndrome children).

Another explanation may be based on the fact that an egg will degenerate in the fallopian tube within about one day if not fertilized. If fertilization should occur toward the close of the period of viability, degenerative changes in the egg will lead to chromosomal abnormalities in zygotene in some animal species, probably including humans. The correlation of trisomy-21 with increased maternal age under this explanation probably is a reflection of less frequent intercourse in older couples. That this is not a *principal* factor is seen in the shape of the curve in Figure 14–13. The increased incidence of Down syndrome in the later maternal years is exponential. In any event, although the causes are not fully clear, the maternal age effect is well established.

Mosaics with Down syndrome, who exhibit the classical symptoms in varying degree, probably arise through failure of the 21s to segregate at mitosis during embryogeny, which produces a fetus with cells of three types: (1) normal 21, (2) triplo-21, and (3) monosomic-21. The latter group's cells appear unable to survive, and thus lead to a disomic–trisomic mosaic. Should mitotic segregation of the 21s fail at the first division of the zygote, however, two daughter cells, one trisomic and the other monosomic, would result. Death of the latter cell would then lead to production of a wholly trisomic embryo.

Examination of parents of translocation Down children almost always discloses that one of them has only 45 chromosomes, including one 21, one 14, and a fused 14/21. Such "carriers" are phenotypically normal, inasmuch as their genetic material is present in the proper amount. More often, the mother of a translocation Down syndrome child is the carrier. The reason why carrier fathers less often produce translocation children is unknown. If the centromere of the maternal translocation chromosome is that of number 14, four kinds of eggs are produced, and this leads apparently to three kinds of children:

Egg	*Sperm*	*Zygote*	*Progeny*
$14\hat{}21$, 21	14, 21	14, $14\hat{}21$, 21, 21	Translocation Down syndrome
$14\hat{}21$	14, 21	14, $14\hat{}21$, 21	Translocation carrier
14, 21	14, 21	14, 14, 21, 21	Normal
14	14, 21	14, 14, 21	Usually lethal

Among cell cultures available to research workers are two 45 XY-21 (that is, monosomy 21) cell lines, one from a five-and-a-half-year-old patient, and the other from a 14-month-old child. Some partial (mosaic) 21 monosomics have been reported.

Edwards' syndrome. First described in 1960 by John H. Edwards and his colleagues, trisomy-18 (Figure 14–14) is now well known in more than 50 published cases. Incidence is about 0.3 per 1,000 births. It is characterized by multiple malformations, primarily low-set ears; small, receding lower jaw; flexed, clenched fingers; cardiac malformations; and various deformities of skull, face, and feet. Harelip and cleft palate often occur. Death takes place generally around 3 to 4 months of age, but may be delayed for nearly 2 years. One mentally retarded female, age 15 has been reported. Triplo-18 children show evidence of severe mental retardation. The defect is about three times more frequent in females; the reasons for this are not clear.

As in Down syndrome, there is a pronounced maternal age effect, although plotting frequency against maternal age produces a bimodal curve (Figure 14–15). The secondary peak in the early twenties reflects the normal maternal age group of maximum births, whereas the pronounced peak from 35 to 45 years is clearly related to increased age of the mother.

Trisomy-13 (Patau syndrome). In 1960, Klaus Patau and associates described a case of multiple malformations in a newborn child, established as a trisomic for one of the D-group of autosomes, now generally agreed on as number 13. Individuals appear to be markedly mentally retarded and very frequently have sloping forehead, harelip, and cleft palate; the last two malformations are so strongly developed that the face has a severely deformed appearance. Polydactyly (both hands and

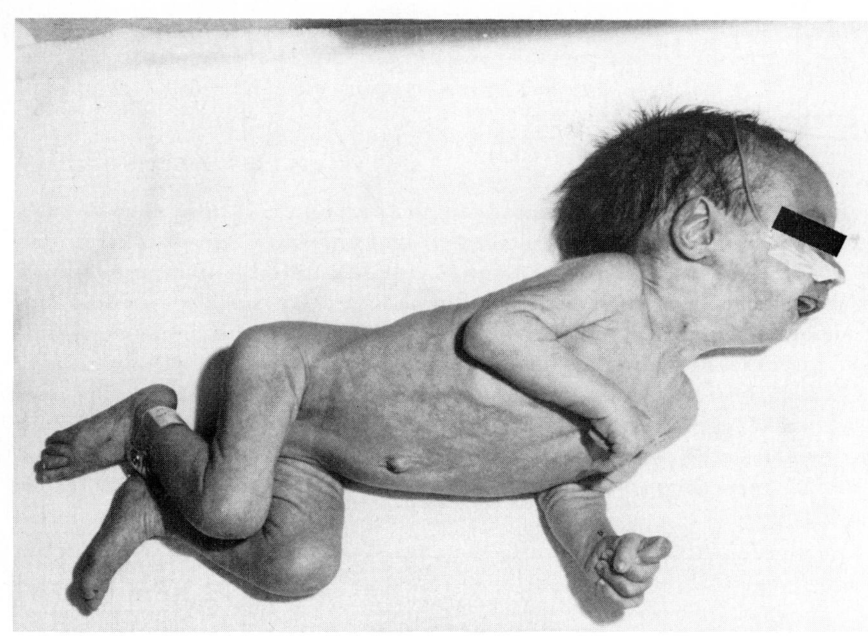

(A)

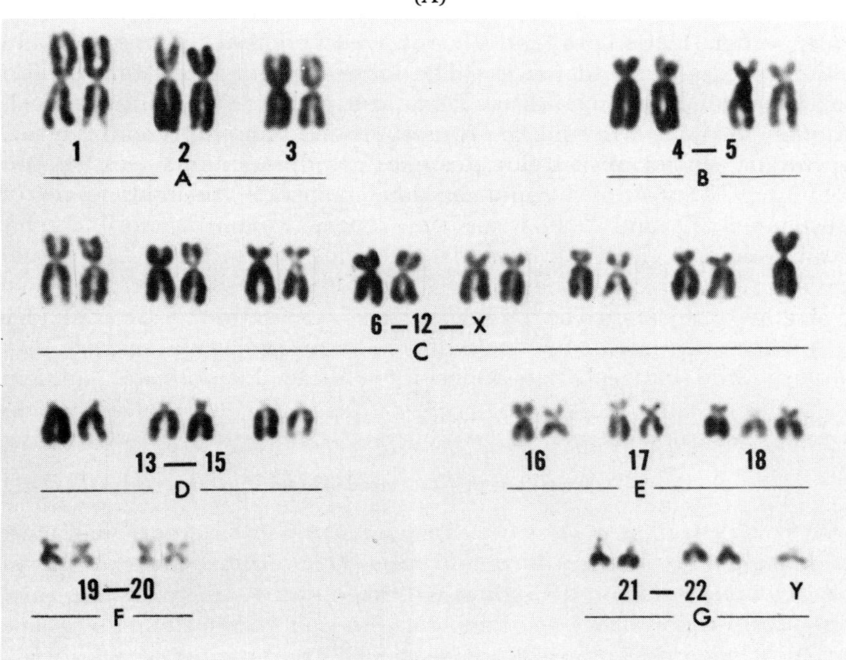

(B)

FIGURE 14–14. *Edwards' syndrome (trisomy-18). (A) Young triplo-18 child, showing low-set ears, small, receding lower jaw, and flexed, clenched fingers. (B) Karyotype of same child. Note the seven E-group autosomes.* (Photo courtesy Dr. Richard C. Juberg used with permission.)

feet) is almost always present; the hands and feet are also characteristically deformed (Figure 14–16). Cardiac and various internal defects (of the kidneys, colon, small intestine) are common.

Death usually occurs within hours or days, but the fetus may abort spontaneously. In an extreme case of longevity, a living triplo-13, age 5 years, was reported. There is a slight excess of affected females. Age of the mother is not as clearly a factor as in Edwards' syndrome, although tabulated cases do form a bimodal curve. However, in one study 40 percent of the triplo-13s were born to mothers over 35, whereas this age group accounted for less than 12 percent of all births. Incidence in the general population is of the order of 0.2 per 1,000 births.

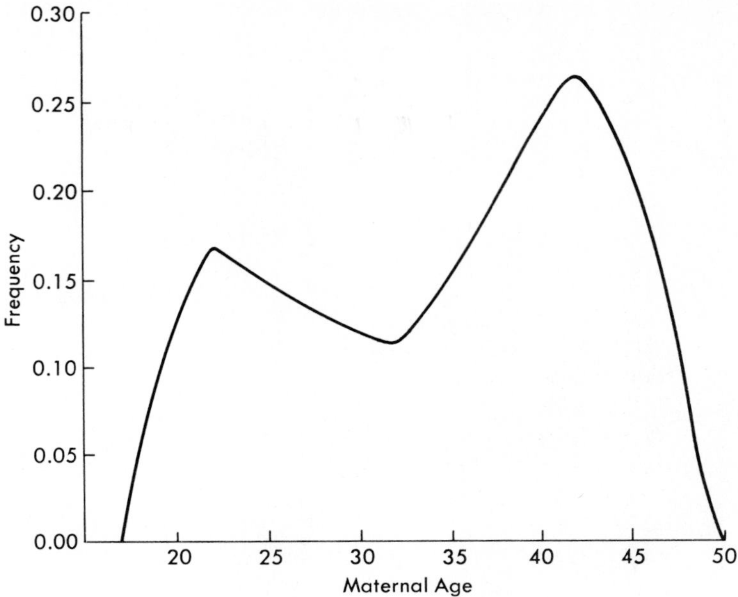

FIGURE 14–15. *Maternal age effect in Edwards' syndrome. The peak between maternal ages 35 and 45 is the significant one here. (See text for details.)*

Trisomy and spontaneous abortion. It appears that trisomy for autosomes other than 13, 18, and 21 may well be lethal. In one study of 1,457 abortuses, 892 (61 percent) had some chromosomal abnormality, including XO, trisomy D, trisomy E, and triploidy. McCreanor and colleagues (1973) reported spontaneous abortion of an empty embryo sac, with wall cells trisomic for chromosome 7, as revealed by Giemsa banding. By 1977 (de Grouchy and Turleau), complete trisomies for chromosomes 5, 12, and 17 had not been reported even in abortuses; for chromosomes 3 and 19 only partial trisomies appear to have been observed. A trisomic for one of the C-group of autosomes is shown in Figure 14–17. Inasmuch as the karyotype shown in that figure antedates

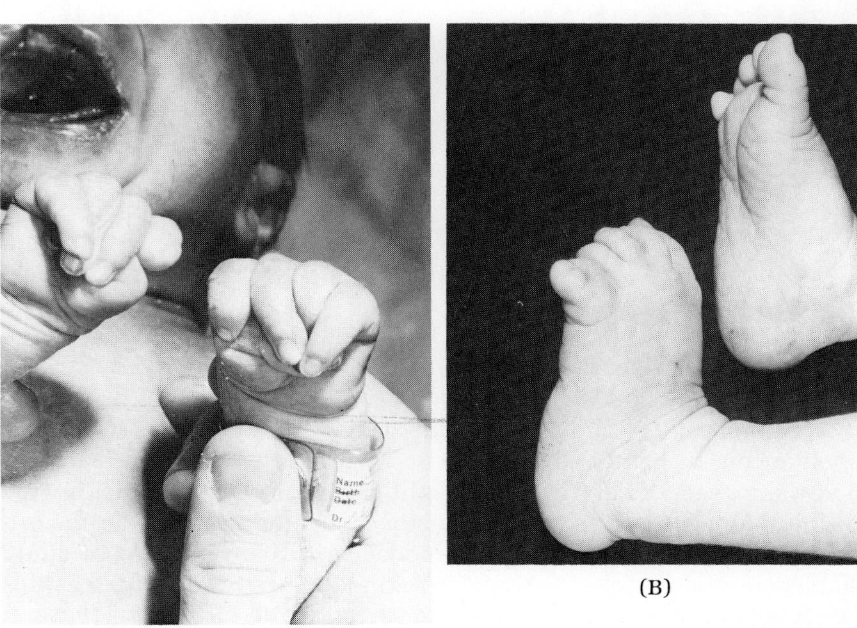

(A) (B)

FIGURE 14–16. *Patau's syndrome (trisomy-13) in a live-born infant. (A) Clenched fists. (B) Deformities of feet. See text for details.* (Photos courtesy Dr. Richard C. Juberg, used with permission.)

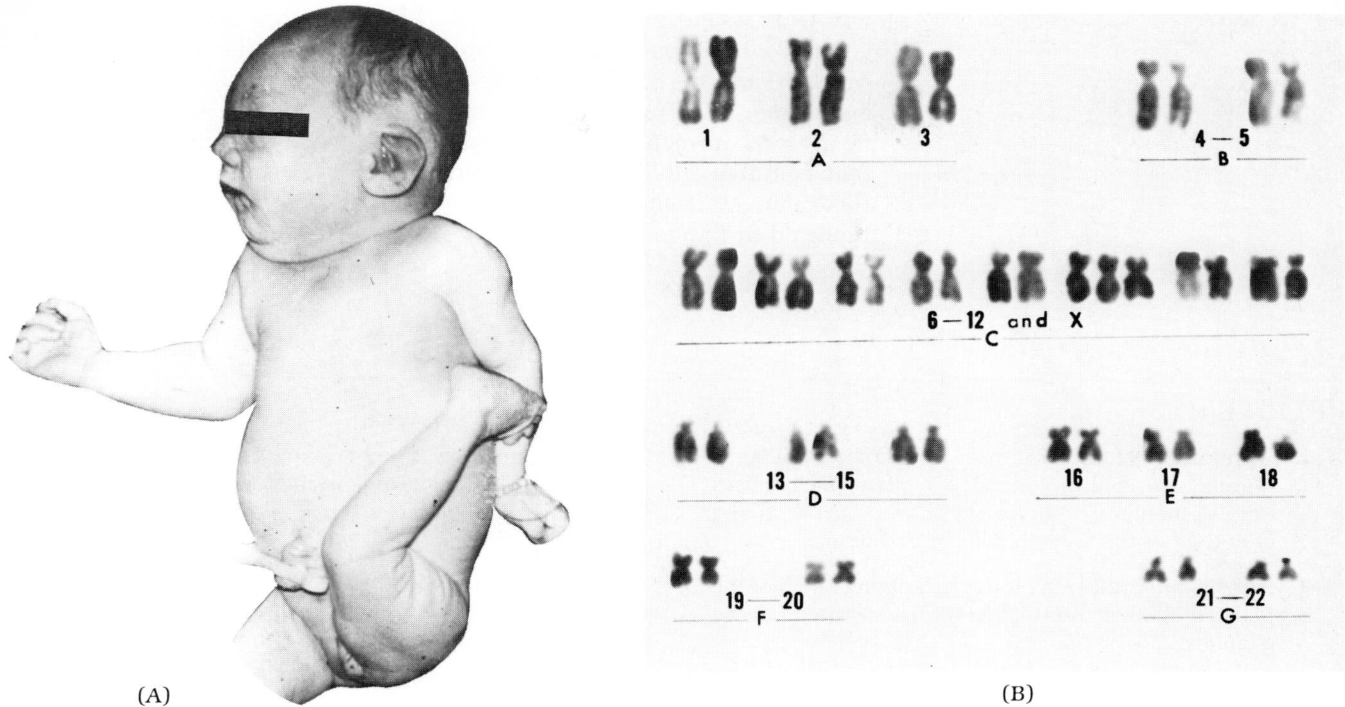

(A) (B)

FIGURE 14–17. *C-trisomy in a female. (A) External features; note ears, jaw, hands and leg position. (B) Karyotype of same child showing 17 chromosomes in the C, X group instead of the normal 16, for a total of 47.* (From R. C. Juberg et al. 1970, used with permission.)

development of banding techniques, no method was available for more precise identification of the extra chromosome.

In a study of a very large number of trisomic spontaneous abortions, Hassold and colleagues (1984) found that 17 of 21 different trisomies had higher mean maternal ages than for normal newborns. For thirteen of the seventeen (chromosomes 2, 7, 9, 10, 13, 14, 15, 16, 17, 18, 20, 21, and 22) the increase in maternal age was statistically significant. The higher maternal ages were found for trisomies involving the small chromosomes. Trisomies 3, 5, 6, 11, and 12 were not associated with statistically significant higher maternal ages, though in these studies fewer than 15 cases were found for each. However, elevated (although not statistically significant) maternal age was found for trisomies 3 and 11. Thus it is very likely that trisomy is associated with increased maternal age in all human cases.

Trisomy in Nonhumans

That trisomy, with phenotypic features similar to trisomic Down humans, may also occur in animals is clear from the work of McClure and colleagues (1969), which described a trisomic young female chimpanzee. This animal displayed clinical and behavioral characteristics similar to those of human Down syndrome cases with scores below normal in a variety of behavioral and postural tests. Chimpanzees have a diploid chromosome number of 48 (Table 4–1) but in the animal in question most of the blood cells examined had 49, with an extra small acrocentric chromosome that matched pair 22 (the second smallest of the autosomal pairs). This same chromosome was identifiable in triplicate even in the few cells where only a total of 46, 47, or 48 chromosomes could be counted. Both parents were cytologically and behaviorally normal. McClure and his colleagues note that "a comparable condition has not been reported in nonhuman primates. The occurrence of this condition in a lower primate again emphasizes the close phylogenetic relation between man and the great apes and may provide a model for studying this relatively frequent human syndrome." Trisomy-21 has also been reported in the gorilla.

PROBLEMS

14–1 Application of colchicine to a vegetative bud of a homozygous tall diploid tomato plant (*DD*) causes development of a tetraploid branch. What is the genotype of the somatic cells of this branch?

14–2 Flowers are produced on the tetraploid branch of the plant in Problem 14–1. What is the genotype of the gametes?

14–3 Pollinating one of the flowers of Problem 14–2 with pollen from a diploid dwarf plant produces embryos of what genotype?

14–4 If the plant in Problem 14–1 had been heterozygous tall (*Dd*), what would be the genotype of the somatic cells of the tetraploid branch?

14–5 Give the gamete genotypes produced on the tetraploid branch of Problem 14–4. In what ratio are they produced?

14–6 (a) Self-pollinating flowers on one of the tetraploid branches referred to in Problem 14–4 would produce embryos with what specific kind of ploidy? (b) What is the probability of a dwarf plant in the progeny?

14–7 If an autotetraploid plant heterozygous for two pairs of genes, for example, of genotype *AAaaBBbb*, is self-pollinated, what is the probability of an *aaaabbbb* plant in the progeny?

14–8 If an autotetraploid plant of genotype *AAaaBBbb CCcc* is self-pollinated, what is the probability of an *aaaabbbbcccc* progeny individual?

14–9 If an autotetraploid plant, heterozygous for n pairs of genes (that is, *AAaaBBbbCCcc* . . . and so on for n pairs of alleles) is self-pollinated, what mathematical expression would you solve to determine the probability of a completely recessive progeny individual?

14–10 If only pairs of alleles that show complete dominance are considered, what is the effect of tetraploidy on phenotypic variability?

14–11 A number of species of the birch tree have a somatic chromosome number of 28. The paper birch (*Betula papyrifera*) is reported to occur with several different chromosome numbers. Individuals with the somatic numbers 56, 70, and 84 are known. With regard to chromosome number, how should the 28, 56, 70, and 84 chromosome individuals be designated?

14–12 The sugar maple (*Acer saccharum*) and the box elder (*Acer negundo*) each have diploid chromosome numbers of 26. Note that they are different species of the same genus. However, hybrids between the two are sterile. What explanation can you offer?

14–13 Should it be possible to secure a fertile hybrid of the cross sugar maple × box elder? How?

14–14 Different species of rhododendron have somatic chromosome numbers of 26, 39, 52, 78, 104, and 156. By what means does evolution appear to be taking place in this genus?

14–15 What appears to be the basic haploid chromosome number in rhododendrons?

14–16 How many sets are represented in the species with 156 chromosomes?

14–17 Referring back to Table 14–1, what was the sex phenotype of the triploid child reported by (a) Schindler and Mikamo, (b) Butler et al.?

14–18 If polyspermy were demonstrated to occur in human beings, what would be the chromosomal complement of an embryo resulting from polyspermy involving an X and a Y sperm?

14–19 Suppose the first mitosis in a normal 46,XY zygote were abnormal so that it became a tetraploid cell. (a) What would be the chromosomal complement of that tetraploid cell? (b) Do you find any evidence in this chapter or its references to suggest that a viable child would result?

14–20 In the Jimson weed what gamete ratios are produced by (a) *Ppp* ♀, (b) *Ppp* ♂, (c) *PPP* ♀, (d) *PP* ♂, (e) *PPp* ♂?

14–21 What is the F_1 phenotypic ratio produced in Jimson weed by crossing (a) purple *Ppp* ♀ × purple *PPp* ♂, (b) *PPp* ♀ × *Pp* ♂, (c) *Ppp* ♀ × *Ppp* ♂?

14–22 "Eyeless" (eyes small or absent) is a recessive character whose gene is located on the small chromosome IV of *Drosophila melanogaster*. Both triplo-IV eggs and triplo-IV sperms are functional. The dominant gene for normal eyes is designated +, the recessive for eyeless *ey*. What is the F_1 phenotypic ratio produced by each of the following crosses: (a) + *ey ey* × + *ey ey*, (b) + + *ey* × + *ey ey*, (c) + + *ey* × + + *ey*?

14–23 The Jimson weed (*Datura*) has 12 pairs of chromosomes, and 12 different viable trisomics are known. How would you explain the fact that the only trisomies reported for human *live* births involve chromosomes 8, 9, 13, 14, 18, 21, 11, X, and Y and that only trisomy 8, X, and Y appear not to reduce life expectancy measurably?

14–24 In a hypothetical study of 5,000 spontaneously aborted human trisomics, none for chromosome 1 is found. Suggest a logical positive reason.

14–25 Describe three ways that a nonhaploid cell can have an odd number of chromosomes.

CHROMOSOMAL ABERRATIONS: STRUCTURAL CHANGES

CHAPTER

15

For all their complexities of structural organization, chromosomes are far from indestructible. Through such agents as radiation and chemicals, breakage can occur, which may result in genetic damage to subsequent generations. Suitable precautions must be taken in X-ray diagnoses, and nuclear weaponry constitutes a definite hazard. A study of 43 Marshall Islanders ten years after accidental exposure to radioactive fallout following testing of a high-yield nuclear device at Bikini disclosed chromosomal breaks in 23 of 43 persons. A wide variety of chemical substances, some formerly or presently used by humans, have been implicated or suspected of inducing chromosomal change.

Breaks may also occur "naturally," for no assignable cause. Any of several cytological and genetic consequences may result. Breaks may be detected in almost any dividing cell, though certain kinds offer special advantage. Particularly useful in this regard are the salivary glands of dipteran insects such as *Drosophila*, even though these cells do not divide. Genetic effects of chromosomal damage or aberrant behavior are detectable as unexpected phenotypes, gross physical malformations, altered enzyme production, altered linkage relationships, reduced fertility (or complete infertility), and an increase in spontaneous abortions.

SALIVARY GLAND CHROMOSOMES

Structural changes in chromosomes that result from breakage are most profitably studied in (1) salivary gland chromosomes of dipteran insects (such as *Drosophila*) and (2) meiocytes during the process of meiosis. Because of their large size and banded structure, salivary gland chromosomes are particularly appropriate for such studies.

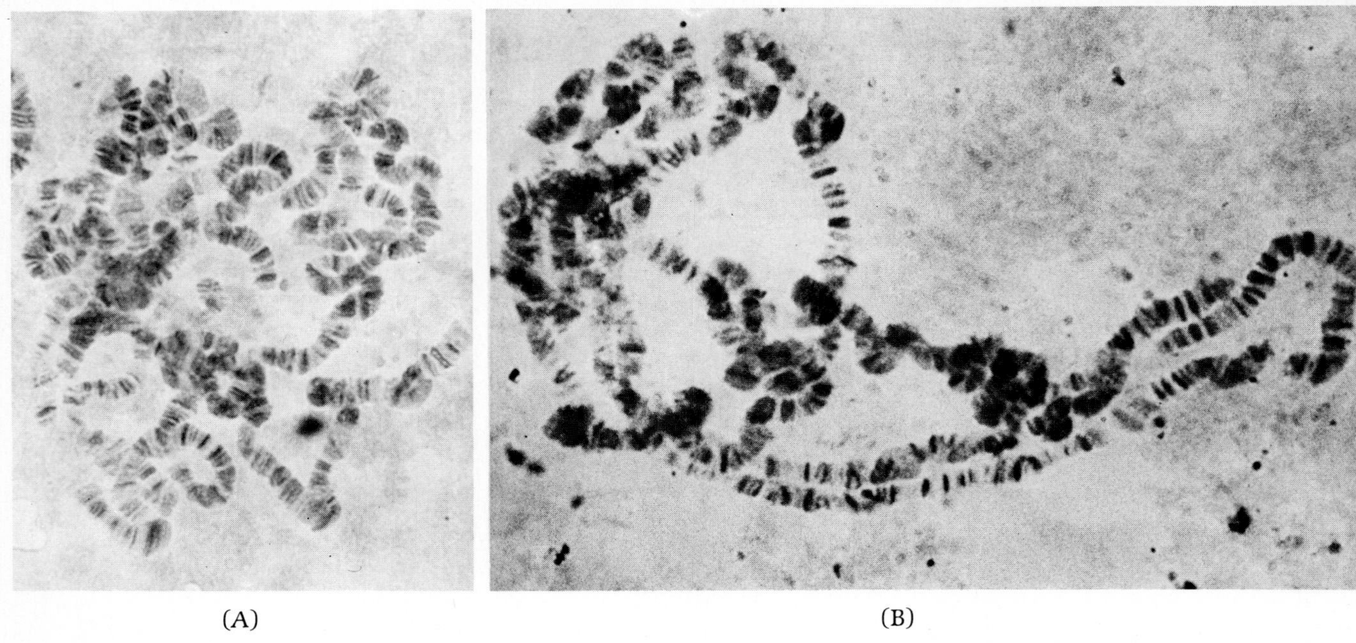

(A) (B)

Figure 15–1. *Salivary gland chromosomes of* Drosophila melanogaster *showing the banding that permits rather accurate cytological mapping. (A) Entire complement.* (Courtesy Miss Chris Arn.) *(B) Partial complement, enlarged.* (Courtesy Mr. Jon Derr.) (Photos used by permission.)

Nuclei of the salivary gland cells of the larvae of dipterans like *Drosophila* have unusually long and wide chromosomes, 100 to 200 times the size of the chromosomes in meiosis or mitosis of the same species (Figure 15–1). This is particularly surprising, because, as noted previously, the salivary gland cells do not divide after the glands are formed, yet their chromosomes replicate and become unusually long (replication is reported to occur 10 times, producing 1,024 side-by-side replicates of each chromosome). Moreover, these chromosomes are marked by numerous cross-bands that are apparent even in unstained nuclei. These bands are quite constant in size and spacing for a given normal chromosome. The multiple chromosomes appear to be in a perpetual prophase and are synapsed. Thus any difference in banding between homologs can be easily compared.

TYPES OF STRUCTURAL CHANGES

A chromosomal break may or may not be followed by a "repair." If segments are rejoined in the original configuration, the break normally passes undetected. Should such repair not be effected, one or more of the aberrations listed in Table 15–1 occur. Note that a deletion in one chromosome may be accompanied by a translocation or a duplication in another. Reciprocal translocations may also occur in which two nonhomologs *exchange* segments that are often unequal in length.

Deletions

Nonhumans. Perhaps the simplest result of breakage is the loss of a part of a chromosome. Portions of chromosomes without a centromere (*acentric fragments*) lag in anaphase movement and are lost from reorganizing nuclei. Such loss of a portion of a chromosome is called **deletion.**

TABLE 15–1 Types of aberrations produced by chromosomal breaks

Type	*Description*	*Gene changes*
Normal	(*ABCDEFGH*)	None
Deletion	No rejoining; chromosomal segment lost	*ABFGH, CDEFGH*, etc.
Inversion	Broken segment reattached to original chromosome in reverse order	*ABFEDCGH*, etc.
Duplication	Broken segment becomes attached to homolog that has experienced a break; homolog then bears one block of genes in duplicate	*ABCDEFGEFGH*
Translocation	Broken segment becomes attached to a nonhomolog resulting in new linkage relations	*LMNOPQRCDEFGH*, etc.

Deletions often make possible *cytological mapping* of chromosomes. For example, if large numbers of fruit flies of autosomal genotype *ABCDEF . . ./ABCDEF . . .* (that is, homozygous dominant for several traits) are subjected to X-irradiation, a few of the individuals may suffer a break and deletion in one of the chromosomes bearing the genes. If a deletion for genes *C* and *D* occurs in one primary spermatocyte, then two of the four sperms produced from it would have an *intercalary deletion* in this particular chromosome:

Primary spermatocyte (irradiated)	*AB—EF . . .* *ABCDEF . . .*
Secondary spermatocytes	*ABEF . . . , ABCDEF . . .*
Spermatids and sperms	*ABEF . . . , ABEF . . . , ABCDEF . . . , ABCDEF . . .*

Mating such males to recessive females, *abcdef . . . /abcdef . . .* will produce offspring, some having the genotype *abcdef . . . /ABEF*. Therefore, these flies will express the recessive phenotype for genes *c* and *d*, whereas those receiving an *ABCDEF . . .* sperm from the male parent will express the dominant phenotype for all six genes. The expression of genes *c* and *d*, which would have been obscured by genes *C* and *D* if these were present, is called **pseudodominance.** This is really not a good term, but it is a useful one and is firmly established in the literature.

Next, flies showing pseudodominance are mated with normal *abcdef . . . /abcdef . . .* individuals. Half the offspring will now exhibit pseudodominance for the same genes. Examination of the salivary gland chromosomes of larvae of this cross will quickly disclose those individuals that are *abcdef . . . /ABEF*. Remember, these chromosomes are banded and pair exactly, band for band. The segment that includes genes *C* and *D* will also include some of the bands, and the giant chromosomes of *abcdef . . . /ABEF . . .* will appear somewhat as shown in Figure 15–2 where the normal, unaltered one shows a characteristic **deletion loop.**

Another strain of flies with an *overlapping deficiency* such as *ABC—F . . ./ABCDEF . . .* might be developed and mated as in the preceding case. Giant chromosomes of *abcdef . . ./ABC-F . . .* individuals will again display the characteristic deletion loop. Comparison with the loops pro-

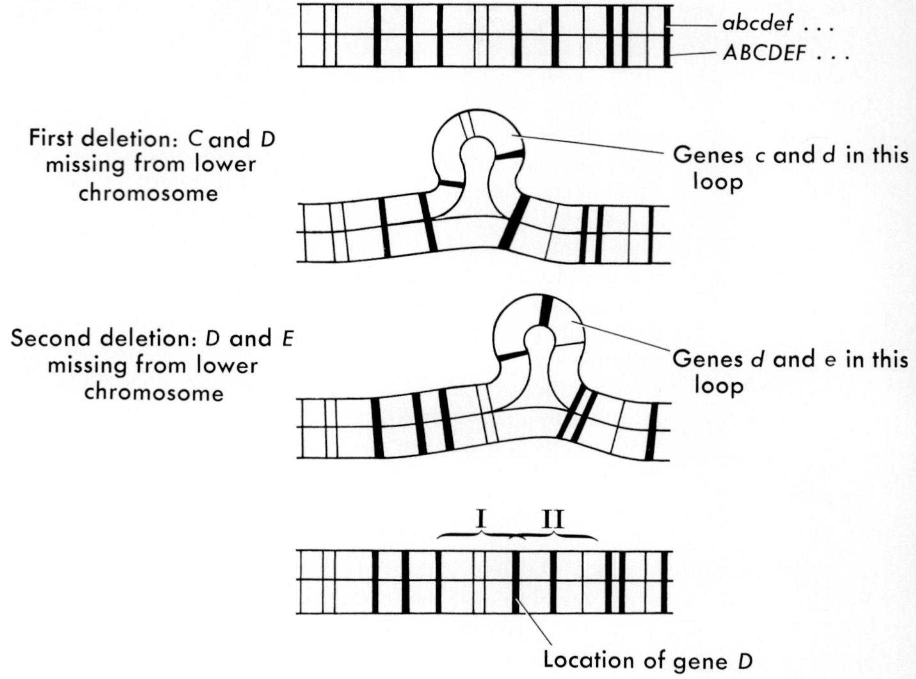

FIGURE 15–2. *Determining which band of salivary gland chromosomes bears a given gene is done by a series of overlapping deficiencies. (See text for details.)*

duced by the *abcdef . . ./ABEF . . .* individuals permits easy detection of the precise chromosomal region that bears gene *D*; its locus can be seen to be associated with a particular band (Figure 15–2).

Detailed *cytological maps* of *Drosophila* have been prepared in just this way. The geneticist now has visible bands within which genes appear to be located, and their distances can be measured in ordinary units. Comparison of cytological and genetic maps confirms the linear sequence given by the latter. Cytological maps, however, show known genes to be more uniformly distributed over the chromosome than is suggested by the genetic map. This results from the greater degree of interference near the centromere and termini of the chromosome. For example, in chromosome II of *Drosophila* the centromere is at locus 55.0. Painter's map (Figure 6–4) shows gene *pr* (purple eyes) to be at locus 54.5, just 0.5 unit to the left of the centromere. This is about 1 percent of the 55 map units between the centromere and the end of the left arm. Yet on the salivary gland map, gene *pr* is about 12 map units to the left of the centromere, or nearly 22 percent of the total distance involved. Figure 15–3 compares genetic and cytological map locations for several genes of chromosome II.

Humans. Probably the best-known disorder to be associated definitely with a deletion in humans is the *cri du chat,* or cat-cry syndrome described by Lejeune and colleagues in 1963 and based on three cases. Manifestations include a characteristic high-pitched, plaintive cry, very

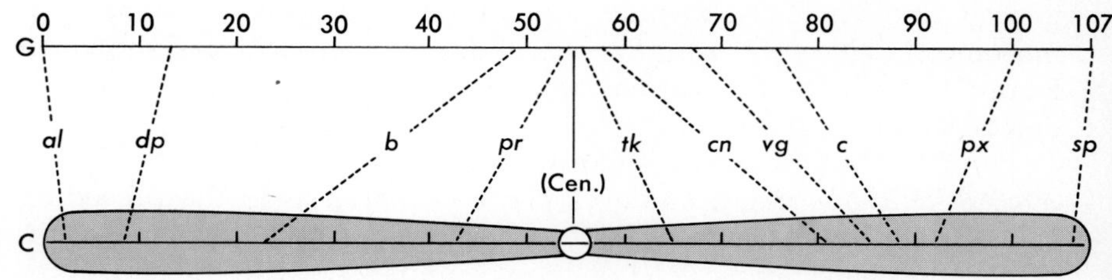

FIGURE 15–3. *Diagrammatic comparison of genetic map (G) and cytological map (C) of chromosome II of* Drosophila melanogaster. *Note particularly the much closer spacing of genes near the centromere on the genetic map, as opposed to the distances between them on the cytological map. This reflects the interference effect of the centromere (cen = centromere).*

similar to that of a kitten in distress. de Grouchy and Turleau (1977) report that acoustic recordings of *cri du chat* infants and of a kitten yield identical tracings. This symptom is associated with a malformation of the larynx. Other characteristic malformations involve the head and face. Mental retardation is severe, with intelligence quotient usually below 20. Other nonspecific malformations and dysfunctions involving the brain, heart, eyes, kidneys, and skeleton may also occur. At least 120 cases have been studied since the initial description. The condition appears not to be significantly life-threatening; many affected individuals attain adulthood.

The karyotype is unvarying; in each case a large part of the smaller arm of one of the number 5 chromosomes is deleted (Figure 15–4). The standard designation for the shorter arm of a nonmetacentric chromosome is *p*; that for the longer arm is *q*. A deletion is indicated by a

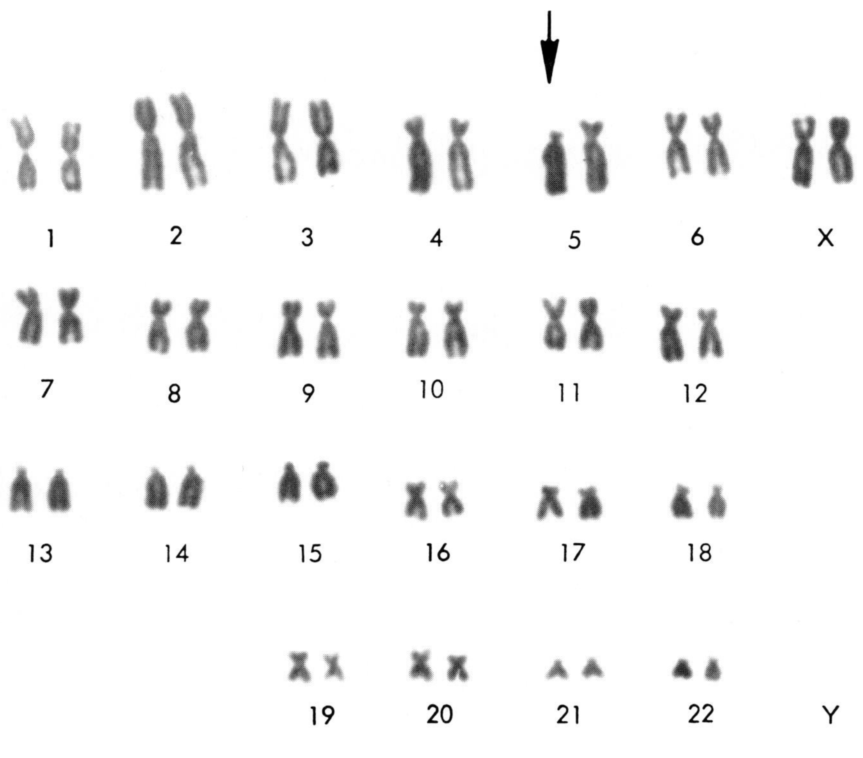

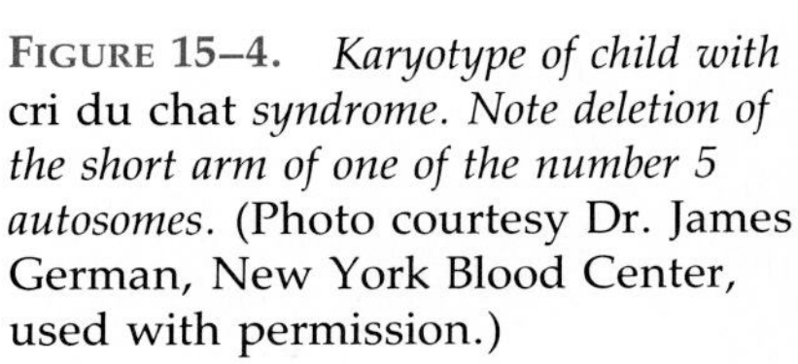
FIGURE 15–4. *Karyotype of child with* cri du chat *syndrome. Note deletion of the short arm of one of the number 5 autosomes.* (Photo courtesy Dr. James German, New York Blood Center, used with permission.)

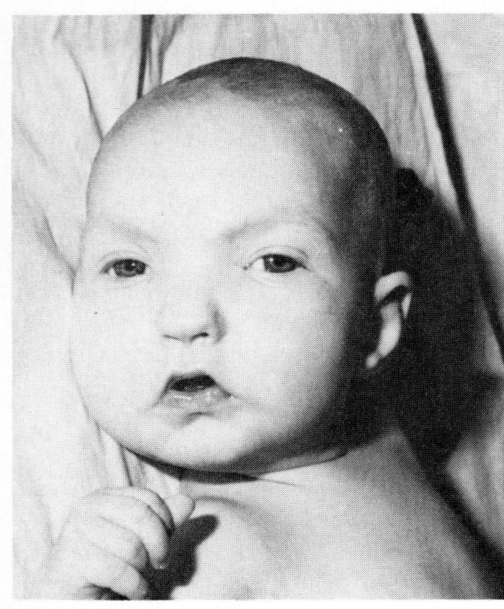

(A)

FIGURE 15–5. *(A) Infant with a deletion of the short arm of one of the number 4 autosomes. The syndrome is described in the text. (B) Karyotype of the child in (A). Note missing segment of the short arm of one of the number 4 autosomes (arrow).* (Photos courtesy Dr. James German, New York Blood Center, used by permission.)

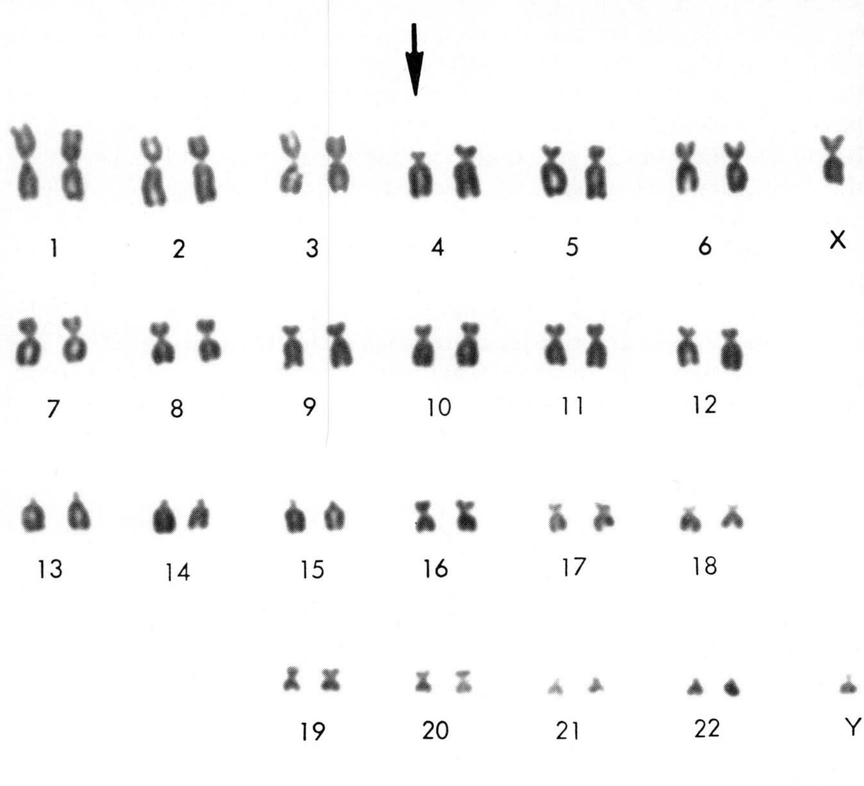

(B)

superscript minus sign, and added segments are indicated by a superscript plus sign. Hence the karyotype of a *cri du chat* patient is $5p^-$. Cytological studies indicate association of major symptoms with deletion of a small portion of band 5*p*14*p*15. In most, if not all cases, the deletion arises during gametogenesis in one parent or the other; both parents of affected children present normal karyotypes. No parental age effect is discernible.

Several other deletions in human beings are well known, and new ones are reported in the literature from time to time. For example, a male infant was found that had a deletion of the short arm of a number 4 chromosome ($4p^-$). Such a child is shown in Figure 15–5A, and its karyotype is shown in Figure 15–5B. The child was described as "unusually small, [with] severe psychomotor retardation, convulsions, a wide flat nasal bridge, prominent forehead, cleft palate, and congenital heart disease." Some mentally defective patients have been found to have other chromosomal deletions; de Grouchy, for example, reported one such child who lacked the short arm of one of its number 18 chromosomes. Notice that, in most departures from the normal human karyotype, the individual is abnormal in one or more aspects.

Deletion of part of the short arm of one of the X chromosomes produces a typical Turner syndrome (see Chapter 7). Deletion of part of the long arm of an X, on the other hand, results in an atypical, Turnerlike syndrome. The latter individuals are of normal height, but have vestigial gonads and are sterile. No other anomalies are consistently associated with this cytological condition.

A computerized technique, termed *automated fast-scanning microscope photometry,* was recently used to identify a rare interstitial deletion of 4*q*. A female infant was born with a number of malformations including large head with wide sutures, broad chest with short sternum, and general shortening of all limbs and digits. Internally there was malrotation of the gut, kidneys showing degeneration, and enlarged liver. The infant died at three months following respiratory arrest. The results of the computerized scan of G-banded chromosomes are found in Figure 15–6. The scan shows a small portion (12 percent) deleted from 4*q*.

Increasing use is being made of cultured amniotic fluid cells obtained by amniocentesis (see Chapter 20) to determine fetal karyotypes. Amniotic cells appear to show a rather high frequency of tetraploidy, unaccompanied by any cytological abnormality of the fetus, but structural aberrations, trisomy, and the like are fairly easy to detect from these cells around the sixteenth week of gestation. This information can be extremely useful to parents and physician in considering a possible termination of the pregnancy.

Translocations

Nonhumans. **Translocations** involve the shift of part of one chromosome to another nonhomologous chromosome. Two major kinds of translocations are recognized:

1. *Simple,* in which, following breaks, a segment of one chromosome is transferred to another nonhomologous chromosome, where it occupies an intercalary location.
2. *Reciprocal* (interchange), in which segments, which need not be of the same size, are exchanged between nonhomologous chromosomes.

Reciprocal translocations are well known in animals and plants that have been studied extensively. They have been produced frequently in *Drosophila* (as well as in other organisms) by radiation and have occurred widely in natural populations of the evening primrose (*Oenothera*), the plant whose resulting phenotypic variability led DeVries to formulate his mutation theory of evolution. Simple translocations have been found in humans.

Two reciprocal translocation types are recognized: homozygous and heterozygous (Figure 15–7). The former may have normal meiosis and, in fact, be difficult to detect cytologically unless morphologically dissimilar chromosomes are involved, or banding patterns differ markedly.

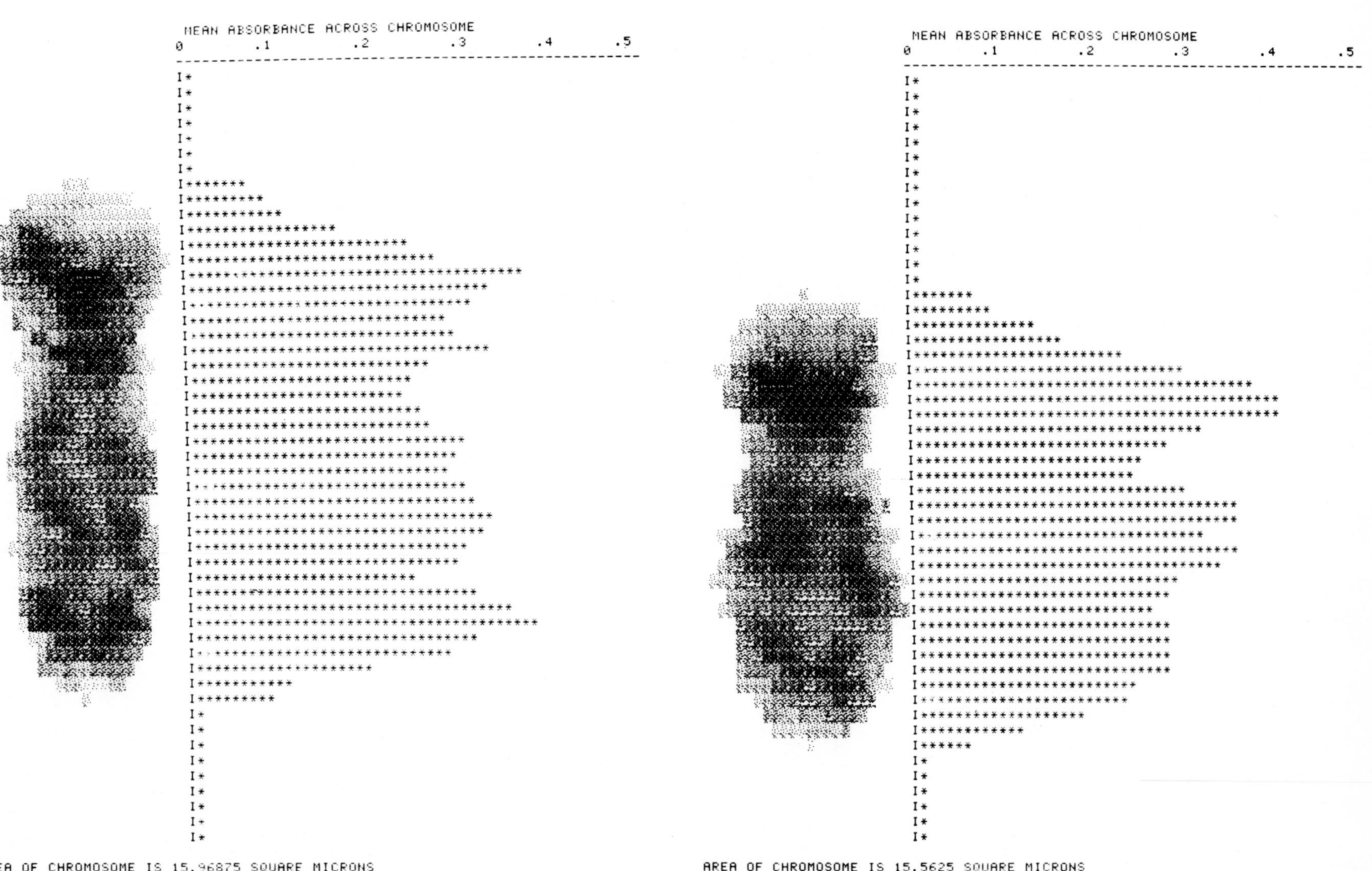

FIGURE 15–6. *Computerized scan of G-banded chromosome number 4 showing the normal (left) and (deleted) chromosome* $4q^-$. *(See text for details on phenotype of individual.)*

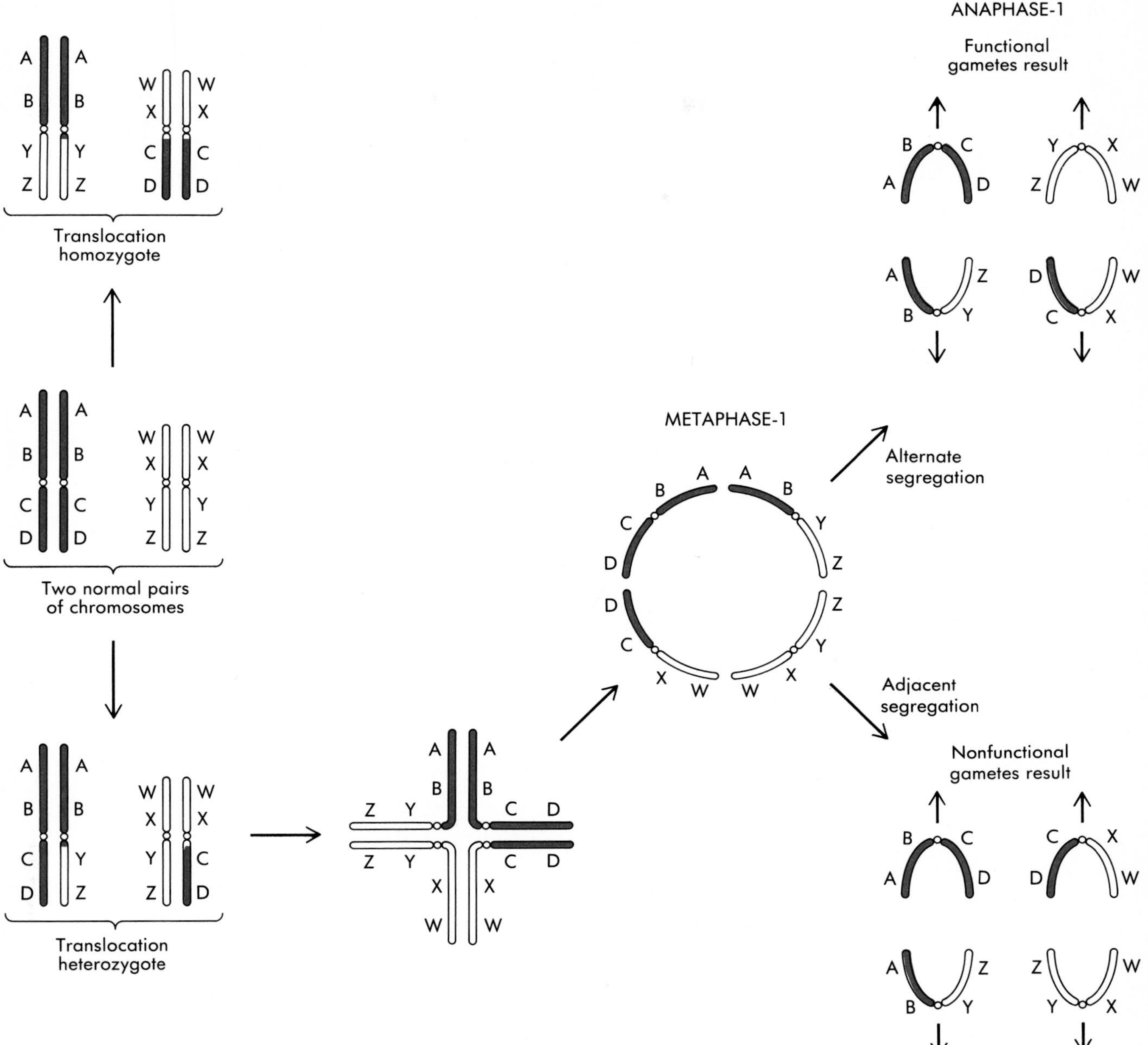

FIGURE 15–7. *Two types of reciprocal translocation. Homozygous, which has a normal meiosis, and heterozygous, which leads to the typical cross-shaped configuration at prophase I, and the ring of four chromosomes at metaphase I. Depending on whether alternate or adjacent segregation occurs from the ring at anaphase I, functional or sterile gametes will result. The translocation heterozygote, therefore, is semisterile.*

Reciprocal translocations are characterized genetically by altered linkage groups and by the fact that a gene with "new neighbors" may produce a somewhat different effect in its new location. Such a **position effect** will be examined shortly.

Translocation heterozygotes, however, are marked by a considerable degree of meiotic irregularity. Peculiar and characteristic formations occur at synapsis because of the difficulty of attaining a pairing of homologous parts. Typically, a cross-shaped formation is seen in prophase-I; this often opens out into a ring (Figures 15–7, 15–8, 15–9). In some cases a ring of chromosomes becomes twisted, with the result that (1) both normal chromosomes move to one pole, and (2) both translocation chromosomes move to the other pole at anaphase-I. In this case functional meiospores (in higher plants) or gametes (animals) will be produced because the meiotic products will have a full gene comple-

FIGURE 15–8. *Ring chromosomes in the tropical plant* Rhoeo discolor. *(A) Ring of 12 chromosomes. (B) Another ring of several chromosomes. Rings form in translocation heterozygotes as a result of terminalization of chiasmata.*

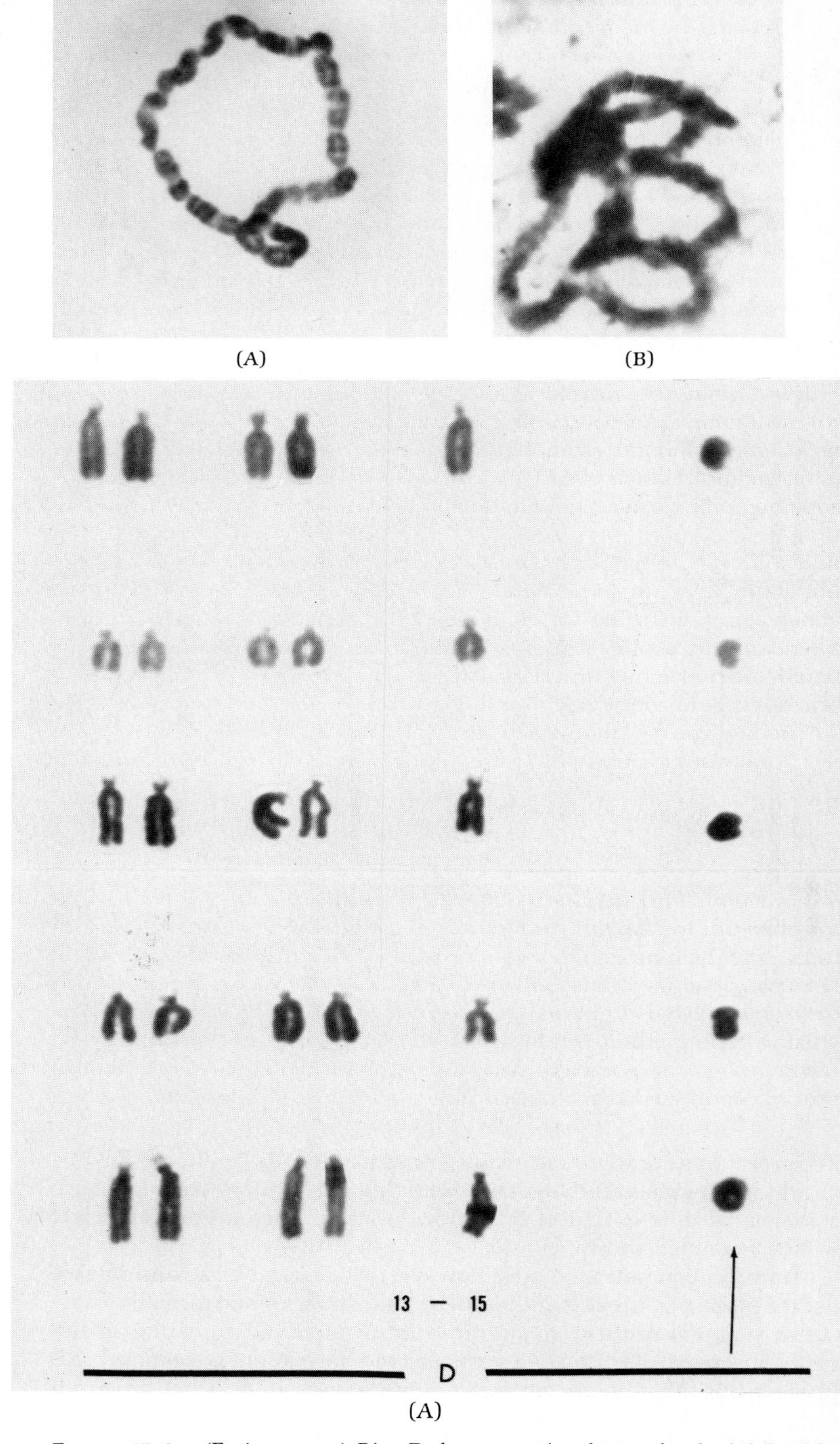

FIGURE 15–9. (Facing pages) *Ring-D chromosome in a human female. (A) Partial karyotype of the D group from five different metaphases showing ring chromosomes at the right. (B) Phenotypic abnormalities associated with the ring-D chromosome.* (Courtesy Dr. Richard C. Juberg, 1969, used by permission.)

ment. This is **alternate segregation** (Figure 15–7). But, if one each of (1) translocation and (2) normal chromosomes move to opposite poles, deletions and duplications will occur in the meiotic products. Meiospores and gametes will be nonfunctional in each case where such **adjacent segregation** has taken place (Figure 15–7). Semisterility resulting from reciprocal translocations is easily observed in such plants as corn. Ears lack about half the kernels, and these are arranged irregularly (Figure 15–10). Abortive pollen is reduced in size.

FIGURE 15–10 (Above) *Partial sterility results from a reduction in functional gametes caused by multiple translocations.* (Courtesy DeKalb Agricultural Association, Inc., DeKalb, Illinois.)

Humans. Translocations in addition to the 14/21 associated with Down syndrome (Chapter 14) have been reported in a few instances for human beings. Those most likely to receive attention involve chromosomal segments present in triplicate. Among other translocations, groups D and G, B-D, and C and E (see Figure 4–15) have been detected. When members of the several groups of autosomes are not easily differentiated, translocations in humans are designated by two letters separated by a slash; the first letter indicates the chromosome or chromosome group that supplied the centromere, and the second indicates the group or chromosome that received the transferred segment. In a reciprocal translocation, one can thus identify both components, for example, 8/14 and 14/8. When break points are right (or perhaps in) in the centromere an arbitrary choice must be made. The 14/21 in Down syndrome is a special case, because only one component (referred to as 14/21) persists. Presumably, the tiny 21/14 was lost generations before.

In most cases there are numerous phenotypic abnormalities, and

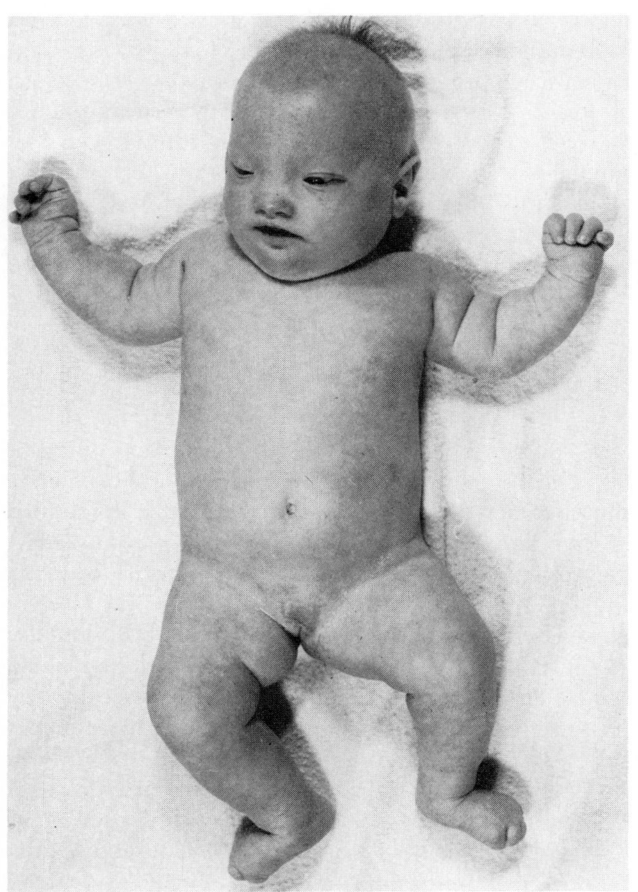

(B)

spontaneous abortion, or death within a few months after birth, usually ensues. Those who do live longer almost always exhibit mental retardation. A probable unbalanced translocation has been reported in which a large part of one of the D-group chromosomes was present in triplicate and a small part of one of the B chromosomes was lacking. There were numerous anatomical abnormalities, and death of the subject occurred at seven months. Forty-six chromosomes were present, including an unusually long one believed to be a B + D. Both parents were cytologically normal.

A C/E translocation with effects resembling Down syndrome was apparently first detected in 1969 at a Boston hospital. The child had one extra long chromosome in its complement of 46, interpreted as an E with added material from a C-group member, with material thus present in triplicate. This diagnosis was made on the basis of the mother's karyotype, which showed one shortened member of group C (believed to be number 8) and a longer than normal E chromosome (probably number 18). The mother was phenotypically normal, as would be expected, because she had the normal amount of chromosomal material.

Patients with chronic myelocytic (myologenous) leukemia display an interesting chromosome abnormality. In the bone marrow and in cells derived from it, a short chromosome, called the Philadelphia (Ph^1) chromosome (after the city where the case was first discovered) is present. Ph^1 persists in bone marrow, and in cells derived from it, even during remission. Detailed cytological study discloses this to be a number 22 chromosome that has lost most of the distal part of its longer arm ($22q^-$). The deleted part of autosome 22 is *translocated* to one of the other autosomes, preferentially to the distal end of $9q$.

Some early confusion over whether 21 or 22 is involved was derived from the morphological similarity between numbers 21 and 22 (see Figure 4–15 for similarity). Moreover, it has not been possible to explain the changed behavior of cells containing the Philadelphia chromosome. The translocation appears to involve no change in the total amount of genetic material. It may be a *position effect* in that genes on $22q$ normally interact with each other, and adjacent ones on an unaltered 22, but cannot do so when the latter genes are translocated so that they are then associated with an entirely different block of genes on chromosome 9. Ph^1 does not appear in gametes because it is not transmitted to offspring of persons with this chromosomal abnormality.

Inversions

Inversions develop most likely when breaks occur at a place where a chromosome forms a tight loop during synapsis. An inversion heterozygote, in which only one of two homologs has a particular inversion, forms a characteristic **inversion loop** in prophase-I of meiosis. With a little care, differences between deletion loops and inversion loops can be distinguished easily.

An inversion requires two breaks in a chromosome, followed by reinsertion of the segment in the opposite direction. A particular block of genes thus occurs in reverse sequence (Table 15–1). Inversions may either include the centromere (*pericentric inversions*) or not include the centromere (*paracentric inversions*). Homologous chromosomes, with identical inversions in each member, pair and undergo normal distribution in meiosis. If a crossover forms within a paracentric inversion, for example, a dicentric bridge is produced at anaphase-I, which results in the loss of an acentric fragment (Figure 15–11). The bridge itself breaks as anaphase-I progresses, and this results in additional aberra-

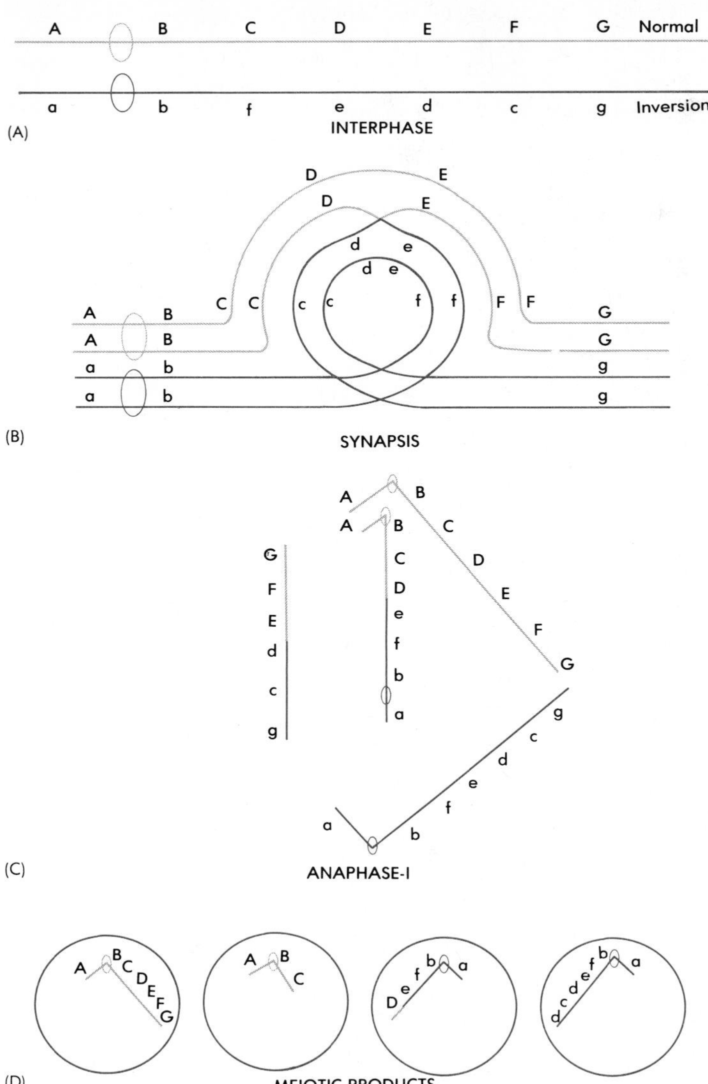

FIGURE 15–11. *Crossing-over within the inversion loop of a paracentric inversion decreases the frequency of functional, chromosomally normal gametes. (A) One pair of chromosomes, one member of which has previously sustained an inversion involving loci* c, d, e, *and* f. *(B) Synapsis of homologous chromosomes showing chromatids and a cross-over between nonsister chromatids within the inversion loop. (C) Anaphase-I. Two normal daughter chromosomes* (ABCDEFG *and* (abcdefg) *one dicentric* (ABCDefba) *and one acentric* (GFEdcg). *The acentric is lost, either in meiosis or in an early subsequent mitosis; the dicentric often breaks (if the centromeres move in opposite directions) randomly. (D) The four meiotic products (for example, meiospores or pollen mother cells in flowering plants or spermatids or ovum and polar bodies) with two normal nuclei (left and right) plus two abnormal nuclei (center).*

tions such as deletions or duplications. Although inversions have been referred to as "crossover suppressors," such a term is inaccurate because crossing-over itself is not suppressed but, rather, *crossover products* are eliminated. Comparison of banded karyotypes of humans and the apes reveals numerous paracentric as well as pericentric inversions in humans as compared to apes.

Duplications and the Position Effect

Duplications occur when a portion of a chromosome is represented more than twice in a normally diploid cell. The extra segment may be attached either to the chromosome whose loci are repeated or to a different linkage group, or it may even be present as a separate fragment. Duplications are useful in studying the quantitative effects of genes normally present only in pairs in diploid cells.

The first duplication to receive critical study was the *bar eye* variant in *Drosophila*. The wild-type eye is essentially oval in shape; the *bar eye* phenotype is characterized by a narrower, oblong, bar-shaped eye with fewer facets. The classical studies of Bridges (1936) showed this trait to

TABLE 15–2. **Comparison of Genotypes and Phenotypes for Bar Eye in *Drosophila* Females**

X chromosomes	*Phenotype*	*Mean number of facets*
A/A	Normal	779
AA/A	Heterozygous bar eye	358
AA/AA	Homozygous bar eye	68
AAA/A	Heterozygous ultrabar	45
AAA/AAA	Homozygous ultrabar	25

A = One section 16A of the X chromosome.

be associated with the duplication of a segment of the X chromosome, called section 16A, as observed in salivary gland chromosomes. Each added section 16A intensifies the bar phenotype. However, the narrowing effect is greater if the duplicated segments are on the same chromosome. If we let A represent one section 16A in a given X chromosome, we can recognize the genotypes and phenotypes listed in Table 15–2. Other arrangements are also possible, but these show clearly that the bar effect of a given number of duplicated 16A sections is intensified if the duplications occur in one X chromosome rather than being divided between the two of the female. Compare heterozygous ultrabar and homozygous bar eyes, for example.

Bar eye is another example of a trait that shows a *position effect*. In bar eye, each added segment narrows the eyes still farther, and this effect is enhanced as more duplications occur in one chromosome. Other duplications are known that produce the opposite effect, counteracting the effect of mutant genes. Moreover, duplications need not always be immediately adjacent to exert this position effect.

Chemical Causes of Chromosomal Aberrations in Humans

Many substances, among them LSD, marijuana, nicotine, cyclamates, DDT, caffeine, asbestos, and some former hair dye ingredients, have been suspected in chromosome damage in humans. These results parallel those with environmental mutagens discussed in more detail in Chapter 13.

In cultured cells of certain individuals specific sites on the chromosomes fail to stain and give the appearance of a constriction or gap. These have been termed **fragile sites.** These sites appear to be places on chromosomes susceptible to *chromosome breakage.* They are not present in normal preparations but must be induced by altering the medium on which the cells are being cultured. There are 17 fragile sites mapped in the human genome that occur in metaphases at exactly the same points in any given individual. The most common group of fragile sites are induced by culture medium low in folic acid and thus are called *folate-sensitive sites.* Nothing is known about the molecular structure of a fragile site. However, the folate-sensitive sites are inherited as Mendelian codominant traits indicating some unique DNA sequence at the fragile site.

Even though fragile sites are assumed to be sites of chromosome breakage, so far only one human disorder, the **fragile X syndrome,** has been associated with the occurrence of a fragile site. The syndrome affects nearly 1 in 2,000 males and is characterized by mental retardation of varying degrees. The syndrome is associated with a fragile site on the long arm of the X chromosome (Xq27); however, 20 percent of males who inherit the fragile site are unaffected. Females who have one fragile X and one normal chromosome show no clear obvious defects. The overall inheritance pattern of this syndrome from pedigree analysis shows some very interesting characteristics. They can be summarized by the following:

1. The sons of carrier females have a 50 percent chance of inheriting the *fragile X mutation,* but only a 40 percent overall risk for mental retardation.
2. The daughter of a carrier female, who has a 50 percent chance of inheriting the fragile X, is found to have a risk of mental retardation dependent on whether her mother is mentally impaired or not. Nearly 28 percent of the daughters of mentally impaired female carriers are also retarded, but only 16 percent of the daughters of mentally normal female carriers are retarded.
3. The daughters of transmitting males are at a 100 percent risk of inheriting the fragile X, whereas daughters of normal carrier females are at a 50 percent risk of inheriting the X chromosome carrying the fragile site from their mother. However, daughters of transmitting males are rarely, if ever, mentally retarded, whereas daughters of normal carrier females have a 16 percent incidence of retardation.
4. There is a distinct difference in the risk for mental retardation in children of mothers of transmitting males when compared to the risk to children born to the daughters of transmitting males.

The fragile X syndrome is the most common inherited form of mental retardation. Some have considered it to be a chromosomal disorder second in occurrence only to Down syndrome, but others suggest it is not a chromosome aberration but rather a single gene mutation associated with a chromosomal structural marker. Clearly, there is still much to be learned about the cause of the syndrome as well as completely understanding its inheritance pattern.

Chromosomal Aberrations and Evolution

Speciation, the evolutionary divergence of segments of a population to the point that they are no longer able to combine their genes through sexual reproduction, is a complex process with no single causative mechanism. The diversification on which evolution is built does, in every case, however, require alterations in the genetic material itself (mutation, Chapter 13) and/or changes in its arrangement (chromosomal aberrations), both of which lead to reproductive incompatibility. The effectiveness of both of these factors is increased if the population is broken into two or more isolated groups. In this way, frequencies of mutant genes may often be spread more rapidly, probability of *different* mutations that arise in separated groups is increased, and differing groups are often prevented from crossing by various isolating mechanisms.

Each of the chromosomal aberrations described in Chapters 14 and 15 is, if not lethal or disabling, related to the development of new taxonomic entities, though each rarely acts alone over time. Thus, trisomy in the Jimson weed leads to morphological differences that are recognizable and forms the basis for an aneuploid series. An aneuploid series, which may arise in a variety of ways, does exist in many plant families and genera and is associated with interspecific morphological differences.

But structural changes are important too in a gradual, step-by-step change. Many inversions are known to differentiate the several species of *Drosophila*. Thus *D. pseudoobscura* and *D. persimilis*, which are morphologically very similar, produce sterile male but fertile female hybrids and differ in four major inversions. On the other hand, *D. pseudoobscura* and *D. miranda* produce completely sterile offspring when crossed. The very complex pairing arrangements assumed by giant chromosomes in hybrid larvae of this interspecific cross indicate repeated and extensive inversions.

Translocations in the evening primrose, *Oenothera*, were largely responsible for the differences on which DeVries based his mutation the-

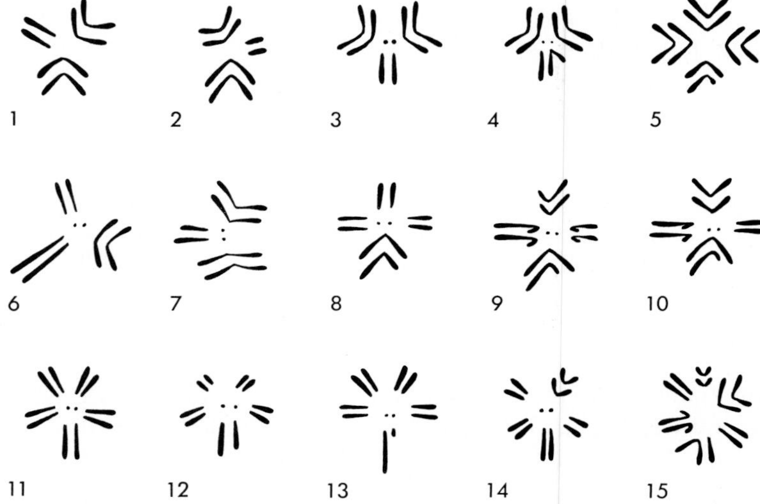

FIGURE 15–12. *Male karyotypes in several species of* Drosophila. *The X and Y chromosomes are at the bottom of each drawing. (1)* D. willistoni; *(2)* D. prosaltans; *(3)* D. putrida; *(4)* D. melanogaster; *(5)* D. ananassae; *(6)* D. spinofemora; *(7)* D. americana; *(8)* D. pseudoobscura; *(9)* D. azteca; *(10)* D. affinis; *(11)* D. virilis; *(12)* D. funebris; *(13)* D. repleta; *(14)* D. montana; *(15)* D. colorata.

ory. A series of many reciprocal translocations in isolated populations may lead to complete reproductive incompatibility, as well as definitive morphological differences, so that speciation may be said to have occurred. Comparison of the complements of several species of *Drosophila* (Figure 15–12) suggests all manner of chromosomal aberrations, including translocations (even unions of whole chromosomes) and deletions, at least. Note the series 3, 4, 5, 6.

On the other hand, the disabling and lethal effects of many types of chromosomal aberrations, especially in higher animals such as humans, remind us that aneuploidy, polyploidy, and structural changes in chromosomes may carry a high negative selection value.

PROBLEMS

15–1 Suggest a meiotic configuration at synapsis for a situation where six chromosomes have undergone reciprocal translocations of about the same length.

15–2 Below are diagrams of a pair of chromosomes (for simplicity chromatids are not shown). Loci are indicated by Arabic numerals along the chromosome. For each of the following structural changes in "chromosome 7" (maternal and paternal homologs are designated A and B, respectively) construct a simple diagram showing synaptic figures assumed in meiosis in each of the cases given below the chromosome diagrams; chromatids need be shown only for part (c):

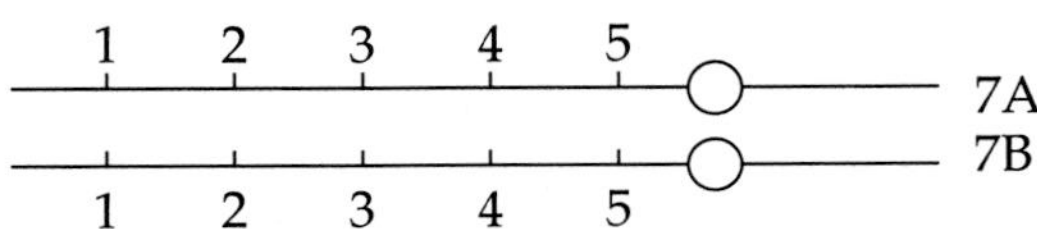

(a) a deletion in 7B involving loci 2 and 3;
(b) an inversion in 7B involving loci 2, 3, and 4;
(c) a reciprocal translocation with chromosome 8B involving loci 1, 2, and 3 of autosome 7B, and loci F, G, and H of chromosome 8B (all of the latter on the long arm of 8B, which has the sequence (centromere) D, E, F, G, H (terminus);
(d) a duplication involving loci 2, 3, and 4 in 7B.

15–3 Let the diagrams immediately below represent a pair of human chromosomes 6s (with A and B representing maternal and paternal homologs, respectively), each of which consists of two sister chromatids. Letters along the chromatids represent several hypothetical loci.

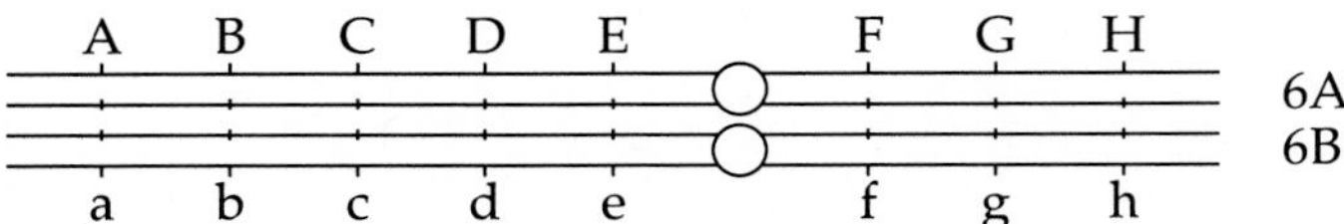

(a) Diagram the synaptic figure assumed by the two homologs (showing chromatids) if chromosome 6 sustains one nonsister chromatid crossover between loci F and G.
(b) Diagram chromosome 6, with loci, as it would appear in each of four sperms derived from the primary spermatocyte.

15–4 Differentiate between paracentric and pericentric inversions.

15–5 Assume a chromosome with the following gene sequence (the o represents the centromere):

A B C D o *E F G H*

You find the following aberrations in this chromosome; for each identify the specific kind of aberration:

(a) *A B C D* o *E F H*
(b) *A D C B* o *E F G H*
(c) *A B C D C D* o *E F G H*

15–6 In relation to the information in Table 15–2, where would *AAA*/*AA* and *AAAA*/*AAA* be properly placed? Note that progressively narrower eyes are shown from the top to the bottom of Table 15–2.

15–7 In *Drosophila*, *e* is a gene at locus 70.7 on chromosome III; *ee* flies have ebony bodies, much darker than wild-type flies of genotype + + or + *e*. If the cross + + × *e e* yields a small percentage of ebony flies, but greater than could be accounted for by the known mutation frequency of this gene, (a) what would you suspect as the cause? (b) What would you look for as confirmation?

15–8 Which term(s) from the list at the right is (are) correctly associated with each term in the list at the left:

(a) Klinefelter	(1) Deletion
(b) Philadelphia chromosome	(2) Duplication
(c) *Cri du chat* syndrome	(3) Inversion
(d) Ring chromosome	(4) Translocation
(e) Bar eye in *Drosophila*	(5) None of the above

15–9 In a hypothetical organism, gene *a* maps genetically three map units from the centromere. Would a cytological map be expected to show this gene farther from, closer to, or the same distance from the centromere? Why?

CYTOPLASMIC GENETIC SYSTEMS

CHAPTER

16

The existence of genes as segments of nucleic acid molecules, located in chromosomes and controlling phenotypes in known and predictable fashion, has been amply demonstrated on sound, observable, verifiable bases. But the firm establishment of such a chromosomal mechanism of inheritance does not necessarily preclude a role for other cell parts (plastids and mitochondria). In fact, when certain mutants demonstrated an inheritance pattern very different from normal Mendelian inheritance patterns, geneticists were forced to look to places other than the nucleus to explain the results. Abnormal segregation patterns showing inheritance from only a single parent, **uniparental inheritance,** or more specifically, **maternal inheritance,** where the genotype of the female parent was preferentially inherited in the progeny, led to considering the cytoplasm as an additional genetic system.

EVIDENCE FOR CYTOPLASMIC GENES

Based on the pattern of hereditary transmission, there are four observations that definitely point to the existence of cytoplasmic genetic systems:

1. Because the female gamete contributes almost all of the cytoplasm to the zygote and the male gamete contributes only a nucleus, an inheritance pattern that differs between reciprocal crosses suggests a cytoplasmic involvement. This is clearly the basis for uniparental or maternal inheritance where the progeny always resemble one parent, most commonly the female parent.

2. Whenever traits fail to demonstrate classical segregation patterns and deviate from standard ratios, the conclusion is again a cytoplasm-based type of inheritance. This is particularly evident when presumptive cytoplasmic markers do not segregate along with known nuclear markers.
3. When the trait fails to show linkage to any known nuclear linkage groups, and assorts independently from nuclear genes, a cytoplasmic mode of inheritance is suggested.
4. Many types of mutants that fit the above criteria will show segregation during mitotic division. This is very common in variegated plants that carry more than one type of plastid (chloroplast) per cell. This leads to variegation, suggesting somatic or vegetative segregation of the plastid types. In addition, it may actually be possible to either observe in the electron microscope or to detect by biochemical methods the actual transmission of the organelles carrying the cytoplasmic markers.

Cytoplasmic inheritance therefore will be understood to be based on cytoplasmically located, independent, self-replicating nucleic acids, which differ from chromosomal genes by their location within the cell, and have their own unique nucleotide sequences. The basis for identifying cytoplasmic transmission in most cases is the differential contribution of cytoplasm of the male and female gametes.

CHLOROPLASTS

FIGURE 16–1. *The four o'clock plant,* Mirabilis jalapa, *green-leaved variety.* (Photo courtesy Burpee Seeds.)

Chloroplast inheritance in the four o'clock plant (*Mirabilis jalapa*; Figure 16–1) is a classic illustration of cytoplasmic inheritance. This example was first studied in 1909 by Carl Correns, one of the rediscoverers of Mendel's original paper. This plant may exist in three forms: normal green, variegated (patches of green and of nongreen tissue), and white (no chlorophyll). Now and then a plant with branches of two or three of these types occurs. Plastids in the white areas have no chlorophyll. Offspring of crosses are phenotypically like those of the pistillate (egg-producing) parent except where the pistillate parent is variegated, in which case the offspring are of all three types, in irregular ratios. Egg cells from green plants carry normal green plastids, those from white plants contain only white plastids, and eggs produced by variegated parents may have both plastid types or just one type. Pollen (which produces the sperm and rarely contains any plastids) has no effect on progeny phenotype.

All that is really involved here is the type of proplastids and plastids present in the egg cytoplasm. If the plastids are defective with regard to chlorophyll synthesis, the F_1 plant will be nongreen; if they are not defective, the F_1 plant will be green. Variegated parents produce eggs that contain both normal and defective plastids. These eggs give rise to plants, whose cells fortuitously receive a majority of green plastids, or whose cells receive larger numbers of white plastids, which results in variegated plants. Recall that division of the cytoplasm following mitosis is not a quantitatively exact process.

In this example, the inheritance pattern is strictly maternal, following the inheritance of plastid type in the egg cytoplasm. A related but different type of cytoplasmic inheritance is demonstrated by the green alga *Chlamydomonas reinhardi*. In this alga (life cycle in Appendix B) the two types of gametes have equal amounts of cytoplasm (isogamy). The motile vegetative cells are haploid; the zygote is the only diploid cell. In germination the zygote undergoes meiosis to form four haploid zoospores that quickly mature into vegetative adults.

Sexual reproduction in most species represents morphological isogamy in which the gametes are morphologically distinguishable. In *Chlamydomonas* physiological anisogamy occurs in that the gametes are differentiated into + and − mating strains that are physically indistinguishable from each other. Mating type depends on nuclear genes; the locus is in linkage group 6 (there are 16 linkage groups in *C. reinhardi*; see Table 4–1). Chemical attraction between gametes of opposite mating strains appears to be exerted through proteins present in the flagella.

Two types of reaction to the antibiotic streptomycin occur in *C. reinhardi*. Vegetative cells of genotype *sr-1* are resistant to 100 μg per milliliter of streptomycin; those of genotype *ss* (sensitives) are killed at this concentration. This pair of alleles shows regular Mendelian segregation. Another genotype, *sr-2*, confers resistance to 500 μg per milliliter of streptomycin; sensitives are again designated as *ss*. Reaction to this higher concentration of the drug is *not* transmitted in Mendelian fashion, but streptomycin response of the progeny is that of the + mating type parent, even though the progeny segregate normally for the nuclear trait, mating type, as seen in this pair of reciprocal crosses:

(a) P: $sr\text{-}2\ mt^{+} \times ss\ mt^{-}$
F$_1$: $\frac{1}{2}\ sr\text{-}2\ mt^{+} + \frac{1}{2}\ sr\text{-}2\ mt^{-}$

(b) P: $sr\text{-}2\ mt^{-} \times ss\ mt^{+}$
F$_1$: $\frac{1}{2}\ ss\ mt^{+} + \frac{1}{2}\ ss\ mt^{-}$

Notice that in each case, mating strain segregates in normal Mendelian fashion (1:1), whereas streptomycin response of the progeny follows that of the (+) parent. More recent evidence suggests that when fertilization occurs, and the single plastid from each parent fuses, the DNA from the (−) mating strain plastid is degraded by an endonuclease from the (+) type so that only the chloroplast DNA of the (+) mating strain remains in the zygote. This clearly explains the inheritance pattern of *sr-2* being associated only with the (+) mating strain parent and the chloroplast DNA as the location of the *sr-2* gene.

MITOCHONDRIA

All living cells except bacteria, cyanobacteria, and mature erythrocytes contain *mitochondria*—small, self-replicating organelles of considerable internal structural complexity, which are centers of aerobic respiration. A few are shown in cross-sectional view in Figure 4–2. Although the shape of mitochondria is not constant, they often appear as elongated, slender rods of rather variable dimensions, and average about 0.5 μm in diameter and 3 to 5 μm in length. Extremes of 0.2 to 7 μm in width and 0.3 to 40 μm in length have been reported. Their number per cell varies considerably, from one in the unicellular green alga *Micrasterias* to as many as about a half million in the giant amoeba *Chaos chaos*.

A simple case of mitochondrial cytoplasmic inheritance is demonstrated by mutants with impaired mitochondrial function. A classic example is the *poky* mutant of the fungus *Neurospora* (see life cycle in Appendix B). Poky is characterized by its very slow growth compared to the normal fungus. This slow growth has been traced to impaired mitochondrial function related to certain cytochromes (proteins essential to electron transport). The pattern of inheritance of poky is strictly maternal and is illustrated in two reciprocal crosses:

Poky (female) × normal (male)
Progeny: all poky

Normal (female) × poky (male)
Progeny: all normal

In each cross the phenotype of the progeny is identical to the phenotype of the female parent.

Studies of respiratory-deficient strains of baker's yeast (*Saccharomyces cerevisiae*) indicate clearly the genetic role of mitochondria. Yeasts are unicellular ascomycete fungi. In the life cycle of some species (Appendix B), diploid and haploid adults alternate; the former reproduces by meiospores called ascospores and the latter reproduces by isogametes. Respiratory-deficient strains are able to respire only anaerobically and, on agar, produce characteristically small colonies known as "petites." Both nuclear and cytoplasmic controls affect this trait.

In one type of petite (*segregational petite*), respiratory deficiencies have been clearly shown to be caused by a recessive nuclear gene. Crosses between haploid petite and normal result in all normal diploid F_1 progeny, but ascospores produced by the latter segregate 1:1 for petite and normal. On the other hand, another strain (*neutral petite*) has defective mitochondrial DNA. More will be said about mitochondrial DNA in the next section. In this strain, the petite character fails to segregate and F_1 and F_2 of the cross petite × normal are all normal; the cytoplasm containing normal mitochondria is incorporated into F_1 zygotes and vegetative cells and distributed to ascospores and haploid vegetative cells of the next generation. In a third type of petite (*suppressive petite*), crosses with normals produce a highly variable fraction of petites in the progeny. It is thought that suppressives have rapidly replicating, abnormal mitochondrial DNA. So petites may lack mitochondrial function because of either mutant nuclear DNA or mutant mitochondrial DNA, or both. Respiratory enzymes (except for cytochrome c) are here under a double genetic control.

An interesting example of a mitochondrially determined trait in some higher plants is found in **cytoplasmic male sterility.** A plant showing male sterility simply does not produce functional pollen. If the trait shows maternal inheritance, it is called cytoplasmic male sterility or CMS. The best example studied so far is in maize. There are three different systems of CMS in maize: CMS-S, CMS-C, and CMS-T. The first two types, S and C, can be reversed by nuclear restorer genes, and the other, CMS-T, cannot. The molecular basis for the CMS trait in maize has been shown to be located in the mitochondria. This CMS trait is used by plant breeders to avoid the time-consuming task of anther removal in order to prevent self pollination. As a result, some form of CMS is found in most economically useful varieties of maize. One type, CMS-T, was present in approximately 80 percent of the breeding lines. In 1970, it was discovered that this type T cytoplasm, while conferring CMS on the plants, also made the plants very susceptible to race T of the pathogen *Helminthosporum maydis*, which causes *corn blight*. This resulted in a major epidemic of the blight, and loss of almost 80 percent of the corn crop that year. The mitochondrial membrane in CMS-T plants is in some yet unknown way sensitive to the toxin produced by the pathogen, thereby causing the disease. Here, as in the case of plastid DNA, there is a degree of cytoplasmic control over a phenotype; however, this control is not entirely independent of nuclear DNA.

MOLECULAR BIOLOGY OF CHLOROPLASTS AND MITOCHONDRIA

In order for these two cellular organelles to contribute to the inheritance of a cell, the first requirement is the presence of DNA, and the second is that the DNA code for organelle-specific proteins. This now has been demonstrated in both cases.

Chloroplasts

DNA is easily isolated from chloroplasts. Figure 16–2 shows the highly supercoiled DNA (cpDNA) from *Acetabularia mediterranea.* A large number of investigations have provided definitive evidence that cpDNA is a closed, circular, double-stranded molecule. In maize, each molecule of cpDNA is about 139,000 base pairs long and is repeated about 50 times per chloroplast. Each molecule is long enough to encode approximately 140 proteins, although very few have actually been identified as yet. Plastids contain their own DNA polymerases and the DNA is replicated in a semiconservative fashion. The RNA polymerase found in plastids most closely resembles RNA polymerase II from the nucleus. This RNA polymerase transcribes several messenger RNAs, unique chloroplast ribosomal RNAs, and a full set of chloroplast tRNAs. Several proteins encoded on the cpDNA are synthesized on the chloroplast ribosomes. A partial list of these proteins is found in Table 16–1.

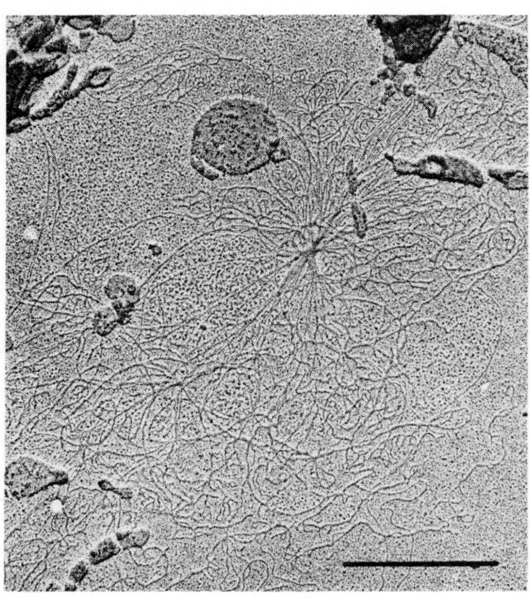

(*A*)

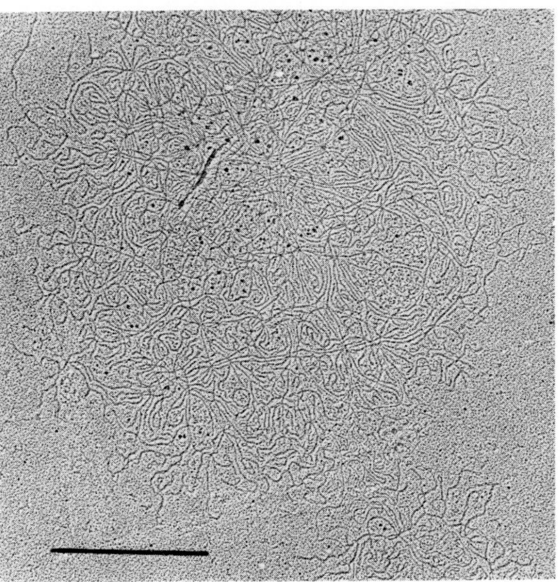

(*B*)

FIGURE 16–2. *Electron micrographs of chloroplast DNA from the green alga* Acetabularia mediterranea. *(A) Lysed chloroplast showing chloroplast DNA fibrils around a "center." (B) Large array of DNA released from a chloroplast. The bar represents 1 μm.* (Courtesy Dr. Beverly R. Green, University of British Columbia, used by permission.)

TABLE 16–1. Chloroplast proteins known to be coded by cpDNA and synthesized in the chloroplast

Proteins	*Function*
Ribulose-biphosphate carboxylase oxygenase (RUBISCO) large subunit	Part of photosynthetic CO_2-fixing enzyme
Subunit of coupling factor 1	Membrane-bound photosynthetic protein
Ubiquinone-binding protein	Membrane-bound protein binds ubiquinone
Light-induced proteins	Unknown functions—but synthesized when plant is transferred from dark to light

So much is now known about the maize cpDNA that a genetic map has been constructed illustrating the location of many genes. This map is found in Figure 16–3. The most striking feature of the map is the large inverted repeat area. This inverted repeat area contains the rRNA genes and some tRNA genes in the same order but in reverse orientation. While there is much to learn, the chloroplast clearly represents a separate genetic system in the cell, but not totally independent from the nucleus.

Mitochondria

A similar case can be made for the mitochondria. The presence of DNA has been clearly established. Figure 16–4 shows the circular mtDNA from mouse fibroblast cells. The interesting finding about mtDNA is its tremendous variation in size. The three major systems studied so far are human, yeast, and higher plant mtDNA. The human mtDNA contains 16,569 base pairs and has been completely sequenced. Yeast mtDNA is about five times larger than human mtDNA, and maize is about five times larger than yeast mtDNA. The mtDNA encodes several

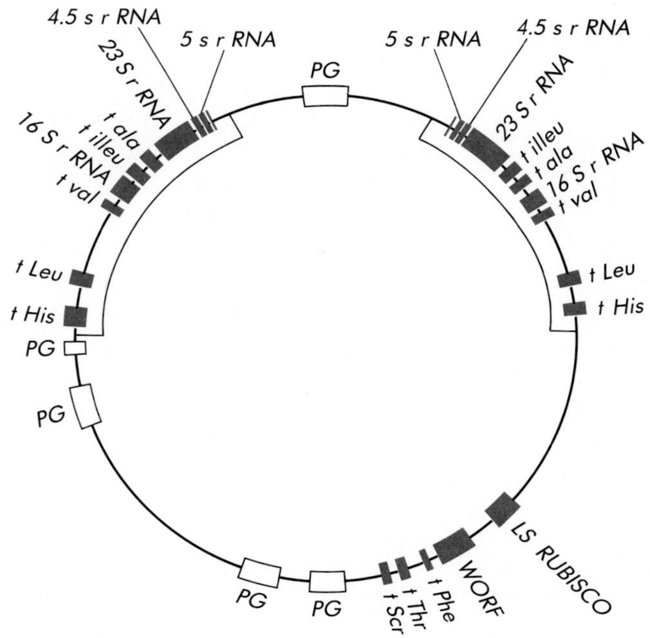

FIGURE 16–3. *A genetic map of the circular maize chloroplast DNA. Note the positions of the two large inverted repeats at the top, containing several ribosomal RNA and transfer RNA genes. The location of several other known tRNA genes are given as well as photogenes (PG, which are induced in the presence of light) and the large subunit gene of RUBISCO. The large UORF (unidentified open reading frame) contains sequences that are probably functional genes, but are as yet unidentified.*

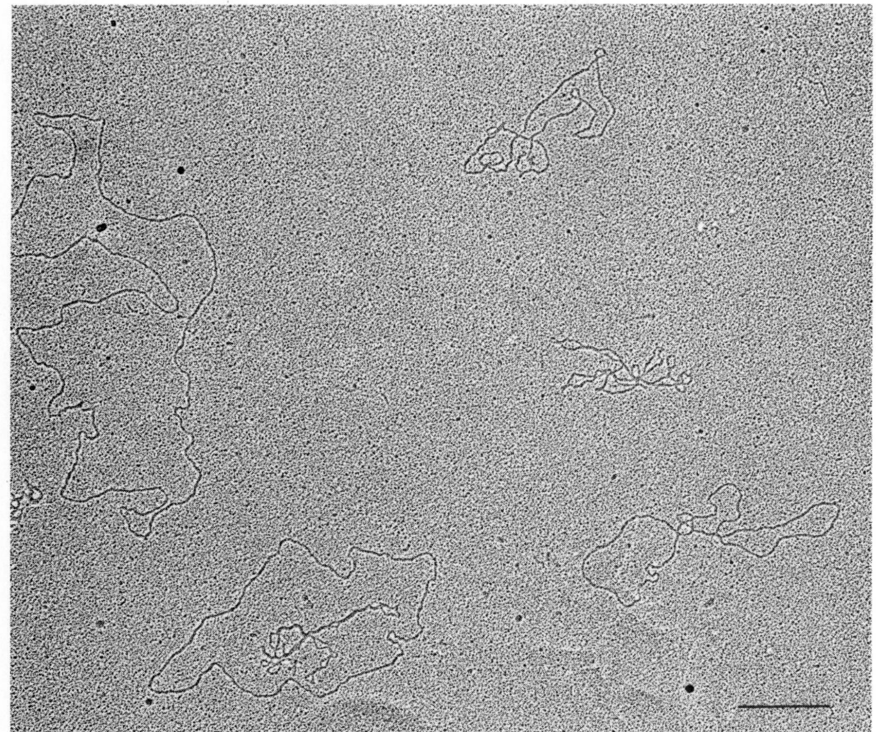

FIGURE 16–4. *Electron micrograph of DNA isolated from mitochondria of mouse fibroblast cells. Note the circularity of these molecules. A highly supercoiled molecule lies inside an open monomer (lower left), and three loosely twisted molecules are located at the right. Bar represents 1 μm.* (Courtesy Dr. Margit M. K. Nass, used by permission.)

proteins as well as rRNA and tRNAs. A list of mitochondrial proteins and RNAs known to be encoded by the mtDNA is found in Table 16–2.

While the genetic map of the human mtDNA is not as far along as the map for maize cpDNA, it is believed that every base pair in the human mtDNA is involved in coding for either a mRNA for a protein, a rRNA, or a tRNA. The genes appear to be very close together, with no noncoding sequences between them. Furthermore, there is only one major promoter on each strand. The primary transcription product is a full-length copy of each strand. The transcript is then cleaved into the various RNA molecules. At least some of the genes contain introns. It appears that tRNA genes flank almost every major gene and are actually used as spacers between these genes. The mtDNA shows the minor variations in the genetic code shown in Table 10-4.

It should be very clear from this discussion that plastids and mitochondria represent separate genetic systems in eukaryotic cells. They have DNA and encode proteins that determine phenotypes whose pattern of inheritance does not follow conventional Mendelian patterns. They are not, however, independent from the nucleus as many of the proteins in plastids and mitochondria are encoded by nuclear genes. Understanding how the nuclear and organellar genetic systems cooperate to produce a normally functioning organelle is a major question of modern research in molecular biology.

TABLE 16–2. **Gene products known to be coded by mtDNA**

Proteins
Cytochrome oxidase subunit I
Cytochrome oxidase subunit II
Cytochrome oxidase subunit III
Cytochrome B
ATPase subunit 6
ATPase subunit 8
ATPase subunit 9
ATPase subunit alpha
var 1 (ribosomal protein)
RNA
rRNA
Mouse 16s, 12s
Yeast 9s, 15s, 21s
Maize 5s, 18s, 26s
tRNA
22 in Mouse
24 in Yeast
3 in Maize

(From Mulligan and Walbot, 1986.)

PLASMIDS

Extrachromosomal genetic elements composed of circular, closed DNA molecules occur in bacteria and have been most clearly characterized in *E. coli.* These are now generally and collectively termed **plasmids,** although the term *episome* was originally introduced by Jacob and his colleagues in 1960 to designate genetic material such as the F (fertility or sex) factor, which can exist either integrated into the bacterial "chromosome" or separately and autonomously in the cytoplasm. The term *plasmid* was reserved for those bits of genetic material that exist *only* extrachromosomally and cannot be integrated into the nucleoid (the DNA). Episomes were said to be integrable, whereas plasmids were not. For simplicity's sake, all bacterial extrachromosomal genetic material, whether integrable or not, will be designated *plasmids*. There are three important classes of plasmids: **F** (fertility) factors, **R** (resistance) factors, and **Col** (colicigenic) factors.

F Factors

Attention has already been directed to the F factor and its important function in bacterial conjugation. Under strict semantic separation, the F factor is an episome, for it is integrable into the nucleoid. When so integrated, the donor bacterial cell is designated a high-frequency recombinant (*Hfr*); when it exists as a separate cytoplasmic organelle, the donor cell is designated F^+. The latter behaves differently from the *Hfr* cell in conjugation. The fertility factor is composed of DNA; in *E. coli* it is composed of about 1×10^5 base pairs. It determines that the cell in which it resides will serve as a donor in conjugation, and by its location (that is, whether integrated or not) whether that cell will be *Hfr* or F^+. (Cells without the F factor are receptors in conjugation.) Therefore, the fertility factor does fulfill the criteria outlined at the opening of this chapter for a true cytoplasmic genetic determinant.

R Factors

The R factors are plasmids; they range in size from an apparent low of 1.5×10^4 to a high of 1×10^5 deoxyribonucleotide pairs. They are most often present in one to three copies per cell, although somewhat higher numbers have been reported for a few R factors.

These plasmids carry genes for resistance to one or more chemicals—for present purposes these include many drugs used to combat infection (for example, chloramphenicol, neomycin, penicillin, streptomycin, sulfonamides, and tetracyclines). Resistance to one or more (usually several) drugs is infectious in the sense that blocks of DNA-carrying resistance markers are often transferred during conjugation.

R factors are composed of two parts: the *resistance transfer factor* (RTF) responsible for the transfer, and the various *drug-resistance* genes. The latter includes all drug-resistance factors that may be present, except for the factor conferring tetracycline resistance, which appears to be in the RTF position. Lewin and others have described the discovery of bacterial drug resistance during treatment of cases of bacillary dysentery (caused by species of the bacterial genus *Shigella*) in Japan in the late 1950s. Four drugs in common use for treatment of this disease are chloramphenicol, streptomycin, sulfonamides, and tetracycline. Strains of the causal organism, isolated from human cases of the disease, were observed to acquire resistance to one or, more commonly, to all four drugs.

Japanese investigators in the late 1950s and early 1960s established that *Shigella* strains with multiple drug resistance arose in the human gut by conjugation with resistant strains of *E. coli*. Drug resistance can be transferred only by cell-to-cell contact but may be passed both from *E. coli* to *Shigella* and from *Shigella* to *E. coli*. Intergeneric transfer of this kind is widespread in bacteria, especially in the members of the family Enterobacteriaceae, to which both *E. coli* and *Shigella* belong. Second, at least a low level of resistant *E. coli* cells is often present in the intestinal tract, and treatment of dysentery with any of the four drugs kills off all the susceptible *E. coli* and *Shigella*. The remaining resistants multiply and "trade" RTFs by conjugation in the gut.

Colicinogenic Factors

The colicinogenic (Col) factors are plasmids that carry genes for production of **colicins,** which are highly specific proteins produced by some strains of intestinal bacteria. Other strains are susceptible if their walls bear specific colicin receptor sites and resistant if the receptors are absent. Colicinogenic cells are resistant to the colicins that they themselves synthesize. There are many, many kinds of colicins, each having its own killing action. One of the most interesting, colicin K, inhibits DNA replication, RNA transcription, and polypeptide synthesis. Many but not all of these plasmids are transferable in conjugation. They are closed, double-stranded, circular DNA molecules ranging roughly between 6,000 and 21,000 base pairs in size.

Detection of Plasmids

The occurrence of plasmids in a given cell line can be detected in several ways. First, and most convincingly, they can be photographed by use of an electron microscope. Second, some are "infectious," that is, they can be transferred in conjugation. This is notably true in the case of drug resistance. Third, because they replicate independently of the chromosome, they may be lost from a cell line in a random pattern. Fourth, they can be genetically mapped, and their loci show no linkage to any markers on the bacterial chromosome.

INFECTIVE PARTICLES

Cytoplasmic inheritance may also be observed in the transmission of invading microorganisms or cytoplasmic particles that have definite phenotypic effects.

Carbon Dioxide Sensitivity in *Drosophila*

A certain strain of *Drosophila melanogaster* shows a high degree of sensitivity to carbon dioxide. Whereas the wild type can be exposed for long periods to pure carbon dioxide without permanent damage, the sensitive strain quickly becomes uncoordinated with even brief exposure to low concentrations. This trait is transmitted primarily, but not exclusively, through the maternal parent. Tests have disclosed that sensitivity is dependent on an infective, viruslike particle, called *sigma,* in the cytoplasm. It is normally transmitted via the eggs' larger amount of cytoplasm but occasionally through the sperms as well. Sensitivity may even be induced by injection of a cell-free extract from sensitives.

Sigma contains DNA and is mutable, but it is clearly an infective, "foreign" particle. Multiplication is independent of any nuclear gene, but the mechanism of its sensitizing action is unknown.

Sex Ratio in *Drosophila*

In another strain of *Drosophila,* almost all male embryos die soon after zygote formation. Studies have disclosed this trait to be dependent on a *spirochete* transmitted in the cytoplasm of the female. The maintenance of the spirochete is dependent on nuclear genes.

Killer Trait in *Paramecium*

Some races ("killers") of the common ciliate *Paramecium aurelia* produce a substance called paramecin that is lethal to other individuals ("sensitives"). Paramecin is water-soluble and diffusable, and depends for its production on particles called **kappa** in the cytoplasm. Kappa contains DNA and RNA and is mutable, but its presence is dependent on the nuclear gene K. Animals of nuclear genotype *kk* are unable to harbor kappa.

K-individuals do not possess kappa unless and until it is introduced through a cytoplasmic bridge during conjugation (Figure 16–5). Nonkiller K-animals may also be derived from killers by decreasing the number of kappa particles. They may be accomplished either by starving the culture or subjecting it to low temperatures or, in contrast, by causing killers to multiply more rapidly than kappa.

Kappa particles can be seen with the microscope. Electron microscopy reveals them to have minute amounts of cytoplasm and to be bounded by a membrane. In addition, kappa contains DNA and protein.

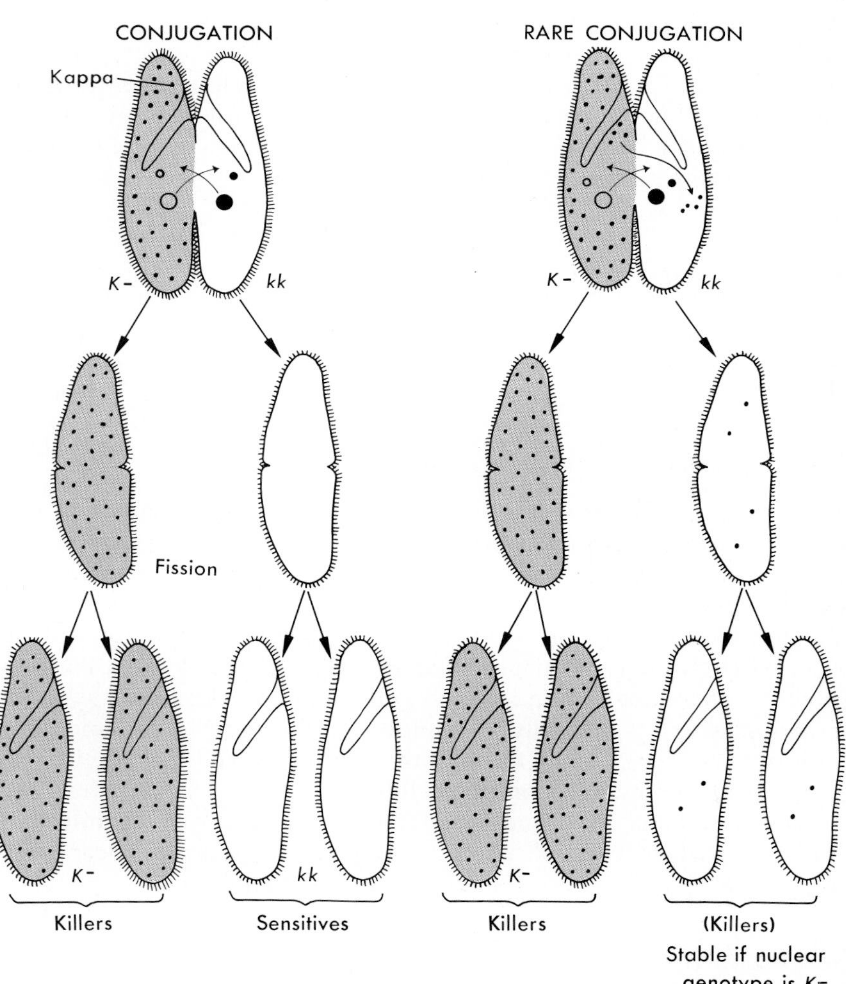

FIGURE 16–5. *Conjugation in* Paramecium *and the killer trait. Progeny of sensitives are killers only in rare situations where conjugation persists for a longer period so that kappa-containing cytoplasm is introduced into the conjugating sensitive. Kappa particles, however, are maintained only in the presence of a K-nuclear genotype.*

The particles can be transferred to other ciliates by feeding. Far from being the illustration of the cytoplasmic gene it was first surmised to be, kappa must be regarded as an infectious organism that has attained a high degree of symbiosis with its host.

Similarly, the mate-killer trait in *Paramecium* is imparted by a *mu* particle that, in turn, exists only in those cells whose micronucleus contains at least one dominant allele of a nuclear gene. *Mu* particles, too, appear to be endosymbiotes.

Milk Factor in Mice

The females of certain lines of mice are highly susceptible to mammary cancer. Results of reciprocal crosses between these and animals of a low-cancer-incidence strain depend on the characteristic of the female parent. Allowing young mice of a low-incidence strain to be nursed by susceptible foster mothers produces a high rate of cancer in these low-incidence young. Apparently this is a case of an infective agent that is transmitted in the milk. This so-called milk factor has many of the characteristics of a virus and has been discovered to be transmissible also by saliva and semen. Its presence in body fluids, moreover, is again dependent on certain nuclear genes.

Situations of the kind involving the milk factor, which often show a maternal inheritance pattern, are certainly not examples of cytoplasmic inheritance as we have defined it. Rather than being components of the individual's genetic mechanism and results of independent, conserved nucleic acids located in the cytoplasm, the components simply acquired infective agents.

MATERNAL EFFECTS

Shell Coiling in the Snail *Limnaea*

The direction of the coiling of the shell in such snails as *Limnaea* illustrates the influence of nuclear genes acting through effects produced in the cytoplasm.

The shells of snails coil either to the right (dextral) or to the left (sinistral), as seen in Figure 16–6. A shell held so that the opening through which the snail's body protrudes is on the right and facing the observer is termed *dextral;* if the opening is on the left, coiling is *sinistral.* Direction of coiling is determined by a pair of nuclear alleles; dextral (+) is dominant to sinistral (*s*). But expression of the trait depends on the maternal genotype. A cross between a dextral female and a sinistral male produces F_1s that are dextral. When the F_1s are selfed (this is possible because the snails are hermaphroditic) all of the F_2 individuals are also dextral even though one-fourth of the progeny are genotypically sinistral. When each of the F_2 genotypes is again self-fertilized, F_3 progeny of the (+ +) and (+*s*) animals are dextral, but those of the *ss* individuals are sinistral.

The explanation for this inheritance pattern is that the coiling phenotype of any individual is determined by the genotype, not the phenotype of the mother. This is because the mother determines the initial direction of the orientation of the spindle in the first mitosis of the zygote. Spindle orientation, in turn, is controlled by the genotype of the oocyte from which the egg develops and appears to be built into the egg before meiosis or syngamy occurs. The exact basis of this rather unusual control is unknown. This type of inheritance is due to a **maternal effect** and is not due to organelle-based genetic systems.

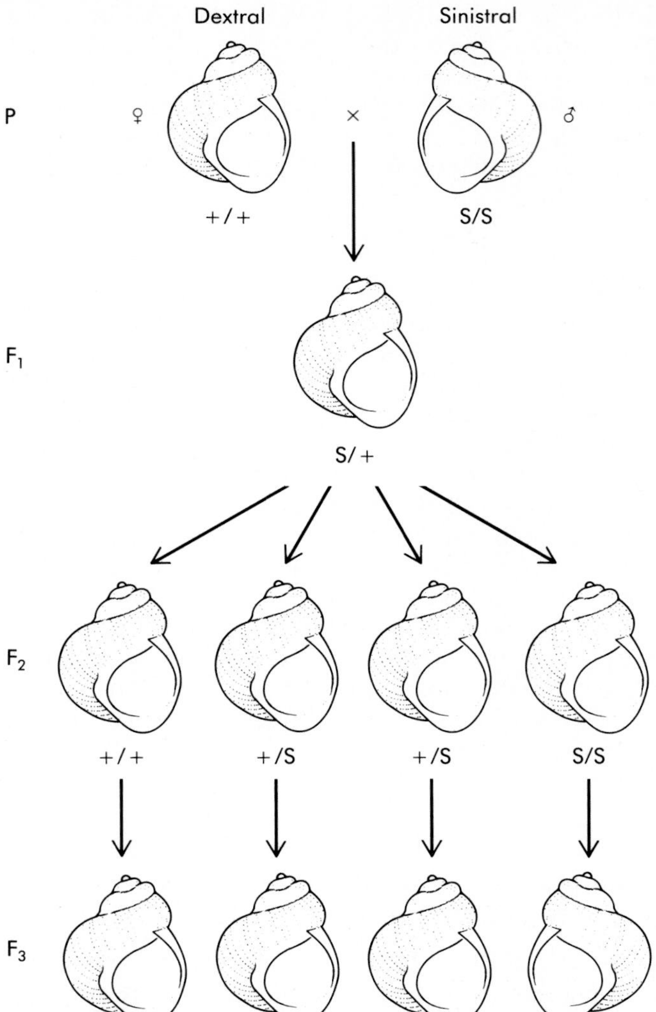

FIGURE 16–6. *Inheritance pattern of dextral (+) and sinistral (s) coiling in the water snail* Limnaea. *The direction of coiling is determined by the genotype of the mother through a maternal effect, and is not due to an organelle-based genetic system.*

Water Fleas and Flour Moths

A similar kind of maternal effect occurs in at least two very different invertebrates, the water flea (*Gammarus*) and the flour moth (*Ephestia*). Pigment production in eyes of young individuals depends on a pair of nuclear alleles, *A* and *a*. The dominant allele directs production of **kynurenine,** a diffusable substance that is involved in pigment synthesis. The cross *Aa* (female) × *aa* (male), for example, produces progeny that have dark eyes while young. Upon reaching the adult stage, half the offspring (those of genotype *aa*) become light-eyed. The explanation is that kynurenine diffuses from the *Aa* mother into all the young, enabling the young to manufacture pigment regardless of their genotype.

The *aa* progeny, however, have no means of continuing the supply of kynurenine, with the result that the eyes eventually become light. This is clearly a cytoplasmic maternal effect, not due to a cytoplasmic genetic mechanism.

ORIGIN OF CYTOPLASMIC GENETIC SYSTEMS

With the occurrence of genetic systems in the cytoplasm, such as plastids and mitochondria, comes the question of their origin. The currently accepted theory is the **endosymbiont theory,** which states that plastids and mitochondria originated with the invasion of a primitive eukaryotic cell by a prokaryotic organism. Eventually, a symbiotic relationship was established with each partner benefiting from the other. In time, they became dependent on each other and could not live separately. Certainly there is enough similarity between prokaryotes and organelles to support the theory. These similarities include:

1. Both have similar genome organization, with circular DNA, 70s ribosomes, absence of histones, and lack of a nuclear membrane.
2. Both respond similarly to antibiotics that inhibit protein synthesis.

Additional support comes from the fact that these organelles code essential cellular proteins not encoded by nuclear genes. This molecular cooperation indicates a symbiotic relationship for many cellular functions. For example, even though respiration occurs in the mitochondria, some respiratory enzymes are encoded by mitochondrial genes and some by nuclear genes. The large subunit of the enzyme **ribulose-bisphosphate carboxylase oxygenase (RUBISCO)** is encoded by a chloroplast gene and the small subunit is encoded by a nuclear gene even though the enzyme functions exclusively in the chloroplast. Also, when comparisons are made between organisms, transitional steps are observed. The subunit 9 protein of the mitochondrial ATPase complex is

WHERE ARE THE STRIPED BASS?

In an effort to explain why the once abundant population of striped bass in the Chesapeake Bay has all but disappeared, marine scientists are even turning to the mitochondrial DNA (mtDNA) for the answer. One of the first questions scientists have to answer pertains to the current location of the bass: Exactly where are the major subpopulations of bass in the bay? One of the most accurate and revealing ways of answering this question comes from analyzing the mtDNA of fish samples from all over the Chesapeake region.

A small sample of mtDNA is isolated from a fish and is cut into specific pieces with restriction enzymes (Chapter 20). A unique fragment pattern is obtained for each group of fish, as identifiable and characteristic as a fingerprint. This mtDNA fingerprint is then matched with the site where the fish was collected. By noting that mtDNA is inherited only through the mother, maternal families with identical fingerprints can be constructed. The results indicate that each family of striped bass can be located in distinct regions of the bay and adjoining areas. The largest family division is between the fish of the Chesapeake Bay and those from the Dan River in North Carolina. The next biggest division occurs between fish from the northern and eastern parts of the bay and fish from the southern and western parts. Even subgroups within a family can be identified. For example, the mtDNA of the Potomac River stripers is distinguishable from the mtDNA of the stripers off the eastern Chesapeake Bay shore, and even differs from the mtDNA collected from bass in the Gulf of Mexico.

These studies using mtDNA are now providing the most accurate picture ever obtained of the striped bass population in the Chesapeake Bay. The next step will be to track the populations to see if they migrate, and to follow their spawning behavior. Eventually, scientists will be able to confidently answer the question: Where have all the striped bass gone? This will be the first step in restoring their populations to the levels of several years ago.

encoded by a nuclear gene in *Neurospora* and mammals, but by the mtDNA in yeast. This type of example has led some to propose gene flow between organelles and the nucleus.

Additional support for this gene flow hypothesis comes from the discovery of **promiscuous DNA** or the occurrence of the same nucleotide sequences in more than one of the three membrane-bound genetic compartments (nucleus, plastids, mitochondira) of eukaryotic cells. For example, sequences that encode the chloroplast 16s rRNA and several chloroplast tRNAs also have been found in the mitochondrial DNA in maize. In yeast, the mitochondrial ATPase is encoded in the mitochondrial DNA, whereas in *Neurospora* the same active gene is in the nucleus; however, the mitochondrial genome of *Neurospora* does contain an inactive copy of the gene.

This phenomenon certainly helps extend and support the conclusions of the endosymbiont theory. No one knows the actual prokaryotic invaders; however, in the case of plastids, a primitive cyanobacterium seems likely and purple sulfur bacteria are thought to be the precursors of mitochondria. In the case of the mitochondria a primitive bacterium is also possible. What appears true for now is that both types of organelles did not have a common origin.

Finally, since mitochondrial DNA apparently has evolved quite rapidly, analysis of this DNA has been used to study possible evolutionary relationships. When human mtDNA was compared with four species of ape (chimpanzee, gorilla, orangutan, and gibbon), it was possible to construct in the fewest possible steps an evolutionary tree with the five groups being derived from a common ancestor. The order of divergence was gibbon, orangutan, human, chimpanzee, and gorilla (Figure 16–7). It was not possible to draw any conclusions using nuclear DNA.

In a similar study, chloroplast DNA has been used by Kung and his co-workers to form evolutionary relationships in the genus *Nicotiana* (tobacco). They found that by using a single restriction enzyme, Sma-I (Chapter 20), they were able to identify each of the species in the genus *Nicotiana*. The results of this work are found in Figure 16–8. This study not only was able to elucidate individual species relationships but also confirmed the origin of present-day cultivated tobacco, *Nicotiana tabacum*, as an interspecific hybrid between *N. sylvestris* and *N. tomentosoformis*. This origin had already been proposed based on other data.

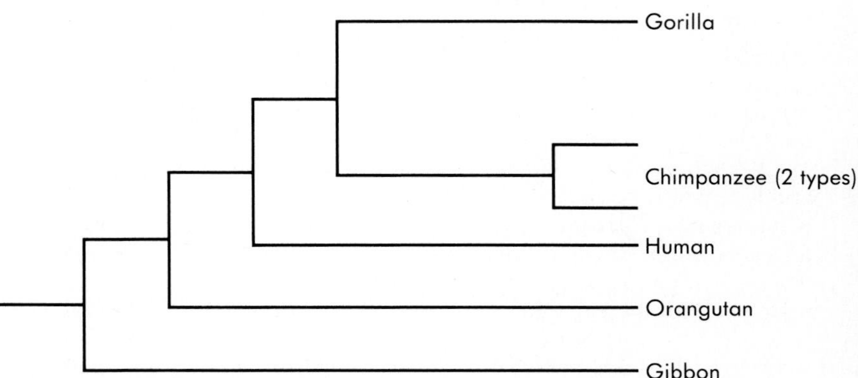

FIGURE 16–7. *Evolutionary tree for mitochondrial DNA of five higher primates. Tree is based on a minimum of 67 mutations at 42 positions in the mtDNA. (See reference by Ferris et al. 1981.)*

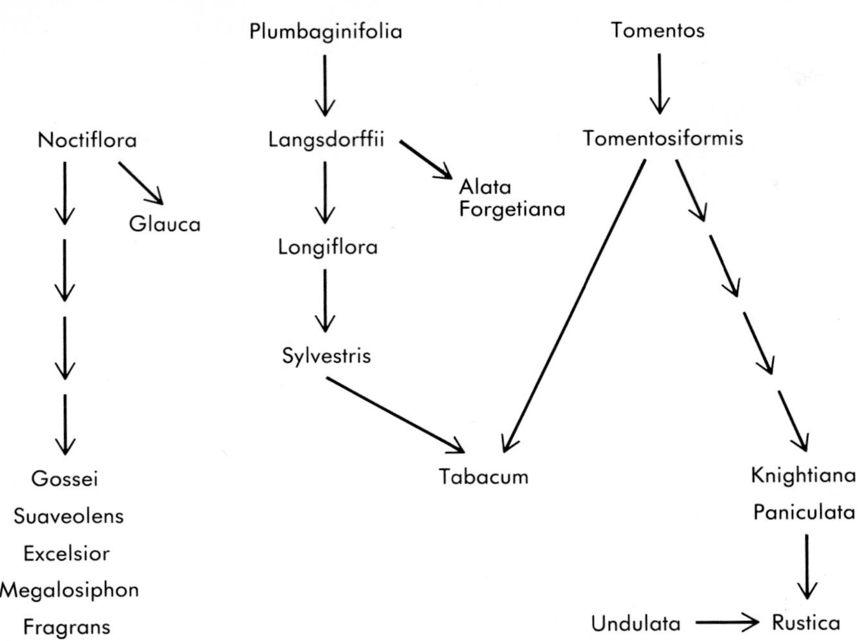

FIGURE 16–8. *Phylogenetic tree for the species of* Nicotiana *based on changes in chloroplast DNA. This information is supported by biochemical and other evidence, and establishes the sequence of evolution. (See reference by Kung et al. 1982)*

PROBLEMS

16–1 You have just discovered a new trait in *Drosophila* and find that reciprocal crosses give different results. How would you determine whether this trait was sex-linked, a purely maternal effect, or the result of an extranuclear genetic system?

16–2 What phenotype would be exhibited by each of the following genotypes in the snail: + *s*, *s* +, *ss*, + +?

16–3 What kind(s) of progeny with respect to eye color result from these crosses in the flour moth (*Ephestia*): (a) light ♀ × homozygous dark ♂, (b) homozygous dark ♀ × light ♂?

16–4 A four-o'clock plant with three kinds of branches (green, variegated, and "white") is used in a breeding experiment. What kinds of progeny are to be expected from each of these crosses: (a) green ♀ × white ♂, (b) white ♀ × green ♂, (c) variegated ♀ × green ♂?

16–5 Green and Burton (1970) were able to secure intact segments of *Acetabularia* cpDNA as long as 419 μm. Such DNA would consist of about 1.23×10^6 deoxyribonucleotide pairs. (a) If it is assumed that all this cpDNA is transcribed, for how many codons could it be responsible? (b) If this entire amount of cpDNA is transcribed and the resulting codons are all translated, how many polypeptide chains could be produced if each consists of 400 amino acid residues?

16–6 Although a neutral petite in yeast has defective mitochondrial DNA, it may have a nuclear gene for normal mitochondrial function. As indicated in the text, a haploid segregational petite carries a recessive nuclear gene for defective mitochondria, but it may possess normal mitochondrial DNA. If such a neutral is crossed with segregational petite of the type described here, what is the phenotype of (a) the diploid F_1, (b) the haploid generation that develops from ascospores produced by these diploid cells?

16–7 Employing different substrains of *Escherichia coli* strain K-12 as *Hfr* conjugants produces different results (see linkage map, Figure 6–7):

K-12 Substrain	Chromosomal genes transferred in conjugation: First	Last
C	*lys + met*	*gal*
H	*pil*	*pyr-B*

For each substrain, give (a) the location of F and (b) the second chromosomal gene that would be transferred.

16–8 For the following crosses in the four o'clock plant give the progeny phenotypes:

Pistillate parent (♀)	*Staminate parent* (♂)
(a) White	Green
(b) Green	White
(c) Green	Variegated
(d) Variegated	Green
(e) Green	Variegated

16–9 In isogamous sexual reproduction in the unicellular green alga *Chlamydomonas* the chloroplast of the minus mating strain is lost, whereas that of the plus parent is retained, becomes the chloroplast of the zygote and, by division, the chloroplasts of the four cells derived by meiosis from the zygote. Isogametes have about equal amounts of cytoplasm and are combined in the zygote. The sm4 strain of this organism *requires* the antibiotic streptomycin for survival. This trait is passed to progeny only through the plus mating strain. Where is the *sm4* gene most likely located?

16–10 Nitrosoguanidine is a potent mutagen. Its application to a culture of *Chlamydomonas* just prior to nuclear division causes mutation in a variety of nuclear genes, but its application to a culture tube of cells of the plus mating strain just prior to their being allowed to mix with and engage in sexual reproduction with untreated minus cells results in many sm4 mutants. On the other hand, application of nitrosoguanidine to a tube of only the minus strain just before sexual reproduction fails to produce any sm4 mutants. What effect do these facts have on your answer to 16–9?

MULTIPLE ALLELES AND BLOOD GROUP INHERITANCE

CHAPTER

17

In discussions so far, a basic assumption has been that a particular chromosomal position, or **locus,** is occupied by either of two alleles. Many instances are known, however, in which a given locus may bear any one of a series of several alleles, so that a diploid individual might possess any two alleles of a series. *When any one of three or more alleles may occupy a given locus, such alleles are said to constitute a series of multiple alleles.*

COAT COLOR IN RABBITS

The coat of the ordinary (wild type) rabbit is referred to as "agouti" or full color, in which individuals have banded hairs, the portion nearest the skin being gray, succeeded by a yellow band, and finally a black or brown tip (Figure 17–1). Albino rabbits, which are totally lacking in pigmentation, have also long been known (Figure 17–2). Crosses of homozygous agouti and albino individuals produce a uniform agouti F_1; interbreeding of the F_1 produces an F_2 ratio of 3 agouti:1 albino. Two-thirds of the F_2 agouti individuals can be shown by testcrosses to be heterozygous. Clearly then, this is a case of monohybrid inheritance, with agouti dominant to albino.

Other individuals, lacking yellow pigment in the coat, have a silvery-gray appearance because of the optical effect of black and gray hairs. This phenotype is referred to as chinchilla (Figure 17–3). Crosses between chinchilla and agouti produce all agouti individuals in the F_1 and

FIGURE 17–1. *Wild-type Agouti rabbit.* (Photo courtesy American Genetic Association.)

FIGURE 17–2. *Albino rabbit. Albino animals are totally lacking in pigment.* (Photo courtesy American Genetic Association.)

FIGURE 17–3. *Chinchilla rabbit.* (Photo courtesy American Genetic Association.)

FIGURE 17–4. *Himalayan rabbit.* (Photo courtesy American Genetic Association.)

a 3 agouti:1 chinchilla ratio in the F_2. Thus genes determining chinchilla and agouti appear to be alleles, with agouti again dominant. If, however, the cross chinchilla × albino is made, the F_1 are all chinchilla, and the F_2 shows 3 chinchilla:1 albino. Therefore, genes for chinchilla and albino are also alleles, and agouti, chinchilla, albino are said to form a multiple allelic series.

Still another phenotype is often encountered in pet shops. This is Himalayan (Figure 17–4), in which the coat is white except for black extremities (nose, ears, feet, and tail). Eyes are pigmented, unlike albino. By appropriate crosses it can be shown that the allele for Himalayan is dominant to that for albino, but recessive to alleles determining agouti and chinchilla. Some of the possible crosses, with F_1 and F_2 progeny, are shown in Figure 17–5. Gene symbols often assigned are c^+ (agouti), c^{ch} (chinchilla), c^h (Himalayan), and c (albino). From the crosses shown in Figure 17–5 the order of dominance can be established:

$$c^+ > c^{ch} > c^h > c$$

It is easy, of course, to predict F_1 and F_2 progeny for two crosses not shown in Figure 17–5, such as agouti × Himalayan or chinchilla × albino.

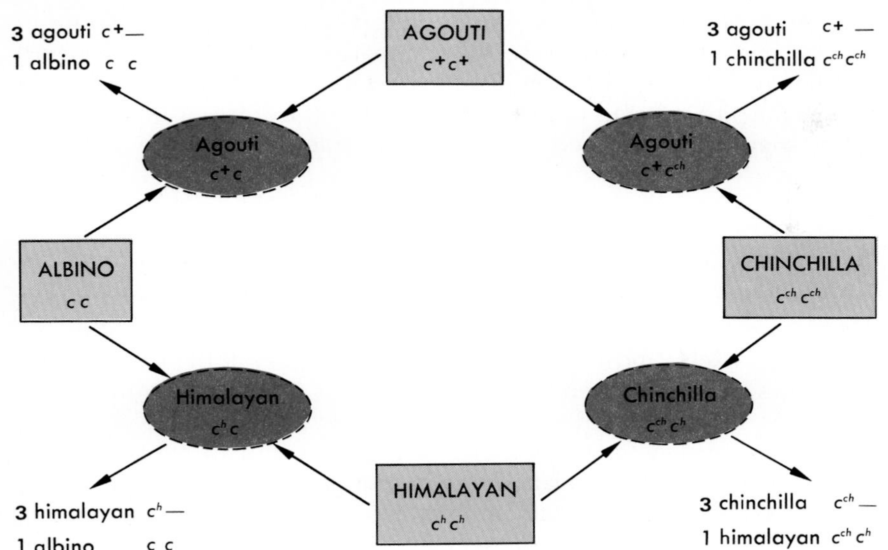

FIGURE 17–5. *Diagram representing several crosses in the multiple-allele coat series in the rabbit. Parental generations in solid rectangles.* F_1 *generations within dashed ovals.* F_2 *generations not enclosed.*

Phenotypes and their associated genotypes, therefore, for this series in rabbit are as follows:

Phenotype	*Genotype*
Agouti	c^+c^+, c^+c^{ch}, c^+c^h, c^+c
Chinchilla	$c^{ch}c^{ch}$, $c^{ch}c^h$, $c^{ch}c$
Himalayan	c^hc^h, c^hc
Albino	cc

Ten different genotypes occur in this series. Earlier (Chapter 2), it was seen that a single pair of alleles at a given locus produces three genotypes. By the same token, a series of three multiple alleles produce six genotypes. Note that as the number of alleles in a series of multiple alleles increases, the variety of genotypes rises still more rapidly:

Number of alleles in series	*Number of genotypes*
2	3
3	6
4	10
5	15
n	$\frac{n}{2}(n + 1)$

With the number of possible genotypes increasing more rapidly than the number of alleles, a considerable increase in genetic variability ensues. Consider, for example, a hypothetical organism having only 100 loci, with a series of exactly four multiple alleles at each locus. Ten genotypes are possible at the first locus; these ten can be combined with any of the ten at the second locus, and so on. The total number of possible genotypes becomes 10^{100}.

Available evidence indicates that a given locus may mutate in several directions many times in the history of a species. The various members of the series in rabbit undoubtedly arose at different times and places as mutations of an ancestral gene, quite possibly c^+. Small wonder, then, that many apparent cases of one-pair differences ultimately turn out to involve series of multiple alleles! Other instances in mammals and plants are referred to in the problems at the close of this chapter, while some interesting series in humans are described in the following sections.

THE ANTIGEN–ANTIBODY REACTION

Blood consists of two principal components: cells (red, white, and subcellular platelets) and liquid (plasma). Plasma, minus the clotting protein fibrinogen, is referred to as serum. In early attempts at transfusion, in fact as long ago as the eighteenth century, death of the recipient sometimes ensued for no determinable reason. But in 1901, Dr. Karl Landsteiner, working in a laboratory in Vienna, observed that red blood cells (erythrocytes) of certain individuals would clump together into macroscopically visible groups when mixed with the serum of some, but not all, persons.

The basis for this clumping is the **antigen–antibody reaction.** Entry of a foreign substance (**antigen**) into the bloodstream of an animal brings about the production of a characteristic and specific **antibody** that reacts with the antigen. Such antibodies are termed *acquired* because their production depends on the entry of the foreign antigen; they are not otherwise produced (Chapter 11). The antigen is usually a protein, at least in part, and may be some plant or animal protein, a bacterial toxin, the protein of a virus coat, or may be derived from pollen. As discussed in Chapter 11, an antibody is highly specific for a particular antigen (though cross-reactions of varying degree may occur between one antibody and other closely similar antigen molecules). Such a system forms the basis of immunization practices as well as of allergic reactions. On the other hand, in a few cases antibodies are produced naturally and normally by the blood, even in the absence of the appropriate antigen. These *natural* antibodies include several involved in human blood groups, particularly the important ABO groups, which will be discussed shortly.

Depending on the nature of the antigen and of its antibody, numerous sorts of antigen–antibody reaction may occur. If, for example, the antigen is a *toxin* (such as produced by typhoid, cholera, staphylococcus, and many other bacteria, or by such substances as snake venom), neutralizing antibodies are called **antitoxins.** If the antigen is cellular in nature, the antibody may be a **lysin,** which lyses or disintegrates the invading cells, or an **agglutinin,** which causes clumping or agglutination of the cells. These are but a few of the recognized antibody types.

Following Landsteiner's discovery of agglutination of red blood cells and an understanding of the antigen–antibody reaction, further studies revealed the occurrence of two natural antibodies in blood serum and two antigens on the surfaces of the erythrocytes. With regard to antigens, an individual may produce either, both, or neither; he or she may produce either, neither, or both antibodies. After some early and confusing multiplicity of nomenclature for these substances, the system in most general use today designates the antigens as A and B, and the corresponding antibodies as anti-A (or α) and anti-B (or β). Chemically, the A and B antigens are mucopolysaccharides, consisting of a protein and a sugar. The protein portion is identical in both antigens; the sugar is the basis for the antigen–antibody specificity. An individual's blood group is designated by the type of antigen he or she produces, as in-

TABLE 17–1. **Antigens and antibodies of the human blood groups**

Blood group	*Antigen on erythrocytes*	*Antibody in serum*
A	A	anti-B
B	B	anti-A
AB	A and B	neither
O	neither	anti-A and anti-B

dicated in Table 17–1; blood tests for each of the four major groups are shown in Figure 17–6.

A rather uncommon subgroup of A was discovered in 1911 so that group A was subdivided into A_1 and A_2. In 1936, a still rarer subgroup, A_3, was found; an even less common subgroup, A_4, is now known. These subgroups of A react weakly with anti-A_1 so that the presence of an A antigen can be recognized. Several slightly different variants of group B are also reported.

Groups O and A are the most common in the United States population (Table 17–2). Group A_1 is by far the most frequently encountered in A and AB persons; for example, of 1,698 genetics students at Ohio Wesleyan University tested over a twelve-year period, only one possessed antigen A_2 and that was in an A_2B person. No A_3 or A_4 antigens have been encountered in this sample.

Although the A, B, AB, and O groups are important in transfusions and other medical situations, subgroups are relevant to certain kinds of

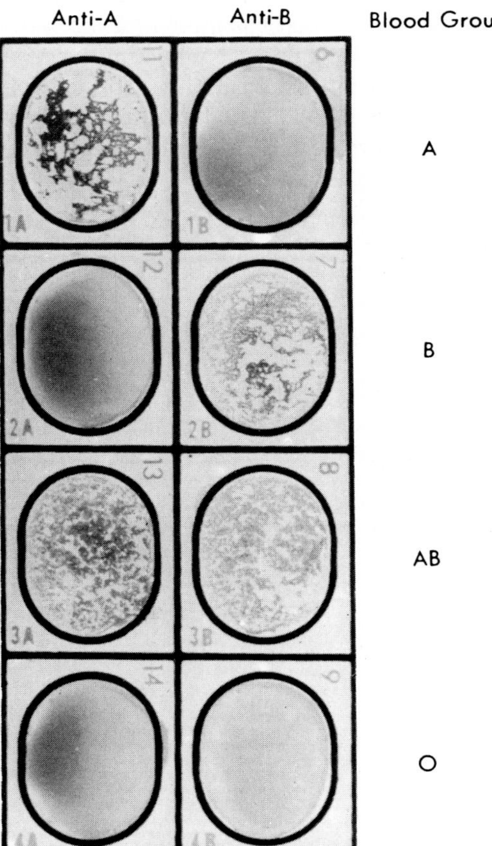

FIGURE 17–6. *The A, B, AB, and O blood group tests. Erythrocytes of group A blood are agglutinated by anti-A serum, of group B by anti-B serum, of group AB by both sera, and of group O by neither serum.* (Photo by Art Green, courtesy Pfizer Diagnostics.)

TABLE 17–2. **Frequencies of the four major blood groups in United States samples**

Sample description	*A*	*B*	*AB*	*O*
Rochester, N.Y. (mixed black and white) (n = 23,787)*	0.418	0.100	0.038	0.444
Blacks (Iowa)†	0.265	0.201	0.043	0.491
Whites (Massachusetts)†	0.397	0.106	0.034	0.463

*From Altman and Dittmer, eds., 1964.
†From Mourant, 1954 (sample size not available).

legal problems. Therefore, attention will be given to the genetics of these four principal groups and some of their subgroups.

INHERITANCE OF A, B, AB, AND O BLOOD GROUPS

Studies of large numbers of human pedigrees have shown that children produce the A antigen only if at least one parent also produced it. Similarly, the B antigen is found only in individuals when at least one parent has it. However, group O individuals may occur in the progeny of A and/or B parents, but O parents have only O children, suggesting recessiveness of the allele for group O. In some cases, marriages of A and B persons produce children having both A and B antigens, indicating that alleles for these latter antigens are codominant. Table 17–3 presents a summary of progeny phenotypes.

Pedigree analysis clearly shows that an individual possesses, in either the homozygous or heterozygous state, any two of a series of *multiple alleles*. Because the antigens involved are of the type known as isoagglutinogens (or isohemagglutinogens), these alleles are often designated as I^A, I^B, and i. Neglecting for the moment the subgroups, dominance relationships of these three alleles may be represented thus: $(I^A = I^B) > i$. Additional studies that take into account the subgroups of the A antigen indicate that I^A may occur in at least four allelic forms. These are symbolized I^{A_1}, I^{A_2}, I^{A_3}, and I^{A_4}. I^{A_1} is dominant to all other I^A alleles, I^{A_2} is recessive to I^{A_1} but dominant to the other two, and so on. Considering four forms of I^A, one of I^B, and one of i, dominance within the series may be shown in this way:

$$[(I^{A_1} > I^{A_2} > I^{A_3} > I^{A_4}) = I^B] > i$$

Omitting the very rare I^{A_4}, this series of multiple alleles produces 15 genotypes and 8 phenotypes:

Genotype	*Phenotype*	*Genotype*	*Phenotype*
$I^{A_1} I^{A_1}$		$I^{A_1} I^B$	A_1B
$I^{A_1} I^{A_2}$			
$I^{A_1} I^{A_3}$	A_1	$I^{A_2} I^B$	A_2B
$I^{A_1} i$		$I^{A_3} I^B$	A_3B
$I^{A_2} I^{A_2}$		$I^B I^B$	
$I^{A_2} I^{A_3}$	A_2	$I^B i$	B
$I^{A_2} i$			
		ii	O
$I^{A_3} I^{A_3}$			
$I^{A_3} i$	A_3		

TABLE 17–3. **Inheritance of blood group phenotypes in humans***

Parental groups ↓→	A_1	A_2	A_3	B	O	A_1B	A_2B	A_3B
A_1	A_1 A_2 A_3 O	A_1 A_2 A_3 O	A_1 A_2 A_3 O	A_1B A_2B A_1 A_3B B A_2 O A_3	A_1 A_2 A_3 O	A_1 A_1B A_2B A_3B B	A_1 A_2 A_1B A_2B A_3B B	A_1 A_2 A_3 A_1B A_2B A_3B B
A_2		A_2 A_3 O	A_2 A_3 O	A_2 A_2B A_3 A_3B B O	A_2 A_3 O	A_1 A_2B A_3B B	A_2 A_2B A_3B B	A_2 A_2B A_3 A_3B B
A_3			A_3 O	A_3 A_3B B O	A_3 O	A_1 A_3B B	A_2 A_3B B	A_3 A_3B B
B				B O	B O	A_1 A_1B B	A_2 A_2B B	A_3 A_3B B
O					O	A_1 B	A_2 B	A_3 B
A_1B						A_1 A_1B B	A_1 A_1B A_2B B	A_1 A_1B A_3B B
A_2B							A_2 A_2B B	A_2 A_2B A_3B B
A_3B								A_3 A_3B B

*Group A is arbitrarily shown as consisting of three subtypes. Phenotypes of parents are listed across the top and at the left; phenotypes of possible children are shown in the body of the table.

Curiously enough, *apparent* changes in one's A-B-O phenotype are associated with certain pathologic conditions. Best known is a weakening of A or B antigenic reaction by a variable proportion of erythrocytes in persons suffering from acute myelocytic leukemia. This is especially well documented in the case of group A persons; in a few instances the diminished response to anti-A antiserum is reported to be accompanied by a weak positive reaction to anti-B antiserum in leukemic persons of group A. During remissions normal antigenic response of the red blood cells returns, only to fall again during relapse. The basis of this change is not known, but it may be related to characteristic chromosome changes (described in Chapter 15) occurring in myelocytic leukemia. In addition to changes associated with leukemia, certain bacterial infections sometimes cause group A cells to acquire a weak response to anti-B serum. This has been explained on the basis of adsorption of a bacterial polysaccharide, chemically similar to the B antigen, on the A red cells (Race and Sanger, 1968). Except for such unusual situations, one's blood antigen–antibody traits are constant and accurate reflections of genotype.

THE H ANTIGEN

Antigens A and B are synthesized from a precursor mucopolysaccharide in the presence of the dominant allele of another pair designated as *H* and *h*. With genotypes *HH* or *Hh*, the precursor is converted to an H antigen, which, in turn, in the presence of I^A and/or I^B, is *partly* converted to antigens A and/or B (Figure 17–7). Gene *h* is termed an amorph because it is responsible for no demonstrable product. So long as individuals are of genotype H−, group A persons produce antigens A and H, group B persons test positively for antigens B and H, and group AB persons produce antigens A, B, and H. Individuals of group O produce only antigen H if of the genotype *iiH*−. On the other hand, blood of persons of genotype − −*hh* (Figure 17–7) does not react with anti-A, anti-B, or anti-H; this is the very rare *Bombay phenotype*, so named because it was first described in a family from that city. A similar case was described in 1955 in an American family of Italian descent, part of whose pedigree is shown in Figure 17–8. Allele *h* has an extremely low frequency; the Bombay phenotype has been described in the literature for fewer than 30 cases. The frequency of the Bombay phenotype has been estimated from large samples as only 1 in 13,000 in the Bombay area, and no cases have been found in England in more than 1 million persons tested.

Because the H antigen is partly converted to antigens A and/or B in the presence of suitable genotypes, as shown in Figure 17–7, it should be expected that not all − −*H*− genotypes would exhibit an equally strong reaction to anti-H serum, and this is true. Blood of persons of genotype *iiH*− (group O) shows the strongest anti-H reaction and blood of AB persons shows the weakest. Within the subtypes of A, A_3 reacts more strongly than A_2, which, in turn, reacts more strongly than A_1. A_1 blood gives a stronger reaction to anti-H than does group B.

MEDICOLEGAL ASPECTS OF THE A-B-O SERIES

From Table 17–3 and a knowledge of the dominance relationships of the multiple allelic series involved, from which the data of the table are derived, applications to cases of disputed parentage can readily be seen. Although mix-ups are rare today in hospitals, situations have occurred in the past in which one or more sets of parents believed they were given someone else's child upon discharge of the mother and new baby from the hospital. For example, a court case of some years ago involved such a situation. Two sets of parents had taken babies home from a

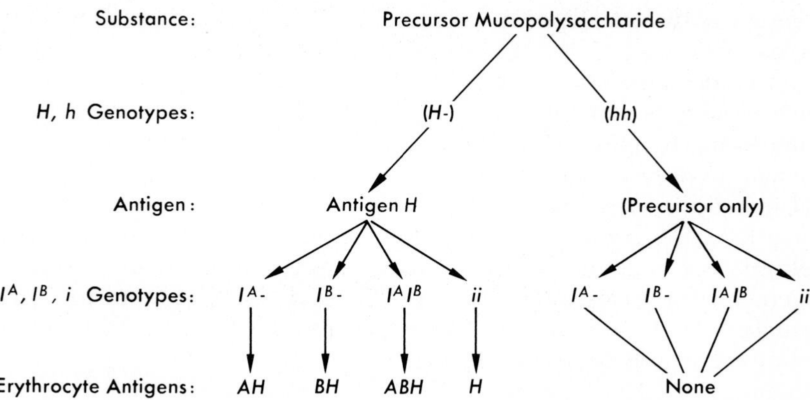

FIGURE 17–7. *Pathways leading to production of antigens on the red blood cells. The precursor mucopolysaccharide is converted into H substance in the presence of genotype H− and is, itself, partly converted into A and/or B antigens in the presence of genes I^A and/or I^B, with I^A1 more effective than I^A2, and I^A3 least effective. The very rare gene h (for lack of H substance) is epistatic to the multiple alleles at the A-B-O locus. Cells of persons of − − h h genotype give no reaction with anti-A or anti-B sera (even though they possess genes IA or IB; this is the rare Bombay phenotype.*

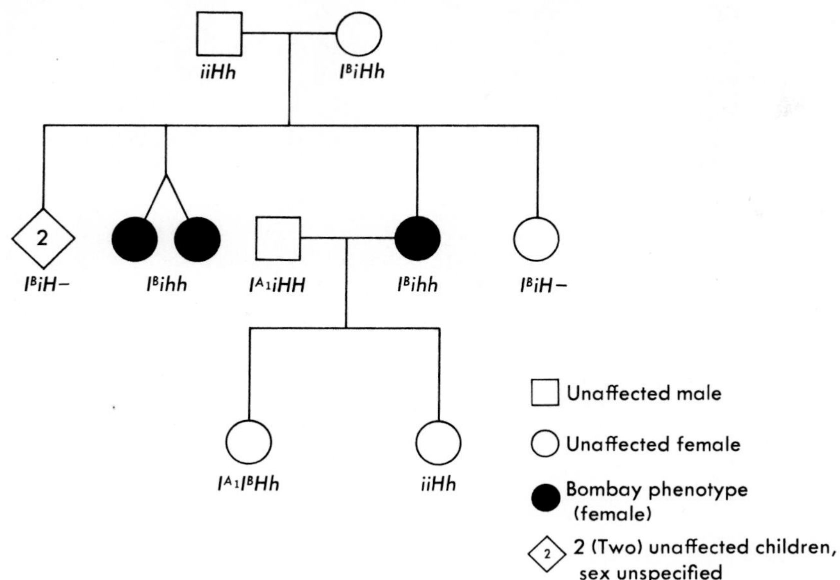

FIGURE 17–8. *Partial pedigree of a family with three children having the Bombay phenotype. The fact that these are all girls is merely coincidental. (Based on work of Levine, Robinson, Celano, Briggs, and Falkinburg, 1955.)*

particular hospital at about the same time; in the process, the identification bracelet of the infant taken by family 1 had become detached. Family 2 soon discovered family 1's name on the child they had received, but the latter family would not agree to an exchange. Fortunately, blood tests quickly demonstrated that neither child could have belonged to the family that had taken it home, but each *could* belong to the other parents:

	Parental blood groups	*Blood group of child taken home*
Family 1	A × AB	O
Family 2	O × O	B

An exchange satisfied both families. Obviously, the tests in this case did not *prove* that the child received by family 1 belonged to family 2, but only that it was possible. Suppose the two families had been A × B and O × B, and that the children given to each had been found to be O and B, respectively. With no additional information from other tests (described later in this chapter) it would be manifestly impossible to make a valid decision, because either family could have produced either child.

Quite clearly, too, blood tests are of considerable value in cases of illegitimacy. Again, tests cannot *prove* a man to be the father, but can in some instances show that he is not the father. Based only on the A-B-O system, probability of exclusion of a wrongly accused man in a paternity case is only about 0.18. Adding tests for MNSs antigens and the Rh system raises this probability to about 0.53. Adding the Kell, Lutheran, Duffy, and Kidd systems, plus the secretor trait (all described later in this chapter), raises the probability to about 0.71. Making use of human leukocyte antigen tests (HLA; these antigens are also discussed later in this chapter) raises the accuracy of exclusion to at least 98 percent. In California, where a battery of genetic tests may be required in paternity suits, the accuracy level is reported as 99.98 percent. In that state, only 1 percent of paternity legal actions go to trial. Now, with the new DNA fingerprinting techniques (see Box in Chapter 20), the accuracy could be raised to 100 percent.

In 1982, Ohio joined the few states with a sound scientific basis for hearing paternity actions. Under that state's "Uniform Parentage Act" the court "may upon its own motion or upon the motion of any party to the action," require a *series* of sophisticated genetic tests, the results of which may be used *either to establish or to exclude* parentage. The wording of the law regarding these tests is unequivocal:

> (Section 3111.09) (E) As used in this chapter, "genetic tests" means a series of serological tests that are either immunological or biochemical or both immunological and biochemical in nature, and that are specifically selected because of their known genetic transmittance. "Genetic tests" include, but are not limited to, tests for the presence or absence of the common blood group antigens, the red blood cell antigens, human leukocyte antigens, serum enzymes, and serum proteins.

The law further stipulates (Section 3111.10) that:

> In an action brought under this chapter, evidence relating to paternity may include:
>
> (A) Evidence of sexual intercourse between the mother and alleged father at any possible time of conception;
>
> (B) An expert's opinion concerning the statistical probability of the alleged father's paternity, which opinion is based upon the duration of the mother's pregnancy;
>
> (C) Genetic test results, weighted in accordance with evidence, if available, of the statistical probability of the alleged father's paternity;
>
> (D) Medical evidence relating to the alleged father's paternity of the child based on tests performed by experts. If a man has been identified as a possible father of the child, the Court may, and upon the request of a party shall, require the child, the mother, and the man to submit to appropriate tests. . . .

Before legislatures and courts had "caught up" with scientific fact, many injustices had occurred. One of the most celebrated cases of this kind occurred in 1944 in California. A widely known movie star was accused by a former starlet-protégé of being the father of the starlet's daughter. The plaintiff's case rested on a remarkable memory for dates and details, a memory so precise that all other potential fathers were eliminated. Three physicians ran blood tests of the alleged father, the mother, and the baby with these results:

	Blood group
Alleged father	O
Mother	A
Daughter	B

A moment's reflection on the genetics of these phenotypes, or reference to Table 17–3, shows clearly that the defendant could not possibly have been the father (barring a highly improbable mutation). Rather, the real father must have belonged either to group B or AB. In spite of such scientific evidence, the jury in a second trial (the first ended in a "hung jury") found the defendant *guilty*! As far as can be determined, the defendant was required to contribute for twenty-one years to the support of a child that was not his own.

Blood tests involving the A-B-O series, as well as others to be described, may, of course, also be used to good advantage in cases of claimants to estates or in certain kinds of criminal proceedings, particularly since blood group may usually be determined from corpses. In fact, blood type can often be determined from mummies, and this has become an important anthropological tool in some investigations. Problems at the end of this chapter explore some hypothetical possibilities.

OTHER BLOOD PHENOTYPES

The Secretor Trait

As study of the A, B, AB, and O blood groups continued, it was noted that in *some* persons whose erythrocytes bear A, B, and/or H antigens, these antigens could also be detected in aqueous secretions such as those from eyes, nose, and salivary glands. Persons with this trait are referred to as *secretors*; they produce water-soluble antigens. Several reports in the literature indicate that about 77 to 78 percent of all persons tested are secretors. Individuals lacking this trait are termed *nonsecretors*, and their antigens are only alcohol-soluble. Note that the secretor-nonsecretor phenotype can be determined only for *HH* or *Hh* genotypes. The blood group of *H*− secretors can be determined even from dried saliva.

Pedigree studies indicate a single pair of alleles, *Se* and *se*, to be responsible. The secretor trait is completely dominant. This pair of alleles markedly increases the number of blood-group phenotypes.

The MNSs System

In the course of their investigations of human blood antigens, Landsteiner and Levine in 1927 discovered two antigens, M and N, which, when injected into rabbits or guinea pigs, stimulated antibody production in the serum of the experimental animal. Because human beings do not produce antibodies for M and N, these antigens are of no importance in transfusion. They are, however, of some interest in genetics, since inheritance of the trait depends on a pair of codominant alleles sometimes referred to as L^M and L^N (for Landsteiner), but more frequently as M and N, producing phenotypes as follows:

Phenotype	*Antigen produced*
M	Antigen M only
MN	Antigens M *and* N
N	Antigen N only

No allele for the absence of either antigen is known. However, a large variety of alleles of the *M-N* locus, varying from uncommon to exceedingly rare, do exist and their antigens can be detected by appropriate antisera.

Another pair of antigens, S and s, with intimate genetic relation to the M-N series, was discovered in 1947. Unfortunately, the designation of antigens has developed rather randomly, so that we have here two sets of antigens, one denoted by two different capital letters (M and N), the other by the same letter in upper- and lowercase (*S* and *s*). Studies show that all human beings produce either antigen S, antigen s, or both;

therefore, another pair of codominant alleles (now generally designated as *S* and *s*) is involved.

As indicated, there is a close genetic relationship between the M-N and *S-s* genes. For example, in families where the parents are phenotypically MNSs and NS, the children, with very rare exceptions, fall into either of two categories: (1) MNS and NSs, or (2) MNSs and NS. Clearly in the first of these two cases, children received genes for either MS or Ns from the heterozygous parent, whereas in the second they received either genes for Ms or NS. Earlier explanations favored a series of four multiple alleles at a single locus: M^S, M^s, N^S, and N^s, and indeed such results as those just cited could rest on a multiple allele mechanism. However, Race and Sanger, in their extensive studies on human blood groups, prefer an alternate explanation, that is, two pairs of very closely linked codominant alleles, *M-N*, and *S-s*. This conclusion is based on rare instances of recombination between the two suggested loci. On the basis of the latter assumption, the two cases described in this paragraph may be diagramed as follows:

Case 1: P $MS/Ns \times NS/NS$
F_1 $\frac{1}{2}\, MS/NS + \frac{1}{2}\, Ns/NS$

Case 2: P $Ms/NS \times NS/NS$
F_1 $\frac{1}{2}\, Ms/NS + \frac{1}{2}\, NS/NS$

Other Antigens

Many other blood antigens have been described in the literature; some are quite rare. These are usually designated by the family name of the individual in whom the antigen or antibody was first demonstrated. Examples include the Kidd factor (about 77 percent of the tested United States population is Kidd-positive) and the Cellano, Duffy, Kell, Lewis, and Lutheran factors (all named for the family in which each was first discovered), to name a few.

These and a large number of additional antigens and/or antibodies produce a great diversity of human blood groups. In one test of 475 persons in London for A_1-A_2-B antigens, the M-N-S-s series, Rh, Kell, Lutheran, and Lewis groups, 269 types were reported, 211 of which included only a single person each. Considering the presently known groups and the fact that additional ones continue to be reported, the already large number of blood phenotypes may some day rise to the point where an individual's blood group may identify him or her as certainly as do fingerprints. This typing has been applied successfully in recent cases of identifying missing children. Only identical twins, triplets, and so forth, who have identical genotypes, would be indistinguishable by appropriate tests.

THE MAJOR HISTOCOMPATIBILITY COMPLEX (MHC) IN HUMANS

Organ Transplants

Since the 1950s, it has been demonstrated that organ transplants (for example, kidney, heart, liver, lung, and pancreas, as well as bone marrow), between animals of the same species, usually are safe and relatively practical. However, although life-span for most recipients is extended beyond that expected with the malfunctioning organ, rarely is the life expectancy equal to that of persons with normally functioning organs of their own.

Kidney transplants are exceptions to the last statement because (1) most recipients still have one functioning kidney, and (2) even when the graft is sooner or later rejected, the patient can go back on dialysis. Rejection rate for kidney transplants is only about 10 percent a year, and rejection (that is, death of the transplanted kidney) does not result in death of the recipient. Kidney transplants are the most successful and lung transplants are the least successful when longevity is the criterion for judgment.

Although the same condition that made the transplant necessary originally often appears later in the transplanted organ, an early rejection occurs unless donor and recipient have a high degree of tissue compatibility. Use of immunosuppressants such as cyclosporine has markedly reduced rejection. This tissue compatibility depends in large part on similarity in the human leukocyte antigens (HLA) of the major histocompatibility complex. This histocompatibility is highest in monozygotic sibs (who have identical genotypes) and lowest in unrelated subjects. In general, tissue compatibility decreases as closeness of relationship decreases.

Genetic Basis

The genetics of histocompatibility is one of the most complex of all fields of human genetics and is by no means fully understood at present. Briefly, tissue compatibility/incompatibility rests on cellular immune reactions that involve lymphocyte (one type of white blood cell) antigens. Such antigens occur on the surfaces of the lymphocytes and enable these cells to "recognize" tissues whose cells carry "foreign" antigens. The lymphocytes then attack and kill foreign tissues or organs, and thus bring about rejection. Histoincompatibility causes rejection about as rapidly as does failure to match donor and recipient for the A-B-O blood groups.

The genetic basis of the immune reaction derives from at least four loci on the sixth human chromosome. After some early disagreement as to the precise location of the HLA genes on chromosome 6, Pearson and colleagues discovered an unusual inversion duplication of chromosome 6, with trisomic codominant expression of the HLA antigens, and presented convincing evidence that the HLA loci are between segment p21 and the terminus of the short arm of the sixth chromosome (that is, 6p21→pter). The four (or more) loci together comprise the **major histocompatibility complex (MHC)** and are designated HLA-A, HLA-B, HLA-C, and HLA-D. At each of these loci there may occur any of a series of codominant multiple alleles, each responsible for producing a particular lymphocyte antigen. These four loci map within an interval of less than 2 map units, and the sequence (reading from the terminus of the short arm of the chromosome toward the centromere) is HLA-A, HLA-C, HLA-B, HLA-D.

Because of the large number of alleles, any of which may occur at each locus, the probability that any two persons (except for monozygotic sibs) have identical MHC genotypes is so low that it is effectively zero. Many million different HLA phenotypic classes are theoretically possible. The MHC loci are so closely linked that crossing-over occurs only rarely. Every individual, therefore, ordinarily inherits from each parent an unaltered number 6 autosome bearing four particular codominant HLA alleles. Each such chromosomal contribution to the next generation is termed a **haplotype.** Everyone thus has two haplotypes, one from each parent and written as, for example, A1B8C2D4/A7B27C4D7.

Assume for the sake of illustration that any of 20 multiple alleles may occur at each MHC locus (at least 20 had been identified at the HLA-A locus, 40 at HLA-B, 8 at HLA-C, and 12 at HLA-D). With this hypothetical figure of 20, the number of possible combinations *within a single haplotype* would be 20^4, or 160,000. This figure, of course, makes some assumptions about the number of multiple alleles at each locus. Although these assumptions are not necessarily accurate, the figure of 160,000 does suggest the very high degree of polymorphism (multiple classes) that exists at the MHC loci. Any of such a large number of haplotypes may be combined with the haplotype from the other parent to yield a tremendously large number of phenotypic classes. Adding tests for MHC antigens to other blood tests for determining paternity would raise the probability of excluding a wrongly accused man to 1 or very close to that value. Unfortunately HLA typing is involved and expensive, and so it is not widely used in most cases of disputed parentage.

One of the most exciting and valuable benefits from HLA research is the discovery that certain of the MHC antigens are associated with particular disorders (Table 17–4). A word of caution is needed in interpreting data such as those in Table 17–4. The occurrence of a given HLA type does not *guarantee* development of a particular disease, but does indicate a statistically significant probability of developing the associated condition. For example, persons with HLA B27 are 87.4 percent more at risk of developing ankylosing spondylitis (an arthritic spinal condition sometimes referred to as "poker spine") than are persons without HLA B27. In one study HLA B27 was present in 90 percent of persons afflicted with ankylosing spondylitis, whereas only 9.4 percent of nonsufferers possessed that antigen.

In Table 17–4 increased risk of developing the particular disease listed in the presence of the HLA antigen ranges from 1.4 × (Hodgkins disease) to 87.4 × (ankylosing spondylitis). An individual with given HLA markers will develop the associated disorder under appropriate dietary, environmental, or infectious conditions. What one inherits, then, is not the disease itself, but a genetic liability to develop it. A determination of the MHC genotype would permit identification of individuals who are genetically *at risk* for a number of serious disabling

TABLE 17–4. **Association of some human diseases with HLA antigens***

Disease	*HLA*	*Increased risk due to presence of HLA antigen*
Hodgkin's disease	A1	1.4×
Ankylosing spondylitis	B27	87.4×
Subacute thyroiditis	B35	13.7×
Psoriasis vulgaris	**Cw6	13.3×
Dermatitis herpetiformis	†D/DR3	15.4×
Coeliac disease	D/DR3	10.8×
Insulin-dependent diabetes	D/DR3, D/DR4, D/DR2	3.3, 6.4, 0.2×
Myasthenia gravis	D/DR3	2.5×
Multiple sclerosis	D/DR2	4.1×
Rheumatoid arthritis	D/DR4	4.2×

*Adapted from Ryder et al. (1980).

**The designation *w* (for example, Cw6) stands for "workshop" and indicates a provisionally identified specificity.

†D and DR designations refer to the method of identification of the antigen; the *R* signifies "related."

or lethal conditions. As noted earlier, the necessary tests are expensive and complex, keeping identification of human leukocyte antigens out of reach for most persons for the present.

DROSOPHILA

Eye Color

In *Drosophila* the wild-type eye is red; this phenotype is produced by a completely dominant allele at about locus 1.5 on chromosome I (the X chromosome; see also Chapter 7). Among the many eye color mutants known in *Drosophila* are several of concern here including coral, cherry, apricot, eosin, ivory, and white. Early work suggested that genes for these and some other eye colors formed a multiple-allelic series, and all mapped at the same locus. Under this hypothesis (+) (or w^+) designated the allele for wild-type red; w^a, apricot; and *w*, white. As data were accumulated for crosses where progeny were expected to be of only two phenotypic classes, apricot and white, an occasional wild-type red appeared. The frequency of such red-eyed flies was low, that is, less than 1 per 1,000, but too high to be explained by mutation.

By using the "marker genes" *y* (yellow body) and *spl* (split thoracic bristles) at loci 0.0 and 3.0, respectively, in the heterozygous condition, it was found that genes for apricot and white were at different, although very close, loci and thus were not part of a multiple-allelic series. The symbol for apricot was therefore changed to *apr*. From this work it was soon determined that (1) *apr* +/+ *w* flies had apricot eyes, (2) + +/*apr w* flies had red eyes, and (3) linkage relationships of *apr* and *w* did not change unless *y* and *spl* did also. Thus + + + +/*y apr w spl* flies were red-eyed, whereas + + *w spl*/*y apr* + + flies, for example, had apricot eyes, demonstrating separate loci for *apr* and *w*.

Frequency of crossing-over between *apr* and *w* was found to be less than 0.001. Loci *apr* and *w* are thus *functionally allelic* in that they both affect the same general phenotypic character (eye color). They are, however, *structurally nonallelic*, that is, they can be shown by recombination to occupy different loci. In the *cis* configuration, the red-eye phenotype is produced, but the *trans* configuration results in a different phenotype, apricot. The two loci exhibit **complementation** in the *cis* position, but **noncomplementation** in the *trans* position (see Chapter 11). This is referred to as a *cis-trans* position effect. If the *cis* and *trans* configurations produce different phenotypes, the two forms are parts of the same gene; but should the *cis* and *trans* configurations produce identical phenotypes, the two are considered to be different genes (Figure 17–9).

The Lozenge Loci

Study of complex loci began in 1940 with work on the so-called lozenge locus in *Drosophila*. In this organism the wild-type eye is broadly oval in shape (Figure 17–10); in the recessive phenotype lozenge the eyes are narrowly ovoid with an irregular surface. The genes responsible are on chromosome I, mapping at about locus 27.7. Subsequent work showed four separate, closely linked loci (Figure 17–11). These loci exhibit the *cis-trans* effect as in the preceding case. For example, the *cis* heterozygote $1z^8$ $1z^k$/+ + has the wild-type phenotype, but the *trans* heterozygote $1z^8$ +/+ $1z^k$ is lozenge. About 20 alleles are known to occur at these four loci.

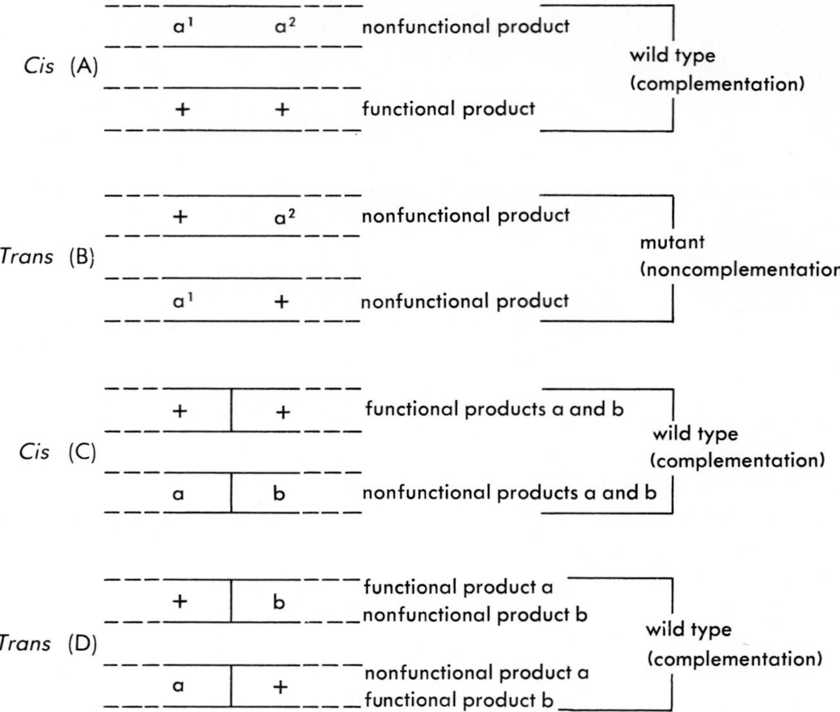

FIGURE 17–9. Cis-trans *effect. A functional product is produced only when the two loci are in the* cis *position (A); in the* trans *position no functional product is produced (B); Loci* a^1 *and* a^2 *exhibit complementation only in the* cis *configuration and are functionally allelic but structurally nonallelic. In (C) and (D) loci a and b exhibit complementation in both the* cis *and* trans *positions and are, therefore, considered to be different genes. See text for further details.*

FIGURE 17–10. *A comparison of the wild-type eye of* Drosophila melanogaster *(left) with the lozenge phenotype (right).*

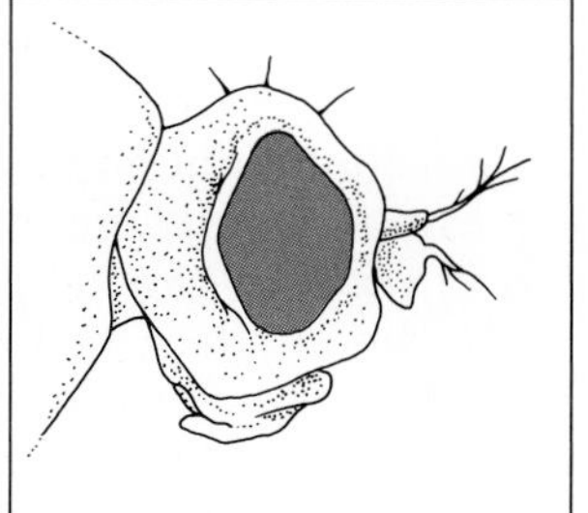

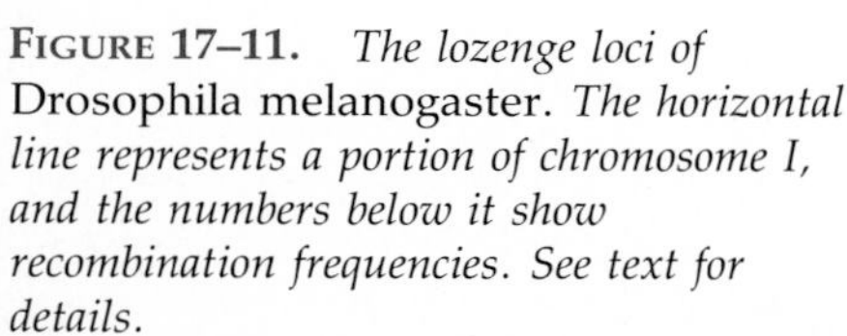

FIGURE 17–11. *The lozenge loci of* Drosophila melanogaster. *The horizontal line represents a portion of chromosome I, and the numbers below it show recombination frequencies. See text for details.*

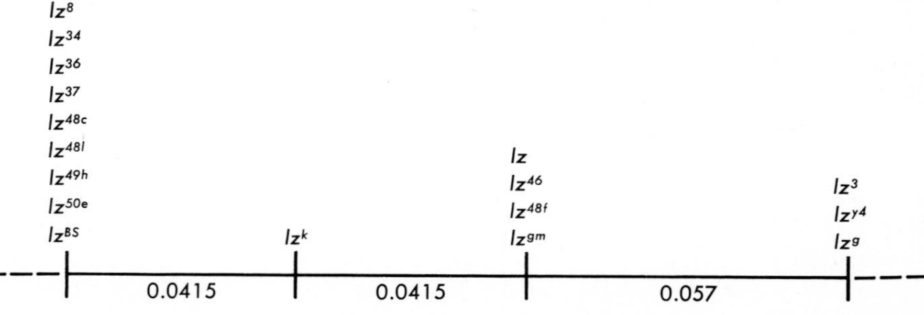

HUMAN BEINGS: THE Rh FACTOR

History

The now well-known Rh factor was discovered in 1940 by Landsteiner and Wiener, who reported that if a rabbit was injected with blood of the *Macaca rhesus* monkey, antibodies were formed by the rabbit. These antibodies would agglutinate the red blood cells of all rhesus monkeys. Thus surfaces of monkey erythrocytes bear a specific antigen, designated as Rh. Tests of human beings show that most persons also produce this antigen; in fact, some 85 percent of white Americans and more than 91 percent of black Americans do. (Reasons for this kind of disparity are explained in Chapter 19.) Persons who produce the Rh antigen are designated as Rh-positive (Rh^+); the much smaller percentage who do not produce the rhesus antigen are termed Rh-negative (Rh−).

Genetic Bases

Evidence almost immediately indicated a genetic basis to the Rh positive phenotype, with Rh^+ as the dominant trait. A single pair of genes, *R* and *r*, was postulated, with Rh^+ persons having genotype *RR* or *Rr* and Rh^- persons having genotype *rr*. Since the work of Landsteiner and Wiener, more Rh antigens have been discovered; the number is now over 30, and the genetics is much more complex than originally believed. Summaries of the complexity of the Rh blood groups are well represented in the literature. Two major explanatory theories have been developed, one by the American investigator Wiener, and a second, in England, by Fisher and others.

The **Wiener** system postulates a series of at least 10 multiple alleles (Table 17–5), some common, others very rare, at a single locus. As noted in Chapter 6 this locus has since been located on chromosome 1. Each, except the completely recessive allele *r*, is responsible for production of one or more of the Rh antigens.

The **Fisher** system, as elaborated by Race and Sanger and others, proposes a group of at least three very closely linked pseudoalleles, *D,d*, *C,c*, and *E,e* (as new antigens and antibodies are discovered, the Fisher system has had to be expanded to include at least two new allele

TABLE 17–5. A comparison of the Wiener and Fisher Rh gene systems and their antigens

Genes			*Antigens* Rh_0	*rh'*	*rh''*	*hr'*	*hr''*	rh^w	*hr*	*rh*			*(Wiener)*
Wiener	*Fisher*	*Frequency**	D	C	E	*c*	*e*	C^w	*ce*	*Ce*	*CE*	*cE*	*(Fisher)*
r	*dce*	0.385	−	−	−	+	+	−	+	−	−	−	
r'	*dCe*	0.007	−	+	−	−	+	−	−	+	−	−	
r''	*dcE*	0.007	−	−	+	+	−	−	−	−	−	+	
r^y	*dCE*	rare	−	+	+	−	−	−	−	−	+	−	
R^o	*Dce*	0.022	+	−	−	+	+	−	+	−	−	−	
R^1	*DCe*	0.405	+	+	−	−	+	−	−	+	−	−	
R^2	*DcE*	0.154	+	−	+	+	−	−	−	−	−	+	
R^z	*DCE*	0.002	+	+	+	−	−	−	−	−	+	−	
R^{1w}	DC^we	0.016	+	+	−	−	+	+	−	+	−	−	
$R^{o''}$	D^uce	rare	weak	−	−	+	+	−	+	−	−	−	

*Approximate, in white populations of western European affinities.

In this table a plus sign indicates production of a given antigen as well as agglutination of red cells by the corresponding antiserum. Some very rare genes are not included here.

Current practice designates as Rh^+ any genotype that includes ability to produce antigen D (Rh_0); all others are designated Rh^-.

Antigen C^w was first described by Callender and Race (1946).

pairs *F,f* and *V,v* not included here). Evidence supports the sequence D-C-E. Instances of compound antigens (for example, *ce*), a single reported case of possible crossing-over (Steinberg), and some apparent cases of position effect are more readily explained than by the pseudoallele concept. For example, Race and Sanger point out that the *cis* heterozygote *Dce/DCE* produces compound antigen *ce* (among others), but in the *trans* configuration (for example, *DCe/DcE*) this antigen is not produced. Loci *c* and *e* thus illustrate the *cis-trans* position effect and are noncomplementary.

The case of possible crossing-over reported by Steinberg and cited above is interesting in that it can be explained either by recombination between closely linked pseudoalleles or by mutation within a single site. In this instance, the husband was determined to have genotype *DCe/dce* and the wife had *dce/dce*. There were eight children, four *dce/dce*, three *DCe/dce*, and one *dCe/dce*. Steinberg was able to eliminate the possibility of illegitimacy of the *dCe/dce* child. Because evidence has become increasingly clear that the correct sequence of loci under the pseudoallele concept must be D-C-E, as first suggested by Fisher, birth of this child must have been preceded by a crossover in the father or by mutation at D in a particular sperm.

Although difficulties are inherent in both the Wiener and Fisher systems, each also has its advantages, and both are in current use. A final judgment between the two cannot yet be made. A comparison of the two systems is shown in Table 17–5. In the Wiener system each gene listed is responsible for formation of more than one antigen.

Antigen D (= Rh_0) is the one that most commonly causes problems in transfusions and in certain pregnancies (see next section); therefore, under ordinary circumstances, persons are typed as Rh^+ if their erythrocytes are agglutinated by anti-D (anti-Rh_0) antiserum. Thus an Rh^- person (for example, of genotype *dce/dce, dCE/dce*, and so on) does not produce antigen D (Rh_0), but does produce other Rh antigens, as shown in Table 17–5. Some interesting results of Rh typing can usually be observed when, say, a genetics class is tested. Erythrocytes of some persons are strongly agglutinated by anti-D antiserum (the one most commonly used, for this purpose), whereas blood of some other persons reacts much more weakly against anti-D. For example, red blood cells of persons of genotype *DCe/dce* (R^1/r) are strongly agglutinated by anti-D, but cells of persons having genotype *DCe/dCe* (R^1/r′) react only weakly. The reason for this sort of difference is unknown; however, under the Fisher system, it can be classed as a position effect. The strong reaction given by *DCe/dce* could then be due to the fact that *dce* are adjacent; the weaker reaction of *DCe/dCe* would result from *d* being next to C. It is not clear how such a position effect operates, although such evidence would support the Fisher concept. No demonstrable antigen is produced by gene *d*.

Some of the commoner genotypes for samples from the American population are listed in Table 17–6. Heiken and Rasmuson have published an extensive list of genotypic frequencies for a large sample of Swedish children (Table 17–7).

Erythroblastosis Fetalis

No cases are known of persons whose blood naturally contains anti-Rh antibodies, though Rh^- individuals can and do develop them if exposed to the corresponding Rh antigen. Such exposure can occur by transfusion, and this, as noted earlier, is the reason the Rh type (ordinarily for

TABLE 17–6. Rh phenotypes and genotypes with percent frequencies in the American population

	Genotypes		$n = 135$	$n = 105$	$n = 766$
Phenotype	*Wiener*	*Fisher*	*Black*	*Oklahoma Indian*	*White*
Rh^+	R^0/r	*Dce/dce*	45.9	2.9	2.2
Rh^+	R^1/R^1	*DCe/DCe*	0.9	34.3	20.9
Rh^+	R^1/r	*DCe/dce*	22.8	5.7	33.8
Rh^+	R^2/R^2	*DcE/DcE*	16.3	17.1	14.9
Rh^+	R^1/R^2	*DCe/DcE*	4.4	36.2	13.9
Rh^+	R^1/R^z	*DCe/DCE*	0.0	2.9	0.1
Rh^-	r/r	*dce/dce*	9.6	0.9	13.9

production or nonproduction of the D, or Rh_0, antigen only) is now routinely determined for blood donors and recipients. But anti-Rh antibody development can also occur in certain pregnancies, often resulting in a fetal condition known as **erythroblastosis.** This is a hemolytic anemia, often accompanied by jaundice, as liver capillaries become clogged with red blood cell remains and bile is absorbed by the blood. The damaged erythrocytes are imperfect oxygen carriers and resemble the immature cells of the marrow, where they are formed. Death may occur before birth or soon after unless appropriate corrective measures are taken.

This disorder occurs only when a number of coincident conditions are met. The mother must be Rh^-, the fetus Rh^+; therefore only marriages of Rh^- women and Rh^+ men are involved. There must also be a placental defect whereby fetal blood, the red cell surfaces of which carry the Rh antigen, passes from the embryo into the maternal circulation. This occurs largely just before or during birth. As a result, Rh

TABLE 17–7. Rh phenotypes, genotypes, and percent frequencies in 8,297 Swedish children

	Representative probable genotype		
Phenotype	*Wiener*	*Fisher*	*Percent of sample*
Rh^+	R^0/r	*Dce/dce*	1.48
Rh^+	R^2/R^2	*DcE/DcE*	3.08
Rh^+	R^2/r	*DcE/dce*	12.50
Rh^+	R^1/r	*DCe/dce*	32.72
Rh^+	R^{1w}/r	DC^we/dce	1.45
Rh^+	R^1/R^2	*DCe/DcE*	14.46
Rh^+	R^z/R^2	*DCE/DcE*	0.05
Rh^+	R^{1w}/R^2	DC^we/DcE	0.66
Rh^+	R^1/R^1	*DCe/DCe*	16.16
Rh^+	R^{1w}/R^z	DC^we/DCE	1.86
Rh^+	R^z/R^1	*DCE/DCe*	0.05
		Total Rh^+	84.47
Rh^-	r/r	*dce/dce*	14.90
Rh^-	r''/r	*dcE/dce*	0.22
Rh^-	r'^w/r	dC^we/dce	0.02
Rh^-	r'/r	*dCe/dce*	0.39
		Total Rh^-	15.53

Based on work of Heiken and Rasmuson, 1966.

antibody concentration is gradually built up in the mother and she becomes *sensitized*. In a second or subsequent pregnancy involving an Rh^+ child, these antibodies may return to the fetus, where they destroy the antigen-carrying red cells. The buildup of antibodies in the mother is gradual; moreover, she is sensitized only at or just before birth of her first Rh^+ child (unless, of course, she previously received an Rh^+ transfusion—not a likely event any longer).

Although fetal erythroblastosis is a severe and tragic condition, occurrence is, fortunately, relatively infrequent. One study in Chicago showed only 92 affected children in 22,742 births in a seven-year period. This is a frequency of 0.004, a fairly average figure for erythroblastosis resulting from Rh incompatibility. The expected frequency of erythroblastotic births, in the absence of any complicating factors, can be calculated from the frequencies of alleles D and d, following the methods introduced in Chapter 5. Disregarding loci C and E, Rh^- persons are of the genotype dd, and in the white American population occur with a frequency of about 0.15. This latter figure is the value b^2 in the expansion of $(a + b)^2$. The frequency of the recessive allele d is then equal to $\sqrt{b^2}$ or $\sqrt{0.15} = 0.39$. Because the example here deals with only one pair of alleles, $(a + b)^2 = 1$, and $a = 1 - b$, or 0.61, which is, then, the frequency of allele D. Only two types of marriages, DD (male) $\times$ dd (female) and Dd (male) $\times$ dd (female), can result in Rh^+ children by Rh^- women (the rare D^u allele is involved here but is not included in the calculations because of its very low frequency). Frequencies of these marriages, based on the expectation of random mating, would be

$$DD \text{ (male)} \times dd \text{ (female)}: a^2 \times b^2 = 0.61^2 \times 0.39^2 = 0.0566$$
$$Dd \text{ (male)} \times dd \text{ (female)}: 2ab \times b^2 = 2(0.61 \times 0.39) \times 0.39^2 = 0.0723$$

In the DD (male) $\times$ dd (female) marriages, all the children are Rh^+, whereas in the second type of marriage the probability of an Rh^+ child is 0.5; therefore, the expected frequency of all erythroblastotic births (if all pregnancies come to term) is $0.0566 + \frac{1}{2}(0.0723) = 0.0927$, or nearly 10 percent of all pregnancies, neglecting the fact that in most cases the first child is unaffected. The 10 percent figure is far greater than the observed frequencies of most studies.

The relative infrequency of erythroblastosis is in part due to other probably genetically based traits, such as the defective placenta. But another important factor is *ABO incompatibility*. ABO-compatible marriages are those in which the husband is of suitable ABO group to donate blood to his wife; incompatible marriages are the reverse, namely, those in which the husband is not of an ABO group suitable for transfusion to his wife, that is,

	♀	♂
Compatible	A	A, O
	AB	A, B, AB, O
	B	B, O
	O	O
Incompatible	A	B, AB
	B	A, AB
	O	A, B, AB

Some studies show far fewer Rh-erythroblastotic children are born to ABO-incompatible marriages than to compatible ones. In ABO-incompatible marriages, fetal erythrocytes that bear an antigen for which the mother's blood contains the corresponding antibody and crosses the placenta will be quickly destroyed before anti-Rh antibody formation can occur.

The discovery of two types of anti-D (anti Rh_0) antibodies almost simultaneously in 1944 by Wiener and by Race has, fortunately, provided a simple preventive measure for Rh-erythroblastosis so that the condition need no longer be a cause of infant death or anemia. These two types of anti-D antibody are (1) *complete,* which agglutinates red cells carrying antigen D, and (2) *incomplete,* which does not. The incomplete antibody, however, attaches to receptor sites on Rh^+ erythrocytes, "sensitizes" them, and prevents them from acting antigenically. Intramuscular injection of human D (Rh_0) immune globulin, containing incomplete anti-D antibody, into Rh^- women within 72 hours after giving birth to a D^- or D^{u-} child effectively blocks the child's red blood cells already in the mother's circulatory system from inducing antibody production on her part. The incomplete antibody is often obtained from Rh^- males who have been injected with Rh^+ blood.

PROBLEMS

17–1 Is it possible to cross two agouti rabbits and produce both chinchilla and Himalayan progeny?

17–2 In the ornamental flowering plant nasturtium, flowers may be either single, double, or superdouble. These differ in number of petals, superdouble having the largest number. Crosses of superdouble × double sometimes yield 1 superdouble:1 double, and sometimes all superdouble. Superdouble × superdouble produces all superdouble, or 3 superdouble:1 double, or 3 superdouble:1 single. Single × single produces only single. (a) How many multiple alleles occur in this series? (b) Arrange the phenotypes in order of relative dominance. (c) Another cross of superdouble × double produces progeny in the ratio of 1 double:2 superdouble:1 single. What do you know about the parental genotypes?

Use the following information in answering the next four problems. In the Chinese primrose the flower has a center, or "eye," of a color different from the remainder of the petals. Normally this eye is of medium size and yellow in color. These variants also occur: very large yellow eye ("Primrose Queen" variety), white eye ("Alexandra"), and blue eye ("Blue Moon"). Results of certain crosses are the following:

P	F_1	F_2
Normal × Alexandra	Alexandra	3 Alexandra:1 Normal
Alexandra × Primrose Queen	Alexandra	(not reported)
Blue Moon × Normal	Normal	3 Normal:1 Blue Moon
Primrose Queen × Blue Moon	Blue Moon	(not reported)

17–3 Arrange these phenotypes in order of relative dominance.

17–4 How many genotypes can produce the "Alexandra" phenotype?

17–5 How many genotypes can produce the "Primrose Queen" phenotype?

17–6 How many different combinations of parental genotypes will produce a progeny ratio of 3 Alexandra:1 normal?

17–7 The frequency of gene I^{A1} was found in one study involving 3,459 persons to be approximately 0.21, of I^{A2} 0.07, and of I^B 0.06. What is the calculated frequency of each of these phenotypes: (a) A_1B; (b) A_2B? (Carry answers to four decimals.)

17–8 How many of the 3,459 persons in the sample of the preceding problem should be A_2B? (Carry answer to the nearest *whole* number.)

17–9 A man has genotype I^AI^BHh, whereas that of his wife is *iihh* (the *ii* was determined by pedigree analysis). What is the probability of their having a child who exhibits the Bombay phenotype?

17–10 Assume that you do not know your M-N phenotype, or whether you are a secretor or a nonsecretor (most people do not know this information about themselves). What is your *most probable* phenotype?

17–11 A hypothetical series of 20 multiple alleles is known for a certain locus. How many phenotypic classes are possible?

17–12 How many different genotypic classes are possible for the locus referred to in Problem 17–11?

17–13 Considering human blood group A to include three subtypes, and groups B and O to include one each, how many phenotypes are included in the A-B-O series?

17–14 Considering three subtypes of group A, which of the blood groups making up your answer to the preceding question would you expect to give the weakest anti-H response?

17–15 A paternity case involves these facts: The woman is A_1, her child is O, and the alleged father is B. Could he be the father? Explain.

17–16 In a paternity case it is determined that the woman is A_1, MS/NS, and a secretor. The man she charges is O, MS/NS, and a secretor. The child is A_2, MS/Ns, and a nonsecretor. Using only this information, can you eliminate the man as the father of this child? Explain.

17–17 In another paternity case it is determined that the woman is group A, MNS, a secretor, and A2B1C2D1/A3B4C3D4; the man is group B, Ms, a secretor, and A1B2C1D2/A4B3C4D3. Her child is group O, MSs, a nonsecretor, and A2B1C2D1/A4B5C6D7. (a) Could he be the father? (b) On which test(s) is the correct answer based?

17–18 In the following case consider that the HLA loci are sufficiently closely linked that crossing-over can be eliminated as a possibility. With regard to the HLA-A and HLA-B loci, a woman's *genotype* is A2B5/A9B12. She and her husband have three children whose *phenotypes* are (1) A1A2 B5B8; (2) A1A9 B8B12; (3) A2A3 B5B7. What is the father's *genotype*?

17–19 The HLA-A and HLA-C loci have been mapped 0.6 map units apart. If an individual's genotype for these two loci is A6B12/A9B5, what would be the expected frequency of A6B5 gametes per thousand produced by that person?

17–20 What is the phenotype of each of the following genotypes in *Drosophila melanogaster:* (a) lz^{BS} + + +/+ lz^k + +; (b) lz^{BS} lz^k + +/+ + + +?

17–21 Recall that in erythroblastosis the important alleles are *D* and *d* and that, therefore, Rh^+ persons are of either genotype *DD* or *Dd*, and that Rh^- persons are *dd*. In the following pedigree the marriage shown is the first for both the man and the woman, and the woman has never had a blood transfusion, nor has she ever received an injection of D immune globulin. Solid symbols represent cases of erythroblastosis fetalis. After examining the pedigree, give the Rh genotype of each member of the family.

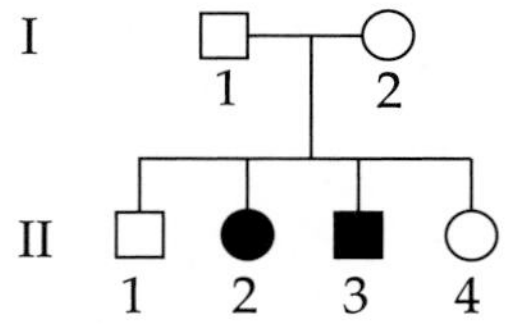

17–22 In the preceding problem why can you be sure of II-1's genotype and phenotype?

17–23 In which of the following marriages is the risk of erythroblastosis fetalis greater: A,Rh − ♀ × O,Rh + ♂, or O,Rh − ♀ × A,Rh + ♂?

17–24 Considering human blood group A to include three subtypes, and groups B and O to include one each, how many phenotypes can be recognized if the A-B-O classes and the two Rh classes are all taken into account?

17–25 A woman of group A_2 charges that a man of group A_2 is the father of her group A_3Rh^+ child. Could he be?

17–26 Both the man and the woman of the preceding problem are Rh^-. Does this change your judgment?

17–27 A couple believes they have brought the wrong baby home from the hospital. The wife is O^+, her husband B^+, and the child O^-. Could the child be theirs?

17–28 An additional test discloses the husband and wife of the preceding problem to be of blood type M, whereas the child is MN. Does this added fact change your judgment?

17–29 In a court action a man claims that some of the six children purportedly belonging to him and his wife are not his children. Blood tests of the husband, wife, and six children gave the following information:

Husband:	O,	*Dce/DcE,*	*MS/Ms*
Wife:	A_1,	*DcE/dce,*	*MS/Ns*
Child 1:	A_1,	*DcE/DcE,*	*MS/MS*
Child 2:	O,	*Dce/dce,*	*MS/Ns*
Child 3:	O,	*DcE/dce,*	*Ms/Ns*
Child 4:	A_1,	*DcE/dce,*	*NS/Ns*
Child 5:	O,	*dce/dce,*	*MS/NS*
Child 6:	A_1B,	*DCe/DcE,*	*MS/NS*

Assuming all were, in fact, born to the wife, could all these children belong to the husband? Explain.

17–30 A student in a genetics class is blood typed as B^+. She reports that her father is B^-, her mother O^+. Using standard gene symbols I^A, I^B, and *i* for the A-B-O system and restricting Rh type to the *D, d* gene pair, give this student's *genotype* for blood group and Rh type.

POLYGENIC INHERITANCE

CHAPTER

18

Of the heritable traits examined thus far, recall that phenotypic classes have always been **discontinuous.** Such traits may be termed **qualitative** ones. Thus *Coleus* leaves have either regular or irregular venation; cattle may or may not have horns, and may be red, roan, or white; rabbits may be distinguished by coat color, as may a host of other animals; people belong to one blood group or another, and so on. This has been the case whether the traits involved form and structure, pigments (that is, an aspect of physiology), antigens and antibodies, and so on; and whether the genes involved show complete dominance, incomplete dominance, or codominance.

Not all inherited traits are expressed in this discontinuous fashion, however. In humans, for example, height is a genetically determined trait. But if an attempt were made to classify a random sample of students on campus according to height, results would quickly disclose a trait showing essentially **continuous** phenotypic variation. It may be asked, then, where does one draw the line between phenotypic classes? Should these classes be one inch apart, one-half inch, one-tenth inch, a millimeter, and so forth? Many other traits are expressed in a similar fashion, including intelligence, skin and eye color in people, color and food yield in various plants, size in many plants and animals, as well as degree of coat spotting in some animals or of seed coat mottling in some plants. The essential difference between continuous and discontinuous inheritance is illustrated with generalized data from crosses of each type compared graphically in Figure 18–1. Clearly, concepts of simple Mendelian inheritance, with the reappearance of the parental phenotypes as distinct and separate classes in the F_1 and F_2 generations, must be modified in order to explain continuous variation in **quantitative** characters, such as height, weight, intelligence, or skin color.

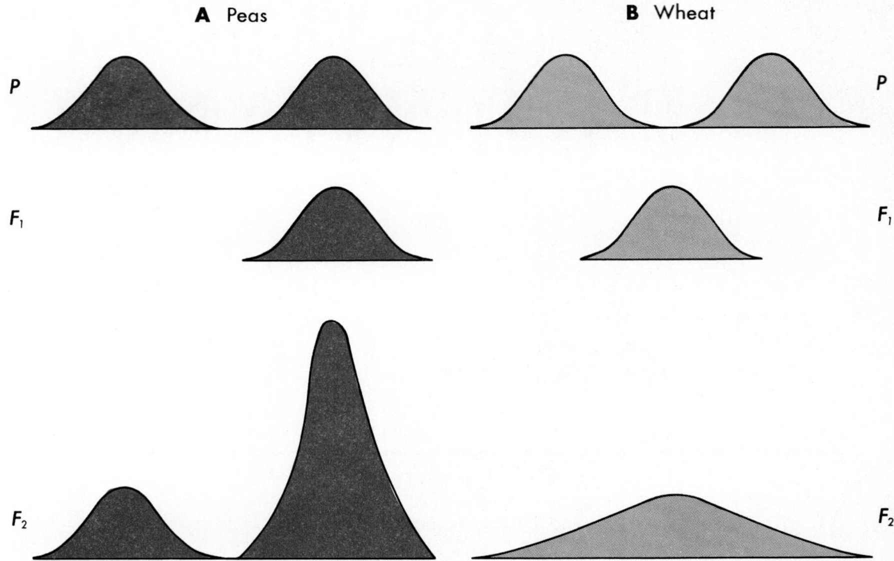

FIGURE 18–1. *Curves comparing results of crosses involving (A) height, a discontinuous trait in peas and (B) kernel color, a continuous trait in wheat, followed for three generations. Ordinates represent the number of individuals; abscissas represent the particular quantitative trait.*

Concern is not just with a person's relative height, but with *how* tall, that is, with *continuous characters of degree rather than discontinuous characters of kind.* Moreover, quantitative inheritance more often deals with a population in which all possible matings occur, and less often with individual matings. This kind of study involves quantification of data and statistical treatment such as means, variances, standard deviations, and others in order to evaluate data properly.

KERNEL COLOR IN WHEAT

Among the earliest investigations that gave a significant clue to the mechanism of quantitative inheritance was the work of Nilsson-Ehle (1909) with wheat. One of his crosses consisted of a red-kerneled variety × a white-kerneled variety. Grain from the F_1 was uniformly red, but of a shade intermediate between the red and white of the parental generation. This might suggest incomplete dominance, but by intercrossing members of the F_1, Nilsson-Ehle produced an F_2 in which five phenotypic classes occurred in a 1:4:6:4:1 ratio. Noting that $\frac{1}{16}$ of the F_2 was as extreme in color as each of the parental plants (that is, as red or as white as the P individuals), he theorized that two pairs of alleles controlling production of red pigment were operating in this cross. The key, of course, is in the fraction ($\frac{1}{16}$) that was as extreme as each of the original parental strains. Symbolizing the genes for red with the capital letters *A* and *B* and their alleles responsible for lack of pigment by *a* and *b*, this cross can be diagramed as follows:

$$
\begin{array}{ll}
\text{P} & AABB \quad \times \quad aabb \\
 & \text{dark red} \qquad \text{white} \\
\text{F}_1 & AaBb \qquad (\times\ AaBb) \\
\text{F}_2 & \frac{1}{16}\,AABB + \frac{2}{16}\,AaBB + \frac{1}{16}\,aaBB \\
 & + \frac{2}{16}\,AABb + \frac{4}{16}\,AaBb + \frac{2}{16}\,aaBb \\
 & + \frac{1}{16}\,AAbb + \frac{2}{16}\,Aabb + \frac{1}{16}\,aabb
\end{array}
$$

Assuming each "dose" of a gene for pigment production increases the depth of color, this F_2 can be sorted phenotypically according to the number of alleles for red in this way:

Genotype	*Number of genes for red*	*Phenotype*	*Fraction of* F_2
AABB	4	dark red	$\frac{1}{16}$
AABb, AaBB	3	medium red	$\frac{4}{16}$
AAbb, aaBB, AaBb	2	intermediate red	$\frac{6}{16}$
aaBb, Aabb	1	light red	$\frac{4}{16}$
aabb	0	white	$\frac{1}{16}$

Alleles symbolized by capital letters, those "*contributing*" to red color in this case, are termed **contributing alleles.** Those that do not "contribute" to red color (here symbolized by the lowercase letters) may be designated **noncontributing alleles.** Some geneticists refer to these as *effective* and *noneffective* alleles, respectively. Here, then, is a polygenic series of as many as four contributing alleles. The term **polygene** was introduced by Kenneth Mather, who has summarized the modern interpretation of quantitative inheritance. This term has since found wide usage and is supplanting the older term *multiple factors.*

In an effort to determine whether the mechanism of polygenic inheritance is the same as that operating in instances of qualitative characters, several simplifying assumptions must be made. *Assume* that

1. There is no dominance; rather, there exist pairs of contributing and noncontributing alleles.
2. Each contributing allele in the series produces an equal effect.
3. Effects of each contributing allele are cumulative or additive.
4. There is no epistasis (masking of the phenotype) among genes at different loci.
5. There is no linkage.
6. Environmental effects are absent or are so controlled that they may be ignored.

Certainly, as the number of pairs of alleles increases, the probability of linkage rises, and the effect of environment can be ignored only in the most closely controlled experiments. Furthermore, some geneticists have found evidence suggesting less than universal operation of the first and/or second of these assumptions. Many polygenic effects, however, do appear to operate in a manner consistent with the first four points, and the task is eased considerably if all six *simplifying assumptions* are made.

In another cross in wheat, reported by Nilsson-Ehle, a different red variety was used, with the result that $\frac{1}{64}$ of the F_2 were as extreme as each parent, with seven classes in a 1:6:15:20:15:6:1 ratio. A little reflection will serve to suggest the operation here of *three* pairs of alleles. From earlier chapters on monohybrid inheritance, recall that, if only one pair of alleles were involved, one fourth of the F_2 should be as extreme as either parent. If information for one, two, and three pairs of polygenes is tabulated, a pattern emerges:

Number of pairs of polygenes in which two parents differ	*Fraction of* F_2 *like either parent*	*Number of genotypic* F_2 *classes*	*Number of phenotypic* F_2 *classes*	F_2 *phenotypic ratio = coefficients of*
1	$\frac{1}{4}$	3	3	$(a + b)^2$
2	$\frac{1}{16}$	9	5	$(a + b)^4$
3	$\frac{1}{64}$	27	7	$(a + b)^6$
n	$(\frac{1}{4})^n$	3^n	$2n + 1$	$(a + b)^{2n}$

Thus with four pairs of polygenes, $\frac{1}{256}$ of the F_2 is as extreme as each parent; with five, only $\frac{1}{1024}$; with 10, the fraction drops to 1/1,048,576; and with 20 pairs only one in 1,099,511,627,776 of the F_2 will have measurements like one parent or the other! The number of genotypic classes increases, of course, with startling rapidity as the number of pairs of polygenes becomes larger: For four pairs of alleles there are 81 F_2 classes; five pairs of alleles produce 243 F_2 genotypes; 10 pairs, 59,049; and 20 pairs, 3,486,784,401! Thus as the number of polygenes governing a particular trait goes up, the progeny very quickly form a continuum of variation in which class distinctions become virtually impossible to make.

Dividing the total quantitative difference by the number of contributing alleles indicates the amount contributed by each effective allele. For example, in the following hypothetical case in pumpkin, how many contributing alleles are operating and how much does each contribute?

P 5-lb fruits × 21-lb fruits
F_1 13-lb fruits
F_2 $\frac{3}{750}$ 5-lb fruits . . . $\frac{3}{750}$ 21-lb fruits

Note that $\frac{3}{750}$, the fraction of the F_2 that has fruits as light or as heavy as the P generation, simplifies to $\frac{1}{250}$. Using the formula $(\frac{1}{4})^n$ (where n = pairs of alleles), note that $(\frac{1}{4})^n = \frac{1}{256}$ if $n = 4$, and $\frac{1}{250}$ is close only to the fraction ($\frac{1}{256}$) in the series of $(\frac{1}{4})^n$. Therefore, four pairs of alleles must be involved, where the plants producing the heaviest fruits have all eight contributing alleles. Since the total weight difference is 16 pounds (21 − 5), $\frac{16}{8} = 2$ pounds are contributed by each effective allele.

Alternatively, $(\frac{1}{2})^n$ could be used to determine directly the numbers of contributing alleles (instead of the number of *pairs*). In this case $(\frac{1}{2})^n = \frac{1}{256}$ only if $n = 8$. In this example with weight of pumpkin fruits, a weight of 5 pounds is termed the *base weight*, and suggests that of all the polygenes that may be involved in fruit weight, the two parental strains were homozygous for all but the four pairs determined by these calculations.

It is also apparent that the *number* of F_2 phenotypic classes follows a pattern, but one that produces a less dramatic increase as the number of pairs of polygenes becomes larger. Thus with one pair, the F_2 of the cross *AA* × *aa* includes three phenotypic classes that correspond to the number of genotypic classes (*AA, Aa, aa*). The two examples from Nilsson-Ehle's work on wheat indicate that the number of F_2 *phenotypic classes* is one more than twice the number of pairs of polygenes, or $2n + 1$.

Note that the *phenotypic ratio* also follows a pattern. One pair gives rise to a 1:2:1 F_2 ratio, two pairs to 1:4:6:4:1, and three pairs to 1:6:15:20:15:6:1. These ratios are the same as the sequence of *coefficients* in binomials of a power equal to *twice* the number of allelic pairs involved. Note the relationship of these coefficients to certain lines in Pascal's triangle, in Chapter 5. Thus expansion of $(a + b)^2$ gives a coefficient sequence of 1:2:1, which is the same as the F_2 phenotypic ratio for one pair of polygenes; expanding $(a + b)^4$ produces the coefficient series 1:4:6:4:1, and so on. So an F_2 phenotypic ratio can be determined for any number of pairs of polygenes by thinking of the sequence of coefficients given by expanding a binomial raised to the power $2n$ where, again, n represents the number of pairs of polygenes.

HUMAN EYE COLOR

People clearly differ in eye color, that is, in the amount of melanin pigment in the iris. Except for albinos, no one is without some eye pigmentation. Those with the least pigment have eyes that appear blue, those with the most have eyes that appear brown. Blue eyes owe their color to the scattering of white light by the nearly colorless superficial cells of the iris. This effect is greatest in the shorter (blue) wavelengths of the visible spectrum, giving the iris its blue appearance. Close inspection of the apparently nonbrown iris of some persons discloses small to very small flecks of brown or somewhat orange coloration, because the small amount of pigment is most abundant in discrete groups of cells. In other persons the pigment is more evenly distributed, and the eyes appear uniformly blue.

Clearly, though, there is a gradation in eye color, ranging from the lightest blue to the darkest brown ("black"). Human beings simply do not fall into "either-or," blue or brown categories, but rather, large samples form a continuum of variation, strongly indicative of polygenic inheritance. The number of phenotypic classes recognized is therefore arbitrary and depends in part on the observational techniques and equipment used, and in part on the observer. Although the inheritance of eye color is complex and only incompletely understood, at least nine phenotypic classes (Davenport and Hughes designate five and seven, respectively) may be recognized as a matter of convenience. In order of increasing amounts of melanin pigmentation, these can be designated as light blue, medium blue, dark blue, gray, green, hazel, light brown, medium brown, and dark brown.

If one considers that the numbers of phenotypic classes is one more than twice the number of pairs of polygenes ($2n + 1$), nine classes would result from the action of four pairs of alleles. Under this hypothesis, a simplified basis for human eye color would be as follows:

Number of contributing alleles in genotype	*Eye color*
0	Light blue
1	Medium blue
2	Dark blue
3	Gray
4	Green
5	Hazel
6	Light brown
7	Medium brown
8	Dark brown

Regardless of the number of pairs of polygenes postulated (and it must be emphasized that studies so far do not permit a definitive determination of the number of pairs), it is clear that eye color is due to polygenes, some of which may interact in poorly understood ways. No conclusive data are yet available on linkage of genes affecting this trait, although some have suggested X-linkage.

OTHER HUMAN TRAITS

Skin color in humans also depends on relative amounts of melanin. Studies on skin color began with C. B. Davenport in 1913, attempting to relate sample frequencies of various degrees of pigmentation to models based on different numbers of pairs of polygenes. His rationale was simply based on a Mendelian dihybrid cross with *AABB* as two gene pairs for black color, and *aabb* as two gene pairs for white color. The actual color was determined by how many *A* or *B* genes a person had, in an additive fashion. Since that time, with better methods to quantitate skin color, from three to six gene pairs are considered responsible for skin color. Davenport's basic conclusion that skin color was a polygenic trait was still correct. Although polygenic inheritance is clearly suggested in humans for many quantitative traits, such as height, intelligence, and hair color (except for red versus nonred), no complete hypothesis setting forth the exact number of pairs of alleles and measurement of their individual and collective effect has yet been developed for any of these traits.

TRANSGRESSIVE VARIATION

Some progeny may be more extreme than either parent or grandparent. Recall examples in humans where, say, some children are shorter or taller than either parent or any of their more remote ancestors. The same phenomenon sometimes occurs, too, with respect to intelligence, skin color, and eye color. Such examples illustrate **transgressive variation.**

One of the earliest instances of transgressive variation to be reported in the literature was one described in 1914 and 1923 by Punnett and Bailey. They crossed the large Golden Hamburg chicken with the smaller Sebright Bantam. The F_1 was intermediate in weight between the parents and fairly uniform, but a few of the F_2 birds were heavier or lighter than either of the parental individuals. Their results suggested to Punnett and Bailey four pairs of alleles, with the Golden Hamburg being of, say, genotype *AABBCCdd* and the Sebright Bantam being of genotype *aabbccDD*. Likewise, children may have darker or lighter eyes than either parent. Some of the problems at the end of this chapter deal with interesting illustrations of transgressive variation.

OTHER ORGANISMS

Instances of polygenic inheritance are known from many other plant and animal species. One of the more suggestive cases was discovered in tomato by Lindstrom. Crosses between the larger-fruited Golden Beauty (average fruit weight 166.6 g) and the smaller-fruited Red Cherry (average fruit weight 7.3 g) varieties produced an F_1 having fruits intermediate in size between the parental types, but distinctly closer to the smaller-fruited variety (average weight 23.9 g). Such results could be explained by assuming dominance or unequal effect among at least some of the noncontributing alleles.

In cattle both solid color and spotting occur. Solid color is due to a dominant allele, *S*, and spotting to its recessive allele, *s*. Studies indicate that the degree of spotting in *ss* individuals depends on a rather large series of polygenes (Figure 18–2). Those cattle that are *S*− will, of course, not be spotted, regardless of the remainder of the genotype with regard to spotting.

C. K. Chai has shown that the difference between high and low leukocyte count in the mouse is a polygenic trait. He suggests dominance, however, for low count and makes the interesting point that "the evidence is accumulating . . . that a quantitative trait is an aggre-

FIGURE 18–2. *Variation in degree of spotting in a herd of cattle. Amount of spotting depends on a series of polygenes, but these are hypostatic to S (solid color).* (Photo courtesy Ayrshire Breeders' Association.)

gate of effects from different biological systems, each of which contributes specific effects that differ in magnitude and biological effect under the control of individual genes. The present results, although not considered as definitive, further indicate this to be the case."

In summary, the basic mechanisms operative in quantitative inheritance appear to be the same as those for qualitative characters. Such studies as are reported here also further emphasize the fact that many traits are the result of *interaction* of one kind or another among several pairs of alleles.

In the case of *quantitative characters* such as those dealt with here, the possible effect of environment must be considered and carefully regulated in any controlled experiment. For example, height in many plants (corn, tomato, pea, marigold, zinnia) is a genetically controlled character, but it is obvious that such environmental factors as soil fertility, texture, water, temperature, duration and wavelength of incident light, occurrence of parasites, to name just a few, also affect height. *Genotype* determines the range an individual will occupy with regard to a given quantitative character; *environment* determines the *point* within the genotypically determined range at which an individual's measurement will fall.

STATISTICAL CONCEPTS

Analysis of the inheritance of polygenes that govern continuously variable, quantitative traits requires application of certain techniques from the branch of mathematics called statistics. Statistics is useful in two fundamental problems commonly encountered in scientific research: (1) What can be learned about a population from measurements of a sample from the population and (2) how much confidence can be placed on conclusions about that population.

There is a clear difference between **population** and **sample**. A population consists of the entire group of individuals, measured for some variable quantitative character. In this context the important aspect of a population is a series of numbers that represents such a variable, quantitative trait. Thus figures showing, for example, gains in weight of adult laboratory rats that have been fed a specific diet for a certain number of days represent a biological population, as do height measurements of college males or females, or weights of ripe pumpkin fruits. The statistician is concerned with these figures and not with the rats, people, and pumpkins themselves. Even a more discrete problem dealing with a population of an endemic species on an isolated small island includes all the individuals of the species on that island. For obvious reasons, it is either impractical or impossible to accumulate measurements for such a group; rarely is the population sufficiently finite for *all* individuals to be measured.

So descriptions of the population must generally be formulated from *samples* of it. To be useful in making estimates of a population, the sample must have been drawn as randomly as possible. In dealing with heights or weights, for example, sample measurements must not be selected more from the taller, or shorter, or heavier, or lighter individuals, but should reflect the same kind and degree of variability as does the population.

Actual values for populations are constants called **parameters;** estimates of populations based on samples are **statistics.** The statistics are subject to some degree of chance error resulting from sampling practices, but once a statistic is determined, it is possible to state the range of the corresponding parameter with a particular degree of confidence. The geneticist needs to be able not only to estimate the parameters of importance but also to determine how likely it is that one is dealing with individuals from either the same or different populations. In the genetic context, then, statistics may be said to provide

1. A concise description of the quantitative characteristics of the sample.
2. An estimate of
   a. The quantitative characteristics of the population from which the sample was drawn.
   b. How well the sample represents that population.
3. An expression of the probability that two samples differ significantly (within limits set by chance) in terms of a particular hypothesis explaining the reason for observed differences.

The five principal statistics that follow will provide the estimates and descriptions just cited.

Mean

A very elementary statistic, and one with which you are undoubtedly familiar, is the average or **mean**. Calculation of the mean, symbolized by $\bar{x}$ ("x-bar"), may be represented by the formula

$$\bar{x} = \frac{\Sigma x}{n} \tag{1}$$

where Σ (the uppercase Greek letter sigma) directs us to sum all following terms, x the individual measurements, and n the number of individuals in the sample. Often, when n is quite large, it becomes convenient to group data by *classes*. Thus, to determine the average grade on a quiz for a large class, it might be more practical to tally the number or *frequency* of individuals scoring between 96 and 100 in one class or group, those between 91 and 95 in another, and so on. A "class-value" midway between the extremes of each class range is also entered. The tabulated data would then be arranged as follows:

Class range	*Class value* x	*Frequency* f	fx
96–100	98	1	98
91–95	93	4	372
86–90	88	8	704
81–85	83	12	996
76–80	78	18	1,404
71–75	73	25	1,825
66–70	68	17	1,156
61–65	63	10	630
etc.	etc.	etc.	etc.

If the data are grouped in this way, the mean will be given by

$$\bar{x} = \frac{\Sigma fx}{n} \tag{2}$$

Equation (2) loses a little in accuracy over Equation (1), but the much simpler arithmetic of its method more than compensates for this slight inaccuracy.

Although the mean is a necessary statistic, it is a rather uninformative one in that a comparison of means of different samples reflects nothing of their spread, or *variability*. Consider as an example three students; the first has grades of 75, 75, 75; the second, 65, 75, 85; and the third 50, 75, and 100. Obviously the mean for each is 75, but the distributions reflect quite different spreads. There is no variability in the first student's record and quite a bit in the last. Furthermore, the mean is greatly affected by a few extreme values.

Variance

Variability of the population is measured by the *variance*, σ^2:

$$\sigma^2 = \frac{\Sigma(x - \mu)^2}{N} \tag{3}$$

where x represents each individual measurement in the population, μ the population mean, and N the number of individuals making up the population. Of course, the population mean and the total number of individuals generally are not determinable because it is impractical or impossible to measure every *individual* in a population. These difficulties are avoided by substituting sample values for those of the population:

$$s^2 = \frac{\Sigma(x - \bar{x})^2}{n} \tag{4}$$

But Equation (4) is biased in the direction of underestimating because, in using the sample mean, the number of independent measurements is $n - 1$. For example, the series of six values 4, 3, 6, 2, 1, 2 has a mean of 3, which is calculated in such a way that one value is fixed by the sum of the others. That is, given

$$\frac{4 + 3 + 6 + 2 + 1 + x}{6} = 3$$

the value of x can only be 2. To obviate the bias in formula (4), it is multiplied by the correction factor

$$\frac{n}{n - 1}$$

giving

$$s^2 = \frac{\Sigma(x - \bar{x})^2 \, n}{n(n - 1)}$$

Canceling n in numerator and denominator gives an unbiased estimate of the sample variance.

$$s^2 = \frac{\Sigma(x - \bar{x})^2}{n - 1} \qquad (5)$$

If data are grouped by classes, as in our calculation of the mean, equation (5) becomes

$$s^2 = \frac{\Sigma f(x - \bar{x})^2}{n - 1} \qquad (6)$$

The sample variance, s^2, provides an unbiased estimate of the population variance (σ^2). But variance is in *squared units,* and we do not express height in square feet, or weight in square pounds. This difficulty is resolved by the next statistic to be described.

Standard Deviation

To avoid expressing variability in squared units of measurement, we simply extract the square root of the variance. This statistic is the **standard deviation,** s, which is given by the formula

$$s = \sqrt{\frac{\Sigma f(x - \bar{x})^2}{n - 1}} \qquad (7)$$

In effect, the standard deviation reflects the extent to which the mean represents the entire sample. If all individuals had exactly the same value, there would be no variability and the mean would represent the sample perfectly. Examination of Equation (7) indicates that the standard deviation would then be *zero.* As the sample becomes more variable, the mean serves progressively less well as an index of the entire sample, and as departures from the mean, class by class, increase, so does the standard deviation. However, extracting the square root of the variance reintroduces bias. Nevertheless, the standard deviation has the necessary advantage of expression in the units of measurement, as well as usefulness in determining other statistics.

To see how this statistic may be calculated and what it discloses regarding the sample, consider some length measurements of 200 hypothetical F_1 plants resulting from a particular cross. Calculation will be facilitated if data are grouped by classes and tabulated as follows:

1	2	3	4	5	6
Class value (cm) x	*Frequency* f	fx	*Deviation from mean* $(x - \bar{x})$	*Squared deviation* $(x - \bar{x})^2$	$f(x - \bar{x})^2$
48	8	384	−4.75	22.56	180.50
50	32	1,600	−2.75	7.56	242.00
52	75	3,900	−0.75	0.56	42.19
54	52	2,808	+1.25	1.56	81.25
56	28	1,568	+3.25	10.56	295.75
58	5	290	+5.25	27.56	137.81
	$n = 200$	$\Sigma fx = 10{,}550$			$\Sigma f(x - \bar{x})^2 = 979.50$

$$\bar{x} = \frac{\Sigma fx}{n} = \frac{10{,}550}{200} = 52.75 \text{ cm}$$

$$s = \sqrt{\frac{\Sigma f(x - \bar{x})^2}{n - 1}} = \sqrt{\frac{979.5}{199}} = \sqrt{4.92} = 2.218 \cong 2.22$$

This calculation provides a mean, plus or minus a standard deviation; that is, $\bar{x} = 52.75 \pm 2.22$. To understand the meaning of this expression and the information it conveys, "curves of distribution" must be examined.

As data for large samples are plotted with a quantitative measurement, such as length, along the abscissa and numbers of individuals (frequency) are plotted along the ordinate, the resulting curve is frequently bell-shaped; the variation is symmetrical about the largest class (or mode), as indicated in Figure 18–3. Such a curve is a **normal curve,** or a curve of normal distribution. If data are carefully plotted and a perpendicular is erected from the abscissa at a value equal to the mean, it will intersect such a curve at the latter's highest point and will divide the area under the curve into two equal parts (Figure 18–3) and, therefore, divide the sample into two groups of equal size. Now if perpendiculars to the abscissa are erected on it at points having values equal to $\bar{x} + s$ and $\bar{x} - s$, the area under the curve between $\bar{x} + s$ and $\bar{x} - s$ is 68.26 percent of the area under the curve (Figure 18–4).

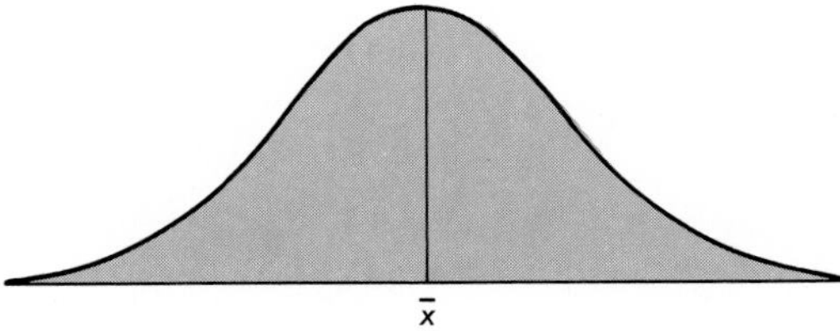

FIGURE 18–3. *The curve of normal distribution. A perpendicular erected from the abscissa at a value equal to the mean intersects the curve at its highest point and divides the area under the curve into areas of equal size.*

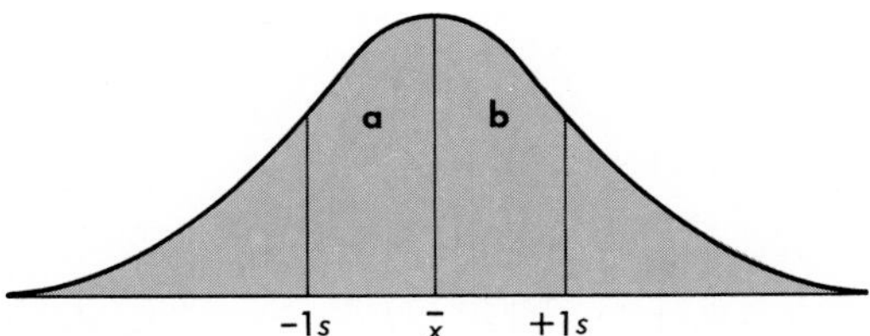

FIGURE 18–4. *The curve of normal distribution with perpendiculars to the abscissa erected at points showing values of* x + s *and* x − s. *Areas a and b each comprise 34.13 percent of the areas under the curve.*

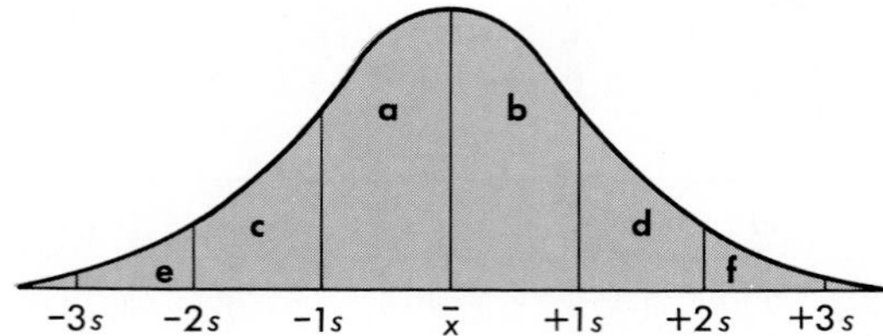

FIGURE 18–5. *Curve of normal distribution with perpendiculars to the abscissa erected at values x +/− 1s, x +/− 2s, and x +/− 3s.* Areas under the curve are as follows: *a + b = 68.26 percent; (a + b) + (c + d) = 95.44 percent; (a + b) + (c + d) + (e + f) =* 99.74 percent.

Similarly, the area under the curve between $\bar{x} - 2s$ and $\bar{x} + 2s$ is 95.44 percent of the total area; for $\bar{x} \pm 3s$, the area included is 99.74 percent of the total (Figure 18–5). This means that, *in a normal distribution,* about 68 percent (or roughly two-thirds) of the individuals will have values between $\bar{x} - s$ and $\bar{x} + s$, about 95 percent will have values between $\bar{x} - 2s$ and $\bar{x} + 2s$, and so on. Therefore, if an individual is chosen *at random* from a normally distributed population, the probability is about 0.68 that it will belong to that part of the population lying in the range $\bar{x} \pm s$. Similarly, there is a 0.95 probability that the individual selected will lie within the limits $\bar{x} \pm 2s$, or, only a 0.05 probability that the randomly chosen individual will lie outside those limits. The percentages of the sample determined by the mean plus or minus different multiples of the standard deviation are shown more fully in Table 18–1.

Thus the standard deviation is a useful description of the variability of the sample, and if the sample is large and randomly chosen, it is a good indicator of the variability of the population. As variability of the sample increases, so does its standard deviation.

TABLE 18–1. **Percentages of the sample falling within given multiples of the standard deviation from the mean**

Mean ± values of s	*Percent of sample included*	*Mean ± values of s*	*Percent of sample included*
0.1	7.96	2.0	95.44
0.2	15.86	2.1	96.42
0.3	23.58	2.2	97.22
0.4	31.08	2.3	97.86
0.5	38.30	2.4	98.36
0.6	45.14	2.5	98.76
0.675	50.00	2.58	99.00
0.7	51.60	2.6	99.06
0.8	57.62	2.7	99.30
0.9	63.18	2.8	99.48
1.0	68.26	2.9	99.62
1.1	72.86	3.0	99.74
1.2	76.98	3.1	99.80
1.3	80.64	3.2	99.86
1.4	83.84	3.3	99.90
1.5	86.64	3.4	99.94
1.6	89.04	3.5	99.96
1.645	90.00	3.6	99.96
1.7	91.08	3.7	99.98
1.8	92.82	3.8	99.98
1.9	94.26	3.9	99.99
1.96	95.00	4.0	99.99

Standard Error of the Sample Mean

If a series of samples were drawn from the same population, their means and standard deviations would probably not be the same. Page 378 carries measurements for a sample of hypothetical plants having a mean of 52.75 cm. Another group of 200 F_2 plants from the same population would be expected, by chance, to have a somewhat different mean. However, if a series of samples of 200 individuals each were drawn at random, we could assume from our knowledge of the laws of probability that a few samples would have relatively low means and a few relatively high, but that most would be intermediate. In fact, if a large number of these successive sample means were plotted, they would be found to form a normal curve, and thus give a fairly clear picture of the population distribution. The population mean and standard deviation could then be calculated. Practical limitations preclude doing exactly this, but the standard deviation of means, or the **standard error of the sample mean,** can be calculated from one representative sample.

The standard error of the sample mean, $s_{\bar{x}}$, represents an estimate of the standard deviation of the means of many samples that might be taken, and is a measure of the closeness with which the sample mean, $\bar{x}$, represents the population mean, μ. The size of the sample used and the variability of the population affect the reliability of $\bar{x}$ as an estimate of μ. The greater the variation in the population, the larger the sample needed to provide an adequate representation of the population. Both these factors are taken into account in the equation for calculating the standard error of the sample mean:

$$s_{\bar{x}} = \frac{s}{\sqrt{n}} \tag{8}$$

where s is the standard deviation of the sample and n is the number of individuals composing the sample.

In the hypothetical sample of 200 F_1 plants (page 378), where $\bar{x}$ = 52.75 cm and s = 2.22, the standard error of the sample mean becomes

$$s_{\bar{x}} = \frac{2.22}{\sqrt{200}} = \frac{2.22}{14.14} = 0.157, \text{ or about } 0.16$$

Because $s_{\bar{x}}$ represents the standard deviation of a *series* of sample means, and recalling that the area relationships of a normal curve (Figure 18–5) indicate that $\bar{x} \pm s$ includes 68.26 percent of the sample, $\bar{x} \pm 2s$ includes 95.44 percent of the sample, etc., the value $s_{\bar{x}} = 0.16$ indicates that there is about a 0.68 probability that μ lies in the range 52.75 cm ± 0.16, that is, between 52.59 cm and 52.91 cm. Likewise, the probability that μ is in the range $\bar{x} \pm 2s_{\bar{x}}$, or between 52.43 and 53.07, is about 0.95. Thus $\bar{x} = \mu \pm s_{\bar{x}}$ 68.26 percent of the time by *chance alone*, $\bar{x} = \mu \pm 2s_{\bar{x}}$ in 95.44 percent of the cases, and $\bar{x} = \mu \pm 3s_{\bar{x}}$ 99.74 percent of the time. If the sample is large (>100), other confidence levels may be determined from Table 18–1. For example, there is a 0.9 probability that μ lies in the range 52.74 ± $1.654s_{\bar{x}}$, or 52.75 ± 0.263 cm; a 0.5 probability that it is within the range 52.75 ± $0.675s_{\bar{x}}$; and so on. Obviously, the smaller the standard error, the more reliable the estimate of the population mean. As sample size, n, increases, the magnitude of the standard error decreases. Therefore, it is desirable to use samples as large as possible in order to determine characteristics of the population.

In cases involving polygenic inheritance, the standard error of the sample mean will give a measure of the population mean. It will also help fix, for example, a parental mean to which a given fraction of the F_2 may be compared. In other words, in the example of pumpkin fruits (page 368), the question could be raised as to what value between 4 and 6 or 19 and 23 is acceptable as representative of the parental strains so that certain F_2 individuals can be designated as being as extreme as each parent. If the standard error of the mean is calculated for each parental sample, a range is arrived at within which there is 0.68, 0.95, or 0.99 confidence where the population mean lies, and therefore this range provides a valid representation of the parental populations against which we can compare the F_2.

Standard Error of the Difference in Means

It is often necessary to determine whether the difference in the means of two samples is statistically significant, that is, to determine the likelihood that two sample means represent genetically different populations rather than chance differences in two samples from the same population. A judgmental answer to this problem is provided by a statistic known as the **standard error of the difference in means** (S_d):

$$S_d = \sqrt{(s_{\bar{x}_1})^2 + (s_{\bar{x}_2})^2} \qquad (9)$$

For example, consider the statistics developed for the hypothetical group of 200 plants ("sample 1") as compared with like statistics for a second hypothetical sample:

sample 1	*sample 2*
$n_1 = 200$	$n_2 = 200$
$\bar{x}_1 = 52.75$	$\bar{x}_2 = 55.87$
$s_1 = 2.218$	$s_2 = 3.150$
$s_{\bar{x}_1} = 0.16$	$s_{\bar{x}_2} = 0.22$

Substituting in equation (9) to determine S_d for these two samples gives

$$\begin{aligned} S_d &= \sqrt{(0.16)^2 + (0.22)^2} \\ &= \sqrt{0.026 + 0.048} \\ &= \sqrt{0.074} \\ &= 0.272 \end{aligned}$$

The meaning of a value of $S_d = 0.272$ can be seen by comparing the difference in sample means, here $\bar{x}_2 - \bar{x}_1$, with the standard error of the difference in means, S_d:

$$\frac{\bar{x}_2 - \bar{x}_1}{S_d} \qquad (10)$$

Substituting values, Equation (10) becomes

$$\frac{3.12}{0.27} = 11.55$$

(Note that the smaller mean is always subtracted from the larger in order to have a positive numerator.)

What, then, does the value of S_d and its relation to the difference in sample means signify? Remember that in a normal curve the area under the curve equal to $\bar{x} \pm 2s$ comprises about 95 percent of the total area, that is, 95 percent of the individuals will have a quantitative value between $\bar{x} - 2s$ and $\bar{x} + 2s$. Thus, for a normal distribution, a standard error represents the standard deviation of a series of means, and a standard error of the difference in sample means deals with this same relationship between $\pm 1s$, $\pm 2s$, and so on. Therefore, *if the difference in sample means is greater than twice the standard error of the difference of the sample means, the difference in sample means is considered significant*. Significance begins whenever $\bar{x}_1 - \bar{x}_2 > 2S_d$. Here *significance* means two different populations.

Would a difference in means of, say, *exactly* twice the value of S_d mean that two different populations are represented by the two samples? No! However, the values of $\bar{x}_1 - \bar{x}_2 = 2S_d$ would give 0.95 *confidence* that two populations are involved, a 0.05 probability that the two samples were taken from the same population. *When the probability that two samples have come from the same population falls below 0.05 the difference in means is considered significant*. Therefore, to be significant, $\bar{x}_1 - \bar{x}_2$ *must exceed* $2S_d$. In the example, here, $\bar{x}_1 - \bar{x}_2 = 11.55\ S_d$; hence $\bar{x}_1 - \bar{x}_2$ is considered highly significant. The probability that only one population is represented is so very small that it is rejected. Note that a significant difference between two sample means is inherently neither "good" nor "bad."

USE OF STATISTICS IN GENETIC PROBLEMS

Two examples of the application of statistics to genetic problems will demonstrate the usefulness of such analysis. In the first example, assume a commercial producer of hybrid seed corn wishes to market grains that will produce plants having ears of very uniform length. He has two varieties, A and B, both of which produce ears of just under 8 in., which is a satisfactory length for his marketing purposes. Although A averages closer to 8 in., it appears to be more variable than B. The producer grows several acres of each variety in as uniform an environment as possible, then analyzes 100 ears from each. Variety A has the following statistics:

$$\bar{x} = 7.95 \text{ in.}$$
$$s = 0.52$$
$$s_{\bar{x}} = 0.05$$

Although the sample mean is very close to the desired length, the standard deviation indicates that two-thirds of the ears of this variety may be expected to vary up to 0.52 in. from the mean of 7.95 in. Of course, one-third will deviate more than this.

This is more variability than the grower would prefer, so, for comparison, 100 ears of variety B are similarly analyzed. This variety has the statistics:

$$\bar{x} = 7.88 \text{ in.}$$
$$s = 0.23$$
$$s_{\bar{x}} = 0.02$$

Although the ears average slightly shorter than those of variety A, B has a much narrower range of variation. Two-thirds of the ears of the latter may be expected to fall within the range 7.65 to 8.11 in., as compared with 7.43 to 8.47 for A. Therefore, the breeder elects to use B.

The standard errors of the two samples are useful in indicating to the grower just how much the mean length of his samples might be expected to vary from the mean lengths of *all* plants of the two varieties. His samples are sufficiently large, and the standard errors are quite small. This shows that the samples do reflect reliably the magnitude of variability in the two varieties.

Another example will show an application of statistics to a more theoretical type of problem (Table 18–2). Data were accumulated on days to maturity for two varieties of tomato (Burpeeana Early Hybrid, P_1, and Burpee Big Boy, P_2) and their hybrids (F_1 and F_2). Maturation time is dependent on both heredity and environment, so environmental differences must be minimized. This is often done in randomized plots, whereby different varieties are distributed randomly in the field. The time elapsing between setting the plants in the field and the ripening of the first fruit was recorded as "days to maturation." Data were re-

TABLE 18–2. **Data for two parental and two progeny strains of tomato based on days required to reach maturity (data shown are numbers of plants)**

Days	P_1	P_2	F_1	F_2
55	1			
56	6			
57	9			1
58	40			1
59	28			2
60	14			3
61	2		3	4
62			8	9
63			20	10
64			31	12
65			19	14
66			10	20
67			8	7
68			1	4
69				3
70				3
71				1
72				2
73				1
74				1
75		4		1
76		12		1
77		20		
78		35		
79		15		
80		10		
81		3		
82		1		
$\bar{x}$	58.38	77.92	64.22	65.14
s	1.14	1.43	1.49	3.43
$s_{\bar{x}}$	0.114	0.143	0.149	0.343
S_d	0.183		0.374	

corded for samples of 100 plants in each of the four varieties, all grown in the same season in randomized plots.

The data in Table 18–2 clearly suggest that P_1 and P_2 represent different populations, and this is confirmed by the standard deviation and standard error of each, as well as by the standard error of the difference in sample means. In fact, the difference in sample means is 19.54 days, which is about 108 times the standard error of the difference in means! Because a difference in sample means of more than $2S_d$ is considered significant, the difference here is highly significant.

The same statistics for the F_1 and F_2 generations also show that although differences in maturation times are considerably less than for the parental strains, they are significantly different from each other. One therefore has more than 95 percent confidence that the F_1 and F_2 samples represent two genetically different populations. The difference in means for these two samples is only 0.92 day, but this is about 2.5 times the standard error of the difference in the sample means.

These data clearly suggest polygenes. The F_1 is intermediate between the parents as is the F_2, but the F_2 has a wider range than the F_1. If P_1 and P_2 are assumed to be completely homozygous, then the variability that each shows must be wholly environmental. The F_1 would then be completely and uniformly heterozygous and its variability would be environmental. The F_2 would be expected to segregate; therefore, genetic variation is superimposed on environmental variation.

The F_2 data may be used to furnish a very rough estimate of the number of contributing alleles. Eleven of the 100 F_2 were as extreme as P_1, and 2 were as extreme as P_2. Now $\frac{11}{100}$ simplifies to about $\frac{1}{9}$, and $\frac{2}{100}$ to $\frac{1}{50}$. The equation for computing numbers of effective alleles from the F_2 data (page 380) shows that $\frac{1}{16}$, which indicates four contributing alleles, is between the two extremes of $\frac{1}{9}$ and $\frac{1}{50}$.

This approach is really too simple and probably gives too low an estimate, for it would take many hundreds or thousands of F_2 individuals to provide a reasonable chance of recovering the maximum extremes possible. Sewall Wright has shown that the number of *pairs of polygenes (n)* can be calculated from the equation

$$n = \frac{R^2}{8(s^2_{F_2} - s^2_{F_2})} \qquad (11)$$

where R is the range (greatest difference) between the mean values of extreme phenotypes (whether found in the parental generations or elsewhere), $s^2_{F_2}$ is the *variance* of the F_2, and $s^2_{F_1}$ is the variance of the F_1. An application of this formula to the data of Table 18–2 with substitutions shows

$$n = \frac{(77.92 - 58.38)^2}{8(11.76 - 2.22)} = \frac{(19.54)^2}{8(9.54)} = \frac{381.81}{76.32} = 5.003$$

or about five pairs of polygenes.

In using this formula the following assumptions have been made: (1) no environmental effect, (2) no dominance, (3) no epistasis, (4) equal, additive contributions by all loci, (5) no linkage, and (6) complete homozygosity in each parent, with complete heterozygosity in the F_1. With these assumptions, the F_1 mean would be expected to fall midway between the parental means. That it does not suggests that not all of those assumptions are completely valid in this case and/or that the sample is too small. Moreover, the fact that a crude estimate of the number of pairs of alleles from the fraction of the F_2 as extreme as each parent is lower than the value given by the Wright formula also illustrates the need for a considerably larger sample.

These are examples of practical contributions of statistical analysis to genetic problems. There are, however, more subtle advantages. In many cases it has been possible to separate genetic mechanisms from sampling errors, and in others to differentiate between genetic and environmental variation. Above all, critical attitudes toward design of experiments and treatment of data have been sharpened.

PROBLEMS

18–1 Which of the following human phenotypes would appear to be based on polygene inheritance: intelligence, absence of incisors, height, **phenylketonuria** (inability to metabolize the amino acid phenylalanine), ability to taste phenylthiocarbamide, skin color, **cryptophthalmos** (failure of eyelids to separate in embryonic development), eye color?

18–2 Show, by means of appropriate genotypes, how parents may have children taller than themselves.

18–3 Suppose another race of wheat is discovered in which kernel color is determined to depend on the action of six pairs of polygenes. From the cross *AABBCCDDEEFF* × *aabbccddeeff*, (a) what fraction of the F_2 would be expected to be like either parent? (b) How many F_2 phenotypic classes result? (c) What fraction of the F_2 will possess any six contributing alleles?

18–4 Two races of corn, averaging 48 and 72 in. in height, respectively, are crossed. The F_1 is quite uniform, averaging 60 in. tall. Of 500 F_2 plants, two are as short as 48 in. and two as tall as 72 in. What is the number of polygenes involved, and how much does each contribute to height?

18–5 In Problems 6–17 to 6–20 it was pointed out that *Pl*— corn plants are purple, whereas *pl pl* individuals are green. Assume now that the F_1 in Problem 18–4 are also *Pl pl*. A breeder wishes to recover a pure-breeding green, 84-in. variety from the cross given in Problem 18–4. What fraction of the F_2 will satisfy his requirement?

18–6 Two 30-in. individuals of a hypothetical species of plant are crossed, producing progeny in the following ratio: one 22-in., eight 24-in., twenty-eight 26-in., fifty-six 28-in., seventy 30-in., fifty-six 32-in., twenty-eight 34-in., eight 36-in., and one 38-in. What are the genotypes of the parents? (Start with the first letter of the alphabet and use as many more letters as needed.)

18–7 Mr. A has dark brown eyes; his wife's eyes are light blue. On the hypothesis that four pairs of polygenes (*A, a; B, b; C, c; D, d*) are responsible for human eye color, give the genotype of (a) Mr. A and (b) his wife. (c) Give the phenotype(s) of children they could have.

18–8 A daughter of Mr. and Mrs. A marries a man (Mr. B) having the same genotype as herself. What is the probability that they could have a (a) dark-brown-eyed child, (b) dark-blue-eyed child, (c) hazel-eyed child?

18–9 (a) What eye color is most likely to occur in any child of Mr. and Mrs. B? (b) What is the probability of a child having this eye color?

18–10 The children referred to in Problem 18–8 illustrate what phenomenon?

18–11 As noted in the text, spotting in certain breeds of cattle is dependent on interaction of *S* (solid color) or *ss* (spotted) and a number of polygenes for degree of spotting. Assume four pairs of the latter (which is almost certainly too low), designated as *A, a; B, b; C, c; D, d*. If data are accumulated from enough *SsAaBbCcDd* × *SsAaBbCcDd* crosses to give a total of, say, 1,024 calves from such matings, how many of these should be unspotted?

18–12 Assume height in a particular plant to be determined by two pairs of unlinked polygenes, each effective allele contributing 5 cm to a base height of 5 cm. The cross *AABB* × *aabb* is made. (a) What are the heights of each parent? (b) What height is to be expected in the F_1 if there are no environmental effects? (c) What is the expected phenotypic ratio in the F_2?

18–13 If each pair of alleles in Problem 18–12 exhibited complete dominance instead of an additive effect, (a) what are the heights of each parent? (b) What height would be expected in the F_1? (c) What is the expected phenotypic ratio in the F_2?

18–14 If you were dealing with a case of polygenic inheritance in corn, and determined that six pairs of polygenes were involved, what binomial would have to be expanded to arrive at the expected F_2 phenotypic ratio (assume the P generation to consist of two genotypes: (1) all homozygous contributing and (2) all homozygous noncontributing)?

The following data were obtained on the weight in pounds of a given sample of pumpkin fruits:

30	26	22	16	24
24	24	30	22	14
28	22	16	26	22
24	20	28	14	28
22	24	22	22	26

18–15 What is the mean to the nearest tenth of a pound?

18–16 What is the variance?

18–17 (a) What is the standard deviation to the nearest tenth of a pound? (b) What does this value tell you about the sample?

18–18 If an individual is selected at random from this sample, what should be the probability that it will weigh more than 18 but less than 28 lb. (Assume that the sample is normally distributed.)

18–19 (a) What is the standard error of the sample mean? (b) What information does this give you?

18–20 What is the probability that the population mean lies between about 21 and 25 lb?

Data on a second sample of 25 individuals were collected and the following statistics determined:

$$\bar{x}_2 = 21.8 \text{ lb}$$

$$s_2 = 4.0$$

$$s_{\bar{x}_2} = 0.8$$

18–21 What is the standard error of the difference in sample means?

18–22 What is the approximate probability that samples 1 and 2 represent two different populations?

18–23 In view of your answer to the preceding question, is the difference in sample means significant? Why?

18–24 Two parental strains with mean heights of 64.29 and 135 cm, respectively, were crossed. The F_1 and F_2 that resulted had the following statistics:

	$\bar{x}$	s
F_1	99.64	10.000
F_2	102.11	13.342

If we neglect any possible environmental effects, dominance, linkage, or epistasis, and assume that each effective allele makes the same additive contribution, how many pairs of polygenes are involved?

POPULATION GENETICS

CHAPTER

19

Population genetics deals with *natural* populations. A natural population consists of all the interbreeding individuals that share a common **gene pool,** which is the total genetic information possessed by the reproductive members of the population. Alleles in the pool interact with each other and with the environment with the result that a selection effect is exerted on those genes. Throughout this text our concern has been largely with results of experimental breeding programs, and family pedigree studies, from which certain familiar genotypic and phenotypic ratios emerge. In producing such F_2 ratios as 1:2:1 and 9:3:3:1, the parental generation was usually of such genotypes $AA \times$ aa, or $AABB$ and $aabb$; that is, the alleles A and a as well as B and b were introduced in *equal frequency*.

Similarly, in all the other experiments dealt with in this text, gene frequency was intentionally included among the controlled factors. But, in natural populations the frequencies of alleles may vary considerably over both space and time. For example, the allele for polydactyly is dominant to its recessive allele, yet the polydactylous phenotype is fairly infrequent among newborn infants, even though the trait is not known to play any part in survival and no more than a very minor role in choosing a mate. Apparently the *frequency of the dominant allele here is lower than that of the recessive allele;* this pair obviously does not exist in the *population* in the 1:1 ratio so commonly encountered in the laboratory where gene frequency and mating pattern are controlled.

In 1908, G. H. Hardy, a British mathematician, and the German physician, W. Weinberg, independently developed a relatively simple mathematical concept, now referred to as the **Hardy-Weinberg principle,** to describe this genetic equilibrium. This principle is the foundation of population genetics and, in essence, states that in the absence of migration, mutation, and selection, gene and genotypic frequencies remain constant within certain narrow, determinable limits, generation after generation, in a large, randomly mating population. The Hardy-Weinberg principle may be used to determine the frequency of each allele of a pair or of a series, as well as frequencies of homozygotes and heterozygotes in the population. This chapter will examine applications of this principle and the forces that alter gene frequencies.

CALCULATING GENE FREQUENCY

The Hardy-Weinberg principle can be used to calculate the gene frequency under a number of conditions. These include codominant genes, completely dominant genes, multiple alleles, and sex-linked genes.

Codominance

The M-N blood type furnishes a useful example of a series of phenotypes due to a pair of codominant alleles. None of the three possible phenotypes, M, MN, and N, appears to have any selective value; in fact, most persons do not know their M-N blood type. Frequencies of the two alleles involved will be calculated for samples from two different groups.

A study of 6,129 white Americans living in New York City, Boston, and Columbus disclosed a genotypic ratio closely approximating 1:2:1. Because the closely related alleles *S* and *s* were not included, these individuals can be grouped according to the M-N system alone:

Genotype	*Number*
MM	1,787
MN	3,039
NN	1,303
	6,129

To calculate frequencies of the two alleles M and N, remember that these 6,129 persons possess a total of 6,129 × 2 = 12,258 alleles. The number of M alleles, for example, is 1,787 + 1,787 + 3,039. Thus, calculation of the frequencies for M and N may be worked out in this way:

$$\text{M} = \frac{1{,}787 + 1{,}787 + 3{,}039}{12{,}258} = \frac{6{,}613}{12{,}258} = 0.5395$$

$$\text{N} = \frac{1{,}303 + 1{,}303 + 3{,}039}{12{,}258} = \frac{5{,}645}{12{,}258} = 0.4605$$

So frequencies of the two alleles in *this* sample are almost equal, and this is reflected in the close approximation to a 1:2:1 ratio so often seen in laboratory results.

Gene frequencies expressed as decimals may be used directly to state probabilities. Assume this sample to be representative of the population; then there is a probability of 0.5395 that of the chromosomes bearing this pair of alleles, any one selected randomly will bear gene M, and a probability of 0.4605 that any one will bear N. Thus, the frequencies of the three phenotypes to be expected in the population are:

Genotype	*Phenotype*	*Phenotypic frequency*
MM	M	$0.5395 \times 0.5395 = 0.2911$
MN / *NM*	MN	$2(0.5395 \times 0.4605) = 0.4968$
NN	N	$0.4605 \times 0.4605 = 0.2121$
		1.0000

Notice that these genotypic frequencies follow a binomial distribution.

The general relationship between gene and genotypic frequencies is described in albegraic terms by the Hardy-Weinberg principle. If we let p represent the frequency of M, and q represent the frequency of N, the total array of probabilities of the three genotypes (MM, MN, and NN) is

$$p^2 + 2pq + q^2$$

which is the expansion of $(p + q)^2$. Note also that in this case $p + q = 1$ and, of course, so does $(p + q)^2$. Thus the probability of a type M individual, for example, under a system of random mating is given by the expression p^2 where the value of p has already been calculated to be 0.5395. Similarly, the probability of MN persons (the heterozygotes) is given by $2pq$, and of N persons by q^2.

Alternatively, the frequencies of M and N could have been calculated by another approach. Recall that an individual of genotype NN, for example, represents the simultaneous occurrence of two events of equal probability, namely the fusion of two gametes, each with the genotype N. Thus q^2 has the value $1{,}303/6{,}129 = 0.2126$, and $q = \sqrt{0.2126}$, or 0.46. Because $p + q = 1$, $p = 1 - q$, or $1 - 0.46 = 0.54$.

For comparison, look briefly at the following sample of 361 Navaho Indians from New Mexico:

Phenotype	*Number*
M	305
MN	52
N	4
	361

This sample is far from a laboratory-type 1:2:1 ratio. Do such frequencies mean that the Navahos do not conform to the same genetic laws as the previous sample? The answer is "no" as soon as it is recalled that the 1:2:1 proportion is based on *equal frequencies* of the alleles in the population. The raw data clearly suggest that M is considerably more frequent in Navaho Indians than is the N allele. Using the same method as was employed for the earlier sample to calculate gene frequencies produces this result:

$$\text{let } p = \text{frequency of M} = \frac{305 + 305 + 52}{722} = 0.9169$$

$$\text{let } q = \text{frequency of N} = \frac{52 + 4 + 4}{722} = \frac{0.0831}{1.0000}$$

Applying these gene frequencies to the sample produces these genotypic frequencies:

Genotype	Gene frequencies	Genotype probability	Genotypes in sample: Expected	Observed
MM	$p^2 = (0.9169)^2$	0.8407	303.5	305
MN	$2pq = 2(0.9169 \times 0.0831)$	0.1524	55.0	52
NN	$q^2 = (0.0831)^2$	0.0069	2.5	4
		1.0000	361.0	361

A chi-square test (see Chapter 5) of the data for the Navaho sample shows $\chi^2 = 1.071$; for the earlier sample of 6,129 persons, $\chi^2 = 0.0237$. So deviation from expectancies based on the calculated gene frequencies is well below the level of significance in each case. Note that although there are three phenotypic classes, there is but one degree of freedom, because only two alleles, M and N, are involved. Therefore, only one phenotypic class can be set at random. For example, in a sample of 400 persons having just 200 M alleles and therefore 600 N alleles, any number of individuals up to 100 may be of genotype MM. If there are, say, 60 persons of type M, the other classes are thereby automatically determined:

Phenotype	Number of persons	Number of alleles: M	N
M	60	120	—
MN	80	80	80
N	260	—	520
Totals	400	200	600

Complete Dominance

An interesting phenotypic trait with no known selective value is the ability or inability to taste the chemical phenylthiocarbamide ("PTC," $C_7H_8N_2S$), also called phenylthiourea. The test is a simple one that can easily be performed by any genetics class. The usual procedure is to impregnate filter paper with a dilute aqueous solution of PTC (about 0.5 to 1 g per liter), allow it to dry, then place a bit of the treated paper on the tip of the tongue . About 70 percent of the white American population can taste this substance, generally as very bitter, rarely as sweetish. Although the physiological basis is unknown, tasting ability does depend on a completely dominant allele that we will designate as T. Thus tasters are $T-$ (that is, TT or Tt); nontasters are tt.

In one study of 280 genetics students, 198 were tasters and 82 were nontasters. From such data the frequencies of genes T and t in the sample may be readily calculated. The 82 nontasters (29.29 percent of the sample) are persons of genotype tt and in the Hardy-Weinberg principle may be represented by q^2. Therefore,

$$q^2 = 0.2929 \quad \text{and} \quad q = \sqrt{0.2929} = 0.5412$$

which is the frequency of gene t. Because only a pair of alleles are involved, frequencies of the two genes must again total 1, $p + q = 1$, and $p = 1 - q$. Therefore, in this example p, the frequency of allele T equals $1 - 0.5412$, or 0.4588. Frequencies of homozygous and heterozygous tasters may now be computed. The binomial expansion $p^2 + 2pq + q^2$ shows the distribution of the three possible genotypes:

$$p^2 = TT = (0.4588)^2 = 0.2105$$
$$2pq = Tt = 2(0.4588 \times 0.5412) = 0.4966$$
$$q^2 = (0.5412)^2 = 0.2929$$

By testing representative samples of different populations, the frequencies of T and t in those groups may be similarly calculated.

Multiple Alleles

The binomial $(p + q)^2$ can be used only when two alleles occur at a particular locus. For cases of multiple alleles, one simply adds more terms to the expression. Recall that the A-B-O blood groups are determined by a series of three multiple alleles, I^A, I^B, and i, if the various subtypes are neglected. Hence in a gene-frequency analysis, let

$$p = \text{frequency of } I^A$$
$$q = \text{frequency of } I^B$$
$$r = \text{frequency of } i$$

and

$$p + q + r = 1$$

Thus genotypes in a population under random mating will be given by $(p + q + r)^2$.

To see how this trinomial is applied, consider the following sample of 23,787 persons from Rochester, New York:

Phenotype	*Number*	*Frequency*
A	9,943	0.418
B	2,379	0.100
AB	904	0.038
O	10,561	0.444
	23,787	1.000

The frequency of each allele may now be calculated from these data, where p, q, and r represent the frequencies of genes I^A, I^B, and i, respectively. The value of r, that is, the frequency of gene i, is immediately evident from the figures given:

$$r^2 = 0.444, \quad \text{hence}$$
$$r = \sqrt{0.444} = 0.6663 \; (= \text{frequency of } i)$$

The sum of A and O phenotypes is given by $(p + r)^2 = 0.418 + 0.444 = 0.862$; therefore,

$$p + r = \sqrt{0.862} = 0.9284$$

so $p = (p + r) - r = 0.9284 - 0.6663 = 0.2621$ (= frequency of I^A). Because $p + q + r = 1$, $q = 1 - (p + r) = 1 - 0.9284 = 0.0716$ (= frequency of I^B); genotypic frequencies, as shown in Table 19–1, can

TABLE 19–1. **Calculation of genotypic frequencies for a sample of 23,787 persons living in Rochester, New York**

Phenotypes	*Genotypes*	*Genotypic frequencies*	*Population probability based on sample*		
O	ii	r^2	0.440	=	0.4440
A	I^AI^A	p^2	0.0687	=	0.4180
	I^Ai	$2pr$	0.3493		
B	I^BI^B	q^2	0.0051	=	0.1005
	I^Bi	$2qr$	0.0954		
AB	I^AI^B	$2pq$	0.0375	=	0.0375
					1.0000

now be calculated. The probability figures arrived at for this large sample check quite closely with those arrived at for other samples of the general United States population.

Once a population has been characterized as to allelic frequencies, the gametes drawn from that population (to lead to the next generation) match the input frequencies of the present generation. The major impact of the Hardy-Weinberg principle is that allelic and genotypic frequencies (and therefore phenotypic frequencies) tend to repeat generation after generation until and unless some (selective) factor acts to change allelic frequencies.

Samples taken from other races or nationalities may, however, show quite different frequencies for the same alleles. In a sample of Navaho Indians, the following allele frequencies were obtained:

$$I^A = 0.1448$$
$$I^B = 0.0020$$
$$i = 0.8532$$

If we assume this sample to be representative of the Navaho population, the percentage of individuals of each genotype and phenotype may be calculated from these frequencies. Thus, one may either calculate allelic frequencies from numbers of each phenotype or compute percentages

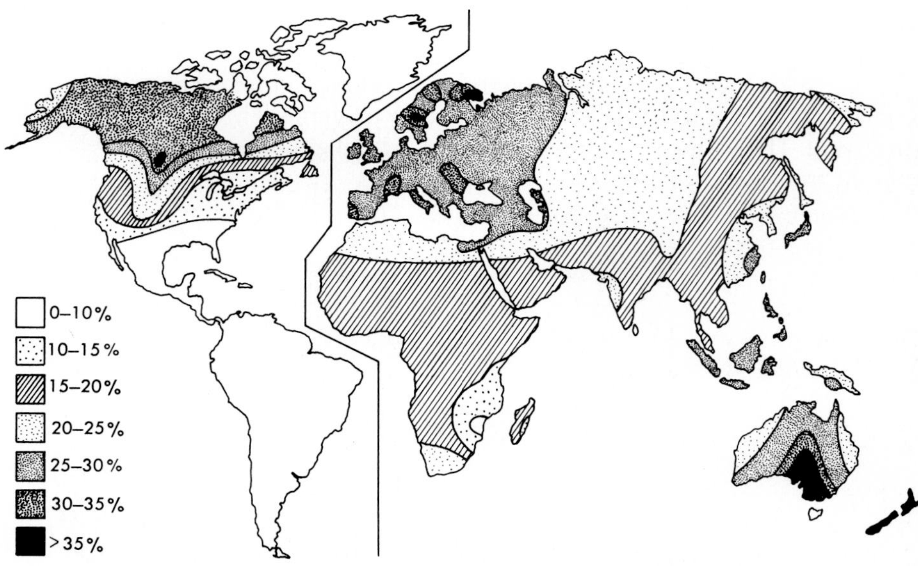

FIGURE 19–1. *Generalized world distribution of I^A for native populations.*

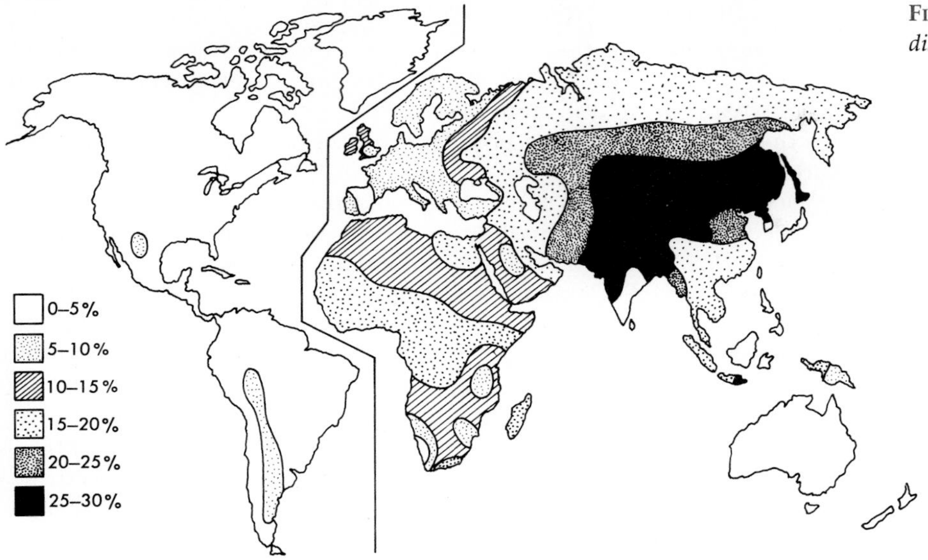

FIGURE 19–2. *Generalized world distribution of I^B for native populations.*

of the population that have each phenotype if allelic frequencies are known.

The two preceding examples clearly demonstrate that different frequencies of genes I^A and I^B, and i may occur in different populations. That this is true, especially for I^A and I^B, is shown in Figures 19–1 and 19–2.

Sex Linkage

Thus far, in considering allele frequencies, only autosomal genes have been discussed. The same techniques, with one modification, may be used in treating sex-linked genes. Because human males have only one X chromosome, they cannot reflect a binomial distribution for random combination of pairs of sex-linked genes as do females. Equilibrium distribution of genotypes for a sex-linked trait, where $p + q = 1$, is given by

$$(\male)\ p + q$$
$$(\female)\ p^2 + 2pq + q^2$$

Consider, for example, red-green color blindness. This trait is due to a sex-linked recessive, as noted in Chapter 7, and may be designated r. About 8 percent of all males suffer from this deutan color blindness. This shows at once that q, the frequency of allele r, is 0.08 and p, the frequency of its normal allele, R, is 0.92. Thus the frequency of color-blind females is expected to be $q^2 = 0.0064$. This is about what is found. Sex-linked dominants may be dealt with in the same way; in the case of normal color vision, with value of $p = 0.92$, the incidence of normal women is $p^2 + 2pq = 0.9936$.

FACTORS AFFECTING GENE FREQUENCY

Isolating Mechanisms

Any mechanism that prevents gene exchange is termed an *isolating mechanism.* Broadly considered, these mechanisms may be either (1) geographic or physical, such as great distances or mountain or oceanic barriers that keep populations apart, or (2) other mechanisms that effectively prevent gene exchange between populations in the same area. Because geographic separation keeps populations physically isolated, there is a question of whether in all cases such groups would remain reproductively isolated if brought together. In fact, it is clear in many cases that physically isolated populations do exchange genes when the isolation is ended.

Whenever individuals move from one population to another and interbreed this is called migration or gene flow. Migration is also used to describe flow of genes between populations that were separated but later became joined. Comparisons of gene frequencies for mixed and unmixed Eskimo and Flathead Indian samples (Table 19–2) shows that gene frequencies in a population may be changed by flow of alleles from other populations. Presumably the "mixing" in these cases is from the western European–American complex. Note how the frequency of the *T* allele has been increased by migration into the Eskimo population. The same sort of change, though in the opposite direction, has occurred by outbreeding of Flathead Indians where the frequency of *T* is high (0.683) in the unmixed group and somewhat lower (0.583) in the mixed population.

Because of the actual or possible exchange of genes between previously separated populations once they are permitted to intermingle, it is often preferred to confine the concept of isolating mechanisms to those that prevent gene exchange between populations that occupy the same area. These include one or more of the following: restriction to quite different habitats (especially in plants), reproductive maturation at different times, mating behavioral differences and/or physical incompatibility of genitalia (animals), destruction of sperm (principally in animals) or lack of development of pollen tubes (plants), and/or death of the zygote or of the embryo. In addition, some populations produce sterile hybrids, as in the mule (donkey × horse) and in many horticultural varieties of plants, effectively preventing establishment of new self-perpetuating genetic lines.

In summary, different populations are often characterized by particular allelic frequencies, which produce phenotypic frequencies that may be expected to fluctuate narrowly around a mean until something occurs

TABLE 19–2. **Percentage of PTC tasters in various human populations**

Population	*Place*	*Sample size*	*Percent tasters*	*Gene frequencies*	
				T	*t*
Welsh	Five towns	237	58.7	0.36	0.64
Eskimo (unmixed)	Labrador and Baffin	130	59.2	0.36	0.64
Arab	Syria	400	63.5	0.40	0.60
American white	Montana	291	64.6	0.41	0.59
Eskimo (mixed)	Labrador and Baffin	49	69.4	0.45	0.55
American white	Columbus, Ohio	3,643	70.2	0.45	0.55
American students	Delaware, Ohio	280	70.7	0.46	0.54
American black	Alabama	533	76.5	0.52	0.48
Flathead Indians (mixed)	Montana	442	82.6	0.58	0.42
Flathead Indians (unmixed)	Montana	30	90.0	0.68	0.32
American black	Ohio	3,156	90.8	0.70	0.30
African black	Kenya	110	91.9	0.72	0.28
African black	Sudan	805	95.8	0.80	0.20
Navaho Indians	New Mexico	269	98.2	0.87	0.13

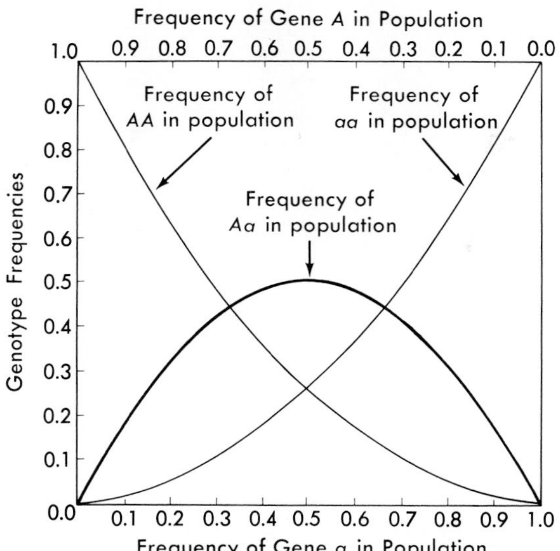

FIGURE 19–3. *The effect of changes in gene frequencies on frequencies of genotypes in a population.*

to alter gene frequencies. Figure 19–3 indicates how frequencies of homozygotes and heterozygotes are changed by a shift in allelic frequencies.

Earlier it was noted that in large populations changes in gene frequency may come about not only through alteration of isolating mechanisms but also through mutation and selection. But in populations of finite size, an additional factor comes into play. This is random fluctuation in gene frequency, or *genetic drift*. Each of these forces as mechanisms of change in gene frequency will now be examined.

Mutation. Basically, a mutation is a sudden, random alteration in the genotype of an individual (see Chapter 13). Strictly speaking, it is a change in the genetic material itself, but the term is often loosely extended to include chromosomal aberrations such as those already considered (Chapters 14 and 15). Its importance in the genetics of populations is to provide new material on which selection can operate, as well as to alter gene frequencies.

If, for example, gene T mutates to t, the relative frequencies of the two alleles are changed. If the mutation $T \rightarrow t$ recurs consistently, T could disappear from the population. But not only is mutation recurrent, it is also reversible, with a known frequency in many cases. These *back mutations* will at least slow the otherwise inexorable shift of T to t and perhaps prevent the total disappearance of T. But notice that the mutation $T \rightarrow t$ inevitably would shift gene frequencies over time (1) unless the rate of back mutation $T \leftarrow t$ equals the rate of forward mutation $T \rightarrow t$, and (2) only if possession of gene t either gives its bearer an advantage or confers no disadvantage. Thus mutation as a force in altering gene frequencies can scarcely be considered apart from the factor of selection, and their *combined* effect will be considered later in this chapter.

Equilibrium resulting from mutation is easily derived algebraically. Because the rate of mutation $T \rightarrow t$ generally does not equal the rate of change $T \leftarrow t$, we can show this relationship as

$$\underset{v}{\overset{u}{T \rightarrow t}} \atop \leftarrow$$

where u is the rate of mutation $T \rightarrow t$, and the back mutation $T \leftarrow t$ occurs at rate v, with $u \neq v$. Generally u will be greater than v.

If q represents the frequency of mutating gene t in any one generation, and p represents the frequency of its mutating allele T (some authors prefer to designate the frequency of the dominant allele as $1 - q$), the change in frequency of t resulting from mutation will be governed by

1. **The addition of *t*, as determined by**
   a. the rate of forward mutation (u), and
   b. the frequency of T (p); that is,
   c. to the limit set by up.
2. **The loss of *t*, as determined by**
   a. the rate of back mutation (v), and
   b. the frequency of t (q); that is,
   c. to the limit set by vq.

Frequencies of T and t will be in equilibrium under mutation when additions and losses balance, that is, when $up = vq$. The rate of change in q under mutation, or q_m, can be stated as

$$\Delta\hat{q}_m = up - vq \qquad (1)$$

Thus, $\hat{q}_m$, the equilibrium frequency of t under mutation, will equal the rate $T \rightarrow t$ divided by the sum of the rates:

$$\hat{q}_m = \frac{u}{u + v} \qquad (2)$$

If, for example, T mutates to t three times as often as t mutates to T, then

$$u = 3v$$

and

$$\hat{q}_m = \frac{3v}{3v + v} = \tfrac{3}{4}$$

So the population will reach equilibrium with regard to the frequencies of the mutating alleles T and t when q, the frequency of t, is 0.75, and p, the frequency of T, is $1 - 0.75$, or 0.25.

Complete Selection. The common colon bacillus (*Escherichia coli*) is usually killed by the antibiotic streptomycin; the normal, wild-type phenotype is thus "streptomycin-sensitive." The alleles for this trait are designated str^s (streptomycin-sensitive) and str^r (streptomycin-resistant). Because vegetative cells of *E. coli* are haploid, any given cell has the genotype str^r or str^s. In a freely growing culture of this organism, about 1 in 10 million cells can be shown by appropriate culture techniques to be streptomycin-resistant. Gene str^s mutates to str^r with a frequency of about 1×10^{-7}. Now in their normal streptomycin-free environment, neither str^s nor str^r confers any advantage or disadvantage. Hence the frequency of str^r would reach equilibrium after a number of generations. However, let streptomycin be introduced into the environment, and possession of str^r becomes a distinct advantage; without it an individual does not survive. So with a specific environmental change, str^r individuals survive and str^s individuals do not. One might say an ordinarily inconsequential gene (str^r) has thus suddenly assumed a very high positive selective value and, at the same time, str^s has acquired a strong negative selective value.

The Tay-Sachs disorder, with symptoms briefly described in Chapter 2, is caused by a recessive autosomal allele. The condition is most common in Ashkenazic Jews. Persistance of this deleterious allele is due primarily to symptomless heterozygous carriers who can, and do, pass on the allele to their offspring, and secondarily to mutation of the normal allele in homozygous normals. Affected children (homozygous recessives) die in infancy or very early childhood so that there is *complete* selection against the recessive allele.

The extreme impact of selection may also be seen in an example from tomato. In this plant several linkage groups contain dominant alleles that confer resistance to different strains of the leaf mold fungus (*Cladosporium*). The disease is not ordinarily a problem in outdoor culture but frequently becomes severe in commercial greenhouses, where temperature and humidity may be consistently high enough to cause severe infection of susceptible plants. In such cases, homozygous recessives, which are susceptible, are quickly killed and may not live to reproductive age. For simplicity, consider this situation as the result of a single pair of alleles, *C* and *c*, with *cc* plants being susceptible to the fungus. A population of plants in which the frequencies of *C* and *c* are each 0.5 at the outset, as well as continuous greenhouse culture in which seed for new plantings is obtained from the crop being grown, will be assumed.

Initially then, the parental generation may be represented as follows:

Frequency of $C = p = 0.5$; frequency of $c = q = 0.5$

P genotypes:	*CC*	*Cc*	*cc*
Genotypic frequencies:	p^2 +	$2pq$ +	q^2
	= 0.25 =	0.50 +	0.25

If *cc* individuals are unable to reproduce in this environment, the breeding population is reduced to *CC* and *Cc* plants, which occur now in the ratio of 1:2 as follows:

$$CC\text{:} \frac{0.25}{0.25 = 0.25} = 0.33 \text{ (new genotypic frequency)}$$

$$Cc\text{:} \frac{0.50}{0.25 = 0.25} = 0.67 \text{ (new genotypic frequency)}$$

This leaves only $C- \times C-$ crosses possible, with consequent changes in phenotypic and genotypic frequencies in the next generation (Table 19–3).

TABLE 19–3.

Crosses	*Frequencies*		*F_1 genotypic frequencies*		
			CC	*Cc*	*cc*
CC × *CC*	$(0.33)^2$	= 0.11	0.11		
CC × *Cc*	2(0.33 × 0.67)	= 0.44	0.22	0.22	
Cc × *Cc*	$(0.67)^2$	= 0.45	0.11	0.23	0.11
			0.44	0.45	0.11

The frequency of c has declined in a single generation from 0.5 to 0.33 ($= \sqrt{0.11}$), and C's frequency has risen from 0.5 to 0.67.

In effect, we have been considering *complete selection* against a recessive lethal. We can represent the frequency of such a gene after any given number of generations as

$$q_n = \frac{q_0}{1 + nq_0} \tag{3}$$

where q_n is the frequency of the recessive lethal after n additional generations, and q_0 is the initial frequency of that gene. Thus, in our tomato example, where $q_0 = 0.5$ and $n = 1$,

$$q_1 = \frac{0.5}{1 + (1 \times 0.5)} = \frac{0.5}{1.5} = 0.33$$

In a closed population, with no admixture from other groups, only mutation can prevent the frequency of an allele so radically selected against from dropping quickly toward zero. The curve for complete selection against a recessive allele, however, is hyperbolic, with the final slope being determined less by mutation than by the fact that the deleterious allele persists in relatively rare heterozygotes that more frequently mate with homozygous dominant individuals. The number of additional generations (n) required to reduce q from a value of q_0 to some desired value, q_n, is given by the equation

$$n = \frac{1}{q_n} - \frac{1}{q_0} \tag{4}$$

Thus, to reduce q from 0.5 ($= q_0$) to 0.33 ($= q_n$), one additional generation is required:

$$n = \frac{1}{0.33} - \frac{1}{0.5} = 3 - 2 = 1$$

With this equation the frequency of a recessive lethal for any number of additional generations can be calculated readily, as shown in Table 19–4.

Note that the frequency of the lethal is reduced by half in generation two, but four more are required to reduce it by half again, and so on. Furthermore, application of Equation (4) shows that the number of generations required to halve the frequency of such an allele is very large when the initial frequency is quite low. On the other hand, had the lethal been a dominant instead of a recessive, it would have been eliminated in one generation in the absence of complicating influences.

TABLE 19–4. **Reduction in frequency (q) of a recessive lethal by generations**

Generation	q
0	0.500
1	0.333
2	0.250
3	0.200
4	0.167
5	0.143
6	0.125
7	0.111
8	0.100
9	0.091
10	0.083
50	0.019
100	0.010
1,000	0.001

Partial selection. More often the genotypes involved differ much less in their relative advantage and disadvantage, and the homozygous recessive, for example, will be eliminated much more slowly than in the illustration for tomato. For instance, assume genotype $A-$ produces 100 offspring, all of which reach reproductive maturity in a given environment, whereas genotype *aa* produces only 80 that do so. The proportion of the progeny of one genotype surviving to maturity (relative to that of another genotype) may be designated as its fitness or *adaptive value*, W. Thus, in this case W_{A-} may be set arbitrarily at 1; W_{aa} is then 0.8. The measure of reduced fitness of a given genotype is referred to

TABLE 19–5. Selection against a recessive allele

	AA	Aa	aa	*Total*
Initial frequency	p^2	$2pq$	q^2	1
Adaptive value (W)	1	1	$1 - s$	
Frequency after selection	p^2	$2pq$	$q^2(1 - s)$	$p^2 + 2pq + q^2 - sq^2 = 1 - sq^2$

as its **selection coefficient** s. The relationship between adaptive value (W) and the selection coefficient (s) can be represented as

$$W = 1 - s, \text{ or } s = 1 - W$$

In the example being considered here, $s = 0$ for genotype $A-$, and $1 - 0.8 = 0.2$ for *aa*.

If p is the frequency of gene A, and q is the frequency of its recessive allele a, selection against the latter is as shown in Table 19–5. By the same method used to calculate frequencies of genes T and t, for example, in the Hardy-Weinberg equilibrium, the information in Table 19–5 gives an equation for determining the frequency of gene a when the value of W_{aa} is known:

$$q_1 = \frac{q_0 - sq_0^2}{1 - sq_0^2} \tag{5}$$

where q_0 again is the frequency of a in a given generation, q_1 is the frequency one generation later under selection, and s is the selection coefficient.

If, for instance, $q = 0.5$ and $s = 0.2$, then substituting in equation (5), $q_1 = 0.4737$. So with *partial selection* against a fully recessive allele, the decrease in its frequency is much less per generation than for complete selection against a recessive lethal. The change in frequency of allele a under selection (Δq_s or $q_1 - q_0$) can be represented by the equation

$$\Delta q_s = \frac{-sq_0^2 p}{1 - sq_0^2} \tag{6}$$

Upon substituting here, $q_s = -0.0263$. If q_0 were very small, as it would ordinarily be in the case of deleterious mutations, the quantity q_s becomes almost equal to $-sq_0^2$.

In time a large and randomly mating population attains a genetic equilibrium that is the resultant between selective forces and rate of mutation. This is apparent in many different species of animals, including humans, as well as plants. But once a population ceases to be isolated, gene frequencies may begin to change if genetic material is contributed by other populations.

Assortive mating. Not all populations are randomly mating. In populations of self-fertilizing plants, for example, given genotypes, in effect, mate only with like genotypes. This is **complete positive genotypic assortive mating** and results in rapid increase in frequencies of homozygotes. For instance, assume a population where genotypic frequencies in generation 0 are $0.25AA + 0.5Aa + 0.25aa$, that is, $p_0 = q_0 = 0.5$. With complete positive genotypic assortative mating, progeny frequencies in generation 1 can be calculated as follows:

Mating	Proportions of matings	Progeny genotypes		
		AA	*Aa*	*aa*
AA × *AA*	0.25	0.25		
Aa × *Aa*	0.5	0.125	0.25	0.125
aa × *aa*	0.25			0.25
Total	1.00	0.375	0.25	0.375

Similarly, progeny in successive generations would occur with the frequencies shown in Table 19–6. With continued inbreeding, frequencies of homozygotes approach the values of p_0 and q_0, respectively. Heterozygotes become so low in frequency that they may easily be eliminated, for example, by an unseasonal freeze (plants) or through predators (animals). Should this happen, the frequencies of *AA* individuals would then equal p_0, and the frequency of *aa* individuals would equal q_0. A similar but less rapid decrease in frequency of heterozygotes will occur in populations where like genotypes, although not "forced" to mate with each other, may mate preferentially (partial positive genotypic assortative mating).

On the other hand, mating may occur preferentially (or even only) between *unlike* genotypes (outbreeding). This is **negative genotypic assortative mating**, and may, of course, be either complete or partial. Negative genotypic assortative mating results in an *increase* in the frequency of heterozygotes.

Similarly, *phenotypic* assortative mating may occur. **Positive phenotypic assortative mating** consists of mating of like phenotypes, that is, $A- \times A-$ and $aa \times aa$; **negative phenotypic assortative mating** involves mating of unlike phenotypes, that is, $A- \times aa$. The effect of the former is to increase homozygosity, and the effect of the latter is to increase heterozygosity.

Combined effects of mutation and selection. In the Tay-Sachs disorder homozygous recessives fail to produce the enzyme hexosaminidase A (see Chapter 2). Mental and motor deterioration rapidly follow the onset of symptoms, accompanied by paralysis and degeneration of the retina, which leads to blindness; the culmination is death, generally

TABLE 19–6. **Progeny genotypic frequencies under complete positive genotypic assortative mating where $p_0 = q_0 = 0.5$**

Generation	Progeny genotypes and frequencies		
	AA	*Aa*	*aa*
0		1	
1	$\frac{1}{4} = 0.25$	$\frac{2}{4} = 0.5$	$\frac{1}{4} = 0.25$
2	$\frac{3}{8} = 0.375$	$\frac{2}{8} = 0.25$	$\frac{3}{8} = 0.375$
3	$\frac{7}{16} = 0.4375$	$\frac{2}{16} = 0.125$	$\frac{7}{16} = 0.4375$
4	$\frac{15}{32} = 0.46875$	$\frac{2}{32} = 0.0625$	$\frac{15}{32} = 0.46875$
5	$\frac{31}{64} = 0.484375$	$\frac{2}{64} = 0.03125$	$\frac{31}{64} = 0.484375$
n	$\frac{2^n - 1}{2^{n+1}}$	$\frac{2}{2^{n+1}}$	$\frac{2^n - 1}{2^{n+1}}$

before the age of four. Frequency of the recessive allele in the Jewish population of New York City is about 0.015, but approximately 0.0015 in non-Jewish individuals.

The selection coefficient for the Tay-Sachs allele is, of course, 1, inasmuch as homozygotes die in infancy or very early childhood. A reasonable estimate of forward mutation (u) appears to be 1×10^{-6}, although some reports range as high as 1.1×10^{-5}. Because of its lethality early in life, the rate of back mutation (v) is effectively zero. In summary, then, for the Jewish population of New York City:

$$\begin{aligned} q &= 0.015 = 1.5 \times 10^{-2} \\ p &= 0.985 = 9.85 \times 10^{-1} \\ u &= 0.000001 = 1 \times 10^{-6} \\ v &= 0 \\ s &= 1 \end{aligned}$$

What, then, is the equilibrium frequency of this recessive lethal in the Jewish population of New York City under the *combined* effect of mutation and selection?

Recall (from page 394) that the rate of change in frequency of a recessive gene under mutation is given by

$$\Delta q_m = up - vq$$

Substituting,

$$\begin{aligned} \Delta q_m &= (1 \times 10^{-6} \times 9.85 \times 10^{-1}) - (0 \times 1.5 \times 10^{-2}) \\ &= (1 \times 10^{-6} \times 9.85 \times 10^{-1}) - 0 \\ &= up = 9.85 \times 10^{-7} \end{aligned}$$

This last figure is approximately equal to u, because $0.000000985 \cong 0.000001$. Hence

$$\Delta q_m \cong u \tag{7}$$

Turning for a moment to frequency change under selection, we use equation (6) from page 397

$$\Delta q_s = \frac{-sq_0^2 p}{1 - sq_0^2}$$

Substituting in this equation,

$$\begin{aligned} \Delta q_s &= \frac{-(0.015)^2 \times 1 \times 0.985}{1 - 1(0.015)^2} \\ &= \frac{-2.25 \times 10^{-4} \times 9.85 \times 10^{-1}}{1 - 2.25 \times 10^{-4}} \\ &= \frac{-2.216 \times 10^{-4}}{9.998 \times 10^{-1}} \\ &= -2.22 \times 10^{-4} \end{aligned}$$

So

$$\Delta q_s \cong -sq_0^2 \tag{8}$$

since $-2.22 \times 10^{-4} \cong -2.25 \times 10^{-4}$.

Thus $\Delta q_m \cong u$, $\Delta q_s \cong -sq^2$, and $\Delta q = \Delta q_s + \Delta q_m = -sq^2 + u$. But, by definition, at equilibrium, $\Delta q = 0$. Therefore at equilibrium,

$$u - sq^2 = 0$$

and

$$u = sq^2$$

Rewriting this latter equation to solve for q^2,

$$q^2 = \frac{u}{s}$$

and therefore

$$\hat{q} = \sqrt{\frac{u}{s}} \tag{9}$$

Substituting in equation (9) for the Tay-Sachs problem,

$$\hat{q} = \sqrt{\frac{1 \times 10^{-6}}{1}} = 1 \times 10^{-3}$$

Hence in the population under consideration here, the equilibrium frequency of the Tay-Sachs gene under the combined effect of mutation and selection is 0.001. The frequency of Tay-Sachs births in this group, then, at equilibrium would be 1×10^{-6}, that is, $(1 \times 10^{-3})^2$, as compared with the present frequency of 2.25×10^{-4}, or $(0.015)^2$.

Random Genetic Drift

From generation to generation the number of individuals carrying a particular allele, either in the homozygous or heterozygous state, may be expected to vary somewhat, so that gene frequencies will fluctuate about a mean. The amplitude of this fluctuation is *random genetic drift* and is also due to the vagaries of chance mating and to the fact that, even in cases where $p = q = 0.5$, theoretical ratios (for example, 3:1, 1:1, 1:2:1) are certainly not always produced.

If a population is large, drift is nondirectional and of small magnitude; it may be expected to vary within rather narrow limits above and below the mean. But in small populations all the progeny might, by chance alone, be of the same genotype with respect to a particular pair of alleles, for example, *Aa*. Their progeny (the F_1) would be expected in the genotypic ratio 1 *AA*:2 *Aa*:1 *aa*. The F_2 is assumed to arise from random mating among members of the F_1. The probability of any of the possible F_1 matings is then a function of the frequency of each F_1 genotype. Under these conditions, **fixation** will occur in one-eighth of the progeny (F_2) per generation:

F_1 mating	Probability of mating	Values of p and q in mating population		Progeny (F_2)
		p	q	
$aa \times aa$	$\frac{1}{4} \times \frac{1}{4} = \frac{1}{16}$	0	1	all aa
$Aa \times aa$	$2(\frac{2}{4} \times \frac{1}{4}) = \frac{4}{16}$	0.25	0.75	1 Aa:1 aa
$Aa \times Aa$	$\frac{2}{4} \times \frac{2}{4} = \frac{4}{16}$	0.5	0.5	1 AA:2 Aa:1 aa
$AA \times aa$	$2(\frac{1}{4} \times \frac{1}{4}) = \frac{2}{16}$	0.5	0.5	all Aa
$AA \times Aa$	$2(\frac{1}{4} \times \frac{2}{4}) = \frac{4}{16}$	0.75	0.25	1 AA:1 Aa
$AA \times AA$	$\frac{1}{4} \times \frac{1}{4} = \frac{1}{16}$	1	0	all AA

The founders principle. The formation of a new population by migration of a sample of individuals may likewise lead to different gene frequencies. Imagine a large population in which $p = 0.4$ and $q = 0.6$. The most probable values of p and q, therefore, in a sample from this population are also 0.4 and 0.6, respectively. Expected deviation from these values is, of course, provided by the standard deviation. The standard deviation for a simple proportionality, such as heads versus tails, is given by formula 10:

$$s = \sqrt{\frac{pq}{n}} \tag{10}$$

AN EXCEPTION TO THE FOUNDERS PRINCIPLE?

According to the founders principle that originated in 1954, if a very small group of individuals from one population leaves to found a new population, the gene frequency and variance of the new population will be quite different compared to the original population. For example, if you have 200 alleles at a locus in a population, and only 2 individuals—a male and a female—begin a new population, the maximum number of alleles in the new population is only 4. This would greatly reduce the variability in the new population.

A recent study on houseflies has challenged this dogma and is causing some interesting discussion on the founders principle. In the housefly study, 1, 4, and 16 pairs of flies were taken from a population and used to found new populations. In the new populations, 3,000 flies were painstakenly measured for several traits. The result was not the expected decrease in variance in the newly developed population, but an increase in variance for some of the traits measured. There was more variability in the new population for wing size and shape than in the original population. Not all the traits showed the same increase in variance.

Most population geneticists are excited about these results, but they are not really sure what the results mean. For example, in nature the kind of trait involved would be very important. It seems that those traits involving fitness, such as body size and reproductive characteristics, did not show an increase in variance. When the overall survival of the species in natural populations is considered, a reduced variance in fitness-related traits may be a serious consequence. Therefore, the observed increase in variance in certain traits should not be considered to be beneficial to the species.

These findings also have a very practical implication. In our attempts to save endangered species, we protect and use the existing limited populations to increase the numbers of the species. Increasing their numbers may not be the complete solution to their survival, however, if their overall fitness is adversely affected.

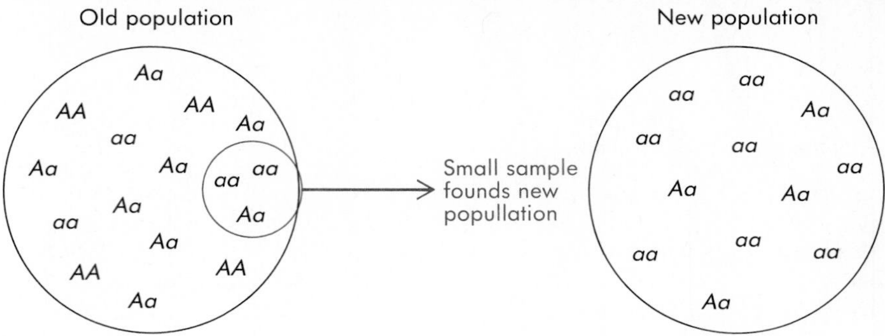

FIGURE 19–4. *Illustration of the founder principle. Gene frequency in the new population can be very different from the old population due to the very small sample.*

where n is the number of observations. But when gene frequencies are calculated from the frequency of homozygous recessive phenotypes, as has been done here, the equation for standard deviation becomes (equation 11):

$$s = \sqrt{\frac{pq}{2N}} \tag{11}$$

where N is the number of (diploid) individuals in the sample.

If, in the large population under consideration, a sample of 50,000 persons is taken,

$$s = \sqrt{\frac{0.24}{100{,}000}} = 0.00155$$

That is, in any sample of 50,000 individuals from this population with $p = 0.4$ and $q = 0.6$, 68 percent of the time p will lie between 0.39845 and 0.40155, that is, 0.4 ± 0.00155; 95 percent of the time it will fall within the range 0.4 ± 0.0031, or 0.3969 to 0.4031. But if the sample were to consist of only 50 persons,

$$s = \sqrt{\frac{0.24}{100}} = 0.049$$

In 68 percent of the samples of this size, p would be expected to fall within the range 0.4 ± 0.049, or between 0.351 and 0.449. Similarly, in 95 percent of the cases of samples of 50, p would be within the range of 0.302 to 0.498. Formation of a new population by emigration of a sample as small as 50 might be expected to lead purely by chance to a very different gene frequency in the next generation. This genetic drift due to a small sample (founders) being sent forth to establish a new population is termed the **founders principle.** This principle is illustrated diagrammatically in Figure 19–4.

A presumed example of such a case of genetic drift has been reported for the Dunkers of Pennsylvania. These are members of a religious sect who migrated from Germany in the early eighteenth century and have remained relatively isolated. The frequency of blood group A in this small group is almost 0.6, whereas it is between 0.40 and 0.45 in German and American populations, and the I^B allele is nearly absent in the Dunkers, whereas group B persons make up 10 to 15 percent of German and American populations.

THE MECHANISM OF EVOLUTION

The principles thus far examined, especially in this chapter and Chapters 14 and 15, constitute the basic mechanisms whereby new species evolve, sometimes slowly, sometimes suddenly, from preexisting ones. A species is more a taxonomic concept than a concrete entity, and its parameters necessarily differ somewhat among various groups of animals and plants. Thus a species in bacteria is quite a different concept from one in birds. Species "boundaries" in viruses and vertebrates have different emphases, as they do even in more closely related units like freely hybridizing, genetically more adaptable groups, such as willows, versus less adaptable groups, such as maples. However, in very general terms, speciation, or the origin of new species, depends largely on such cytological and genetic mechanisms as:

1. Chromosomal aberrations, especially translocations and allopolyploidy (in plants)
2. Addition of new genetic material by mutation (although most are either neutral or disadvantageous in an existing environment).
3. Changes in gene frequencies:
   a. By mutation
   b. Through genetic drift
   c. By environmental selection
   d. By migration or the breakdown of isolating barriers.

These mechanisms may sooner or later break up larger populations into smaller units that (1) develop genetically determined, distinctive morphological and/or physiological characters that differ from those of other such units, and (2) become reproductively isolated from related groups, and develop thereby into "Mendelian populations" that have their own gene pools. How, in what direction(s), and to what extent evolution will develop depends on an intricate interaction among a variety of influences. But basically only those changes in the genetic material that can be passed on to succeeding generations and do not disappear from the gene pool can serve as agents of evolution in living organisms.

PROBLEMS

19–1 If you consider the data set forth in this chapter for M, MN, and N persons in this country and assume that you do not know the genotype of either yourself or your parents: (a) What genotype are you most likely to have? (b) What genotype are you least likely to have?

19–2 A sample of 1,000 persons tested for M-N blood antigens was found to be distributed: M, 360; MN, 480; N, 160. What is the frequency of alleles *M* and *N*?

19–3 A sample of 1,522 persons living in London disclosed 464 of type M, 733 of type MN, and 325 of type N. Calculate gene frequencies of *M* and *N*.

19–4 A sample of 200 persons from Papua (southeast New Guinea) showed 14 M, 48 MN, and 138 N. Calculate frequencies for alleles *M* and *N*.

19–5 A sample of 100 persons disclosed 84 PTC tasters. Calculate gene frequencies of *T* and *t*.

19–6 (a) How many heterozygotes should there be in the sample of Problem 19–5? (b) How many *TT* persons? (c) What is the probability that a taster in this sample is heterozygous (*Tt*)?

19–7 Albinism is the phenotypic expression of a homozygous recessive genotype. One source estimates the frequency of albinos in the American population as 1 in 20,000. What percentage of the population is heterozygous for this gene?

19–8 Alcaptonuria, which results from the homozygous expression of a recessive autosomal allele, occurs in about 1 in 1 million persons. What is the proportion of heterozygous "carriers" in the population?

19–9 A sample of 1,000 hypothetical persons in the United States showed the following distribution of blood groups: A, 450; B, 130; AB, 60; O, 360. Calculate the frequencies of genes I^A, I^B, and i.

19–10 Another sample of 1,000 hypothetical persons had these blood groups: A, 320; B, 150; AB, 40; O, 490. What is the frequency in this sample of each of the following genotypes: I^Ai^A, I^Ai, I^BI^B, I^Bi, I^AI^B, ii?

19–11 Assume the data of Table 19–1 to represent accurately the distribution of A, B, AB, and O blood groups in the United States population. Using the chi-square test, determine whether the sample of Problem 19–9 represents a significant deviation. (Round the frequencies of Table 19–1 to the whole numbers 440, 420, 100, and 40, respectively, to obtain calculated values for computing chi-square.)

19–12 A sample of 429 Puerto Ricans showed the following allelic frequencies: I^A, 0.24; I^B, 0.06; i, 0.70. Calculate the percentage of persons in this sample with A, B, AB, and O blood.

19–13 What percentage of the sample in the preceding problem is (a) homozygous A, (b) heterozygous B?

19–14 If one man in 25,000 suffers from hemophilia A, what is the frequency of gene h in the population?

19–15 One *man* in 100 exhibits a trait that results from a certain sex-linked recessive gene. What is the frequency of (a) heterozygous *women*, (b) homozygous recessive *women*?

19–16 Data represented in Table 19–4 show a frequency of 0.64 for gene i and 0.61 for the Rh allele D in the U.S. white population. If only these two alleles are considered, what would be the frequency of O^+ persons in that population?

19–17 Increasing use of amniocentesis (see Chapter 20) is being made to detect a Tay-Sachs fetus *in utero*, being carried by mothers who are at risk. Some parents in cases of a finding of *positive* for Tay-Sachs elect to have the fetus aborted. What effect would the latter practice have over time on the frequency of the Tay-Sachs gene?

19–18 Great effort is being made to eliminate the muscular dystrophy caused by allele d. If we are someday successful in eliminating the *effect* of this gene in *treated* individuals, will this result in genetic improvement or deterioration in the world population?

19–19 Use the tabulated data (Table 19-3) concerning frequency of gene c in tomato to determine the frequency of this allele after one more generation of inbreeding.

19–20 (a) If we consider the tabulated data (Table 19-3) concerning frequency of allele c in tomato, what would be the frequency of this gene in the twentieth generation of inbreeding? (b) If the parental generation is considered as generation 1, in what generation would the frequency of c be reduced to exactly 0.005?

19–21 Allele f is an autosomal recessive lethal that kills ff individuals before they reach reproductive age. If, in an isolated population, this gene had a frequency of 0.4 in the P generation, what would be its frequency in the F_2?

19–22 For a certain pair of alleles, completely dominant allele A has an initial frequency of 0.7 and an adaptive value of 1. Its recessive allele has a frequency of 0.3 and an adaptive value of 0.5. What is the frequency of a in the next generation?

19–23 Allele A mutates to its recessive allele a four times more frequently than a mutates to A. What will be the equilibrium frequency of a under mutation?

19–24 If allele B has a forward mutation rate of 5×10^{-6} and a back mutation rate of 1×10^{-6}, what is the equilibrium frequency of allele b under mutation?

19–25 In a given population allele C has a frequency of 0.2. If 50 individuals from that population are sampled, there is a 0.68 probability that the frequency of C will be no more than how much above or below 0.2 in that sample?

19–26 Persons with cystic fibrosis of the pancreas occur with a frequency of about 0.0004. This condition is due to a recessive autosomal allele and is fatal in childhood. (a) Give the frequency of this allele in the present population. (b) If the rate of forward mutation is assumed to be 4×10^{-6}, what is the equilibrium frequency of the allele for cystic fibrosis under the combined effect of mutation and selection?

19–27 Based on your answer to part (a) of the preceding problem, and assuming a human generation to be 30 years, (a) how many years would be required to reduce this frequency to 0.01? (b) Is this likely to occur? Why?

19–28 Let p represent the frequency of dominant autosomal allele A and q the frequency of its recessive allele a. For a randomly mating population, give a mathematical expression for (a) the frequency of an AA individual, (b) the probability of an $AA \times AA$ mating, (c) the frequency of an Aa individual, (d) the probability of an $Aa \times Aa$ mating, (e) the probability of an $Aa \times aa$ mating, (f) the total of all possible matings.

19–29 Give a mathematical expression for the frequency of (a) dominant and (b) recessive progeny phenotypes from a *single* $Aa \times aA$ mating.

19–30 For the *entire* randomly mating population, with all possible matings equally free to occur, give (a) a mathematical expression for the frequency of recessive progeny in the population resulting from such random matings and (b) the numerical value of this expression if $p = q = 0.5$.

19–31 Using the information in Table 19–6, calculate the frequency of heterozygotes in generation 10.

19–32 In view of your answer to the preceding question, what would be the frequency of each of the two homozygous genotypes in generation 10?

19–33 Phenylketonuria (PKU) is a condition characterized by such low intelligence that persons not diagnosed at birth and *put on immediate treatment* almost never reproduce. This condition is produced by an inherited inability to metabolize the essential amino acid phenylalanine and is caused by a completely recessive autosomal allele. PKUs have an incidence of about 3.6×10^{-6}. (a) Assuming lack of diagnosis in time to treat the condition effectively, what is the frequency of the PKU gene at present? (b) What is the incidence of homozygous normal persons? (c) What is the incidence of heterozygotes? (d) If u is assumed to be 4×10^{-6}, what is the allele's equilibrium frequency under the combined effect of mutation and selection? (e) At this value of u, what is the probability of two homozygous normal persons producing a PKU child?

19–34 Assume a certain dominant allele to have a forward mutation rate of 2×10^{-6} and its recessive allele to have a frequency now of 0.015 and a selection coefficient of 0.5. What is the equilibrium frequency of the recessive allele under the combined effect of mutation and selection?

19–35 As noted in the text, some reports of the rate of forward mutation for the recessive Tay-Sachs allele are as high as 1.1×10^{-5}. Calculate the equilibrium frequency of this allele under the combined effect of mutation and selection using this higher rate of forward mutation.

19–36 Lengths of index finger relative to that of the fourth finger is thought to be due to a sex-influenced autosomal allele, with allele F (shorter index finger) dominant in males and recessive in females. A class of 85 genetics students consisted of 32 males with short index finger, 8 males with long index finger, 27 females with short index finger, and 18 females with long index finger. Based on the hypothesis of pair of sex-influenced alleles, calculate the frequency of (a) allele F in the females and (b) allele f in the males. (c) How many of the males in this sample should be of genotype FF?

19–37 Tyrosinemia is a recessive disorder in which a liver enzyme (parahydroxyphenylpyruvic acid) is not produced, with the result that toxins that cause cancer and cirrhosis of the liver accumulate. It affects 1 in 100,000 children in the United States. These children usually live less than one year (although a few may survive to age 5) unless a successful liver transplant can be made. (a) What is the frequency of children with this disorder (expressed as a decimal)? (b) What is the frequency of heterozygotes (to four decimal places)? (c) What is the probability of two heterozygotes marrying (to five decimal places)? (d) What is the probability that persons referred to in part (c) will have an affected child (to five decimal places)? (e) If the dominant normal gene for production of the enzyme mutates to the recessive allele for tyrosinemia at the rate of 1×10^{-18}, what is this recessive allele's equilibrium frequency under the combined effect of mutation and selection (express as a power of 10)?

RECOMBINANT DNA: NEW GENETICS AND THE FUTURE

CHAPTER

20

In the early 1970s, a group of scientists met and decided to declare a moratorium on the particular type of experiments they were doing that involved the use of a newly developed technique for gene manipulation. The deferral of these experiments would extend until an assessment could be made of the potential dangers of this research. The ultimate finding was that these dangers were minimal. The technique of concern centered around the ability to construct new combinations of DNA molecules that do not exist naturally. These molecules are generally referred to as **recombinant DNA** molecules. This chapter examines the basis for the research and its possible applications and concerns, as well as other new developments in genetics and medicine.

RECOMBINANT DNA

The joining together of DNA sements derived from biologically different sources is the technology referred to as **recombinant DNA** technology or **genetic engineering.** Advances in this field have made many experiments routine that were not even possible a few years ago. Out of these techniques is springing an industry in **biotechnology,** or the use of biological processes for the production of useful substances. The overall process centers on three main experimental approaches:

1. Generating a series of fragments of DNA from different sources containing gene sequences of interest.

2. Joining these segments to a DNA molecule that is able to replicate—usually a bacterial plasmid called a vector.
3. Transforming bacterial cells with the recombinant molecule so that they are first replicated (cloned) and then expressed.

Restriction Enzymes

The basis for the recombinant DNA technique is the use of **restriction endonucleases.** These are bacterial, DNA cleaving enzymes that recognize short sequences of bases and make a double-stranded cut in the DNA molecule. Their function in the bacterial cell is to destroy foreign DNA that might enter the cell. The enzyme recognizes the foreign DNA and cuts it at several sites along the molecule. Each bacterium has its own unique restriction enzymes and each enzyme recognizes only one type of sequence. The sequences recognized by restriction enzymes are called **palindromes.** Palindromes are base sequences that read the same on the two strands but in opposite directions. For example, if the sequence on one strand is G A A T T C read in the 5′→3′ direction, the sequence on the opposite strand is C T T A A G read in the 3′→5′ direction, but when both strands are read in the 5′→3′ direction the sequence is the same. The palindrome appears accordingly:

5′ G A A T T C 3′
3′ C T T A A G 5′

In addition, there is a point of symmetry within the palindrome. In our example, this point is in the center between the AT/TA. The value of restriction enzymes is that they make cuts in the DNA molecule around this point of symmetry. Some cut straight across the molecule at the symmetrical axis producing blunt ends. Of more value, however, are the restriction enzymes that cut between the same two bases away from the point of symmetry on the two strands thus producing a staggered break:

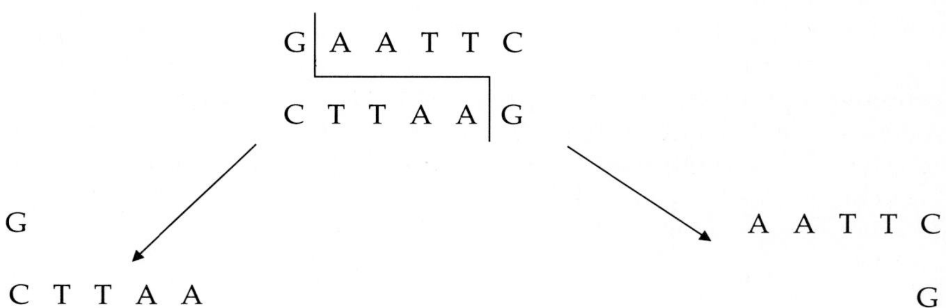

The example just used is the palindrome sequence recognized by one of the most popular restriction enzymes **Eco RI** from *E. coli.* In Table 20–1 additional restriction enzymes and their sites of DNA cleavage are listed. There are now several hundred known restriction enzymes.

The most useful aspect of restriction enzymes is that each enzyme recognizes the same unique base sequence regardless of the source of the DNA. It furthermore means that these enzymes establish fixed landmarks along an otherwise very regular DNA molecule. This allows dividing a long DNA molecule into fragments that can be separated from each other by size with the technique of gel electrophoresis. Each fragment is thus also available for further analysis, including the sequencing discussed in Chapter 8.

TABLE 20–1. Restriction enzymes, sources, and cleavage sites

Restriction enzyme	*Source*	*Cleavage site*
Bam H1	*Bacillus amyloliquefaciens H*	↓ G G A T C C C C T A G G ↑
Hind III	*Haemophilus influenzae*	↓ A A G C T T T T C G A A ↑
Sal I	*Streptococcus albus G*	↓ G T C G A C C A G C T G ↑
Pst I	*Providencia stuartii*	↓ C T G C A G G A C G T C ↑

One value in cutting a DNA molecule up into discrete fragments is being able to locate a particular gene on the fragment where it resides. This is done by the general technique of **Southern blotting.** This technique is illustrated in Figure 20–1. In this technique, a DNA molecule is cut into discrete fragments by a restriction enzyme. It is then electrophoresed through an agarose gel. This separates the various fragments according to size. The DNA is then denatured into single strands by exposing the gel to NaOH. A few pieces of filter paper soaked in buffer are placed under the gel. A large piece of nitrocellulose paper is layed over the agarose gel, followed by several layers of absorbent material such as filter paper. This dry absorbent material pulls the buffer

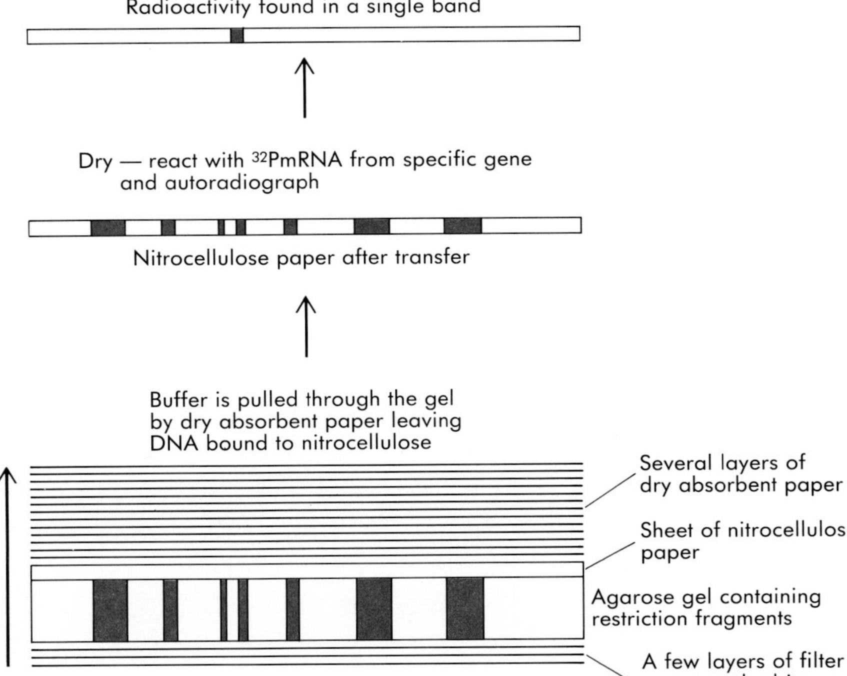

FIGURE 20–1. *Procedure for Southern blotting. Restriction fragments are transferred from agarose gel to nitrocellulose paper where radioactively labeled mRNA reacts only with complementary DNA sequences in one band corresponding to a single fragment size.*

up through the gel from the lower layer. This washes the DNA off the gel and onto the filter, where it covalently binds to the filter.

The position of the DNA molecules on the filter are identical to their position in the gel. The nitrocellulose filter containing the DNA is first dried and then exposed to a solution of ^{32}P labeled mRNA from the gene to be isolated. The radioactive mRNA hybridizes (hydrogen bonds) only with the single-stranded DNA in restriction fragments that contain complementary sequences. The nitrocellulose filter is then removed and placed in contact with photographic film that when developed will reveal the fragments from the original gel containing complementary sequences to the mRNA used in the assay. This procedure allows specific identification of restriction fragments containing DNA sequences to specific RNA molecules.

In addition, these fragments of DNA molecules, produced by the same enzyme from the same or different sources, contain identical staggered cut ends. Since these ends can overlap and complementary base pairing can occur, fragments from different sources may join together to form hybrid (recombinant) molecules. If the joined DNA fragments are treated with DNA ligase they will be permanently joined together. A diagram of this process is shown in Figure 20–2.

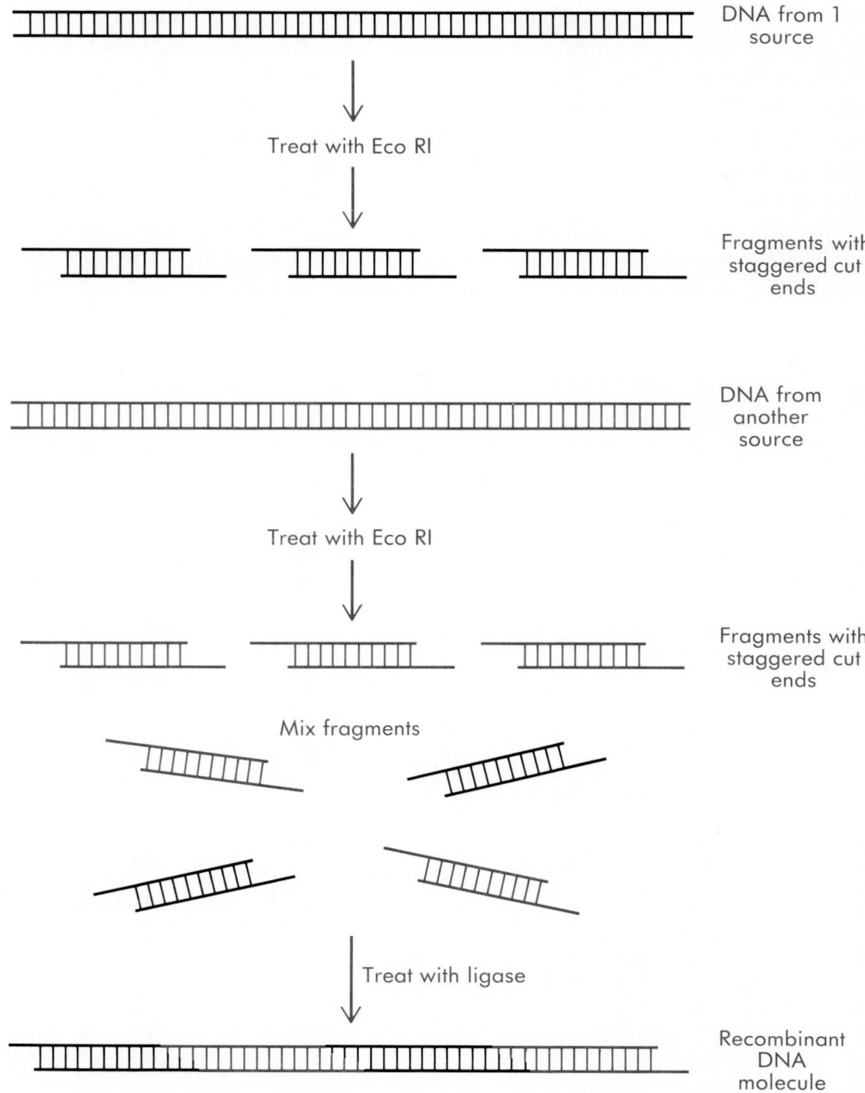

FIGURE 20–2. *Construction of a recombinant DNA molecule using a restriction enzyme.*

Molecular Cloning

In order to construct a useful hybrid DNA molecule using restriction endonucleases, the fragments from one source are joined to a molecule that is able to replicate when reintroduced into a bacterial cell. This is usually a bacterial plasmid (see Chapter 16). Plasmids, with their intimate relationship with the bacterial cells in which they exist, carry genes that confer advantages on their host cell. The host cell in return provides conditions necessary for limited replication of the plasmid, and its transmission to daughter cells whenever the cell divides, thus allowing the plasmid to spread and exist in the entire bacterial population.

Plasmids are easily isolated in large numbers from bacterial cells, and usually have a small limited number of restriction sites. When a plasmid is cut with a restriction enzyme, and DNA ligase is used to join the fragments with a foreign DNA fragment produced by the same enzyme, a hybrid (recombinant) DNA molecule results. The plasmid carrying a piece of foreign DNA is referred to as a **vector.** Phage vectors, which are also used in this way, have the advantage of being able to carry a larger fragment of foreign DNA. The plasmid or phage vector provides a means of establishing and expressing the foreign DNA fragment in a bacterial cell. The method of joining a piece of foreign DNA to a plasmid vector is found in Figure 20–3.

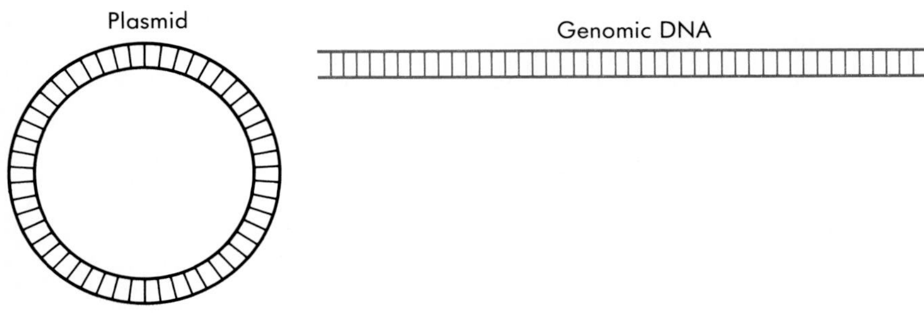

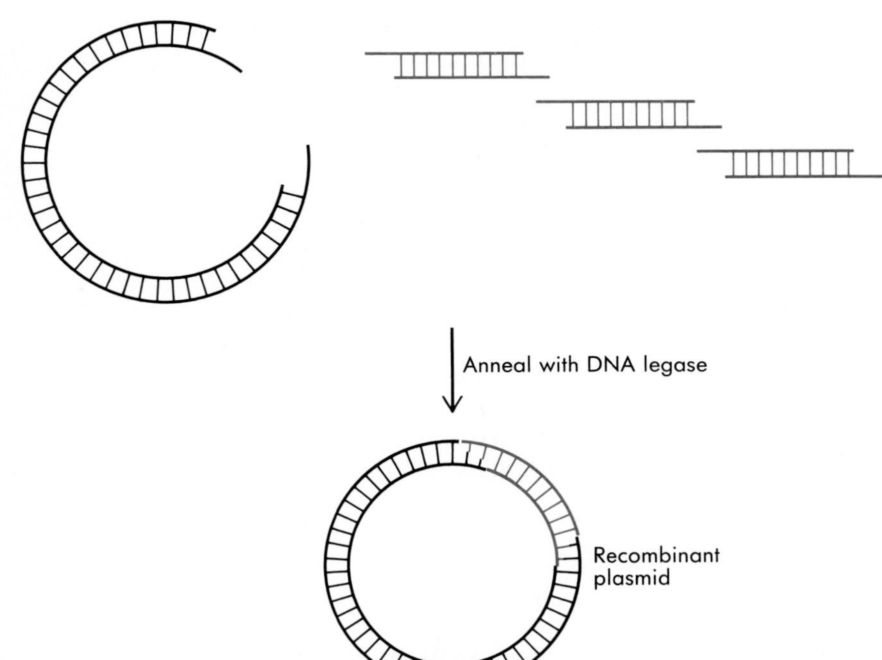

FIGURE 20–3. *Construction of a recombinant plasmid between a bacterial plasmid and genomic DNA from another organism.*

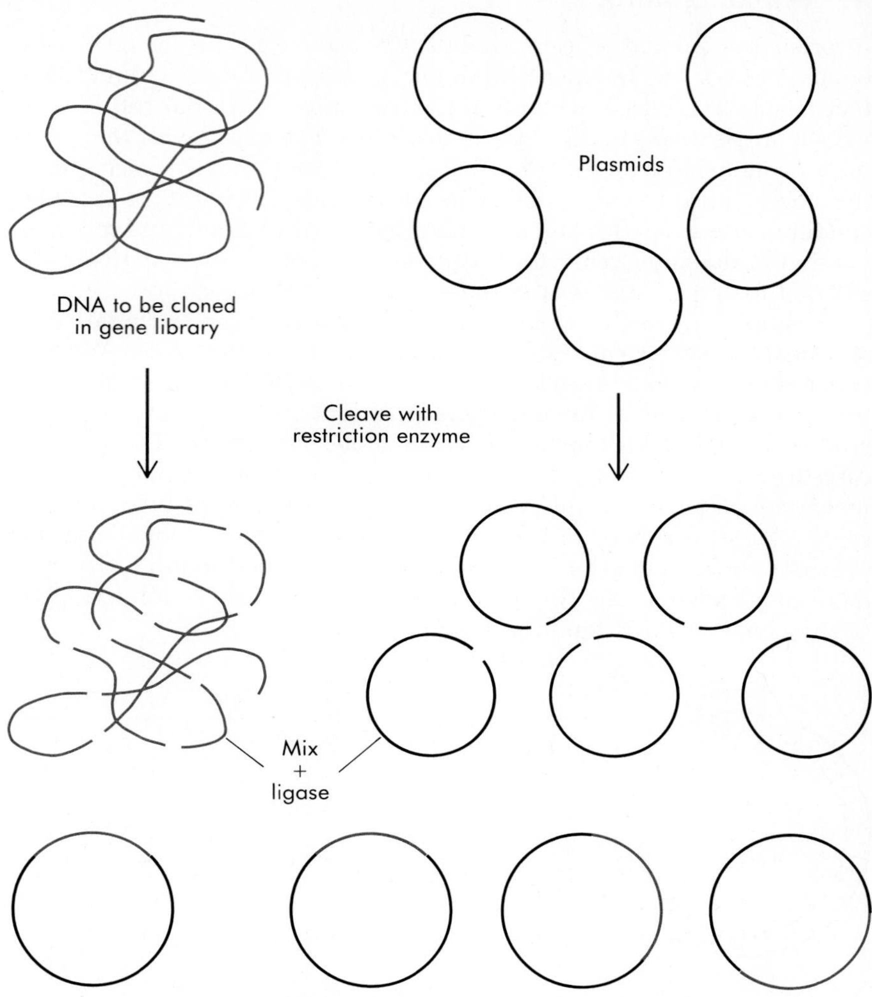

FIGURE 20–4. *Procedure for "shotgun" cloning and creation of a gene library.*

The recombinant plasmid is introduced into a bacterial cell as in transformation. This is easily done by adding the vector to bacterial cells in the presence of $CaCl_2$. Once inside the bacterial cell the plasmid is replicated and the number increased. The plasmid vector usually carries one or more antibiotic resistance markers so that when cultured on a medium containing the antibiotic, only bacterial cells containing the recombinant plasmid vector will survive and grow. Each hybrid molecule results in a bacterial cell population with the same foreign DNA fragment being present in all the cells. In this context, this piece of foreign DNA is referred to as being cloned. This constitutes the **molecular cloning** process.

A major application of gene cloning is to have DNA fragments representing the entire genome of an organism cloned on plasmids. This constitutes a so-called **gene library.** It represents a collection of plasmid-containing fragments that is large enough so that each segment of the entire genomic DNA is represented at least once. A general method for producing a gene library is called "shotgun" cloning. An outline of the procedure is found in Figure 20–4.

The DNA is isolated from a particular source and cleaved into fragments with a restriction enzyme. Likewise a plasmid-cloning vector is also cleaved with the same enzyme. When the fragments are mixed and ligated, a plasmid population is constructed that contains a random array of the fragments. Based on the size of the genome, the number of independent clones, N, required in order to insure complete representation of the entire genome in the library can be calculated as follows:

$$N = \ln (1 - P)/\ln (1 - f)$$

where P is the probability of a given nucleotide sequence being in the library (a normal acceptable value is 99 percent), and f is the fraction of the genomic DNA represented by each independent clone.

Clones of these plasmids are then maintained in the bacterial population. Various techniques are available to probe the library for specific bacterial strains containing a fragment with a particular sequence. One strategy involves the isolation of the mRNA transcript from tissue actively making large quantities of the protein whose gene is being sought. This mRNA is radioactively labeled and reacted through Southern blotting with DNA isolated from the various library clones. This isolated RNA will hybridize only to those clones carrying complementary DNA sequences.

Another approach can be used to identify clones carrying specific genes. This is called cDNA cloning. The mRNA transcript from the gene of interest can be isolated and a DNA copy of this RNA can be made with reverse transcriptase. This DNA is called copy DNA or cDNA. The cDNA can be inserted into a plasmid and cloned in a bacterium. This cloned cDNA is used as the probe to identify library clones carrying that particular gene. If the protein whose gene is sought is not present in large quantity, a small amount of the protein is purified, and the amino acid sequence of some portion of it is determined. A piece of cDNA corresponding to this sequence can be synthesized. This human-made fragment is used as the probe for the clone of interest. Finally, if enough protein is produced to obtain antibodies, the various clones may be screened with antibody to identify clones in which the protein is being made. Through these techniques and others it is now possible to isolate and clone any gene whose protein product is known.

Expression of Cloned Genes

An immediate application of recombinant DNA and cloning technology is to use bacteria containing cloned genes as factories to produce a specific gene product. This requires that a foreign gene be transcribed and translated in a bacterial cell. Many problems are encountered in trying to accomplish this goal, especially if the foreign gene is eukaryotic. These include:

1. The bacterial RNA polymerase may not recognize the eukaryotic promoters accompanying the gene.
2. If the gene is transcribed, it may lack the prokaryotic sequences necessary to attach the message to the bacterial ribosome.
3. If the eukaryotic gene contains introns, the bacterial cell will not have the enzymes for processing the RNA.
4. The bacterial cell may not be able to process the translation products to make them functional.
5. Bacterial proteases can recognize the proteins as foreign and degrade them before they can be isolated from the bacterial cell.

Some of these difficulties can be overcome by using a bacterial or yeast promoter and cDNA that does not contain introns. Once these obstacles are overcome, and a gene is redesigned for high-level expression in a bacterium or yeast system, proteins that are normally synthesized only in small amounts can be produced in very large quantities. In addition, as will be seen with insulin, a processing step of the gene product outside the bacterium may still be required. This overall approach allows the very inexpensive growth of bacteria containing cloned genes in large-scale fermentation chambers to be used for production of valuable proteins in very large quantity.

APPLICATIONS OF RECOMBINANT DNA TECHNOLOGY

There are several applications of this recombinant DNA technology presently being pursued. These include: (1) engineering of bacteria to carry out specific processes or to produce important molecules such as hormones or antibiotics, (2) altering the genotypes of plants as an aid in plant breeding, and (3) altering genotypes of animals to correct genetic defects.

Engineering Bacteria

Perhaps one of the first applications of recombinant DNA techniques was the combination of several genes for metabolizing petroleum on a single plasmid and introducing this plasmid into a marine bacterium. This organism can be used to clean up oil spills in the oceans. Another application is the construction of bacteria with genetic information to convert biological waste into ethyl alcohol, which is directly applicable for use as a fuel. Several protein products are currently being produced in engineered bacteria. These include insulin for treating diabetes, interferon for controlling infections and possibly tumors, urokinase and plasminogen activator for dissolving blood clots, tumor necrosis factor for providing cancer therapy, and cellulase for producing sugar from plant cellulose.

Two successful examples of engineered bacteria being used as factories to produce human proteins—the human growth hormone somatostatin, and insulin—on a commercial basis will be discussed. The tricks employed to get bacterial cells to make these proteins are fascinating in themselves. The gene for the 14-amino acid peptide somatostatin was attached to the beginning of the lac Z gene that codes for beta galactosidase (see Chapter 12). This gene is inserted into a plasmid with the lac promotor and used to transform *E. coli.* When the lac operon is induced by lactose, the protein made contains the first few amino acids of beta galactosidase, and then somatostatin. The protein is isolated from the bacterium, is purified, and the beta galactosidase amino acids are cleaved off, leaving the somatostatin peptide intact.

The situation is only a little more complex for insulin, the 51-amino-acid hormone that regulates sugar metabolism in humans. The major source of insulin for the treatment of diabetes is cattle or pig pancreas. However, some diabetics are allergic to this insulin. Recombinant DNA techniques have been used to successfully produce human insulin in bacteria. Normally insulin is synthesized as an 86-amino-acid molecule called preproinsulin (Figure 20–5) and is the product of a single gene. The molecule folds back on itself, forms disulfide bonds, is processed first into proinsulin, and finally processed into the 51-amino-acid hormone insulin. The functional hormone consists of a 30-amino-acid B chain and a 21-amino-acid A chain.

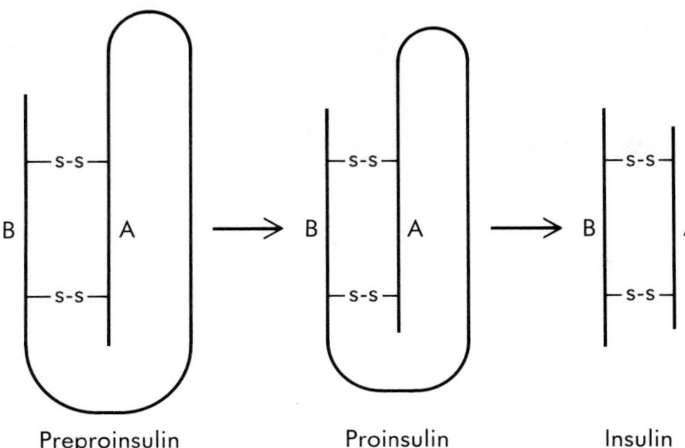

FIGURE 20–5. *Processing of preproinsulin into proinsulin and then insulin.*

Insulin has been produced in *E. coli* in a similar way to somatostatin. DNA segments encoding the human A and B chains are separately cloned in two different bacteria within the beta galactosidase Z gene. When the two proteins are synthesized in *E. coli* they are isolated and purified. The chains are mixed and the bonds holding the chains together can form. Human insulin produced using recombinant DNA technology is presently available on a commercial basis (Figure 20–6).

Altering the Genotypes of Plants

Recombinant DNA technology is presently being applied to problems of gene transfer and directed changes in the genotype of important crop plants. The main vector of interest is the Ti plasmid of the crown gall tumor-inducing bacterium *Agrobacterium tumefaciens*. The tumor-inducing ability of *A. tumefaciens* resides in the Ti plasmid. When dicotyledonous plants are infected with the bacterium, a portion of the Ti plasmid, the **T-DNA,** is transferred to the genome of the plant. The T-DNA stably integrates into the plant's DNA. Because the T-DNA carries genes for production of the plant hormones, auxin and cytokinin, the

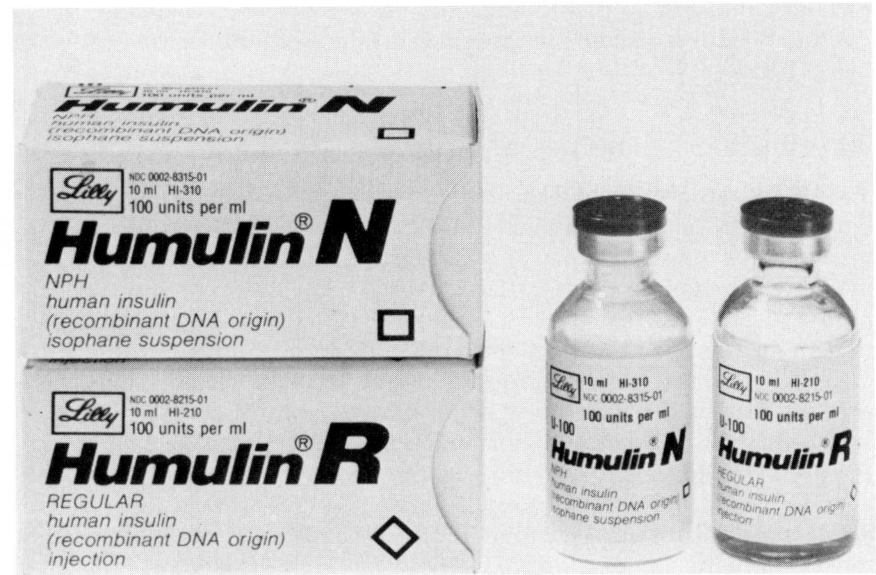

FIGURE 20–6. *Commercially available insulin produced by recombinant DNA techniques.* (Photo courtesy Mary P. Grein, Eli Lilly Co.)

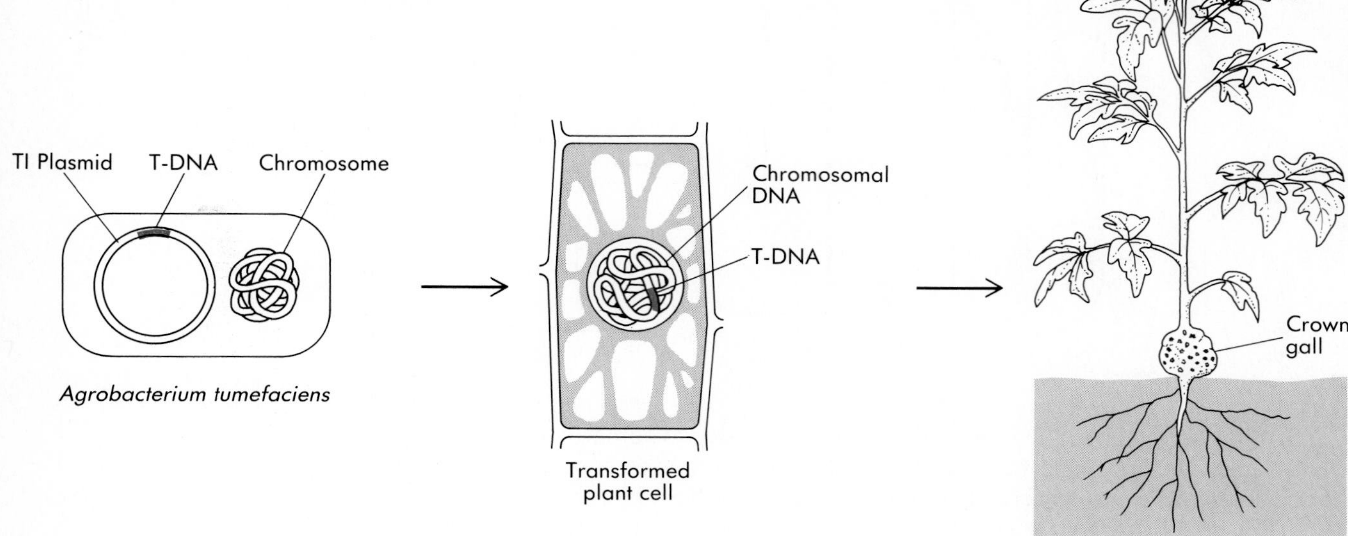

FIGURE 20–7. *Transforming activity of Ti plasmid of* Agrobacterium tumefaciens. *T-DNA from the Ti plasmid is transferred to host cell DNA. The T-DNA contains genes for hormone biosynthesis that interrupts normal hormone balances and causes uncontrolled cell growth into a crown gall tumor on the plant.*

normal balance in the plant is upset, causing the infected plant cells to divide out of control. The result is a crown gall tumor (Figure 20–7).

Scientists have found that they can delete the hormone genes from the T-DNA and introduce foreign genes in their place. When a plant is infected with the bacterium, the foreign genes are transferred to the plant cell where they are expressed. When whole plants are regenerated from the cells, they are transformed for the trait. An outline of this procedure is found in Figure 20–8. Thus far only a few genes have been successfully transferred in this manner. Most of these genes have been concerned with antibiotic resistance, but very recently **Petunia** plants have been successfully transformed for tolerance to the herbicide glyphosate, and tomato plants have been made insect-resistant by introduction of a bacterial gene whose protein is toxic to the larvae of several insect pests.

Altering the Genotypes of Animals

The use of recombinant DNA techniques is also in progress to alter the heredity of animals including—some day—humans. Two different approaches have been successful so far. First, the structural gene for human growth hormone was attached to the regulatory region from a mouse gene. The "fusion genes" as they are called were microinjected into mouse fertilized eggs. Some of the progeny that developed produced the human growth hormone naturally. What is even more impressive is that the mice grew substantially more than controls in response to the human growth hormone. Figure 20–9 shows a comparison between the mice that did and did not receive the transferred gene. Furthermore, the increased growth characteristics were transmitted to and expressed in half the progeny of these mice. All of the mice that inherited the fusion gene grew two to three times faster than their litter mates that did not inherit the gene.

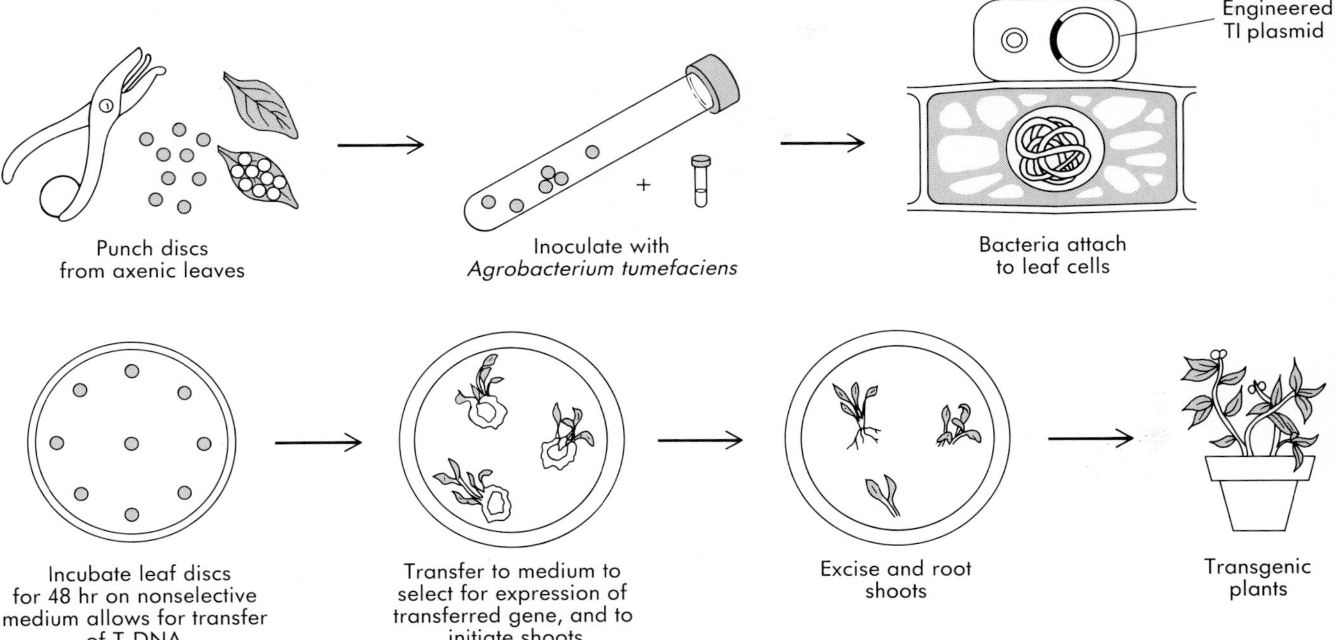

FIGURE 20–8. *Outline for transformation of plants using Ti plasmid of* Agrobacterium tumefaciens.

A slightly different approach was used in an experiment in *Drosophila*. Flies homozygous for rosy (*ry*) have a reddish-brown eye color, lacking the enzyme xanthine dehydrogenase necessary to form the dark red color of wild type. The "rosy (+)" allele (ry^+) that encodes xanthine dehydrogenase and permits the distinctive wild type color was inserted into a *Drosophila P* transposable element (see Chapter 13) and cloned in a bacterium. The cloned elements containing "rosy (+)" were microinjected into fertilized *ry/ry Drosophila* eggs. Almost 40 percent of the progeny obtained from the microinjected eggs exhibited the ry^+ phenotype.

FIGURE 20–9. *Comparison between mouse siblings 24 weeks old. The mouse on the right received human growth hormone gene and the mouse on the left did not.* (Photo courtesy Dr. R. L. Brinster, University of Pennsylvania, used by permission.)

COMPLICATIONS OF GENETIC SCREENING

Recent applications of molecular biological techniques have resulted in the ability to determine the presence of genes responsible for a number of human disorders. The essence of the approach is the same as that used in DNA fingerprinting—**DNA probes.** The DNA probes are radioactively labeled DNA sequences that can be used to search through the entire genome of a cell to find a specific target sequence. The target sequence is a piece of DNA immediately adjacent to the disease gene that is of interest. When the probe binds to the target sequence, its location can be determined by the radioactive label. If the target sequence is always inherited by victims of the disease, then the defective gene must be near the target sequence.

The question now is whether to use these probes to screen people who do not yet have a disease, but who may develop one later in life. An example is the screening of normal persons to see if they carry the gene for Huntington's disease. Everyone who inherits the gene sooner or later develops this devastating, progressive, irreversible, and ultimately fatal disease. Since there is no means of prevention, should persons at a high risk be automatically screened for the gene, or should they be allowed to decide for themselves whether or not they want to know if they have the gene? In one study 96 percent of children of Huntington's patients said that they thought the test should be made available, but only 66 percent said they would be tested.

Another problem area is the potential practice of screening employees for these genetic disorders. Employers may not want to hire or promote someone who might develop a debilitating genetic disease in the near future. Furthermore, insurance companies may require screening of new or present employees for certain disorders. The first problem is maintaining secrecy of the test results in order to protect the employee's rights. Second, there is a difference depending on whether health insurance or life insurance companies are doing the screening. Is it fair to let a health insurance company screen persons for genetic diseases over which they have no control? Nevertheless, the potential costs to the health insurance company could be very great. Life insurance companies may want to exclude persons because of poor health or who are at a high risk of dying. Finally, if the insurance company pays for the test, which is expensive, who has the right to the see the results?

The controversy sounds very similar to that presently involved with AIDS screening. In New York, insurance companies who require AIDS testing cannot give the results to the employer. However, we do not even have a national policy regarding AIDS testing. Presently, there are no easy solutions that balance individual rights against the rights of companies and employers.

The second generation showed inheritance of ry^+ in a manner expected for a stable dominant allele. Apparently the *P* transposable element containing the ry^+ allele jumped from the plasmid to the *Drosophila* genome. In about one-half of the transformed flies, the ry^+ allele had been inserted in the correct chromosomal location for this gene.

RESULTS OF THE RECOMBINANT DNA DEBATE

Many applications of recombinant DNA have been developed since the original moratorium in 1974 on recombinant DNA research. In 1976, a set of detailed guidelines was published by the National Institutes of Health (NIH) governing all government-supported research. The guidelines specified certain types of experiments that should not be done, and recommended rules on those being carried out. Depending on the hazard, the guidelines specify levels of **physical containment** for experiments to prevent accidental escape of organisms from the laboratory and levels of **biological containment** to restrict the growth of an organism if it does escape.

In the interval of time since the publishing of the original NIH guidelines, as more has been learned, there has been a general relaxation of some of the restrictions. Since no health hazard has resulted from recombinant DNA research, the current questions center mainly around commercial applications and patenting of techniques, control over the release of engineered organisms into the environment, and ways that this technology can best be put to use for the benefit of humankind.

OTHER TECHNOLOGIES IN THE NEW GENETICS

It appears that the original concern over the hazards of recombinant DNA research were basically unfounded. However, when combined with additional contemporary technologies such as **amniocentesis** and ***in vitro* fertilization,** possible genetic intervention into human reproduction can become a potential problem. So far, amniocentesis and in vitro fertilization have been employed mainly as diagnostic tools for genetic disease.

Amniocentesis (Figure 20–10) is a technique that involves removing amniotic fluid, culturing fetal cells contained within the fluid, and screening the cultured cells for chromosome aberrations or other genetic disorders. A combination of this technique with analysis of restriction enzyme fragments now enables scientists to do prenatal screening for sickle-cell anemia. The single nucleotide substitution in the gene for the beta hemoglobin chain eliminates a restriction site in that gene. As a result a larger fragment is produced upon enzyme digestion of DNA obtained from cultured cells from the amniotic fluid. This diagnosis could not be made until several months after birth, as the gene is inactive before that time. With the diagnosis of certain genetically inherited diseases such as Down syndrome a mother may elect to abort the fetus after being counseled.

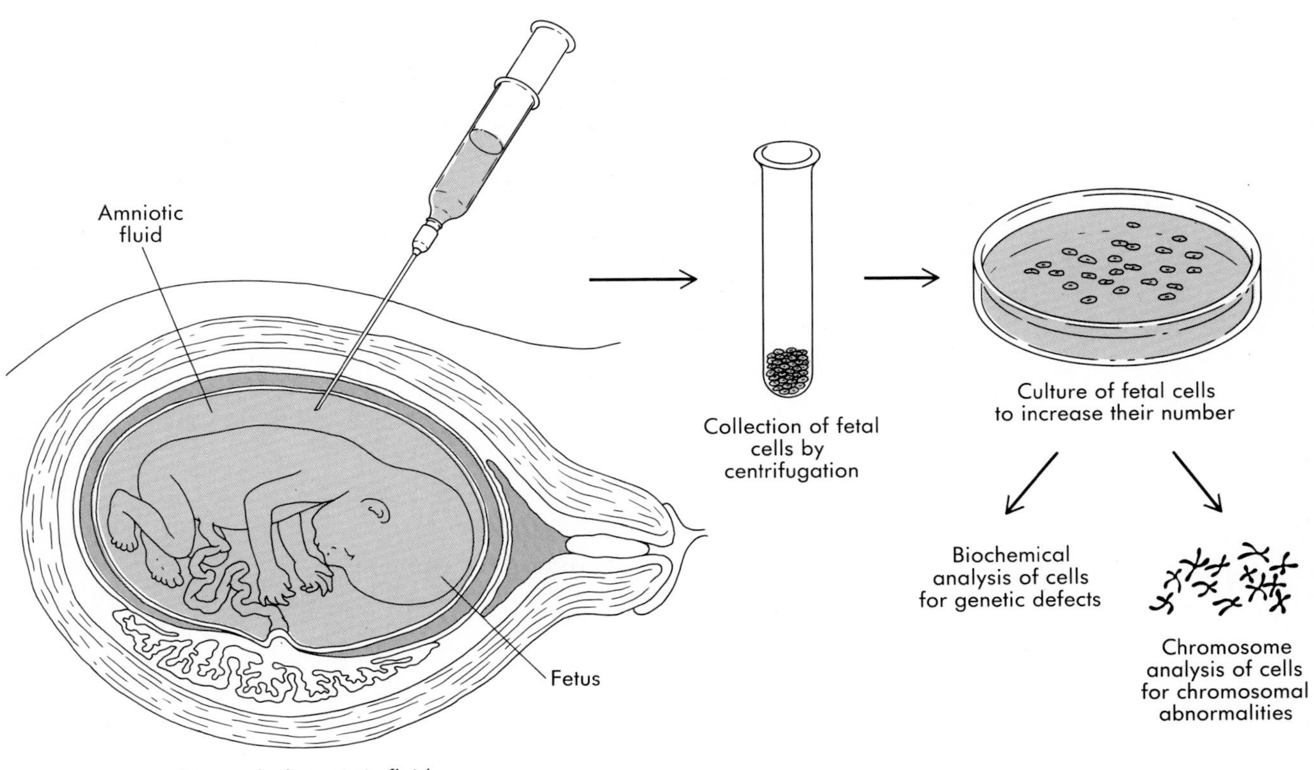

FIGURE 20–10. *Procedure for amniocentesis and screening of fetal cells for genetic defects.*

In vitro **fertilization** is a technique whereby a human egg is removed from the female and fertilized with sperm from the husband or other donor. After the embryo begins to develop, it is reimplanted into the uterus of the woman. This procedure is being successfully applied to women who have physical disorders that prevent normal fertilization. If this approach is combined with recombinant DNA techniques, the genotype of the future fetus could be altered. A retrovirus or transposable element engineered to contain a specific gene could be injected into the egg before or after fertilization, and the egg could be implanted back into the female. The transferred gene might be a normal copy of a gene known to be defective in the mother. This procedure is called **gene therapy.** Thus far only a small number of human genes have been isolated and cloned for possible use in such experiments. These are listed in Table 20–2.

To be successful the newly introduced gene would have to be expressed and regulated only in those tissues where the gene is normally expressed, and it should confer some selective advantage on a cell with the new gene.

Presently research with gene therapy uses cultured bone marrow cells (a somatic-cell type, since experiments with germ-line cells are presently banned) and an engineered retrovirus as the vector. After transfer of the normal gene into the cells and expression in those cells, the bone marrow cells are reimplanted in the animal. While these experiments have been successful in correcting an enzyme defect in the cultured cells, the expression of the new gene upon transfer of the cells back into the animal is so low that it would not result in correction of the defect in the animal. This approach has great potential because of the ability to permanently correct certain genetic defects, such as replacing a defective beta hemoglobin gene with a normal one, or correcting diabetes by replacing the defective insulin gene with the normal gene. One major concern with the use of the retrovirus as a vector is that insertion of the new gene may be near an oncogene (see Chapter 13) that could become activated, thus causing the cell to become cancerous.

One technique that could be classed as gene therapy involves the

TABLE 20–2. **List of human genetic diseases for which genes have been cloned**

alpha-Thalassemia
beta-Thalassemia
Collagen disorder
Dwarfism
Emphysema
Lesch-Nyhan syndrome
Phenylketonuria
Christmas disease
Hemophilia
Glucose-6-phosphate dehydrogenase deficiency
Ornithine transcarbamoylase deficiency
Antithrombin-3 deficiency
Cholesterol metabolism/heart disease
Cancer
Leukemia, lymphomas
Immune deficiencies

From Ellis and Davies (1985).

reinduction of normal fetal genes, which are turned off during development, to replace the activity of a defective adult gene. This is the case with beta-thalassemia and homozygous sickle-cell disease. In these diseases, the defective gene is expressed only in adults. It has been found that 5′-aza-cytidine can be used to reactivate the normal fetal globin gene that has been turned off during development. The fetal gene supplies the normal protein in place of the defective protein from the adult gene.

ETHICS OF THE NEW GENETICS

There are, of course, many puzzling issues in genetic engineering, whether it involves recombinant DNA, *in vitro* fertilization and implantation, or gene therapy. Like so many advances, the "new genetics" is viewed with alarm by some, but by others as potentially of great benefit to the human species. It should be only a short time until several more important human proteins are produced by vats of microorganisms carrying human genes, which will be proven safe for use by human beings and available on the pharmaceutical shelf. These techniques are also being investigated in numerous laboratories for faster, more controllable, and more specific improvements in farm animals and crop plants. Many new commercial laboratories, funded by both private and public investment, have been formed to pursue these objectives. Some of these goals have already been realized; others are still in the experimental stage and give promise of soon being commercially feasible. However, others (for example, development of nitrogen-fixing corn, wheat, and other crop plants) appear to most investigators to be farther in the future. There are a few concerns of the public, such as the environmental impact of the release of a genetically engineered organism into the environment, or the possible escape from industrial production of an engineered organism.

There is legitimate concern over some of the ethical issues raised by continued application to *Homo sapiens*, but a great lack of understanding exists among lay persons. Some heat, but little light, is produced by such magazine-cover scare headlines as "What's Science Doing to the Human Race? A Shocking Report on Genetic Experimentation" (*Parents Magazine*, May 1981). As pointed out earlier, medical science is already using a number of techniques in humans (for example, sperm storage, artificial insemination, superovulation and *in vitro* fertilization, and prenatal screening by amniocentesis). Further possibilities are sex selection by sperm typing (X versus Y), parthenogenesis (either by nuclear transfer or egg fusion), and obviation of genetic disorders by recombinant DNA techniques.

Our increasing control over genotype and phenotype and altering those properties in directed fashion is, for some, an exciting challenge, for others a source of vague uneasiness, and for still others an alarming threat. Of course, this growing capability can be viewed as usurpation of the responsibility that has been historically entrusted to forces other than *Homo sapiens* for maintaining our genetic integrity. On the other hand, it can also be seen in a positive light as a means of improving our lot, both now and in the future.

Religious arguments have been made on both sides of this question. Pope John Paul II has held genetic engineering to be in opposition to natural law. On the other hand, a prominent Catholic philosopher, Robert Francoeur, has written:

DNA FINGERPRINTING: THE ULTIMATE IDENTIFICATION TEST

Is the drop of blood found at the crime scene from the suspect on trial? Who is the child's father? Every year in court cases all over the world the ability to establish a person's identity is essential to a just decision. Until recently, there was no foolproof test. In a criminal case, if there were no identifiable fingerprints left behind at the crime scene, there was no case. Blood tests can only determine who isn't the parent, not who is. Other tests are only about 90 percent accurate. A test has now been developed that provides 100 percent positive identification. The test is called **DNA fingerprinting.** This test can show conclusively whether the genetic material in a drop of blood matches that of the suspect, or it can be used to solve paternity cases.

The technique of DNA fingerprinting relies on developments from recombinant DNA technology and allows an examination of each individual's unique genetic blueprint-DNA. The technique was discovered in England by Alec Jeffreys. It is based on the fact that the DNA of each individual is interrupted by a series of identical DNA sequences called repetitive DNA or tandem repeats. The pattern, length, and number of these repeats is unique for each individual. Jeffreys developed a series of DNA probes, which are short pieces of DNA that seek out any specific sequence they match, and base pair with that sequence. The probes are used to detect the unique repetitive DNA patterns characteristic of each individual.

The procedure is outlined in the figure below. DNA is purified from a small sample of blood, semen, or other DNA-bearing cells, and digested into smaller fragments with restriction endonucleases. The fragments are separated by agarose gel electrophoresis, a technique that was described for DNA sequencing in Chapter 8. The separated fragments are transferred to a nylon membrane by the technique of Southern blotting (Figure 20–1). The DNA probes labeled with radioactive material are added to a solution containing the nylon membrane. Wherever the probes fit a band containing repetitive DNA sequences, they attach. X-ray film is pressed against the nylon filter and exposed only at bands carrying

THE DNA FINGERPRINTING PROCESS

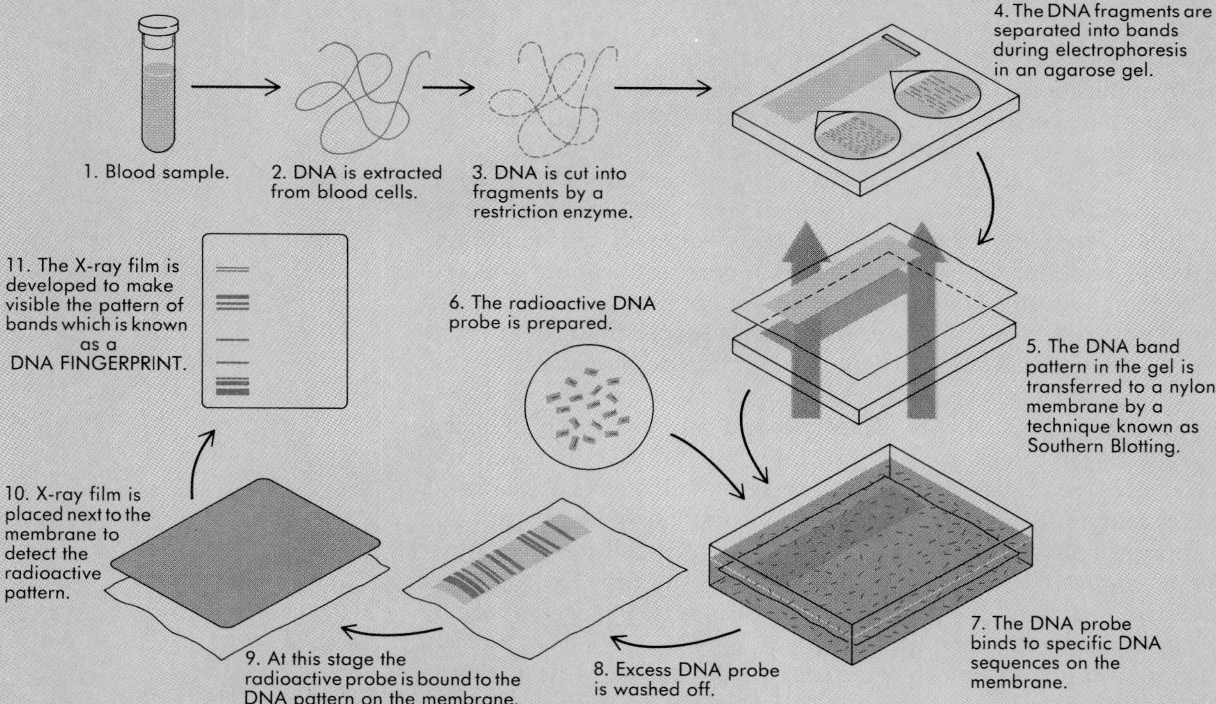
1. Blood sample.
2. DNA is extracted from blood cells.
3. DNA is cut into fragments by a restriction enzyme.
4. The DNA fragments are separated into bands during electrophoresis in an agarose gel.
5. The DNA band pattern in the gel is transferred to a nylon membrane by a technique known as Southern Blotting.
6. The radioactive DNA probe is prepared.
7. The DNA probe binds to specific DNA sequences on the membrane.
8. Excess DNA probe is washed off.
9. At this stage the radioactive probe is bound to the DNA pattern on the membrane.
10. X-ray film is placed next to the membrane to detect the radioactive pattern.
11. The X-ray film is developed to make visible the pattern of bands which is known as a DNA FINGERPRINT.

the radioactive probes attached to the fragments. The pattern of bands obtained on the film is 100 percent unique for each person, except for identical twins who would have the same pattern.

The forensic application of the technique involves a comparison between the DNA fingerprint obtained from cells at a crime scene with a DNA fingerprint from cells provided by the suspect. If the DNA patterns match exactly, certain identification is made. For paternity determination, DNA fingerprints of the mother, child, and alleged father are compared. In this case, one-half the bands in the child come from the mother and the other half from the father. All of the paternal bands in the child's DNA fingerprint must match with the alleged father for positive paternity identification. A third and very important use of this technique will be to identify missing children, as well as to use the information as part of a national data base for identifying missing or kidnapped children.

> Man has played God in the past, creating a whole new artificial world for his comfort and enjoyment. Obviously we have not always displayed the necessary wisdom and foresight in that creation; so it seems to me a waste of time for scientists, ethicists, and laymen alike to beat their breasts today, continually pleading the question of whether or not we have the wisdom to play God with human nature and our future.

The Office of Technology assessment of the 97th Congress of the United States, in a report entitled "Impacts of Applied Genetics," makes this significant statement:

> Genetics thus poses social dilemmas that most other technologies based in the physical sciences do not. Issues such as sex selection, the abortion of a genetically defective fetus, and *in vitro* fertilization raise conflicts between individual rights and social responsibility, and they challenge the religious or moral beliefs of many.

By such techniques and procedures as have been outlined, then, almost any faulty or "undesirable" gene could perhaps be changed, first in the somatic cells, and next in the reproductive cells. Modification of the genetic material of somatic cells can, of course, have no effect on future generations; to exercise quality control over human evolution, modification would have to be made in the sex cells, either directly by genetic engineering or by the more inefficient and problem-laden recombination route.

We are still left, however, with several grave questions:

1. To what extent will diminution of negative selection pressures (for example, keeping alive those whose genotypes are lethal today) "pollute" the human gene pool?
2. How is "desirability" in the genome to be defined; that is, desirability from whose viewpoint?
3. What presently unforeseen dangers lie ahead in genetic engineering experimentation, even with microorganisms?
4. Knowing that some applications of genetic engineering are, even now, saving some lives or at least improving the quality of life for some, should research in genetic engineering be halted either permanently or until our ability to use this new knowledge has matured?
5. What is the responsibility of the scientist in helping the policy makers and the public sort out and understand the complex issues involved?
6. How can ethical and legal questions be answered?

Scientists themselves are devoting more and more thought to these and similar questions, and, as might be surmised, agreement is less than complete.

There is consensus that medical progress, in keeping alive to reproductive age persons whose genotypes presently doom them to death or incapacitation before reproducing, will have only a small quantitative effect on gene frequencies. These deleterious genes have, for the most part, very low frequencies in the large human gene pool. The number of generations required to double the number of heterozygotes, for example, depends on both the present frequency of the gene under consideration and the number of births. Dr. Arthur Steinberg has arrived at the figures shown in Table 20–3, where q_0 represents the present frequency of the (recessive) deleterious allele and n represents the number of children per family. If we assume a human generation of 30 years, and only two children per family, the doubling time for heterozygote frequency ranges from 600 years if $q_0 = 0.1$ to nearly half a million years if $q_0 = 0.0001$. Even without future medical progress to alleviate lethal genetic disorders (which is certainly not to be expected), the effect on the species is far from great. It is only when one focuses on a particular family that concerns become more immediate.

TABLE 20–3. **Number of generations required to double the frequency of heterozygotes**

n	q_0			
	0.1	0.01	0.001	0.0001
2	20	156	1,517	15,127
4	14	110	1,070	10,675

Data of Steinberg.

Definitions of "desirability" are simply not possible for at least two reasons. First, human history has been one of using discoveries for evil as well as for good; atomic fission is a recent illustration, and many other examples may occur to you. Second, future applications of seemingly harmless or even beneficial discoveries cannot always be anticipated. Signer cites a case in point. The Ph.D. research of Arthur Galston disclosed that 2,3,5-triiodobenzoic acid increases the number of flowers and fruits in soybean, which results in greater yield per acre—an important discovery in an age of food shortages. However, governmental chemical warfare research teams found that higher concentrations of this substance produce defoliation. This discovery was followed by development of defoliants widely used in Vietnam. One of these, "Agent Orange," which is a mixture of 2,4-dichlorophenoxyacetic acid and 2,4,5-trichlorophenoxyacetic acid, accounted for 58 percent of the spraying in Vietnam, where it was applied at a rate of 20 to 30 pounds per acre—10 times the amount used in the United States. It produces malformations in fetuses of rats and mice and possibly induced cancer in American soldiers exposed to the spraying. Unusual increases in stillbirths and in birth defects (for example, cleft palate and spina bifida—a defect in the walls of the spinal canal resulting in tumor formation) were reported in Tay Ninh Provincial Hospital and in Saigon in 1971 after heavy and widespread spraying by the United States.

CONCLUSIONS

Although the successful application of genetic engineering still lies largely in the future, the future quickly becomes the present, and it is not at all unlikely that it will one day soon be possible in human beings. As Friedmann (1979) has pointed out:

> The scientific dreams and expectations of one generation become the techniques that later generations take for granted. It has always been so, and it should be no surprise that startling new techniques continue to appear, making possible work that was previously unimaginable . . . Already these techniques have shed light on the mechanisms of gene control in prokaryotic and eukaryotic systems, and they promise to be used at increasing rates for the study of gene expression, for potential treatment of human disease, for agriculture, and all other fields of genetics.

The new genetics—recombinant DNA—prologue or epilogue? With our past record of inept handling of new technologies, the outlook may not be so good, but with wisdom and sober, logical regard to the ethical issues involved, the new genetics offers the human race an exciting prologue. For you and your children to be part of that stimulating future, however, you must become informed of the facts and weigh the ethical issues with care and logic. If you adopt an infomed, unprejudiced position, you will be a witness, possily even a participant, in perhaps humanity's greatest achievement of all time.

QUESTIONS FOR REFLECTION

(No claim is made for originality in the following questions—many writers have raised similar ones, implicitly or explicitly. Moreover, few of them can be given absolute answers at our present stage of understanding. Most of them are opinion questions to which, it is hoped, your genetics course will have contributed some knowledge as well as a sense of priorities and values. Therefore, no answers will be found in Appendix A. However, all these questions are important to you personally.)

20–1 In a recent case of child abuse, the child was so severely beaten by the parents that it suffered permanent brain damage and was crippled for life. Do you recommend taking children away from such parents?

20–2 A couple has a child who develops the Tay-Sachs syndrome; tests confirm that both husband and wife are heterozygous. Does the probability in this case indicate they should "try again"?

20–3 The couple in the preceding question does "try again"; amniocentesis discloses that the second child will also have the Tay-Sachs syndrome. Should they abort?

20–4 How would you answer the two preceding questions if one member of the couple is yourself?

20–5 Suppose a child of yours is born with the Tay-Sachs syndrome, but needs a respirator to sustain life during the period shortly after birth. Would you want the respirator used?

20–6 You are one of approximately 3 in 100 who is heterozygous for cystic fibrosis. In this condition, which is due to a recessive autosomal gene, affected persons are unable to digest food properly and are highly subject to infection; their ligaments do not form properly, and the lungs fill with fluid that has to be removed (sometimes painfully) almost daily. Death often occurs in the teens. (a) Would you marry? (b) If so, would you elect to have children of your own? (c) If you did, and amniocentesis disclosed the fetus to be homozygous recessive (which, of course, discloses your spouse also to be heterozygous), would you opt for an abortion?

20–7 Like most persons you are heterozygous for several lethal genes. Would you like to subject yourself to genetic engineering by transducing viruses in order to change your genotype with respect to those genes? Explain the bases for your choice.

20–8 Some people are unable to synthesize arginase and consequently have high blood levels of the amino acid arginine. As a result, they suffer spastic paraplegia, epileptic seizures, and severe mental retardation. This condition, called arginemia, is caused by a recessive autosomal gene. Infection with the Shope virus, which carries a gene for arginase synthesis, has resulted in elevated levels of the enzyme in both normal and affected persons. The virus produces skin cancer in rabbits, though its carcinogenic effect on humans has not been established. If you had a child born with arginemia, would you want it treated with the Shope virus?

20–9 Should parents in general have freedom to choose whether or not to have children of their own (a) if both are heterozygous for several different lethal or disabling genes, (b) if both are heterozygous for the same lethal or disabling gene?

20–10 If you knew you and your spouse were both heterozygous for the same recessive lethal gene (which, when homozygous, causes intense physical suffering, then death between the ages of 5 and 10), (a) would you want the freedom to make your own choice as to whether or not to have children? (b) Would you want the decision to be made for you, say, by a government commission?

20–11 A woman is carrying fraternal (dizygotic) twins; amniocentesis discloses one of the fetuses to be homozygous for a lethal gene that will kill the child some time between the ages of 5 and 10 after intense physical suffering. Assume the other fetus to be normal. Assuming that an abortion generally cannot be selective in aborting only the defective fetus, what option would you accept if (a) the woman were unknown to you, (b) the woman were your sister, (c) the woman were yourself or your spouse?

20–12 Would you like to be able to choose the sex of your offspring?

20–13 If it were possible to choose the sex of one's offspring with near certainty, can you foresee any possible disadvantages for the human species?

20–14 Would you like to be able to choose the intelligence range of your children within, say, about 10 I.Q. points?

20–15 Would you like to have one or more clonants? Give the bases for your answer.

20–16 Do you feel that we should try to alter human genotypes (a) now or (b) in the future? Give the rationale for your viewpoint.

20–17 What is recombinant DNA and how is it made?

20–18 What potential benefits and/or dangers for the human race do you see in recombinant DNA? Evaluate those benefits and/or dangers critically.

20–19 Do you feel that research on recombinant DNA should be (a) carried out without interruption but with all possible precaution against contamination of the human gene pool, (b) temporarily and voluntarily halted until risks and advantages can be more clearly evaluated, (c) permanently and voluntarily halted, or (d) permanently halted by statutory fiat? Explain your view in the light of your present genetic knowledge and your assessment of the likelihood of future progress and its potential uses.

20–20 Assuming that this planet can no longer sustain its population, that agriculture-related remedies have failed, and that time has run out, do you recommend (a) mass starvation, (b) starvation of selected populations, or (c) the elimination of "substandard" or noncontributing members of the human race? Do you have any other options to suggest?

ANSWERS TO PROBLEMS

APPENDIX

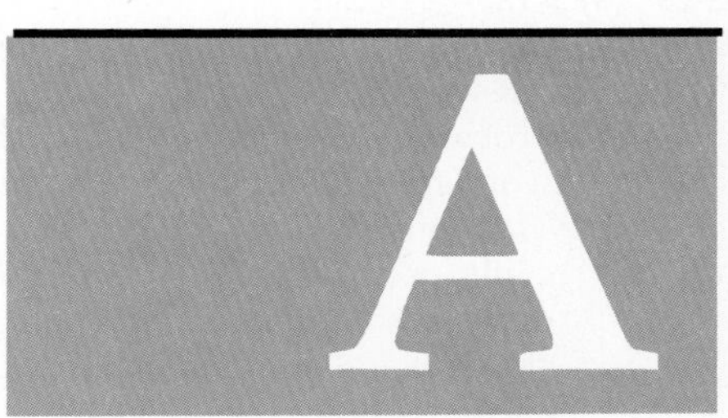

CHAPTER 2

2–2 *ww*.

2–3 (a) 1. *aa*; 2. *Aa*; 3. *Aa*; 4. *aa*; 5. *A*−.
(b) 1:1.

2–4 Incomplete dominance.

2–5 (a) 25%.
(b) 50%.
(c) *Probably zero; evidence suggests* woman is homozygous normal.

2–6 (a) Incompletely dominant as relates to chloride excretion.
(b) Recessive lethal.

2–7 (a) None.
(b) Yes; this couple's children have a 50% probability of being heterozygotes. If one of such heterozygotes marries another, *their* children (grandchildren of the original couple) have a 25% chance of having the disease.

2–8 (a) Purple is heterozygous; blue is homozygous.
(b) Purple.

2–9 Hornless is the dominant character. Hornless animals producing horned offspring are heterozygotes; any that do so should not be bred. Because he needs to get rid of a recessive, and cattle usually produce but one offspring per year, the problem is not going to be solved quickly. On the other hand, red animals are homozygous, so roans and whites can be excluded from the breeding program.

2–10 (a) *hh*. (e) *h*.
(b) *HH*. (f) *Hh*.
(c) *H*. (g) *Hhh*.
(d) *h*. (h) *hh*.

2–11 Curly is the heterozygous expression of a recessive lethal; 341:162 is a close approximation of a 2:1 ratio.

2–12 *Ff*; *Ff*; *ff*; *Ff*; *F*—; *ff*; *ff*; *Ff*.

2–13 Testcross.

2–14 *Aa*; *Aa*; *aa*.

2–15 On the basis of her daughter, III-4, who has to be *aa*, inheriting one *a* from each parent.

2–16 Rh positive.

2–17 (a) Both completely recessive.
(b) Both are heterozygous.

2–18 Codominance.

2–19 Incomplete dominance, but with the gene for thalassemia major recessive with regard to lethality.

2–20 (a) $\frac{1}{2}$. (b) $\frac{1}{4}$.

2–21 No; the transfused blood does not affect the recipient's genotype.

2–22 (a) Recessive.
(b) It is maintained in and transmitted by heterozygotes, some of whose children are PKUs. Mutation is another factor, and probably balances the occasional loss of recessive genes in PKUs.

2-23 (a) $\frac{2}{3}$ (*not* $\frac{1}{2}$, because his normal phenotype eliminates the possibility that he might be homozygous recessive for PKU).
(b) Probably none; it appears highly likely that the woman is homozygous dominant for normal.
(c) That any heterozygous child of theirs (which could occur if the husband is heterozygous) has one chance in four of having a PKU child if he or she marries another heterozygote. This eventuality is more likely if such a heterozygous child marries a relative, such as a cousin.

2-24 (a) $(\frac{2}{3})^2 = \frac{4}{9}$. (b) $(\frac{1}{9})$. (c) $\frac{1}{4}$.

2–25 (a) Completely dominant.
(b) 1:1.
(c) Testcross.

2–26 Genes for antigens A and B are codominant, and both are dominant to the gene that results in production of neither antigen.

CHAPTER 3

3–1 6 (i.e., *AA* × *AA*; *AA* × *Aa*; *AA* × *aa*; *Aa* × *Aa*; *Aa* × *aa*; *aa* × *aa*).

3–2 (a) $\frac{3}{16}$. (c) $\frac{2}{16}$.
(b) $\frac{3}{16}$. (d) $\frac{1}{16}$.

3–3 (a) 8.
(b) 2^{12}.

3–4 (a) 1:1.
(b) 1:1:1:1:1:1:1:1.

3–5 16.

3–6 (a) 16.
(b) 81.

3–7 256.

3–8 (a) 4. (c) 16.
(b) 8. (d) 2^n.

3–9 $(\frac{1}{4})^n$.

3–10 (a) $\frac{1}{64}$.
(b) $\frac{1}{16}$.

3–11 (a) 24.
(b) $\frac{1}{256}$.
(c) $\frac{1}{128}$.

3–12 Letting *Y* represent red and *y* yellow, the parental genotypes are $Yyh^1h^2 \times Yyh^1h^2$ (red, scattered hairs).

3–13 (a) Cream.
(b) $\frac{2}{16}$.

3–14 3:6:3:1:2:1.

3–15 (a) Two pairs, both incompletely dominant.
(b) Broad red, narrow red, broad white, narrow white.

3–16 9 red:3 pink:4 white.

3–17 9 normal:7 deaf.

3–18 (a) 9:7. (c) *AaBb*.
(b) *A* – *B* –. (d) *AAbb* × *aaBB*.

3–19 (a) *aabb*.
(b) *AaBb*.
(c) Anything *except aabb*.

3–20 9 black:3 brown:4 white.

3–21 12 white:3 black:1 brown.

3–22 12 white:3 brown:1 black.

3–23 9 brown:3 black:4 white.

3–24 (a) 2.
(b) *AaBb*.
(c) Purple, *A* – *B* –; red, *A* – *bb*; white, *aa* – –, if one assumes gene *A* is responsible for the enzyme converting colorless precursor to cyanidin, and *B* for the enzyme converting cyanidin to dephinidin.

3–25 9 both enzymes:3 enzyme number 1 only:3 enzyme number 2 only:1 neither enzyme.

3–26 *aaBB* × *AAbb*; *aaBb* × *Aabb*.

3–27 Four: red long, red round, white long, and white round.

3–28 (a) 3:6:3:1:2:1.
(b) 1:2:1.

3–29 (a) Let *A* represent a color inhibitor gene, *a* the gene for color, *B* yellow, *b* green. Then *A* – – – is white, *aaB* – yellow, and *aabb* green.
(b) P: *AAbb* (white) × *aaBB* (yellow)
F_1: *AaBb*
F_2: 9 *A* – *B* – } white
3 *A* – *bb* } white
3 *aaB* – yellow
1 *aabb* green.

3–30 (a) Disk *C* – *D* –, sphere *C* – *dd* and *ccD* –; elongate *ccdd*.
(b) P: *CCdd* × *ccDD*
F_1: *CcDd*
F_2: 9 *C* – *D* – disk
3 *C* – *dd* } sphere
3 *ccD* – } sphere
1 *ccdd* elongate.

3–31 (a) 8.
(b) 1.
(c) 24.

3–32 (a) 9.
(b) $\frac{108}{256}$.

3–33 (a) red $R - S -$; sandy $rrS -$ and $R - ss$; white $rrss$.
(b) case 1 $RRSS \times RRSS$.
case 2 $RrSS \times RrSS$, or $RRSs \times RRSs$, or $RrSs \times RrSS$, or $RrSs \times RRSs$.
case 3 $RRSS \times rrss$.
case 4 $rrSS \times RRss$.
case 5 $rrSs \times Rrss$.

3–34

	Genotypic:	*Phenotypic:*
(a)	1:2:1:2:4:2:1:2:1	9:3:3:1
(b)	1:2:1:2:4:2:1:2:1	3:6:3:1:2:1
(c)	1:2:1:2:4:2:1:2:1	1:2:1:2:4:2:1:2:1
(d)	1:2:1:2:4:2	3:1
(e)	1:2:1:2:4:2	1:2:1
(f)	1:2:2:4	all alike.

3–35 3 red:6 purple:3 blue:4 white.

3–36 (a) 4; reduced penetrance. (c) 1; epistasis.
(b) 3; lethal gene. (d) 4; reduced penetrance.

4–12 $(\frac{1}{2})^7$.

4–13 (a) 1. (d) 4.
(b) 2. (e) 2.
(c) 2. (f) 32.

4–14 (a) 2.
(b) *AB* and *ab*.

4–15 (a) 4. (c) 0.1 each.
(b) *AB*, *ab*, *Ab*, *aB*. (d) 0.4 each.

4–16 (a) Yes.
(b) Crossing-over between the two pairs of alleles.

4–17 (a) 0.000,005 m. (c) 5,000 nm.
(b) 0.005 mm. (d) 50,000 Å.

4–18 Increases variability by recombining characters from two parents.

4–19 May give rise to deleterious gene combinations in progeny or disrupt gene combinations (supergenes) that confer selective advantage on the organism.

4–20 Chromosomes usually occur in pairs in diploid organisms. So, regardless of the number of *pairs* (odd or even), the total number of chromosomes (= number of pairs × 2) will be an even number except in a few cases like the male grasshopper.

CHAPTER 4

4–1 (a) 48. (e) 48.
(b) 24. (f) 24.
(c) 48. (g) 12.
(d) None. (h) 12.

4–2 (a) 20. (d) None.
(b) 40. (e) 40.
(c) 40.

4–3 (a) 20. (e) 10.
(b) 10. (f) 10.
(c) 30. (g) 20.
(d) 10.

4–4 (a) 40.
(b) 40.
(c) 40.

4–5 160.

4–6 (a) 80.
(b) 160.

4–7 (a) $(\frac{1}{2})^{30}$.
(b) $(\frac{1}{4})^{30}$.

4–8 (a) 33.
(b) Irregularities of pairing at synapsis lead to defective gametes with more or less than one complete set of chromosomes.

4–9 All *Aa*.

4–10 Two *A* and two *a*.

4–11 Prophase longest, next telophase, next metaphase, with anaphase shortest; why?

CHAPTER 5

5–1 (a) $(\frac{1}{2})^3$, or $\frac{1}{8}$.
(b) $\frac{3}{8}$

5–2 $\frac{1}{2}$.

5–3 $\frac{1}{2}$.

5–4 $(\frac{1}{2})^8 = \frac{1}{256}$.

5–5 (a) $(\frac{1}{2})^5 = \frac{1}{32}$.
(b) $\frac{1}{16}$.

5–6 (a) $\frac{1}{6}$.
(b) $\frac{1}{36}$.
(c) $\frac{1}{6}$.

5–7 (a) $\frac{3}{4}$.
(b) $\frac{1}{4}$.

5–8 (a) $\frac{4}{16}$, or $\frac{1}{4}$.
(b) $\frac{1}{16}$.
(c) $\frac{9}{16}$.
(d) $\frac{3}{16}$.

5–9 (a) $28a^6b^2$.

5–10 $\frac{18}{256}$.

5–11 (a) $\frac{1}{8}$.
(b) $\frac{1}{4}$.
(c) 6.

5–12 $\frac{270}{32,768}$.

5–13 (a) $\frac{243}{32,768}$.
(b) No, because other phenotypes are also possible.

5–14 0.2.

5–15 $\frac{36}{4,096}$, or $\frac{9}{1,024}$.

5–16 $\frac{24}{81}$.

5–17 (a) 0.02.
(b) 0.0392.

5–18 See Table 5–2. The expectancy with the lower value of chi-square is the preferred interpretation.

5–19 Table 5–2 shows that chi-square equals zero when there is *no* deviation from the calculated or expected ratio.

5–20 P = 1.0.

5–21 (a) 1.
(b) Yes, for the 2:1 expectancy.
(c) A 3:1 expectancy.

5–22 Chi-square for a 13:3 ratio is 0.993.

5–23 (a) Chi-square for a 1:1 expectancy is 2.0.
(b) Chi-square for a 9:7 expectancy is 0.127.
(c) No.
(d) Unless a larger sample can be obtained, accept the expectancy that has the lower chi-square value; in this case this is the 9:7 expectancy.

5–24 (a) Chi-square for a 1:1 expectancy is 20.0.
(b) Chi-square for a 9:7 expectancy is 1.27.
(c) The deviation is not significant for the 1:1 expectancy only.
(d) The larger the sample, the greater the usefulness of the chi-square test.

5–25 (a) Chi-square equals 0.015.
(b) Chi-square equals 0.415.
(c) Chi-square equals 0.563.
(d) Chi-square equals 0.618.
None of the values of chi-square is significant.

5–26 Very nearly 2% (actually, 0.198, or 1.98%).

CHAPTER 6

6–1 (a) *R* and *ro* each 0.4375; *R*, *ro*, and *rRo* each 0.0625.
(b) *RRo/RRo* progeny are calculated to be 0.875.
(c) Purple long progeny are calculated to have an expected frequency of 0.0625.

6–2 15 map units.

6–3 (a) El_1d/el_1D.
(b) Trans.

6–4 (a) 0.485.
(b) 0.015.

6–5 (a) *wo-dil-o-aw* (or the reverse order).
(b) Because of inability to detect double crossovers in the map distance arrived at in part (a).

6–6 (a) *jvl-fl-e* (or the reverse).
(b) Because double crossovers are missed in two-pair crosses involving *jvl* and *e*.

6–7 Locus *obt* could be 19 map units either side of *jvl*.

6–8 The sequence is now confirmed as *jvl-fl-e-obt*.

6–9 (a) Yes.
(b) 0.77%

6–10 4.

6–11 12.

6–12 (a) 12.
(b) 12.
(c) 23.
(d) 24.

6–13 Because genes for *some* of the traits he followed are on different chromosome pairs; linked traits with which he worked (seed color/flower color; pod contour/plant height, among others) are so far apart on their chromosomes that genes for those traits appear to segregate randomly. In one other case Mendel did not make the cross that would have revealed linkage.

6–14 (a) Yes.
(b) No. (See text.)

6–15 (a) 4.
(b) 2.

6–16 (a) 8.
(b) 2.

6–17 (a) Peach, round and smooth, ovate.
(b) The pistillate parent is given as heterozygous and, with the progeny given, the pistillate plant must be *PO/po*. The staminate parent's, given as peach, ovate, must be *po/po*. Therefore, the parental cross is *PO/po* ♀ × *po/po* ♂.
(c) *Cis*.
(d) Pistillate parental gamete genotype frequency should be *PO* and *po* 0.44 each, and crossover gametes 0.06 each.
(e) 12 map units.

6–18 Yes, they are linked. Normal beaked plants are doubly heterozygous in this case, and should produce about 6 percent (= $\frac{1}{16}$) of the progeny with this phenotype. The number actually observed is about 3.68 times greater than is to be expected with unlinked genes.

6–19 (a) *Cis* configuration.
(b) 4 map units.

6–20 0.4 *PlPy* + 0.4 *plpy* + 0.1 *Plpy* + 0.1 *plPy*.

6–21 16%.

6–22 9%.

6–23 "True-breeding" plants must be homozygous dominant or homozygous recessive for each of the two pairs of linked genes. The percentage of "true-breeding" plants here is 34%.

6–24 (a) Progeny having maternal chromosome + *ss* + or *cu* + *sr* total 88.2%.
(b) Progeny having maternal chromosome + + + or *cu ss sr* would total 0.2%.

6–25 (a) Sequence of genes is *h fz eg*. The *h-fz* distance is 14 map units; the *fz-eg* distance is 6 map units.
(b) 0.238.

6–26 0.5 (rounded from 0.502).

6–27 (a) *d* + + and + *m p*.
(b) *p d m* (or the reverse order).
(c) *p d*, 4.5 map units; *d m*, 4.5 map units.
(d) 0.5.

6–28 (a) Noncrossover types are *b* + + and + *cn vg* and total 81.8%.
(b) Double crossover types are *b n* + and + + *vg*, totaling 0.3 percent.
(c) *Trans.*
(d) *cn.*
(e) *cn b*, 9 map units; *vg cn*, 9.5 map units.
(f) Yes.
(g) 0.35.

6–29 (a) 0.85.
(b) 0.05.
(c) 0.1.
(d) None (zero).

6–30 (a) Noncrossovers, 0.4275 each.
Single crossovers between pg_{12} and gl_{15}, 0.0225 each.
Single crossovers between gl_{15} and bk_2, 0.0475 each.
Double crossovers 0.0025 each.
(b) Noncrossovers, 0.42625 each.
Single crossovers pg_{12} to gl_{15} 0.02375 each.
Single crossovers gl_{15} to bk_2, 0.04875 each.
Double crossovers, 0.00125 each.

CHAPTER 7

7–1 (a) Metafemale.
(b) Metamale.
(c) Metafemale.
(d) Intersex.
(e) Tetraploid female.
(f) Metamale.

7–2 (a) ♀.
(b) ♂.
(c) ♀.
(d) ♂.
(e) ♂.

7–3 9 monoecious:3 staminate:4 pistillate.

7–4 $\frac{1}{4}$.

7–5 3♂:1♀.

7–6 (a) *bW* (or *b*).
(b) *B*−.
(c) *BW* (or *B*).
(d) *bb*.

7–7 (a) $\frac{1}{4}$ each of the following: barred ♀, nonbarred ♀, barred ♂, nonbarred ♂.
(b) All ♂s barred, ♀s 1 barred:1 nonbarred.

7–8 $\frac{1}{3}$ ♂, $\frac{2}{3}$♀.

7–9 0.1.

7–10 45% (that is, 20% XX + 25% XXY).

7–11 (a) $\frac{1}{4}$.
(b) $\frac{1}{4}$.
(c) $\frac{1}{3}$.

7–12 (a) White (X^w O).
(b) Red ($\widehat{X^+X^w}$Y).

7–13 Fluorescence pattern, especially the bright longer arm, is the best single identification; the longer arms are closer together and satellites are absent. Its length relative to members of the F and G groups of autosomes is more variable.

7–14 From the theoretical standpoint, several alternative explanations are possible. Among the most likely are
(a) Nondisjunction of Y in the second meiotic division in spermatogenesis;
(b) Nondisjunction in the first meiotic division in spermatogenesis, giving rise to an XY sperm;
(c) First or second division in either spermatogenesis or oogenesis, producing either O sperm or egg, which then fuses with an X gamete from either sex. The XXY condition could also arise through nondisjunction in the first cleavage division of a normal XY zygote, whereby one daughter cell receives XXY and the other OY (the latter is nonviable).

7–15 Most likely by the lagging of an X chromosome in early mitoses of an XX zygote.

7–16 The X chromosome carries a large number of genetic loci, most or all of which appear to be necessary for normal development.

7–17 Humans. Single genes (*Asparagus*) are subject to mutation that could result in a sex imbalance in small populations or even a lethal condition, but in humans there appear to be many loci governing sex on at least the X and Y chromosomes. Furthermore, the occasional "male × male" crosses that occur in *Asparagus* increase the likelihood of homozygosity of deleterious genes in the progeny. Also a 1:1 sex ratio in *Asparagus* occurs only in populations where "males" are heterozygous.

7–18 (a) XX. (c) XY.
(b) XY. (d) XX.

7–19 Because of irregularities in meiosis.

7–20 Maternal.

7–21 Because the Xg^a − phenotype is recessive, the genotypes must be: mother, *Xg Xg*, father Xg^a Y, daughter, *Xg* O.

7–22. (a) Mother.
(b) Father.

7–23 Mother, *Xg Xg*; father Xg^a Y; son, *Xg Xg* Y.

7–24 All ♀s "bent," all ♂s normal.

7–25 A 50% probability of a girl with slight nystagmus, and a 50% chance of a boy with severe nystagmus. The likelihood of normal children of either sex is 50%.

7–26 All ♀ progeny are normal, and all ♂ offspring are deranged.

7–27 All ♂s are barred, rose (*BbRr*), and all ♀s are non-barred, rose (*bWRr*).

7–28 The F_1 includes $\frac{3}{16}$ *BbR*– males and $\frac{3}{16}$ *BWR*– for $\frac{6}{16}$ barred, rose, equally divided between the two sexes.

7–29 Progeny $\frac{1}{4}$ *RR* + $\frac{2}{4}$ *Rr* (= $\frac{3}{4}$ rose) + $\frac{1}{4}$ *rr*.

7–30 There is no chance that any children of the young man will develop the disease as long as he marries a + + young lady; each woman has a probability of 0.5 of being heterozygous and therefore transmitting the trait to half her sons.

CHAPTER 8

8–1 Dominance, epistasis, sex linkage.

8–2 3′TTGCATGACG5′ because of the pairing qualities of deoxynucleotides.

8–3 4^n.

8–4 $L_{\mu m} = 3.4 \times 10^{-4}P$ where $L_{\mu m}$ represents the length in micrometers, and P represents the number of nucleotide pairs.

8–5 68 μm.

8–6 (a) 1.38×10^{10}, assuming the molecular weight of a deoxyribonucleotide pair to be 650.
(b) 4.7×10^6.
(c) Length in inches (L_{in}) = 185.

8–7 (a) 7.7×10^8.
(b) 2.6×10^5.
(c) 10.2.

8–8 (a) 135.
(b) 126.
(c) 111.
(d) 151

8–9 (a) 251.
(b) 242.
(c) 227.
(d) 267.

8–10 (a) 329.
(b) 320.
(c) 305.
(d) 345.

8–11 649.5 (but round off to 650 for greater ease in calculating; this figure loses an inconsequential level of accuracy).

8–12 1.3×10^8.

8–13 4.15×10^6.

8–14 One per minute.

8–15 At the rate of one per minute.

8–16 A, 20 percent; T, 20 percent; G, 30 percent; C, 30 percent.

8–17 No. Because of the pairing qualities of the bases, the total amount of thymine + cytosine should equal the amount of adenine + guanine. The four bases would then total only 50% instead of 100%.

8–18 (a) A/T = 0.99, G/C = 1.00.
(b) That it is double stranded.
(c) That the DNA in Table 8–1 is the single-stranded replicative form.

8–19 This suggests that gene function is nuclear in nature. The matter is explored in Chapter 16.

8–20 31.26%.

8–21 (a) None.
(b) Two of the four will have one ^{15}N strand and one ^{14}N.
(c) The other two strands will have incorporated only ^{14}N.

8–22 Transformation involves naked DNA from one cell becoming incorporated into another's DNA, whereas in transduction a virus serves as the vector transferring DNA derived from one cell into another.

8–23 (a) $L_{in} = \dfrac{L_{\mu m}}{2.54 \times 10^4}$. The denominator represents the number of micrometers per inch.
(b) $L_{in} = \dfrac{3.4 \times 10^{-4} P}{2.54 \times 10^4}$ or $1.3386 \times 10^{-8} P$.

CHAPTER 9

9–1 No. Why?

9–2 Yes. Why?

9–3 No. Why?

9–4 Two different pairs of alleles (Figure 9–1) are involved at two different points in the synthesis of melanin. If, for example, parents were *AAa′a′* and *aaA′A′*, they would be albinos, but their children would be *AaA′a′*, hence normally pigmented.

9–5 Strain 1.

9–6 Strain 2 grows only if thiamine or thiazole is supplied; it therefore cannot produce enzyme "*a*." If thiazole, for example, is supplied exogenously, strain 2 grows because it does produce enzymes "*b*" and "*c*."

9–7 Strain 4 cannot produce either enzyme "*a*" or "b".

9–8 Genotype of each strain is: 1. + + +; 2. *a* + +; 3. + *b* +; 4. *a b* +.

9–9 (a) Strain 2 fails to produce enzyme "*a*."
(b) Strain 3 is incapable of producing enzyme "*b*."
(c) Strain 1 is incapable of producing enzyme "*c*."

9–10 (a) 3′AUCUUUACGCUA5′.
(b) Adenylic acid (A).
(c) Uridylic acid (U).

9–11 200.

9–12 (a) $L_{\text{Å}} = 3.4\ P = 3.4 \times 100 = 340$. For a nucleotide of such a small number as 101, use $n - 1$ internucleotide spaces for greater accuracy.
(b) $L_{\mu m} = 3.4 \times 10^{-4} = 3.4 \times 10^{-4} \times 100 = 3.4 \times 10^{-2} = 340$. Again, use $n - 1$ internucleotide spaces for greater accuracy.

9–13 Use 337 as the average molecular weight of a ribonucleotide, a useful figure but accurate only if the four ribonucleotides are present in equal number. Thus one calculates as follows:
(a) 80.
(b) $L_{\mu m} = 0.027$.
(c) tRNA, based on its number of nucleotides (which, in turn, determines its molecular weight).

9–14 2,967.

9–15 5′UAGCCAAUC3′.

9–16 (a) By definition, codons occur only in mRNA.
(b) The anticodon is complementary, base for base, with the codon; hence is UAC.
(c) By definition, anticodons occur only in tRNA.
(d) TAC (complementary to the *codon* AUG).

9–17 No explanation beyond Chapter 9 and Appendix A is needed.

9–18 Chapter 9 and Appendix A contain sufficient information.

9–19 Longer; why?

CHAPTER 10

10–1 From Table 10–2,
(a) A triplet, nonoverlapping code would produce these sequences of mRNA codons and amino acid residues:

(DNA)	3′ TAC CGG AAT TGC 5′
(mRNA)	5′ AUG GCC UUA ACG 3′
(amino acids)	met ala leu thr

(b) A triplet code, overlapping by two bases would (Table 10–2) produce these sequences of codons and amino acid residues:

mRNA	5′	AUG	UGG	GGC	GCC	
amino acids		met	trp	gly	ala	
		CCU	CUU	UUA	UAA	
		pro	leu	leu	stop	
		AAC	ACG	CG-		3′

(AAC ACG CG-: not read.)

(c) The trinucleotide TAC transcribes into AUG (met), which serves as a polypeptide chain start signal.

10–2 Deletion of the second C in DNA here produces:

DNA	3′ TAC GGA ATT GC- 5′
mRNA	5′ AUG CCU UAA CG- 3′
amino acids	met pro stop not read →

10–3 (a) With the second C deleted and a T inserted after the GG sequence,

DNA	3′ TAC GGT AAT TGC 5′
mRNA	5′ AUG CCA UUA ACG 3′
amino acids	met pro leu thr

(b) Missense in the second amino acid residue (pro substituted for ala).

10–4 (a) $\frac{6}{216}$, or $\frac{1}{36}$. With the relative proportions of 3U:2G:1A, the probability that a given ribonucleotide in the synthetic mRNA is: U, $\frac{3}{6}$; G, $\frac{2}{6}$; A, $\frac{1}{6}$. Therefore, $\frac{3}{6} \times \frac{2}{6} \times \frac{1}{6} = \frac{6}{216}$ of the triplets would be expected to be UGA.
(b) For UUU the probability is $(\frac{3}{6})^3 = \frac{27}{216} = \frac{1}{8}$.

10–5 (a) 423.
(b) 0.14 μm.
(c) Too low for the real gene because of introns, plus start and stop signals; correct only for the exons.

10–6 At GCU, the mRNA codon for alanine. The mRNA-tRNA pairing is determined solely by the codon-anticodon complementarity; the amino acid bound to the tRNA has nothing to do with process.

10–7 Tryptophan, which has only one codon. Arginine has six.

10–8 The A-14 mutant has undergone a substitution in the second base of the isoleucine codon (AUU, AUC, or AUA) to AUG (met). The Ni-1055 mutant has undergone a substitution in the third base of an isoleucine codon (AUU, AUC, or AUA) to AUG, the methionine codon as shown by Table 10–2.

10–9 In A-446, UAU or UAC (tyrosine) has undergone a substitution in the third base to UGU or UGC (cysteine). In mutant A-187, a glycine codon (GG-) has undergone a substitution in the second base of GG- to GU-, which codes for valine.

10–10 288.

10–11 (a) Degeneracy.
(b) Ambiguity.

10–12 (a) Nonsense; UGA is a stop signal and does not translate.
(b) Missense; GGA codes for glycine, GAA for glutamic acid.
(c) Degeneracy; both GGA and GGC code for glycine.
(d) Missense; GGA codes for glycine, CGA for arginine.

10–13 (a) 402.
(b) No, because they are "clipped out" in processing.

CHAPTER 11

11–1 (a) 1,500.
(b) Too high; why?

11–2 267.

11–3 801.

11–4 (a) 5×10^9.
(b) 5×10^9.
(c) 1.7×10^6.
(d) 8.33×10^8.
(e) 2.78×10^6.

11–5 Depending on the amino acid position involved, some missense mutations result in substitutions that do not materially affect the functioning of the resulting protein but, for example, they may result in peptides with different electrophoretic mobilities because of charge differences.

11–6 By accumulation of a series of missense mutations affecting different amino acid positions in a given polypeptide; those not lethal or disabling might be expected to be perpetuated and passed on to later generations. This is the basis of genetic polymorphism.

11–7 *Cistron:* a segment of DNA specifying one polypeptide chain.
Muton: the smallest segment of DNA that can be changed and thereby bring about a mutation; it can be as small as one deoxyribonucleotide pair. *Recon:* the smallest segment of DNA that is capable of recombination; it can be as small as one deoxyribonucleotide pair.

11–8 *Complementation:* the ability of linearly adjacent segments of DNA to supplement each other in phenotypic effect.
Recombination: A new association of genes in a recombinant individual, arising from (1) independent assortment of unlinked genes, (2) crossing over between linked genes, (3) intracistronic crossing over, (4) insertion or deletion of transposons.

11–9 In the hemoglobin example, as in many others, there are distinct protein molecules, each having the same function prevalent at different stages of development of the organism. The differences in the proteins are only a few amino acids. Each form of the protein is encoded by a distinct gene active at particular stages of development. A collection of such genes would be termed a developmentally regulated gene family.

11–10 An embryonic cell contains many base sequences coding for all antibodies. These sequences are contiguous in the DNA. During development reciprocal recombination occurs, joining distant sequences and at the same time removing large blocks of DNA that include the adjacent sequences. There are many different blocks that can be removed, so that an infinite number of combinations of coding sequences can remain after the recombinational event. One event occurs in a single cell and its cellular progeny with a unique coding sequence that enables the clone of cells to make a specific antibody.

11–11 The constant region has nearly the same amino acid sequence in all antibody molecules, whereas the amino acid sequence in the variable region is different in each type of antibody molecule.

CHAPTER 12

12–1 They are similar in that each is composed of a given segment of deoxyribonucleotides and each participates in regulatory control over cistrons. On the other hand, although the operator is transcribed in large part, it is not translated. The regulator is responsible for production of a regulatory protein (either a repressor or an activator); no such product has been demonstrated for the operator.

12–2 The repressor is a protein, the product of a regulator site, which inhibits the action of an operator site so that cistrons of the operon are "turned off." An effector is any substance, often the substrate of the enzyme(s) for which the cistron(s) of the operon are responsible; it binds to the repressor protein, which thereby undergoes a change in shape so that it can no longer bind to the operator.

12–3 Both involve control by a regulatory protein, which binds to an operator. In positive control the regulator protein serves as an activator, permitting translation of the cistrons of the operon by binding to the operator. In negative control the regulatory protein represses the operator, either in the absence of an effector (*lac* operon) or in the presence of an effector (*his* operon).

12–4 Transcriptional level.

12–5 (a) Inductive.
(b) Constitutive.
(c) Constitutive.
(d) Absent.
(e) Constitutive.

12–6 (a) Not produced.
(b) Beta-galactosidase constitutive, beta-galactoside permease inductive, and thiogalactoside transacetylase constitutive.

12–7 (a) Presence of glucose interferes with production of cyclic adenosine phosphate (cAMP) so that the cAMP-CRP complex cannot be formed, thereby halting transcription of the lactose cistrons (even though lactose is present).
(b) Although, in the absence of glucose, the cAMP-CRP complex is formed, absence of lactose allows an active repressor to bind to the operator, thus preventing transcription of the lactose cistrons.
(c) In the absence of glucose, the cAMP-CRP complex is produced, lactose binds to the repressor, inactivating it and allowing the cistrons to be transcribed.

12–8 (a) No; why?
(b) Neither will be produced. Why?

12–9 All three are produced constitutively; why?

12–10 Translation; a nonsense mutation causes both chain termination and release (assuming release factors to be present). Termination and release function in translation, not in transcription.

12–11 (a) Negative; why?
(b) Produced; why?

12–12 Bacterial gene expression is usually turned on and off many times throughout the lifetime of the cell, whereas in a differentiated cell the gene is usually turned on and never turned off.

12–13 Eukaryotic genes are not organized into operons, and because of the presence of introns in many eukaryotic genes, only a single protein is made from any primary transcript.

12–14 Heterochromatin is more densely staining than euchromatin; therefore, whatever picks up stain must be more available in heterochromatin than euchromatin or there must be more of it due to dense packaging of heterochromatin over euchromatin.

12–15 A promotor mutation, which would prevent transcription, or a mutation in the start codon, which would prevent initiation.

12–16 First, it would have to pass through the cell membrane and cytoplasm, then it must pass through the nuclear membrane. Then it would have to disrupt the nucleosome and remove histones before it can interact with the DNA.

12–17 Labeling occurs only in the chromosomal regions of a puff where the genes are actively transcribing RNA and therefore incorporating the radioactive uracil into RNA.

12–18 The pattern of expression of these genes is clearly influenced by the temperature. Whatever the temperature at the second instar stage determines the pattern of development. This pattern cannot be reversed by changing the temperature. After the second instar the cells are destined to form whatever structure had been determined at that time.

CHAPTER 13

13–1 For the environment of a given species, those mutations that have either a positive or a neutral selection value would be expected to increase in frequency, although deleterious recessive mutations may be expected to persist at a low frequency in heterozygotes. In addition, many (but not all) mutations result in proteins of lowered functional capability; such mutant individuals are usually at a disadvantage in survival and reproduction.

13–2 See Figure 13–13 and accompanying text.

13–3 Short-term effects include radiation sickness, surface and deep tissue burning and destruction, loss of hair, and so on. Long-term effects include an increased incidence of leukemia (although radiation is employed, along with chemotherapy, to slow or arrest progress of the disease), and a variety of mutations. Increases in incidence of birth defects in children and miscarriages in women exposed to radiation also show a probable positive correlation.

13–4 Figure 13–2 shows no such threshold dose.

13–5 Recessive mutations are more easily detected in homozygous males.

13–6 Haploid greatest, polyploid least. Most mutations are recessive, and recessives have only a very low probability of being expressed in polyploids with their multiple sets of chromosomes bearing normal (dominant) alleles. Both dominant and recessive mutations are expressed at once in haploids.

13–7 *Avena brevis* ($2n = 14$) is a diploid, *A. barbata* ($2n = 28$) is a tetraploid, and *A. sativa* ($2n = 42$) is a hexaploid. Because most mutations are recessive, frequency of *detectable* mutations may be expected to be inversely related to ploidy.

13–8 *Triticum monococcum* is the only diploid among the species of wheat listed in Table 4–1.

13–9 (a) Two; why?
(b) Because $\frac{9}{16}$ of the progeny must have at least one dominant of each pair (*A—B—*) and are blue, blue must be the result of the second enzyme-controlled process. Inasmuch as white is given as the colorless precursor, red must be the phenotype resulting from at least one dominant of one pair, but homozygous recessive for the other (for example, $\frac{3}{16}$ *A-bb*). White then results from either $\frac{3}{16}$ *aaB—* or $\frac{1}{16}$ *aabb* genotypes. Thus, the sequence is white → red → blue.
(c) (1) Parents were *AaBb* × *AaBb*; these are the only parental genotypes that can produce a 9:3:4 progeny phenotypic ratio. (2) *A—B—*.
(d) (1) *A—bb* (or *aaB—*) ($\frac{3}{16}$). (2) *aa— —*(or — — *bb* if *aaB —* if *aaB —* was chosen for (d) (1).
(e) The *A,a* pair (or *B,b* pair if the alternatives were chosen in part (d).

13–10

DNA:	3′AAA5′	changed to	3′AAC5′
mRNA:	5′UUU3′	changed to	5′UUG3′
amino acids:	phe	changed to	leu.

13–11 A transition because it pairs with guanine, which replaces an adenine, both of which are purines.

13–12 A78 is a transversion; A58 is a transition:

	A78	wild	A58
Residue 233:	Cysteine	Glycine	Aspartic acid
DNA:	ACG	CCG	CTG
mRNA codon:	UGC	GGC	GAC
Change:	C→A		C→T
	(pyr—pur)		(pyr—pyr)

13–13 The single-stranded molecule would form a so-called stem and loop structure with a short base-paired stem region formed from the inverted repeat ends of the transposable element, and a large single-stranded loop region corresponding to the coding region of the transposable element.

13–14 Somatic mutation.

13–15 Three, as any other number of deletions (or additions) would cause a frameshift and other amino acid changes.

CHAPTER 14

14–1 *DDDD*. Explain.

14–2 *DD*, assuming normal disjunction.

14–3 This cross is *DD* egg X *d* sperm, giving rise to *DDd*.

14–4 The diploid becomes *DDdd*, assuming normal replication.

14–5 Gamete ratio:1 *DD*:4 *Dd*:1 *dd*.

14–6 (a) Autotetraploidy.
(b) A dwarf autotetraploid results only if a *dd* egg is fertilized by a *dd* sperm. Each of these gametic genotypes has a $\frac{1}{6}$ chance of being produced by the respective parent (see the preceding problem); the probability of two such gametes fusing is $(\frac{1}{6})^2 = \frac{1}{36}$.

14–7 Probability of an *aabb* gamete from *each* parent is $\frac{1}{6} \times \frac{1}{6} = \frac{1}{36}$; the probability of their fusing is $(\frac{1}{36})^2 = \frac{1}{1,296}$.

14–8 $(\frac{1}{6})^6 = \frac{1}{46,656}$.

14–9 For each heterozygous allelic pair in an autotetraploid (*AAaa*) the probability of an *aa* gamete is $\frac{1}{6}$, and of their fusing $(\frac{1}{6})^2$. For two pairs (*AAaaBBbb*) the probability of an *aabb* gamete is $(\frac{1}{6})^2$ and of their fusing $(\frac{1}{6})^4$. Thus, for n pairs of heterozygous genes in an autotetraploid, the probability of a wholly recessive individual in the progeny of a cross between two heterozygous autotetraploids is $(\frac{1}{6})^{2n}$

14–10 Reduces it; why?

14–11 The basic haploid set here is 14 chromosomes; hence diploids have 28 chromosomes (2 × 14), tetraploids 56 (4 × 14), pentaploids 70 (5 × 14), and hexaploids 84 (6 × 14).

14–12 A different series of chromosomal aberrations in each species, so that normal pairing is impossible. Translocation (see Chapter 15) is probably the most frequent of these aberrations.

14–13 Yes, by creating an allotetraploid hybrid.

14–14 Euploidy; the different chromosome numbers are, respectively, $2n$, $3n$, $4n$, $6n$, $8n$, and $12n$.

14–15 13 (half the diploid number of 26).

14–16 12 sets (12 × 13).

14–17 (a) ♂, triploid 69,XXY.
(b) ♀, triploid 69,XXX.

14–18 69,XXY. There would be 3 sets of chromosomes (66). 1 X from the mother and 1 X plus 1 Y from the father.

14–19 (a) 92,XXYY, or four of every autosome, plus 2 X and 2 Y sex chromosomes.
(b) No. Such embryos are usually aborted early, often before the woman knows she is pregnant.

14–20 Eggs bearing two like chromosomes (for example, number 9) are functional, but such sperms are not in this plant. Therefore, in jimson weed,
(a) 1 *P* + 2 *p* + 2 *Pp* + 1 *pp*.
(b) Only 1 *P* + 2 *p*.
(c) 1 *P* + 1 *PP*.
(d) all *P*.
(e) 2 *P* + 1 *p*.

14–21 (a) $\frac{2}{18}$ *PP* + $\frac{4}{18}$ *Pp* + $\frac{4}{18}$ *PPp* + $\frac{2}{18}$ *Ppp* + $\frac{1}{18}$ *Pp* + $\frac{2}{18}$ *pp* + $\frac{2}{18}$ *Ppp* + $\frac{1}{18}$ *ppp*. All individuals having at least one *P* allele are purple, hence this cross yields a 15:3 (or 5:1) phenotypic ratio.
(b) Progeny: $\frac{2}{12}$ *PP* + $\frac{1}{12}$ *Pp* + $\frac{1}{12}$ *PPP* + $\frac{2}{12}$ *PPp* + $\frac{2}{12}$ *Pp* + $\frac{1}{12}$ *pp* + $\frac{1}{12}$ *PPp* + $\frac{2}{12}$ *Ppp* for an 11 purple:1 white ratio.
(c) Progeny: $\frac{1}{18}$ *PP* + $\frac{2}{18}$ *PPp* + $\frac{1}{18}$ *Ppp* + $\frac{2}{18}$ *Pp* + $\frac{2}{18}$ *Pp* + $\frac{4}{18}$ *Ppp* + $\frac{2}{18}$ *ppp* + $\frac{4}{18}$ *pp* for a 12 purple:6 white (= 2:1).

14–22 (In *Drosophila* genetics it is customary to use a + sign for the dominant allele, and a letter or letters for the recessive.)

"Eyeless" (eyes small or absent) is a recessive trait whose allele is located on the small chromosome IV of *D. melanogaster*. Both triplo-IV eggs and triplo-IV sperms are functional in this insect.
(a) $\frac{1}{4}$ + +, $\frac{1}{4}$ + (*ey*)*ey*, $\frac{1}{4}$ + *ey* (*ey*), $\frac{1}{4}$ *ey ey*, for a 3:1 normal:1 eyeless ratio.
(b) 11 normal:1 eyeless.
(c) 35 normal:1 eyeless.

14–23 Autosomal monosomy probably constitutes a lethal genic imbalance.

CHAPTER 15

15–1 See drawing in text.

15–2 See drawing in text.

15–3 See drawing in text.

15–4 By definition a paracentric inversion does not include the centromere, whereas a pericentric one does include the centromere.

15–5 (a) Deletion.
(b) Inversion.
(c) Duplication.

15–6 *AAA*/*AA* between heterozygous ultrabar and homozygous ultrabar. *AAAA*/*AAA* below homozygous ultra bar. Each added 16A section intensifies the bar phenotype by decreasing the number of facets. The narrowing effect is greater if the added segments are on the same chromosome.

15–7 (a) A deletion.
(b) A deletion loop in the appropriate salivary gland chromosome (number III).

15–8 Farther from the centromere, which has an interfering effect on crossing over.

15–9 See text for justification of answers.

CHAPTER 16

16–1 Sex-linked recessive traits show a characteristic inheritance sequence from affected father to "carrier" daughter to about half her sons; sex-linked dominants are transmitted by an affected mother (× normal father) to about half her sons and half her daughters or by an affected father (× normal mother) to all his daughters and none of his sons. Purely maternal effects are transmitted from mother to all her progeny but do not persist in certain nuclear genotypes. Extranuclear genetic systems would ordinarily operate through the maternal line. If the trait is repeatedly transmitted through backcrosses of F_1 individuals with maternal parent but not with paternal parent, an extranuclear genetic system may be involved. It should be identified and located, and such guiding criteria as those listed at the outset of this chapter applied.

16–2 Any genotype will have the phenotype determined by the maternal genotype (a maternal effect).

16–3 (a) This cross is *aa* ♀ × *AA* ♂; the progeny are all *Aa*. Due to maternal effect, the young are light-eyed, but the dominant *A* gene directs the production of kynurine, which darkens the eyes by the young adult stage.
(b) This cross is *AA* ♀ × *aa* ♂; progeny are all *Aa* and are dark-eyed, regardless of age. Young have dark eyes because of diffusion of kynurine from the *AA* mother into the young. Because the progeny are *Aa*, by the time maternally derived pigment has disappeared, the *Aa* progeny are producing their own kynurine. Hence, progeny are all dark-eyed at both young and adult stages.

16–4 (a) All green progeny because of chloroplast-containing eggs from the pistillate parent.
(b) All "white" because of all colorless (defective) plastids in the eggs of the pistillate parent.
(c) Green, variegated, and "white" in irregular ratio, reflecting the three kinds of eggs produced by the pistillate parent.

16–5. (a) $P/3$, where P represents the number of deoxyribonucleotide pairs, and 3 is the number of nucleotides per codon. Thus, $1.23 \times 10^6/3 = 4.1 \times 10^5$.
(b) Divide the number of codons by the number of amino acid residues per polypeptide chain; thus, $4.1 \times 10^5/400 = 1{,}025$.

16–6 (a) The diploid F_1 will all be heterozygous normal-petite (hence of normal phenotype) because the gene for normal is dominant; here it is transmitted by the neutral petite. The cytoplasm of the F_1 individuals contains normal mitochondria derived from the segregational petite.
(b) 1:1 petite:normal. The cytoplasm of all members of the succeeding haploid generation contain normal mitochondria contributed by the diploid F_1. However, in the meiotic division, which produces the haploid progeny, the nuclear genes segregate, with the result that half the progeny haploid cells receive the normal allele, and half the recessive allele for petite.

16–7 Progeny phenotypes in all four o'clock crosses given are identical to those of the pistillate parents because of the kind of plastid contained in the egg. The green, variegated, and "white" of part (d) occur in irregular ratio.

16–8 In the chloroplast DNA; the chloroplast of the minus mating strain is lost, and the chloroplast of each of the four progeny cells resulting from meiosis is derived by division from the plus mating strain. The only DNA-containing structure shared alike by all four meiotic products is the chloroplast derived from the + mating strain.

16–9 This behavior confirms the location of the sm4 gene in the cpDNA. Only the cpDNA of the plus strain is affected by the mutagen.

CHAPTER 17

17–1 No, because of the dominance relationships of the four multiple alleles: $c^+ > c^{ch} > c^h > c$. The cross $c^+ c^{ch} \times c^+ c^{ch}$, for example, produces agouti and chinchilla in a 3:1 phenotypic ratio, and the cross $c^+ c^h \times c^+ c^h$ yields agouti and Himalayan in a 3:1 phenotypic ratio. Depending on parental genotypes, it is possible to secure agouti and *either* chinchilla *or* Himalayan in the progeny, but not both in the same mating.

17–2 Parental genotypes had to include genes c^{ch}, c^h, and c, with each parent heterozygous for either c^{ch} or c^h, with c; thus, parental genotypes were $c^{ch}\,c$ and $c^h\,c$. The 1:2:1 progeny ratio also indicates heterozygosity of the parents.

17–3 (a) 3 multiple alleles, one for each phenotype.
(b) Superdouble > double > single. Dominance of superdouble over both its alleles is indicated by the occurrence of superdouble in all, $\frac{3}{4}$, or $\frac{1}{2}$ of the progeny, depending on parental genotypes. Whenever superdouble occurs in a parent, the trait also appears in the progeny. The cross single × single produces only single-flowered offspring, indicating that single is recessive to both of its alleles.
(c) The 1:2:1 progeny ratio indicates heterozygosity of both parents, one for superdouble and single, the other for double and single. Parental heterozygosity is also indicated by the occurrence of single in the progeny.

17–4 Alexandra > Normal > Blue Moon > Primrose Queen.

17–5 4.

17–6 1.

17–7 3.

17–8 (a) 0.0252. (You may wish to review Chapter 5.)
(b) 0.0084.

17–9 29.

17–10 0.5

17–11 MN, secretor, because MN and the secretor trait are the most common throughout the world population.

17–12 20.

17–13 210.

17–14 8.

17–15 A_1B. See text.

17–16 Yes, *if* the woman is I^A *i* and the man I^B *i*.

17–17 Yes, he is exonerated by the MNSs test (only). Neither the mother nor the alleged father could have contributed Ns to the child.

17–18 (a) No.
(b) The HLA test. The haplotype *A4B5C6D7* could not have been contributed by the alleged father. The other tests do not eliminate him.

17–19 *A1B8/A3B7*. The first and second children's *A1B8*, and the third child's *A3B7* had to be contributed by the father.

17–20 *A6B5* gametes, one of the two possible crossover combinations and, therefore, would here be expected in a frequency of 0.6/2 = 0.3.

17–21 (a) Lozenge; this is a *trans* heterozygote.
(b) Wild; this is a *cis* heterozygote.

17–22 Yes, the *c-d* pair, because of the *cis-trans* effect.

17–23 The occurrence of erythroblastosis in some of the children indicates the woman (I-2) is Rh−, hence of genotype *dd* and the man (I-1) is Rh+. The latter appears to be of genotype *Dd* because, from the data given, the last child (II-4) is most likely Rh− (*dd*) and she was born of an already sensitized mother. Child II-1 must be of genotype *Dd* (Rh+) and have sensitized the mother. Second and third children (II-2 and II-3) are identifiable as Rh+ because of their erythroblastosis, and each has the genotype *Dd*.

17–24 The first child is Rh+.

17–25 Had an injection been given the mother within 72 hours of the birth of II-1 and II-2, no cases of erythroblastosis would have occurred in this family.

17–26 The A Rh− ♀ × O Rh+ ♂ is A-B-O compatible; therefore, the risk of erythroblastosis is higher in that marriage. The other marriage is A-B-O incompatible and thus is at lower risk.

17–27 16.

17–28 48.

17–29 162 (sic).

17–30 Yes, if both parents are heterozygous I^{A2} I^{A3}.

17–31 Yes, two Rh− parents cannot have an Rh+ child.

17–32 Yes, if the husband is I^B *i Dd* and his wife is also *Dd*. She is necessarily *ii*.

17–34 No, it is not valid. How do you know?

17–35 Children 4, 5, and 6 cannot belong to the husband, number 4 on the basis of *NS/Ns*, number 5 on the basis of *dce/dce* and NS/Ns, number 6 on the basis of all three tests, even allowing for a very low probability of recombination. Moreover, it appears that more than one man fathered the last three children (4, 5, 6); numbers 4 and 5 *could* have been fathered by the same man (but not the husband), and number 6 by still a third man.

17–36 I^B *i Dd*. The parental genotypes are *iiD-* ♀ × I^B — *dd* ♂.

CHAPTER 18

18–1 All traits showing practically continuous variation are likely to be due to polygenes. Here, the traits of intelligence, height, skin color, and eye color probably involve polygenes. As will be seen later in this chapter, some of these furnish excellent indication of polygene action.

18–2 Any parental genotypes that can produce at least some progeny with a greater number of contributing alleles than they themselves have are possible, for example, *AaBbCcDd* × *AaBbCcDd*, *AaBbCcDd* × *aabbCcDd*, etc.

18–3 (a) $\frac{1}{4,096}$.
(b) 13.
(c) $\frac{924}{4,096}$.

18–4 4 pairs of polygenes, or a total of 8. The contribution of each effective allele is 8 inches, assuming the conditions set forth in this chapter.

18–5 Averaging the progeny extremes as $\frac{3}{3,100} = \frac{1}{1,033}$. This latter fraction satisfactorily approximates $\frac{1}{1,024}$, which here equals $(\frac{1}{4})^5$, indicating 10 polygenes for a contribution of 4.8 inches per contributing allele.

18–6 $\frac{1}{4,096}$.

18–7 The frequencies of progeny as extreme as either parent is $\frac{1}{256}$, or $(\frac{1}{2})^4$, which indicates 4 *pairs* of polygenes. Therefore, each parent is *AaBbCcDd*.

18–8 Any in which each parent is homozygous for all four pairs of polygenes and in which two *different* pairs are homozygous contributing, as *AABBccdd* × *aabbCCDD*.

18–9 (a) *AABBCCDD;* see text.
(b) *aabbccdd;* see text.
(c) Only green-eyed (*AaBbCcDd* in this case).

18–10 This cross is *AaBbCcDd* × *AaBbCcDd*.
(a) $\frac{1}{256}$.
(b) $\frac{28}{256}$.
(c) $\frac{56}{256}$.

18–11 (a) Green; see text.
(b) $\frac{70}{256}$.

18–12 Transgressive variation.

18–13 769 = $\frac{3}{4}$ of 1,024 which are *S*- - - - - - - - -, plus 1 which is *ssaabbccdd* and, therefore, also unspotted.

18–14 8.

18–15 (a) 25 and 5 cm.
(b) 15 cm.
(c) 1:4:6:4:1.

18–16 (a) 15 cm for *AABB*, 5 cm for *aabb*.
(b) 15 cm.
(c) 9 (15):6 (10):1 (5).

18–17 $(a + b)^{2n}$, or, in this case, $(a + b)^{12}$.

18–18 23.04, or rounded to 23.

18–19 20.04.

18–20 4.477, or approximately 4.5.

18–21 0.6826, or approximately $\frac{2}{3}$.

18–22 (a) 0.895.
(b) That there is a 0.6826 probability that μ, the population mean, lies between the limits 23 ± 0.895, and so forth.

18–23 The limits of 21 and 25 are approximately 23 ± 2 or $\bar{x} \pm 2\, s_{\bar{x}}$. Therefore, there is a probability of 0.9544 that μ lies between 21 and 25.

18–24 1.2.

18–25 0.6826; that is, only a 0.6826 probability that two different populations are involved, because 0.6826 only *equals* S_d.

18–26 $x_1 - x_2$ is *not* significant.

18–27 The number of polygenes (n) equals 8.

CHAPTER 19

19–1 (a) *MN*; why?
(b) *NN*; why?

19–2 0.6.

19–3 The frequency of *M* is 0.546. The frequency of *N* is 1 − 0.546 = 0.454.

19–4 The frequency of allele *M* is 0.19, that of allele *N* is 1 − 0.19 or 0.81.

19–5 The frequency of *T* is 0.6, and of *t* 0.4.

19–6 (a) The frequency of heterozygotes is 0.48, or 48 of the 100 persons tested.
(b) The frequency of *TT* persons is 0.36, or 36 of the 100 persons tested.

19–7 1.4%.

19–8 One in 500 persons is a "carrier."

19–9 The frequencies of the three alleles is: I^A, 0.3; I^B, 0.1; *i*, 0.6.

19–10 I^Ai^A, 0.04; I^Ai, 0.28; I^BI^B, 0.01; I^Bi, 0.14; I^AI^B, 0.04; *ii*, 0.49.

19–11 Deviation is highly significant; chi-square value equals 35.69.

19–12 A = 39.36 percent; B = 8.76 percent; AB = 2.88 percent; O = 49.0 percent.

19–13 (a) Homozygous A, 5.76 percent.
(b) Heterozygous B, 8.4 percent.

19–14 The frequency of allele *his* is $\frac{1}{25{,}000}$ or 0.00004.

19–15 (a) The frequency of heterozygous women is 0.0198, or approximately 0.02.
(b) The frequency of homozygous women here would be 0.0001, or 1 in 10,000.

19–16 The frequency of O+ persons in this sample is 0.3473, or, rounded, 35%.

19–17 None, because the allele owes its continuing presence in the human population almost entirely to perpetuation in heterozygotes.

19–18 Deterioration.

19–19 0.25.

19–20 (a) 0.045.
(b) 198.

19–21 0.22.

19–22 0.267.

19–23 0.8.

19–24 8.83.

19–25 0.04.

19–26 (a) 0.02.
(b) 2×10^{-3}.

19–27 (a) 1,500 years.
(b) Probably not, because of mutation from dominant normal to recessive lethal.

19–28 (a) p^2.
(b) p^4.
(c) $2pq$.
(d) $4p^2q^2$.
(e) $4\,pq^3$.
(f) $p^4 + 4p^3q + 6p^2q^2 + 4pq^3 + q^4$.

19–29 (a) $p^2 + 2pq$.
(b) q^2.

19–30 (a) $p^2q^2 + 2pq^3 + q^4$.
(b) 0.0625 + 0.125 + 0.0625 = 0.25.

19–31 0.0009765.

19–32 1 − 0.0009765 (from preceding problem), dividing by 2, equals 0.49995117.

19–33 (a) 6×10^{-3}.
(b) 0.988.
(c) 0.012.
(d) 2×10^{-3}
(e) $(4 \times 10^{-6})^2 = 1.6 \times 10^{-11}$.

19–34 2×10^{-3}.

19–35 3.317, or just over 0.003.

19–36 (a) 0.775.
(b) 0.447.
(c) 12 (rounded from 12.232).

19–37 (a) $\frac{1}{100,000}$, or 0.00001.
(b) 0.0063.
(c) 0.00004.
(d) 0.00001.
(e) 1×10^{-9}.

SELECTED LIFE CYCLES

APPENDIX

B

1. BACTERIA

Bacteria reproduce most frequently by asexual cell division, but at least certain genera, notably *Escherichia* and closely related genera, may also engage in a type of reproduction called conjugation. Conjugants are of two general types, F^- ("female" or, better, **recipient** cells) and either F^+ or *Hfr* ("male" or, better, **donor** cells). Donor cells possess the fertility factor *F*; in F^+ cells *F* is an independent cytoplasmic plasmid, whereas in *Hfr* cells *F* is integrated into the cell's large DNA molecule (its "chromosome"). Conjugation between an *Hfr* (**high frequency recombinant**) and an F^- cell includes the following steps:

a. Chance contact between *Hfr* and F^- cells in pairs.
b. Formation of a conjugation tube connecting the two conjugants.
c. One strand of the double-stranded DNA "chromosome" of the *Hfr* cell is cut enzymatically at a point within *F*. DNA replication occurs by the rolling circle method and a single strand with its 5′ and enters the F^- cell through the conjugation bridge.
d. The integrated *F* is at the trailing end (3′) of the strand being transferred.
e. Cell-to-cell contract is usually broken by external forces before most of the 5′, 3′ strand is transferred from donor to recipient.
f. Synapsis between donated segment and the homologous section of the F^- "chromosome" occurs.

g. Donor strand segment is integrated into the recipient "chromosome"; the displaced single-stranded segment from the recipient cell is enzymatically degraded after its excision from the F^- "chromosome."

h. Replication of DNA strand complementary to the now integrated donor strand occurs. Recipient cell is now a *recombinant* if the donated DNA segment bore alleles of those of the recipient's original "chromosome" (e.g., *donor:* leu^+ pan^+ $met\ D^-$ $pro\ A^-$; and *recipient:* leu^- pan^- $met\ D^+$ $pro\ A^+$ and so forth).

i. The *F* factor sometimes carries a variable number of chromosomal cistrons in its own circular form. This type of fertility factor is designated F'. The recipient cell in conjugation with an F' cell can and usually does become diploid (a **merozygote**) for those chromosomal genes that are included within the fertility factor. Conjugation involving F' donor cells is called **sexduction.**

In the case of $F^+ \times F^-$ conjugation almost always (>99.99 percent of the cases) only the plasmid *F* is transferred to the recipient, which thereby becomes an F^+ cell. The F^+ donor remains F^+ because plasmid *F* replicates just prior to the actual conjugation process. On the other hand, in $Hfr \times F^-$ conjugation, the *Hfr* donor remains *Hfr* except in very rare instances when all the donor strand (i.e., including *F*) is transferred.

2. NEUROSPORA

Like most fungi, *Neurospora* produces large numbers of asexual spores (here called conidia) but also reproduces sexually if + and − mating strains come into contact. In *Neurospora,* as in many fungi, pairing of nuclei of sex cells does not result in immediate syngamy. The nuclei that ultimately fuse are daughter nuclei of the original pairing nuclei. The essential steps are diagramed in Figure B–1 and include:

a. Contact between filaments (hyphae) of + and − strains.

b. Pairing (not fusion) of + and − nuclei.

c. Development of the "fruiting body" (ascocarp), called a perithecium, which consists of n^+, n^-, and dikaryon (n^+/n^-) hyphae.

d. Development of large numbers of elongate, saclike sporangia called asci (sing., ascus) in the perithecium.

e. Fusion in the young asci of a + and a − nucleus that were derived through several mitoses from the original pairing nuclei, thus forming a diploid zygote.

f. Meiosis of zygote soon after formation, in the developing ascus, to form four meiospores.

g. Mitosis of the four meiospores to form eight (monoploid) spores called ascospores. Four of these will give rise to + mating strain plants, the other four to − strain plants.

h. Release of ascospores and their germination to form new adults.

3. SACCHAROMYCES (YEAST)

Yeasts are one-celled ascomycete fungi. Multiplication is by "budding," in which the nucleus divides by mitosis. In the full life cycle, however, morphologically identical diploid and monoploid generations alternate, as shown in Figure B–2. The life history consists of the following steps:

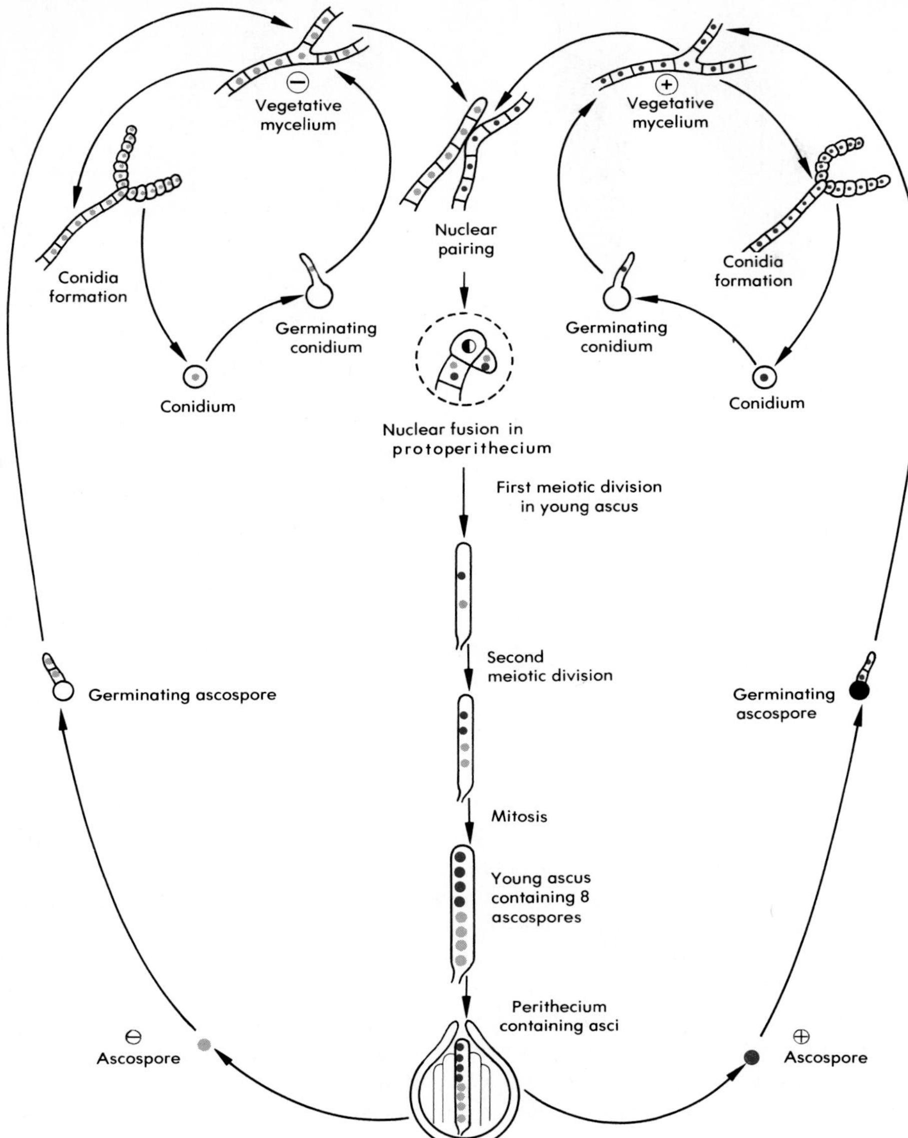

FIGURE B–1

a. Diploid adults multiply by "budding" but, under certain environmental conditions, undergo meiosis to form four monoploid meiospores within the old cell wall.
b. Maturation of the four meiospores to become four ascospores (baker's yeast) or, in some other species, mitosis of each of the four meiospores to form eight ascospores.
c. Liberation of ascospores from ascus (old vegetative cell wall).
d. Germination of ascospores to form haploid adult cells. Half the ascospores from any one ascus give rise to + mating strain adults and half to − mating strain.
e. Multiplication of haploid + and − adults by "budding."
f. Contact between + and − cells.
g. Formation of intercellular cytoplasmic bridge.
h. Fusion of + and − nuclei (each haploid cell in the pair furnishes a single gamete).
i. Formation of diploid vegetative cell, often from the cytoplasmic bridge.

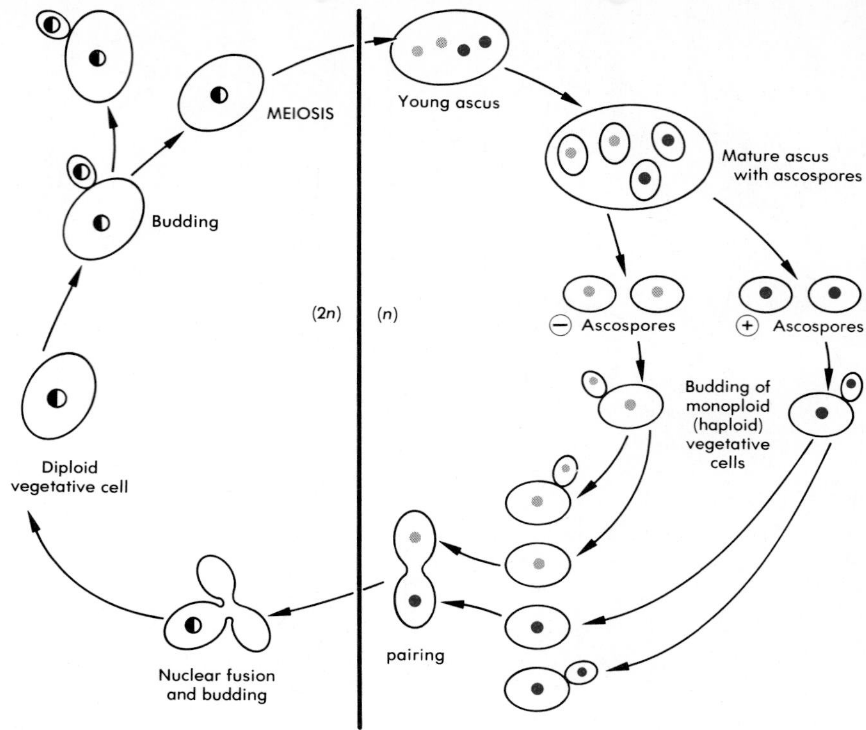

FIGURE B–2

4. CHLAMYDOMONAS

The small, unicellular, motile green alga *Chlamydomonas* reproduces freely by cell division (mitosis and cytokinesis). The vegetative cells are monoploid (haploid). In sexual reproduction, the vegetative cell functions as a gametangium; its protoplast divides mitotically to produce 4, 8, 16, or 32 gametes. These sex cells are morphologically similar to the vegetative cells but smaller in size. In many species the gametes are identical in appearance, hence may be referred to as isogametes. In other species varying degrees of morphological differentiation of gametes occur. Chemical differences among gametes, and the cells producing them, occur and mating strains are designated as + and −. Gametes of opposite mating strain come into contact at their flagellar ends; the protoplasts fuse to form a four-flagellate zygote. The zygote soon loses its flagella, develops a wall, and becomes dormant. Germination of the zygote begins with meiosis of its diploid nucleus and ends with liberation of biflagellate zoospores from the old zygote wall. In some species only four zoospores are thus formed, but in others meiosis is followed by one or more mitoses so that 8, 16, or more zoospores are produced. Zoospores resemble the vegetative cells into which they will develop; and of the number produced from a single zygote, half are of each mating strain. The process is diagramed in Figure B–3.

5. SPHAEROCARPOS (LIVERWORT)

Vegetative plants are small, thin, and lobed; these are monoploid (haploid), unisexual gametophytes (gamete-producing plants). Males produce motile sperms in antheridia; females develop one egg in each of several archegonia. Syngamy occurs when liquid water (from rain or dew) is present, and thus allows sperms to swim to the archegonia. The resulting zygote develops into a small, multicellular sporophyte (spore-

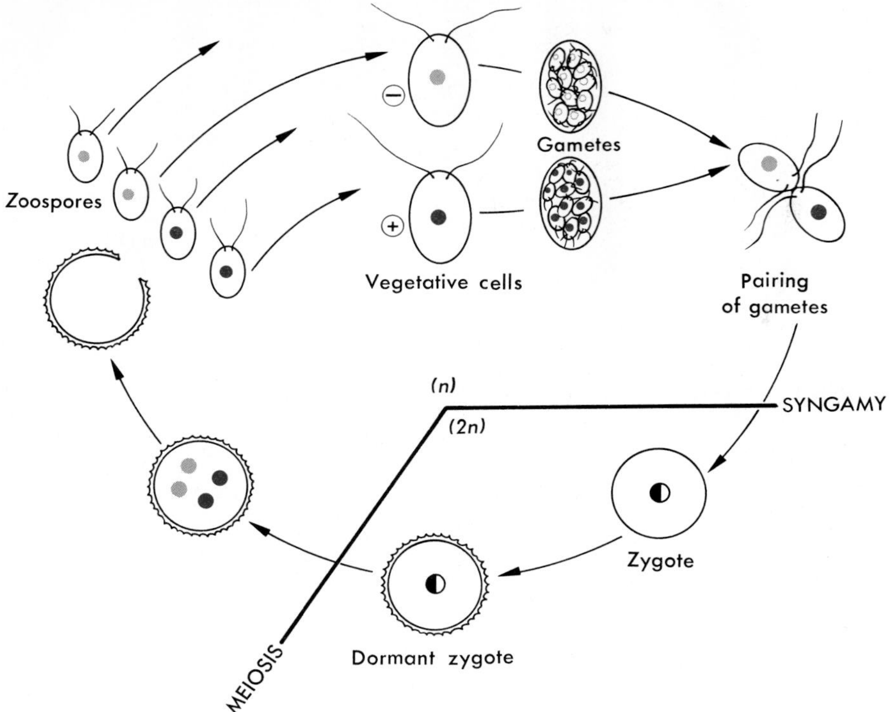

FIGURE B–3

bearing plant), which remains permanently attached to the parent gametophyte. It ultimately protrudes from the remains of the archegonium and produces internally a large number of diploid sporocytes that undergo meiosis to produce four meiospores each. Two of each of these will produce male gametophytes, and two female. The life cycle is diagramed in Figure B–4.

6. FLOWERING PLANTS (CLASS ANGIOSPERMAE)

The plant we recognize by name in the angiosperms is the diploid sporophyte. The monoploid (haploid) gametophyte is microscopic and contained almost entirely within various floral structures. A (usually) triploid food storage tissue, the endosperm, also occurs and is cytologically unique to the angiosperms. The life cycle of a representative angiosperm is diagramed in Figure B–5 and consists of the following structures and steps:

a. Production of flowers by the sporophyte.
b. Meiosis of microsporocytes (pollen mother cells) in anthers of stamens to form four functional, uninucleate microspores each.
c. Development of young male gametophyte by mitosis of the microspore nucleus within the microspore wall, inside the anther. This two nucleate structure (tube nucleus and generative nucleus) is sometimes called a pollen grain.
d. Transfer of pollen grains to stigmas of the pistils, where each grain produces a tubular outgrowth, the pollen tube, which grows down through structures of the pistil to the ovule, which it enters via the micropyle.

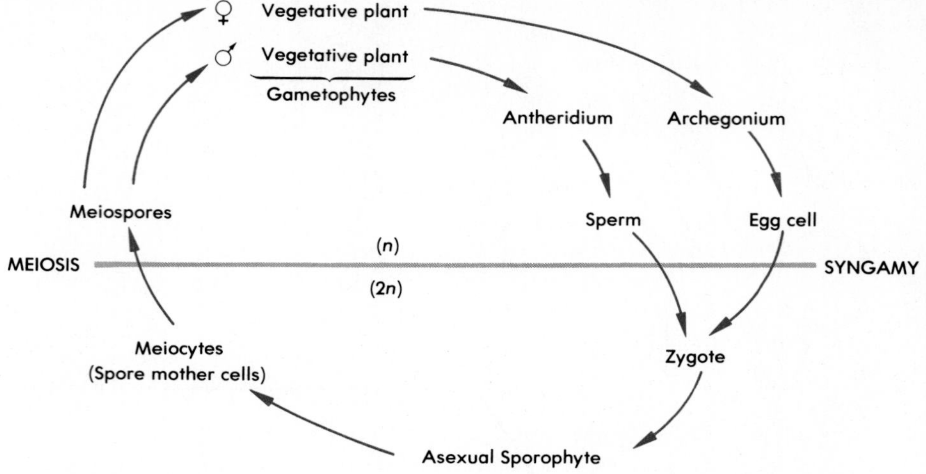

FIGURE B–4

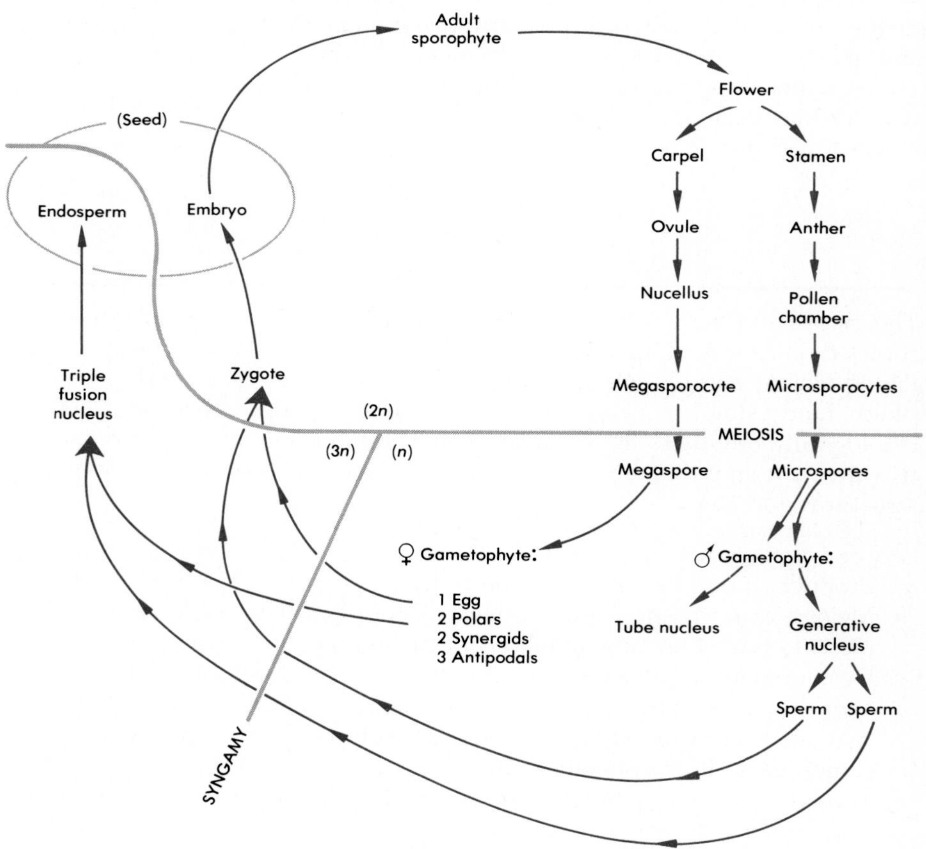

FIGURE B–5

e. Development in each ovule of a megasporocyte, which then undergoes meiosis to form four megaspores, three of which degenerate.
f. Development of the female gametophyte by mitosis from the one functional megaspore. In the classical case the mature female gametophyte consists of eight nuclei (one egg, two polars, two synergids, three antipodals) in a common cytoplasm within each ovule, and contained within the ovary of the pistil.
g. Mitosis of generative nucleus to form two sperms in the pollen tube.
h. Entry of the pollen tube into the embryo sac.
i. Fusion of egg and sperm to form the zygote.
j. Fusion of the second sperm with the two polars to form the triploid triple fusion nucleus.
k. Degeneration of synergids and antipodals.
l. Development of multicellular embryo from the zygote.
m. Development of endosperm from the triple fusion nucleus. In some plants this tissue is absorbed by the cotyledons of the embryo during the latter's development.
n. Development of seed coat, primarily from the integuments of the ovule.
o. Development of a fruit from the ovary.

7. PARAMECIUM

Paramecia are elongate ciliates of the phylum Protozoa. Each animal contains a large macronucleus, which exerts phenotypic control for that individual, and two micronuclei, which function in the sexual process. The macronucleus is polyploid, the micronucleus diploid in the vegetative animal. Paramecia increase in number only by fission; other processes, important in genetics, also occur and are as follows:

Fission

a. Mitosis of micronuclei.
b. Constriction of macronucleus to form two.
c. Movement of one macronucleus and one micronucleus to each end of the animal, which then constricts in the middle to form two new individuals.

Conjugation (Figure B–6)

a. Pairing of two animals (conjugants) and formation of intercellular bridge.
b. Disintegration of macronucleus.
c. Meiosis of each of the two micronuclei.
d. Disintegration of seven of the eight products of meiosis.
e. Mitosis of the remaining haploid nucleus to form two.
f. One of the two haploid nuclei of each conjugant passes through the connecting bridge to the other animal (reciprocal transfer).
g. Fusion of the two haploid nuclei in each conjugant, to restore the diploid condition.
h. Two mitoses of the fertilization nuclei, which results in four diploid nuclei per conjugant.
i. Separation of the conjugants, which are now genetically alike.
j. Two of the four nuclei in each ex-conjugant become macronuclei, two become micronuclei.
k. Distribution of two macronuclei to each daughter cell at next fission.

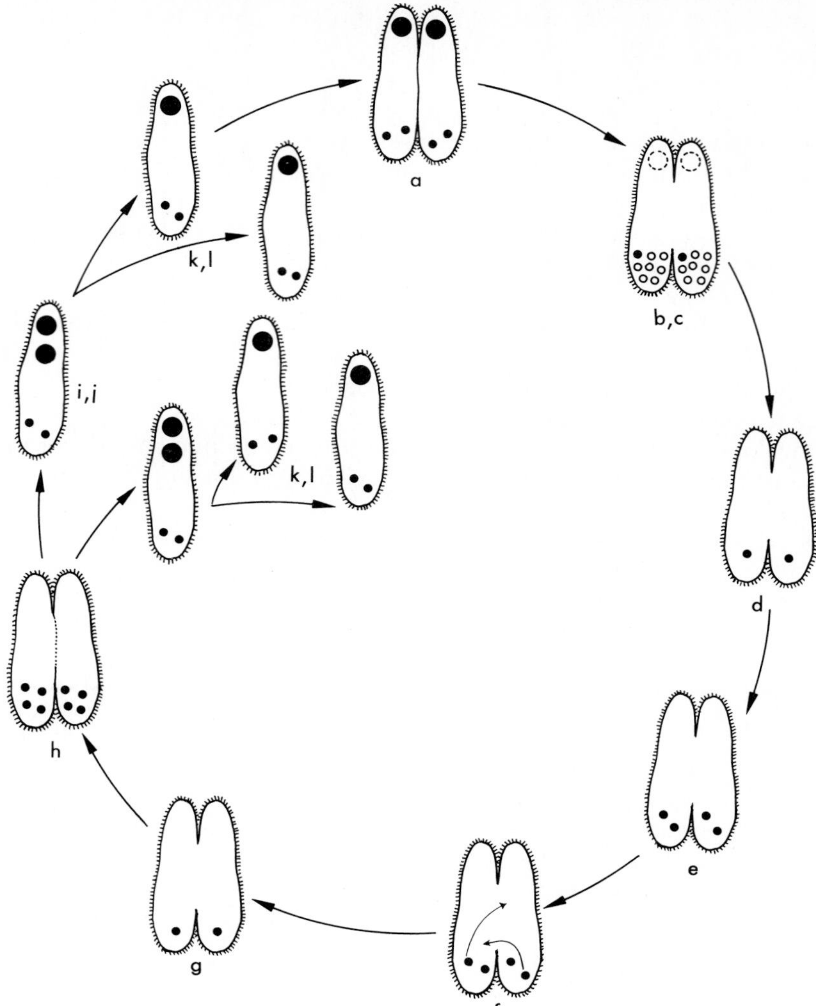

FIGURE B–6

l. Mitosis of the two micronuclei at next fission, two being distributed to each daughter cell.
m. Conjugation is usually a short-term event with little cytoplasmic transfer. Under certain conditions the intercellular connection may persist for a longer period, and allow exchange of considerable cytoplasm.

Autogamy (Figure B–7)

This is a type of internal self-fertilization that resembles somewhat the events of conjugation but involves only a single animal. It results in homozygosity of the individual undergoing the process.

a. Macronucleus behaves as in conjugation.
b. Meiosis of the two micronuclei and disintegration of seven of the eight resulting nuclei as in conjugation.
c. Mitosis of the remaining haploid nucleus to form two.
d. Fusion of the two haploid nuclei resulting from step (c).
e. Restoration of micronuclei and the macronucleus as for conjugation.

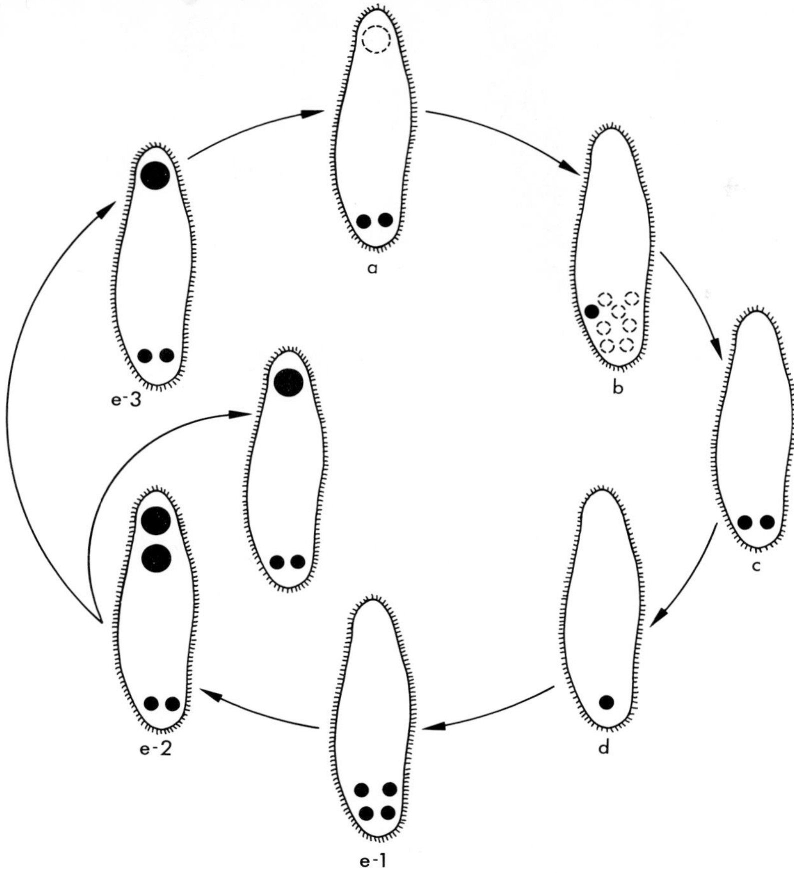

FIGURE B–7

8. MAMMALS

In mammals, the somatic cells (except for some, such as liver cells, which may be polyploid) are diploid; meiosis immediately precedes the formation of gametes. The diploid condition is restored at syngamy (Figure B–8). The following steps are involved in the male:

a. Development of diploid primary spermatocytes.
b. Meiosis. The two haploid products of the first meiotic division are called secondary spermatocytes. The four cells resulting from the second meiotic division are called spermatids.
c. Maturation of spermatids into sperms.

In the female:

a. Development of diploid primary oocytes.
b. Meiosis. The first division produces two unequal cells, a smaller first polar body and a larger secondary oocyte. The second division produces two second polar bodies from the first polar body and, from the secondary oocyte, a third second polar body and a larger ootid.
c. Maturation of one egg from the ootid; degeneration of the three second polar bodies.

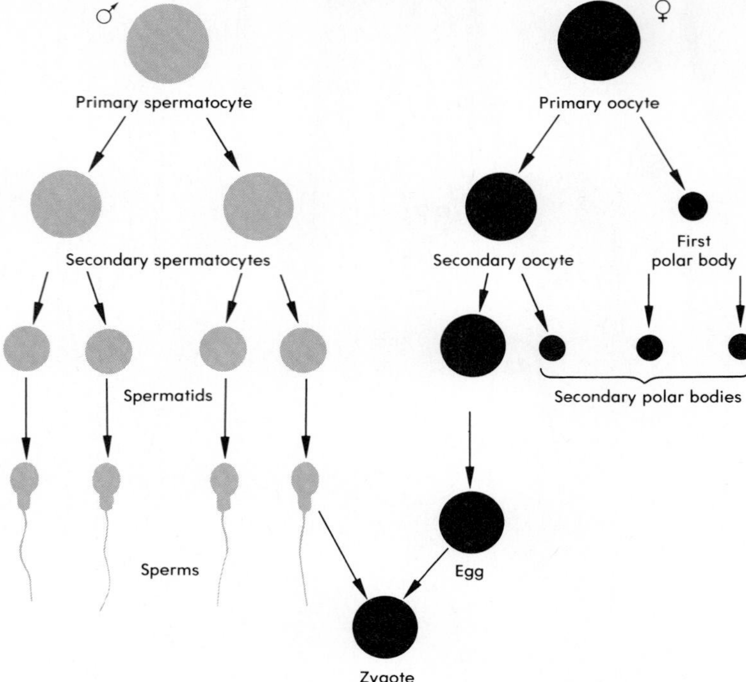
♂
♀
Primary spermatocyte
Primary oocyte
Secondary spermatocytes
Secondary oocyte
First
polar body
Spermatids
Secondary polar bodies
Sperms
Egg
Zygote

FIGURE B–8

USEFUL FORMULAS, RATIOS, AND STATISTICS

APPENDIX

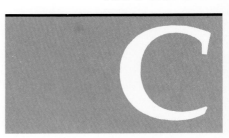

3:1	Monohybrid phenotypic ratio produced by *Aa* × *Aa*.
1:2:1	Monohybrid genotypic ratio produced by *Aa* × *Aa;* monohybrid phenotypic and genotypic ratio produced by $a^1a^2 \times a^1a^2$.
1:1	Monohybrid testcross phenotypic and genotypic ratio produced by *Aa* × *aa*.
2:1	Monohybrid lethal genotypic and phenotypic ratio produced by $a^1a^2 \times a^1a^2$ where either a^1a^1 or a^2a^2 is lethal; also sex ratio produced by *Aa* × *AY* where *a* is a sex-linked recessive lethal.
"1:0"	Monohybrid lethal phenotypic ratio produced by *Aa* × *Aa* where *aa* or *A* − is lethal; also monohybrid testcross phenotypic and genotypic ratio produced by *AA* × *aa*.
9:3:3:1	Dihybrid phenotypic ratio produced by *AaBb* × *AaBb* where phenotypes may be represented as 9 *A* − *B* −, 3 *A* − *bb*, 3 *aaB* −, 1 *aabb;* note epistatic possibilities producing "condensations" of this ratio (e.g., 9:7, 9:6:1).

1:1:1:1	Dihybrid testcross genotypic and phenotypic ratio produced by *AaBb* × *aabb* where there is no linkage.
3:6:3:1:2:1	Dihybrid phenotypic ratio produced by $Aab^1b^2 \times Aab^1b^2$. Note that dihybrid and polyhybrid ratios are products of their component monohybrid ratios and may be combined in any way.
2^n	Number of gamete genotypes and progeny phenotypes where n = the number of pairs of heterozygous genes with complete dominance.
3^n	Number of zygote genotypes under the preceding conditions.
4^n	Number of possible zygote combinations under the preceding conditions, which yield 3^n zygote genotypes.
$\frac{n}{2}(n + 1)$	The chance of selecting at random any two items in pairs (e.g., the number of possible genotypes for n multiple alleles in a series).
$(a + b)^{2n}$	Progeny phenotypic ratios in polygene crosses are given by the *coefficients* of the expansion of $(a + b)^{2n}$ where n = the number of *pairs* of polygenes.
$(a + b)^n$	Expansion of the binomial provides probability determinations where n = the number of independent events and the choices, represented by a and b, respectively, are two.
$(p + q)^2 = 1$	Binomial whose expansion permits calculation of the frequency of each member of a pair of alleles.
$(p + q + r)^2 = 1$	Trinomial whose expansion permits calculation of the frequency of each of three multiple alleles.
$(\frac{1}{4})^n$	In polygene cases, the fraction of the F_2 like either P, is given by this expression, where n = the number of pairs of genes in which the parents (P) differ.
$(\frac{1}{2})^n$	In polygene cases, the fraction of the F_2 like either P is given by this expression, where n = the number of effective or contributing alleles.
$\chi^2 = \Sigma \left[\frac{(o - c)^2}{c}\right]$	Consult tables of chi-square for levels of significance (where degrees of freedom equal one less than the number of classes). In general, a value of chi-square ≥ than that for $P = 0.05$ is regarded as significant; i.e., there is significant evidence against the hypothesis. A value of chi-square showing a level of $P = 0.05$, for example, does *not* mean that a deviation as large or larger will *not* occur by chance alone

under the hypothesis adopted; but, it is likely to in only five trials out of 100. This is considered too few; at this level, the chance of rejecting a right hypothesis is only one in 20.

$$\bar{x} = \frac{\Sigma fx}{n} \text{ or } \frac{\Sigma x}{n}$$

The sample mean is self-explanatory.

$$s^2 = \frac{\Sigma f(x - \bar{x})^2}{n - 1}$$

The variance (s^2) provides an unbiased estimate of the population variance (σ^2).

$$s = \sqrt{\frac{\Sigma f(x - \bar{x})^2}{n - 1}}$$

The standard deviation measures the variability of the sample; in a normal distribution, 68.26 percent of the sample will lie in the range $\bar{x} \pm s$, and 95.44 percent will fall in the range $\bar{x} \pm 2s$. Used with normal distributions.

$$s_{\bar{x}} = \frac{s}{\sqrt{n}}$$

The standard error of the sample mean indicates the degree of correspondence between $\bar{x}$ and μ; there is 68.26 percent confidence that $\mu = \bar{x} \pm s_{\bar{x}}$ by chance alone, and 95.44 percent confidence that $\mu = \bar{x} \pm 2s_{\bar{x}}$ by chance alone.

$$S_d = \sqrt{(s_{\bar{x}_1})^2 + (s_{\bar{x}_2})^2}$$

The standard error of the difference in means is useful in comparing two samples to determine whether the difference in their means is significant. If $(\bar{x}_1 - \bar{x}_2) > 2S_d$, the difference in sample means is considered significant and the two samples represent two different populations.

$$s = \sqrt{\frac{pq}{n}}$$

The standard deviation (s) of a simple proportionality, such as heads (p) versus tails (q) for n trials.

$$s = \sqrt{\frac{pq}{2N}}$$

The standard deviation of gene frequencies where N represents the number of *diploid* individuals, and p and q represent the frequency of each of a pair of alleles.

$$n = \frac{R^2}{8(s^2_{F_2} - s^2_{F_1})}$$

The number of pairs of polygenes (n) is calculated from this equation where R is the maximum quantitative range between phenotypes, $s^2_{F_2}$ is the variance of the F_2, and $s^2_{F_1}$ is the variance of the F_1.

$$q_n = \frac{q_0}{1 + nq_0}$$

The frequency of a recessive lethal (q_n) after n additional generations equals the initial frequency of the gene (q_0) divided by 1 plus the product of the number of additional generations (n) times the initial frequency of the gene.

$$n = \frac{1}{q_n} - \frac{1}{q_0}$$

The number of additional generations (n) required to reduce the frequency of a gene from its initial value (q_0) to any particular value (q_n) is given by this equation.

$W = 1 - s$, and $s = 1 - W$

The adaptive value of a genotype (W) is 1 minus its selection coefficient(s), and the selection coefficient is 1 minus the adaptive value of the genotype.

$$q_1 = \frac{q_0 - s(q_0)^2}{1 - s(q_0)^2}$$

The equation for determining the frequency of a gene after one generation under selection (q_1) when its selection coefficient (s) and its intial frequency (q_0) are known.

$$\Delta q_s = \frac{-sq_0^2 p}{1 - sq_0^2}$$

Change in frequency of a recessive gene under selection in one generation.

$$L_{\mu m} = 3.4 \times 10^{-4} P$$

Length in micrometers ($L_{\mu m}$) of double-stranded DNA equals 0.00034 times the number of deoxyribonucleotide pairs (P).

$$P = \frac{M}{650}$$

The number of deoxyribonucleotide pairs (P) equals the molecular weight (M) of a DNA molecule divided by 650.

$$M = 650P$$

The molecular weight of a DNA molecule (M) equals 650 times the number of deoxyribonucleotide pairs (P) of which it consists.

$$P = \frac{L_{\mu m}}{3.4 \times 10^{-4}}$$

The number of deoxyribonucleotide pairs (P) in a DNA molecule equals its length in micrometers ($L_{\mu m}$) divided by 0.00034.

$$\Delta q_m = up - vq$$

The rate of change in recessive gene frequency under mutation alone equals the rate of forward mutation times the frequency of the dominant allele minus the quantity the rate of back mutation times the frequency of the recessive allele. See page 399.

$$\hat{q}_m = \frac{u}{u + v}$$

The equilibrium frequency of a recessive allele under mutation alone equals the rate of forward mutation divided by the sum of the rates of forward mutation plus back mutation.

$$\hat{q} = \sqrt{\frac{u}{s}}$$

The equilibrium frequency of a recessive allele under the combined effect of mutation and selection equals the square root of the quantity rate of forward mutation divided by the selection coefficient.

USEFUL METRIC VALUES

APPENDIX

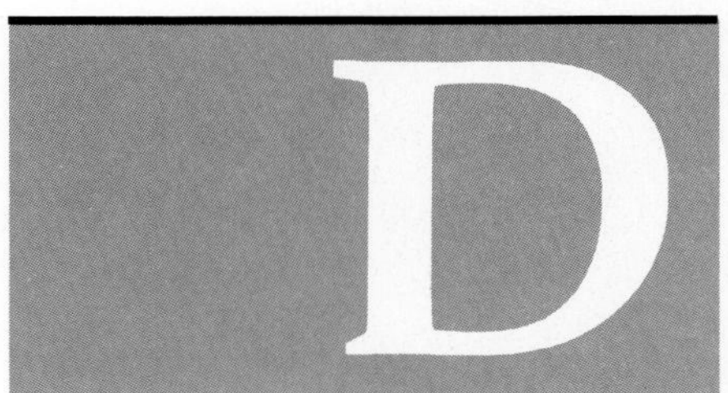

Name	*Numerical value (m)*	*Power of 10*	*Symbol*	*Synonym*
Meter	1.0		m	
Decimeter	0.1	10^{-1}	dm	
Centimeter	0.01	10^{-2}	cm	
Millimeter	0.001	10^{-3}	mm	
Micrometer	0.000 001	10^{-6}	μm	micron (μ)
Nanometer	0.000 000 001	10^{-9}	nm	millimicron (mμ)
	0.000 000 000 1	10^{-10}		Angstrom unit (Å)

Note: 1 Å = 0.0001 or 1×10^{-4} μm; 1 μm = 10,000 or 1×10^{4} Å.

JOURNALS AND REVIEWS

APPENDIX

Advances in Genetics
Advances in Human Genetics
American Journal of Human Genetics
Annals of Human Genetics
Annual Review of Biochemistry
Annual Review of Genetics
Annual Review of Microbiology
Biochemical Genetics
BioEssays
Bio/Technology
Cell
Chromosoma
Clinical Genetics
Cold Spring Harbor Symposia in Quantitative Biology
Cytogenetics
Genetica
Genetical Research
Genetics
Genetics Abstracts
Hereditas
Heredity
Human Genetics
Journal of Bacteriology
Journal of Genetics
Journal of Heredity
Journal of Medical Genetics
Journal of Molecular Biology
Journal of Virology
Lancet
Molecular and General Genetics
Nature
New England Journal of Medicine
Plant Molecular Biology
Proceedings of the National Academy of Science (U.S.)
Science
Scientific American
Theoretical and Applied Genetics
Trends in Biochemical Sciences
Trends in Biotechnology
Trends in Genetics

REFERENCES

CHAPTER 1

To celebrate the 100th anniversary of Mendel's discovery, a number of books appeared reviewing the historical developments of this field of genetics. These are recommended reading for both scientists and historians of science.

CARLSON, E. A., 1966. *The Gene: A Critical History.* W. B. Saunders Co., Philadelphia.

DUNN, L. C., 1965. *A Short History of Genetics.* McGraw Hill, New York.

OLBY, R. C., 1966. *Origins of Mendelism.* Shocken Books, New York.

STURTEVANT, A. H., 1965. *A History of Genetics.* Harper & Row, New York.

CHAPTER 2

ABBOT, U. K., R. M. CRAIG, and E. B. BENNETT, 1970. Sex-linked coloboma in the chicken. *J. Heredity* 61:95–102.

CUENOT, L., 1904. L'Heredite de la pigmentation chex les souris 3me Note. *Arch Zool. Exp. et Gen.,* 3me Serie 10:Notes et Revues, 27–30.

CUENOT, L., 1905. Les races pures et leur combinaisons chex les souris. *Arch. Zool. Exp. et Gen.* 3:123–132.

EATON, G. J., and M. M. GREEN, 1962. Implantation and lethality of the yellow mouse. *Genetica* 33:106–112.

HENAULT, R. E., and R. CRAIG, 1970. Inheritance of plant height in the Geranium. *J. Heredity* 61:75–78.

KRETCHMER, N., 1972. Lactose and lactase. *Sci. American* 227:(4)71–78.

ROBERTSON, G. G., 1942. An analysis of the development of homozygous yellow mouse embryos. *J. Exp. Zool.* 89:197–231.

ROSENFELD, A., 1981. The heartbreak gene. *Science* 812:46–50.

CHAPTER 3

ATWOOD, S. S., and J. T. SULLIVAN, 1943. Inheritance of a cyanogenic glucoside and its hydrolyzing enzyme in *Trifolium repens. J. Heredity* 34:311–320.

STEWARD, R. N., and T. ARISUMI, 1966. Genetic and histogenic determination of pink bract color in Poinsettia. *J. Heredity* 57:217–220.

CHAPTER 4

ALBERTS B., D. BRAY, J. LEWIS, M. RAFF, K. ROBERTS, and J. D. WATSON, 1985. *Molecular Biology of the Cell.* Garland Publishing, New York.

BAHR, G. F., 1977. Chromosomes and chromatin structure. In J. J. Yunis, ed., *Molecular Structure of Human Chromosomes.* Academic Press, New York.

CASPERSSON, T., I. ZECH, and C. JOHANSSON, 1970a. Differential binding of alkylating fluorochromes in human chromosomes. *Exp. Cell Res.* 60:315–319.

CASPERSSON, T., L. ZECH, and C. JOHANSSON, 1970b. Analysis of the human metaphase chromosome set by aid of DNA-binding fluorescent agents. *Exp. Cell Res.* 62:490–492.

CASPERSSON, T., L. ZECH, C. JOHANSSON, and E. J. MODEST, 1970c. Identification of human chromosomes by DNA-binding fluorescing agents. *Chromosoma* 30:215–227.

CHAMBON, P., 1978. Summary: The molecular biology of the eukaryotic genome is coming of age. *Cold Spring Harbor Symposia Quant. Biol.* 42:1209–1234.

FELSENFELD, G., 1978. Chromatin. *Nature* 271:115–122.

KEDES, L. H., 1979. Histone genes and histone messengers. *Ann. Rev. Biochemistry* 48:837–870.

KING, R. D., and W. STANSFIELD, 1985. *A Dictionary of Genetics,* 3rd ed. Oxford University Press, New York.

MAZIA, D., 1974. The cell cycle. *Sci. American* 230:(1)54–64.

PARDEE, A. B., R. DUBROW, J. L. HAMLIN, and R. F. KLETZIEN, 1978. Animal cell cycle. *Ann. Rev. Biochemistry* 47:715–750.

RIS, H., and D. F. KUBAI, 1970. Chromosome structure. *Ann. Rev. Genet.* 4:263–294.

SWANSON, C. P., T. MERZ, and W. J. YOUNG, 1981. *Cytogenetics,* 2nd ed. Prentice-Hall, Englewood Cliffs, N.J.

TSO, P. O. P., ED., 1977. *The Molecular Biology of the Mammalian Genetic Apparatus,* vol. 1. Elsevier North-Holland, New York.

VON WETTSTEIN, D., S. W. RASMUSSEN, and P. B. HOLM, 1984. The synaptonemal complex in genetic segregation. *Ann. Rev. Genet.* 18:331–413.

WRAY, W., M. MACE, JR., Y. DASKAL, and E. STUBBLEFIELD, 1978. Metaphase chromosome architecture. *Cold Spring Harbor Symposia Quant. Biol.* 42:361–365.

YUNIS, J. J., ED., 1977. *Molecular Structure of Human Chromosomes.* Academic Press, New York.

YUNIS, J. J., and D. PRAKASH, 1982. The origin of man: A chromosomal pictorial legacy. *Science* 215:1525–1530.

CHAPTER 5

SNEDECOR, G. W., and W. G. COCHRAN, 1980. *Statistical Methods,* 7th ed., Iowa State University Press, Ames, Iowa.

SOKAL, R. R., and F. J. ROHLF, 1981. *Biometry—The Principles and Practices of Statistics in Biological Research,* 2nd ed. W. H. Freeman, San Francisco.

CHAPTER 6

BACHMAN, B. J., and K. B. LOW, 1980. Linkage map of *Escherichia coli* K-12, Edition-6. *Microbiol. Rev.* 44(1):1–56.

BATESON, W., and R. C. PUNNETT, 1905–1908. Experimental studies in the physiology of heredity. Reports to the Evolution Committee of the Royal Society, 2, 3, and 4. Reprinted in J. A. Peters, ed. 1959. *Class Papers in Genetics.* Prentice-Hall, Englewood Cliffs, New Jersey.

BATESON, W., E. R. SAUNDERS, and R. C. PUNNETT, 1905. Experimental studies in the physiology of heredity. Reports to the Evolution Committee of the Royal Society, II, 1–55 and 80–99.

BEREGSMA, D., ED. 1974. *Human Gene Mapping.* Intercontinental Medical Book Corp., New York.

BLIXT, S., 1975. Why didn't Gregor Mendel find linkage? *Nature* 256:206.

CONNEALLY, P. M., and M. L. RIVAS, 1980. Linkage analysis in man. *Adv. Genet.* 10:209–266.

CREIGHTON, H. S., and B. MCCLINTOCK, 1931. A correlation of cytological and genetical crossing-over in *Zea mays. Proc. Nat. Acad. Sci. (U.S.)* 17:492–497. Reprinted in J. A. Peters, 1959. *Classic Papers in Genetics.* Prentice-Hall, Englewood Cliffs, New Jersey.

HUTCHISON, C. B., 1922. The linkage of certain aleurone and endosperm factors in maize, and their relation to other linkage groups. *Cornell Agr. Exp. Sta. Mem.,* 60.

MCKUSICK, V. A., 1971. The mapping of human chromosomes. *Sci. American* 224:(4)104–113.

MCKUSICK, V. A., 1986. *Mendelian Inheritance in Man,* 7th ed. The Johns Hopkins Press, Baltimore.

MCKUSICK, V. A., 1986. The human gene map. *Clin. Genet.* 29:545–588.

MCKUSICK, V. A., and F. H. RUDDLE, 1977. The status of the gene map of the human chromosomes. *Science* 196:390–405.

MORGAN, T. H., 1910a. Sex-limited inheritance in *Drosophila. Science* 32:120–122.

MORGAN, T. H., 1910b. The method of inheritance of two sex-limited characters in the same animal. *Proc. Soc. Exp. Biol. Med.* 8:17.

MORGAN, T. H., 1911a. The application of the conception of pure lines to sex-limited inheritance and to sexual dimorphism. *Am. Naturalist* 45:65.

MORGAN, T. H., 1911b. Random segregation versus coupling in Mendelian inheritance. *Science* 34:384.

O'BRIEN, S. J., ED., 1984. *Genetic Maps.* Cold Spring Harbor Press, New York.

PAINTER, T. S., 1934. Salavary gland chromosomes and the attack on the gene. *J. Heredity* 25:465–476.

RENWICK, J. H., 1971. The mapping of human chromosomes. *Ann. Rev. Genet.* 5:81–120.

SUTTON, W. S., 1903. The chromosomes in heredity. *Biol. Bull.* 4:231–251. Reprinted in J. A. Peters, ed., 1959. *Classic Papers in Genetics.* Prentice-Hall, Englewood Cliffs, N.J.

CHAPTER 7

BRAZZEL, J., T. TEGENKAMP, and R. ASHCOM, 1978. Pregnancy in a bisexually active true hermaphrodite, 46,XX. *Program and Abstracts,* The American Society of Human Genetics, 29th Annual Meeting, The University of Chicago Press, Chicago.

BRIDGES,C. B. 1916a. Nondisjunction as proof of the chromosome theory of heredity. *Genetics* 1:1–52.

BRIDGES, C. B., 1916b. Nondisjunction as proof of the chromosome theory of heredity (concluded). *Genetics* 1:107–163.

BRIDGES, C. B., 1925. Sex in relation to chromosomes and genes. *Am. Naturalist* 59:127–137.

BUHLER, E. M., 1980. A synopsis of the human Y chromosome. *Human Genet.* 55:145–175.

CURTISINGER, J. W., and M. W. FELDMAN, 1980. Experimental and theoretical analysis of the sex-ratio polymorphism in *Drosophila pseudoobscura. Genetics* 94:445–466.

de Grouchy, J., and C. Turleau, 1977. *Clinical Atlas of Human Chromosomes.* John Wiley, New York.

Eicher E. M., and L. L. Washburn, 1986. Genetic control of primary sex determination in mice. *Ann. Rev. Genet.* 20:327–360.

Fichman, K. R., B. R. Migeon, and C. J. Migeon, 1980. Genetic disorders of male sexual differentiation. In H. Harris and K. Hirschhorn, eds. *Advances in Human Genetics,* vol. 10. Plenum Press, New York.

German, J., J. L. Simpson, R. S. K. Chaganti, R. L. Summit, L. B. Reid, and I. R. Markats, 1978. Genetically determined sex-reversal in 46,XY humans. *Science* 202:53–56.

Lubs, H. A., and F. H. Ruddle, 1970. Chromosomal abnormalities in the human population: Estimation of rates based on New Haven newborn study. *Science* 169:495–497.

Lyon, M. F., 1962. Sex chromatin and gene action in mammalian X-chromosomes. *Am. J. Human Genet.* 14:135–148.

Lyon, M., B. M. Cattanach, and H. M. Charlton, 1981. Genes affecting sex determination in animals. In C. R. Auston and R. G. Edwards, eds. *Mechanisms of Sex Determination in Animals and Man.* Academic Press, New York.

McKusick, V. A., 1986. *Mendelian Inheritance in Man,* 7th ed. The Johns Hopkins University Press, Baltimore.

McKusick, V. A., 1986. The human gene map. *Clin. Genet.* 29:545–588.

Morgan, T. H., 1910. Sex limited inheritance in *Drosophila. Science* 32:120–122. Reprinted in J. A. Peters, ed., 1959. *Classic Papers in Genetics.* Prentice-Hall: Englewood Cliffs, N.J.

Morreale, S. J., G. J. Ruiz, J. R. Spotila, and E. A. Standora, 1982. Temperature-dependent sex determination. *Science* 216:1245–1247.

Nothiger, R., and M. Steinmann-Zwicky, 1985. Sex determination in *Drosophila. Trends in Genet.* 1:209–214.

Ohno, S., 1976. Major regulatory genes for mammalian sexual development. *Cell* 7:315–321.

Rick, C. M., and G. C. Hanna, 1943. Determination of sex in *Asparagus officinalis L. Am. J. Botany* 33:711–714.

Simpson, E., 1982. Sex reversal and sex determination. *Nature* 300:404.

Simpson, J. L., 1982. Abnormal sexual differentiation in humans. *Ann. Rev. Genet.* 16:193–224.

Summitt, R. L., and D. Bergsma, eds., 1978. *Sex Differentiation and Chromosomal Abnormalities.* Annual Review of Birth Defects, 1977. Proceedings of the 1977 Memphis Birth Defects Conference, Part C. Alan R. Liss, New York.

Warmke, H. W., 1946. Sex determination and sex balance in *Melandrium. Am. J. Botany* 33:648–660.

CHAPTER 8

Avery, O. T., C. M. MacLeod, and M. McCarty, 1944. Studies on the chemical nature of the substance inducing transformation of Pneumococcal types. *J. Exp. Med.* 79:137–158. Reprinted in J. A. Peters, ed., 1959. *Classic Papers in Genetics.* Prentice-Hall, Englewood Cliffs, N.J.

Bahr, G. F., 1977. Chromosomes and chromatin structure. In J. J. Yunis, ed., *Molecular Structure of Human Chromosomes,* Academic Press, New York.

Burlingame R. W., W. E. Love, B-C Wang, R. Hamlin, N-H. Xuong, and E. N. Moudrianakis, 1985. Crystallographic structure of the octameric histone core of the nucleosome at a resolution of 3.3 Å. *Science* 228:546–553.

Denhardt, D., and E. A. Faust, 1985. Eukaryotic DNA replication. *BioEssays* 2:148–153.

DuPraw, E. J., 1970. *DNA and Chromosomes.* Holt, Rinehart, Winston, New York.

Edenberg, H. J., and J. A. Huberman, 1975. Eukaryotic chromosome replication. *Ann. Rev. Genet.* 9:245–284.

Geis, I., 1983. Visualizing the anatomy of A, B and Z-DNAs. *J. Biomol. Struc. Dynam.* 1:581–591.

Freifelder, D., 1987. *Molecular Biology, A Comprehensive Introduction to Prokaryotes and Eukaryotes,* 2nd ed. Jones and Bartlett, Boston.

Gilbert, W., 1981. DNA sequencing and gene structure. *Science* 214:1305–1312.

Griffith, F., 1928. The significance of Pneumococcal types. J. Hygene 27:113–156.

Hershey, A. D., and M. Chase, 1952. Independent functions of viral protein and nucleic acid in growth of bacteriophage. *J. Gen. Physiol.* 36:39–56. Reprinted in G. S. Stent, ed., 1965. *Papers on Bacterial Viruses,* 2nd ed. Little, Brown, Boston.

Hotchkiss, R. D., and M. Gabor, 1970. Bacterial transformation, with special reference to recombination process. *Ann. Rev. Genet.* 4:193–224.

Judson, H. F., 1979. *The Eighth Day of Creation. The Makers of the Revolution in Biology.* Simon and Schuster, New York.

Kornberg, A., 1984. DNA replication. *Trends Biochem. Sci.* 9:122–124.

Kriegstein, H. J., and D. S. Hogness, 1974. Mechanism of DNA replication in *Drosophila* chromosomes: Structures of replication forks and evidence for bidirectionality. *Proc. Natl. Acad. Sci. (USA)* 71:135–139.

Lehman, I. R., 1974. DNA ligase: Structure, mechanism, and function. *Science* 186:790–797.

Leng, M., 1985. Left-handed Z-DNA. *Biochim. Biophys. Acta* 825:339–344.

Lewin, B., 1974. *Gene Expression, Bacterial Genomes,* vol 1. John Wiley, New York.

Lewin, B., 1978. *Gene Expression, Plasmids and Phages,* vol 3. John Wiley, New York.

Lewin, B., 1980. *Gene Expression, Eukaryotic Chromosomes,* vol 2. John Wiley, New York.

Lewis, B. J., J. W. Abrell, R. G. Smith, and R. C. Gallo, 1974. Human DNA polymerase III (R-DNA polymerase): Distinction from DNA polymerase I and reverse transcriptase. *Science* 183:867–869.

MACE, M. L., JR., Y. DASKAL, H. BUSCH, V. P. WRAY, and W. WRAY, 1977. Isolated metaphase chromosomes: Scanning electron microscope appearance of salt-extracted chromosomes. *Cytobios* 19:27–40.

MESELSON, M. S., and F. W. STAHL, 1958. The replication of DNA in *Escherichia coli. Proc. Nat. Acad. Sci. (USA)* 44:671–682.

OGAWA, T., and T. OKAZAKI, 1980. Discontinuous DNA replication. *Ann. Rev. Biochemistry* 49:421–457.

OZEKI, H., and H. IKEDA, 1968. Transduction mechanisms. *Ann. Rev. Genet.* 2:245–278.

SANGER, F., 1981. Determination of nucleotide sequences of DNA. *Science* 214:1205–1210.

SAYRE, A., 1975. *Rosalind Franklin and DNA.* W. W. Norton & Co, New York.

SHEININ, R., J. HUMBERT, and R. E. PEARLMAN, 1978. Some aspects of eukaryotic DNA replication. *Ann. Rev. Biochemistry* 47:277–316.

THOMASZ, A., 1969. Some aspects of the competent state in genetic transformation. *Ann. Rev. Genet.* 3:217–232.

WATSON, J. D., 1968. *The Double Helix.* Atheneum, New York.

WATSON, J. D., 1971. The regulation of DNA synthesis in eukaryotes. In D. M. Prescott, L. Goldstein, and E. McConkey, eds. *Advances in Cell Biology.* Appleton-Century-Crofts, New York.

WATSON, J. D., and F. H. C. CRICK, 1953. Molecular structure of nucleic acids. A structure for deoxyribonucleic acid. *Nature* 171:737–738. Reprinted in J. H. Taylor, ed., 1965. *Selected Papers on Molecular Genetics.* Academic Press, New York.

WATSON, J. D., and F. H. C. CRICK, 1953. Genetic implications of the structure of deoxyribonucleic acid. *Nature* 171:964–969. Reprinted in J. H. Taylor, ed., 1965. *Selected Papers on Molecular Genetics.* Academic Press, New York.

WATSON, J. D., N. H. HOPKINS, J. W. ROBERTS, J. A. STEITZ, and A. M. WEINER, 1987. *Molecular Biology of the Gene,* 4th ed. The Benjamin/Cummings Publishing Co., Inc., Menlo Park, CA.

WICKNER, S. H., 1978. DNA replication proteins of *Escherichia coli. Ann. Rev. Biochemistry* 47:1163–1191.

YUNIS, J. J., 1977. *Molecular Structure of Human Chromosomes.* Academic Press, New York.

ZINDER, N. D., and J. LEDERBERG, 1952. Genetic exchange in *Salmonella. J. Bacteriology* 64:679–699.

CHAPTER 9

ABELSON, J., 1979. RNA processing and the intervening sequence problem. *Ann. Rev. Biochemistry* 48:1035–1069.

BEADLE, G. W., and E. L. TATUM, 1941. Genetic control of biochemical reactions in *Neurospora. Proc. Natl. Acad. Sci. (USA)* 27:499–506. Reprinted in J. A. Peters, ed., 1959. Classic Papers in Genetics. Prentice-Hall, Englewood Cliffs, N.J.

BIRNSTIEL, M., M. BUSSLINGER, and K. STRUB, 1985. Transcription termination and 3′ processing: The end is in site. *Cell* 41:349–359.

BITTAR, E. E., ED., 1973. *Cell Biology in Medicine.* Wiley (Interscience Division), New York.

BRIMACOMBE, R., and W. STEGE, 1985. Structure and function of ribosomal RNA. *Biochem. J.* 229:1–17.

BRIMACOMBE, R., R. G. STOFFLER, and H. G. WITTMAN, 1978. Ribosome structure. *Ann. Rev. Biochemistry* 47:217–249.

BUSCH, H., F. HIRSCH, K. K. GUPTA, M. RAO, W. SJLOHN, and B. C. WU, Structural and functional studies on the "5′-Cap": A survey method for mRNA. In W. E. Cohn and E. Volkin, eds. 1976. *Progress in Nucleic Acid Research and Molecular Biology,* vol. 19. Academic Press, New York.

CASKEY, C. TH., 1980. Polypeptide chain termination. *Trends Biochem. Sci.* 5:234–237.

CLARK, B., 1980. The elongation step of protein synthesis. *Trends Biochem. Sci.* 5:207–210.

CLARK-WALKER, G. D., 1973. Translation of messenger RNA. In P. R. Stewart and D. S. Letham, eds. *The Ribonucleic Acids.* Springer-Verlag, New York.

DOOLITTLE, R. F., 1985. Proteins. *Sci. American* 253:(4)88–99.

FREIFELDER, D., 1987. *Molecular Biology—A Comprehensive Introduction to Prokaryotes and Eukaryotes,* 2nd ed. Jones and Bartlet Publishers, Boston.

FURUICHI, Y., S. MUTHUKRISHNAN, J. TOMASZ, and A. J. SHATKIN, 1976. The 5′-terminal sequence ("Cap") of mRNAs. In W. E. Cohen and E. Volkin, *Progress in Nucleic Acid Research and Molecular Biology,* vol. 19. Academic Press, New York.

GREEN, M., 1986. Pre-mRNA splicing. *Ann. Rev. Genet.* 20:671–708.

HAMKALO, B., 1985. Visualizing transcription in chromosomes. *Trends Genet.* 1:255–260.

HARRIS, H., 1975. *The Principles of Human Biochemical Genetics,* 2nd ed. American Elsevier, New York.

HASELKORN, R., and L. B. ROTHMAN-DENES, 1973. Protein synthesis. *Ann. Rev. Biochemistry* 42:397–438.

HOWELLS, A. J., 1973. Messenger RNA. In P. R. Stewart and D. S. Letham, eds. *The Ribonucleic Acids.* Springer-Verlag, New York.

HUNT, T., 1980. The initiation of protein synthesis. *Trends Biochem. Science* 5:207–210.

KABACK, M. M., ED., 1977. *Tay-Sachs Disease: Screening and Prevention. Progress in Clinical and Biological Research,* vol. 18. Alan R. Liss, New York.

KIM, S. H., F. L. SUDDATH, G. J. QUIGLEY, A. MCPHERSON, J. L. SUSSMAN, A. H. J. WANG, N. C. SEEMAN, and A. RICH, 1974. Three-dimensional tertiary structure of yeast phenylalanine transfer RNA. *Science* 185:435–440.

LAKE, J., 1981. The ribosome. *Sci. American* 245:(2)84–97.

MADISON, J. T., G. A. EVERETT, and H. KUNG, 1966. Nucleotide sequence of a yeast tyrosine transfer RNA. *Science* 153:531–534.

MILLER, O. L., JR., and B. A. HAMKALO, 1972. Visualization of Genetic Transcription. In M. Sussman, ed. *Molecular*

Genetics and Developmental Biology. Prentice-Hall, Englewood Cliffs, N.J.

PATWARDHAN, S., G. KALTWASSER, P. R. DIMALRIA, and C. J. GOLDENBERG, 1985. Splicing of messenger RNA precursors. *BioEssays* 2:205–208.

POLYA, G. M., 1973. Transcription. In P. R. Stewart and D. S. Letham, eds. *The Ribonucleic Acids*. Springer-Verlag, New York.

RICH, A., and S. H. KIM, 1978. The three-dimensional structure of transfer RNA. *Sci. American* 238:(1)52–62.

RICH, A., and L. RAJBHANDARY, 1976. Transfer RNA: Molecular structure, sequence, and properties. *Ann. Rev. Biochemistry* 45:805–860.

SHARP, P. A., 1987. Splicing of messenger RNA precursors. *Science* 235:766–771.

CHAPTER 10

ADAMS, J. M., and M. R. CAPECCH, 1965. N-formylmethionine-RNA as the initiator of protein synthesis. *Proc. Nat. Acad. Sci. (USA)* 55:147–155.

BEAUDET, A. L., and C. T. CASKEY, 1972. Polypeptide chain termination. In L. Bosch, ed. *The Mechanism of Protein Synthesis and Its Regulation*. American Elsevier, New York.

BRENNER, S., A. O. W. STRETTON, and S. KAPLAN, 1965. Genetic code: The "nonsense" triplets for chain termination and their suppression. *Nature* 206:994–998.

COREY, S., K. R. MARCKER, S. K. DUBE, and B. F. C. CLARK, 1968. Primary structure of a methionine transfer RNA from *Escherichia coli*. *Nature* 220:1039–1040.

CRICK, F. H. C., 1963. On the genetic code. *Science* 139:461–464.

CRICK, F. H. C., 1966. Codon-anticodon pairing: The wobble hypothesis. *J. Molec. Biol.* 19:548–555.

CRICK, F. H. C., L. BARNETT, S. BRENNER, and R. J. WATTS-TOBIN, 1961. General nature of the genetic code for proteins. *Nature* 192:1227–1232.

DUBE, S. K., K. R. MARKER, B. F. C. CLARK, and S. CORY, 1968. Nucleotide sequence of N-formyl-methionine-transfer RNA. *Nature* 218:232–233.

FREIFELDER, D., 1987. *Molecular Biology—A Comprehensive Introduction to Prokaryotes and Eukaryotes*, 2nd ed. Jones and Bartlett Publishers, Boston.

FRISCH, L., ED., 1967. The genetic code. *Cold Spring Harbor Symposia Quant. Biol.* 31 (1966).

GAREN, A., 1968. Sense and nonsense in the genetic code. *Science* 160:149–159.

GRUNBERG-MANAGO, M., and S. OCHOA, 1955. Enzymatic synthesis and breakdown of polynucleotides: Polynucleotide phosphorylase. *J. Am. Chem. Soc.* 778:3165–3166.

HOSMAN, D., D. GILLESPIE, and H. F. LODISH, 1972. Removal of formyl-methionine residue from nascent bacteriophage f2 protein. *J. Molec. Biol.* 65:163–166.

HOYER, B. H., B. J. MCCARTHY, and E. T. BOLTON, 1964. A molecular approach in the systematics of higher organisms. *Science* 144:959–967.

JUKES, T. H., 1983. Evolution of the amino acid code. In *Evolution of Genes and Proteins*, M. Nei and R. K. Koehn eds. Sinauer Associates Inc, Sunderland, Mass.

LEDER, P., and M. W. NIRENBERG, 1964. RNA code words and protein synthesis III. On the nucleotide sequence of a cysteine and a leucine RNA code word. *Proc. Natl. Acad. Sci. (USA)* 52:1521–1529.

NIRENBERG, M., and P. LEDER, 1964. RNA code words and protein synthesis. *Science* 145:1319–1407.

NIRENBERG, M. W., and J. H. MATTHAEI, 1961. The dependence of cell-free protein synthesis in *E. coli* upon naturally occurring or synthetic polyribonucleotides. *Proc. Nat. Acad. Sci. (USA)* 47:1588–1602.

REVEL, M., 1972. Polypeptide chain initiation: The Role of ribosomal protein factors and ribosomal subunits. In L. Bosch, ed. *The Mechanism of Protein Synthesis and Its Regulation*. American Elsevier, New York.

SARABHAI, A. S., A. O. W. STRETTON, and S. BRENNER, 1964. Co-linearity of the gene with the polypeptide chain. *Nature* 201:13–17.

SHAW, D. C., J. E. WALKER, F. D. NORTHROP, B. G. BARRELL, G. N. GODSON, and J. C. FIDDERS, 1978. Gene K, a new overlapping gene in bacteriophage G4. *Nature* 272:510–515.

SIBLEY C. G., and J. E. AHLQUIST, 1984. The phylogeny of the hominoid primates, as indicated by DNA-DNA hybridization. *J. Mol. Evol.* 20:2–15.

WHITFIELD, H., 1972. Suppression of nonsense, frameshift, and missense mutation. In L. Bosch, ed. *The Mechanism of Protein Synthesis and Its Regulation*. American Elsevier, New York.

WOESE, C. R., 1967. *The Genetic Code—the Molecular Basis for Genetic Expression*. Harper and Row, New York.

WOESE, C. R., 1970. The problem of evolving a genetic code. *BioScience* 20:471–485.

YANOFSKY, C. 1967. Structural relationships between gene and protein. *Ann. Rev. Genet.* 1:117–138.

YANOFSKY, C., G. R. DRAPEAU, J. R. GUEST, and B. C. CARLTON, 1967. The complete amino acid sequence of the tryptophan synthetase A protein (a subunit) and its colinear relationship with the genetic map of the A gene. *Proc. Natl. Acad. Sci. (USA)* 57:296–298.

ZINDER, N. D., D. L. ENGLEHARDT, and R. E. WEBSTER, 1966. Punctuation in the genetic code. *Cold Spring Harbor Symposia Quant. Biol.* 31:251–256.

CHAPTER 11

BENZER, S., 1955. Fine structure of a genetic region in bacteriophage. *Proc. Natl. Acad. Sci. (USA)* 41:344–354.

BENZER, S., 1961. On the topography of the genetic fine structure. *Proc. Natl. Acad. Sci. (USA)* 47:403–415.

BENZER, S., 1962. The fine structure of the gene. *Sci. American* 206:(1)70–84.

Benzer, S., and E. Freese, 1958. Induction of specific mutations with 5-bromouracil. *Proc. Natl. Acad. Sci. (USA)* 44:112–119. Reprinted in G. S. Stent, ed., 1965. *Papers on Bacterial Viruses,* 2nd ed. Little Brown, Boston.

Freifelder, D., 1987. *Molecular Biology, A Comprehensive Introduction to Prokaryotes and Eukaryotes,* 2nd ed. Jones and Bartlett Publishers, Boston.

Gilbert, W., 1985. Genes-in-pieces revisited. *Science* 228:823–824.

Lewin, B., 1987. *Genes III.* John Wiley, New York.

Manitais, T., E. F. Fritsch, J. Lauer, and R. M. Lawn, 1980. The molecular genetics of human hemoglobins. *Ann. Rev. Genet.* 14:145–178.

Stadler, D. R., 1973. The mechanism of intragenic recombination. *Ann. Rev. Genet.* 7:113–127.

Stamatoyannopoulos, G., 1972. The molecular basis of hemoglobin disease. *Ann. Rev. Genet.* 6:47–70.

Tonegawa, S., 1985. The molecules of the immune system. *Sci. American* 253:(4)122–131.

CHAPTER 12

Amabis, J. M., and D. Cabral, 1970. RNA and DNA puffs in polytene chromosomes of *Rhynchosciara:* Inhibition by extirpation of prothorax. *Science* 169:692–694.

Ames, B. N., and P. E. Hartman, 1963. The histidine operon. *Cold Spring Harbor Symposia Quant. Biol.* 28:349–365. Reprinted in E. A. Adelberg, ed., 1966. *Papers on Bacterial Genetics.* Little, Brown, Boston.

Angelier, N., M. Paintrand, A. Lavaud, and J. P. Lechaire, 1984. Scanning electron microscopy of amphibian lampbrush chromosomes. *Chromosoma* 89:243–253.

Beckwith, J. R., 1967. Regulation of the lac operon. *Science* 156:596–604.

Beckwith, J., and P. Rossow, 1974. Analysis of genetic regulatory mechanisms. *Ann. Rev. Genet.* 8:1–13.

Beckwith, J. R., and D. Zipser, eds., 1970. *The Lactose Operon.* Cold Spring Harbor Laboratory, Cold Spring Harbor, N.Y.

Dickson, R. C., J. Abelson, W. M. Barnes, and W. S. Reznikoff, 1975. Genetic regulation: The lac control region. *Science* 187:27–35.

Englesberg, E., and G. Wilcox, 1974. Regulation: Positive control. *Ann. Rev. Genet.* 8:219–242.

Gehring, W. J., 1985. The molecular basis of development. *Sci. American* 253:(4)153–162.

Gehring, W. J., and Y. Hiromi, 1986. Homeotic genes and the homeobox. *Ann. Rev. Genet.* 20:147–174.

Jacob, F., and J. Monod, 1961. Genetic regulatory mechanisms in the synthesis of proteins. *J. Molec. Biology* 3:318–356.

Jacob, F., D. Perrin, C. Sanchez, and J. Monod, 1960. The operon: A group of genes whose expression is coordinated by an operator. *Compt. Rend. Acad. Sci.* 250:1727–1729.

Lewin, B., 1987. *Genes III.* John Wiley, New York.

Maclean, N., 1976. *Control of Gene Expression.* Academic Press, New York.

O'Malley, B. W., H. C. Towle, and R. J. Schwartz, 1977. Regulation of gene expression in eukaryotes. *Ann. Rev. Genet.* 11:239–275.

Pardee, A. B., F. Jacob, and J. Monod, 1959. The genetic control and cytoplasmic expression of "inducibility" in the synthesis of B-galactosidase by *E. coli. Jour. Molec. Biol.* 1:165–178.

Reznikoff, W. S., 1972. The operon revisited. *Ann. Rev. Genet.* 6:133–156.

Ruddle, F. H., C. P. Hart, and W. McGinnis, 1985. Structural and functional aspects of the mammalian homeobox sequences. *Trends Genet.* 1:48–50.

von Holt, C., 1985. Histones in perspective. *BioEssays* 3:120–124.

Wang, T. Y., N. C. Kostraba, and R. S. Newman, 1976. Selective transcription of DNA mediated by nonhistone proteins. In W. E. Cohn and E. Volkin, eds., *Progress in Nucleic Acid Research and Molecular Biology.* Academic Press, New York.

Wilcox, G., K. J. Clemetson, P. Clearya, and E. Englesberg, 1974. Interaction of the regulatory gene product with the operator site in the L-arabinose operon of *Escherichia coli. J. Molec. Biology* 85:589–602.

CHAPTER 13

Ames, B. N., 1979. Identifying environmental chemicals causing mutations and cancer. *Science* 204:587–593.

Bishop, J. M., 1982. Oncogenes. *Sci. American* 246:(1)80–93.

Bishop, J. M., 1985. Trends in oncogenes. *Trends Genet.* 1:245–249.

Bishop, J. M., 1987. The molecular genetics of cancer. *Science* 235:305–311.

Cole, M., 1986. The myc oncogene: Its role in transformation and differentiation. *Ann. Rev. Genet.* 20:361–384.

Croce, C. M., and G. Klein, 1985. Chromosome translocations and human cancer. *Sci. American* 252:(3)54–73.

Crow, J. F., and C. Denniston, 1985. Mutation in human populations. *Adv. Human Genet.* 14:59–216.

Doring, H.-P. 1985. Plant transposable elements. *Bioessays* 3:164–166.

Doring, H.-P., and Starlinger P., 1986. Molecular genetics of transposable elements in plants. *Ann. Rev. Genet.* 20:175–200.

Drake, J. W., 1970. *The Molecular Basis of Mutation.* Holden-Day, San Francisco.

Fedoroff, N., 1984. Transposable genetic elements in maize. *Sci. American* 250:(6)84–98.

Hollaender, A., 1971. *Chemical Mutagens.* Plenum Press, New York.

Howard-Flanders, P., 1981. Inducible repair of DNA. *Sci. American* 245:(5)72–80.

HUNTER, T., 1984. The proteins of oncogenes. *Sci. American* 251:(2)70–79.

KLECKNER, N., 1981. Transposable genetic elements. *Ann. Rev. Genet.* 15:341–404.

KNUDSON, A. G., 1986. Genetics of human cancer. *Ann. Rev. Genet.* 20:327–360.

MCCLINTOCK, B., 1951. Chromosome organization and genetic expression. *Cold Spring Harbor Symp. Quant. Biol.* 16:13–47.

MCCLINTOCK, B., 1965. The control of gene action in maize. *Brookhaven Symp. Biol.* 18:162–184.

MCKUSICK, V. A., 1986. *Mendelian Inheritance in Man,* 7th ed. Johns Hopkins University Press, Baltimore.

MULLER, H. J., 1927. Artificial transmutation of the gene. *Science* 66:84–87.

REIF, A. E., 1981. The causes of cancer. *Am. Scientist* 69:437–447.

STADLER, L. J., 1928. Mutations in barley induced by X rays and radium. *Science* 68:186–187.

WEINBERG, R. A., 1983. A molecular basis of cancer. *Sci. American* 249:(5)126–142.

WHITKIN, E. M., 1976. Ultraviolet mutagenesis and inducible DNA repair in *Escherichia coli. Bact. Rev.* 40:869–907.

CHAPTER 14

BURNS, G. W., 1942. The taxonomy and cytology of *Saxifraga pensylvanica* L. and related forms. *Am. Midl. Nat.* 28:127–160.

BUTLER, L. J., P. E. CONEN, and B. ERKMAN-BALIS, 1966. Frequency and occurrence of chromosomal syndromes. 1. D-trisomy, 2. E-trisomy. *Am. J. Human Genet.* 18:374–398.

DAY, R. W. 1966. The epidemiology of chromosome aberrations. *Am. J. Human Genet.* 18:70–80.

ERIKSON, J. D., 1979. Paternal age and Down's syndrome. *Am. J. Human Genet.* 31:489–497.

EDWARDS, J. H., D. G. HARNDEN, A. H. CAMERON, V. M. CROSSE, and O. H. WOLFF, 1960. A new trisomic syndrome. *Lancet* 1:787–789.

GERMAN, J., 1970. Studying human chromosomes today. *Am. Scientist* 58:182–201.

GROUCHY, J. DE, and C. TURLEAU, 1977. *Clinical Atlas of Human Chromosomes.* John Wiley, New York.

GROUCHY, J. DE, C. TURLEAU, M. ROUBIN, and F. CHAVIN-COLIN, 1973. Chromosomal evolution of man and the primates *(Pan troglodytes, Gorilla gorilla, Pongo pygmaeus). Nobel Symposium 23: Chromosome Identification.* Academic Press, New York.

GUPTA, P. K., and P. M. PRIYADARSHAN, 1982. Triticale: Present status and future prospects. *Adv. Genet.* 21:256–346.

HAMERTON, J. L., 1971. *Human Cytogenetics,* vol 2. Academic Press, New York.

HASSOLD, T., P. JACOBS, J. KLEIN, Z. STEIN, and D. WARBURTON, 1980. Effect of maternal age on autosomal trisomies. *Ann. Human Genet.* 44:29–36.

HASSOLD, T. J., and P. JACOBS, 1984. Trisomy in man. *Ann. Rev. Genet.* 18:69–97.

HASSOLD, T., D. WARBURTON, J. KLINE, and Z. STEIN, 1984. The relationship of maternal age and trisomy among trisomic spontaneous abortions. *Am. J. Human Genet.* 36:1349–1356.

HOOK, E. B., R. LEHRKE, A. ROESNER, and J. J. YUNIS, 1965. Trisomy-18 in a 15-year old female. *Lancet* 2:910–911.

HUSKINS, C. L., 1930. The origin of *Spartina townsendii. Genetica* 12:531–538.

JUBERG, R. C., E. F. GILBERT, and R. S. SALISBURY, 1970. Trisomy C in an infant with polycystic kidneys and other malformations. *J. Pediatrics* 76:598–603.

KAJII, T., N. HIIKAWA, A. FERRIER, and H. TAKAHARA, 1973. Trisomy in abortion material, *Lancet* 2:1214.

KARPECHENKO, G. D., 1928. Polyploid hybrids of *Raphanus sativus* L. × *Brassica oleracea* L. *Ztschr. Ind. Abst. Vererb.* 48:1–83.

LEWIS, W. H., ED., 1980. *Polyploidy: Biological Relevance.* Plenum Press, New York.

MCCLURE, H. M., K. H. BELDEN, W. A. PIEPER, and C. B. JACOBSEN, 1969. Autosomal trisomy in a chimpanzee: Resemblance to Down's syndrome. *Science* 165:1010–1011.

MCCREANOR, H. R., F. M. O'MALLEY, and R. A. REID, 1973. Trisomy in abortion material. *Lancet* 2:972–973.

MANING, C. H., and H. O. GOODMAN, 1981. Parental origin of chromosomes in Down's syndrome. *Human Genet.* 59:101–103.

MARCHANT, C. J., 1963. Corrected chromosome numbers for *Spartina* × *townsendii* and its parent species. *Nature* 199:929.

MIOLA, E. S., 1987. Down syndrome: Update for practitioners. *Ped. Nursing* 13:6–10.

PANTELAKIS, S. N., O. M. CHRYSSOSTOMIDOU, D. ALEXIOU, T. VALAES, and S. A. DOXIADIS, 1970. Sex chromatin and chromosome abnormalities among 10,412 liveborn babies. *Arch. Dis. Childhood* 45:87–92.

PATAU, K., D. W. SMITH, E. THERMAN, S. L. INHORN, and H. P. WAGNER, 1960. Multiple congenital anomaly caused by an extra autosome. *Lancet* 1:790–793.

PENROSE, L. S., and J. D. A. DELHANATY, 1961. Triploid cell cultures from a macerated fetus. *Lancet* 1:1261–1262.

SCHINDLER, A.-M., and K. MIKAMO, 1970. Triploidy in man: Report of a case and a discussion in etiology. *Cytogenetics* 9:116–130.

SIMMONDS, N. W., ED., 1976. *Evolution in Crop Plants.* Longman, London.

STEBBINS, G. L., 1966. Chromosome variation and evolution. *Science* 152:1463–1469.

STEBBINS, G. L., 1971. *Chromosome Evolution in Higher Plants.* Addison-Wesley, Reading, Mass.

SWANSON, C. P., T. MERZ, and W. J. YOUNG, 1981. *Cyto-*

genetics. The Chromosome in Division, Inheritance, and Evolution, 2nd ed. Prentice-Hall, Englewood Cliffs, N.J.

YUNIS, J. J., ED., 1977. *New Chromosomal Syndromes.* Academic Press, New York.

CHAPTER 15

BENEDICT, W. F., 1976. Morphological transformation and chromosome aberrations produced by two hair dye components. *Nature* 260:368–369.

BRIDGES, C. B., 1936. The bar "gene," a duplication. *Science* 83:210–211. Reprinted in J. A. Peters, ed., 1959. *Classic Papers in Genetics,* Prentice-Hall, Englewood Cliffs, N.J.

BUTLER, L. J., A. V. PALMER, T. SPENCER, R. TABIOS-BROADWAY, and W. J. WALL, 1987. A new interstitial deletion of chromosome No. 4 del(4) (q22:q25). *Clin. Genet.* 31:199–205.

DISHOTSKY, N. I., W. D. LOUGHMAN, R. E. MOGAR, and W. R. LIPSCOMB, 1971. LSD and genetic damage. *Science* 172:431–440.

GROUCHY, J. DE, and C. TURLEAU, 1977. *Clinical Atlas of Human Chromosomes.* John Wiley, New York.

GROUCHY, J. DE, C. TURLEAU, M. ROUBIN, and F. CHAVIN-COLIN, 1973. Chromosomal evolution of man and the primates *(Pan troglodytes, Gorilla gorilla, Pongo pygmaeus). Nobel Symposium 23: Chromosome Identification,* Academic Press, New York.

HSU, L. Y., L. STRAUSS, and K. HIRSCHORN, 1970. Chromosome abnormality in offspring of LSD user. *J. Am. Med. Assn.* 211:987–990.

HUNGERFORD, D. A., K. M. TAYLOR, C. SHAGASS, G. U. LABADIE, G. B. BALABAN, and G. R. PATON, 1968. Cytogenetic effects of LSD therapy in man. *J. Am. Med. Assn.* 206:2287–2291.

JOHNSON, G. A., and S. M. JALAL, 1973. DDT-induced chromosome damage in mice. *J. Heredity* 64:7–8.

LEUCHTENBERGER, C., R. LEUCHTENBERGER, U. RITTER, and N. INNUI, 1973. Effects of marijuana and tobacco smoke on DNA and chromosomal complement in human lung explants. *Nature* 242:403–404.

LEUCHTENBERGER, C., R. LEUCHTENBERGER, and A. SCHNEIDER, 1973. Effects of marijuana and tobacco smoke on human lung physiology. *Nature* 241:137–139.

LONG, S. Y., 1972. Does LSD induce chromosomal damage and malformations? A review of the literature. *Teratology* 6:75–90.

NUSSBAUM, R. L., and D. H. LEDBETTER, 1986. Fragile × syndrome: A unique mutation in man. *Ann. Rev. Genet.* 20:109–145.

SINCOCK, A., and M. SEABRIGHT, 1975. Induction of chromosome changes in chinese hamster cells by exposure to asbestos fibers. *Nature* 257:56–58.

SWANSON, C. P., T. MERZ, and W. J. YOUNG, 1981. *Cytogenetics: The Chromosome in Division, Inheritance and Evolution,* 2nd ed. Prentice-Hall, Englewood Cliffs, N.J.

CHAPTER 16

BIRKY, C. W., 1978. Transmission genetics of mitochondria and chloroplasts. *Ann. Rev. Genet.* 12:471–512.

BOGORAD, L., ET AL., 1983. The organization and expression of maize plastid genes. In L. D. Owens, ed., *Genetic Engineering: Applications to Agriculture.* Granada Press, London.

BUCHER, T., W. NEUPERT, W. SEBALD,and S. WERNER, 1976. *Genetics and Biogenesis of Chloroplasts and Mitochondria.* North-Holland Publishing, New York.

CHIANG, K. S., 1976. On the search for a molecular mechanism of cytoplasmic inheritance: Past controversy, present progress and future outlook. In T. Bucher, W. Neupert, W. Sebald, and S. Werner, eds., *Genetics and Biogenesis of Chloroplasts and Mitochondria.* North-Holland Publishing, New York.

ELLIS, R. J., 1981. Chloroplast proteins: Synthesis, transport and assembly. *Ann. Rev. Plant Physiology* 32:111–137.

FALKOW, S., E. M. JOHNSON, and L. S. BARON, 1967. Bacterial conjugation and extrachromosomal elements. *Ann. Rev. Genet.* 1:87–116.

FERRIS, S. D., A. C. WILSON, and W. M. BROWN, 1981. Evolutionary tree for apes and humans based on cleavage maps of mitochondrial DNA. *Proc. Natl. Acad. Sci. (USA)* 78:2432–2436.

GILLHAM, N. W., 1978. *Organelle Heredity.* Raven Press, New York.

GREEN, B. R., and H. BURTON, 1970. *Acetabularia* chloroplast DNA: Electron microscopic visualization. *Science* 168:981–982.

GREEN, B. R., and M. P. GORDON, 1966. Replication of chloroplast DNA of tobacco. *Science* 152:1071–1074.

GRIVELL, L., 1983. Mitochondrial DNA. *Sci. American* 248:(3)78–89.

KUNG, S. D., Y. S. ZHU, and G. F. SHEN, 1982. *Nicotiana* chloroplast genome III. Chloroplast DNA evolution. *Theor. Appl. Genet.* 61:73–79.

LEWIN, B., 1977. *Gene Expression III Plasmids and Phages.* John Wiley, New York.

MULLIGAN, R. M., and V. WALBOT, 1986. Gene expression and recombination in plant mitochondrial genomes. *Trends Genet.* 2:263–266.

PALMER, J. D., 1985. Comparative organization of chloroplast genomes. *Ann. Rev. Genet.* 19:325–354.

PREER, J. R., JR., 1971. *Extrachromosomal inheritance. Ann. Rev. Genet.* 5:361–406.

SAGER, R., 1976. The circular diploid model of chloroplast DNA in *Chlamydomonas.* In T. Bucher, W. Neupert, W. Sebald, and S. Werner, eds., *Genetics and Biogenesis of Chloroplasts and Mitochondria.* North-Holland Publishing, New York.

SAGER, R., 1985. Chloroplast genetics. *BioEssays* 3:180–184.

SAGER, R., and Z. RAMANIS, 1963. The particulate nature of nonchromosomal genes in *Chlamydomonas. Proc. Natl. Acad. Sci. (USA)* 50:260–268.

CHAPTER 17

ALTMAN, P. L., and D. S. DITTMER, EDS., 1962, 1972, *Biology Data Book*. Federation of American Societies for Experimental Biology, Bethesda, Maryland.

AMOS, D. B., and D. D. KOSTYU, 1980. HLA-A central immunological agency of man. In H. Harris and K. Hirshhorn, eds., *Advances in Human Genetics*, vol. 10. Plenum Press, New York.

BACH, F. H., 1976. Genetics of transplantation: The major histocompatibility complex. *Ann. Rev. Genet.* 10:319–339.

ERSKINE, A. G., and W. W. SOCHA, 1978. *Principles and Practices of Blood Grouping*, 2nd ed. Mosby, St. Louis.

FISHER, R. A., 1947. The rhesus factor: A study in scientific method. *Am. Scientist* 35:95–103.

GREEN, M. M., 1961. Phenogenetics of the lozenge locus in *Drosophila melanogaster*. II. Genetics of lozenge-krivshenko (lzk). *Genetics* 46:1169–1176.

GREEN, M. M., 1963. Pseudoalleles and recombination in *Drosophila*. In W. J. Burdette, ed. *Methodology in Basic Genetics*. Holden-Day, San Francisco.

GREEN, M. M., and K. C. GREEN, 1949. Crossing over between alleles at the lozenge locus in *Drosophila melanogaster*. *Proc. Natl. Acad. Sci. (USA)* 35:586–591.

HEIKEN, A., and M. RASMUSSON, 1966. Genetical studies on the Rh blood group system. *Hereditas* 55:192–212.

LANDSTEINER, K., 1901. Uber Agglutinationserscheinungen Normlalen Menschlichen Blutes. *Wien Klin. Wochenschr.* 14:1132–1134. Reprinted (in English) in S. H. Boyer, ed., 1963. *Papers on Human Genetics*, Prentice-Hall, Englewood Cliffs, N.J.

LANDSTEINER, K., and P. LEVINE, 1927. A new agglutinable factor differentiating individual human bloods. *Proc. Soc. Exp. Biol. N.Y.* 24:600–602.

LANDSTEINER, K., and A. S. WIENER, 1940. An agglutinable factor in human blood recognized by immune sera from Rhesus blood. *Proc. Soc. Exp. Biol. Med. N.Y.* 43:223.

LEWIS, E. B., 1952. The pseudoallelism of white and apricot in *Drosophila melanogaster*. *Proc. Natl. Acad. Sci. (USA)* 38:953–961.

MOURANT, A. E., 1954. *The Distribution of Human Blood Groups*. Blackwell Scientific Publications, Oxford, England.

PEARSON, G., J. D. MANN, J. BENSEN, and R. W. BULL, 1979. Inversion duplication of chromosome 6 with trisomic codominant expression of HLA antigens. *Am. J. Human Genet.* 31:29–34.

RACE, R. R., and R. SANGER, 1968. *Blood Groups in Man*, 5th ed. F. A. Davis Company, Philadelphia.

RYDER, L. P., A. SVEJGAARD, and J. DAUSSET, 1980. Genetics of HLA disease association. *Ann. Rev. Genet.* 15:169–187.

STEINBERG, A. G., 1965. Evidence for a mutation or crossing-over at the Rh locus. *Vax Sang.* 10:721.

WALSH, R. H., and C. MONTGOMERY, 1947. A new isoagglutinin subdividing the MN blood group system. *Nature* 160:504.

WATKINS, W. M., 1980. Biochemistry and genetics of the ABO, Lewis, and P blood group systems. In H. Harris and K. Hirschhorn, eds., *Advances in Human Genetics*, vol 10. Plenum Press, New York.

YOSHIDA, A., 1982. Biochemical genetics of the human blood group ABO system. *Am. J. Human Genet.* 34:1–14.

CHAPTER 18

BODMER, W. F., and L. L. CAVALLI-SFORZA, 1970. Intelligence and race. *Sci. American* 223:(4)19–29.

BRUES, A. M., 1946. A genetic analysis of human eye color. *Am. J. Phys. Anthropology* (new series) 4:1–36.

CHAI, C. K., 1970. Genetic basis of leukocyte production in mice. *J. Heredity* 61:61–71.

DAVENPORT, C. B., 1913. Heredity of skin color in negro–white crosses. *Carnegie Inst. Washington Pub.* 188:1–106.

DAVENPORT, G. C., and C. B. DAVENPORT, 1910. Heredity of skin pigmentation in man. *Am. Naturalist* 44:641–672.

FALCONER, D. S., 1981. *Introduction to Quantitative Genetics*, 2nd ed. Longman, New York.

HUGHES, B. O., 1944. The inheritance of eye color—brown and nonbrown. Contributions from the Laboratory of Vertebrate Biology, University of Michigan, No. 27:1–10.

LINDSTROM, E. W., 1924. A genetic linkage between size and color factors in the tomato. *Science* 60:182–183.

LINDSTROM, E. W., 1926. Hereditary correlation of size and color characters in tomatoes. *Iowa Agr. Exp. Sta. Research Bull. 93*.

LINDSTROM, E. W., 1929. Linkage of qualitative and quantitative genes in maize. *Am. Naturalist* 63:31–327.

MATHER, K., 1954. The genetic units of continuous variation. *Proc. IX International Cong. Genet.* Part I:106–123.

MATHER, K., and J. L. JINKS, 1971. *Biometrical Genetics*. Chapman, London.

NILSSON-EHLE, H., 1909. *Lunds Univ. Arsskrift N. F. Avd. 2: Bd. 5*.

PUNNETT, R. C., 1923. *Heredity in Poultry*. Macmillan, New York.

THOMPSON, J. N., JR., and J. M. THODAY, EDS., 1979. *Quantitative Genetic Variation*. Academic Press, New York.

CHAPTER 19

BLAKESLEE, A. F., and T. N. SALMON, 1935. Genetics of sensory thresholds: Individual taste reactions for different substances. *Proc. Natl. Acad. Sci. (USA)* 21:84–90.

CAVALLI-SFORZA, L. L., 1974. The genetics of human populations. *Sci. American* 231:(3)80–89.

GLASS, H. B., M. S. SACKS, E. F. JAHN, and C. HESS. 1952. Genetic drift in a religious isolate: An analysis of the causes of variation in blood group and other gene frequencies in a small population. *Am. Naturalist* 86:145–159.

HENDRICK, P. W., 1983. *Genetics of Populations.* Van Nostrand, New York.

KABACK, M. M., ED., 1977. *Tay-Sachs Disease and Prevention.* Alan R. Liss, New York.

LEWONTIN, R. C., 1973. Population genetics. *Ann. Rev. Genet.* 7:1–17.

LEWONTIN, R. C., 1974. *The Genetic Basis of Evolutionary Change.* Columbia University Press, New York.

REED, T. E., 1969. Caucasian genes in American Negroes. *Science* 165:762–768.

SCHULL, W. J., and J. W. MACCLUER, 1968. Human genetics: Structure of population. *Ann. Rev. Genet.* 2:279–304.

STERN, C., 1943. The Hardy-Weinberg law. *Science* 97:137–138.

VOLPE, E. P., 1984. *Understanding Evolution.* Wm. C. Brown Publishers, Dubuque, Iowa.

WHITE, M. J. D., 1969. Chromosomal rearrangements and speciation in animals. *Ann. Rev. Genet.* 3:75–98.

CHAPTER 20

ABELSON, J., 1980. Recombinant DNA. A revolution in biology. *Science* 209:1319–1321.

ANDERSON, W. F., and E. G. DIACUMAKOS, 1981. Genetic engineering in mammalian cells. *Sci. American* 245(1):106–121.

BERG, P., 1981. Dissections and reconstructions of genes and chromosomes. *Science* 213:296–303.

CAVALIERI, L. F. *The Double-Edged Helix: Science in the Real World.* Columbia University Press, New York.

CHILTON, M. D., 1983. A vector for introducing new genes into plants. *Sci. American* 248:(6)50–59.

COHEN, S. N., 1975. The manipulation of genes. *Sci. American* 233:(1)24–33.

ELLIS, K. P., and K. E. DAVIES, 1985. An appraisal of the application of recombinant DNA techniques to chromosomal defects. *Biochem. J.* 226:1–11.

FREIFELDER, D., 1987. *Molecular Biology, A Comprehensive Introduction to Prokaryotes and Eukaryotes.* 2nd ed. Jones and Bartlett Publishers, Boston.

HOPWOOD, D. A., 1981. Genetic programming of industrial microorganisms. *Sci. American* 245:(3)91–102.

JACKSON, D. A., and S. P. STICH, EDS., 1979. *The Recombinant DNA Debate.* Prentice-Hall, Englewood Cliffs, N.J.

MOTULSKY, A. G., 1983. Impact of genetic manipulation on society and medicine. *Science* 219:135.

NATHANS, D., 1979. Restriction endonucleases, simian virus 40, and the new genetics. *Science* 206:903–909.

PALMITER, R. D., and R. L. BRINSTER, 1985. Transgenic mice. *Cell* 41:343–345.

PALMITER, R. D., and R. L. BRINSTER, 1986. Germ-line transformation of mice. *Ann. Rev. Genet.* 20:465–499.

PALMITER, R. D. ET AL., 1983. Metallothionein-human GH fusion genes stimulate growth of mice. *Science* 222:809–814.

RODRIGUES, R. L., and R. C. TAIT, 1983. *Recombinant DNA Techniques: An Introduction.* Addison-Wesley, Reading, Mass.

RUBIN, G. M., and A. C. SPRADLING, 1982. Genetic transformation of *Drosophila* with transposable element vectors. *Science* 218:348–353.

SEEBERG, P. H., ET AL., 1978. Synthesis of growth hormone by bacteria. *Nature* 276:795–798.

SHAH, D. M., ET AL., 1986. Engineering herbicide tolerance in transgenic plants. *Science* 233:478–481.

SINSHEIMER, R. L., 1977. Recombinant DNA. *Ann. Rev. Biochem.* 46:415–438.

SMITH, D. H., 1979. Nucleotide sequence specificity of restriction endonucleases. *Science* 205:455–462.

SPRADLING, A. C., and G. M. RUBIN, 1982. Transposition of cloned P elements into *Drosophila* germ line chromosomes. *Science* 218:341–347.

TORREY, J. G., 1985. The development of plant biotechnology. *Am. Scientist* 73:354–363.

WEINBERG, R. A., 1985. The molecules of life. *Sci. Am.* 253:(4)48–57.

WILLIAMSON, B., 1982. Gene therapy. *Nature* 298:416–418.

WU, R., ED., 1979. Recombinant DNA. *Methods in Enzymology,* vol. 68. Academic Press, New York.

GLOSSARY

Abortus. An aborted fetus.

Acentric. A chromatid or chromosome that lacks a centromere.

Acquired character. An alteration in function or form resulting from a response to environment; it is not heritable.

Acridine dye. Organic molecules that produce mutations by binding to DNA and causing insertions or deletions of bases.

Acrocentric. A chromosome with a nearly terminal centromere.

Activator. An autonomous controlling element that has the ability to excise itself from one chromosomal site and transpose to another site.

Adaptation. Adjustment in some way to the environment.

Adaptive value. The proportion of the progeny of a given genotype surviving to maturity (relative to that of another genotype) is its adaptive value (also called *fitness*). Expressed as a pure number between 0 and 1. Thus if only 60 percent of the progeny of genotype *aa* survive, whereas 100 percent of those of genotype *A*— survive, the adaptive value of *aa* is 0.6.

Adenine. A purine base occurring in DNA and RNA. Pairs normally with thymine in DNA.

Adenosine. The ribonucleoside containing the purine adenine.

Adenylic acid. The ribonucleotide containing the purine adenine.

Agglutinin. An antibody that produces clumping (agglutination) of the antigenic structure.

Albino. An individual characterized by absence of pigment. Ordinarily applied to animals whose skin, hair, and eyes are unpigmented and whose skin and eyes are pinkish because of blood vessels showing through. Also used to describe plants in which a particular pigment (usually the chlorophylls) is absent.

Aleurone layer. Outermost layer of endosperm in some seeds, rich in aleurone (proteinaceous) grains; typically triploid cells.

Allele (also **allelomorph**; adj., **allelic** or **allelomorphic**). One member of a pair or a series of genes that can occur at a particular locus on homologous chromosomes.

Allopolyploid. A polyploid having whole chromosome sets from different species.

Allosteric. Applied to any enzyme (or protein) having two or more nonoverlapping receptor (or attachment) sites.

Alpha helix. A typical helical secondary structure found in many proteins. One turn of the helix occurs every 3.6 amino acids.

Alternation of generations. The regularly occurring alternation of haploid (gametophyte) and diploid (sporophyte) phases in the life cycle of sexually reproducing plants.

Amber. The nonsense codon UAG.

Ambiguity. The coding for more than one amino acid by a given codon. See also *genetic code* and *codon.*

Ames test. A bacterial test for mutagenicity, used to screen suspected chemicals for their carcinogenicity.

Amino acid. A class of chemicals containing an amino (NH_2) group and a carboxyl (COOH) group, plus a side chain; the basic constructional unit of proteins.

Amino group. The $-NH_2$ chemical group.

Amniocentesis. A method for testing the genotype of an unborn child by withdrawal of a small amount of fluid from the amniotic sac; it contains sloughed fetal cells.

Amphidiploid. A tetraploid individual, that is, having two sets of chromosomes from each of two known ancestral species. An allotetraploid in which the species source of the two different genomes is clearly known. See also *allopolyploid.*

Anaphase. The stage of nuclear division characterized by movement of chromosomes from spindle equator to spindle poles. It begins with separation of the centromeres and closes with the end of poleward movement of chromosomes.

Aneuploidy. Variation in chromosome number by whole chromosomes, but less than an entire set; e.g., $2n + 1$ (trisomy), $2n - 1$ (monosomy).

Angstrom unit (Å). A measurement of length or distance, often used in describing intra- or intermolecular dimensions; equal to 1×10^{-10} meter and 1×10^{-1} nanometer (which see).

Anisogamy. Sexual reproduction involving gametes similar in morphology but differing in size.

Anther. The microsporangium of flowering plants; distal part of stamen in which microspores, and later pollen grains, are produced.

Antibiotic. Any chemical substance, elaborated by a living organism (usually a microorganism), that kills or inhibits growth of bacteria.

Antibody. A protein produced by B-lymphocytes (plasma cells) in response to a foreign antigen, which is capable of combining with the antigen.

Anticodon. The group of three nucleotides in transfer RNA that pairs complementarily with three nucleotides of messenger RNA during protein biosynthesis.

Antigen. Any substance, usually a protein, that causes antibody production when introduced into a living organism.

Antiparallel. A term used to refer to the opposite but parallel arrangement of the two sugar-phosphate strands in double-stranded DNA; the 5′-3′ orientation of one such strand is aligned along the 3′-5′ orientation of the other strand.

Ascospore. One of the asexual, monoploid (haploid) spores contained in the ascus of ascomycete fungi, such as *Neurospora* and the yeasts.

Ascus (pl., **asci**). The generally elongate, saclike meiosporangium in which ascospores are produced in ascomycete fungi.

Asexual reproduction. Any process of reproduction that does not involve fusion of cells. Vegetative reproduction, by portions of the vegetative body, is sometimes distinguished as a separate type.

Assortive mating. The situation where mating combinations between males and females is not random, but involves the tendency of males of one particular type to breed preferentially with females of a particular type.

ATP. Adenosine triphosphate, an energy-rich compound participating in energy-storing and energy-using reactions in the cell.

Attached-X. A strain of *Drosophila* in which the two X chromosomes of the female are permanently attached (symbolized $\widehat{XX}$) so that only XX and O eggs are produced.

Autoimmunity. The production of antibodies against one's own tissues.

Autonomous element. A transposable element capable of independent excision and transposition.

Autopolyploid. A polyploid all of whose sets of chromosomes are those of the same species.

Autosome. A chromosome not associated with the sex of the individual and therefore possessed in matching pairs by diploid members of both sexes.

Auxotroph. An individual that, unable to carry on some particular synthesis, requires supplementing of minimal medium by some growth factor.

B cells. White blood cells produced by the bone marrow, which produce antibodies.

Backcross. The cross of a progeny individual with one of its parents. See also *testcross.*

Bacteriophage. See *phage.*

Balbiani ring. An RNA-producing puff in the polytene chromosomes (which see) of the dipteran *Chironomus.*

Barr body. The inactive, densely staining, condensed X chromosome, generally found next to the nuclear membrane, in nuclei of somatic cells of XX females. The number of Barr bodies in such nuclei is one less than the total number of X chromosomes.

Base analog. A slightly modified purine or pyrimidine molecule that may substitute for the normal base in nucleic acid molecules.

Bivalent. A pair of synapsed homologous chromosomes.

Carboxyl group. An acidic chemical group,—COOH.

Carcinogen. Any physical or chemical agent that causes cancer.

Carotenoid. Any of a group of yellow, orange, or red pigments, in plants associated with chlorophylls, and in animal fat.

Carrier. A heterozygous individual. Term ordinarily used in cases of complete dominance.

C-bands. A method of staining chromosomes that involves treating with alkali and controlling the hydrolysis with buffered salt solutions. The technique emphasizes the centromeric regions.

Cell culture. A growth of cells *in vitro.*

Cell-free extract. A fluid extract of soluble materials of cells obtained by rupturing the cells and discarding particulate materials and any intact cells.

Centimeter (cm). 1×10^{-2} meter.

Centimorgan. A unit of relative distance between linked genes; equal to 1 percent crossing-over. See also *Morgan unit.*

Centriole. The central granule in the *centrosome* (which see).

Centromere. A specialized, complex region of the chromosome, consisting of one kinetochore (which see) for each sister chromatid.

Centrosome. A self-propagating cytoplasmic body present in animal cells and some lower plant cells, consisting of a centriole and astral rays at each pole of the spindle during nuclear division.

Chiasma, (pl., **chiasmata**). The visible connection or crossover between two chromatids seen during prophase-I of meiosis.

Chi-square test. A statistical test for determining the probability that a set of experimentally obtained values will be equaled or exceeded by chance alone for a given theoretical expectation.

Chloroplast. Cytoplasmic organelle containing several pigments, particularly the light-absorbing chlorophylls, and also DNA and polysomes (which see).

Chromatid. One of the two identical longitudinal halves of a chromosome, which shares a common centromere with a sister chromatid; this results from the replication of chromosomes during interphase.

Chromatin. Nuclear material comprising the chromosomes; the DNA-protein complex or nucleoprotein.

Chromomere. Small, stainable thickenings arranged linearly along a chromosome.

Chromosome. Nucleoprotein structures, generally more or less rodlike during nuclear division, the physical sites of nuclear genes arranged in linear order. Each species has a characteristic number of chromosomes, although individuals with fewer or more than this characteristic number occur, especially in plants.

Cis arrangement. Linkage of the dominants of two or more pairs of alleles on one chromosome and the recessives on the homologous chromosome.

Cis-trans test. A test to determine whether two mutant sites are in the same or different cistrons. In the cis configuration both mutations are on the same chromosome, and both wild-type sites are on the homolog. In the trans configuration, each homolog has one mutant and one wild type (e.g., $a^1\ a^+\ /\ a^+\ a^2$). In the cis configuration the wild-type phenotype occurs; in the trans, the mutant phenotype occurs.

Cistron. A segment of DNA specifying one polypeptide chain in protein synthesis. Under the concept of a triplet code, one cistron must contain three times as many nucleotide pairs as amino acids in the chain it specifies.

Clone. A group of cells or organisms, derived from a single ancestral cell or individual and all genetically alike.

Code. See *genetic code.*

Coding sequence. That portion of a eukaryotic gene separated by intervening sequences, which actually codes for the protein.

Codominance. The condition in heterozygotes where both members of an allelic pair contribute to phenotype, which is then a *mixture* of the phenotypic traits produced in either homozygous condition. In cattle the cross of red × white produces roan offspring whose coat consists of both red hairs and white hairs. Codominance differs from incomplete dominance.

Codon. A set of nucleotides that is specific for a particular amino acid in protein synthesis; generally agreed to consist of three nucleotides, the last of which, in the case of some amino acids, may be any of the four nucleotides.

Coincidence. The observed frequency of double crossovers, divided by their calculated or expected frequency. Expressed as a pure number; a measure of interference. In *positive* interference the coincidence is < 1; in *negative* interference the coincidence is > 1.

Col factors. Plasmids (which see) in bacteria conferring ability to produce colicins (antibiotic lipocarbohydrate-proteins).

Colinearity. Said of a genetic code (which see) in which the sequence of nucleotides corresponds to the sequence of amino acid residues in a polypeptide.

Commaless. Said of a genetic code (which see) in which successive codons (which see) are contiguous and not separated by noncoding bases or groups of bases.

Complementary base pairing. Base pairing in which A is always paired with T and G is always paired with C.

Complementation. The ability of linearly adjacent segments of DNA to supplement each other in phenotypic effect. Complementary genes, when present together, interact to produce a different expression of a trait.

Conditional lethal mutation. A mutation that is lethal under some conditions and not under other conditions.

Conjugation. Side-by-side association of two bodies, as of synapsed chromosomes in meiosis or of two organisms during sexual reproduction.

Consensus sequence. A sequence of nucleotides most often present in a DNA segment of interest.

Constant region. Part of the antibody molecule that has the same amino acid sequence among a large group of antibody types.

Constitutive enzyme. An enzyme that is constantly produced independent of environmental conditions.

Controlling element. A type of transposable element which cause highly unstable mutations.

Corepressor. A small molecule that inhibits transcription of genes in an operon by binding to the regulatory protein. The corepressor is usually the end product in a pathway.

cpDNA. The DNA located in chloroplasts.

Crossing-over. A process whereby genes are exchanged between nonsister chromatids of homologous chromosomes. *Unequal crossing-over* may occur, with the result that one chromatid receives a given gene twice, whereas the other chromatid lacks that gene entirely. Chiasmata (which see) are visible evidences of crossing-over.

Crossover. Said of a chromatid or gamete resulting from crossing-over.

Crossover unit. A crossover value of 1 percent between linked genes. See also *map unit.*

Cross-reacting material (CRM). A defective protein produced by a mutant gene, enzymatically inactive but antigenically similar to the wild-type protein.

C-terminus. That end of a peptide chain carrying the three alpha carboxyl group of the last amino acid in the sequence; written at the right end of the structural formula.

Cytidine. The ribonucleoside containing the pyrimidine cytosine.

Cytidylic acid. The ribonucleotide containing the pyrimidine cytosine.

Cytogenetics. Study of the cellular structures and mechanisms associated with genetics.

Cytokinesis. The division of the cytoplasm during cell division.

Cytology. The study of the structure and function of cells.

Cytosine. A pyrimidine base that occurs in DNA and RNA. Pairs with guanine in DNA.

Dalton. A unit that equals the mass of the hydrogen atom (1.67×10^{-24} g).

Deficiency. See *deletion*.

Degeneracy. Said of a genetic code (which see) in which a particular amino acid is coded for by more than one codon (which see).

Deletion. The loss of a part of a chromosome, usually involving one or more genes (rarely a portion of one gene).

Deletion loop. The loop formed by the nonsynapsing portion of an unaltered chromosome, caused by a deletion in the homologous chromosome.

Deoxyadenosine. The deoxyribonucleoside containing the purine adenine.

Deoxyadenylic acid. The deoxyribonucleotide containing the purine adenine.

Deoxycytidine. The deoxyribonucleoside containing the pyrimidine cytosine.

Deoxycytidylic acid. The deoxyribonucleotide containing the pyrimidine cytosine.

Deoxyguanosine. The deoxyribonucleoside containing the purine guanine.

Deoxyguanylic acid. The deoxyribonucleotide containing the purine guanine.

Deoxyribonucleic acid (DNA). A usually double-stranded, helically coiled, nucleic acid molecule, composed of deoxyribose-phosphate "backbones" connected by paired bases attached to the deoxyribose sugar; the genetic material of all living organisms and many viruses.

Deoxyribonucleoside. Portion of a DNA molecule composed of one deoxyribose molecule plus either a purine or a pyrimidine.

Deoxyribonucleotide. Portion of a DNA molecule composed of one deoxyribose phosphate bonded to either a purine or a pyrimidine.

Deoxyribose. The 5-carbon sugar of DNA.

Deviation. A departure from the expected or from the norm.

Dicentric. Said of a chromosome or chromatid with two centromeres.

Dihybrid. An individual heterozygous for two pairs of alleles; also said of a cross between individuals differing in two gene pairs.

Dioecious. Individuals producing either sperm or egg, but not both. In dioecious species, the sexes are separate. Compare with *monoecious*.

Diploid. An individual or cell with two complete sets of chromosomes.

Disjunction. The separation of homologous chromosomes during anaphase-I of meiosis.

DNA ligase. An enzyme catalyzing the covalent union of segments of an interrupted sugar-phosphate strand in double-stranded DNA.

DNA polymerase. Any one of a system of enzymes that catalyze the formation of DNA from deoxyribonucleotides, using one strand of DNA as a template.

DNase I. An endonuclease that digests either single-stranded or double-stranded DNA.

DNA topoisomerase. A class of enzymes that produce single- or double-stranded breaks in DNA during replication to release tension brought about by the separating of the double strands.

Domain. A sequence of amino acids within a protein that can be identified with a particular function.

Dominance. That situation in which one member of a pair of allelic genes expresses itself in whole (complete dominance) or in part (incomplete dominance) over the other member.

Dominant. Pertaining to that member of a pair of alleles that expresses itself in heterozygotes to the complete exclusion of the other member of the pair. Also, the trait produced by a dominant gene.

Drift. See *genetic drift*.

Duplication. A chromosomal aberration in which a segment of the chromosome bearing specific loci is repeated.

Dysgenic. Any effect or situation that is or tends to be harmful to the genetics of future generations.

Effector (molecule). A substance that combines with regulatory proteins, either activating or inactivating them with respect to their ability to bind to an operator site.

Embryo sac. The female gametophyte of a flowering plant. A large, thin-walled cell within the ovule, containing the egg and several other nuclei, within which the embryo develops after fertilization of the egg.

Endonuclease. An enzyme that breaks the internal phosphodiester bonds in a DNA molecule.

Endoplasmic reticulum. A double membrane system in the cytoplasm, continuous with the nuclear membrane and bearing numerous ribosomes.

Endosperm. A polyploid (in many species, triploid) food-storage tissue in many angiosperm seeds formed by fusion of two (or more) female cells and a sperm.

Enzyme. Any substance, protein in whole or in part, that regulates the rate of a specific biochemical reaction in living organisms.

Episome. A closed, circular molecule of DNA that may be present in a given (bacterial) cell, either separately in the cytoplasm or integrated into the chromosome.

The *F*, or fertility factor, in *Escherichia coli* is an example. See also *plasmid*.

Epistasis. The masking of the phenotypic effect of either or both members of one pair of alleles by a gene of a different pair. The masked gene is said to be hypostatic.

Equatorial plate. The figure formed at the spindle equator in nuclear division.

Equilibrium frequency. In a population, that gene frequency that varies nondirectionally about a mean by an amount described by the standard deviation (which see) under conditions of unchanging selection pressure and mutation rate, with no intermixing from other populations.

Euchromatin. Chromatin (which see) that is noncondensed during interphase and condensed during nuclear division, reaching a maximum in metaphase. The banded segments of the polytene chromosomes of *Drosophila* larval salivary glands contain euchromatin.

Eukaryote (eucaryote). Any organism or cell with a structurally discrete nucleus. Contrast with *prokaryote*.

Euploidy. Variation in chromosome number by whole sets or exact multiples of the monoploid (haploid) number, e.g., diploid, triploid. Euploids above the diploid level may be referred to collectively as polyploids.

Excision repair. Repair of DNA lesions by removal of the damaged segment and replacement with a newly synthesized corrected segment.

Exon. DNA sequences of a cistron that *are* transcribed into mRNA. See also *intron*.

Expressivity. The degree of phenotypic expression within one phenotype under a variety of environmental conditions.

F^- cell. A bacterial cell lacking the fertility factor (F); it acts as the receptor cell in conjugation.

F^+ cell. A bacterial cell having the fertility factor (F) as a plasmid in the cytoplasm.

F' cell. A bacterial cell, the fertility factor of which induces some chromosomal genes.

F factor. The fertility factor in the bacterium *Escherichia coli;* it is composed of DNA and must be present for a cell to function as a donor in conjugation. See also *episome*.

F_1. The first filial generation; the first generation resulting from a given cross.

F_2. The second filial generation; the generation resulting from interbreeding or selfing members of the F_1.

Fertility factor. See *F factor*.

Fitness. See *adaptive value*.

Fixation. Attainment of a gene frequency of 1.0.

Founder effect. Change in gene frequency in a new population founded by a very small sample from an existing population. In this case, the founding group does not carry the representative gene frequency of the original population.

Fragile site. A nonstaining gap in a chromosome, inherited in a Mendelian fashion, and susceptible to breakage.

Frame shift (reading frame shift). The shift in code reading that results from addition or deletion of nucleotides in any number other than 3 or multiples thereof.

Gamete. A protoplast that, in the process of sexual reproduction, fuses with another protoplast.

Gametophyte. In plants, the phase of the life cycle that reproduces sexually by gametes and characterized by having the reduced, usually haploid, chromosome number.

G-bands. Stained bands in chromosomes produced by treatment with trypsin and stained with Giemsa. Euchromatin stains lightly and heterochromatin darkly producing the characteristic G-bands.

Gene. The particulate determiner of a hereditary trait; a particular segment of a DNA molecule, generally located in the chromosome. See also *cistron, muton,* and *recon*.

Gene frequency. The proportion of one allele of a pair or series present in the population or a sample thereof; that is, the number of loci at which a gene occurs, divided by the number of loci at which it could occur, expressed as a pure number between 0 and 1.

Gene interaction. The coordinated effect of two or more genes in producing a given phenotypic trait.

Gene library. A large collection of cloning vectors containing a complete set of fragments of the genome of an organism.

Gene pool. The total of all genes in a population.

Generalized transduction. Phage-induced transduction whereby the phage can insert anywhere in the host chromosome.

Genetic code. The collection of base triplets of DNA and RNA carrying the genetic information by which proteins are synthesized in the cell.

Genetic drift. A change in gene frequency in a population. *Steady drift* is a directed change in frequency toward either greater or lower values; *random drift* is random fluctuation of gene frequencies caused by chance in mating patterns or to sampling errors.

Genetic equilibrium. Constancy of a particular gene frequency through successive generations.

Genetic load. The proportional reduction in average fitness, relative to an optimal genotype; the average number of lethals per individual in a population.

Genetic polymorphism. The continued occurrence in a population of two or more discontinuous genetic variants in frequencies that cannot be accounted for by recurrent mutation.

Genome. A complete set of chromosomes, or of chromosomal genes, inherited as a unit from one parent, or the entire genotype of a cell or individual.

Genotype (adj., **genotypic**). The genetic makeup or constitution of an individual, with reference to the traits under consideration, usually expressed by a symbol, e.g., "+," "D," "Dd," "str." Individuals of the same genotype breed alike. See *phenotype.*

Guanine. A purine base occurring in DNA and RNA. Pairs normally with cytosine in DNA.

Gynandromorph; gynander. An individual part of whose body exhibits male sex characters and part female characters.

Haplo-. A prefix before a chromosome number denoting an individual whose somatic cells lack one member of that chromosome pair.

Haploid. An individual or cell with a single complete set of chromosomes. Synonym, *monoploid.*

Hardy-Weinberg equilibrium. Occurrence in a population of three genotypes, two homozygous and one heterozygous (e.g., *AA, Aa,* and *aa*), where the frequencies of the three genotypes are p^2 for homozygous dominants, $2pq$ for heterozygotes, and q^2 for homozygous recessives and p = frequency of A and q = frequency of a. At these frequencies there is no change in genotypic frequencies as long as isolation mechanisms, selection mutation frequencies, and drift are constant.

HeLa cells. Cells originally (1952) from a carcinoma of the cervix, maintained as a standard human cell line for tissue and genetic experimentation. "HeLa" is derived from the pseudonym (Helen Lane) of the original patient source (Henrietta Lacks).

Helicase. An enzyme that unwinds the DNA double helix in front of DNA polymerase III.

Helix-destabilizing protein. Proteins that bind the single-stranded portion of the DNA molecule, preventing it from assuming the double-stranded conformation during replication.

Hemizygous. An individual having but one of a given gene; or a gene present only once. Designates either a haploid organism, a sex-linked gene in XY males (or an XY individual with regard to a particular X-linked gene), or an individual heterozygous for a given chromosomal deletion.

Hemoglobin. An iron-protein pigment of blood functioning in oxygen-carbon dioxide exchange of living cells.

Hemophilia. A metabolic disorder characterized by free bleeding from even slight wounds because of the lack of formation of clotting substances. It is associated with a sex-linked recessive gene.

Hermaphrodite. An individual with both male and female reproductive organs.

Heterochromatin. Condensed chromatin (which see). *Constitutive heterochromatin* occurs in the centromeric region of chromosomes and in various intercalary positions characteristic of a given species; it includes certain important, genetically active regions. *Facultative heterochromatin* is that chromatin that makes up the genetically inactive whole chromosomes (e.g., the X chromosome in human females) or parts of certain other chromosomes.

Heteroduplex. A double-stranded nucleic acid in which the two strands have different origins and therefore are not perfectly complementary.

Heterogametic sex. That sex with either only one or two different sex chromosomes, as XO, XY, or ZW. The heterogametic sex thus produces two kinds of gametes with respect to sex chromosomes, e.g., X and Y sperms in human beings.

Heterogeneous nuclear RNA (hnRNA). Various unprocessed RNA molecules located in the nucleus.

Heterospory. In higher plants the production of two kinds of meiocytes (megasporocytes and microsporocytes), which gives rise to two different kinds of meiospores. Compare with *homospory.*

Heterozygote (adj., **heterozygous**). An individual whose chromosomes bear unlike genes of a given allelic pair or series. Heterozygotes produce more than one kind of gamete with respect to a particular locus.

Histone. Any of several basic, low molecular weight proteins that can complex with DNA.

HLA. Human leukocyte antigens; antigens located on the surfaces of the leukocytes and which affect tissue compatibility in organ transplants and skin grafting. These antigens occur in quite a large number of types and are produced by at least four closely linked loci (in the sequence *D B C A*) on autosome 6. Collectively, they make up the *major histocompatibility complex* (MHC).

Hogness box. A sequence approximately 30 base pairs upstream from a eukaryotic gene to which RNA polymerase II binds.

Holandric gene. A gene located only on the Y chromosome in XY species.

Homeobox. A highly conserved segment of DNA found in all homeotic genes. This sequence may code for the part of a regulatory protein that binds to DNA.

Homeotic mutation. A mutation in which one developmental pattern is replaced by another. The genes in which the mutations occur appear to be involved in regulation of developmental patterns possibly by coding for proteins that bind to DNA. The antennapedia and bithorax mutations are homeotic mutations.

Homogametic sex. That sex possessing two identical sex chromosomes (XX or ZZ). Gametes produced by the homogametic sex are all alike with respect to sex chromosome constitution.

Homolog. See *homologous chromosomes.*

Homologous chromosomes. Chromosomes that occur in pairs (in diploids), one derived from each of two parents; Normally (except for chromosomes associated with sex), these chromosomes are morphologically alike and bear the same gene loci. Each member of such a pair is the *homolog* of the other.

Homospory. In plants, the production of but one kind of meiocyte, which gives rise to meiospores morphologically indistinguishable from each other.

Homozygote (adj., **homozygous**). An individual whose chromosomes bear identical genes of a given allelic pair or series. Homozygotes produce only one kind of gamete with respect to a particular locus and therefore "breed true."

H-Y antigen. A histocompatibility factor determined by the Y chromosome. It is thought to be the major male-determining factor in mammals. In humans the gene for the H-Y antigen is on the short arm of the Y chromosome.

Hybrid. An individual that results from a cross between two genetically unlike parents.

Hybridoma. A cell hybrid between an antibody-producing B cell and a tumor cell, which divides indefinitely and produces a single antibody (monoclonal antibody) in culture.

Hypha (pl. **hyphae**). One of the filaments of a mycelium in fungi.

Immunoglobulin. One of several types of antibodies secreted by plasma cells.

Imperfect flower. One lacking either stamens or pistil.

Incomplete dominance. The condition in heterozygotes where the phenotype is intermediate between the two homozygotes. In some plants the cross of red × white produces pink-flowered progeny. Incomplete dominance differs from codominance.

Incompletely sex-linked genes. Genes located on homologous portions of the X and Y (or Z and W) chromosomes.

Independent segregation. The random or independent behavior of genes on different pairs of chromosomes.

Inducer. See *effector.*

Inducible enzyme. An enzyme synthesized only in the presence of an effector (which see).

Interference. The increase (*negative interference*) or decrease (*positive interference*) in likelihood of a second crossover closely adjacent to another. In most organisms interference increases with decreased distance between crossovers. See *coincidence.*

Interphase. The stage of cell life during which that cell is not dividing.

Intersex. An individual showing secondary sex characters intermediate between male and female or some of each sex.

Intron (intervening sequence). A sequence of nucleotides in DNA that does not appear in mRNA; probably excised from hnRNA in its processing. Its function is unknown; polypeptide synthesis proceeds unimpeded in vitro without the introns.

Inversion. Reversal of the order of a block of genes in a given chromosome. *PQUTSRVWX* would represent an inversion of genes *RSTU* if the normal order is alphabetical.

Inversion loop. The loop configuration in a pair of synapsed, homologous chromosomes, caused by an inversion in one of them.

Inverted repeat. Two copies of the same DNA sequence oriented in opposite directions on the same molecule. Commonly found at the ends of transposons and the T-DNA of *Agrobacterium tumefaciens.*

In vitro. Experimentally induced biological processes outside the organism (literally, "in glass").

In vivo. Experimentally induced biological processes within the organism.

Isogametes. Gametes not sexually differentiated. Fusing isogametes are physically similar but in some species are chemically differentiated.

Kappa. A particle present in the cytoplasm of some Paramecia (killers) that secretes a substance that kills Paramecia not possessing the kappa particle.

Karyokinesis. The division of the nucleus during cell division.

Karyolymph. A clear fluid material within the nuclear membrane; "nuclear sap."

Karyotype. The somatic chromosome complement of an individual, usually as defined at mitotic metaphase by morphology (including centromere location and often by special staining techniques, as fluorescence banding) and number, arranged in a sequence that is standard for that organism.

Kilobase (abbr., **KB.**). One thousand nucleotides or nucleotide pairs in sequence. May be used to pertain to either DNA or RNA.

Kinetochore. The attachment region (within the centromere) of the microtubules of the spindle in cells undergoing mitosis or meiosis.

Klinefelter syndrome. A genetic disease due to the XXY karyotype. It produces sterile males with some mental retardation.

Lampbrush chromosome. A chromosome that has paired loops extending laterally, and occurs in primary oocyte nuclei; they represent sites of active RNA synthesis.

Lethal gene. A gene whose phenotypic effect is sufficiently drastic to kill the bearer. Death from different lethal genes may occur at any time from fertilization of the egg to advanced age. Lethal genes may be dominant, incompletely dominant, or recessive.

Life cycle. The entire series of developmental stages undergone by an individual from zygote to maturity and death.

Linkage. The occurrence of different genes on the same chromosome. They show nonrandom assortment at meiosis. See also *synteny.*

Linkage group. All of the genes located physically on a given chromosome.

Linkage map. A scale representation of a chromosome that shows the relative positions of all its known genes.

Locus (pl., **loci**). The position or place on a chromosome occupied by a particular gene or one of its alleles.

Lysis (n.) (v.i. or v.t., **lyse**). Disintegration or dissolution; usually, the destruction of a bacterial host cell by infecting phage particles.

Lysogenic bacteria. Living bacterial cells that harbor temperate phages (viruses).

Major histocompatibility complex. See *HLA*.

Map unit. A distance on a linkage map represented by 1 percent of cross-overs (recombinants), that is, by a recombination frequency of 1 percent.

Maternal inheritance. Phenotypic differences due to factors such as chloroplasts and mitochondria transmitted by the female gamete.

Mean. The arithmetic average; the sum of all values for a group, divided by the number of individuals. Symbolized by $\bar{x}$ (sample) or by μ (population).

Median. The middle value of a series of readings arranged serially according to magnitude.

Megasporocyte. In plants, the meiocyte that is destined to produce megaspores by meiosis. Synonym, *megaspore mother cell*.

Meiocyte. Any cell that undergoes meiosis.

Meiosis. Nuclear divisions in which the diploid or somatic chromosome number is reduced by half. In the first of the two meiotic divisions, homologous chromosomes first replicate, then pair (synapse), and finally separate to different daughter nuclei, which thus have half as many chromosomes as the parent nucleus, and one of each kind instead of two. A second division, in which chromosomal replicates separate into daughter nuclei, follows so that meiosis produces four monoploid daughter nuclei from one diploid parent nucleus.

Meiospore. In plants, one of the asexual reproductive cells produced by meiosis from a meiocyte.

Meristem. An undifferentiated cellular region in plants characterized by repeated cell division.

Merozygote. A partially diploid receptor bacterial cell that results from conjugation.

Messenger RNA. Ribonucleic acid conferring amino acid specificity on ribosomes; complementary to a given DNA cistron.

Metacentric. A chromosome with a centrally located centromere.

Metafemale (also **superfemale**). Abnormal females in *Drosophila*, usually sterile and weak, with an overbalance of X chromosomes with respect to autosomes; X/A ratio greater than 1.0.

Metamale (also **supermale**). Abnormal males in *Drosophila* with an overabundance of autosomes to X chromosomes; X/A ratio less than 0.5.

Metaphase. That stage of nuclear division in which the chromosomes are located in the equatorial plane of the spindle prior to centromere separation.

MHC. Major histocompatibility complex. See *HLA*.

Micrometer (μm). A commonly employed unit of measurement in microscopy, it equals 1×10^{-6} meter or 1×10^{-3} millimeter. Synonym, *micron* (μ) in older usage.

Micron. See *micrometer*.

Microspore. In plants, the meiospore that gives rise to the male gametophyte and sperms.

Microsporocyte. In plants, a cell that is destined to produce microspores by meiosis. Synonyms, *microspore mother cell* and *"pollen mother cell."*

Microtubule. A hollow tubular cytoplasmic component that forms the spindle of the mitotic apparatus (which see), and is found especially in motile cells. It has an outside diameter about 15 to 30 nm.

Millimeter (mm). A unit of distance measurement equal to 1×10^{-3} meter.

Missense. A mutation by which a particular codon is changed to incorporate a different amino acid, which often results in an inactive protein. A *missense codon* results when one or more bases of a *sense codon* (which see) are changed so that a different amino acid is coded for.

Mitochondrion (pl., **mitochondria**). Small cytoplasmic organelle where cellular respiration occurs.

Mitosis. Nuclear division in which a replication of chromosomes is followed by separation of the products of replication and their incorporation into two daughter nuclei. Daughter nuclei are normally identical with each other and with the original parent nucleus in both kind and number of chromosomes. Mitosis may or may not involve cytoplasmic division.

Mitotic apparatus. The collection of cytoplasmic structures that are present in mitotic cells and consists of the asters (if present) surrounding each centrosome and the spindle of microtubules.

Mode. The numerically largest class or group in a series of measurements or values.

Monoclonal antibody. A specific antibody produced by a hybridoma cell line.

Monoecius. Individuals producing both sperm and egg.

Monohybrid. The offspring of two homozygous parents that differ in only one gene locus or in which only one such locus is under consideration.

Monohybrid cross. A cross between two parents that differ in only one heritable character or in which only one such character is under consideration.

Monoploid. An individual with a single complete set of chromosomes. Also the fundamental number of chromosomes comprising a single set. Synonym, *haploid*.

Monosomic. An individual lacking one chromosome of a set ($2n - 1$).

Morgan unit (morgan). A unit of relative distance between genes on a given chromosome; it equals a cross-over value of 100 percent.

Morphology. The study of form and structure in organisms.

Mosaic. An individual part of whose body is composed of tissue genetically different from another part.

mtDNA. DNA that occurs as a normal component within mitochondria.

Multiple alleles. A series of three or more alternative alleles, any one of which may occur at a particular locus on a chromosome.

Multiple genes. See *polygenes.*

Mutagen. Any agent that brings about a mutation.

Mutation. A sudden change in genotype having no relation to the individual's ancestry. Used for changes in a single gene itself ("point mutations") and for chromosomal aberrations.

Muton. The smallest segment of DNA or subunit of a cistron that can be changed and thereby bring about a mutation; can be as small as one nucleotide pair.

Mycelium. The threadlike filamentous vegetative body of many fungi.

Nanometer (nm). A unit of distance equal to 1×10^{-9} meter. Synonym, *millimicron* (mμ).

Negative control. Repression of an operator site (which see) by a regulatory protein that is produced by a regulator site (which see).

Neoplasm. A localized population of rapidly dividing cells whose growth is not subject to usual growth control mechanisms.

Nonautonomous element. A type of controlling element dependent on another autonomous element for its transposition. The Ds element in the Ac-Ds system in maize is such an example.

Nondisjunction. The failure of homologous chromosomes to separate at anaphase-I of meiosis. *Primary nondisjunction* may occur in an XX female, and lead to production of XX or O eggs (in addition to the normal X eggs); or it may occur (first division) in an XY male, and result in XY and O sperms or (second division) in XX and O or YY and O sperm (in addition to normal Y or X sperms). *Secondary nondisjunction* may occur in an XXY female, giving rise to eggs with XX, XY, X, or Y chromosomal combinations.

Nonhistone protein. A large class of chromosomal proteins, not of the histone class, possibly involved in specific gene regulation.

Nonsense. A codon that does not specify any amino acid in the genetic code.

Normal curve. A smooth, symmetrical, bell-shaped curve of distribution.

N-terminus. The amino ($—NH_2$) end of a peptide chain, by convention written as the left end of the structural formula.

Nucleolus. Deeply staining body containing rRNA; multiple copies of DNA that code for rRNA, and proteins occur in the nucleus.

Nucleoside. Portion of a DNA or RNA molecule composed of one deoxyribose molecule (in DNA), or ribose (in RNA), plus a purine or a pyrimidine.

Nucleosome. Short, disk-shaped cylinders, composed of nucleoproteins and spaced at roughly 100 Å intervals on chromosomes. Histones (which see) H2A, H2B, H3, and H4 are complexed with DNA in the nucleosomes.

Nucleotide. A monomeric unit from which DNA and RNA is constructed. It consists of a purine or pyrimidine base, a pentose, and phosphoric acid.

Ocher. The nonsense codon UAA.

Okazaki fragment. Segments of approximately 1,000 to 2,000 nucleotides of the daughter strand of replicating DNA. These are joined in succession by the enzyme DNA ligase to form a normal double-stranded DNA molecule.

Oligonucleotide. A linear sequence of a few (generally not over 10) nucleotides.

Oncogene. A gene that induces uncontrolled cellular proliferation. Oncogenes may be either cellular or viral in origin.

Ontogeny. The complete development of the individual from zygote, spore, and so on, to adult form.

Oocyte. The diploid cell that will undergo meiosis to form an egg.

Oogenesis. Egg formation.

Opal. The nonsense codon UGA.

Operator site. A segment of DNA in an operon that affects activity or nonactivity of associated cistrons; it may be combined with a repressor and thereby "turn off" the associated cistrons.

Operon. A system of cistrons, operator and promoter sites, by which a given genetically controlled, metabolic activity is regulated.

Organelle. A specialized cytoplasmic structure with a particular function (e.g., chloroplast, mitochondrion).

Ovary. The female gonad in animals or the ovule-containing portion of the pistil of a flower.

Ovule. Structure within the ovary of the pistil of a flower that becomes a seed. It represents a megasporangium, together with some overgrowing tissue (integuments), and ultimately contains the female gametophyte (embryo sac), one of whose nuclei is an egg.

P. The parental generation in a given cross.

Palindrome. A sequence of DNA base pairs that reads the same on the complementary strands. For example, 5′GAATTC3′ on one strand, and 5′CTTAAG3′ on the other strand. Palindrome sequences are recognized by restriction endonucleases.

Paracentric. Refers to an inversion that does not involve the centromere but lies entirely within one arm of the chromosome.

Parameter. Actual value of some quantitative character for a population. Compare *statistic.*

Parthenogenesis. Development of a new individual from an unfertilized egg.

Pedigree. The ancestral history of an individual; a chart showing such history.

Penetrance. The proportion of individuals of a particular genotype that show the expected phenotype under a certain set of environmental conditions.

Peptide bond. A chemical bond (CONH) linking amino acid residues together in a protein.

Perfect flower. One with both stamen(s) and pistil(s).

Pericentric. Refers to an inversion that does include the centromere, hence involves both arms of a chromosome.

Petite. A slow-growing strain of yeast (*Saccharomyces*) that lacks certain respiratory enzymes and forms unusually small colonies on agar. *Segregational petites* bear mutant nuclear gene(s), whereas *neutral petites* (sometimes called *vegetative petites*) bear mutant mitochondrial DNA.

Phage. A virus that infects bacteria.

Phenotype (adj., **phenotypic**). The appearance or discernible character of an individual, which is dependent on its genetic makeup, usually expressed in words, e.g., "tall," "dwarf," "wild type," "prolineless." Identical phenotypes may not necessarily breed alike.

Phosphodiester bond. A bond between the sugar and a phosphate group in the sugar-phosphate strand of DNA.

Photoreactivation. A light-induced reversal of ultraviolet light causing injury to cells.

Phylogeny. The evolutionary development of a species or other taxonomic group.

Pilus (pl., **pili**). One of the filamentous projections from the surface of a bacterial cell. One kind of pilus serves as the conjugation tube in bacteria.

Pistil. The entire part of the flower that produces megaspores (which, in turn, produce eggs). Sometimes, and *incorrectly,* referred to as a female floral part.

Plaque. Clear area on culture plate of bacteria where these have been killed by phages.

Plasmagene. A self-replicating, cytoplasmically located gene.

Plasmid. Originally, a closed, circular DNA molecule of restricted size (a few tens of thousands of nucleotide pairs), existing only in the cytoplasm (of bacteria); incapable of integration into the bacterial "chromosome." See also *episome; plasmid* is often used to include both episomes and plasmids.

Pleiotropy. The influencing of more than one trait by a single gene.

Polar body. One of the three very small cells produced during meiosis of an oocyte, which contains a haploid nucleus but little cytoplasm. It is nonfunctional in reproduction.

Pollen. The young male gametophyte of a flowering plant, surrounded by the microspore wall.

Poly-A tail. The initially long sequence of adenine nucleotides at the 3′ end of mRNA; added after transcription.

Polygenes. Two or more different pairs of alleles, with a presumed cumulative effect that governs such quantitative traits as size, pigmentation, intelligence, among others. Those contributing to the trait are termed contributing (effective) alleles; those appearing not to do so are referred to as noncontributing or noneffective alleles.

Polymer. A chemical compound composed of two or more units of the same compound.

Polynucleotide. A linear sequence of many nucleotides.

Polynucleotide phosporylase. An enzyme that can link ribonucleotides together in a random fashion, which can be used to construct an artificial mRNA molecule.

Polypeptide. A compound containing amino acid residues joined by peptide bonds. A protein may consist of one or more specific polypeptide chains.

Polyploid. An individual having more than two complete sets of chromosomes, e.g., triploid ($3n$), tetraploid ($4n$).

Polyribosome. See *polysome.*

Polysome. A group of ribosomes joined by a molecule of messenger RNA.

Polytene chromosome. Many-stranded giant chromosomes produced by repeated replication during synapsis in certain dipteran larval tissues. Synonym, *giant chromosome.*

Population. An infinite group of individuals, measured for some variable, quantitative character, from which a sample is taken.

Position effect. A phenotypic effect dependent on a change in position on the chromosome of a gene or group of genes.

Positive control. Activation of an operator site (which see) by a regulatory protein that is produced by a regulator site (which see).

Pribnow box. A DNA sequence upstream from the start codon in a eukaryotic gene to which RNA polymerase binds.

Probability. The likelihood of occurrence of a given event. Usually expressed as a number between 0 (complete certainty that the event will *not* occur) and 1 (complete certainty that the event *will* occur).

Progeny. Offspring individuals.

Prokaryote. A cell or organism lacking a discrete nuclear body (also *procaryote*).

Promiscuous DNA. The occurrence of the same DNA sequences in more than one cellular compartment. It suggests that DNA may have been exchanged between organelles, or between organelles and the nucleus.

Promotor. A specific site on DNA to which RNA polymerase binds and initiates transcription.

Prophage. Phage nucleoprotein (DNA or RNA, depending on the phage) integrated into the bacterial DNA.

Prophase. The first stage of nuclear division, including all events up to (but not including) arrival of the chromosomes at the equator of the spindle.

Protoplast. A structural unit of protoplasm; all the living (protoplasmic) material of a cell. The two principal parts are nucleus and cytoplasm.

Prototroph. An individual able to carry on a given synthesis; a wild-type individual able to grow on minimal medium.

Pseudoalleles. Nonalleles so closely linked that they are often inherited as one gene, but shown to be separable by crossover studies.

Pseudodominance. The expression (apparent dominance) of a recessive gene at a locus opposite a deletion.

Punnett square. A "checkerboard" grid designed to determine all possible genotypes produced by a given cross. Genotypes of the gametes of one sex are entered across the top, those of the other down one side. Zygote genotypes produced by each possible mating are then entered in the appropriate squares of the grid.

Pure line (pure breeding line). A strain of individuals homozygous for all genes being considered.

Purine. Nitrogenous base occurring in DNA and RNA; these are adenine and guanine.

Pyrimidine. Nitrogenous base occurring in DNA (thymine and cytosine) or RNA (uracil and cytosine).

Pyrimidine dimer. The compound that results when UV radiation induces the covalent joining of two thymine residues, two cytosine residues, or a thymine–cytosine pair.

Q-bands. The banding pattern observed when chromosomes are stained with the fluorochrome quinicrine mustard and viewed under ultraviolet light. This technique is particularly useful for identifying the Y chromosome.

r. That portion of the *R* plasmid carrying genes for all drug resistance, with the probable exception of tetracycline resistance. See also *R* factors and *RTF*.

R factors. Bacterial *plasmids* (which see) carrying genes for drug resistance. See also *r* and *RTF*.

Rad. Term meaning radiation-absorbed dose; it is the amount of ionizing radiation that liberates 100 ergs of energy in one gram of matter.

Reading frame. The codon sequence resulting from mRNA "reading" the DNA nucleotides in groups of three.

Recessive. An adjective applied to the member of a pair of genes that fails to express itself in the presence of its dominant allele. The term is also applicable to the trait produced by a recessive gene. Recessive genes express themselves ordinarily only in the homozygous state.

Reciprocal cross. A second cross of the same genotypes in which the sexes of the parental generation are reversed. The cross *AA* (♀) × *aa* (♂) is the reciprocal of the cross *aa* (♀) × *AA* (♂).

Reciprocal translocation. The exchange of segments between two nonhomologous chromosomes.

Recombinant. An individual derived from a crossover gamete.

Recombinant DNA. A DNA molecule (in practice generally a bacterial plasmid) which has been enzymatically cut and DNA of another individual of the same or a different species inserted in the space so produced, then reannealed to the closed, circular form. Many recombinant DNA molecules, therefore, can be so constructed that an individual (e.g., a bacterium) has a genetic capability not previously associated with that organism.

Recombination. The new association of genes in a recombinant individual; this association arises from independent assortment of unlinked genes, from crossing-over between linked genes, or from intracistronic crossing-over.

Recon. The smallest segment of DNA or subunit of a cistron that is capable of recombination; may be as small as one deoxyribonucleotide pair.

Reduction division. See *meiosis*.

Regulator site. The specific segment of DNA responsible for production of a regulatory protein that may serve as a repressor in some cases or as an activator in others. See also *negative control* and *positive control*.

Relational coiling. The loose coiling of chromatids about each other.

Replicate. To form replicas from a model or template; applies to synthesis of new DNA from preexisting DNA as part of nuclear division.

Replicon. A sequentially replicating segment of a nucleic acid, controlled by a subsegment known as a replicator. A single replicator is present in the bacterial "chromosome," whereas the chromosomes of eukaryotes bear large numbers of replicons in series.

Repressor. A protein produced by a regulator gene that can combine with and repress action of an associated operator gene.

Resistance transfer factor. See *RTF*.

Restriction endonuclease. Any of a group of enzymes that break internal bonds of DNA at highly specific points.

Restriction site. The base sequence at which a restriction endonuclease cuts the DNA molecule, usually a point of symmetry within a palindrome sequence.

Retrovirus. RNA viruses that replicate with reverse transcriptase. Since the enzyme produces a DNA copy of the viral RNA that is the reverse of transcription, the name *retro*, suggesting backward transcription, is used.

Reverse transcriptase. An enzyme, carried within the coat of the retroviruses, that makes double-stranded DNA from a single-stranded RNA template.

Ribonucleic acid (RNA). A single-stranded nucleic acid molecule, synthesized principally in the nucleus from deoxyribonucleic acid, composed of a ribose-phosphate backbone with purines (adenine and guanine) and pyrimidines (uracil and cytosine) attached to the sugar ribose. RNA is of several kinds of functions to carry the "genetic message" from nuclear DNA to the ribosomes.

Ribonucleoside. Portion of an RNA molecule composed of one ribose molecule plus either a purine or a pyrimidine.

Ribonucleotide. Portion of an RNA molecule composed of one ribosephosphate unit plus a purine or a pyrimidine.

Ribose. The 5-carbon sugar of ribonucleic acid.

Ribosomal RNA. That ribonucleic acid incorporated into ribosomes; it is nonspecific for amino acids.

Ribosome. Cytoplasmic structure, usually adherent to the endoplasmic reticulum, which is the site of protein synthesis. In bacteria the ribosomes are free in the cytoplasm. See also *polysome.*

RNase. An enzyme that hydrolyzes RNA.

RNA-dependent DNA polymerase. A group of enzymes that catalyze formation of DNA molecules from RNA templates. These occur in some viruses (e.g., those that produce tumors). See also *reverse transcriptase.*

RNA polymerase. An enzyme that catalyzes the formation of RNA from ribonucleotide triphosphates, using single-stranded DNA as a template.

RNA splicing. The removal of large noncoding sequences (introns) from the primary RNA transcript followed by rejoining of the nonadjacent coding sequences (exons) to produce the functional mRNA.

Roentgen (R). The amount of ionizing radiation that produces about two ion pairs per cubic micrometer of matter, or about 1.6×10^{12} ion pairs per cubic centimeter.

RTF (resistance transfer factor). That part of the *R* plasmid responsible for transfer of the *R* factor during bacterial conjugation. It appears to include the cistron for tetracycline resistance, whereas other drug resistance factors occur in the *r* portion of the *R* factor.

SI nuclease. A nuclease specific for single-stranded nucleic acids.

Satellite DNA. Any fraction of DNA from a eukaryotic cell that differs significantly in base composition from the majority of the DNA so that it produces separate bands of DNA in a CsCl density gradient. If the satellite DNA is rich in A + T, it is lighter than the main DNA, and if it is rich in C + G, it is heavier than the main DNA. Satellite DNA sequences are usually highly repetitious.

Sedimentation coefficient. The rate of sedimentation of a solute in an appropriate solvent under centrifugation. An *s* value of 1×10^{-13} second is one *Svedberg unit.*

Selection coefficient. The measure of reduced fitness of a given genotype; it equals one minus the adaptive value (which see) of that genotype.

Self-fertilization. Functioning of a single individual as both male and female parent. Plants are "selfed" if sperm and egg are supplied by the same individual.

Semiconservative. Replication of DNA in which the two sugar-phosphate "backbones" become separated; each is conserved as one of the two strands of two new DNA molecules.

Sense codon. A codon (which see) specifying a particular amino acid in protein synthesis.

Sex chromosomes. Heteromorphic chromosomes that do not occur in identical pairs in both sexes in diploid organisms; in humans and fruit flies these are designated as X and Y chromosomes, respectively; in fowl, as the Z and W chromosomes.

Sexduction. Incorporation of bacterial chromosomal genes in the fertility plasmid, with subsequent transfer to a recipient cell in conjugation.

Sex-influenced trait. One in which dominance of an allele depends on sex of the bearer; e.g., pattern baldness in humans is dominant in males, recessive in females.

Sex-limited trait. One expressed in only one of the sexes; e.g., cock feathering in fowl is limited to normal males.

Sex-linked gene. A gene located only on the X chromosome in XY species (or on the Z chromosomes in ZW species).

Shine-Delgarno sequence. The mRNA sequence preceding the start codon, which serves as the binding site for ribosomes.

Siblings (also **sibs**). Individuals with the same maternal and paternal parents; brother-sister relationship.

Sickle-cell anemia. Anemia in humans inherited as an autosomal recessive and due to a single amino acid substitution in the beta-hemoglobin chain.

Significance. In statistical treatments, probability values of >0.05 are termed *not significant,* those ≤ 0.05 but >0.01 are *significant,* those ≤ 0.01 but >0.001 are *highly significant,* and those ≤ 0.001 are *very highly significant.* Significant probability values indicate that the results, although possible, deviate too greatly from the expectancy to be acceptable when chance alone is operating.

Soma (adj., **somatic**). The body, cells of which in mammals and flowering plants normally have two sets of chromosomes, one derived from each parent.

Specialized transduction. A type of transduction where only a few bacterial genes are transferred because the phage has only specific sites of integration on the host chromosome.

Spermatid. The haploid cells, resulting from meiosis of a primary spermatocyte that will mature into sperms.

Spermatocyte. The cell that undergoes meiosis to produce four spermatids.

Spermatogenesis. Development of sperms.

Spore. An asexual reproductive protoplast capable of developing into a new individual. See also *meiospore.*

Sporocyte. In plants, a meiocyte.

Sporogenesis. Formation of spores.

Sporophyte. In plants, the phase of the life cycle that reproduces asexually by meiospores and is characterized by having the double (usually diploid) chromosome number.

Stamen. That part of a flower that produces microspores (which, in turn, produce sperms). Sometimes, and *incorrectly,* referred to as a male floral part.

Standard deviation. A measure of the variation in a sample. Symbolized by s.

Standard error of difference in means. Measure of the significance of the difference in two sample means. Symbolized by S_d.

Standard error of sample mean. An estimate of the standard deviation of a series of hypothetical sample means that serves as a measure of the closeness with which a given sample mean approximates the population mean. Symbolized by $s_{\bar{x}}$.

Statistic. Actual value of some quantitative character for a sample from which estimates of parameters may be made.

Stigma. The pollen-receptive portion of the pistil of a flower.

Structural gene. See *cistron.*

Submetacentric. A chromosome where the centromere is nearer one end than the other, resulting in the arms not being of equal length.

Supercoiling. The coiling of a covalently closed circular DNA molecule whereby the helix is coiled upon itself so that it crosses its own axis.

Svedberg unit. See *sedimentation coefficient.*

Synapsis (v.i., **synapse**). The pairing of homologous chromosomes that occur in prophase-I of meiosis.

Synaptonemal complex. A structure that forms between the paired homologous chromosomes at the pachytene stage, consisting of a tripartite ribbon of parallel, dense, and lateral elements surrounding a medial complex.

Syngamy. The union of the nuclei of sex cells (gametes) in reproduction.

Synteny. The occurrence of two or more genetic loci on the same chromosome. Depending on intergene distance(s) involved, they may or may not exhibit nonrandom assortment at meiosis.

TaTa box. The same thing as the Hogness and Pribnow box. Site of attachment of RNA polymerase.

Tautomer. An alternate molecular form of a compound, characterized by a different arrangement of its electrons and protons as compared with the common form of the molecule.

Taxon (pl., **taxa**). A taxonomic group of any rank.

Taxonomy. The study of describing, naming, and classifying living organisms, and the bases on which resultant classification systems rest.

Telocentric. A chromosome with a terminal centromere.

Telophase. The concluding stage of nuclear division, characterized by the reorganization of interphase nuclei.

Temperate phage. A phage (bacterium-infecting virus) that invades and multiplies in but does not ordinarily lyse its host.

Template. A model, mold, or pattern; DNA acts as a template for RNA synthesis.

Testcross. The cross of an individual (generally of dominant phenotype) with one having the recessive phenotype. Generally used to determine whether an individual of dominant phenotype is homozygous or heterozygous, or to determine the degree of linkage.

Tetrad. The four monoploid (haploid) cells arising from meiosis of a megasporocyte or microsporocyte in plants; also, a group of four associated chromatids during synapsis.

Tetraploid. A polyploid cell, tissue, or organism with four sets of chromosomes ($4n$). See also *polyploid, allopolyploid,* and *autopolyploid.*

Three-point cross. A trihybrid testcross (e.g., *ABC/abc* × *abc/abc,* etc.) used primarily in chromosome mapping.

Thymidine. The deoxyribonucleoside that contains the pyrimidine thymine.

Thymidylic acid. The deoxyribonucleotide that contains the pyrimidine thymine.

Thymine. A pyrimidine base occurring in DNA. Pairs normally with adenine.

Totipotency. The property of a cell (or cells) whereby it develops into a complete and differentiated organism.

Trans arrangement. Linkage of the dominant allele of one pair and the recessive of another on the same chromosome.

Transcription. Synthesis of messenger RNA from a DNA template.

Transduction. Recombination in bacteria whereby DNA is transferred by a phage from one cell to another. *Generalized* transduction involves phages that have incorporated a segment of bacterial chromosome during packaging of the phage; *specialized* transduction involves temperate phages that are always inserted into the bacterial chromosome at a site specific for that phage.

Transfer RNA. Amino acid-specific RNA that transfers activated amino acids to mRNA, where polypeptide synthesis takes place.

Transformation. Genetic recombination, particularly in bacteria, whereby naked DNA from one individual becomes incorporated into that of another.

Transgressive variation. Appearance in progeny of a more extreme expression of a trait than occurs in the parents. Assumed to result from cumulative action of polygenes, but careful testing of variation in parental lines is necessary for verification.

Transition. The substitution in DNA or RNA of one purine for another or of one pyrimidine for another.

Translation. The process by which a particular messenger RNA nucleotide sequence is responsible for a specific amino acid residue sequence of a polypeptide chain.

Translocation. The shift of a portion of a chromosome to another part of the same chromosome or to an entirely different chromosome. (See also *reciprocal translocation.*)

Transposon (Transposable element). A segment of DNA, which generally consists of more than 2,000 nucleotides, and is capable of moving into and out of a chromosome or plasmid, and/or from place to place within the chromosome in both prokaryotes and eukaryotes. It is capable of turning genes "on" or "off."

Transversion. The substitution in DNA or RNA of a purine for a pyrimidine or vice versa.

Trihybrid. An individual heterozygous for three pairs of alleles.

Triplet. A group of three successive nucleotides in RNA (or DNA) that, in the genetic code (which see), specifies a particular amino acid in the synthesis of polypeptide chains.

Triplo-. Prefix denoting a trisomic individual where the identity of the extra chromosome is known, e.g., triplo-IV *Drosophila*, or triplo-21 human beings.

Triploid. A polyploid cell, tissue, or organism with three sets of chromosomes ($3n$).

Trisomic. An individual with one extra chromosome of a set ($2n + 1$).

Turner syndrome. A series of abnormalities in humans due to monosomy for the X chromosome. Individuals are phenotypically female, but are sterile.

Uniparental inheritance. Inheritance pattern where the offspring have received certain phenotypes from only one parent. This inheritance pattern is due to transmission of DNA containing cytoplasmic particles.

Universal donor. A person with group O blood, whose erythrocytes therefore bear neither A nor B antigens, and whose blood can be donated to members of groups O, A, B, and AB if necessary.

Universal recipient. A person of blood group AB who can receive blood from members of groups AB, A, B, or O if necessary.

Unwinding proteins. Proteins that bind to, and unwind, the DNA helix at the replicating fork.

Uracil. A pyrimidine base occurring in RNA.

Uridine. The ribonucleoside that contains the pyrimidine uracil.

Uridylic acid. The ribonucleotide that contains the pyrimidine uracil.

Variable region. The portion of an antibody molecule that varies greatly in amino acid sequence. This is the part of the antibody that reacts with the antigen.

Variance. A statistic providing an unbiased estimate of population variability; it is the square of the standard deviation (which see).

Vector. A DNA molecule capable of replication into which a gene is inserted by recombinant DNA techniques.

Virulence. The ability to produce disease.

Virulent phage. A phage (virus) that destroys (lyses) its host bacterial cell.

Wild type. The most frequently encountered phenotype in natural breeding populations; the "normal" phenotype.

Wobble hypothesis. The partial or total lack of specificity in the third base of some triplet codons (which see) whereby two, three, or four codons differing only in the third base may code for the same amino acid.

Xeroderma pigmentosum. A genetic disorder in which the skin is extremely sensitive to sunlight and death usually occurs from skin cancer. It is inherited as an autosomal recessive.

X-linkage (sex linkage). Refers to genes located on the X chromosome.

Y-linkage (holandric genes). Refers to genes located on the Y chromosome.

Zygote. The protoplast resulting from the fusion of two gametes in sexual reproduction; a fertilized egg.

INDEX

INDEX

A

B

E

F

G

H

I

J

K

L

M

N

O

P

R